AF353123

Paleontology in the 21st Century

Paleontology in the 21st Century

Editors

Mary H. Schweitzer
Ferhat Kaya

MDPI • Basel • Beijing • Wuhan • Barcelona • Belgrade • Manchester • Tokyo • Cluj • Tianjin

Editors
Mary H. Schweitzer
North Carolina State
University
USA

Ferhat Kaya
University of Helsinki
Finland

Editorial Office
MDPI
St. Alban-Anlage 66
4052 Basel, Switzerland

This is a reprint of articles from the Special Issue published online in the open access journal *Biology* (ISSN 2079-7737) (available at: https://www.mdpi.com/journal/biology/special_issues/paleontology_in_21st_century).

For citation purposes, cite each article independently as indicated on the article page online and as indicated below:

LastName, A.A.; LastName, B.B.; LastName, C.C. Article Title. *Journal Name* **Year**, *Volume Number*, Page Range.

ISBN 978-3-0365-7572-8 (Hbk)
ISBN 978-3-0365-7573-5 (PDF)

Cover image courtesy of Mary Higby Schweitzer

© 2023 by the authors. Articles in this book are Open Access and distributed under the Creative Commons Attribution (CC BY) license, which allows users to download, copy and build upon published articles, as long as the author and publisher are properly credited, which ensures maximum dissemination and a wider impact of our publications.

The book as a whole is distributed by MDPI under the terms and conditions of the Creative Commons license CC BY-NC-ND.

Contents

Editorial

Paleontology in the 21st Century

Mary H. Schweitzer [1,2,3,4]

1 Department of Biological Sciences, North Carolina State University, Raleigh, NC 27606, USA;
   schweitzer@ncsu.edu
2 North Carolina Museum of Natural Sciences, Raleigh, NC 27601, USA
3 Department of Geology, Lund University, 223 62 Lund, Sweden
4 Museum of the Rockies, Montana State University, Bozeman, MT 59717, USA

Citation: Schweitzer, M.H. Paleontology in the 21st Century. *Biology* **2023**, *12*, 487. https://doi.org/10.3390/biology12030487

Received: 15 March 2023
Accepted: 19 March 2023
Published: 22 March 2023

Copyright: © 2023 by the author. Licensee MDPI, Basel, Switzerland. This article is an open access article distributed under the terms and conditions of the Creative Commons Attribution (CC BY) license (https://creativecommons.org/licenses/by/4.0/).

For much of its 300+ year history, "modern" paleontology has been a descriptive science, firmly housed within geological sciences. The application of rigorous phylogenetic methods [1–3] to extinct organisms was a driver of a movement that began in the 1980s to bring paleontology, and particularly deep time paleontology (i.e., >1 Ma), more firmly into the biological sciences. Paleohistology—the investigation of the microstructure of fossil bones initiated by Enlow [4,5] and expanded upon by De Ricqlés and colleagues—(e.g., [5–7]) also influenced how dinosaurs and other fossils would be studied, as biological organisms rather than as geological oddities. The microscopic examination of ancient bone—even dinosaur bone—has revealed the presence of osteons, osteocyte lacunae, and vascular canals in these ancient specimens. Rather than being obliterated by fossilization processes, the retention of these features in bone recovered from Mesozoic and earlier sediments has allowed us to discern biological information from these once-living organisms, such as comparative growth rates and estimates of ontogenetic stages [7–9]. Phylogenetic and histological analyses forever changed the direction of paleontological studies.

However, the application of most molecular biological tools to elucidate evolutionary processes and timing was, until recently, reserved for extant species. The genomic revolution fostered by the advent of a polymerase chain reaction (PCR) and next gen technologies was limited in application to only living or very recently extinct organisms, and was not rigorously and regularly applied to deep time fossils, Jurassic Park notwithstanding.

Although DNA technologies were more readily accepted and proved useful for archaeological materials (see, e.g., [10–13]), the assumption that original, informative organic components were lost at some point during the transition from the biosphere to the geosphere [14,15], and the proposal of a predictable, and short, half-life for DNA and other biomolecules effectively slowed the application of these methods to all but the most recent fossil materials. What else older fossils may be telling us, and what else was possible to know, were questions that were not widely asked.

This is beginning to change, as illustrated in this Special Issue, because it is widely recognized that 1. some endogenous structures and the molecules comprising them are retained in ancient specimens; 2. morphological studies alone have failed to adequately account for convergence or parallel evolution; and 3. fossils are absolutely necessary to incorporate into any studies seeking to determine the patterns, processes, and timing of evolution deep in the Earth's history.

Just as technology in the life sciences has expanded at a record pace, the application of new technologies to paleontological specimens, though slower in coming, is resulting in an explosion in the type of data recoverable from fossils, as well as the type and age of fossils to which these can be applied. For example, the upper limit of DNA preservation, just 20 years ago, was proposed to be 100,000 years [10], but this limit has been repeatedly pushed back in time, most recently to >2 Ma with the recovery of environmental DNA from Greenland [16]. Protein sequence data, on the other hand, has been reported from 3 Ma eggshells [17] to multi-million-year-old dinosaurs [18,19], and immunological evidence for such preservation, suggested as far back as the 1950s [20,21], is becoming more prevalent.

This Special Issue sheds light on the quantum leap in understanding the Earth's rich biological history over the last 4 billion years, brought about by the application of new technologies to old fossils. It highlights new questions we can ask of the fossil record, and the expansion of fossils we can interrogate to yield robust high-resolution data. The authors within this Issue discuss the rapid development of new (or, new to paleontology) methods, once limited to extant organisms, that are now becoming broadly applied to fossils, including those from deep time.

This Issue contains several broad review articles. Tihelka et al. [22] discuss the possibility that animals, in the form of arthropods, invaded the land several million years before what was previously assumed, yielding a terrestrial "Cambrian explosion" of these widespread and taxonomically diverse invertebrates. This hypothesis is supported by combining molecular clock data with fossil evidence. The discrepancies between molecular clock and fossil data are addressed, and a method combining molecular and morphological data to estimate lineage divergence in a total evidence framework is proposed.

The value of fossils for phylogenetic studies has always been recognized, but the value of fossils for molecular studies has recently, and increasingly, been elucidated by authors contributing here [23] and elsewhere. Torres et al. [24] review the long history of paleoimmunological approaches to paleontology, and apply these methods to fossils from the 1.3 Ma Venta Micena site, whereas Tahoun et al. [25] provide an overview of organic molecules recovered from non-avian dinosaurs and contemporary organisms, including pigments and various proteins. They highlight the effort within the paleontological community to understand the mechanisms of such preservation, contributing to the emerging disciplines of molecular paleontology, and molecular taphonomy/diagenesis.

López-Antoñanzas and colleagues [26] further emphasize the need for incorporating fossil data to better understand evolutionary rates and relationships in their review of new, integrative phylogenetic methods. They stress the need for a unified approach to improve accuracy in modeling evolutionary processes and diversity distributions in deep time, and highlight new methods that combine morphological data from living and extinct groups with available molecular data to achieve more accurate evolutionary syntheses. Some of these include combining geometric morphometric and phylogenetic methods, incorporating stratigraphic data into parsimony analyses, and various statistical methods to calibrate a "morphological clock" that uses morphological data from both extant and extinct species.

Tamborini [27] reviews the changing role of paleobiology between the 20th and 21st centuries, pointing to the necessity of fossils in elucidating "deep time patterns and processes". He contrasts the historical differences between paleontology, a geology-based discipline, and the more biological approaches of paleobiology. He focuses his discussion first on the role of paleocolor as a testable hypothesis, stimulated by integrating more data with more technology; second, the search for endogenous organics in fossil materials that has driven the application of new (to paleontology) technologies, such as high-resolution, high-mass accuracy tandem mass spectrometry to address and characterize organic molecules in fossils; third, the integration of morphology and evolutionary theory in investigations of locomotion and mastication via the new role of robotics in 21st century; fourth, the relationship between evolution as expressed in fossils and the development of living organisms ('evo-devo'), resulting in a broader integration between paleontology and biology; and finally, the examination of both biotic and abiotic factors in shaping organisms through evolution. He ends his review by encouraging a new synthesis of knowledge brought about by new data, new fossils, new technologies, and the deeper integration of these disciplines.

Zhou [28] reviews a century of development in paleontology in China, identifying the influences of foreign collectors shaping the first part of this century-wide overview, and notes their influence of both geology and paleontology in China for decades. He notes not only the increasing participation of Chinese paleontologists to the discipline, but also the role of Chinese fossils, from early hominins (e.g., Peking Man, *Homo erectus*) to early feathered dinosaurs, in shaping a much broader picture of Earth's history. Native-born Chinese paleontologists greatly expanded the breadth of paleontology in the 20th century and continue this trend today, both within and outside of China. The role of international collaborations, increased funding from government sources, and, of course, the incredibly rich paleontological flora and fauna in Chinese deposits has greatly expanded paleontology of all kinds, and will, no doubt, continue this trend into the foreseeable future.

Monson and colleagues [23] highlight various modern approaches to traditional paleontological questions, pointing out the expansion of data derived from CT imaging (including synchrotron imaging) to achieve 3D reconstruction of fossil data, and non-destructive means to study fossil histology as well. These authors describe how combining quantitative genetics and developmental biology approaches allows us to incorporate genotype:phenotype mapping to address morphological variation and its significance in informing phylogenetic history and the role of selection, and provide a case study using these tools.

The presence of endogenous molecules, including protein, DNA, various pigments, and biomarkers, is becoming an increasingly important aspect of paleontological studies, and new technology continues to push back the time limit for the preservation of these important molecules. New methods are being used to test hypotheses rooted in analyses of DNA. Churchill et al. [29] seeks to test the idea of interbreeding between Neandertal and modern humans in Europe and Asia, proposed in earlier DNA studies [30,31] by comparing facial size and shape parameters that may reflect the expression of Neandertal genes, using morphometric techniques.

A wide array of these new technologies are applied to address biomolecular remnants in bones as young as 350 years [32] and as old as the Carboniferous [33]. Phylogenetically and physiologically informative tissues were probed by synchrotron [34] to support the previous identification of reproductive tissues in dinosaurs [35,36]. Technologies continue to broaden not only the type of questions to be asked, but the type of fossils we can analyze, from coprolites [33], teeth [37], and invertebrates [22,38,39] to dinosaurs [25,34,40–44], mammals [45], and our own lineage [29,32,46].

Finally, taphonomic reconstructions remain an important part of paleobiology, and this Special Issue includes multi-dimensional studies on taphonomy using rare earth element studies (REE) to trace the movement of pore waters through bone during fossilization to elucidate the mechanisms contributing to molecular preservation in various dinosaur bone and other fossils [40–43]. However, the recovery of proteins also requires a better understanding of taphonomic modifications, as noted in [37] and previously discussed in [47].

Actualistic taphonomy experiments inform on the modifications introduced during diagenesis, but also informs on possible preservation conditions, recognizing that Mesozoic conditions were very different than today, and may have facilitated preservation through an elevated microbial response to high atmospheric CO_2 [48]. Additionally, although exceptionally preserved tissues have long been the target of molecular studies on fossils, there is clearly more to the story, as illustrated by Colleary et al. [44], as not all exceptional fossils preserve endogenous biomolecules.

This Special Issue briefly reviews these cutting-edge technologies and their applications to fossil data in various case studies, indicating a new 'fossil renaissance' for understanding life on this planet, yielding robust data that may be applied to understanding where we are going by better understanding from where we have come.

Conflicts of Interest: The author declares no conflict of interest.

References

1. Padian, K.; Chiappe, L.M. The origin and early evolution of birds. *Biol. Rev.* **1998**, *73*, 1–42. [CrossRef]
2. Gauthier, J. Saurischian monophyly and the origin of birds. In *The Origin of Birds and the Evolution of Flight*; Padian, K., Ed.; Memoirs of the California Academy of Sciences; The Academy: San Francisco, CA, USA, 1986; Volume 8, pp. 1–56.
3. Padian, K.; Lindberg, D.R.; Polly, P.D. Cladistics and the fossil record: The uses of history. *Ann. Rev. Earth Planet. Sci.* **1994**, *22*, 63–91. [CrossRef]
4. Enlow, D.H.; Brown, S.O. A comparative histological study of fossil and recent bone tissues, Part I. *Tex. J. Sci.* **1958**, *7*, 187–230.
5. Reid, R.E.H. On supposed Haversian bone from the hadrosaur Anatosaurus, and the nature of compact bone in dinosaurs. *J. Paleontol.* **1985**, *59*, 140–148.
6. De Ricqles, A.J.; Squelettiques, E.F.; Ura, W.R.S.I.; Via, U.P.; Padian, K.; Horner, J.R.; Francillon-vieillot, H. Palaeohistology of the bones of pterosaurs and biomechanical implications. *Zool. J. Linn. Soc.* **2000**, *129*, 349–385. [CrossRef]
7. Horner, J.R.; Padian, K.; de Ricqlès, A.; de Ricqles, A. Comparative osteohistology of some embryonic and perinatal archosaurs: Developmental and behavioral implications for dinosaurs. *Paleobiology* **2001**, *27*, 39–58. [CrossRef]
8. De Ricqles, A. Some Remarks on Palaeohistology from a Comparative Evolutionary Point of View. In *Histology of Ancient Human Bone*; Grupe, G., Garland, A.N., Eds.; Springer: Berlin/Heidelberg, Germany, 1993; pp. 37–77.
9. De Ricqles, A.J. Tissue structures of dinosaur bone: Functional significance and possible relation to dinosaur physiology. In *A Cold Look at the Warm-Blooded Dinosaurs*; Thomas, D.K., Olson, E.C., Eds.; Westview Press: Boulder, CO, USA, 1980; pp. 104–140.
10. Höss, M. Neanderthal population genetics. *Nature* **2000**, *404*, 453–454. [CrossRef]
11. Noro, M.; Masuda, R.; Dubrovo, I.A.; Yoshida, M.C.; Kato, M. Molecular phylogenetic inference of the woolly mammoth Mammuthus primigenius, based on complete sequences of mitochondrial cytochrome b and 12S ribosomal RNA genes. *J. Mol. Evol.* **1998**, *46*, 314–326. [CrossRef] [PubMed]
12. Knapp, M.; Hofreiter, M. Ancient Human DNA: Phylogenetic Applications. *eLS* **2014**, 1–4. [CrossRef]
13. Hofreiter, M.; Collins, M.; Stewart, J.R. Ancient biomolecules in Quaternary palaeoecology. *Quat. Sci. Rev.* **2012**, *33*, 1–13. [CrossRef]
14. Lindahl, T. Instability and Decay of the Primary Structure of DNA. *Nature* **1993**, *362*, 709–715. [CrossRef] [PubMed]
15. Lindahl, T. Facts and artifacts of ancient DNA. *Cell* **1997**, *90*, 1–3. [CrossRef] [PubMed]
16. Kjaer, K.H.; Pedersen, M.W.; De Sanctis, B.; De Cahsan, B.; Michelsen, C.S.; Sand, K.K.; Jelavić, S.; Ruter, A.H.; Bonde, A.M.Z.; Kjeldsen, K.K.; et al. A 2-Million-year-old ecosystem in Greenland uncovered by Environmental DNA. *Nature* **2022**, *612*, 283–291. [CrossRef] [PubMed]
17. Demarchi, B.; Hall, S.; Roncal-Herrero, T.; Freeman, C.L.; Woolley, J.; Crisp, M.K.; Wilson, J.; Fotakis, A.; Fischer, R.; Kessler, B.M.; et al. Protein sequences bound to mineral surfaces persist into deep time. *Elife* **2016**, *5*, e17092. [CrossRef]
18. Cleland, T.P.; Schroeter, E.R.; Zamdborg, L.; Zheng, W.; Lee, J.E.; Tran, J.C.; Bern, M.; Duncan, M.B.; Lebleu, V.S.; Schweitzer, M.H.; et al. Mass spectrometry and antibody-based characterization of blood vessels from Brachylophosaurus canadensis. *J. Proteome Res.* **2015**, *14*, 5252–5262.
19. Schroeter, E.R.; Dehart, C.J.; Cleland, T.P.; Zheng, W.; Thomas, P.M.; Kelleher, N.L.; Bern, M.; Schweitzer, M.H. Expansion for the Brachylophosaurus canadensis Collagen i Sequence and Additional Evidence of the Preservation of Cretaceous Protein. *J. Proteome Res.* **2017**, *16*, 920–932. [CrossRef] [PubMed]
20. Abelson, P.H. Paleobiochemistry: Organic constituents of fossils. *Carnegie Inst. Wash. Yearb.* **1955**, *54*, 107–109.
21. Abelson, P.H. Amino acids in fossils. *Science* **1954**, *119*, 576.
22. Tihelka, E.; Howard, R.J.; Cai, C.; Lozano-Fernandez, J. Was There a Cambrian Explosion on Land? The Case of Arthropod Terrestrialization. *Biology* **2022**, *11*, 1516. [CrossRef]
23. Monson, T.A.; Brasil, M.F.; Mahaney, M.C.; Schmitt, C.A.; Taylor, C.E.; Hlusko, L.J. Keeping 21st Century Paleontology Grounded: Quantitative Genetic Analyses and Ancestral State Reconstruction Re-Emphasize the Essentiality of Fossils. *Biology* **2022**, *11*, 1218. [CrossRef] [PubMed]
24. Torres, J.M.; Borja, C.; Gibert, L.; Ribot, F.; Olivares, E.G. Twentieth-Century Paleoproteomics: Lessons from Venta Micena Fossils. *Biology* **2022**, *11*, 1184. [CrossRef] [PubMed]
25. Tahoun, M.; Engeser, M.; Namasivayam, V.; Sander, P.M.; Müller, C.E. Chemistry and Analysis of Organic Compounds in Dinosaurs. *Biology* **2022**, *11*, 670. [CrossRef] [PubMed]
26. López-Antoñanzas, R.; Mitchell, J.; Simões, T.R.; Condamine, F.L.; Aguilée, R.; Peláez-Campomanes, P.; Renaud, S.; Rolland, J.; Donoghue, P.C.J. Integrative Phylogenetics: Tools for Palaeontologists to Explore the Tree of Life. *Biology* **2022**, *11*, 1185. [CrossRef] [PubMed]
27. Tamborini, M. A Plea for a New Synthesis: From Twentieth-Century Paleobiology to Twenty-First-Century Paleontology and Back Again. *Biology* **2022**, *11*, 1120. [CrossRef] [PubMed]
28. Zhou, Z. The Rising of Paleontology in China: A Century-Long Road. *Biology* **2022**, *11*, 1104. [CrossRef]
29. Churchill, S.E.; Keys, K.; Ross, A.H. Midfacial Morphology and Neandertal–Modern Human Interbreeding. *Biology* **2022**, *11*, 1163. [CrossRef] [PubMed]
30. Krause, J.; Briggs, A.W.; Kircher, M.; Maricic, T.; Zwyns, N.; Derevianko, A.; Pääbo, S. A Complete mtDNA Genome of an Early Modern Human from Kostenki, Russia. *Curr. Biol.* **2010**, *20*, 231–236. [CrossRef]
31. Gokcumen, O. Archaic hominin introgression into modern human genomes. *Am. J. Phys. Anthr.* **2019**, *171*, 60–73. [CrossRef]

32. Unterberger, S.H.; Berger, C.; Schirmer, M.; Pallua, A.K.; Zelger, B.; Schäfer, G.; Kremser, C.; Degenhart, G.; Spiegl, H.; Erler, S.; et al. Morphological and Tissue Characterization with 3D Reconstruction of a 350-Year-Old Austrian Ardea purpurea Glacier Mummy. *Biology* **2023**, *12*, 114. [CrossRef]
33. Tripp, M.; Wiemann, J.; Brocks, J.; Mayer, P.; Schwark, L.; Grice, K. Fossil Biomarkers and Biosignatures Preserved in Coprolites Reveal Carnivorous Diets in the Carboniferous Mazon Creek Ecosystem. *Biology* **2022**, *11*, 1289. [CrossRef]
34. Anné, J.; Canoville, A.; Edwards, N.P.; Schweitzer, M.H.; Zanno, L.E. Independent Evidence for the Preservation of Endogenous Bone Biochemistry in a Specimen of Tyrannosaurus rex. *Biology* **2023**, *12*, 264. [CrossRef] [PubMed]
35. Schweitzer, M.H.; Zheng, W.; Zanno, L.; Werning, S.; Sugiyama, T. Chemistry supports the identification of gender-specific reproductive tissue in Tyrannosaurus rex. *Sci. Rep.* **2016**, *6*, 23099. [CrossRef] [PubMed]
36. Schweitzer, M.H.; Wittmeyer, J.L.; Horner, J.R. Gender-specific reproductive tissue in ratites and Tyrannosaurus rex. *Science* **2005**, *308*, 1456–1460. [CrossRef]
37. Pessoa-Lima, C.; Tostes-Figueiredo, J.; Macedo-Ribeiro, N.; Hsiou, A.S.; Muniz, F.P.; Maulin, J.A.; Franceschini-Santos, V.H.; de Sousa, F.B.; Barbosa, F.; Line, S.R.P.; et al. Structure and Chemical Composition of ca. 10-Million-Year-Old (Late Miocene of Western Amazon) and Present-Day Teeth of Related Species. *Biology* **2022**, *11*, 1636. [CrossRef] [PubMed]
38. Pakhnevich, A.; Nikolayev, D.; Lychagina, T. Crystallographic Texture of the Mineral Matter in the Bivalve Shells of Gryphaea dilatata Sowerby, 1816. *Biology* **2022**, *11*, 1300. [CrossRef]
39. Heingård, M.; Sjövall, P.; Schultz, B.P.; Sylvestersen, R.L.; Lindgren, J. Preservation and Taphonomy of Fossil Insects from the Earliest Eocene of Denmark. *Biology* **2022**, *11*, 395. [CrossRef]
40. Ullmann, P.V.; Ash, R.D.; Scannella, J.B. Taphonomic and Diagenetic Pathways to Protein Preservation, Part II: The Case of Brachylophosaurus canadensis Specimen MOR 2598. *Biology* **2022**, *11*, 1177. [CrossRef]
41. Ullmann, P.V.; Macauley, K.; Ash, R.D.; Shoup, B.; Scannella, J.B. Taphonomic and diagenetic pathways to protein preservation, part I: The case of tyrannosaurus rex specimen mor 1125. *Biology* **2021**, *10*, 1193. [CrossRef]
42. Voegele, K.K.; Boles, Z.M.; Ullmann, P.V.; Schroeter, E.R.; Zheng, W.; Lacovara, K.J. Soft Tissue and Biomolecular Preservation in Vertebrate Fossils from Glauconitic, Shallow Marine Sediments of the Hornerstown Formation, Edelman Fossil Park, New Jersey. *Biology* **2022**, *11*, 1161. [CrossRef]
43. Schroeter, E.R.; Ullmann, P.V.; Macauley, K.; Ash, R.D.; Zheng, W.; Schweitzer, M.H.; Lacovara, K.J. Soft-Tissue, Rare Earth Element, and Molecular Analyses of Dreadnoughtus schrani, an Exceptionally Complete Titanosaur from Argentina. *Biology* **2022**, *11*, 1158. [CrossRef]
44. Colleary, C.; O'Reilly, S.; Dolocan, A.; Toyoda, J.G.; Chu, R.K.; Tfaily, M.M.; Hochella, M.F.; Nesbitt, S.J. Using Macro- and Microscale Preservation in Vertebrate Fossils as Predictors for Molecular Preservation in Fluvial Environments. *Biology* **2022**, *11*, 1304. [CrossRef]
45. Cirilli, O.; Machado, H.; Arroyo-cabrales, J.; Barr, C.I.; Davis, E.; Jass, C.N.; Jukar, A.M.; Landry, Z.; Mar, A.H.; Pandolfi, L.; et al. Evolution of the Family Equidae, Subfamily Equinae, in North, Central and South America, Eurasia and Africa during the Plio-Pleistocene. *Biology* **2022**, *11*, 1258. [CrossRef] [PubMed]
46. Garralda, M.D.; Weiner, S.; Arensburg, B.; Maureille, B.; Vandermeersch, B. Dental Paleobiology in a Juvenile Neanderthal (Combe-Grenal, Southwestern France). *Biology* **2022**, *11*, 1352. [CrossRef] [PubMed]
47. Cleland, T.P.; Schroeter, E.R.; Colleary, C. Diagenetiforms: A new term to explain protein changes as a result of diagenesis in paleoproteomics. *J. Proteom.* **2021**, *230*, 103992. [CrossRef] [PubMed]
48. Schweitzer, M.H.; Zheng, W.; Equall, N. Environmental Factors Affecting Feather Taphonomy. *Biology* **2022**, *11*, 703. [CrossRef] [PubMed]

Disclaimer/Publisher's Note: The statements, opinions and data contained in all publications are solely those of the individual author(s) and contributor(s) and not of MDPI and/or the editor(s). MDPI and/or the editor(s) disclaim responsibility for any injury to people or property resulting from any ideas, methods, instructions or products referred to in the content.

biology

Article

Keeping 21st Century Paleontology Grounded: Quantitative Genetic Analyses and Ancestral State Reconstruction Re-Emphasize the Essentiality of Fossils

Tesla A. Monson [1], Marianne F. Brasil [2,3], Michael C. Mahaney [4], Christopher A. Schmitt [5], Catherine E. Taylor [3] and Leslea J. Hlusko [3,6,*]

[1] Department of Anthropology, Western Washington University, 516 High Street, Bellingham, WA 98225, USA
[2] Berkeley Geochronology Center, 2455 Ridge Road, Berkeley, CA 94709, USA
[3] Human Evolution Research Center, Valley Life Sciences Building, University of California Berkeley, MC-3140, Berkeley, CA 94720, USA
[4] Department of Human Genetics, South Texas Diabetes and Obesity Institute, University of Texas Rio Grande Valley School of Medicine, Brownsville, TX 78520, USA
[5] Department of Anthropology, Boston University, 232 Bay State Road, Boston, MA 02115, USA
[6] National Center for Research on Human Evolution (CENIEH), Paseo Sierra de Atapuerca 3, 09002 Burgos, Spain
[*] Correspondence: hlusko@berkeley.edu

Citation: Monson, T.A.; Brasil, M.F.; Mahaney, M.C.; Schmitt, C.A.; Taylor, C.E.; Hlusko, L.J. Keeping 21st Century Paleontology Grounded: Quantitative Genetic Analyses and Ancestral State Reconstruction Re-Emphasize the Essentiality of Fossils. *Biology* **2022**, *11*, 1218. https://doi.org/10.3390/biology11081218

Academic Editors: Mary H. Schweitzer and Ferhat Kaya

Received: 21 July 2022
Accepted: 10 August 2022
Published: 13 August 2022

Publisher's Note: MDPI stays neutral with regard to jurisdictional claims in published maps and institutional affiliations.

Copyright: © 2022 by the authors. Licensee MDPI, Basel, Switzerland. This article is an open access article distributed under the terms and conditions of the Creative Commons Attribution (CC BY) license (https://creativecommons.org/licenses/by/4.0/).

Simple Summary: Over the last two decades of biological research, our understanding of how genes determine dental development and variation has expanded greatly. Here, we explore how this new knowledge can be applied to the fossil record of cercopithecid monkeys. We compare a traditional paleontological method for assessing dental size variation with measurement approaches derived from quantitative genetics and developmental biology. We find that these new methods for assessing dental variation provide novel insight to the evolution of the cercopithecid monkey dentition, different from the insight provided by traditional size measurements. When we explore the variation of these traits in the cercopithecid fossil record, we find that the variation is outside the range predicted based on extant variation alone. Our 21st century biological approach to paleontology reveals that we have even more to learn from fossils than previously recognized.

Abstract: Advances in genetics and developmental biology are revealing the relationship between genotype and dental phenotype (G:P), providing new approaches for how paleontologists assess dental variation in the fossil record. Our aim was to understand how the method of trait definition influences the ability to reconstruct phylogenetic relationships and evolutionary history in the Cercopithecidae, the Linnaean Family of monkeys currently living in Africa and Asia. We compared the two-dimensional assessment of molar size (calculated as the mesiodistal length of the crown multiplied by the buccolingual breadth) to a trait that reflects developmental influences on molar development (the inhibitory cascade, IC) and two traits that reflect the genetic architecture of postcanine tooth size variation (defined through quantitative genetic analyses: MMC and PMM). All traits were significantly influenced by the additive effects of genes and had similarly high heritability estimates. The proportion of covariate effects was greater for two-dimensional size compared to the G:P-defined traits. IC and MMC both showed evidence of selection, suggesting that they result from the same genetic architecture. When compared to the fossil record, Ancestral State Reconstruction using extant taxa consistently underestimated MMC and PMM values, highlighting the necessity of fossil data for understanding evolutionary patterns in these traits. Given that G:P-defined dental traits may provide insight to biological mechanisms that reach far beyond the dentition, this new approach to fossil morphology has the potential to open an entirely new window onto extinct paleobiologies. Without the fossil record, we would not be able to grasp the full range of variation in those biological mechanisms that have existed throughout evolution.

Keywords: primates; Cercopithecidae; monkeys; genotype:phenotype mapping; evolution; dentition; phylogeny

1. Introduction

The most essential, core moment in paleontology is when someone notices a fossil as something other than a rock and collects it for scientific study. This event is often just a person walking across the landscape, scanning the ground for evidence of past life. While this simple act has been fundamentally the same for generations of paleontologists, the lead-up to that moment and the science that follows have evolved dramatically. The technological advances that have taken us from landline telephones to smartphones have similarly altered how the science of paleontology is conducted. We can see this in the way scientists discover fossil sites. Where fossiliferous sediments were once identified mostly by happenstance, aerial photography, then satellite imagery, and now remote sensing are common tools for field paleontologists [1–3]. As well, our protocols for the collection, inventory, and organization of fossils now rely on fine resolution GIS [4] and remote access to the internet [5].

The laboratory side of the science is also remarkably different from 20th century paleontology. Fossils are now imaged by laser scanners as well as through photography [6,7]. Quantification of those scanned surfaces can be performed in three-dimensions with thousands of points, opening the door for new analytical approaches to morphological variation [8,9] and enabling the digital reconstruction of crushed fossils [10]. With the application of computed tomography (CT), paleontologists can more readily study internal bony structures [11,12], giving them the ability to reconstruct soft-tissue anatomies [13,14]. CT scans have become an essential tool in the description of new fossils [15]. With a synchrotron, we can even see fossilized histology without mechanically damaging specimens [16,17]. Advances in geochemistry provide new insight into the evolution of dietary niches [18–21] and life history [22], not to mention the ability to geologically date fossils [23]. As well, of course, advances in artificial intelligence and machine learning have forever changed taphonomy [24,25], approaches to fieldwork [26,27], and trait analysis [28–30].

Paleontologists have also incorporated new knowledge from biology and genomics. As genomic sequencing became increasingly possible for a wide range of organisms, paleontologists began to combine morphological evidence from fossils with genomic data to reconstruct phylogenetic relationships [31–33].

Alongside the genomic revolution, there is another discipline in biology with significant implications for paleontology: elucidating the relationship between genotype and phenotype, often referred to as genotype:phenotype (G:P)-mapping. The insight that comes from G:P-mapping will fundamentally alter how we approach fossil morphologies in the 21st century and, consequently, improve our knowledge of the evolutionary past. To demonstrate this point, we investigated the insight that G:P-mapped dental traits bring to the African fossil record of monkeys (Primates: Cercopithecidae). We first used quantitative genetic analyses to assess the heritability and covariate effects on traditional measurements of tooth size and two types of G:P-mapped traits, one derived from developmental biology and the other from quantitative genetic analyses. We then compared how these traits vary across extant cercopithecids to test Hypothesis 1: G:P-mapped dental traits can provide evidence of phylogenetic history and selection, and therefore, are useful in paleontological investigations. We then focused on the traits defined through our quantitative genetic approach and explored how they vary in the fossil record to test Hypothesis 2: G:P-mapped traits reveal a range of morphological variation that cannot be predicted solely through extant variation.

2. Background: Traditional and G:P-Mapped Dental Traits

Paleontologists have long relied on the size of the postcanine teeth (especially the molars) to serve as a proxy for body size, to provide essential insight into taxonomy, and to observe patterns of evolution [34–37]. Tooth size is traditionally defined as the two-dimensional occlusal area of the crown, calculated by multiplying the mesiodistal length by the buccolingual breadth (Figure 1C). This trait has long been, and still is, an essential trait in mammalian paleontology.

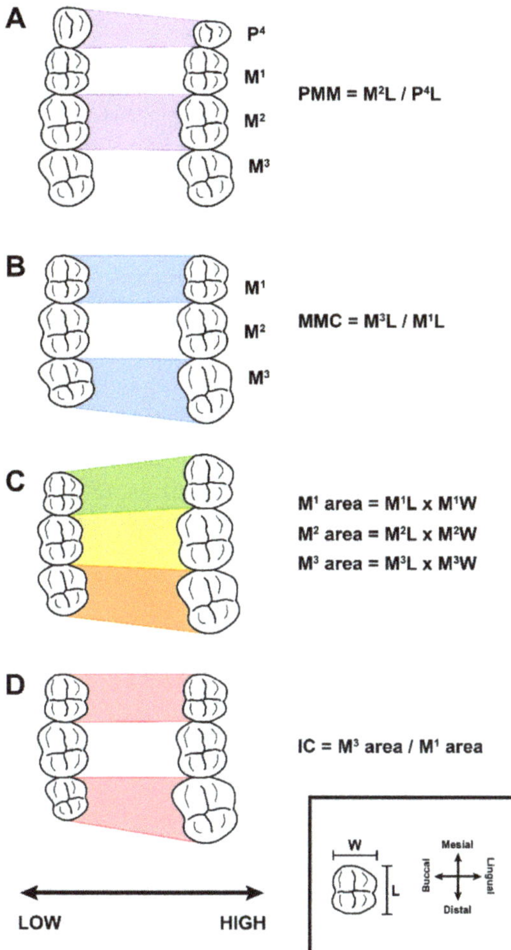

Figure 1. Illustration of the four maxillary dental traits investigated in this analysis. Each panel shows the right maxillary occlusal view of two extremes for one of the traits. Mesial is to the top, distal to the bottom, lingual to the right, and buccal to the left. The axis at the bottom of the figure orients the reader to how the morphology varies according to low and high values of the trait. (Panels (**A**,**B**)) demonstrate the two traits defined through quantitative genetic analyses, ratios that reflect the relative size variation between the premolar and molar genetic modules (PMM; panel (**A**)) and the relative sizes of the molars within the molar module (MMC; panel (**B**)). (Panel (**C**)) shows the traditional method for studying molar size variation within paleontology, by calculating a two-dimensional area of the occlusal view of the crown. (Panel (**D**)) shows the "inhibitory cascade" (IC) trait, defined through developmental gene expression studies of mice. See text for more detailed descriptions.

Over the last couple of decades, technological advances in the biological sciences have enabled scientists to probe the genetic influences on tooth size variation. There are two main avenues for G:P-mapping of dental variation: quantitative genetics and developmental biology. Quantitative genetic analyses approach the G:P-map through phenotypic variation, investigating how anatomical variation is inherited through family lineages. So long as the family structure within a population is known, any taxon can be studied, including large-bodied and long-lived animals such as primates. Because quantitative genetics reveals the genetic contributions to phenotypic variation within a population, this approach is particularly informative for Neogene paleontology, as population-level variation is most applicable to micro-evolutionary questions [38,39]. In contrast, developmental approaches involve the manipulation of embryogenesis and organogenesis to gain insight into the formation of the dentition from a fertilized egg. Consequently, experimental developmental biology is limited to animals that are amenable to being raised in a laboratory setting, who have short generation times, and/or for whom organs can be grown in culture, such as mice.

While there is a deep history of quantitative genetic research on the dentition [40], results from recent analyses have clarified that individual teeth are not genetically or developmentally independent structures, and that different aspects of a tooth are underlain by different genetic and non-genetic influences. For example, minor shape variants on the crown are genetically independent of tooth size [41]. Looking along the dental arcade, we see that the size of the incisors is genetically independent from the size of the premolars and molars (in baboons [42]; and macaques [43]; with some suggestive evidence in humans [44,45]; but see tamarins [46,47], and a different study on humans [48]), yet there is significant pleiotropy between postcanine teeth [42,43,46–48]. Evidence of pleiotropy indicates a genetic correlation, meaning that a significant proportion of the residual phenotypic variance in the two traits is due to the shared additive effects of the same gene or set of genes. Thus, evidence of pleiotropy helps elucidate the underlying genetic architecture. Shared genetic effects are not just limited to within the dentition. In baboons, for example, we also discovered that molar width is genetically correlated with body size (with more than 20% of the additive genetic covariance between these traits estimated to be due to the same gene or set of genes), but in surprising contrast, molar length is not [49]. While this exact correlation has not yet been explored in other primates, variation in crown area for humans has a positive correlation with the length of the dental arch, and a negative correlation with arch width, suggesting that tooth area and size dimensions within human dentitions are similarly not uniform [48]. Based on this genetic evidence, we now know that variation in the 2D occlusal area (as studied by paleontologists) reflects a range of underlying genetic effects related to body size and sex in addition to the genetic effects that pattern dental variation.

In order to make this quantitative genetic evidence translatable to paleontological research, Hlusko and colleagues [38] developed two dental traits that reflect the genetic architecture of the baboon dentition: the molar module component (MMC) and the premolar-molar module (PMM). Both traits are based on our quantitative genetic analyses of baboon mandibular dental variation. These analyses revealed that the mesiodistal lengths of the first, second, and third molars share a genetic correlation that is essentially 100%, indicating that first, second, and third molars are, genetically speaking, not the separate, independent structures that anatomists have long viewed them to be, but rather, one organ [42,50,51]. Consequently, the relative mesiodistal lengths of the first, second, and third molars represent components within one genetic module. As mentioned previously, molar buccolingual width has significant pleiotropic effects on body size [49]. Therefore, Hlusko et al. [38] proposed the ratio of the mesiodistal length of the third molar divided by the mesiodistal length of the first molar as a trait (MMC) that captures the genetic variation influencing tooth size variation within the molar module without the genetic effects that also influence body size (Figure 1B). Consequently, MMC is a more direct reflection of the underlying genetic architecture influencing molar size variation than two-dimensional crown area (length × width) because 2-dimensional crown area results from a combination of genetic effects that include those that influence body size.

We also defined PMM as a ratio that reflects the genetic correlation between the size of the fourth premolar relative to the size of the molar module [38]. Previous analyses demonstrated that the mesiodistal length of the fourth premolar has an overlapping, but not complete genetic correlation with the mesiodistal length of the molars [42,50,51]. PMM is the mesiodistal length of the second molar divided by the mesiodistal length of the fourth premolar (Figure 1A). As with MMC, we focused on the mesiodistal lengths in order to avoid conflating the genetic effects on body size with those that influence dental patterning.

The mandibular versions of MMC and PMM were first identified for cercopithecid monkeys and then expanded to apes, revealing an episode of selection during the Late Miocene [38]. While we do not yet know the genetic mechanisms that underlie PMM and MMC, we do know that these two ratios reflect a genetic architecture that does not simultaneously influence body size or sex, and that appears to primarily influence variation in the relative sizes of teeth in the postcanine dentition of catarrhine primates [38,52] and many other mammals [53,54].

The influence of developmental mechanisms on two-dimensional molar size variation has also been explored. Kavanagh and colleagues [55] reported evidence of an inhibitory cascade within the molar teeth of mice that can explain variation in the relative sizes of the first, second, and third molars. Through experimental manipulation of cultured tooth germs, they found that the timing of first molar initiation influences the initiation time and ultimate size of the second and third molars. For example, the removal of the first molar bud led to earlier initiation of the second and third molars, and these later-forming teeth grew larger. Kavanagh and colleagues [55] observed that across murine rodents, the size of the second molar always accounts for approximately one-third of the two-dimensional size of the molar row in occlusal view, and that the relative sizes of the first and third molar vary around this. From these observations, they [55] proposed that evolution follows this rule of one-third, and that first and third molar size can be predicted from each other. This model is referred to as the inhibitory cascade (IC) model. The model fits well with the phenotypic variation observed across murines [55] and has been supported in a range of other mammals (e.g., early mammaliaforms [56]; kangaroos [57]; many but not all South American ungulates [58]; and many but not all rodents [59]). However, the IC model does not fit the patterns of variation observed for anthropoid primates [60,61], humans [62], and some earlier hominids [63].

For Hypothesis 1, we explore both types of G:P-mapped traits in the maxillary dentitions, the IC (from developmental biology), and the MMC and PMM (from quantitative genetics). For Hypothesis 2, we focus on the quantitative genetics-derived traits, complementing the previously published investigation of the mandibular versions of PMM and MMC with the maxillary analyses.

3. Materials and Methods

Our analyses rely on dental linear metrics from three different samples described in detail in the following paragraphs. The quantitative genetic analyses were performed on data from 611 individuals within a captive pedigreed population of *Papio hamadryas* baboons. The extant, neontological analyses were performed using data from 825 museum skeletal specimens representing 13 genera within Cercopithecidae. Finally, we augmented the data we collected from museum specimens with data culled from the published scientific literature to create a fossil dataset of 1,436 individuals from 17 genera representing the last 20 million years of cercopithecid evolution in Africa.

Sample 1, quantitative genetics: The baboons from which dental data used in our quantitative genetic analyses were obtained are members of a large, six-generation pedigree ($n = 2426$), developed and maintained at the Southwest National Primate Research Center (SNPRC) at the Texas Biomedical Research Institute (Texas Biomed) in San Antonio, Texas. The pedigree was genetically managed to minimize inbreeding, and ascertainment of animals for this study was random with respect to phenotype. We analyzed linear crown metric data for the maxillary fourth premolar and first, second, and third molars obtained from 611 members of the single, large, six-generation pedigree. The female to male sex ratio was approximately 2:1 and the mean age of the sample was approximately 16 years, with ages ranging from 8 to 32 years. All procedures involving animals were reviewed and approved by Texas Biomed's Institutional Animal Care and Use Committee. SNPRC facilities and animal use programs at Texas Biomed are accredited by the Association for Assessment and Accreditation of Laboratory Animal Care International, comply with all National Institutes of Health and U.S. Department of Agriculture guidelines, and are directed by Doctors of Veterinary Medicine.

Sample 2, extant variation: Our comparative sample of extant taxa includes 825 individuals (Table 1). Most of the extant comparative data were collected by the authors and have been included in previously published research [64]. This dataset builds on the published dataset [65].

Table 1. Taxonomic composition of the extant comparative dataset.

Subfamily	Tribe	Genus	Species	Number of Individuals
Cercopithecinae	Cercopithecini	*Cercopithecus*	*albogularis*	1
		Cercopithecus	*campbelli*	9
		Cercopithecus	*mitis*	95
		Chlorocebus	*aethiops*	28
		Erythrocebus	*patas*	2
	Papionini	*Cercocebus*	*atys*	4
		Cercocebus	*galeritus*	1
		Cercocebus	*torquatus*	20
		Lophocebus	*albigena*	3
		Macaca	*fascicularis*	98
		Macaca	*mulatta*	76
		Mandrillus	*leucophaeus*	1
		Mandrillus	*sphinx*	17
		Papio	*hamadryas*	127
		Theropithecus	*gelada*	10
Colobinae	Colobini	*Colobus*	*guereza*	125
		Nasalis	*larvatus*	30
		Piliocolobus	*badius*	15
	Presbytini	*Presbytis*	*melalophos*	83
		Presbytis	*rubicunda*	80
	TOTAL			825

Sample 3, extinct variation: Our comparative sample of fossil taxa includes 1436 individuals (Table 2). Fossil data include measurements collected by the authors, culled from published sources, and downloaded from PRImate Morphometrics Online (PRIMO). Data sources for each sample are specified in Table 2.

Table 2. Taxonomic composition of the fossil comparative dataset *.

Subfamily	Tribe	Genus	Species	Number of Individuals	Source
Cercopithecinae	Cercopithecini	*Cercopithecus*	sp. (Andalee)	30	1
		Cercopithecus	sp. (Upper Andalee)	5	1
		cf. *Chlorocebus*	Asbole	13	2
		cf. *Chlorocebus*	sp. (Chai Baro)	105	3
		cf. *Chlorocebus*	sp. (Faro Daba)	223	3
	Papionini	*Papio*	*hamadryas angusticeps*	12	4, 5
		Papio	*hamadryas robinsoni*	29	4, 6
		Papio	*hamadryas* ssp. (Asbole)	10	2
		Papio	*hamadryas* ssp. (Chai Baro)	143	3
		Papio	*hamadryas ursinus*	1	4
		Papio	*izodi*	7	4, 6, 7
		Parapapio	*broomi*	34	4, 6, 7
		Parapapio	*jonesi*	12	4, 7
		Parapapio	*whitei*	16	4, 6, 7
		Pliopapio	*alemui*	5	8
		Procercocebus	*antiquus*	8	6, 7
		Soromandrillus	*quadratirostris*	11	9
		Theropithecus	*oswaldi* cf. *darti*	124	10
		Theropithecus	*oswaldi darti*	4	7
		Theropithecus	*oswaldi leakeyi*	12	2, 11
		Theropithecus	*oswaldi oswaldi*	8	4
Colobinae	Colobini	*Cercopithecoides*	*kimeui*	12	9, 12
		Cercopithecoides	*meaveae*	2	9
		Cercopithecoides	*williamsi*	91	9
		Colobus	cf. *guereza* (Faro Daba)	360	3
		Colobus	sp. (Andalee)	31	1
		Colobus	sp. (Asbole)	47	2
		Colobus	sp. (Upper Andalee)	4	1
		Kuseracolobus	*aramisi*	5	8
		Kuseracolobus	*hafu*	14	13
		Libypithecus	*markgrafi*	3	9
		Microcolobus	*tugenensis*	1	9
		Paracolobus	*chemeroni*	1	9
		Paracolobus	*enkorikae*	8	14
		Paracolobus	*mutiwa*	22	9
		Rhinocolobus	*turkanaensis*	23	9
Victoriapithecinae		*Victoriapithecus*	*macinnesi*	40	15
			TOTAL	1436	

* Data sources: 1 [66]; 2 [67]; 3 (Authors measured at the National Museum of Ethiopia); 4 (Authors measured at the Ditsong Museum of Natural History); 5 [68]; 6 (Authors measured at the University of California Museum of Paleontology); 7 (Authors measured at University of the Witswatersrand); 8 [69]; 9 (PRIMO); 10 [70]; 11 [71]; 12 [72]; 13 [73]; 14 [74]; 15 [75].

Data collection: Tooth dimensions for the SNPRC baboons are described in Hlusko et al. [76]. For the other two samples, mesiodistal length and buccolingual breadth measurements were collected from the maxillary fourth premolar (P4) and the three maxillary molars (M1, M2, and M3) for each individual, for both left and right sides, following standard protocols (see [64]). For the measurements collected by our research team, we did not account for interstitial wear. For the data culled from other publications, we refer to those publications, noting that some authors do not explicitly state how they measured mesiodistal length on teeth with significant interstitial wear. We used these two linear measurements, mesiodistal length (L) and buccolingual breadth (W) (see inset of Figure 1), to calculate 2-dimensional occlusal area, MMC, PMM, and the IC (see Figure 1 for equations).

Abbreviations: Premolars are abbreviated as P, molars as M. The letter for the tooth (P or M) is followed by a number indicating tooth position. For example, M2 refers to the second molar. We are primarily focused on a discussion of maxillary molars in this manuscript. We specifically indicate if a measurement or tooth is from the mandibular dental arch in the text rather than through abbreviations.

Overview: In order to test Hypothesis 1, we first established that a significant proportion of the phenotypic variation in all of the six traits is attributable to the effects of genes, i.e., that all the traits are heritable. To do this, we estimated the heritability of the traits in the SNPRC baboons. We then assessed the variation of all six traits across a sample of extant cercopithecid monkeys and considered how they vary within a phylogenetic context through a phylogenetic ANOVA. We followed the ANOVA with an analysis to test whether the traits are phylogenetically conserved or show evidence of selection. For the test of Hypothesis 2, we focused on the two traits derived from quantitative genetics: PMM and MMC. We first reconstructed ancestral states (ASR) based on the phylogenetic relationships of the extant genera analyzed for Hypothesis 1. We then compared the ASR trait values derived from the extant taxa to the PMM and MMC values observed in the fossil record.

Quantitative genetic analyses: We conducted statistical genetic analyses using a maximum likelihood-based variance decomposition approach implemented in the computer package SOLAR ([77]; v 8.1.1, www.solar-eclipse-genetics.org). This approach partitions the observed covariance between individuals into genetic and environmental components. The variance components are additive, with the phenotypic variance (σ_P^2) being the sum of the genetic (σ_G^2) and environment (σ_E^2) variances. Estimates of heritability (h^2), the proportion of the phenotypic variance attributable to additive genetic effects, were obtained as:

$$h^2 = \sigma_G^2 / \sigma_P^2$$

Unless otherwise noted, all quantitative genetic analyses were conducted following inverse gaussian normalization of the residuals (trait values were adjusted for the mean effects of sex and/or age, the latter a rough proxy for wear, if significant). Significance of the maximum-likelihood estimates for heritability and other parameters was assessed by means of likelihood ratio tests [78]. The maximum likelihood for a general model in which all parameters were estimated was compared to that for restricted models in which the value of the parameter to be tested was held constant (value dependent on null hypothesis). Twice the difference in the log-likelihoods of the two models compared is distributed asymptotically approximately as either a $1/2:1/2$ mixture of χ^2 with a point mass at zero for tests of parameters such as h^2 for which a fixed value of zero in a restricted model is at a boundary of the parameter space or a χ^2 variate for tests of covariates for which zero is not a boundary value [79]. In both cases, degrees of freedom are obtained as the difference in the number of estimated parameters in the two models [79]. However, in tests of parameters such as h^2, where values may be fixed at a boundary of their parameter space in the null model, the appropriate significance level is obtained by halving the p-value [80].

Descriptive statistics: Statistical analyses were completed in the R statistical environment v3.2.2 [81]. We first calculated univariate descriptive statistics for the two-dimensional areas, IC, MMC, and PMM values for all taxa included in the study, using built-in functions in R. Kurtosis was calculated using the *moments* package in R [82]. We visualized the distribution of the MMC and PMM traits across taxa in R using the package *ggplot2* (v1.0.1; [83]).

Phylogenetic ANOVA: We conducted a phylogenetic ANOVA to investigate variation across cercopithecid genera using the aov.phylo function in *geiger* [84]. The phylogenetic ANOVA uses average species data to compare traits across genera. Analyses were run on left side maxillary data. When no left side data were available, the right side was included. All dental areas were geometric mean size-corrected prior to analysis. All other dental traits are unit-free ratios.

Phylogenetic analyses: For all phylogenetic analyses, we used a consensus molecular chronogram based on a Bayesian phylogenetic analysis of genetic data downloaded from the 10kTrees v.3 database, built using data from six autosomal genes and 11 mitochondrial genes sampled from GenBank [85]. *Presbytis rubicunda* is not available in the 10kTrees database, and so we added this taxon manually to the phylogeny in R using a branch length split age of 1.3 million years from *Presbytis melalophos* [38,86].

Test of phylogenetic signal and selection: We tested the phylogenetic signal of the dental traits with a Blomberg's K analysis using phylosignal in *picante* [87]. Blomberg's K tests whether a trait is present in closely related taxa more frequently than would be expected by Brownian motion [88]. The K value for a trait can be either less than 1, equal to 1, or greater than 1. A K value > 1 is generally interpreted as more phylogenetically conserved than expected under neutral Brownian motion, while a K value of 1 generally indicates Brownian evolution of the trait under drift. In contrast, K < 1 is generally interpreted as a trait that is phylogenetically conserved, although less so than expected under a Brownian model, suggesting that selection pressures may be influencing the distribution of the trait in ways that deviate from the pattern expected based on phylogeny (with K = 0 implying that a trait varies in a pattern completely unrelated to phylogeny). However, heterogeneous rates of genetic drift or rapid divergence between species can also result in low K values [88,89]. We used summary trait values for each species and compared average species values across genera.

Ancestral state reconstruction: To investigate how dental traits have evolved in cercopithecids, we generated a series of ancestral state reconstructions (ASR) using contMap in *phytools* [90], which maps continuous variables across a phylogeny. We quantified the estimated values at internal nodes using fastAnc in *phytools* [90], a function that generates maximum likelihood ancestral states for continuous traits.

4. Results

4.1. Test of Hypothesis 1: G:P-Mapped Dental Traits Can Provide Evidence of Phylogeny and Selection

The results of the quantitative genetic analyses are presented in Table 3. Statistically significant residual h^2 estimates, ranging from 0.611 to 0.728, were obtained for five of six two-dimensional areas, two on the left side and three on the right. Both sex and age exerted significant mean effects on the two left side 2-dimensional areas, while only sex influenced the three right side traits. These covariate effects were substantive, accounting for approximately 28% to 51% of the total phenotypic variance in these five 2-dimensional areas. These same analyses returned significant h^2 estimates (range: 0.491–0.604) for three of the six G:P-mapped traits: right IC, and right and left PMM, with sex being the lone significant covariate, accounting for approximately 2% to 9% of their total phenotypic variance.

The analyses did not return statistically significant heritability estimates for four phenotypes, three on the left side of the arch (M3 2D area, IC, MMC) and one on the right (MMC). Derivation of these traits was based on data from comparatively small numbers of animals: i.e., only 140 to 221 individuals of the more than 600 pedigreed baboons from which data were obtained for this study.

Extant variation descriptive statistics: Univariate statistics for the two-dimensional areas of M1, M2, and M3, and the G:P-mapped traits (IC, PMM, and MMC) are reported in Tables 4 and 5. These are based on the phenotypic observations of the taxa listed in Table 1. See Supplementary Table S1 for more detailed descriptive statistics (Table S1).

Table 3. Residual heritability estimates for the three types of maxillary dental traits: two-dimensional area, IC, MMC, and PMM *.

Trait	h^2	h^2 se	*p*-Value	Number of Individuals	Proportion of Variance Due to Covariates
LM1 2D area	0.611	0.110	<0.001	461	0.353 **
LM2 2D area	0.728	0.091	<0.001	537	0.413 **
LM3 2D area	0.261	0.190	0.260	221	0.490 *
RM1 2D area	0.703	0.132	<0.001	440	0.281 *
RM2 2D area	0.681	0.103	<0.001	531	0.366 *
RM3 2D area	0.726	0.275	0.004	171	0.508 *
L IC	0.181	0.156	0.100	170	0.103 *
R IC	0.604	0.333	0.006	127	0.094 *
L MMC	0.001	0.141	0.496	191	0.082 **
L PMM	0.491	0.093	<0.001	402	0.022 *
R MMC	0.238	0.228	0.096	140	0.044 *
R PMM	0.527	0.114	<0.001	380	0.030 *

* L = left; R = right; 2D = 2-dimensional; M1, 2 or 3 = first, second, or third molar; IC = inhibitory cascade trait; MMC = molar module component ratio; PMM = premolar-molar module ratio. * indicates sex only is a significant covariate. ** indicates sex and age are significant covariates. Shaded rows are statistically non-significant at $p < 0.05$.

Table 4. Descriptive statistics for the two-dimensional area traits *.

	M1 Area		M2 Area		M3 Area	
	Mean (n)	StDv	Mean (n)	StDv	Mean (n)	StDv
Extant Genera						
Cercocebus	59.17 (15)	4.79	70.55 (15)	6.91	65.30 (15)	8.64
Cercopithecus	32.75 (91)	5.75	40.44 (96)	5.16	31.59 (88)	4.90
Chlorocebus	34.56 (27)	9.13	42.58 (27)	10.88	35.08 (25)	12.47
Colobus	45.31 (110)	5.11	55.19 (119)	5.97	51.88 (105)	6.26
Erythrocebus	39.29 (1)	-	48.16 (1)	-	47.06 (1)	-
Lophocebus	47.32 (6)	2.21	58.06 (6)	3.86	49.32 (5)	4.45
Macaca	44.26 (174)	8.30	58.91 (174)	12.73	55.94 (148)	12.40
Mandrillus	95.96 (18)	10.64	130.4 (18)	12.86	132.03 (18)	12.29
Nasalis	49.24 (30)	5.49	59.35 (30)	4.58	54.79 (30)	5.36
Papio	161.85 (84)	23.28	230.63 (106)	34.04	242.68 (104)	40.53
Piliocolobus	39.36 (15)	2.42	44.78 (15)	2.56	45.32 (15)	3.23
Presbytis	32.11 (151)	2.39	34.24 (153)	2.70	30.86 (145)	3.05
Theropithecus	96.15 (9)	11.11	140.87 (9)	11.61	145.15 (9)	11.66
Fossil Genera						
Cercopithecoides	78.92 (22)	16.74	104.41 (25)	22.32	98.85 (22)	19.59
Cercopithecus	34.59 (7)	3.41	45.77 (7)	3.58	36.00 (3)	10.11
cf. Chlorocebus	30.61 (48)	2.43	38.59 (45)	4.01	30.83 (37)	3.77
Colobus	43.59 (85)	5.23	51.64 (73)	7.44	48.50 (59)	6.38
Kuseracolobus	95.88 (1)	-	103.4 (1)	-	76.54 (1)	-
Libypithecus	48.90 (2)	2.97	58.93 (2)	3.78	65.95 (2)	0.49
Papio	95.81 (37)	13.39	139.68 (42)	22.23	137.82 (33)	29.59
Paracolobus	84.95 (4)	27.92	127.95 (9)	31.10	129.37 (8)	39.24
Parapapio	97.06 (20)	16.92	135.16 (20)	24.73	127.91 (28)	23.93
Pliopapio	54.02 (1)	-	73.1 (1)	-	66.75 (1)	-
Procercocebus	97.83 (4)	7.80	134.44 (5)	10.40	120.29 (5)	11.99
Rhinocolobus	85.70 (4)	8.13	104.31 (3)	2.92	114.32 (4)	13.56
Soromandrillus	121.41 (4)	12.21	198.79 (6)	28.07	205.49 (6)	30.60
Theropithecus	1.34 (6)	0.03	1.34 (6)	0.03	1.34 (6)	0.03
Victoriapithecus	42.19 (8)	1.30	57.45 (9)	4.58	44.55 (5)	4.15

* M1, M2, and M3 areas refer to the two-dimensional area of the tooth in occlusal view, calculated as the mesiodistal length multiplied by the buccolingual breadth. See text for details and definitions. StDv = standard deviation. See Supplementary Table S1 for more extensive descriptive statistics.

Table 5. Descriptive statistics for the Genotype:Phenotype (G:P)-mapped traits *.

	MMC		PMM		IC	
	Mean (n)	**StDv**	**Mean (n)**	**StDv**	**Mean (n)**	**StDv**
Extant Genera						
Cercocebus	1.05 (15)	0.05	1.42 (15)	0.06	1.10 (15)	0.10
Cercopithecus	0.95 (92)	0.06	1.49 (103	0.08	0.97 (76)	0.09
Chlorocebus	0.96 (24)	0.07	1.48 (26)	0.10	1.02 (24)	0.13
Colobus	1.08 (118)	0.06	1.47 (120)	0.07	1.15 (93)	0.10
Erythrocebus	1.04 (1)	-	1.48 (1)	-	1.20 (1)	-
Lophocebus	0.98 (5)	0.05	1.49 (6)	0.03	1.04 (5)	0.11
Macaca	1.12 (149)	0.08	1.59 (174)	0.09	1.26 (148)	0.14
Mandrillus	1.16 (18)	0.05	1.45 (18)	0.08	1.38 (18)	0.12
Nasalis	1.09 (30)	0.05	1.57 (29)	0.08	1.12 (30)	0.10
Papio	1.22 (86)	0.07	1.65 (99)	0.09	1.46 (71)	0.13
Piliocolobus	1.08 (15)	0.05	1.47 (15)	0.04	1.15 (15)	0.09
Presbytis	0.97 (151)	0.05	1.41 (160)	0.07	0.96 (138)	0.08
Theropithecus	1.25 (8)	0.05	1.88 (10)	0.06	1.53 (8)	0.06
Fossil Genera						
Cercopithecoides	1.18 (13)	0.15	1.64 (18)	0.15	1.27 (7)	0.25
Cercopithecus	0.92 (2)	0.03	1.51 (4)	0.15	0.85 (2)	0.04
cf. Chlorocebus	0.98(26)	0.07	1.55 (40)	0.11	1.01 (22)	0.09
Colobus	1.07 (55)	0.09	1.44 (66)	0.08	1.12 (46)	0.10
Kuseracolobus	1.17 (1)	-	1.65 (2)	0.14	-	-
Libypithecus	1.14 (2)	0.06	1.41 (2)	0.10	1.35 (2)	0.09
Papio	1.20 (32)	0.10	1.72 (36)	0.14	1.42 (21)	0.24
Paracolobus	1.16 (3)	0.20	1.54 (4)	0.10	1.26 (3)	0.23
Parapapio	1.18 (23)	0.13	1.74 (22)	0.11	1.34 (19)	0.21
Pliopapio	1.20 (1)	-	1.72 (1)	-	1.24 (1)	-
Procercocebus	1.13 (4)	0.03	1.61 (5)	0.03	1.42 (2)	0.19
Rhinocolobus	1.19 (3)	0.07	1.43 (3)	0.07	1.26 (2)	0.16
Soromandrillus	1.28 (6)	0.05	1.78 (6)	0.13	1.55 (4)	0.10
Theropithecus	1.34 (6)	0.03	1.34 (6)	0.03	1.62 (5)	0.06
Victoriapithecus	1.01 (4)	0.05	1.59 (2)	0.08	1.05 (3)	0.12

* MMC = molar module component; PMM = premolar-molar module; IC = inhibitory cascade. See text for details and definitions. StDv = standard deviation. See Supplementary Table S1 for more extensive descriptive statistics.

Phylogenetic ANOVA: Results from the phylogenetic ANOVA are presented in Table 6. The summary *p*-values indicate that all six traits differ significantly across the genera included in the analyses. The *p*-values for each genus are also presented. For two-dimensional areas, *Nasalis, Colobus, Macaca, Lophocebus,* and *Erythrocebus* are not different from the pooled value of the trait across all the extant genera. *Piliocolobus* is only statistically different for the M2. *Chlorocebus* is only statistically different for the M2 and M3 two-dimensional areas. IC and MMC results are identical, demonstrating that *Cercopithecus, Mandrillus, Papio,* and *Theropithecus* are statistically significantly different from the pooled values of IC and MMC. PMM differentiates most of the papionins (*Macaca, Papio,* and *Theropithecus*) as well as the colobine *Nasalis* from the other genera.

Table 6. Phylogenetic ANOVA results for extant genera *.

		M1 2D Area	M2 2D Area	M3 2D Area	IC	MMC	PMM
Summary *p*-Value		0.0004	<0.0001	<0.0001	0.003	0.004	0.009
	Genera:						
Colobines	*Presbytis*	**0.027 ***	**0.004 ****	**0.003 ****	0.078	0.100	0.765
	Nasalis	0.560	0.381	0.436	0.967	0.385	**0.029 ***
	Piliocolobus	0.153	**0.041 ***	0.091	0.663	0.503	0.270
	Colobus	0.346	0.206	0.277	0.659	0.510	0.252
Papionins	*Macaca*	0.233	0.310	0.474	0.075	0.088	**0.006 ****
	Papio	**<0.0001 *****	**<0.0001 *****	**<0.0001 *****	**0.007 ****	**0.009 ****	**0.005 ****
	Theropithecus	**0.006 ****	**0.0002 *****	**<0.0001 *****	**0.003 ****	**0.005 ****	**0.0002 *****
	Lophocebus	0.380	0.227	0.123	0.361	0.251	0.112
	Mandrillus	**0.008 ****	**0.0003 *****	**<0.0001 *****	**0.002 ****	**0.004 ****	0.288
Cercopith-ecins	*Chlorocebus*	0.077	**0.029 ***	**0.017 ***	0.308	0.111	0.238
	Erythrocebus	0.151	0.068	0.123	0.371	0.899	0.174
	Cercopithecus	**0.019 ***	**0.004 ****	**0.002 ****	**0.039 ***	**0.026 ***	0.091

* M1, M2, M3 refer to the first, second, and third molars. 2D refers to the two-dimensional area of the tooth crown in occlusal view, calculated as the mesiodistal length multiplied by the buccolingual breadth. IC is the 2-d area of the M3 divided by the 2D area of the M1. MMC is the mesiodistal length of the M3 divided by the mesiodistal length of the M1. PMM is the mesiodistal length of the M2 divided by the mesiodistal length of the P4 (fourth premolar). All area traits were geometric mean size-corrected before analysis. * indicates significance at $p < 0.05$. ** indicates significance at $p < 0.01$. *** indicates significance at $p < 0.001$.

Phylogenetic signal: Blomberg's K-values for the six traits are reported in Table 7. These all range between 0.625 and 0.673. Statistically non-significant *p*-values indicate that the trait is evolving neutrally under Brownian motion. IC is marginally significant at the $p = 0.05$ level, and therefore may indicate that IC variation observed across these extant taxa is the result of selection. MMC is statistically significant at the $p = 0.05$ level, providing a clear indication that selection has likely been operating on the relative mesiodistal lengths of the molars. Blomberg's K is a conservative test that is sensitive to sample size [88]. Additionally, variation in sample sizes across taxa, as well as variation in sample source populations within taxa, have been demonstrated to skew mean trait values used in these analyses, which can in turn skew results [91]. Sampling more extensively within sparsely sampled taxa, and across a broader range of primate taxa, may reveal stronger phylogenetic signal for these traits.

Table 7. Blomberg's K for the dental traits *.

Trait	K-Value	K *p*-Value
M1A	0.6595	0.070
M2A	0.6727	0.058
M3A	0.6606	0.055
IC	**0.6251**	**0.045**
MMC	**0.6324**	**0.035**
PMM	0.6379	0.059

* M1A, M2A, and M3A = two-dimensional area estimates for first, second, and third maxillary molars; MMC = molar module component; PMM = premolar-molar module; IC = inhibitory cascade. See text for trait definitions. Statistically significant estimates are in bold text. K-values greater than 1 indicate a strong phylogenetic signal. Non-significant *p*-values are interpreted as evolution under neutral genetic drift. For K-values that are significant at $p < 0.05$, the trait is interpreted to show evidence of selection.

4.2. Test of Hypothesis 2: G:P-Mapped Traits Reveal a Range of Morphological Variation That Cannot Be Predicted Solely through Extant Variation

Ancestral State Reconstruction (ASR): ASR estimates based on the extant genera listed in Table 1 are presented in Table 8, with nodes defined on the molecular phylogeny shown in Figure 2.

Table 8. Comparison of trait values from the Ancestral State Reconstruction (ASR) and Possible Fossil Representatives *.

ASR Node	ASR MMC	ASR PMM	Molecular Divergence	Possible Fossil Representative	MMC Value	PMM Value	Geological Age
20	1.07	1.52	16 Ma	*Victoriapithecus*	1.03	1.59	19–12.5 Ma
28	1.14	1.45	5 Ma	*Procercocebus*	1.13	1.62	2.5 Ma
28	1.14	1.45	5 Ma	*Soromandrillus*	1.28	1.78	2–3 Ma
29	1.10	1.43	2 Ma	*Procercocebus*	1.13	1.62	2.5 Ma
30	1.18	1.45	2.5 Ma	*Soromandrillus*	1.28	1.78	2–3 Ma
31	1.17	1.66	2 Ma	*Parapapio*	1.18	1.76	2–5 Ma
31	1.17	1.66	2 Ma	*Pliopapio*	1.20	1.72	4.4 Ma
35	1.07	1.49	<7.5 Ma	*Paracolobus*	1.16	1.50	2–6 Ma
35	1.07	1.49	<7.5 Ma	*Cercopithecoides*	1.18	1.67	2–5 Ma
35	1.07	1.49	<7.5 Ma	*Kuseracolobus*	1.17	1.75	4–4.4 Ma
35	1.07	1.49	<7.5 Ma	*Libypithecus*	1.14	1.41	5 Ma
ASR Tip	**MMC**	**PMM**	**Molecular Divergence**	**Possible Fossil Representative**	**MMC Value**	**PMM Value**	**Geological Age**
Chlorocebus aethiops	0.96	1.48	1 Ma	cf. *Chlorocebus* (Ethiopia)	0.98	1.56	100–600 ka
Colobus guereza	1.08	1.47	<1.6 Ma	*Colobus* sp. (Ethiopia)	1.06	1.44	100–600 ka

* Ma = million years ago; ka = thousand years ago; MMC and PMM are defined in the text; Molecular divergence estimates: Node 20 [92]; Node 28–31 [93]; Node 35 [94]. Geological dates for the fossils: *Victoriapithecus, Parapapio, Paracolobus, Cercopithecoides* [95]; *Procercocebus* [96]; *Soromandrillus* [97]; *Pliopapio* [66]; *Kuseracolobus* [73]; *Libypithecus* [98]; cf. *Chlorocebus* (authors, unpublished data); *Colobus* (authors, unpublished data).

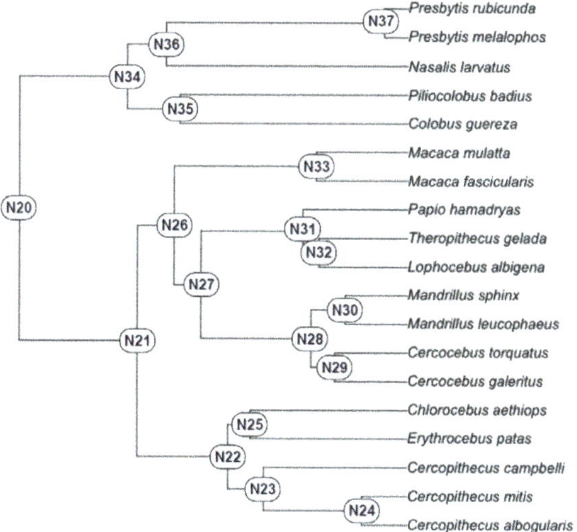

Figure 2. Molecular phylogeny of the extant cercopithecid genera included in this analysis with ASR nodes indicated. See Table 8 for ASR MMC and ASR PMM estimates.

Comparison to fossil data: In order to compare the ASR trait values to the anatomical variation observed in the fossil record, we compiled data for 17 fossil genera (Table 2) that could possibly be a fossil representative for one of the ASR nodes (Table 8). We include the molecular divergence date estimates that correspond to each node in the phylogeny. Next to these data, we list the possible fossil representative genus, along with the MMC and PMM values associated with that genus and the associated geological age range. Note that some fossil genera are potentially associated with more than one node. We present these data visually in Figure 3, along with the extant data for comparison. The averages for the fossil genera are indicated with a skull icon. Each fossil data point is linked with a double-ended arrow to the ASR node/estimate it may potentially represent, highlighting the difference between them. For both the PMM and MMC, the ASR estimates are usually lower than the values observed in the fossils. We present the absolute value of the difference between the ASR trait estimate and the fossil trait in Figure 4. Absolute value of the average difference between ASR MMC and fossil MMC is 0.066. Absolute value of the average difference between ASR PMM and fossil PMM is 0.162. At all of the time points represented by these data, the difference between the ASR value and the fossil value is most distinct for PMM.

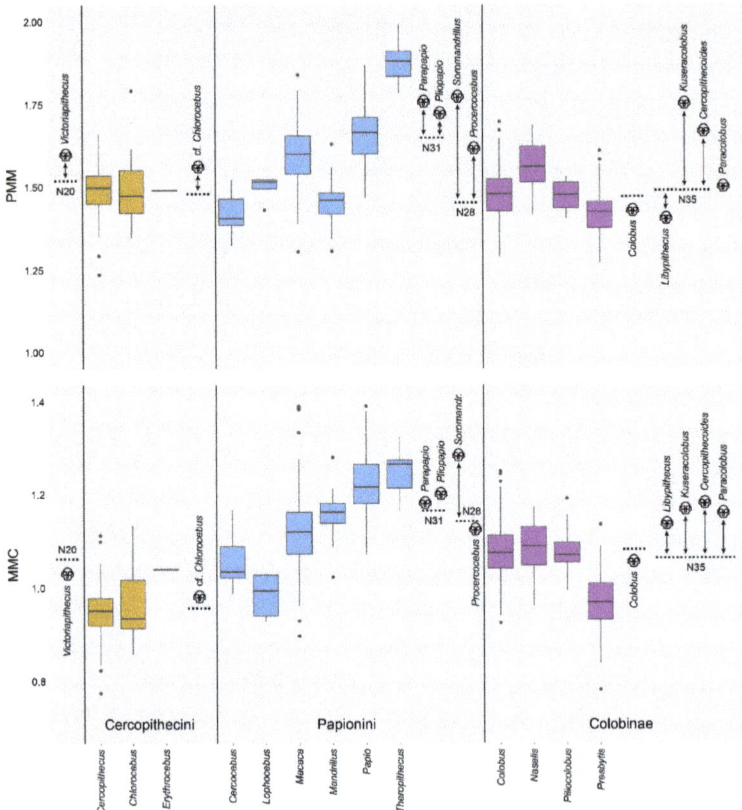

Figure 3. Box and whisker plots showing the range of variation for PMM and MMC within the sampled extant genera (labeled at the bottom of the figure). The genera are color-coded, with tribe Cercopithecini in gold, tribe Papionini in blue, and the subfamily Colobinae in purple. In addition to the extant data, we plot trait estimates for the Ancestral State Reconstruction (ASR) nodes as horizontal dotted lines, labeled with N and the number of the node. The possible fossil representatives for these nodes are plotted within the tribe or subfamily to which the fossil belongs. *Victoriapithecus*, on the far left, is widely thought to be ancestral to the split between the Colobinae and the Cercopithecinae (which includes Cercopoithecini and Papionini, shown here) [99]. Notice that for all but two of the PMM ASR-fossil pairs, the ASR estimate is lower than the observed fossil values. Similarly, for all but two of the MMC ASR-fossil pairs, the ASR estimate is also lower than the observed values. These differences are shown quantitatively in Figure 4.

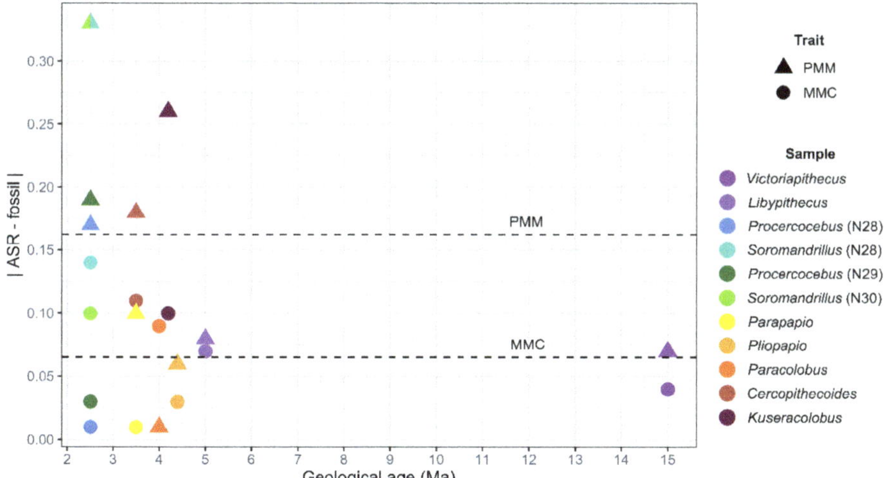

Figure 4. Bivariate plot of the difference between ASR trait values and fossil evidence for PMM and MMC. Geological age is shown on the *X*-axis. On the *Y*-axis, we report the absolute value of the difference between the ASR-estimated trait value for each node (molecular divergence) and the trait values observed for the African cercopithecid fossil genera in the same Tribe living near the time of the molecular divergence. The genera are shown in separate colors, defined in the key to the right. Triangles represent the PMM trait, and circles represent the MMC trait. The average difference for PMM is indicated by the top dashed line. The average difference for MMC is indicated by the lower dashed line. *Procercocebus* and *Soromandrillus* are included twice, as they could represent the ancestral morphology for nodes 28, 29, and 30.

5. Discussion

As advances in genetics and developmental biology make it possible to elucidate the relationship between genotype and phenotype (G:P), paleontologists are able to modify their approaches to anatomical variation accordingly. Our aim in this study was to understand how the method of trait definition influences the ability to reconstruct phylogenetic relationships and evolutionary history inCercopithecidae, the Linnaean Family of monkeys currently living in Africa and Asia. We compared one of the most classic traits in primate paleontology, two-dimensional occlusal tooth size (calculated as the mesiodistal length of the crown multiplied by the buccolingual breadth), to a trait that reflects developmental influences on molar development (the inhibitory cascade, IC [55]) and two traits that reflect the genetic architecture of postcanine tooth size variation defined through quantitative genetic analyses: MMC and PMM [38].

We first established that our maxillary trait types are highly heritable (albeit sensitive to low sample sizes), indicating that variation in tooth size, however it is assessed, is significantly influenced by genetic variation. This result was expected, as it builds on many decades of quantitative genetic analyses of dental variation demonstrating that tooth size is one of the most heritable phenotypes (e.g., [40]). At first glance, there are two caveats to this conclusion. First, while the right IC heritability estimate is significant, the left is not. We know from past analyses that antimeres (left and right side corresponding traits) generally return genetic correlations of one, indicating that they are influenced by identical genetic effects [41,42,50,51,100]. Therefore, we are confident that the left IC is also heritable, similarly to the right, and that our analysis is just underpowered by the small sample size. The second caveat is that we found that both left and ride side maxillary MMC traits returned non-significant heritability estimates. This was not unexpected given the small number of individuals (*n* = 191 for the left and 140 for the right) with data available. We are confident that this non-significant result is due to the analysis being underpowered rather than a true biological signal, given that the component dimensions when analyzed individually are highly heritable [42,50,51], and that the mandibular homologue of this trait is significantly heritable [38]. However, that said, further analyses with larger sample sizes are clearly needed.

These quantitative genetic analyses provide a good example of how challenging this approach can be, and why this type of research within evolutionary biology is only now becoming more common. Sampling is a significant challenge. For example, in our data set for the SNPRC baboons, composite traits reduce the number of individuals that can be included by a remarkable degree, especially for traits that include measurements of the third molar. We see this data reduction because the SNPRC measurements were collected from dental casts made of living animals. Consequently, the gumline often obscures the back edges of the third molar. Therefore, in a sample of 611 animals within the SNPRC colony, we only have M3 mesiodistal lengths for 140 (right side) and 191 (left side) individuals. Another significant factor in the success of quantitative genetic analyses is the location of the individuals within the pedigree. For example, even though we have more SNPRC baboon individuals available for the analysis of the left IC ($n = 170$) compared to the right ($n = 127$), only the right value returned a significant heritability estimate for IC. This is likely the result of where those individuals with data fall in the pedigree rather than evidence of a different biological signal. We are currently in the process of expanding the SNPRC dental data set and anticipate revisiting these analyses with a larger sample size.

Ever since Darwin [101], biologists have recognized that the heritable nature of phenotypic variation is central to the theory of evolution by natural selection. While all paleontologists appreciate this fact, ascertaining heritability is not simple. Even though the fundamental concept of quantitative genetics originated with Mendel, the ability to analyze the inheritance of normal, continuously varying traits across complex pedigrees was not possible until recently, as the algorithms are computationally intense and require modern computing technologies (for a history of approaches to dental variation: [40]). The modern concepts of evolutionary quantitative genetics were developed almost forty years ago [102–105], but it has been over the last 20 years that there has been an incredible expansion of quantitative genetic analyses being applied to evolutionary questions (examples of this research using primate models: [38,43,46–49,100,106–110]).

In addition to the high heritability estimates, we also find that G:P-mapped traits are phylogenetically conserved and show evidence of selection. ANOVA indicates that all six traits vary significantly across the cercopithecid clade, however, there are interesting differences in how variation in these traits is distributed across the Linnaean families, tribes, and genera. Within the colobines, *Presbytis* is significantly different in terms of two-dimensional molar size from other colobines, but not for the G:P-mapped traits. Previous researchers noted that the maxillary M3 morphology and eruption sequence of *Presbytis* sets it apart from other Asian colobines [111,112]. The lack of significant variation in the G:P-mapped traits for *Presbytis* poses the hypothesis that the distinct M3 morphology of *Presbytis* compared to other Asian colobines is not due to variation in the dental genetic architecture of PMM, MMC, or IC. Perhaps the unusual *Presbytis* dental morphology is related to body size, as the two-dimensional areas that are significantly different have pleiotropic effects with body size variation, possibly related to degrees of evolutionary dwarfism in this genus [64,113].

The ANOVA also revealed a distinct separation of three of the papionin genera: *Papio*, *Theropithecus*, and *Mandrillus*. These three genera are derived among the cercopithecids in having elongated muzzles, which is well-known to demonstrate positive allometry [114–117]. Looking more closely, we see that *Papio* and *Theropithecus* differ from the other genera in all six dental traits. However, *Mandrillus* differs in the two-dimensional area traits and the IC and MMC, but not PMM. Given that *Mandrillus* may be in a clade more closely related to *Macaca* than *Papio/Theropithecus/Lophocebus* [93,118], our results suggest that the phenotypic expression of MMC and IC are convergent in these two clades, and that the expressions of PMM differ despite the similarity in overall muzzle elongation. Previous in-depth analysis of the morphological variation of the faces of *Mandrillus* and *Papio* supports the interpretation that their elongated muzzles are convergent [115]. Our G:P analysis offers the first glimpse into the possible genetic mechanisms that may have been co-opted in this example of parallel evolution.

As described in the Introduction, the MMC and the IC are similar conceptually but distinct in their implementation and aims. The "inhibitory cascade" is a model proposed to explain the pattern of molar size variation observed across murines [55]. The IC model is based on the observation that the timing of initiation of the posterior molars is modulated by the growth of the first molar [55], confirming previous research. Lumsden and Osborn [119] and Lumsden [120] observed that all three molars develop from the ectopic transplantation of just the mouse M1 germ. By measuring the daily growth of mouse molars from 14 to 23 days post-fertilization, Sofaer [121] found compensatory changes in growth rate that seem to result from "some kind of competitive interaction" between the molars [121]. Lucas et al. [122] also observed that for 67 primate species, the size of the maxillary M2 is stable in accounting for 33–40% of the size of the molar row, with the M1 and M3 varying around the M2 in a compensatory manner. Kavanagh et al. [55] provided more experimental evidence for the mechanism first identified by the earlier investigators, gave it a name, and tested the model across the dental variation within Murinae. Since then, the authors have extended it to be a "simple rule govern[ing] the evolution and development of hominin tooth size" [61,123].

When the MMC and PMM were first proposed, we described the variation captured by MMC as likely due to the same developmental mechanisms underlying the IC [38]. However, we named our measurement in terms of the anatomical structures being assessed (the components of the molar genetic module) rather than by a hypothetical developmental mechanism [55], the genetics of which have not yet been established to our knowledge. Therefore, the MMC is not a developmental model likethe IC (contra [124]), but rather a measurement protocol for assessing molar size variation. As Lucas et al. [122] noted, the M1/M3 ratio is a measure of the shape of the tooth row. IC and MMC both capture this shape variation through ratios, but with a distinct difference. The IC is based on the two-dimensional size of M3 divided by the two-dimensional size of M1 (the traditional anatomical assessment of tooth size). In contrast, the MMC is the ratio of the length of M3 divided by the length of M1, which focuses the ratio on the genetic effects that result in variation in the relative lengths of the molars, separatefrom the genetic effects that influence molar width and also body size [38,49]. This distinction between the genetic architecture of length and width dimensions accords with Sofaer et al. [125]'s conclusion that mesiodistal lengths and buccolingual widths are influenced by different genetic and environmental effects, as well as Marshall and Corrucini [126]'s observation that molar lengths change much more slowly than widths in marsupial lineages with evolutionary dwarfing. Based on all of this evidence, the MMC is in all likelihood a more precise reflection of the genetic patterning mechanism that influences molar size proportions in cercopithecids, if not primates and other mammals more generally, compared to the IC.

Our analyses presented here further support the interpretation that the IC and MMC overlap in the genetic influences on molar size variation that they capture. For example, in the quantitative genetic analyses, the IC, MMC, and PMM all have much smaller covariate effects compared to two-dimensional areas (0.05 on average compared to 0.38 on average, respectively). Additionally, the IC and MMC have the same pattern of significance across genera in our ANOVA. This molar module pattern is distinct from the PMM, providing additional evidence that PMM is capturing a genetic mechanism distinct from that of the MMC (and IC). Our estimation of Blomberg's K also reveals similarities between MMC and IC. However, the results presented here suggest that our measurement protocol for MMC may well be a more specific reflection of the underlying genetic mechanism influencing molar proportions in cercopithecids compared to IC, given that we removed the known pleiotropic effects with body size. Further genetic analyses are needed to explore this with more certainty.

There has been a lot of enthusiasm for what G:P-mapped dental traits might offer for oral health [127] as well as paleontology (e.g., [128,129]). Evans (of [123]) even suggested that for hominids "This pattern is so strong, we can predict the size of the remaining four teeth without even finding the fossils!" (http://evomorph.org/inhibitory-cascade, accessed on 17 July 2022). With evolutionary biologists expressing this type of sentiment about the utility of fossils, it would not be unreasonable for funding agencies and budding scientists to ask if field paleontology is a thing of the past. Does the future of paleontology need new fossils?

In light of this question, our second major aim was to investigate G:P-mapped traits within the fossil record. For this, we focused on the maxillary MMC and PMM and compared computer-generated estimates of ancestral traits to the traits observed on fossils. We want to be up front about there being no clear consensus on direct ancestor-descendant relationships among cercopithecids over the last five million years, as the African cercopithecids from the Plio-Pleistocene are remarkably different from extant monkeys [95,99]. Consequently, new approaches are clearly needed, and G:P-mapped traits might offer novel insight into this murky evolutionary history.

Our comparisons of the ASR estimates with the fossil values unequivocally demonstrate that ASR based on extant data is compromised by the phenomenon of "the tyranny of the present". The lure of the extant comparative data available in museum collections unintentionally limits our expectations for what ancestral morphologies could have been. For example, we find that both MMC and PMM ASR estimates return values lower than what is observed in the fossil record penecontemporaneous with the ancestral nodes (Figure 3). PMM is underestimated twice as much as is MMC (Figure 4). Anecdotally, Figure 3 shows that ASR essentially averages the observed variation and is therefore unable to predict a wider range of variation than that of the input. While paleontologists are sometimes able to input fossil morphologies into their analyses to avoid this bias (e.g., [130]), this requires a high degree of confidence in the ancestor-descendant relationships, something we do not have for the Cercopithecidae. For monkeys, the modern bias in ASR would lead to the interpretation of the PMM of *Papio* and *Theropithecus* as newly derived, when we see that they actually have quite similar PMM values to early papionin genera such as *Parapapio*, *Pliopapio* and *Soromandrillus*. The high MMC values of the Miocene and Pliocene colobines also change how we view the evolutionary relationship of the African and Asian colobines. Knowing that earlier colobines in Africa had higher MMC values than both extant African and Asian colobines suggests that the African and Asian colobines evolved along the same MMC trajectory (reducing the MMC over time). None of these trends are visible when just size alone is considered.

The next step is to figure out what genetic mechanisms MMC (and IC) and PMM capture. We have a few hints. Previous analyses have shown that mandibular MMC is likely more evolutionarily conserved than PMM within catarrhine primates [38], across Boreoeutheria [53], between the different genera of megabats [54], and in the fossil record of the hominids [52]. Our results here for cercopithecids similarly demonstrate that the genetic mechanism captured by maxillary PMM appears to be more evolutionarily labile than maxillary MMC. We report elsewhere that variation in MMC may covary with prenatal growth rates [131], and therefore, MMC, a dental trait, may actually reflect life history variation rather than mastication and diet. If future analyses bolster this conclusion, G:P-mapping of dental variation opens a new window to the paleobiologies preserved in fossil morphology. But without the fossil evidence, we will never fully understand the range of variation that has existed over the evolutionary history of the Cercopithecidae. Therefore, the discovery of new fossils is not only still relevant, but even more revelatory as we apply 21st century methods to this most ancient data set.

Supplementary Materials: The following supporting information can be downloaded at: https://www.mdpi.com/article/10.3390/biology11081218/s1, Table S1: Descriptive Univariate Statistics.

Author Contributions: Concept development. Performed ANOVA, phylogenetic, and ASR analyses. Drafted an earlier version of the study. Edited the manuscript, T.A.M.; Concept development. Compiled fossil data. Created all figures. Edited the manuscript, M.F.B.; Concept development. Performed quantitative genetic analyses. Edited the manuscript, M.C.M.; Concept development. Ran phylogenetic analyses and consulted with T.A.M. Edited the manuscript, C.A.S.; Ran all univariate descriptive statistics. Created Table 4. Edited the manuscript, C.E.T.; Concept development. PI for the NSF grants that funded much of this research. Organized and wrote the manuscript, L.J.H. All authors have read and agreed to the published version of the manuscript.

Funding: Data collection was conducted over the last 24 years and supported variously by funding from the University of California Berkeley (through the Human Evolution Research Center, Department of Integrative Biology, Museum of Vertebrate Zoology, and Museum of Paleontology), Palaeontological Scientific Trust (PAST), Swiss National Science Foundation, The Leakey Foundation, University of Illinois Urbana-Champaign, Western Washington University, and U.S. National Science Foundation grants (BCS 0500179, 0616308, and 0130277) to L.J.H. and NSF DGE 1752814 to C.E.T. M.F.B. is supported by the John Templeton Foundation.

Institutional Review Board Statement: All procedures involving animals were reviewed and approved by Texas Biomed's Institutional Animal Care and Use Committee. SNPRC facilities and animal use programs at Texas Biomed are accredited by the Association for Assessment and Accreditation of Laboratory Animal Care International, comply with all National Institutes of Health and U.S. Department of Agriculture guidelines, and are directed by Doctors of Veterinary Medicine.

Informed Consent Statement: Not applicable.

Data Availability Statement: Data are available through the publications cited herein and the published dataset [65].

Acknowledgments: We are grateful to Mary Schweitzer and Ferhat Kaya for the invitation to contribute to this special issue, giving us the opportunity to place our research in this broader paleontological context. We thank the repositories and their personnel who provided access to extant and fossil specimens included in this study: American Museum of Natural History, Anthropology Department at the University of Zurich, Cleveland Museum of Natural History, National Museum of Ethiopia (Addis Ababa), University of California Museums of Paleontology (UCMP) and Vertebrate Zoology (MVZ), and the Smithsonian Institution's National Museum of Natural History. We thank the following people for assistance with data collection and/or project development: Julia Addiss, Stephen Akerson, Sarah Amugongo, Liz Bates, Josh Carlson, Selene Clay, Josh Cohen, Theresa Grieco, Anne Holden, Michaela Huffman, Daniel Lopez, Kurtis Morrish, Jackie Moustakas, Alicia Murua-Gonzalez, Danelle Pillie, Whitney Reiner, Oliver Rizk, Antoine Souron, Risa Takenaka, Kara Timmins, Mallory Watkins, Andrew Weitz, Madsen H. White, Tim White, Jeffrey Yoshihara, Sunwoo Yu, and Arta Zowghi. We thank the Texas Biomedical Research Institute (Texas Biomed) in San Antonio, Texas, supported by NIH National Center for Research Resources Grant P51 RR013986, for access and assistance with the quantitative genetic analysis of baboon dental variation. We also thank Eric Delson and colleagues for access to data downloaded from PRIMO, the NYCEP PRImate Morphology Online database (http://primo.nycep.org accessed on 1 November 2021).

Conflicts of Interest: The authors declare no conflict of interest.

References

1. Stucky, R.K.; Krishtalka, L. The application of geologic remote sensing to vertebrate biostratigraphy: General results from the Wind River basin, Wyoming. *Mt. Geol.* **1991**, *28*, 75–82.
2. Anemone, R.; Emerson, C.; Conroy, G. Finding fossils in new ways: An artificial neural network approach to predicting the location of productive fossil localities. *Evol. Anthropol. Issues News Rev.* **2011**, *20*, 169–180. [CrossRef] [PubMed]
3. Anemone, R.L.; Conroy, G.C. (Eds.) *New Geospatial Approaches to the Anthropological Sciences*; University of New Mexico Press: Albuquerque, NM, USA, 2018.
4. Hlusko, L.J. Geospatial approaches to hominid paleontology in Africa: What is old, what is new, and what doesn't change. In *New Geospatial Approaches in Anthropology*; Anemone, R., Conroy, G., Eds.; SAR Press: Santa Fe, NM, USA, 2018; pp. 39–58.

5. Reed, D.; Barr, W.A.; Mcpherron, S.P.; Bobe, R.; Geraads, D.; Wynn, J.G.; Alemseged, Z. Digital data collection in paleoanthropology. *Evol. Anthropol. Issues News Rev.* **2015**, *24*, 238–249. [CrossRef] [PubMed]

6. Cunningham, J.A.; Rahman, I.A.; Lautenschlager, S.; Rayfield, E.J.; Donoghue, P.C. A virtual world of paleontology. *Trends Ecol. Evol.* **2014**, *29*, 347–357. [CrossRef] [PubMed]

7. Otero, A.; Moreno, A.P.; Falkingham, P.L.; Cassini, G.; Ruella, A.; Militello, M.; Toledo, N. Three-dimensional image surface acquisition in vertebrate paleontology: A review of principal techniques. *Publ. Electrón. Asoc. Paleontol. Argent.* **2020**, *20*, 1–14. [CrossRef]

8. Rohlf, F.J.; Marcus, L.F. A revolution morphometrics. *Trends Ecol. Evol.* **1993**, *8*, 129–132. [CrossRef]

9. Mitteroecker, P.; Schaefer, K. Thirty years of geometric morphometrics: Achievements, challenges, and the ongoing quest for biological meaningfulness. *Yearb. Biol Anthr.* **2022**, *178*, 181–210. [CrossRef]

10. Cirilli, O.; Melchionna, M.; Serio, C.; Bernor, R.L.; Bukhsianidze, M.; Lordkipanidze, D.; Rook, L.; Profico, A.; Raia, P. Target deformation of the *Equus stenonis* holotype skull: A virtual reconstruction. *Front. Earth Sci.* **2020**, *8*, 247. [CrossRef]

11. Morimoto, N.; Kunimatsu, Y.; Nakatsukasa, M.; Ponce de Leon, M.S.; Zollikofer, C.P.; Ishida, H.; Sasaki, T.; Suwa, G. Variation of bony labyrinthine morphology in Mio— Plio— Pleistocene and modern anthropoids. *Am. J. Phys. Anthropol.* **2020**, *173*, 276–292. [CrossRef]

12. Monson, T.A.; Fecker, D.; Scherrer, M. Neutral evolution of human enamel–dentine junction morphology. *Proc. Natl. Acad. Sci. USA* **2020**, *117*, 26183–26189. [CrossRef]

13. Lautenschlager, S. Digital reconstruction of soft-tissue structures in fossils. *Paleontol. Soc. Pap.* **2016**, *22*, 101–117. [CrossRef]

14. Kochiyama, T.; Ogihara, N.; Tanabe, H.C.; Kondo, O.; Amano, H.; Hasegawa, K.; Suzuki, H.; Ponce de León, M.S.; Zollikofer, C.P.; Bastir, M.; et al. Reconstructing the Neanderthal brain using computational anatomy. *Sci. Rep.* **2018**, *8*, 6296. [CrossRef] [PubMed]

15. Haile-Selassie, Y.; Melillo, S.M.; Vazzana, A.; Benazzi, S.; Ryan, T.M. A 3.8-million-year-old hominin cranium from Woranso-Mille, Ethiopia. *Nature* **2019**, *573*, 214–219. [CrossRef] [PubMed]

16. Tafforeau, P.; Boistel, R.; Boller, E.; Bravin, A.; Brunet, M.; Chaimanee, Y.; Cloetens, P.; Feist, M.; Hoszowska, J.; Jaeger, J.J.; et al. Applications of X-ray synchrotron microtomography for non-destructive 3D studies of paleontological specimens. *Appl. Phys. A* **2006**, *83*, 195–202. [CrossRef]

17. Smith, T.M.; Tafforeau, P. New visions of dental tissue research: Tooth development, chemistry, and structure. *Evol. Anthropol. Issues News Rev.* **2008**, *17*, 213–226. [CrossRef]

18. Kimura, Y.; Jacobs, L.L.; Cerling, T.E.; Uno, K.T.; Ferguson, K.M.; Flynn, L.J.; Patnaik, R. Fossil mice and rats show isotopic evidence of niche partitioning and change in dental ecomorphology related to dietary shift in Late Miocene of Pakistan. *PLoS ONE* **2013**, *8*, e69308. [CrossRef]

19. Levin, N.E.; Haile-Selassie, Y.; Frost, S.R.; Saylor, B.Z. Dietary change among hominins and cercopithecids in Ethiopia during the early Pliocene. *Proc. Natl. Acad. Sci. USA* **2015**, *112*, 12304–12309. [CrossRef]

20. Ma, J.; Wang, Y.; Jin, C.; Yan, Y.; Qu, Y.; Hu, Y. Isotopic evidence of foraging ecology of Asian elephant (*Elephas maximus*) in South China during the Late Pleistocene. *Quat. Int.* **2017**, *443*, 160–167. [CrossRef]

21. Souron, A. Morphology, diet, and stable carbon isotopes: On the diet of *Theropithecus* and some limits of uniformitarianism in paleoecology. *Am. J. Phys. Anthropol.* **2018**, *166*, 261–267. [CrossRef]

22. Smith, T.M.; Austin, C.; Green, D.R.; Joannes-Boyau, R.; Bailey, S.; Dumitriu, D.; Fallon, S.; Grün, R.; James, H.F.; Moncel, M.H.; et al. Wintertime stress, nursing, and lead exposure in Neanderthal children. *Sci. Adv.* **2018**, *4*, eaau9483. [CrossRef]

23. Reynard, B.; Balter, V. Trace elements and their isotopes in bones and teeth: Diet, environments, diagenesis, and dating of archeological and paleontological samples. *Palaeogeogr. Palaeoclimatol. Palaeoecol.* **2014**, *416*, 4–16. [CrossRef]

24. Courtenay, L.A.; Huguet, R.; González-Aguilera, D.; Yravedra, J. A hybrid geometric morphometric deep learning approach for cut and trampling mark classification. *Appl. Sci.* **2019**, *10*, 150. [CrossRef]

25. Domínguez-Rodrigo, M.; Cifuentes-Alcobendas, G.; Jiménez-García, B.; Abellán, N.; Pizarro-Monzo, M.; Organista, E.; Baquedano, E. Artificial intelligence provides greater accuracy in the classification of modern and ancient bone surface modifications. *Sci. Rep.* **2020**, *10*, 18862. [CrossRef]

26. Martín-Perea, D.M.; Courtenay, L.A.; Domingo, M.S.; Morales, J. Application of artificially intelligent systems for the identification of discrete fossiliferous levels. *PeerJ* **2020**, *8*, e8767. [CrossRef] [PubMed]

27. Gorur, K.; Kaya Ozer, C.; Ozer, I.; Can Karaca, A.; Cetin, O.; Kocak, I. Species-level microfossil prediction for *Globotruncana* genus using machine learning models. *Arab. J. Sci. Eng.* **2022**, 1–18. [CrossRef]

28. Monson, T.A.; Armitage, D.W.; Hlusko, L.J. Using machine learning to classify extant apes and interpret the dental morphology of the chimpanzee-human last common ancestor. *PaleoBios* **2018**, *35*, 1–20. [CrossRef]

29. Püschel, T.A.; Marcé-Nogué, J.; Gladman, J.T.; Bobe, R.; Sellers, W.I. Inferring locomotor behaviours in Miocene New World monkeys using finite element analysis, geometric morphometrics and machine-learning classification techniques applied to talar morphology. *J. R. Soc. Interface* **2018**, *15*, 20180520. [CrossRef]

30. Millar, M. Predicting Theropod Hunting Tactics using Machine Learning. *Open Sci. J.* **2019**, *4*. [CrossRef]

31. Wiens, J.J. Paleontology, genomics, and combined-data phylogenetics: Can molecular data improve phylogeny estimation for fossil taxa? *Syst. Biol.* **2009**, *58*, 87–99. [CrossRef]

32. Geisler, J.H.; McGowen, M.R.; Yang, G.; Gatesy, J. A supermatrix analysis of genomic, morphological, and paleontological data from crown Cetacea. *BMC Evol. Biol.* **2011**, *11*, 112. [CrossRef]

33. O'Leary, M.A.; Bloch, J.I.; Flynn, J.J.; Gaudin, T.J.; Giallombardo, A.; Giannini, N.P.; Goldberg, S.L.; Kraatz, B.P.; Luo, Z.X.; Meng, J.; et al. The placental mammal ancestor and the post–K-Pg radiation of placentals. *Science* **2013**, *339*, 662–667. [CrossRef] [PubMed]
34. Weidenreich, F. *The Dentition of Sinanthropus Pekinensis: A Comparative Odontography of the Hominids*; Palaeontologia Sinica: New Series D, No. 1; Geological Survey of China: Beijing, China, 1937; Volume 101, pp. 1–180.
35. Gould, S.J. On the scaling of tooth size in mammals. *Am. Zool.* **1975**, *15*, 353–362. [CrossRef]
36. Fleagle, J.G.; McGraw, W.S. Skeletal and dental morphology supports diphyletic origin of baboons and mandrills. *Proc. Natl. Acad. Sci. USA* **1999**, *96*, 1157–1161. [CrossRef]
37. Gingerich, P.D. Environment and evolution through the Paleocene–Eocene thermal maximum. *Trends Ecol. Evol.* **2006**, *21*, 246–253. [CrossRef]
38. Hlusko, L.J.; Schmitt, C.A.; Monson, T.A.; Brasil, M.F.; Mahaney, M.C. The Integration of quantitative genetics, paleontology, and neontology reveals genetic underpinnings of primate dental evolution. *Proc. Natl. Acad. Sci. USA* **2016**, *113*, 9262–9267. [CrossRef]
39. Hlusko, L.J. Recent insights into the evolution of quantitative traits in non-human primates. *Curr. Opin. Genet. Dev.* **2018**, *53*, 15–20. [CrossRef]
40. Rizk, O.T.; Amugongo, S.; Mahaney, M.C.; Hlusko, L.J. The quantitative genetic analysis of primate dental variation: History of the approach and prospects for the future. In *Technique and Application in Dental Anthropology*; Irish, J.D., Nelson, G.C., Eds.; Cambridge University Press: Cambridge, CA, USA, 2008; ISBN 13: 9780521870610.
41. Hlusko, L.J.; Mahaney, M.C. Genetic contributions to expression of the baboon cingular remnant. *Arch. Oral Biol.* **2003**, *48*, 663–672. [CrossRef]
42. Hlusko, L.J.; Sage, R.D.; Mahaney, M.C. Evolution of modularity in the mammalian dentition: Mice and monkeys share a common dental genetic architecture. *J. Exp. Zool. Part B Mol. Dev. Evol.* **2011**, *316*, 21–49. [CrossRef]
43. Hardin, A.M. Genetic correlations in the rhesus macaque dentition. *J. Hum. Evol.* **2020**, *148*, 102873. [CrossRef]
44. Cunha, A.S.; Dos Santos, L.V.; Marañón-Vásquez, G.A.; Kirschneck, C.; Gerber, J.T.; Stuani, M.B.; Matsumoto, M.A.N.; Vieira, A.R.; Scariot, R.; Küchler, E.C. Genetic variants in tooth agenesis–related genes might be also involved in tooth size variations. *Clin. Oral Investig.* **2021**, *25*, 1307–1318. [CrossRef] [PubMed]
45. Stojanowski, C.M.; Paul, K.S.; Seidel, A.C.; Duncan, W.N.; Guatelli-Steinberg, D. Heritability and genetic integration of tooth size in the South Carolina Gullah. *Am. J. Phys. Anthropol.* **2017**, *164*, 505–521. [CrossRef] [PubMed]
46. Hardin, A.M. Genetic contributions to dental dimensions in brown-mantled tamarins (*Saguinus fuscicollis*) and rhesus macaques (*Macaca mulatta*). *Am. J. Phys. Anthropol.* **2019**, *168*, 292–302. [CrossRef] [PubMed]
47. Hardin, A.M. Genetic correlations in the dental dimensions of *Saguinus fuscicollis*. *Am. J. Phys. Anthropol.* **2019**, *169*, 557–566. [CrossRef]
48. Hardin, A.M.; Knigge, R.P.; Duren, D.L.; Williams-Blangero, S.; Subedi, J.; Mahaney, M.C.; Sherwood, R.J. Genetic influences on dentognathic morphology in the Jirel population of Nepal. *Anat. Rec.* 2022; *Early View*. [CrossRef]
49. Hlusko, L.J.; Lease, L.R.; Mahaney, M.C. The evolution of genetically correlated traits: Tooth size and body size in baboons. *Am. J. Phys. Anthropol.* **2006**, *131*, 420–427. [CrossRef] [PubMed]
50. Hlusko, L.J.; Mahaney, M.C. Of mice and monkeys: Quantitative genetic analyses of size variation along the dental arcade. In *Dental Perspectives on Human Evolution: State of the Art Research in Dental Paleoanthropology*; Bailey, S., Hublin, J.-J., Eds.; Springer: Dordrecht, The Netherlands, 2007; pp. 237–245.
51. Hlusko, L.J.; Mahaney, M.C. The baboon model for dental development. In *The Baboon in Biomedical Research*; Springer: New York, NY, USA, 2009; pp. 207–223.
52. Brasil, M.F.; Monson, T.A.; Schmitt, C.A.; Hlusko, L.J. A genotype:phenotype approach to testing taxonomic hypotheses in hominids. *Sci. Nat.* **2020**, *107*, 40. [CrossRef] [PubMed]
53. Monson, T.A.; Boisserie, J.-R.; Brasil, M.F.; Clay, S.; Dvoretzky, R.; Ravindramurthy, S.R.; Schmitt, C.A.; Souron, A.; Takenaka, R.; Ungar, P.; et al. Evidence of strong stabilizing effects on the evolution of boreoeutherian (Mammalia) dental proportions. *Ecol. Evol.* **2019**, *9*, 7597–7612. [CrossRef] [PubMed]
54. Zuercher, M.E.; Monson, T.A.; Dvoretzky, R.R.; Ravindramurthy, S.; Hlusko, L.J. Dental variation in megabats (Chiroptera: Pteropodidae): Tooth metrics correlate with body size and tooth proportions reflect phylogeny. *J. Mamm. Evol.* **2021**, *28*, 543–558. [CrossRef]
55. Kavanagh, K.D.; Evans, A.R.; Jernvall, J. Predicting evolutionary patterns of mammalian teeth from development. *Nature* **2007**, *449*, 427–432. [CrossRef]
56. Halliday, T.J.; Goswami, A. Testing the inhibitory cascade model in Mesozoic and Cenozoic mammaliaforms. *BMC Evol. Biol.* **2013**, *13*, 79. [CrossRef] [PubMed]
57. Couzens, A.M.; Evans, A.R.; Skinner, M.M.; Prideaux, G.J. The role of inhibitory dynamics in the loss and reemergence of macropodoid tooth traits. *Evolution* **2016**, *70*, 568–585. [CrossRef] [PubMed]
58. Wilson, L.; Madden, R.; Kay, R.; Sánchez-Villagra, M. Testing a developmental model in the fossil record: Molar proportions in South American ungulates. *Paleobiology* **2012**, *38*, 308–321. [CrossRef]
59. Labonne, G.; Laffont, R.; Renvoise, E.; Jebrane, A.; Labruère, C.; Chateau-Smith, C.; Navarro, N.; Montuire, S. When less means more: Evolutionary and developmental hypotheses in rodent molars. *J. Evol. Biol.* **2012**, *25*, 2102–2111. [CrossRef]
60. Carter, K.E.; Worthington, S. The evolution of anthropoid molar proportions. *BMC Evol. Biol.* **2016**, *16*, 110. [CrossRef]

61. Roseman, C.C.; Delezene, L.K. The inhibitory cascade model is not a good predictor of molar size covariation. *Evol. Biol.* **2019**, *46*, 229–238. [CrossRef]

62. Bermúdez de Castro, J.M.; Modesto-Mata, M.; García-Campos, C.; Sarmiento, S.; Martín-Francés, L.; Martínez de Pinillos, M.; Martinón-Torres, M. Testing the inhibitory cascade model in a recent human sample. *J. Anat.* **2021**, *239*, 1170–1181. [CrossRef]

63. Bermúdez de Castro, J.M.; Modesto-Mata, M.; Martín-Francés, L.; García-Campos, C.; Martínez de Pinillos, M.; Martinón-Torres, M. Testing the inhibitory cascade model in the Middle Pleistocene Sima de los Huesos (Sierra de Atapuerca, Spain) hominin sample. *J. Anat.* **2021**, *238*, 173–184. [CrossRef]

64. Grieco, T.M.; Rizk, O.T.; Hlusko, L.J. A modular framework characterizes micro- and macroevolution of Old World monkey dentitions. *Evolution* **2013**, *67*, 241–259. [CrossRef]

65. Grieco, T.M.; Rizk, O.T.; Hlusko, L.J. Data from: A Modular Framework Characterizes Micro- and Macroevolution of Old World Monkey Dentitions. Dryad, Dataset. 2012. Available online: https://datadryad.org/stash/dataset/doi:10.5061/dryad.693j8 (accessed on 1 January 2022). [CrossRef]

66. Frost, S.R. Fossil Cercopithecidae from the Afar Depression, Ethiopia: Species Systematics and Comparison to the Turkana Basin. Doctoral Dissertation, City University of New York, New York, NY, USA, 2001.

67. Frost, S.R.; Alemseged, Z. Middle Pleistocene fossil Cercopithecidae from Asbole, Afar Region, Ethiopia. *J. Hum. Evol.* **2007**, *53*, 227–259. [CrossRef] [PubMed]

68. Freedman, L. The fossil Cercopithecoidea of South Africa. *Ann. Transvaal Mus.* **1957**, *23*, 121–262.

69. Frost, S.R.; Haile-Selassie, Y.; Hlusko, L.J. Chapter 6 Cercopithecidae. In *Ardipithecus kadabba: Late Miocene Evidence from Middle Awash, Ethiopia*; Haile-Selassie, Y., WoldeGabriel, G., Eds.; University of California Press: Berkeley, CA, USA, 2009.

70. Frost, S.R.; Jablonski, N.G.; Haile-Selassie, Y. Early Pliocene Cercopithecidae from Woranso-Mille (Central Afar, Ethiopia) and the origins of the *Theropithecus oswaldi* lineage. *J. Hum. Evol.* **2014**, *76*, 39–53. [CrossRef] [PubMed]

71. Frost, S.R. Fossil Cercopithecidae from the Middle Pleistocene Dawaitoli Formation, Middle Awash Valley, Afar Region, Ethiopia. *Am. J. Phys. Anthropol.* **2007**, *134*, 460–471. [CrossRef]

72. Frost, S.R.; Marcus, L.F.; Bookstein, F.L.; Reddy, D.P.; Delson, E. Cranial allometry, phylogeography, and systematics of large-bodied papionins (Primates: Cercopithecinae) inferred from geometric morphometric analysis of landmark data. *Anat. Rec.* **2003**, *275A*, 1048–1072. [CrossRef]

73. Hlusko, L.J. A new large Pliocene colobine species (Mammalia: Primates) from Asa Issie, Ethiopia. *Geobios* **2006**, *39*, 57–69. [CrossRef]

74. Hlusko, L.J. A new late Miocene species of *Paracolobus* and other Cercopithecoidea (Mammalia: Primates) fossils from Lemudong'o, Kenya. *Kirtlandia* **2007**, *56*, 72–85.

75. Benefit, B.K. The permanent dentition and phylogenetic position of *Victoriapithecus* from Maboko Island, Kenya. *J. Hum. Evol.* **1993**, *25*, 83–172. [CrossRef]

76. Hlusko, L.J.; Weiss, K.M.; Mahaney, M.C. Statistical genetic comparison of two techniques for assessing molar crown size in pedigreed baboons. *Am. J. Phys. Anthropol.* **2002**, *117*, 182–189. [CrossRef]

77. Almasy, L.; Blangero, J. Multipoint quantitative-trait linkage analysis in general pedigrees. *Am. J. Hum. Genet.* **1998**, *62*, 1198–1211. [CrossRef]

78. Edwards, A.W.F. *Likelihood*; Johns Hopkins University Press: Baltimore, MD, USA, 1992.

79. Hopper, J.L.; Mathews, J.D. Extensions to multivariate normal models for pedigree analysis. *Ann. Hum. Genet.* **1982**, *46*, 373–383. [CrossRef] [PubMed]

80. Boehnke, M.; Moll, P.P.; Kottke, B.A.; Weidman, W.H. Partitioning the variability of fasting plasma glucose levels in pedigrees. Genetic and environmental factors. *Am. J. Epidemiol.* **1987**, *125*, 679–689. [CrossRef] [PubMed]

81. R Core Team. R: A Language and Environment for Statistical Computing. Available online: https://www.R-project.org/ (accessed on 1 January 2022).

82. Komsta, L.; Novomestky, F. Moments: Moments, Cumulants, Skewness, Kurtosis and Related Tests. R. R Package Version 0.14.1. Available online: https://cran.r-project.org/web/packages/moments/index.html (accessed on 1 January 2022).

83. Wickham, H. *ggplot2: Elegant Graphics for Data Analysis*; Springer: New York, NY, USA, 2009.

84. Harmon, L.; Weir, J.; Brock, C.; Glor, R.; Challenger, W.; Hunt, G.; FitzJohn, R.; Pennell, M.; Slater, G.; Brown, J.; et al. Package 'Geiger'. Analysis of Evolutionary Diversification. Available online: https://cran.r-project.org/web/packages/geiger/geiger.pdf (accessed on 1 January 2022).

85. Arnold, C.; Matthews, L.J.; Nunn, C.L. The 10kTrees Website: A New Online Resource for Primate Phylogeny. *Evol. Anthr.* **2010**, *19*, 114–118. [CrossRef]

86. Meyer, D.; Rinaldi, I.D.; Ramlee, H.; Perwitasari-Farajallah, D.; Hodges, J.K.; Roos, C. Mitochondrial phylogeny of leaf monkeys (genus *Presbytis*, Eschscholtz, 1821) with implications for taxonomy and conservation. *Mol. Phylogenet. Evol.* **2011**, *59*, 311–319. [CrossRef]

87. Kembel, S.W.; Cowan, P.D.; Helmus, M.R.; Cornwell, W.K.; Morlon, H.; Ackerly, D.D.; Webb, C.O. *Picante*: R tools for integrating phylogenies and ecology. *Bioinformatics* **2010**, *26*, 1463–1464. [CrossRef]

88. Blomberg, S.P.; Garland, T., Jr.; Ives, A.R. Testing for phylogenetic signal in comparative data: Behavioral traits are more labile. *Evolution* **2003**, *57*, 717–745. [CrossRef]

89. Revell, L.J.; Harmon, L.J.; Collar, D.C. Phylogenetic signal, evolutionary process, and rate. *Syst. Biol.* **2008**, *57*, 591–601. [CrossRef]

90. Revell, L.J. Phytools: An R package for phylogenetic comparative biology (and other things). *Methods Ecol. Evol.* **2012**, *3*, 217–223. [CrossRef]
91. Garamszegi, L.Z.; Møller, A.P. Effects of sample size and intraspecific variation in phylogenetic comparative studies: A meta-analytic review. *Biol. Rev.* **2010**, *85*, 797–805. [CrossRef]
92. Raaum, R.L.; Sterner, K.N.; Noviello, C.M.; Stewart, C.B.; Disotell, T.R. Catarrhine primate divergence dates estimated from complete mitochondrial genomes: Concordance with fossil and nuclear DNA evidence. *J. Hum. Evol.* **2005**, *48*, 237–257. [CrossRef]
93. Liedigk, R.; Roos, C.; Brameier, M.; Zinner, D. Mitogenomics of the Old World monkey tribe Papionini. *BMC Evol. Biol.* **2014**, *14*, 176. [CrossRef] [PubMed]
94. Ting, N. Mitochondrial relationships and divergence dates of the African colobines: Evidence of Miocene origins for the living colobus monkeys. *J. Hum. Evol.* **2008**, *55*, 312–325. [CrossRef] [PubMed]
95. Jablonski, N.G. Fossil Old World monkeys: The late Neogene radiation. In *The Primate Fossil Record*; Cambridge University Press: Cambridge, UK, 2002; pp. 255–299.
96. Gilbert, C.C. Craniomandibular morphology supporting the diphyletic origin of mangabeys and a new genus of the *Cercocebus/Mandrillus* clade, *Procercocebus. J. Hum. Evol.* **2007**, *53*, 69–102. [CrossRef] [PubMed]
97. Gilbert, C.C. Cladistic analysis of extant and fossil African papionins using craniodental data. *J. Hum. Evol.* **2013**, *64*, 399–433. [CrossRef]
98. Frost, S.R.; Gilbert, C.C.; Nakatsukasa, M. The Colobine Fossil Record. *Colobines Nat. Hist. Behav. Ecol. Divers.* **2022**, *3*, 13.
99. Jablonski, N.G.; Frost, S.R. Cercopithecoidea. In *Cenozoic Mammals of Africa*; Werdelin, L., Sanders, W.J., Eds.; University of California Press: Berkeley, CA, USA, 2010; pp. 393–428.
100. Koh, C.; Bates, E.; Broughton, E.; Do, N.T.; Fletcher, Z.; Mahaney, M.C.; Hlusko, L.J. Genetic integration of molar cusp size variation in baboons. *Am. J. Phys. Anthropol.* **2010**, *142*, 246–260. [CrossRef]
101. Darwin, C. *The Origin of Species by Natural Selection: Or, the Preservation of Favored Races in the Struggle for Life*; John Murray: London, UK, 1859.
102. Lande, R. Quantitative genetic analysis of multivariate evolution, applied to brain: Body size allometry. *Evolution* **1979**, *33*, 402–416. [CrossRef]
103. Lande, R.; Arnold, S.J. The measurement of selection on correlated characters. *Evolution* **1983**, *37*, 1210–1226. [CrossRef]
104. Cheverud, J.M. Quantitative genetics and developmental constraints on evolution by selection. *J. Theor. Biol.* **1984**, *110*, 155–171. [CrossRef]
105. Roff, D.A. *Evolutionary Quantitative Genetics*; Springer Science & Business Media: Berlin, Germany, 2012.
106. Hlusko, L.J.; Suwa, G.; Kono, R.; Mahaney, M.C. Genetics and the evolution of primate enamel thickness: A baboon model. *Am. J. Phys. Anthropol.* **2004**, *124*, 223–233. [CrossRef]
107. Marroig, G.; Cheverud, J.M. Size as a line of least evolutionary resistance: Diet and adaptive morphological radiation in New World monkeys. *Evolution* **2005**, *59*, 1128–1142. [CrossRef] [PubMed]
108. Stojanowski, C.M.; Paul, K.S.; Seidel, A.C.; Duncan, W.N.; Guatelli-Steinberg, D. Quantitative genetic analyses of postcanine morphological crown variation. *Am. J. Phys. Anthropol.* **2019**, *168*, 606–631. [CrossRef] [PubMed]
109. Paul, K.S.; Stojanowski, C.M.; Hughes, T.E.; Brook, A.H.; Townsend, G.C. Patterns of heritability across the human diphyodont dental complex: Crown morphology of Australian twins and families. *Am. J. Phys. Anthropol.* **2020**, *172*, 447–461. [CrossRef] [PubMed]
110. Paul, K.S.; Stojanowski, C.M.; Hughes, T.; Brook, A.H.; Townsend, G.C. Genetic correlation, pleiotropy, and molar morphology in a longitudinal sample of Australian twins and families. *Genes* **2020**, *13*, 996. [CrossRef]
111. Willis, M.S.; Swindler, D.R. Molar size and shape variations among Asian colobines. *Am. J. Phys. Anthropol. Off. Publ. Am. Assoc. Phys. Anthropol.* **2004**, *125*, 51–60. [CrossRef]
112. Monson, T.A.; Hlusko, L.J. Breaking the rules: Phylogeny, not life history, explains dental eruption sequence in primates. *Am. J. Phys. Anthropol.* **2018**, *167*, 217–233. [CrossRef]
113. Frazier, B.C. The Cranial Morphology of Dwarf Primate Species. Ph.D. Thesis, The Pennsylvania State University, State College, PA, USA, 15 December 2010.
114. Huxley, J.S. *Problems of Relative Growth*; Methuen: London, UK, 1932.
115. Collard, M.; O'Higgins, P. Ontogeny and homoplasy in the papionine monkey face. *Evol. Dev.* **2001**, *3*, 322–331. [CrossRef]
116. Monson, T.A.; Brasil, M.F.; Stratford, D.J.; Hlusko, J. Patterns of craniofacial variation and taxonomic diversity in the South African Cercopithecidae fossil record. *Palaeontol. Electron.* **2017**, *20*, 1–20. [CrossRef]
117. Monson, T.A. Patterns and magnitudes of craniofacial covariation in extant cercopithecids. *Anat. Rec.* **2020**, *303*, 3068–3084. [CrossRef]
118. Perelman, P.; Johnson, W.E.; Roos, C.; Seuánez, H.N.; Horvath, J.E.; Moreira, M.A.M.; Rumpler, Y. A molecular phylogeny of living primates. *PLoS Genet.* **2011**, *7*, e1001342. [CrossRef]
119. Lumsden, A. Pattern formation in the molar dentition of the mouse. *J. Biol. Buccale* **1979**, *7*, 77–103.
120. Lumsden, A.; Osborn, J. Development of mouse dentition in culture. *J. Dent. Res.* **1976**, *55*, 136.
121. Sofaer, J.A. Co-ordinated growth of successively initiated tooth germs in the mouse. *Arch. Oral Biol.* **1977**, *22*, 71–72. [CrossRef]
122. Lucas, P.W.; Corlett, R.T.; Luke, D.A. Postcanine tooth size and diet in anthropoid primates. *Z. Morphol. Anthropol.* **1986**, *76*, 253–276. [CrossRef]

123. Evans, A.R.; Daly, E.S.; Catlett, K.K.; Paul, K.S.; King, S.J.; Skinner, M.M.; Nesse, H.P.; Hublin, J.J.; Townsend, G.C.; Schwartz, G.T.; et al. A simple rule governs the evolution and development of hominin tooth size. *Nature* **2016**, *530*, 477–480. [CrossRef]

124. Vitek, N.S.; Roseman, C.C.; Bloch, J.I. Mammal molar size ratios and the inhibitory cascade at the intraspecific scale. *Integr. Org. Biol.* **2020**, *2*, obaa020. [CrossRef]

125. Sofaer, J.A.; Bailit, H.L.; Maclean, J. A developmental basis for differential tooth reduction during hominoid evolution. *Evolution* **1971**, *25*, 509–517. [CrossRef]

126. Marshall, L.G.; Corrucini, R.S. Variability, evolutionary rates, and allometry in dwarfing lineages. *Paleobiology* **1978**, *4*, 101–119. [CrossRef]

127. Townsend, G.; Bockmann, M.; Hughes, T.; Brook, A. Genetic, environmental and epigenetic influences on variation in human tooth number, size and shape. *Odontology* **2012**, *100*, 1–9. [CrossRef]

128. Polly, P.D. Development with a bite. *Nature* **2007**, *449*, 413–414. [CrossRef]

129. Polly, P.D. Quantitative genetics provides predictive power for paleontological studies of morphological evolution. *Proc. Natl. Acad. Sci. USA* **2016**, *113*, 9142–9144. [CrossRef] [PubMed]

130. Finarelli, J.A.; Flynn, J.J. Ancestral state reconstruction of body size in the Caniformia (Carnivora, Mammalia): The effects of incorporating data from the fossil record. *Syst. Biol.* **2006**, *55*, 301–313. [CrossRef] [PubMed]

131. Monson, T.A.; Coleman, J.; Hlusko, L.J. Craniodental allometry, prenatal growth rates, and the evolutionary loss of the third molars in New World monkeys. *Anat. Record* **2019**, *302*, 1419–1433. [CrossRef] [PubMed]

biology

MDPI

Article

Morphological and Tissue Characterization with 3D Reconstruction of a 350-Year-Old Austrian *Ardea purpurea* Glacier Mummy

Seraphin H. Unterberger [1], Cordula Berger [2], Michael Schirmer [3], Anton Kasper Pallua [4], Bettina Zelger [5], Georg Schäfer [5], Christian Kremser [6], Gerald Degenhart [6], Harald Spiegl [7], Simon Erler [7], David Putzer [8], Rohit Arora [8], Walther Parson [2,9] and Johannes Dominikus Pallua [2,5,8,*]

[1] Material-Technology, Leopold-Franzens University Innsbruck, Technikerstraße 13, 6020 Innsbruck, Austria
[2] Institute of Legal Medicine, Medical University of Innsbruck, Muellerstraße 44, 6020 Innsbruck, Austria
[3] Department of Internal Medicine, Clinic II, Medical University of Innsbruck, Anichstrasse 35, 6020 Innsbruck, Austria
[4] Former Institute for Computed Tomography-Neuro CT, Medical University of Innsbruck, Anichstraße 35, 6020 Innsbruck, Austria
[5] Institute of Pathology, Neuropathology, and Molecular Pathology, Medical University of Innsbruck, Muellerstrasse 44, 6020 Innsbruck, Austria
[6] Department of Radiology, Medical University of Innsbruck, Anichstraße 35, 6020 Innsbruck, Austria
[7] WESTCAM Datentechnik GmbH, Gewerbepark 38, 6068 Mils, Austria
[8] University Hospital for Orthopaedics and Traumatology, Medical University of Innsbruck, Anichstraße 35, 6020 Innsbruck, Austria
[9] Forensic Science Program, The Pennsylvania State University, State College, PA 16801, USA
*   Correspondence: johannes.pallua@i-med.ac.at

Simple Summary: As glaciers disappear, animal mummies preserved in ice for centuries are released. Depending on the preservation method, residual soft tissues may differ in their biological information content. Paleoradiology, including micro-computed tomography (micro-CT) and magnetic resonance imaging (MRI), is the method of choice for the non-destructive analysis of mummies. A 350-year-old Austrian *Ardea purpurea* glacier mummy from the Öztal Alps was identified with micro-CT, MRI, histo-anatomical analyses, and DNA sequencing.

Citation: Unterberger, S.H.; Berger, C.; Schirmer, M.; Pallua, A.K.; Zelger, B.; Schäfer, G.; Kremser, C.; Degenhart, G.; Spiegl, H.; Erler, S.; et al. Morphological and Tissue Characterization with 3D Reconstruction of a 350-Year-Old Austrian *Ardea purpurea* Glacier Mummy. *Biology* **2023**, *12*, 114. https://doi.org/10.3390/biology12010114

Academic Editors: Mary H. Schweitzer and Ferhat Kaya

Received: 22 September 2022
Revised: 6 January 2023
Accepted: 9 January 2023
Published: 11 January 2023

Copyright: © 2023 by the authors. Licensee MDPI, Basel, Switzerland. This article is an open access article distributed under the terms and conditions of the Creative Commons Attribution (CC BY) license (https://creativecommons.org/licenses/by/4.0/).

Abstract: Glaciers are dwindling archives, releasing animal mummies preserved in the ice for centuries due to climate changes. As preservation varies, residual soft tissues may differently expand the biological information content of such mummies. DNA studies have proven the possibility of extracting and analyzing DNA preserved in skeletal residuals and sediments for hundreds or thousands of years. Paleoradiology is the method of choice as a non-destructive tool for analyzing mummies, including micro-computed tomography (micro-CT) and magnetic resonance imaging (MRI). Together with radiocarbon dating, histo-anatomical analyses, and DNA sequencing, these techniques were employed to identify a 350-year-old Austrian *Ardea purpurea* glacier mummy from the Ötztal Alps. Combining these techniques proved to be a robust methodological concept for collecting inaccessible information regarding the structural organization of the mummy. The variety of methodological approaches resulted in a distinct picture of the morphological patterns of the glacier animal mummy. The BLAST search in GenBank resulted in a 100% and 98.7% match in the cytb gene sequence with two entries of the species Purple heron (*Ardea purpurea*; Accession number KJ941160.1 and KJ190948.1) and a 98% match with the same species for the 16 s sequence (KJ190948.1), which was confirmed by the anatomic characteristics deduced from micro-CT and MRI.

Keywords: magnetic resonance imaging; micro-computed tomography; radiocarbon dating; DNA barcoding

1. Introduction

Glaciers release animal mummies preserved in the ice for centuries or millennia due to climate changes. These mummies are of inestimable value from an archaeological and biological point of view. Glacier animal mummies are relatively recent, as global temperature evolution has shown pronounced warming over the past 150 years. The location of such glacier animal mummies can be natural habitats, hiding places from predators, migration paths, or transport areas through updrafts [1]. Such animal mummies provide the unique opportunity to evaluate and compare different procedures for the analysis of glacier animal mummies, which can then be applied to human glacier mummies. The histological analysis of soft tissues may further expand the information content [2]. Internal organs, such as those comprising the digestive system, are often entirely decomposed. Organs may be shrunken and challenging to identify. The most oft-preserved soft tissues are those with a high collagen content, such as the dermis, muscle fasciae, and tendons [3].

The presence of skin may give essential clues regarding pathology and trauma. Good soft tissue preservation may also indicate good DNA preservation, allowing for genomic species identification [4]. Ancient DNA studies have proven the possibility of extracting and analyzing DNA preserved in skeletal remains and sediments for hundreds of thousands of years [1–4]. Such discoveries assist scientists in answering questions about extinct species and their relationships to others, including animals alive today. However, the very presence of soft tissue, especially the skin, also makes it challenging to examine the body in a non-destructive manner. Many studies focus on developing and applying non-destructive methods for analyzing mummies, for which paleoradiology is the method of choice [5]. Paleoradiologic analyses enable mummies to remain intact, protecting a valuable archaeological resource. These analyses could disclose information on the nature of the skeletal remains and the mummification process.

Generally speaking, computed tomography (CT) is the gold standard diagnostic method for mummy studies [6]. In addition, magnetic resonance imaging (MRI) has successfully been applied to ancient specimens [7]. The soft tissues that are found in mummified remains display radio-anatomical characteristics that are different from those known from clinical data. In ancient mummies, post-mortem dehydration and decomposition often lead to skin folding, and soft tissues appear radio-opaque on CT [8]. The usual lack of moisture in historical material makes MRI exceedingly tricky. MRI has successfully been applied in ancient dry soft tissues after invasive, morphology-alternating rehydration [9,10] and by using MRI settings with ultra-short echo time sequences [7,11]. However, paleoradiologic diagnostic accuracy and spatial tissue differentiation in historic mummies are not satisfactory due to tissue alterations. Therefore, the desire to achieve a high degree of diagnostic sensitivity and specificity is crucial in choosing any methodological approach in studies on glacier animal mummies. The discovery of one of the best-preserved human glacier mummies in the Ötztal Alps laid the foundations for scientific endeavors to diagnose glacier-bearing objects [12–16]. Due to the rarity of these findings, there is no standardized process for investigation. The handling and examination of glacier mummies are complex due to their rare occurrence and the associated lack of experience.

Micro-CT is an imaging procedure that is based on the same physical and technical bases as CT. These devices are primarily a miniaturized form of volume- or cone-beam CT scanners and can be used for non-invasive, three-dimensional investigations in pre-clinical research on bones, teeth, and small animals. A significant advantage of using micro-CT compared to clinical CT is a considerably higher spatial resolution with significantly better visualization of anatomical structures [17–19]. In vivo measurements with a spatial resolution of 10 µm are possible [20,21]. Thus, micro- and nano-CTs comprise an essential non-destructive tool to study internal structures in various disciplines, including biology [22–27], paleontology [28], geology [29], thermochronology [30], hydrology [31], soil science [32–35], materials science [36,37], and medicine [38–43].

In order to conduct a comprehensive non-destructive investigation, MRI was added. MRI provides a non-invasive tool to investigate the internal anatomy and physiology of living organisms and exploits the phenomenon of nuclear magnetic resonance. With this method, atomic nuclei exposed to a strong magnetic field absorb and reemit electromagnetic waves at characteristic frequencies, providing information on the structural and biochemical properties of the tissue [44,45]. Therefore, micro-CT and MRI analyses might offer a valuable and novel extension to conventional methods for glacier mummy research. Ideal tools for the morphological and histomorphological evaluation of mummy specimens include fixation, embedding, cutting, mounting on slides, staining, and examination using microscopic techniques such as optical, electron, and fluorescence microscopy. These methods require substantial preparation procedures and interpretation by experts. Micro-CT and MRI represent complementary tools, enabling the investigation with or without sample processing before image acquisition [46]. These imaging techniques are powerful tools, with several advantages in the structural characterization of biological systems: nearly no damage to samples occurs, and repeated scanning of the same sample is possible [47]. Moreover, imagery from micro-CT and MRI measurements can also be viewed and reviewed in 2D or 3D, and objects of interest can be segmented from the images as digital surfaces or isosurfaces to analyze complex structures [30].

A glacier mummy was found at the Gurgler Ferner on North Tyrolean territory near Ötztal, Tyrol, Austria. This study's primary goal was to apply paleoradiological imaging in the form of micro-CT and MRI analyses, followed by radiocarbon dating and DNA analyses, to this animal glacier mummy. Thus, modern methods for investigating glacier mummies are explored for potential use in human glacier mummies.

2. Materials and Methods

2.1. Find Spot

A glacier mummy was found at 3004 m height on 3 August 2015 by Franz Scheiber and Josef Klotz in the area of the Hochwildehaus towards Hochwilde and the Annakegele at the Gurgler Ferner (degree of latitude: 46.785946, degree of longitude: 11.003878) on North Tyrolean territory near Ötztal, Tyrol, Austria (see Figure 1A). The Gurgler Ferner in Tyrol is one of the largest glaciers in the Ötztal Alps. With an area of 9.58 km^2, it is now the third-largest glacier in the Austrian province of Tyrol [48]. Due to the demarcation of the border, which in this area is not always oriented to the ice or watershed, smaller parts of the glacier are also located on Italian territory and are protected in the South Tyrolean nature park Texelgruppe. The Gurgler Ferner is a typical valley glacier and flows from the Gurgler ridge, which is part of the main alpine ridge, almost eight kilometers to the north into the Gurgler valley [49]. The Gurgler Ferner is embedded between the Ramolkamm with the Schalfkogel in the west and the Schwärzenkamm in the east. On the orographic right bank of the glacier lies the Hochwildehaus. In this area, on 6 August 2015, Franz Scheiber, Josef Klotz, Judith Unterberger, and Seraphin Unterberger collected the mummy parts (see Figure 1B). All parts were sealed in bags and boxes. The head was separated from the body, and the plumage was packed separately in a plastic bag with undefinable parts when recovered (see Figure 1C,D). These boxes and bags were frozen at −18 °C until the start of the investigation.

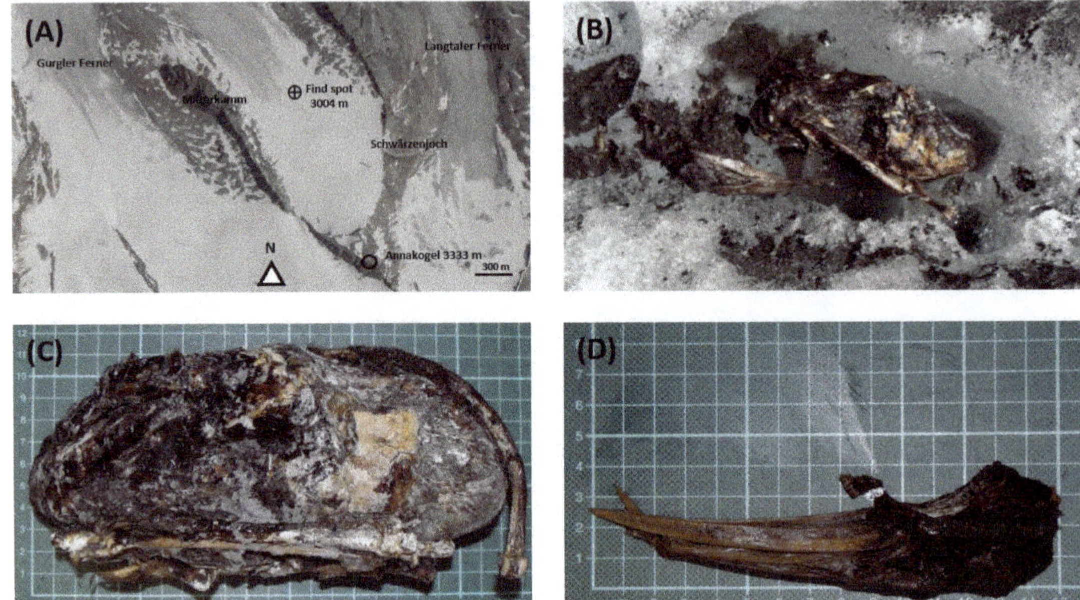

Figure 1. (**A**) Location of sample collection of the glacier mummy on the Gurgler Ferner using tirisMaps (https://maps.tirol.gv.at, accessed on 22 November 2022). This location is west of the sample collection of Ötzi, about 11.16 km apart. (**B**) Presentation of the glacier mummy at first finding with (**C**) focus on the corpus and the (**D**) detailed skull.

2.2. Micro-Computed Tomography (Micro-CT)

Micro-CT measurements were performed on a vivaCt40 and an XtremeCT II (ScancoMedical AG, Brüttisellen, Switzerland). Due to the geometrical dimensions, the corpus of the mummy was scanned only in the XtremeCT II. The mummy's skull was further analyzed in the vivaCT40 due to its higher resolution. The settings for the XtremeCT II experiments were a 30.5 µm isotropic voxel size with a 68 kV, 1470 µA tube setting and 650 ms exposure time. The image matrix was 4096 × 4096, with a 16-bit grey-value resolution. The settings for the micro-CT experiment were a 10.5 µm isotropic voxel size with a 70 kV, 114 µA tube setting, 6500 ms exposure time, 1000 projections, and 2048 samples. The image matrix was 2048 × 2048 with a 16-bit grey-value resolution. The micro-CT data were evaluated by an experienced radiologist and summarized. The reconstructions were carried out with Analyze 14.0 (Analyze Direct Inc., Overland Park, KS, USA) software. The following two-dimensional display formats were used for the reconstructions:

- Planar CT slice images: reconstructed CT slice images along axial, sagittal, or coronal planes.
- Multi-planar reconstruction: reconstructed CT slice images along planes of freely selectable position and angle.
- Curved planar reconstruction: reconstructed CT slice images along planes of arbitrary orientation.

Various three-dimensional reconstructions were also used:

- Maximum intensity projection: the voxel with the highest intensity is displayed along a specific projection through the volume dataset.
- Surface Rendering: according to a defined mean value, the surface is rendered along a certain projection of the volume data set.

- Volume Rendering: assignment of a color value to a voxel according to its X-ray density. It is then possible to make certain regions transparent.
- Segmentation: semi-automatic segmentation tools such as Threshold Volume, Region Grow, and Object Extractor were used to segment the calcifications and internal organs. The procedure consists of selecting a pixel within an area called a seed. Neighboring pixels of similar density are automatically added or connected. A density threshold is chosen to capture the total volume in the pmCT layer. This process is repeated for each layer, and the total volume is automatically added.

For the comparative morphological study, 12 individual bones (sternum, coracoid, scapula, furcula, humerus, radius, ulna, carpometacarpus, pelvis, femur, tibiotarsus, and tarsometatarsus) of the postcranial skeleton were used and compared with published data [50]. The measurements were carried out as described in [50].

2.3. Magnetic Resonance Imaging (MRI) Data Acquisition and Processing

The MRI experiments were performed on a 3-T MR-Scanner (Siemens Magnetom Skyra, Siemens Healthineers, Erlangen, Germany) using a standard 12-channel head coil. Magnetization prepared T1 weighted ultrashort echo time imaging using PETRA ("Pointwise Encoding Time Reduction with Radial Acquisition") was used [51] with TR = 3.32 ms, TE = 0.07 ms, TI = 1300 ms, flip angle: $6°$, receive bandwidth: 400 Hz/pixel, number of radial views: 60,000, FOV: 251 mm, image matrix: 320×320, voxel size: 0.78 mm $\times 0.78$ mm $\times 0.78$ mm. In addition, a multi-slab T2-weighted turbo spin-echo sequence was acquired with TR = 6680 ms, TE = 103 ms, echo train length: 15, slice thickness: 2 mm, spacing between slices: 2.2 mm, acquisition matrix: 448×314, FOV: 140 mm, voxel size: 0.31 mm $\times 0.31$ mm $\times 2$ mm, number of slabs to cover the whole bird: 4 with 25 images per slab. Data processing and analyses were performed using Syngo.Via (Siemens Healthcare, Erlangen, Germany).

2.4. Sample Collection and Tissue Specimens

First, the animal glacier mummy was macroscopically examined. After using the paleoradiological methods, two independent biopsies were taken from different organs. As these did not provide conclusive results, the mummy was then sequentially sliced from caudal to cranial at 3–5 mm intervals, fixed in formalin, and embedded in paraffin as whole-mount sections according to the European standards of Biobanking CEN/TS and the ISO standards ISO 20166-1:2018, ISO 20166-2:2018, and ISO 20166-3:2018 on the pre-examination process for molecular diagnostics [52,53]. This technique assures the preservation of tissues for future histological and biomolecular analyses. Before formalin fixation, samples were taken for radiocarbon dating and DNA sequencing. Figure 2 presents the macroscopical inspection and sampling for DNA sequencing and radiocarbon dating.

2.5. Radiocarbon Dating

Radiocarbon dating was routinely performed at the Ion Beam Physics, ETH Zurich Laboratory. Two samples (20 mg and 200 mg) consisting of feather, skin, bone, and tissue were used. Sample treatment [54,55], reporting [56,57], and reporting ^{14}C ages [58] were conducted according to the cited literature.

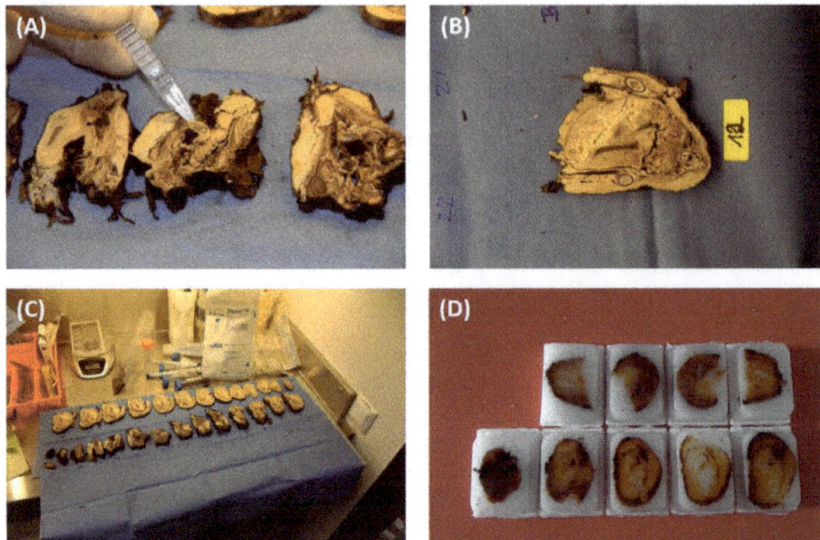

Figure 2. Sample collection. (**A**) Sample for DNA sequencing; (**B**) 3–5 mm slice of the mummy; (**C**) overview of the sequentially sliced mummy from caudal to cranial at 3–5 mm intervals; (**D**) paraffin-embedded tissue blocks.

2.6. DNA Analysis

Two tissue punches (size 1.5 × 2 mm) and one bone sample from the glacier mummy were used for the DNA analysis. The tissue punches were lysed, and DNA was extracted using the EZ-1 and the MagAttract DNA kit (all Qiagen, Hilden, Germany) following the manufacturer's recommendations.

Physical and chemical cleaning of the bone surface: The mechanical and chemical processing of the samples was performed with the necessary care required for forensically relevant samples containing only minute amounts of DNA exposed to potential superficial contamination [59,60]. One bone sample was taken from the unknown animal glacier mummy and subjected to mechanical surface cleaning with sterile scalpel blades. The sample was then bathed in sodium hypochlorite ($\geq$4% active chlorine, Sigma Aldrich, St. Louis, MO, USA) at room temperature for 15 min, washed twice in purified water (DNA/RNA free), and rinsed in absolute ethanol for 5 min. Samples were dried in a closed laminar flow cabinet overnight, UV irradiated for 10 min (λ = 254 nm), and then powdered using a vibrating ball mill (Mixer Mill MM400, Retsch, Haan, Germany). Grinding with the ball mill was performed in cycles of 60 s, with a grinding rate of 25 Hz, followed by 60 s cooling steps. A minimum of two grinding cycles were completed, and the abrasive product was visually evaluated for homogeneity.

DNA extraction of the bone sample: The bone powder was subjected to lysis, and DNA was extracted according to the modified Dabney method as described in Xavier et al., 2021 [61].

Mitochondrial DNA typing: Both the mitochondrial (mt)DNA cytochrome b gene [62], as well as the 16 sRNA [63], were amplified and sequenced on an ABI 3500 Sequencer, and the sequences were aligned using Sequencher (GeneCodes, Ann Arbour, MI, USA). The consensus sequence was BLASTed at NCBI GenBank to recover the closest neighbors for species identification.

3. Results

3.1. Radiocarbon Dating

Data of the radiocarbon dating of the *Ardea purpurea* glacier mummy are shown in Figure 3. The sample indicates the presence of "bomb peak ^{14}C" (post 1950 AD) 1 sigma range BC/AC Lower 1642 Upper 1665, 2 sigma range BC/AC Lower 1529 Upper 1799. All calibrated intervals listed below need to be taken into account. In some cases, due to the shape of the calibration curve in the region of interest, the sample's age falls into a period when precise information about the true age range cannot be provided. Therefore, radiocarbon dating defined the mummy's age as 350 years.

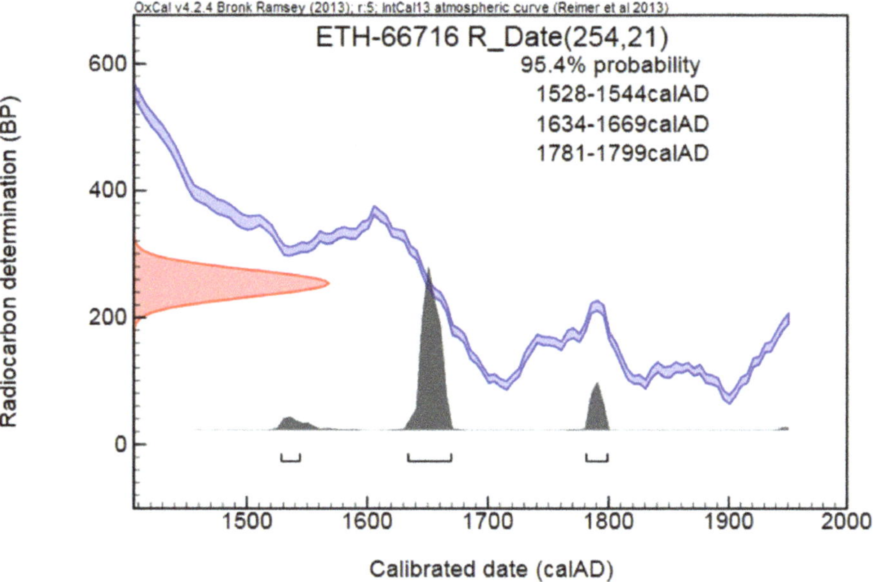

Figure 3. Radiocarbon dating of the *Ardea purpurea* glacier mummy. ^{14}C age (BP)—delta C13 corrected radiocarbon age based on concentration of ^{14}C measured in sample. BP = before present (before 1950 AD) [56,57].

3.2. Morphological Analysis via Micro-CT and MRI

Before further destructive analyses, micro-CT and MRI were used with photographs of anatomical sections to study anatomy. The micro-CT and MRI scans were correlated with three photograph images of the anatomical section (see Figure 4) to identify relevant structures along the trunk from the crop to the end of the thoracoabdominal cavity. A localization image for micro-CT and MRI is shown in Figure 4, with the transverse planes and corresponding anatomical sections shown in lines. Micro-CT, MRI, and anatomical section examination revealed the presence of remains of internal organs, many of which appeared to be lytic. Brain tissue remains were not visible on the inside of the calotte. The remains of the lungs, heart, stomach, and other internal organs were also visible. The lungs are collapsed, and their outlines were still recognizable. The alveolar and bronchial structures were still clearly visible. The remains of all major muscle groups were also preserved. Fat deposits were still visible. The examination of the remaining structures revealed no pathological modifications, and the specific cause of death could not be conclusively determined. There were no signs of artificial body mummification (e.g., no puncture channels, no opening of the body cavities, no removal of organs, and no introduction of foreign material).

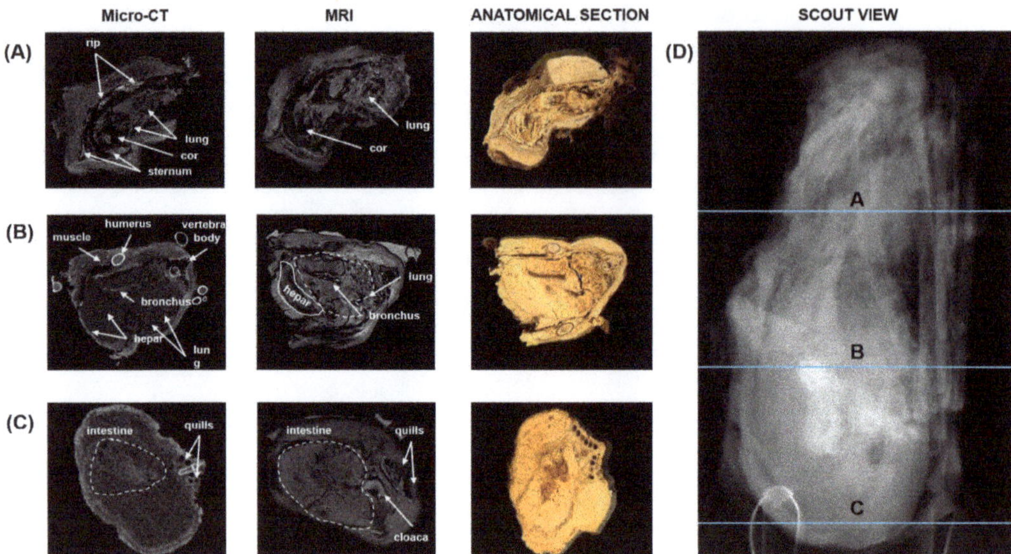

Figure 4. Transversal micro-CT, MRI, and anatomical section images of the trunk at different levels. (**A**) Anatomical section of the trunk at the level of the cor and lung. (**B**) Anatomical section of the trunk at the level of the hepar and lung. (**C**) Anatomical section of the trunk at the level of the cloaca and intestine. (**D**) Scout view of the trunk with lines representing the locations of different levels.

Three-dimensional reconstructions of the skull and the postcranial skeleton based on the micro-CT data of the *Ardea purpurea* glacier mummy were performed for comparative morphological studies on single bones. All bones were in their anatomical position and completely preserved (see Figure 5). The skull was heavily pneumatized, and the posterior skullcap was large. Prominent blood–brain conductors and the bone spur were still present. The beak was deformed but not bent (see Figure 5A,B). Based on the skull, no immediate species identification could be performed due to the severe deformations of the skull. The skeleton showed no fractures, which is demonstrated by the maximum intensity projection (Figure 5C), volume rendering bone of the body (Figure 5D), and volume rendering soft tissue of the body (Figure 5E).

As a result, most of the investigated bone elements fell within the size range of modern *Ardea purpurea*. The dimensions of the coracoid, scapula, humerus, radius, ulna, carpometacarpus, pelvis, femur, tibiotarsus, and tarsometatarsus showed the smallest measured value compared to modern *Ardea purpurea*.

The furcula, coracoid, and pelvis are described as follows in more detail.

Furcula: The characteristic of the furcula of herons is a thorn-like process on the hypocleidium, which extends dorsally at the bifurcation point of the two furcula branches. The hypocleidium bears a distinct suture bar on the caudodorsal side. The best way to distinguish species is the hypocleidium. It is more general in the *Ardea purpurea* and ends, tapering evenly, with two small bone serrations close to each other (see Figure 6).

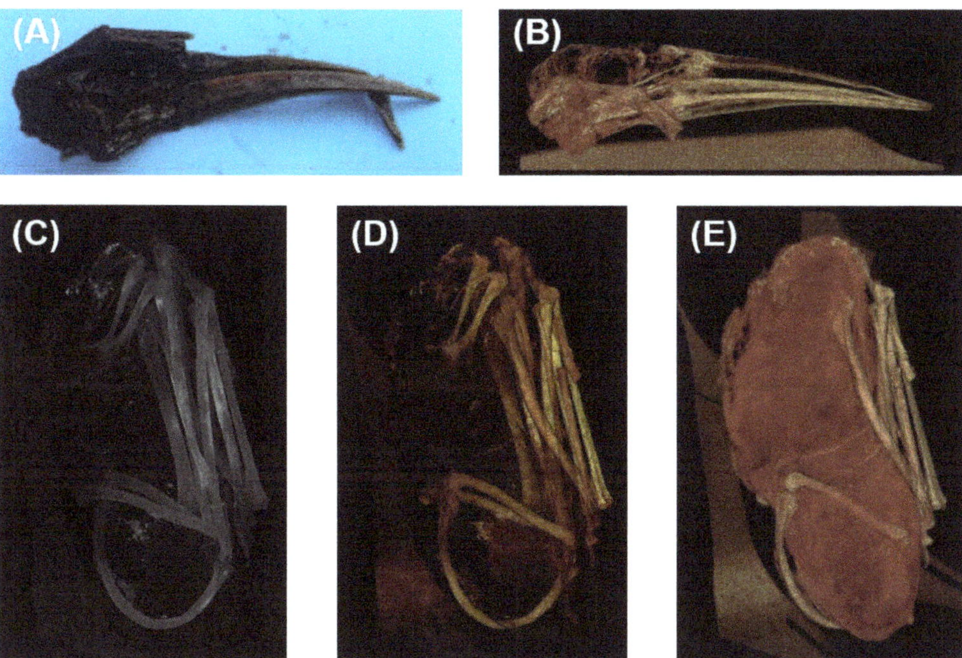

Figure 5. Three-dimensional reconstruction of the skull and the trunk of the *Ardea purpurea* glacier mummy. (**A**) Photograph of the skull. (**B**) Volume rendering bone of the skull. (**C**) Maximum intensity projection of the body. (**D**) Volume rendering bone of the body. (**E**) Volume rendering soft tissue of the body.

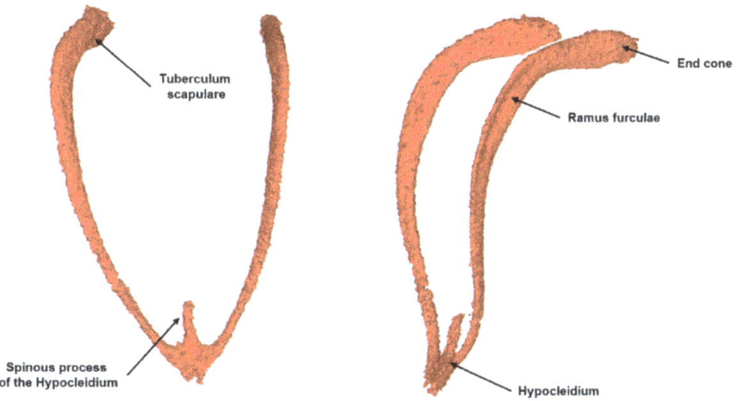

Figure 6. Three-dimensional reconstruction of the furcula of the *Ardea purpurea* glacier mummy.

Coracoid: Compared to other herons, *Ardea purpurea* has a long and slender coracoid with a high process scapularis and a deep concave margo lateralis. The acrocoracoid, when viewed cranially, is narrow as in *Nycticorax nycticorax*. All herons have a well-developed processus scapularis that curls medially with a broad base. In the *Ardea purpurea*, this tip is drawn more craniodorsally than in other herons. The processus lateralis is bent up in the shape of a hook; together with the edge drawing from it to the apex lateralis, the margo lateralis is an essential distinguishing feature. The edge of the small hook, viewed cranially or caudally, drops vertically to bend into a concave edge. Here, the arcuate notch extends deeper into the coracoid plate. At about two-thirds of the total length, the coracoid shaft widens towards the lateral process. The labium articulare sternale extends evenly in a slight arc from the medial margin to the apex lateralis. The asymmetrical design of the articular grooves at the anterior margin of the sternum causes a different form of the ventral end of the coracoids on both sides.

For this reason, the coracoids of both sides were measured. The ventral section of the right coracoid is slightly wider than that of the left, expressed in the measurements broad basal (BB) and broad of the facies articularis basalis (BF). Additionally, the greatest diagonal length (acrocoracoid–apex lateralis) and medial length (acrocoracoid–apex medialis) were determined (see Figure 7).

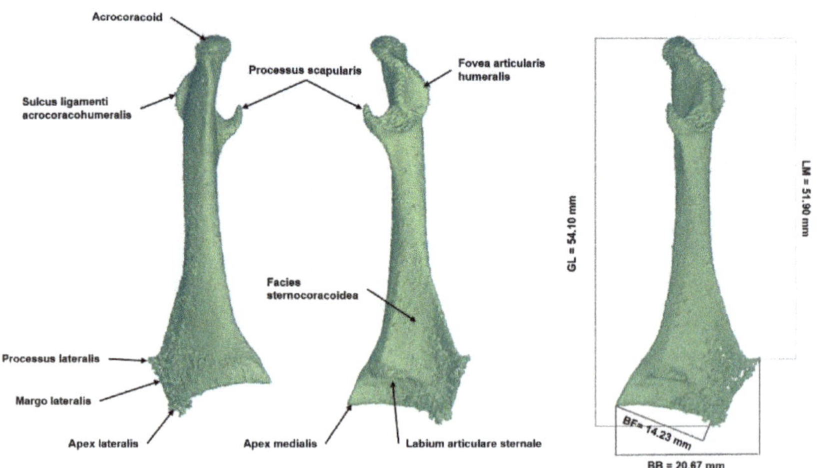

Figure 7. Three-dimensional reconstruction of the coracoid of the *Ardea purpurea* glacier mummy.

BB; broad basal = 20.67 mm.
BF; broad of the facies articularis basalis = 14.23 mm.
GL; greatest diagonal length (acrocoracoid–apex lateralis) = 54.10 mm.
LM; length medial (acrocoracoid–apex medialis) = 51.90 mm.

Pelvis: In herons, the praeacetabular portion of the pelvis is about as long as the postacetabular portion. The pelvis of *Ardea purpurea* is small and narrow. The muscle lines further caudally than in *Ardea cinerea* and ends on a small scale of bone. The cranial tip of the crista spinalis protrudes cranially beyond the pars glutaea ossis ilium. The foramina intertransversaria medialia are less numerous and smaller. The 3D reconstruction of the pelvis in Figure 8 clearly shows taphonomic bone loss.

The sternum could not be reconstructed three-dimensionally and therefore could not be measured correctly. The comparative description of the bones and the differentiation criteria published by Kellner [50] could be confirmed using the 3D reconstructions of the bones. Based on the anatomic characteristics published by Kellner [50], the glacier mummy was determined to be an adult male *Ardea purpurea*.

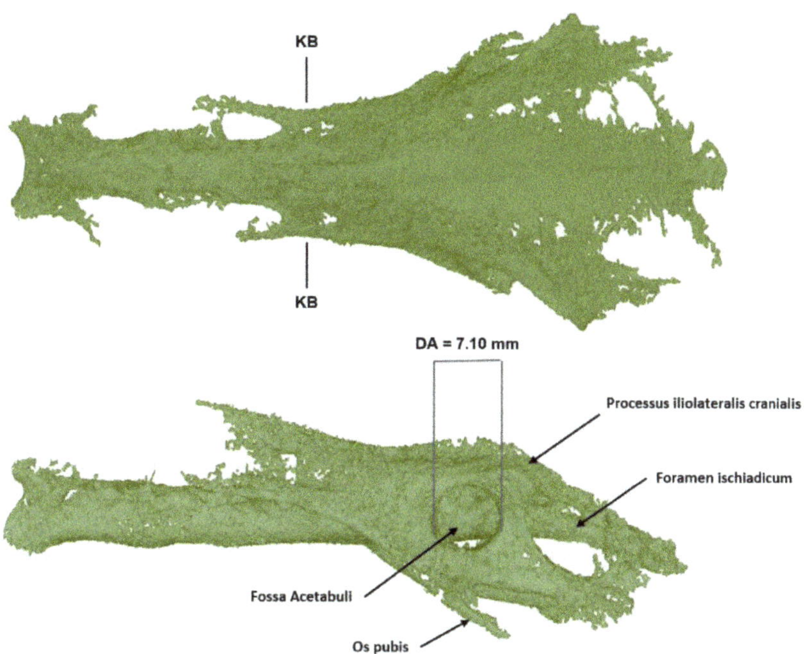

Figure 8. Three-dimensional reconstruction of the pelvis of the *Ardea purpurea* glacier mummy. KB: smallest width of the partes glutaeae = 12.45 mm; DA: diameter of the acetabulum (greatest distance) = 7.10 mm.

3.3. DNA Analysis

The bone sample was successfully examined in the cytb gene region (358 bp (13,733–14,090) and 16 s region (604 bp, 1943–2545—primer sequences included, positions according to reference sequence KJ190948.1). The BLAST search in GenBank resulted in a 100% and 98.7% match of the cytb gene sequence with two entries of the species Purple heron (*Ardea purpurea*; Accession number KJ941160.1 and KJ190948.1) and a 98% match with the same species for the 16 s sequence (KJ190948.1). The two tissue samples did not result in usable sequences.

4. Discussion

This workup of a serendipitously found glacier mummy from the Gurgler Ferner on the North Tyrolean territory near Ötztal (Tyrol, Austria) allowed for the identification of a male *Ardea purpurea*. This identification was confirmed by DNA analyses and anatomical observations with high matches of the cytb gene sequence and the 16 s sequence using a BLAST search in GenBank. Radiocarbon dating defined the mummy's age as 350 years. Then, paleoradiological techniques, including micro-CT and MRI, allowed for a 3D reconstruction showing the mummy's skull and body for further comparison with previously described anatomical structures [50].

Since mummies are potentially precious relics of past times, non-invasive techniques are preferred, and CT imaging is currently the most common method. The multiple planes with optimized settings for different tissues in micro-CT and MRI images have allowed species identification based on osteoanatomical observations for about 40 years [64]. Micro-CT and MRI images allowed for the three-dimensional reconstruction, especially of the skull and pelvis for sex-specific features and of the sternum, coracoid, scapula, furcula, humerus, radius, ulna, carpometacarpus, pelvis, femur, tibiotarsus, and tarsometatarsus for bone identification, as previously shown for *Ardea purpurea* [50]. Only the sternum and parts of the pelvis were severely degraded in the glacier mummy (e.g., leaching of calcium from the bones) and could not be fully morphologically assessed. As a result of the microenvironment, demineralization may not be uniform within the skeletal system or bone. The bone may appear patchy despite being morphologically intact.

Moreover, other tissues seem to become more radio-dense (i.e., the attenuation of X-ray beams increases). In particular, this affects ligaments, fasciae, and the subcutis [3]. This may be due to the deposition of mineral salts (containing metals such as iron) in collagenous tissues. Organs may require manual segmentation in several slices due to the changes caused by water loss. Therefore, the images need to be post-processed by segmenting, delineating, and extracting specific anatomical structures.

Complementing CT images with MRI data is desirable due to the lack of contrast in soft tissues found in CT images. Mummified tissues are invisible to standard MRI techniques due to their dehydration, short T2 relaxation times, and special acquisition, and thus strategies such as ultrashort echo time sequences (UTE) have to be used [65]. The magnetic resonance effect is observed for any atomic nucleus with special magnetic properties (magnetic moment), but hydrogen nuclei (protons) in particular have very advantageous properties. Therefore, protons (i.e., hydrogen) are usually most relevant for MRI, and images show the tissue's water content or the properties of water within the tissue. Water content is higher in living tissues than in bone and enamel, which are highly mineralized, with little water content. In addition to the short T2 relaxation times already mentioned, this also explains the reduced value of MRI in mummies. However, in the glacier mummy studied in this work, the water content was high enough to allow for MRI (Figure 5).

As a limitation of the imaging methods, there is a certain degree of subjectivity in the segmentation process. Detailed anatomical knowledge is necessary. Overall, CT and MR scanning images and 3D renderings of internal structures and tissues should not be viewed as objective and "true" representations. Visualizing the skeleton results in many points of reference, allowing for a more accessible assessment of the remains of internal organs and structures. Pathological processes can be falsely attributed to diagenetic processes and vice versa [64]. In mummy MRIs, the greatest challenge is the extensive dehydration of the tissues, since dehydrated tissues lack the hydrogen (H) in mobile water required for standard MRI signals [65]. MRI imaging is less informative than micro-CT imaging in mummies without sufficient water content. This can be circumvented by rehydrating tissues and organs [10] (but this is an invasive procedure and may not always be possible) or by applying the technique to mummies that are not entirely dehydrated, such as this glacier mummy [66]. Accordingly, reasonable images using MRI have been reported for Ötzi the Tyrolean Iceman, Lindow Man from a bog, and a corpse from a sealed medieval Korean tomb [65].

Several confounding factors, including taphonomic processes and preservation always limit the study of mummies. We estimate that over a hundred mummies have been CT-scanned and reported at this point [64]. None of these data have been synthesized into meaningful work beyond a single individual scan. Often, insufficient imaging data on ancient tissues prevents conclusions from being drawn. Future research is needed to determine the difference between antemortem and postmortem findings in micro-CT and MRI images in long-term observations to eventually produce the most accurate results with optimized segmentation procedures. Comparisons between image-based and specimen-based information are necessary. Mummified tissue biobanks with asservation of tissue specimens have already been proposed [3]. Thus, new trends in mummy research emphasize establishing guidelines and ensuring proper scientific methodology regarding analytic methods.

5. Conclusions

Applying the methodological concept of micro-CT and MRI imaging in combination with invasive but established techniques such as radiocarbon dating and DNA analyses can support the identification of animal species, as in the case of this glacier mummy. Three-dimensional digitization and interactive visualization of micro-CT and MRI allowed us to conduct digital autopsies and to provide a highly detailed 3D reconstruction of the *Ardea purpura* mummy.

Author Contributions: Conceptualization, S.H.U., C.B., M.S., A.K.P., B.Z., G.S., C.K., G.D., H.S., S.E., D.P., R.A., W.P. and J.D.P.; methodology, S.H.U., C.B., M.S., A.K.P., B.Z., G.S., C.K., G.D., H.S., S.E., D.P., R.A., W.P. and J.D.P.; software, S.H.U., C.B., M.S., A.K.P., B.Z., G.S., C.K., G.D., H.S., S.E., D.P., R.A., W.P. and J.D.P.; validation, S.H.U., C.B., M.S., A.K.P., B.Z., G.S., C.K., G.D., H.S., S.E., D.P., R.A., W.P. and J.D.P.; formal analysis, S.H.U., C.B., M.S., A.K.P., B.Z., G.S., C.K., G.D., H.S., S.E., D.P., R.A., W.P. and J.D.P.; investigation, S.H.U., C.B., M.S., A.K.P., B.Z., G.S., C.K., G.D., H.S., S.E., D.P., R.A., W.P. and J.D.P.; resources, S.H.U., C.B., M.S., A.K.P., B.Z., G.S., C.K., G.D., H.S., S.E., D.P., R.A., W.P. and J.D.P.; data curation, S.H.U., C.B., M.S., A.K.P., B.Z., G.S., C.K., G.D., H.S., S.E., D.P., R.A., W.P. and J.D.P.; writing—original draft preparation, S.H.U., C.B., M.S., A.K.P., B.Z., G.S., C.K., G.D., H.S., S.E., D.P., R.A., W.P. and J.D.P.; writing—review and editing, S.H.U., C.B., M.S., A.K.P., B.Z., G.S., C.K., G.D., H.S., S.E., D.P., R.A., W.P. and J.D.P.; visualization, S.H.U., C.B., M.S., A.K.P., B.Z., G.S., C.K., G.D., H.S., S.E., D.P., R.A., W.P. and J.D.P.; supervision, J.D.P.; project administration, J.D.P. All authors have read and agreed to the published version of the manuscript.

Funding: This work was supported by the Verein zur Förderung der Hämatologie, Onkologie und Immunologie, Innsbruck, Austria.

Institutional Review Board Statement: Not applicable.

Informed Consent Statement: Not applicable.

Data Availability Statement: Not applicable.

Acknowledgments: The authors thank Franz Scheiber and Josef Klotz for finding and recovering the mummy. The authors thank Richard Scheithauer, head of the Institute of Forensic Medicine (Medical University of Innsbruck), for the ongoing support. The authors would like to thank Christiane Böhm, Armin Landmann (both Innsbruck), Peter de Knijff (Leiden), Gerhard Forstenpointer (Vienna), and Julia Waldhart for their fruitful discussions. Thanks to Sabine Jöbstl, Inge Jehart, and Sarah Peer for their excellent technical support.

Conflicts of Interest: The authors declare no conflict of interest.

Ethical Statement: For such studies without risk of pain for the animal, neither the Ethics Committee nor the Animal Welfare Committee of the Medical University of Innsbruck considered themselves responsible for this project. The Advisory Board on Ethical Issues in Scientific Research and representatives of the animal protection committee of the Leopold Franzens University, as well as colleagues from the University of Veterinary Medicine in Vienna, further concluded that there are no known legal or ethical standards for projects on animal mummies in this country.

References

1. Hafner, A. Archaeological discoveries on Schnidejoch and at other ice sites in the European Alps. *Arctic* **2012**, *65*, 189–202. [CrossRef]
2. Aufderheide, A.C. *The Scientific Study of Mummies*; Cambridge University Press: Cambridge, UK, 2003.
3. Lynnerup, N. Methods in mummy research. *Anthropol. Anz.* **2009**, *67*, 357–384. [CrossRef] [PubMed]
4. Nystrom, K.C. Advances in paleopathology in context: A focus on soft tissue paleopathology. *Int. J. Paleopathol.* **2020**, *29*, 16–23. [CrossRef] [PubMed]
5. Licata, M.; Tosi, A.; Larentis, O.; Rossetti, C.; Lorio, S.; Pinto, A. Radiology of Mummies. *Semin. Ultrasound CT MR* **2019**, *40*, 5–11. [CrossRef] [PubMed]
6. Ruhli, F.J.; Chhem, R.K.; Boni, T. Diagnostic paleoradiology of mummified tissue: Interpretation and pitfalls. *Can. Assoc. Radiol. J.* **2004**, *55*, 218–227.
7. Ruhli, F.J. Short Review: Magnetic Resonance Imaging of Ancient Mummies. *Anat. Rec.* **2015**, *298*, 1111–1115. [CrossRef]
8. Baldock, C.; Hughes, S.W.; Whittaker, D.K.; Taylor, J.; Davis, R.; Spencer, A.J.; Tonge, K.; Sofat, A. 3-D reconstruction of an ancient Egyptian mummy using X-ray computer tomography. *J. R. Soc. Med.* **1994**, *87*, 806–808.

9. Sebes, J.I.; Langston, J.W.; Gavant, M.L.; Rothschild, B. MR imaging of growth recovery lines in fossil vertebrae. *AJR Am. J. Roentgenol.* **1991**, *157*, 415–416. [CrossRef]
10. Piepenbrink, H.; Frahm, J.; Haase, A.; Matthaei, D. Nuclear magnetic resonance imaging of mummified corpses. *Am. J. Phys. Anthropol.* **1986**, *70*, 27–28. [CrossRef]
11. Ohrstrom, L.M.; von Waldburg, H.; Speier, P.; Bock, M.; Suri, R.E.; Ruhli, F.J. Scenes from the past: MR imaging versus CT of ancient Peruvian and Egyptian mummified tissues. *Radiographics* **2013**, *33*, 291–296. [CrossRef]
12. William, A.; Murphy, J.; Nedden, D.z.; Gostner, P.; Knapp, R.; Recheis, W.; Seidler, H. The Iceman: Discovery and Imaging. *Radiology* **2003**, *226*, 614–629. [CrossRef]
13. Keller, A.; Graefen, A.; Ball, M.; Matzas, M.; Boisguerin, V.; Maixner, F.; Leidinger, P.; Backes, C.; Khairat, R.; Forster, M.; et al. New insights into the Tyrolean Iceman's origin and phenotype as inferred by whole-genome sequencing. *Nat. Commun.* **2012**, *3*, 698. [CrossRef]
14. Callaway, E. Famous ancient iceman had familiar stomach infection. *Nature* **2016**. [CrossRef]
15. Callaway, E. Iceman's DNA reveals health risks and relations. *Nature* **2012**. [CrossRef]
16. Coia, V.; Cipollini, G.; Anagnostou, P.; Maixner, F.; Battaggia, C.; Brisighelli, F.; Gómez-Carballa, A.; Destro Bisol, G.; Salas, A.; Zink, A. Whole mitochondrial DNA sequencing in Alpine populations and the genetic history of the Neolithic Tyrolean Iceman. *Sci. Rep.* **2016**, *6*, 18932. [CrossRef]
17. Engelke, K.; Karolczak, M.; Lutz, A.; Seibert, U.; Schaller, S.; Kalender, W. Micro-CT. Technology and application for assessing bone structure. *Radiologe* **1999**, *39*, 203–212. [CrossRef]
18. Cavanaugh, D.; Johnson, E.; Price, R.E.; Kurie, J.; Travis, E.L.; Cody, D.D. In vivo respiratory-gated micro-CT imaging in small-animal oncology models. *Mol. Imaging* **2004**, *3*, 55–62. [CrossRef]
19. Kalender, W.A.; Deak, P.; Kellermeier, M.; van Straten, M.; Vollmar, S.V. Application- and patient size-dependent optimization of x-ray spectra for CT. *Med. Phys.* **2009**, *36*, 993–1007. [CrossRef]
20. Brouwers, J.E.; van Rietbergen, B.; Huiskes, R. No effects of in vivo micro-CT radiation on structural parameters and bone marrow cells in proximal tibia of wistar rats detected after eight weekly scans. *J. Orthop. Res.* **2007**, *25*, 1325–1332. [CrossRef]
21. Brouwers, J.E.; Lambers, F.M.; Gasser, J.A.; van Rietbergen, B.; Huiskes, R. Bone degeneration and recovery after early and late bisphosphonate treatment of ovariectomized wistar rats assessed by in vivo micro-computed tomography. *Calcif. Tissue Int.* **2008**, *82*, 202–211. [CrossRef]
22. Stock, S.R.; Nagaraja, S.; Barss, J.; Dahl, T.; Veis, A. X-ray microCT study of pyramids of the sea urchin Lytechinus variegatus. *J. Struct. Biol.* **2003**, *141*, 9–21. [CrossRef] [PubMed]
23. Bogart, S.J.; Spiers, G.; Cholewa, E. X-ray microCT imaging technique reveals corm microstructures of an arctic-boreal cotton-sedge, Eriophorum vaginatum. *J. Struct. Biol.* **2010**, *171*, 361–371. [CrossRef] [PubMed]
24. Schulz-Mirbach, T.; Hess, M.; Metscher, B.D.; Ladich, F. A unique swim bladder-inner ear connection in a teleost fish revealed by a combined high-resolution microtomographic and three-dimensional histological study. *BMC Biol.* **2013**, *11*, 75. [CrossRef] [PubMed]
25. Faulwetter, S.; Vasileiadou, A.; Kouratoras, M.; Thanos, D.; Arvanitidis, C. Micro-computed tomography: Introducing new dimensions to taxonomy. *ZooKeys* **2013**, *263*, 1–45. [CrossRef] [PubMed]
26. Dinley, J.; Hawkins, L.; Paterson, G.; Ball, A.D.; Sinclair, I.; Sinnett-Jones, P.; Lanham, S. Micro-computed X-ray tomography: A new non-destructive method of assessing sectional, fly-through and 3D imaging of a soft-bodied marine worm. *J. Microsc.* **2010**, *238*, 123–133. [CrossRef]
27. Schulz-Mirbach, T.; Metscher, B.; Ladich, F. Relationship between swim bladder morphology and hearing abilities—A case study on Asian and African cichlids. *PLoS ONE* **2012**, *7*, e42292. [CrossRef]
28. Collareta, A.; Landini, W.; Lambert, O.; Post, K.; Tinelli, C.; Di Celma, C.; Panetta, D.; Tripodi, M.; Salvadori, P.A.; Caramella, D.; et al. Piscivory in a Miocene Cetotheriidae of Peru: First record of fossilized stomach content for an extinct baleen-bearing whale. *Die Nat.* **2015**, *102*, 70. [CrossRef]
29. Ketcham, R.A.; Carlson, W.D. Acquisition, optimization and interpretation of X-ray computed tomographic imagery: Applications to the geosciences. *Comput. Geosci.* **2001**, *27*, 381–400. [CrossRef]
30. Evans, N.J.; McInnes, B.I.; Squelch, A.P.; Austin, P.J.; McDonald, B.J.; Wu, Q. Application of X-ray micro-computed tomography in (U–Th)/He thermochronology. *Chem. Geol.* **2008**, *257*, 101–113. [CrossRef]
31. Wildenschild, D.; Vaz, C.; Rivers, M.; Rikard, D.; Christensen, B. Using X-ray computed tomography in hydrology: Systems, resolutions, and limitations. *J. Hydrol.* **2002**, *267*, 285–297. [CrossRef]
32. Elliot, T.R.; Heck, R.J. A comparison of 2D vs. 3D thresholding of X-ray CT imagery. *Can. J. Soil Sci.* **2007**, *87*, 405–412. [CrossRef]
33. Elliot, T.R.; Heck, R.J. A comparison of optical and X-ray CT technique for void analysis in soil thin section. *Geoderma* **2007**, *141*, 60–70. [CrossRef]
34. Jassogne, L.; McNeill, A.; Chittleborough, D. 3D-visualization and analysis of macro-and meso-porosity of the upper horizons of a sodic, texture-contrast soil. *Eur. J. Soil Sci.* **2007**, *58*, 589–598. [CrossRef]
35. Torrance, J.; Elliot, T.; Martin, R.; Heck, R. X-ray computed tomography of frozen soil. *Cold Reg. Sci. Technol.* **2008**, *53*, 75–82. [CrossRef]

36. Wang, Y. Morphological Characterization of Wood Plastic Composite (WPC) with Advanced Imaging Tools: Developing Methodologies for Reliable Phase and Internal Damage Characterization. Masters Thesis, Oregon State University, Corvallis, OR, USA, 2007.
37. Tondi, G.; Blacher, S.; Léonard, A.; Pizzi, A.; Fierro, V.; Leban, J.; Celzard, A. X-ray microtomography studies of tannin-derived organic and carbon foams. *Microsc. Microanal.* **2009**, *15*, 384. [CrossRef]
38. Cnudde, V.; Masschaele, B.; De Cock, H.E.; Olstad, K.; Vlaminck, L.; Vlassenbroeck, J.; Dierick, M.; Witte, Y.D.; Van Hoorebeke, L.; Jacobs, P. Virtual histology by means of high-resolution X-ray CT. *J. Microsc.* **2008**, *232*, 476–485. [CrossRef]
39. Schnabl, J.; Glueckert, R.; Feuchtner, G.; Recheis, W.; Potrusil, T.; Kuhn, V.; Wolf-Magele, A.; Riechelmann, H.; Sprinzl, G.M. Sheep as a large animal model for middle and inner ear implantable hearing devices: A feasibility study in cadavers. *Otol. Neurotol.* **2012**, *33*, 481–489. [CrossRef]
40. Maret, D.; Peters, O.A.; Dedouit, F.; Telmon, N.; Sixou, M. Cone-Beam Computed Tomography: A useful tool for dental age estimation? *Med. Hypotheses* **2011**, *76*, 700–702. [CrossRef]
41. Jones, A.C.; Milthorpe, B.; Averdunk, H.; Limaye, A.; Senden, T.J.; Sakellariou, A.; Sheppard, A.P.; Sok, R.M.; Knackstedt, M.A.; Brandwood, A.; et al. Analysis of 3D bone ingrowth into polymer scaffolds via micro-computed tomography imaging. *Biomaterials* **2004**, *25*, 4947–4954. [CrossRef]
42. Hanson, N.A.; Bagi, C.M. Alternative approach to assessment of bone quality using micro-computed tomography. *Bone* **2004**, *35*, 326–333. [CrossRef]
43. Schmidt, V.-M.; Zelger, P.; Woess, C.; Pallua, A.K.; Arora, R.; Degenhart, G.; Brunner, A.; Zelger, B.; Schirmer, M.; Rabl, W.; et al. Application of Micro-Computed Tomography for the Estimation of the Post-Mortem Interval of Human Skeletal Remains. *Biology* **2022**, *11*, 1105. [CrossRef]
44. Sliva, D. Cellular and physiological effects of Ganoderma lucidum (Reishi). *Mini Rev. Med. Chem.* **2004**, *4*, 873–879. [CrossRef]
45. Deng, J.Y.; Chen, S.J.; Jow, G.M.; Hsueh, C.W.; Jeng, C.J. Dehydroeburicoic acid induces calcium- and calpain-dependent necrosis in human U87MG glioblastomas. *Chem. Res. Toxicol.* **2009**, *22*, 1817–1826. [CrossRef] [PubMed]
46. Steppe, K.; Cnudde, V.; Girard, C.; Lemeur, R.; Cnudde, J.P.; Jacobs, P. Use of X-ray computed microtomography for non-invasive determination of wood anatomical characteristics. *J. Struct. Biol.* **2004**, *148*, 11–21. [CrossRef] [PubMed]
47. Dutilleul, P.; Lontoc-Roy, M.; Prasher, S.O. Branching out with a CT scanner. *Trends Plant Sci.* **2005**, *10*, 411–412. [CrossRef] [PubMed]
48. Baumeister, A. Kapitel 5 | Das Potential historischer Karten zur Rekonstruktion des Gletscherrückgangs im Gurgler Tal. *Alp. Forsch. Obergurgl Band* **2013**, *3*, 95.
49. Meixner, W.; Siegl, G. Kapitel 1 | Historisches zum Thema Gletscher, Gletschervorfeld und Obergurgl. *Alp. Forsch. Obergurgl Band* **2010**, *1*.
50. Kellner, M. *Vergleichend Morphologische Untersuchungen an Einzelknochen des Postkranialen Skelettes in Europa Vorkommender Ardeidae*; Universitat Tierarztliche Fakultat: Munich, Germany, 1986.
51. Grodzki, D.M.; Jakob, P.M.; Heismann, B. Ultrashort echo time imaging using pointwise encoding time reduction with radial acquisition (PETRA). *Magn. Reson. Med.* **2012**, *67*, 510–518. [CrossRef]
52. Dagher, G.; Becker, K.-F.; Bonin, S.; Foy, C.; Gelmini, S.; Kubista, M.; Kungl, P.; Oelmueller, U.; Parkes, H.; Pinzani, P.; et al. Pre-analytical processes in medical diagnostics: New regulatory requirements and standards. *New Biotechnol.* **2019**, *52*, 121–125. [CrossRef]
53. Stumptner, C.; Sargsyan, K.; Kungl, P.; Zatloukal, K. Crucial role of high quality biosamples in biomarker development. In *Handbook of Biomarkers and Precision Medicine*; Chapman and Hall/CRC: London, UK, 2019; pp. 128–134.
54. Hajdas, I. Radiocarbon dating and its applications in Quaternary studies. *E&G Quat. Sci. J.* **2008**, *57*, 2–24. [CrossRef]
55. Hajdas, I.; Bonani, G.; Furrer, H.; Mäder, A.; Schoch, W. Radiocarbon chronology of the mammoth site at Niederweningen, Switzerland: Results from dating bones, teeth, wood, and peat. *Quat. Int.* **2007**, *164–165*, 98–105. [CrossRef]
56. Reimer, P.J.; Bard, E.; Bayliss, A.; Beck, J.W.; Blackwell, P.G.; Ramsey, C.B.; Buck, C.E.; Cheng, H.; Edwards, R.L.; Friedrich, M.; et al. IntCal13 and Marine13 Radiocarbon Age Calibration Curves 0–50,000 Years cal BP. *Radiocarbon* **2013**, *55*, 1869–1887. [CrossRef]
57. Ramsey, C.B.; Lee, S. Recent and Planned Developments of the Program OxCal. *Radiocarbon* **2013**, *55*, 720–730. [CrossRef]
58. Stuiver, M.; Polach, H.A. Discussion Reporting of 14C Data. *Radiocarbon* **1977**, *19*, 355–363. [CrossRef]
59. Tully, G.; Bar, W.; Brinkmann, B.; Carracedo, A.; Gill, P.; Morling, N.; Parson, W.; Schneider, P. Considerations by the European DNA profiling (EDNAP) group on the working practices, nomenclature and interpretation of mitochondrial DNA profiles. *Forensic Sci. Int.* **2001**, *124*, 83–91. [CrossRef]
60. Gilbert, M.T.; Bandelt, H.J.; Hofreiter, M.; Barnes, I. Assessing ancient DNA studies. *Trends Ecol. Evol.* **2005**, *20*, 541–544. [CrossRef]
61. Xavier, C.; Eduardoff, M.; Bertoglio, B.; Amory, C.; Berger, C.; Casas-Vargas, A.; Pallua, J.; Parson, W. Evaluation of DNA Extraction Methods Developed for Forensic and Ancient DNA Applications Using Bone Samples of Different Age. *Genes* **2021**, *12*, 146. [CrossRef]
62. Parson, W.; Pegoraro, K.; Niederstätter, H.; Föger, M.; Steinlechner, M. Species identification by means of the cytochrome b gene. *Int. J. Leg. Med.* **2000**, *114*, 23–28. [CrossRef]
63. University of Hawaii. *The Simple Fool's Guide to PCR*; University of Hawaii: Honolulu, HI, USA, 2002.
64. Cox, S.L. A Critical Look at Mummy CT Scanning. *Anat. Rec.* **2015**, *298*, 1099–1110. [CrossRef]

65. Baadsvik, E.L.; Weiger, M.; Froidevaux, R.; Rösler, M.B.; Brunner, D.O.; Öhrström, L.; Rühli, F.J.; Eppenberger, P.; Pruessmann, K.P. High-resolution MRI of mummified tissues using advanced short-T2 methodology and hardware. *Magn. Reson. Med.* **2021**, *85*, 1481–1492. [CrossRef]
66. Lynnerup, N. Mummies. *Am. J. Phys. Anthropol.* **2007**, *134*, 162–190. [CrossRef] [PubMed]

Disclaimer/Publisher's Note: The statements, opinions and data contained in all publications are solely those of the individual author(s) and contributor(s) and not of MDPI and/or the editor(s). MDPI and/or the editor(s) disclaim responsibility for any injury to people or property resulting from any ideas, methods, instructions or products referred to in the content.

biology

MDPI

Article

Dental Paleobiology in a Juvenile Neanderthal (Combe-Grenal, Southwestern France)

María Dolores Garralda [1,*], Steve Weiner [2], Baruch Arensburg [3], Bruno Maureille [4] and Bernard Vandermeersch [4]

[1] Departamento de Biodiversidad, Ecología y Evolución, Facultad de CC. Biológicas, Universidad Complutense de Madrid, 28040 Madrid, Spain
[2] Department of Chemical and Structural Biology, Weizmann Institute of Science, 234 Herzl Street, Rehovot 76100, Israel
[3] Department of Anatomy, Faculty of Medicine, Tel-Aviv University, Tel Aviv 39040, Israel
[4] University Bordeaux, CNRS, MC, PACEA UMR5199, F-33600 Pessac, France
* Correspondence: mdgarral@ucm.es

Simple Summary: Numerous prehistoric sites in Europe and the Near East provided bones and dental remains of the populations of the past. One of them is the Combe-Grenal Cave (SW France), where fossils of children and adults represent the Neanderthals who lived there more than 60 ky ago, during a harsh period of the last glaciation. In this paper, we analyze a sample of the tartar of a juvenile individual. The numerous bacteria forming the plaque are compared to those of one adult from Israel, Kebara 2, revealing the differences between the most common bacteria in a young and an older individual, probably because of their immunological systems, and the different living conditions of the human groups they represented.

Citation: Garralda, M.D.; Weiner, S.; Arensburg, B.; Maureille, B.; Vandermeersch, B. Dental Paleobiology in a Juvenile Neanderthal (Combe-Grenal, Southwestern France). *Biology* 2022, 11, 1352. https://doi.org/10.3390/biology11091352

Academic Editors: Mary H. Schweitzer and Ferhat Kaya

Received: 1 August 2022
Accepted: 9 September 2022
Published: 14 September 2022

Publisher's Note: MDPI stays neutral with regard to jurisdictional claims in published maps and institutional affiliations.

Copyright: © 2022 by the authors. Licensee MDPI, Basel, Switzerland. This article is an open access article distributed under the terms and conditions of the Creative Commons Attribution (CC BY) license (https://creativecommons.org/licenses/by/4.0/).

Abstract: Combe-Grenal site (Southwest France) was excavated by F. Bordes between 1953 and 1965. He found several human remains in Mousterian levels 60, 39, 35 and especially 25, corresponding to MIS 4 (~75–70/60 ky BP) and with Quina Mousterian lithics. One of the fossils found in level 25 is Combe-Grenal IV, consisting of a fragment of the left corpus of a juvenile mandible. This fragment displays initial juvenile periodontitis, and the two preserved teeth (LLP4 and LLM1) show moderate attrition and dental calculus. The SEM tartar analysis demonstrates the presence of *cocci* and filamentous types of *bacteria*, the former being more prevalent. This result is quite different from those obtained for the two adult Neanderthals Kebara 2 and Subalyuk 1, where more filamentous *bacteria* appear, especially in the Subalyuk 1 sample from Central Europe. These findings agree with the available biomedical data on periodontitis and tartar development in extant individuals, despite the different environmental conditions and diets documented by numerous archeological, taphonomical and geological data available on Neanderthals and present-day populations. New metagenomic analyses are extending this information, and despite the inherent difficulties, they will open important perspectives in studying this ancient human pathology.

Keywords: Neanderthal; Combe-Grenal; juvenile; mandible; periodontitis; tooth; tartar; SEM analysis

1. Introduction

The Combe-Grenal site is located east of the Domme village (Dordogne), on the right side of the valley of a small Dordogne dried river. The site, facing south-west, corresponds to a probably very deep and large rock-shelter, naturally formed within Cogniacian limestone. Only a small part of the rock-shelter remains, preserving only a narrow and very small cave at its northeastern angle. Such geomorphology describes the site name "grotte de Combe-Grenal" since at least 1817. Combe-Grenal rock-shelter was first excavated by D. Peyrony in 1929, who identified three Mousterians layers, but the most important investigations were those led by F. Bordes, between 1953 and 1965 [1–4].

Since 2014, new scientific fieldwork is in progress under the direction of J.-Ph. Faivre (PACEA, Bordeaux).

Considering human fossil remains, several pieces were found by Bordes in Mousterian levels 60, 39, 35 and 25. Geological and faunal studies conducted by Guadeli and Laville [5] attribute level 60 to MIS 6, level 39 to MIS 5a, while levels 35 to 25 are related to MIS 4. Most of the Combe-Grenal fossils (those from layers 35, and especially 25, should be placed chronologically at the beginning of MIS 4 (~70 to 60 ky) and assigned to its coldest period (~70 to 65 ky). Paleoenvironmental and chronostratigraphic data document climatic changes toward colder conditions, first humid and later increasingly drier, as well as a progression of the open Arctic milieu fauna, confirming the cold and harsh environment in which people then lived [5].

The anthropological fossils found at Combe-Grenal were the objects of detailed morpho-anatomical descriptions and analyses [6–9]. The whole sample is presently preserved at the Musée National de Préhistoire at Les Eyzies (Southwestern France).

Most of the remains were found in level 25, where several young adult males and females (MNI ~8) of different ages were identified [6]. Bordes' unpublished data demonstrate the dispersion of the fossils in several excavation grid squares, very close to one another, located at the center of the back part of the rock-shelter [6]. All the human fossils were fragmented and randomly mixed with abundant faunal remains and lithics. There were no traces of deliberate burials, but cut marks were identified on several fragments [6,7,9]. Morphological and anatomical analyses of the Combe-Grenal fossils have led to their assignment to Neanderthals.

The aim of this contribution is the study of a tartar sample obtained from the mandibular fragment Combe-Grenal IV. We will briefly summarize the interest in dental calculus analyses, followed by the main morphological characteristics of the fossil, the oral pathology and the results of the tartar SEM analysis in comparison with the previously published data from other Neanderthals and new methods of analyses.

2. Brief Reminder of Tartar (Dental Calculus) Etiology

Tartar or dental calculus is a form of hardened dental plaque, caused by the precipitation of minerals from saliva and gingival crevicular fluid in the tooth's plaque. Such a process kills the bacterial cells within the dental plaque, forming a rough and hardened surface ideal for further plaque formation, namely tartar [10].

Two types of dental calculus have been described. Supragingival tartar affects the gums along the gumline, while subgingival tartar forms within the narrow sulcus existing between the teeth and the gingiva [10]. Dental calculus formation is associated with several clinical manifestations, including receding gums and chronically inflamed gingiva.

According to Lang et al. [10], tartar is composed of both inorganic (mineral) and organic (cellular and extracellular matrix) components. The cells within the dental calculus are primarily bacterial, but also include at least one species of *Archaea* (usually called "*cocci*") and several species of yeast. Trace amounts of dietary and environmental micro debris or plant DNA have also been found.

The processes of dental calculus formation are not well understood. Tartar forms in incremental layers, but the timing and triggering of these events are poorly understood and vary widely among individuals, probably related to age, gender, diet, etc. [10]. Supragingival tartar is more abundant on the buccal surfaces of the upper maxillary molars and the lingual surfaces of the mandibular molars, while subgingival tartar forms below the gumline and is typically dark in color due to the presence of black-pigmented bacteria.

Dental calculus has been described in animals (e.g., [11]) and documented in various human groups and individuals from Prehistory to present times. Sometimes, even if exceptional, tartar can be present as a very thick deposit, such as on the T15 individual from the Medieval cemetery from Clarensac (Gard, Southeastern France; [12,13]). Concerning human fossil teeth, unfortunately, in the past, many of them have been excessively cleaned, destroying and removing tartar deposits. However, we do have a few with preserved tartar deposits. This is the case for several Early Upper Pleistocene Neanderthals, whose analyses offer new data on their biology and expand the knowledge of their hunter–gatherer population behaviors.

3. Materials and Methods

3.1. Materials

Combe-Grenal IV is a fragment of the left side mandibular corpus corresponding to the upper part of the left mandibular body (Figure 1), ranging from the distal margin of the left lower second premolar (from here LLP3) alveolus to the mesial septum of the left lower second molar (LLM2; [6]). It preserves two lower teeth, the lower left second premolar (LLP4) and the first molar (LLM1), both with tartar deposits (Figure 1, red arrows) around the crowns and interproximal facets with the LLP3 and the LLM2 (both absent), which indicate that all four teeth had emerged and were functional.

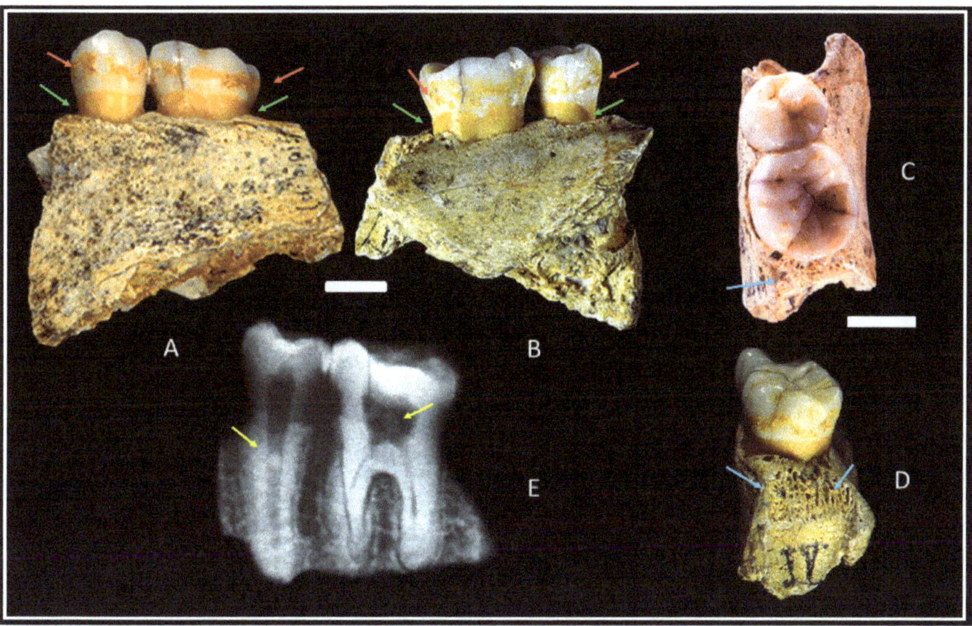

Figure 1. Combe-Grenal IV: external (**A**), internal (**B**), occlusal (**C**), distal side of the LLM1and the alveolus of the LLM2 (**D**), X-ray (**E**) external side. Scales = 10 mm.

On this mandibular fragment, Combe-Grenal IV, we can also observe the alveolus of the LLM2 (Figure 1C,D). It has highly visible osseous trabeculae, and the alveolar ridge is altered by gingivitis (Figure 1 blue arrows), indicating that the LLM2 had fully erupted and was functional long before the individual's death, causing the distal interproximal facet on the LLM1. The LLP4 has the mesiodistal axis slightly oblique in comparison to that of the molar. The crown has traces of occlusal attrition (degree 1; [14]) and two interproximal facets, while the mesial one has a deep vertical sulcus [6].

Few age markers can be considered regarding this incomplete fossil, such as the close apex of the two preserved teeth (Figure 1E), their weak attrition (degree 1 from Murphy; [14]) or the small interproximal facet caused on the LLM1 distal side (Figure 1D) by the (absent) LLM2. It is possible to estimate the age at death of this individual by using the charts on dental eruption published for modern children, although they are based on samples with very different biological and environmental conditions. Thus, according to the Ubelaker [15] schemes and the AlQahtani et al. [16] atlas, the age was

15 ± 3 years and 15.5 years, respectively. Consequently, Combe-Grenal IV can be assigned to a juvenile individual.

The alveolar arch (Figure 1 green arrows) displays a slight degree of resorption and is separated from the tooth cement–enamel junction by 2.5/3.02 mm. The loss of osseous mass and the alveolar destruction can also be seen in the radiograph (Figure 1E), showing a pathology that seems to be the result of incipient periodontal disease. This slight exposure of roots, especially if we consider the lack of conjunctive tissue, varies between 1.0 and 1.5 mm in height [17]. Both preserved teeth on Combe-Grenal IV show supragingival tartar deposits (Figure 1 red arrows) forming a wide band around the crowns, separated by 5/7 mm from the alveolar margin. As in degree 2 on the Brothwell [18] scale, and, according to the classification of periodontal diseases [19], they correspond to stage I (early–mild) and nearly stage II (moderate).

Two pulp stones (pulpoliths) of different sizes (that rattle when shaking the fragment) also appear in the pulp chamber of those teeth (Figure 1E, yellow arrows). Such pathology has been documented in other Neanderthals, as in the case with Combe-Grenal X and 29 [6] and Kebara 2 [20,21].

3.2. Methods

A small sample of supragingival dental calculus was detached from the lingual surface of the Combe-Grenal IV LLM1 and processed for a scanning electron microscope study at the Weizmann Institute of Science at Rehovot (Israel).

We measured the *bacteria* directly from the micrographs, keeping in mind that these are only estimates. This is because we cannot know whether we are viewing the real length or diameter, given that they were partially embedded in the calcified matrix. The average dimensions were photographed in three pictures, with the different magnifications indicated in Tables 1 and 2. We measured the *bacteria* appearing more complete (avoiding the empty cavities). Each crown measurement was repeated on three different days by two of the authors (S.W. and B.A.), and the inter- and intra-observer error between measurements was <4%.

Table 1. Measurements in μm (diameter and length) of the Combe-Grenal IV tartar *bacteria*, taken from the SEM pictures indicated.

	Measurements	N	Mean	Std. Dev.	Std. Error	Minimum	Maximum	Variance
C. Grenal 304 (10.000 M)	Diameter	16	0.493	0.109	0.027	0.360	0.700	0.012
	Length	7	0.791	0.262	0.099	0.590	1.340	0.069
C.-Grenal 306 (3.500 M)	Diameter	31	0.601	0.138	0.025	0.310	0.850	0.019
	Length	10	1.608	0.472	0.149	1.050	2.550	0.222
C.-Grenal 308 (20.000 M)	Diameter	15	0.477	0.165	0.043	0.330	0.840	0.027
	Length	7	1.499	0.376	0.142	1.160	2.200	0.141
Total C.-Grenal IV	Diameter	62	0.543	0.149	0.019	0.330	0.850	0.022
	Length	24	1.338	0.522	0.107	0.590	2.550	0.272

On the obtained images, the bacteria identification and measurements were performed by Image Tools 3 "UTHSCA" analysis, and the results are expressed as means $\pm$ SE (standard error). The analyses included a breakdown and one-way ANOVA tests. The *p*-values indicated the post hoc significance levels for the respective pairs of means, and a *p*-value of <0.05 was considered significant. The calculations were performed using the SPSS statistical package (1990) and the STATISTICA package (StatSoft Inc., Tulsa, OH, USA, 1995), and their results are given in Table 3.

Table 2. The measurements in µm (diameter and length) of the Kebara 2 tartar bacteria, taken from the indicated SEM pictures.

Pictures	Measurements	N	Mean	Std. Dev.	Std. Error	Minimum	Maximum	Variance
Kebara 601 (10.000 M)	Diameter	20	0.718	0.161	0.036	0.350	0.940	0.026
	Length	18	1.241	0.422	0.100	0.830	2.360	0.178
Kebara 603 (5.000 M)	Diameter	17	0.825	0.154	0.037	0.570	1.140	0.024
	Length	16	3.116	0.719	0.180	1.940	4.310	0.517
Kebara 606 (5.000 M)	Diameter	22	0.555	0.142	0.030	0.400	0.810	0.020
	Length	0						
Total Kebara	Diameter	59	0.687	0.187	0.024	0.400	1.140	0.035
	Length	34	2.123	1.109	0.190	0.830	4.310	1.229

Table 3. Statistical analysis of the total lengths and diameters of the tartar *bacteria* from Combe-Grenal IV and Kebara 2.

Individuals	Traits	N	Mean	Std. Error	P
Total Kebara 2	Diameter	59	0.687	0.187	0.0001
Total Combe-Grenal IV		62	0.543	0.149	
Total Kebara 2	Length	34	2.123	1.109	0.02
Total Combe-Grenal IV		24	1.338	0.522	

4. Results

The two preserved teeth on Combe-Grenal IV, left LLP4 and LLM$_1$, show supragingival dental calculus deposits (Figure 1 red arrows) forming a wide band around the crowns, separated 5/7 mm from the alveolar border, as in degree 2 on the Brothwell [18] scale and I/II of the recent classification [19].

On the surface of the Combe-Grenal IV tartar sample, fine crystal dental calculus deposits appear in the macro photographs at different magnifications. They reveal alternate layers running from the first calcified plaque directly covering the enamel surfaces to the outermost and final calcified layer, indicating various stages in the dental calculus formation of this individual. Magnifications of 3500 µm (Figure 2A), 10,000 µm (Figure 2B) and 20,000 µm (Figure 3A) clearly show both empty bacterial cavities and complete bacteria embedded in the calcified matrix. The bacteria present are *cocci* and filamentous types, although it is not possible to recognize the specific fossilized micro-organisms among the ~325 species that could be present in the oral cavity [22].

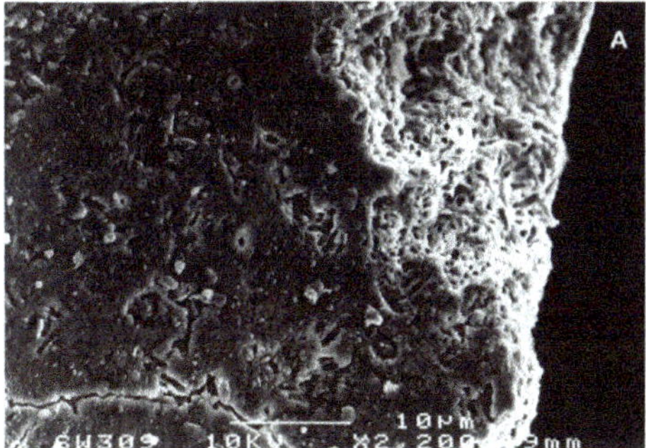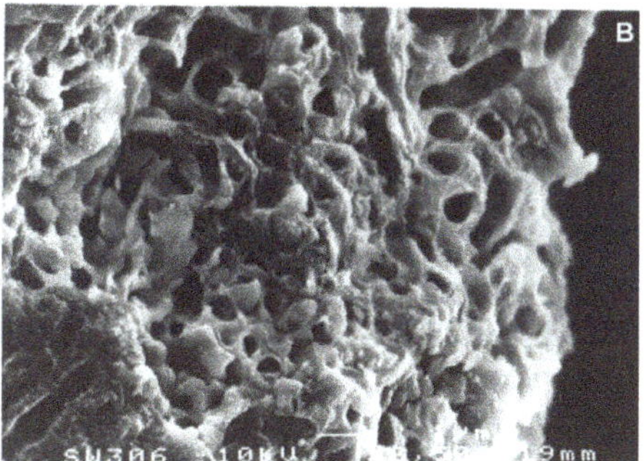

Figure 2. Combe-Grenal IV. (**A**): (picture 306), tartar *bacteria* at 3500 SEM magnification. (**B**): (picture 304): tartar *bacteria* at 10,000 SEM magnification, white scale = 1 µm.

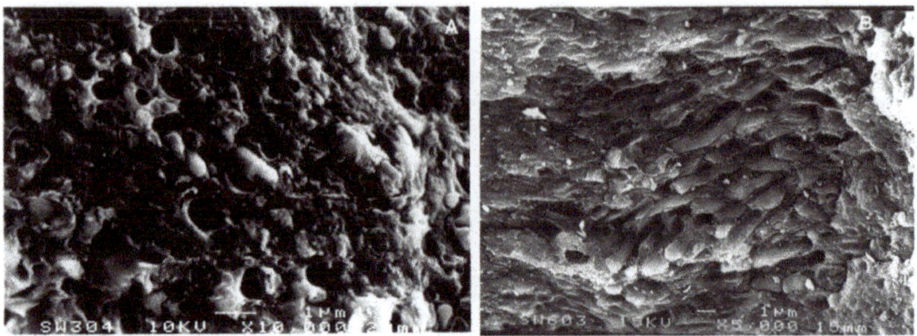

Figure 3. Combe Grenal IV. (**A**): (picture 308), tartar *bacteria* at 20,000 SEM magnification. (**B**): Kebara 2 (picture 603), tartar *bacteria* at 5000 SEM magnification.

We used the SEM images to compare the distribution of these microbiotas in two different Neanderthal fossils: the juvenile Combe-Grenal IV and the adult male Kebara 2 (Israel), for which we have the analysis of a sample also from the calculus of his LLM1 [20].

As can be observed in Figure 3B, the Levantine fossil contains numerous *cocci* bacterial types [20], and rods are more frequent than in the young Combe-Grenal individual.

There are also some differences in the size of the identified bacteria between both individuals. In Tables 1 and 2, the parameters corresponding to the length and diameter (in µm) of the *bacteria* in the compared fossils cited above are given. They were measured on the SEM microphotographs corresponding to the indicated numbers (Combe-Grenal IV: 304, 306, and 308; Kebara 2: 601, 603, and 606).

In Figure 4, the "Box and Whisker Plot" shows the differences obtained for the total dimensions of Combe-Grenal IV and Kebara 2 *bacteria*, with the former (Figure 4, left) corresponding to the diameter and the latter (Figure 4, right) to the length. On both graphs appear the mean values and the variation range of ± 1 and 1.96 standard errors. The differences between the total values obtained for both fossil individuals (Table 3) are statistically significant, particularly those comparing the diameters ($p < 0.001$), indicating larger dimensions of the bacterial flora on the adult male Kebara 2 than on the juvenile Combe-Grenal IV.

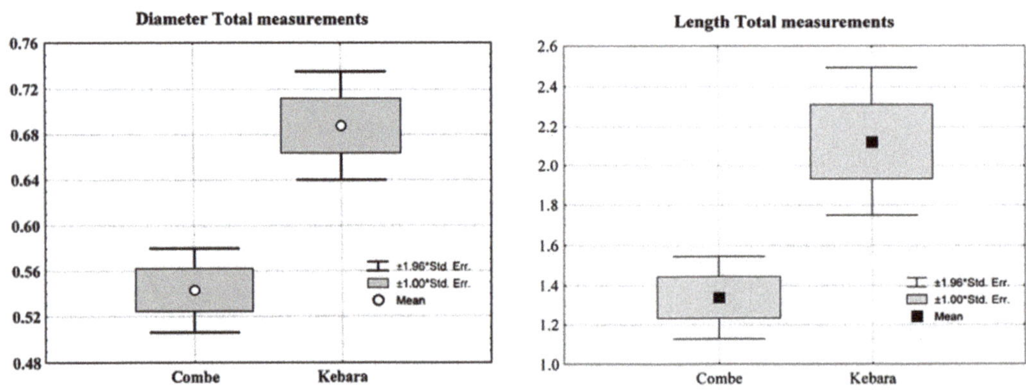

Figure 4. Comparison (Box and Whisker Plot) of the total diameter (**left**) and length (**right**) measurements (in µm) of the Combe-Grenal IV and Kebara 2 bacteria. The mean and the ranges of variation of ±1 and 1.96 error standard deviations are indicated.

These results agree with the available information on the older adult Subalyuk 1 (Hungary), which—following Pap et al. [23]—shows the presence of more filamentous type *bacteria* than on Kebara 2 [20]. As their dimensions were not given, no statistical comparison with the other two Neanderthals can be made.

5. Discussion

Tartar, being a mineralized form of dental plaque adhering to the surface of the tooth, can be preserved, and the study of this durable material using SEM provides information on the microbial flora responsible for the periodontal disease of ancient hominid fossils. It is well known that the presence of periodontitis, in general, is the result of a dense accumulation of micro-organisms on the tooth surface and the host response (innate and acquired immunity) of each individual [24].

The most ancient case among hominids is that described by Ripamonti [25] on a juvenile *Australopithecus*. Concerning the Neanderthals, the publications on the *bacteria* found in the dental calculus of Kebara 2 [20,21], Subalyuk 1 [23] and the present paper demonstrate the rich oral flora present in ancient human populations.

The results of the tartar macroanalyses of these three Neanderthals, Combe-Grenal IV, Kebara 2 and Subalyuk 1, also indicate differences in the oral flora causing supragingival dental calculus and periodontitis among them. When interpreting these findings, it is important to consider not only the chronology and environment in which they lived, but also the individual's age at death (more *cocci* on the juvenile) and the available data on diets. The latter is difficult to obtain when dealing with sites excavated long ago. Another influencing factor could be immunological differences between individuals, because each one of the three Neanderthals studied could have had different reactions to similar stress, illness or diet. This hypothesis is, however, impossible to verify.

As previously indicated, the three considered Neanderthals have similar antiquity, related to MIS 4 (both European fossils) or the beginning of MIS 3 (Kebara 2), meaning that they lived in a cold and dry environment, undoubtedly colder in Europe than in the Levant. In the case of Combe-Grenal level 25, where the studied fossil was found, sedimentological analyses indicate very cold and dry weather conditions with an open Arctic milieu fauna [5]. Layer 11 from Subalyuk Cave displayed similar conditions, according to Bartucz et al. [26]. In Kebara Cave, the analyses demonstrated that Unit XII (where the burial of the individual Kebara 2 appeared) was formed in a dry and cold environment, but not as extreme as that known for the European sites [27].

None of the three individuals under consideration have been analyzed to obtain the "carbon and nitrogen isotopic signature" used to evaluate aspects of the diet, but some data are available for other Neanderthals found, for example, in Scladina [28], Marillac [29], Vindija [30], Saint-Césaire [31,32], Jonzac [33] or Troisième caverne from Goyet [34]. The results obtained from these fossils indicate that animal tissue must have been a very important source of food, regardless of their different chronologies and environments (e.g., the reindeer so frequent in Marillac, and absent from the Kebara environment). Recently, Ca isotopes were used to assess aspects of the diet of the MIS 5 Neanderthal Regourdou 1 [35]. This study demonstrates the carnivore-like diet of the fossil and, also, the ingestion of a small percentage of bone, probably during the consumption of bone fat and red marrow.

The evaluation of the possible ingestion of vegetables in Paleolithic times is often difficult because of the absence of such remains in many sites, or the imitations imposed by the archeological data sets. Nevertheless, several well-documented excavations, such as those for the Kebara Cave, document not only ephemeral seasonal hunting during late spring–summer, to intensive winter to early spring hunting, but also archeobotanical remains indicating that plant foods appear to have been gathered during the autumn (October–December) and early spring (March–April) and remained more or less constant throughout the analyzed sequence [36,37]. Hardy and colleagues' [38] research on dental calculus from five El Sidrón Neanderthals has demonstrated the ingestion by these individuals of different kinds of probably cooked plants. Moreover, based on nitrogen isotope analyses for the Spy Neanderthal bone collagen, it was hypothesized that plant consumption accounted for up to 20% of the sources of their diet [39].

The studies on Combe-Grenal IV, Kebara 2 and Subalyuk 1 also allow us to identify the presence of mild periodontitis in Combe-Grenal IV [6], and a more advanced process with thicker dental calculus deposits in the adults Kebara 2 [40] and Subalyuk 1 [23]. The presence of such pathologies on a juvenile specimen is not an exception in Middle Paleolithic times; the Combe-Grenal I (~7 years old) child's mandible shows a slight periodontosis and a thin tartar line on the two preserved deciduous molars [6]. Genetic factors and living conditions are highly correlated to the development of periodontal disease. Note, also, that the marked interproximal facets described on the juvenile Combe-Grenal IV LLP4 and LLM1 [6] reflect strong masticatory forces, which developed daily, perhaps not only due to diet but also due to paramasticatory activities.

6. Conclusions

The study of the mandibular fragment Combe-Grenal IV (MIS 4, ~ 70 ky) allows the assignment of this fossil to a relatively robust juvenile Neanderthal individual (~15 years old). Two teeth are preserved (LLP4 and LLM_1), and show moderate attrition, pulpoliths and initial root hypercementosis. Mild periodontitis affected the alveolar region, and both teeth display almost continuous 1 mm high strip dental calculus on their buccal crown lengths.

The SEM analysis of the LLM1 dental calculus demonstrated the prevalence of *cocci*-type *bacteria*, which is usual in juvenile individuals. This group of microbiota is less prevalent in the tartar from the adult male Kebara 2, where "rods" are more frequent, although less so than in the Subalyuk 1 older adult. In agreement with these observations, the measurements made on the SEM pictures from Combe-Grenal IV and Kebara 2 show larger dimensions of the *bacteria* in the latter individual, with statistically significant differences being found between the means of the lengths and, particularly, between the diameters.

It is interesting to remark that tartar was preserved in the studied fossils, while other individuals, such as the late Neanderthal maxillary CF-1 from Cova Foradà [41], was assigned to an old individual who suffered advanced periodontal disease, but no tartar has been described. In other fossils, such as Krapina-J, the modern manipulations that removed the dental plaque are visible.

Over the last two decades, scientists have relied increasingly on analyses of stable carbon, nitrogen and oxygen isotopes, as well as strontium and other trace elements, in bone, tooth enamel and dentine in order to determine the role of plants, animal tissues and fish in past human diets. Various studies indicate that one can differentiate between the consumption of C3 and C4 plants and trace the exploitation of terrestrial and marine mammals and fish, e.g., [42–44]. Even with the use of these techniques, we still cannot determine the ratios of animal tissues versus plant foods when analyzing Neanderthal remains [45–47]. However, testing the correlation between ethnographic information and stable isotopes on samples of more recent chronology has demonstrated good agreement [48].

At present, new possibilities are opening with metagenomic analyses which can enable the detailed study of the microbial genomes preserved in the dental calculus. The first results demonstrated, as expected, the regional differences in ecology among the several Neanderthal specimens considered, and consequently differences in diet and paleogenetics [49], and some of them are questionable [50,51]. Similar results have been published by studying the faunal remains found on numerous sites (e.g., [50] in Europe and the Near East, often allowing for the determination of the seasonal periods of hunting, as in the cited case for Kebara [34,35] or Marillac [52]). The macromorphological tartar studies that we present in this contribution for Combe-Grenal IV, or those on the Kebara 2 or Subalyuk 1's tartar, reflect the development of the microbiota related to the different diets and ages at death of these Neanderthals, evolving through life to include more rods, as is known for extant populations [10].

Author Contributions: Conceptualization, M.D.G. and B.A.; methodology and software, S.W. and B.A.; validation, formal analysis and investigation, all the authors; writing—original draft preparation, all the authors; writing—review and editing, all the authors; supervision, M.D.G. All authors have read and agreed to the published version of the manuscript.

Funding: M. D. Garralda received a Professor Research Grant from the Spanish Ministry of Education and Research to study the Combe-Grenal Neanderthal series in collaboration with the Laboratoire d'Anthropologie (Bordeaux University, France). B. Maureille works on Quina Neanderthals benefit of funding from the European Research Council (ERC) under the European Union's Horizon 2020 research and innovation program (grant agreement No 851793). This research also benefited from the scientific framework of the University of Bordeaux's IdEx "Investments for the Future" program/GPR "Human Past".

Institutional Review Board Statement: Not applicable.

Informed Consent Statement: Not applicable.

Data Availability Statement: Not applicable.

Acknowledgments: We thank B. Alonso, Faculty of Odontology of the Universidad Complutense de Madrid, for discussion of the results and bibliographical references, E. Lundin for the revision of the manuscript and P. Billet for the radiographies. Special thanks are due to D. de Sonneville-Bordes for her important help regarding F. Bordes Combe-Grenal excavation documents and to J. J. Cleyet-Merle, Director of the Musée National de Préhistoire des Eyzies, who gave us the permission to study the Combe-Grenal anthropological remains.

Conflicts of Interest: The authors declare that there is no conflict of interest.

References

1. Bordes, F. La stratigraphie de la grotte de Combe-Grenal (Dordogne). Note préliminaire. *Bull. Soc. Préhistorique Française* **1955**, *52*, 426–429. [CrossRef]
2. Bordes, F. *A Tale of Two Caves*; Harper & Row: New York, NY, USA, 1972.
3. Bordes, F.; Prat, F. Observations sur les faunes du Riss et du Würm I en Dordogne. *l'Anthropologie* **1965**, *69*, 31–46.
4. Bordes, F.; Laville, H.; Paquerau, M.M. Observations sur le Pléistocéne supérieur du gisement de Combe-Grenal (Dordogne). *Actes Soc. Linnéenne Bordx.* **1966**, *103*, 3–19.
5. Guadelli, J.L.; Laville, H. L'environnement climatique de la fin du Moustérien à Combe-Grenal et à Camiac. Confrontation des Données Naturalistes et Implications. In *Paléolithique Moyen Récent et Paléolithique Superieur Ancien en Europe*; Farizy, C., Ed.; Mém. du Musée de Préhistoire d'Ile de France: Nemours, France, 1990; pp. 43–48.
6. Garralda, M.D.; Vandermeersch, B. Les Néandertaliens de la Grotte de Combe-Grenal (Domme, France). *Paléo Rev. D'archéologie Préhistorique* **2000**, *12*, 213–259.
7. Garralda, M.D.; Giacobini, G.; Vandermeersch, B. Neanderthal cutmarks: Combe-Grenal and Marillac (France). A SEM analysis. *Anthropologie* **2005**, *43*, 189–197.
8. Maureille, B.; Garralda, M.D.; Madelaine, S.; Turq, A.; Vandermeersch, B. Le plus ancien enfant d'Aquitaine, Combe-Grenal 31 (Domme, France). *PALEO. Rev. D'archéologie Préhistorique* **2011**, *21*, 189–202. [CrossRef]
9. Gómez-Olivencia, A.; Garralda, M.D.; Vandermeersch, B.; Madelaine, S.; Arsuaga, J.-L.; Maureille, B. Two newly identified mousterian human rib fragments from Combe-Grenal. *PALEO. Rev. D'archéologie Préhistorique* **2013**, *24*, 229–235. [CrossRef]
10. Lang, N.P.; Mombelli, A.; Attström, R. Dental Plaque and Calculus. In *Clinical Periodontology and Implant Dentistry*, 3rd ed.; Lindle, J., Karring, T., Lang, N.P., Eds.; Munsksgaard: Copenhagen, Denmark, 1998; pp. 102–137.
11. Ottoni, C.; Guelli, M.; Ozga, A.T.; Stone, A.C.; Kersten, O.; Bramanti, B.; Porcier, B.; Van Neer, W. Metagenomic analysis of dental calculus in ancient Egyptian baboons. *Sci. Rep.* **2019**, *9*, 19637. [CrossRef]
12. Gleize, Y.; Castex, D.; Duday, H.; Chapoulie, R. Analyse préliminaire et discussion sur la nature d'un dépôt dentaire très particulier. *Bull. Mémoires Société d'Anthropologie Paris. BMSAP* **2005**, *17*, 5–12. [CrossRef]
13. Verger-Pratoucy, J.-C. Commentaires sur l'article: Analyse préliminaire et discussion sur la nature d'un dépôt dentaire très particulier. *Bull. Mémoires Société d'Anthropologie Paris. BMSAP* **2007**, *19*, 95–102.
14. Smith, B.H. Patterns of Molar Wear in Hunter-Gatherers and Agriculturalists. *Amer. J. Phys. Anthropol.* **1984**, *63*, 39–56. [CrossRef]
15. Ubelaker, D. *Human Skeletal Remains*, 2nd ed.; Taraxacum Press: Washington, DC, USA, 1989.
16. AlQahtani, S.J.; Hector, M.P.; Liversidge, H.M. The London Atlas of Human Tooth Development and Eruption. *Am. J. Phys. Anthropol.* **2010**, *142*, 481–490. [CrossRef] [PubMed]
17. Lavigne, S.E.; Molto, J.E. System of measurement of the severity of periodontal disease in Past Populations. *Int. J. Osteoarch.* **1995**, *5*, 265–273. [CrossRef]
18. Brothwell, D. *Digging Up Bones*, 3rd ed.; Cornell University Press: Ithaca, NY, USA, 1981.
19. Dietrich, P.; Ower, P.; Tank, M.; West, N.X.; Walter, C.; Needleman, I.; Hughes, F.J.; Wadia, R.; Milward, M.R.; Hodge, P.J.; et al. Periodontal diagnosis in the context of the 2017 classification system of periodontal diseases and conditions—Implementation in clinical practice. *Br. Dent. J.* **2019**, *226*, 16–22. [CrossRef] [PubMed]
20. Vandermeersch, B.; Arensburg, B.; Tillier, A.-M.; Rak, Y.; Weiner, S.; Spiers, M.; Aspillaga, E. Middle Paleolithic dental Bacteria from Kebara, Israel. *Comptes Rendus L Acad. Sci. Ser. Ii* **1994**, *319*, 727–731.

21. Littner, M.; Tillier, A.-M.; Arensburg, B.; Kaffe, I. A Middle Paleolithic mandible from Kebara Israel (60.000 BP) in View of Oral Health of Modern Humans. *J. Isr. Dent. Assoc.* **1996**, *13*, 1.

22. Lindhe, J.; Karring, T.; Lang, N.P. *Clinical Periodontology and Implant Dentistry*, 3rd ed.; Munksgaard: Copenhagen, Denmark, 1998.

23. Pap, I.; Tillier, A.-M.; Arensburg, B.; Weiner, S.; Chech, M. First scanning electron microscope analysis of dental calculus from European Neanderthals: Subalyuk (Middle Paleolithic, Hungary). *Bull. Mémoires Société d'Anthropologie Paris* **1995**, *7*, 69–72. [CrossRef]

24. Hillson, S. *Dental Anthropology*; Cambridge University Press: Cambridge, UK, 1996.

25. Ripamonti, U. Paleopathology in *Australopithecus africanus*: A suggested case of a 3 million-year-old prepubertal periodontitis. *Amer. J. Phys. Anthropol.* **1988**, *76*, 197–210. [CrossRef]

26. Bartucz, L.; Dancza, J.; Hollendonner, F.; Kadic, O.; Mottl, M.; Pataki, V.; Palosi, E.; Szabo, J.; Vendl, A. *Die Mussolini-Höhle (Subalyuk) bei Cserépfalu*; Editio Instituti Regni Hungarici Geologici: Budapest, Hungary, 1939; Volume 14.

27. Bar-Yosef, O. Middle Paleolithic Adaptations in the Mediterranean Levant. In *The Evolution and Dispersal of Modern Humans in Asia*; Akazawa, T., Aoki, K., Kimura, T., Eds.; Hokusen-Sha: Tokyo, Japan, 1992; pp. 189–216.

28. Bocherens, H.; Billiou, D.; Patou-Mathis, M.; Otte, M.; Bonjean, D.; Toussaint, M.; Mariotti, A. Palaeoenvironmental and palaeoditary implications of isotopic biogeochemistry of late interglacial Neandertal and mammal bones in Scladina Cave (Belgium). *J. Archeol. Sci.* **1999**, *26*, 599–607. [CrossRef]

29. Bocherens, H. Isotopic biogeochemistry as a marker of Neandertal diet. *Anthrop. Anz.* **1997**, *55*, 101–120. [CrossRef]

30. Richards, M.P.; Pettitt, P.B.; Trinkaus, E.; Smith, F.H.; Paunovic, M.; Karavanic, I. Neanderthal diet at Vindija and Neanderthal predation: The evidence from stable isotopes. *Proc. Natl. Acad. Sci. USA* **2000**, *97*, 7663–7666. [CrossRef] [PubMed]

31. Balter, V.; Simon, L. Diet and behavior of the Saint-Césaire Neanderthal inferred from biogeochemical data inversion. *J. Hum. Evol.* **2006**, *51*, 329–338. [CrossRef] [PubMed]

32. Bocherens, H.; Drucker, D.; Billiou, D.; Pathou-Mathis, M.; Vandermeersch, B. Isotopic evidence for diet and subsistence pattern of the Saint-Césaire 1 Neanderthal review and use of a multi-source mixing model. *J. Hum. Evol.* **2005**, *49*, 71–87. [CrossRef] [PubMed]

33. Richards, M.P.; Taylor, G.; Steele, T.; McPherron, S.; Soressi, M.; Jaubert, J.; Orschiedt, J.; Mallye, J.-B.; Rendu, W.; Hublin, J.-J. Isotopic analysis of a Neanderthal and associated fauna from the site of Jonzac (Charente-Maritime, France). *J. Hum. Evol.* **2008**, *55*, 179–185. [CrossRef] [PubMed]

34. Wibing, C.; Rougier, H.; Baumann, C.; Comeyne, A.; Crevecoeur, I.; Drucker, D.; Gaudzinski-Windheuser, S.; Germonpré, M.; Gomez-Olivencia, A.; Krause, J.; et al. Stable isotopes reveal patterns of diet and mobility in the last Neanderthls and first modern humans in Europe. *Sci. Rep.* **2019**, *9*, 4433. [CrossRef]

35. Dodat, P.-J.; Tacail, T.; Albalat, E.; Gómez-Olivencia, A.; Couture-Veschambre, C.; Holliday, T.; Madelaine, S.; Martin, J.E.; Rmoutilova, R.; Maureille, B.; et al. The dietary reconstruction of the Regourdou 1 Neandertal (MIS 5, France) using bone calcium isotopes. *J. Hum. Evol.* **2021**, *10*, 21230.

36. Bar-Yosef, O. Eat what is there: Hunting and Gathering in the World of Neanderthals and their Neighbours. *Int. J. Osteoarch.* **2004**, *14*, 333–342. [CrossRef]

37. Lev, E.; Kislev, M.E.; Bar-Yosef, O. Mousterian vegetal food in Kebara Cave, Mt. Carmel. *J. Archaeol. Sci.* **2005**, *32*, 475–484. [CrossRef]

38. Hardy, K.; Buckley, S.; Collins, M.J.; Estalrrich, A.; Brothwel, D.; Copeland, L.; García-Tabernero, A.; García-Vargas, S.; de la Rasilla, M.; Lalueza-Fox, C.; et al. Neanderthal medics? Evidence for food, cooking, and medicinal plants entrapped in dental calculus. *Naturwissenschaften* **2012**, *99*, 617–626. [CrossRef]

39. Naito, Y.I.; Chikaraishi, Y.; Drucker, D.G.; Ohkouchi, N.; Semal, P.; Wißing, C.; Bocherens, H. Ecological niche of Neanderthals from Spy Cave revealed by nitrogen isotopes of individual amino acids in collagen. *J. Hum. Evol.* **2016**, *93*, 82–90. [CrossRef]

40. Duday, H.; Arensburg, B. La Pathologie. In *Le Squelette Moustérien de Kébara 2*; Bar-Yosef, O., Vandermeersch, B., Eds.; Cahiers de Paléoanthropologie, CNRS: Paris, France, 1991; pp. 179–193.

41. Lozano, M.; Subirà, M.E.; Aparicio, J.; Lorenzo, C.; Gómez-Merino, G. Toothpicking and periodontal disease in a Neanderthal specimen from Cova Foradà site (Valencia, Spain). *PLoS ONE* **2013**, *8*, e76852. [CrossRef] [PubMed]

42. van der Merwe, N.J.; Williamson, R.F.; Pfeiffer, S.; Thomas, S.C.; Allegretto, K.O. The Moatfield ossuary: Isotopic dietary analysis of an Iroquoian community, using dental tissue. *J. Anthropol. Archaeol.* **2003**, *22*, 245–261. [CrossRef]

43. Schoeninger, M.J.; Reeser, H.; Hallin, K. Paleoenvironment of *Australopithecus anamensis* at Allia Bay, East Turkana, Kenya: Evidence from mammalian herbivore enamel stable isotopes. *J. Anthropol. Archaeol.* **2003**, *22*, 200–207. [CrossRef]

44. Lee-Thorp, J.; Sponheimer, M. Three case studies used to reassess the reliability of fossil bone and enamel isotope signals for paleodietary studies. *J. Anthropol. Archaeol.* **2003**, *22*, 208–216. [CrossRef]

45. Richards, M.P.; Schmitz, R.W. Isotope evidence for the diet of the Neanderthal type specimen. *Antiquity* **2008**, *82*, 553–559. [CrossRef]

46. Bocherens, H. Diet and Ecology: Implications from C and N isotopes. In *Neanderthal Lifeways, Subsistence and Technology*; Conard, N., Richter, N.H., Eds.; Vertebrate Paleobiology and Paleoanthropology Series; Springer: Dordrecht, The Netherlands, 2011. [CrossRef]

47. Bocherens, H.; Díaz-Zorita, M.; Daujeard, C.; Fernandes, P.; Raynal, J.-P.; Moncel, M.-H. Direct Isotopic evidence for subsistence variability in Middle Pleistocene Neanderthals (Payre, southeastern France). *Quat. Sci. Rev.* **2016**, *154*, 226–236. [CrossRef]

48. Harrison, R.G.; Katzenberg, M.A. Paleodiet studies using stable carbon isotopes from bone apatite and collagen: Examples from Southern Ontario and San Nicolas Island, California. *J. Anthropol. Archaeol.* **2003**, *22*, 227–244. [CrossRef]
49. Weyrich, L.S.; Duchene, S.; Soubrier, J.; Arriola, L.; Llamas, B.; Breen, J.; Morris, A.G.; Alt, K.W.; Caramelli, D.; Dresely, V.; et al. Neanderthal behaviour, diet, and disease inferred from ancient DNA in dental calculus. *Nature* **2017**, *554*, 357. [CrossRef]
50. Charlier, P.; Gaultier, F.; Héry-Arnaud, G. Interbreeding between Neanderthals and modern humans: Remarks and methodological dangers of a dental calculus microbioma analysis. *J. Hum.Evol.* **2019**, *126*, 124–126. [CrossRef]
51. Mann, A.E.; Fellows Yates, J.A.; Fagernäs, Z.; Austin, R.M.; Nelson, E.A.; Hofman, C.A. Do I have something in my teeth? The trouble with genetic analyses of diet from archaeological dental calculus. *Quat. Int.* **2022**; *in press*. [CrossRef]
52. Costamagno, S.; Meignen, L.; Cédric, B.; Vandermeersch, B.; Maureille, B. Les Pradelles (Marillac-le-Franc, France): A Mousterian reindeer hunting camp? *J. Anthrop. Archaeol.* **2006**, *26*, 466–484. [CrossRef]

Article

Structure and Chemical Composition of ca. 10-Million-Year-Old (Late Miocene of Western Amazon) and Present-Day Teeth of Related Species

Caroline Pessoa-Lima [1,*,†], Jonas Tostes-Figueiredo [1,†], Natalia Macedo-Ribeiro [1], Annie Schmaltz Hsiou [2], Fellipe Pereira Muniz [2], José Augusto Maulin [3], Vinícius H. Franceschini-Santos [4,‡], Frederico Barbosa de Sousa [5], Fernando Barbosa, Jr. [6], Sergio Roberto Peres Line [7], Raquel Fernanda Gerlach [1] and Max Cardoso Langer [2]

[1] Department of Basic and Oral Biology, FORP, University of Sao Paulo, Av. Café, Ribeirao Preto 14040904, SP, Brazil
[2] Laboratory of Paleontology, Department of Biology, FFCLRP, University of São Paulo, Ribeirao Preto 14040900, SP, Brazil
[3] Laboratory of Electron Microscopy, Department of Cell and Molecular Biology and Biopathogenic Agents, FMRP, University of Sao Paulo, Ribeirao Preto 14040900, SP, Brazil
[4] Department of Biology, FFCLRP, University of São Paulo, Ribeirao Preto 14040900, SP, Brazil
[5] Departament of Morfology, CCS, Universidade Federal da Paraiba, Cidade Universitaria, Joao Pessoa 58051900, PB, Brazil
[6] Laboratory of Toxicology and Essentiality of Metals, FCFRP, University of Sao Paulo, Ribeirao Preto 14040900, SP, Brazil
[7] Department of Biosciences, FOP, University of Campinas, Piracicaba 13414903, SP, Brazil
* Correspondence: caroline.pessoa.lima@usp.br
† These authors contributed equally to this work.
‡ Current address: Division of Gene Regulation, Netherlands Cancer Institute, 1066 CX Amsterdam, The Netherlands.

Citation: Pessoa-Lima, C.; Tostes-Figueiredo, J.; Macedo-Ribeiro, N.; Hsiou, A.S.; Muniz, F.P.; Maulin, J.A.; Franceschini-Santos, V.H.; de Sousa, F.B.; Barbosa, F., Jr.; Line, S.R.P.; Gerlach, R.F.; Langer, M.C. Structure and Chemical Composition of ca. 10-Million-Year-Old (Late Miocene of Western Amazon) and Present-Day Teeth of Related Species. *Biology* **2022**, *11*, 1636. https://doi.org/10.3390/biology11111636

Academic Editor: Zhifei Zhang

Received: 9 August 2022
Accepted: 6 October 2022
Published: 8 November 2022

Publisher's Note: MDPI stays neutral with regard to jurisdictional claims in published maps and institutional affiliations.

Copyright: © 2022 by the authors. Licensee MDPI, Basel, Switzerland. This article is an open access article distributed under the terms and conditions of the Creative Commons Attribution (CC BY) license (https://creativecommons.org/licenses/by/4.0/).

Simple Summary: The dental enamel is the most mineralized tissue of vertebrates, and its preservation in fossil records is important to better understand the ancient life and environment on Earth. However, the association of morphological features with the mineral and organic information of this tissue is still poorly understood. This study aims to compare morphological features and chemical composition of dental enamel of extinct and extant species of alligators and rodents. Organic, mineral, and water content were obtained on ground sections of four teeth, resulting in a higher organic volume than previously expected (up to 49%). It is observed that both modern and fossil enamel exhibit the same major constituents: 36.7% calcium, 17.2% phosphorus, and 41% oxygen, characteristic of hydroxyapatite, the biomineral of vertebrates. Twenty-seven microelements were measured from superficial enamel. Zinc was the most abundant microelement, followed by lead, iron, magnesium, and aluminium. Semiprismatic enamel was observed in the alligator fossil. The fossilized enamel was in an excellent state for microscopic analyses. Results show that all major dental enamel's physical, chemical, and morphological features are present both in extant and extinct fossil tooth enamel (>8.5 Ma) in both taxa.

Abstract: Molecular information has been gathered from fossilized dental enamel, the best-preserved tissue of vertebrates. However, the association of morphological features with the possible mineral and organic information of this tissue is still poorly understood in the context of the emerging area of paleoproteomics. This study aims to compare the morphological features and chemical composition of dental enamel of extinct and extant terrestrial vertebrates of Crocodylia: *Purussaurus* sp. (extinct) and *Melanosuchus niger* (extant), and Rodentia: *Neoepiblema* sp. (extinct) and *Hydrochoerus hydrochaeris* (extant). To obtain structural and chemical data, superficial and internal enamel were analyzed by Scanning Electron Microscopy (SEM) and Energy Dispersive Spectroscopy (SEM-EDS). Organic, mineral, and water content were obtained using polarizing microscopy and microradiography on ground sections of four teeth, resulting in a higher organic volume than previously expected (up to 49%). It is observed that both modern and fossil tooth enamel exhibit the same major constituents: 36.7% Ca, 17.2% P, and 41% O, characteristic of hydroxyapatite. Additionally, 27 other elements were measured from superficial enamel by inductively coupled mass spectrometry (ICP-MS). Zinc was the most abundant microelement detected, followed by Pb, Fe, Mg, and Al. Morphological features observed include enamel rods in the rodent teeth, while incremental lines and semiprismatic enamel were observed in the alligator species. The fossil enamel was in an excellent state for microscopic analyses. Results show that all major dental enamel's physical, chemical, and morphological features are present both in extant and extinct fossil tooth enamel (>8.5 Ma) in both taxa.

Keywords: teeth; enamel; microanalysis; *Purussaurus*; *Neoepiblema*; miocene; fossils; ICP-MS; MEV; microscopy

1. Introduction

Historically, advances in paleontology and the classification of extinct species have relied mainly on the comparative analysis of morphological aspects of fossilized material, especially in calcified structures such as bones and teeth. These analyses, however, may be constrained for some extinct species as the identification may be based on small fragments of fossilized material. The anatomical similarities between species may not directly result from shared phylogenetic history. Evolutionary convergence is often regarded as a relevant factor in morphological phylogenetics [1]. Paleogenomics and paleoproteomics have recently appeared as essential tools that may assist in classifying extinct species.

Among calcified tissues, tooth enamel has the highest mineral content [2], corresponding to higher density, smaller pores (nanopores) [3], and a unique set of proteins, whose recovery from superficial enamel etches started only in 2006 [4]. This innovation was based on the "enamel biopsy", a superficial enamel etching used to obtain ultraclean samples. Such samples enabled the comparison of the exposure of children to lead, a neurotoxin [5–7]. The superficial enamel sample had two main advantages: teeth did not need to be ground/destroyed to be studied anymore, and the spatial information on the number of chemical elements on the outer enamel (where many elements with high affinity for hydroxyapatite accumulate) would not be lost. The need to develop ultraclean techniques to be able to obtain reliable results on the superficial etchings was a key factor that later resulted in excellent results from the mass spectrometry analysis of biopsy samples [3,8,9].

Nowadays, enamel proteins recovered from ancient species are considered particularly useful for comparing divergent species [10–12]. As a "pre-fossilized" extracellular matrix, dental enamel is superior to other calcified tissues in retaining the features of living specimens [13,14]. The inorganic composition of fossil teeth makes it possible to understand aspects such as diet, species' migration patterns, and paleoenvironmental interpretation [15,16]. Enamel peptides have been recovered from fossil enamel, indicating that the unique properties of this tissue act as a protective barrier preventing protein degradation [17].

Many gaps in knowledge exist, as expected in a very new field of research. To date, information on the association of morphological features of dental enamel with the composition (quantity and diversity) of proteins in forming enamel is still poorly understood. In recent years, this information has been obtained primarily from mouse models [18–21]. Little is known about other species, since the protein information is gained while enamel is forming, and this usually happens in early life. Such information would be very beneficial for the analyses of peptides in different species. Since proteins are the molecules that form the scaffold that organizes crystal growth, it is intuitive to assume that different amounts (quantitative changes) of proteins will determine different phenotypes of dental enamel, and not only the mutations accumulated over time (qualitative changes). So, it may be that both different amounts of the different enamel proteins and different enamel protein sequences will have an influence on the physical and chemical properties of the enamel. This seems particularly important in light of the tooth enamel phenotype observed in transgenic and enamel proteins of knock-out mice, which offers the chance to look into experiments that nature might have tested over large scales of time.

Therefore, it seems that it might be useful to try to associate morphological features with the possible mineral and organic information of this tissue. This information contributes to better interpreting molecular information that is now being gathered in several excellent research centers, where the new area of paleoproteomics is being developed.

In this study, we compare the morphological features and the chemical composition of dental enamel of extinct and extant terrestrial vertebrates belonging to two groups. Crocodylia was represented by *Purussaurus* sp. (fossil) and *Melanosuchus niger* (extant), while Rodentia was represented by *Neoepiblema* sp. (fossil) and *Hydrochoerus hydrochaeris* (extant). We combined several techniques adapted to study developing enamel [22–24], and we used polarized light microscopy and birefringence to determine organic content [25]. The mineral composition was studied with ICP-MS, as opposed to other chemical analyses that are less sensitive [26]. This study shows for the first time the mineral, organic, and water contents of the enamel of extinct and extant species from the same taxonomic families, and some of the species show high organic content.

2. Materials and Methods

2.1. Institutional Abbreviations

UFAC—Universidade Federal do Acre, Rio Branco, AC, Brazil.

LIRP—Laboratorio de Ictiologia de Ribeirao Preto, FFCLRP, University of Sao Paulo, Ribeirao Preto, SP, Brazil.

2.2. Fossil Specimens and Provenance

Both fossil specimens used in this study came from the Acre Basin, a retroarc foreland basin related to the Andean orogenesis [27]. It is the westernmost of a series of interior sag/fracture basins along the Solimões/Amazonas rivers, at the westernmost portion of the Brazilian Amazon, neighboring Peru and Bolivia [28–30]. They were collected in the sedimentary deposits of the Late Miocene Solimões Formation, which is mostly exposed along riverbanks in the states of Acre and Amazonas. Such sediments were deposited mainly under fluvio-lacustrine conditions, both in river channels and in floodplains/lakes [31,32]. A diverse vertebrate fauna has been reported in the Solimões Formation, including cartilaginous and bony fishes, frogs, turtles, birds, crocodylians, lizards, snakes, and mammals (e.g., [33–37]).

UFAC-7226 (Figure 1)—the first sample corresponds to a *Purussaurus* sp. tooth. It was collected in the locality known as "Niterói", which is located on the right embankments of the Acre River, Senador Guiomar municipality (UTM 19L 629983 E/8879539 S, datum WGS84). The fossil comes from the same bone-beds U-Pb dated by Bissaro-Júnior et al. [30], from detrital zircon, with a maximal weighted-mean age of 8.5 ± 0.5 Ma (Tortonian Stage, Late Miocene). *Purussaurus* was one of the largest crocodiles, reaching more than 10 m in length [38], with records in the Miocene of Brazil, Colombia, Peru, Venezuela, and Panama. *Purussaurus* is regarded as a semiaquatic top predator, placed in recent phylogenetic studies within Alligatoridae and Caimaninae, forming, with other taxa, the sister clade to Jacarea [39].

UFAC-7227 —the other fossil sample analyzed consists of a *Neoepiblema* sp. partial lower incisor. It was collected in the locality known as "Talismã", which is located on the right embankments upstream of the Purus River, State of Amazonas, between the Manuel Urbano and the Iaco River mouth (UTM 19L 510475 E/9029741 S, datum WGS84). The fossil comes from the same bone-beds U-Pb dated by Bissaro-Júnior et al. [30], from detrital zircon, with a maximal weighted-mean age of 10.89 ± 0.13 Ma (Tortonian Stage, Late Miocene). *Neoepiblema* is a large-sized, caviomorph rodent, with records in the Miocene of Brazil, Peru, Venezuela, and Argentina. Along with other neoepiblemids, it is more closely related to chinchillids than to other caviomorphs [40].

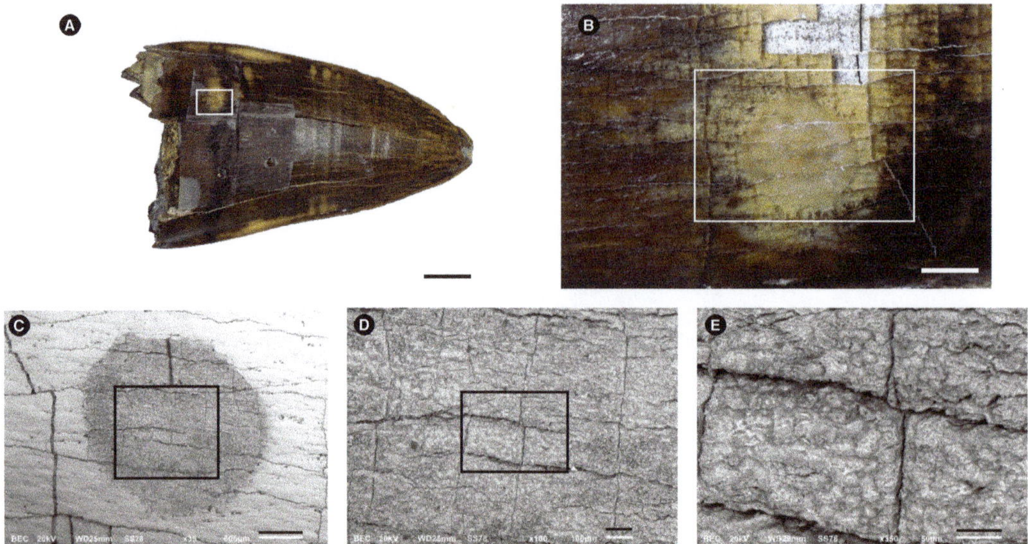

Figure 1. Images of a *Purussaurus* sp.(Crocodylia, Alligatoridae) tooth used to obtain superficial enamel acid etch samples. Images were made with photograph cameras (**A**,**B**) and with a Scanning Electron Microscope using Back-scattered-electrons (BSE) mode (**C–E**). Rectangles are always amplified in the next figure of the series. (**A**). *Purussaurus* sp. (Crocodylia, Alligatoridae) tooth picture taken without magnification lens. The white arrow indicates enamel cracks. Bar = 1 cm. (**B**). Details of the white rectangle depicted in (**A**). Bar = 1 mm. In the lower area of the figure, some "squares" have fallen, showing the dentine below the enamel. (**C**). Amplification of the white rectangle shown in (**B**). The circular area (gray) was etched with the acid. Bar = 500 μm. (**D**). Amplification of the rectangle shown in (**C**). Bar = 100 μm. (**E**). Amplification of the rectangle shown in (**D**). Bar = 50 μm.

2.3. Modern Specimens

Teeth of two extant species, *M. niger* and *H. hydrochaeris*, were analyzed to compare the structure and chemical composition of enamel of the modern and fossil samples. *M. niger* is a South-American alligator with geographical distribution in the north of the continent. *H. hydrochaeris* is the largest living rodent and has a widespread geographical occurrence in South America. The specimens' teeth were obtained from LIRP's zoological collection. The *M. niger* specimen was from "Reserva Mamirauá", Amazon State, and the *H. hydrochaeris* specimen was from Ribeirao Preto, São Paulo State.

2.4. SEM-EDS Analysis

Analyses by Scanning Electron Microscopy (SEM) using conventional and Back-scattered-electron mode (BSE) were performed on a scanning electron microscope (Jeol JSM–5600LV, Tokyo, Japan) [22,41]. EDS signals were captured using a standard setup on BSE mode. The EDS signals were obtained using the following parameters: Samples were positioned at a 10 mm distance from the emitter, with 15 to 25 kV of emission. Spot size ranged from 69 to 80 nm. Measurements were made on the surface of the dental enamel. For the SEM-EDS analysis, internal multielement standards were used for calibration according to the supplier (Oxford Instruments, Scotts Valley, CA, USA). Based on those standards, the maps of the most abundant elements and quantifications were made [42]. The *Purussaurus* tooth have three different superficial colors. Thus, three regions on the tooth surface were chosen for analysis: (1) a black area (visible in lines), (2) a brown area (most abundant), and (3) a yellow area. The images were taken in both the conventional and Back-scattered-electron (BSE) mode in different magnifications. Multielement maps were produced automatically by the Aztec program (Oxford Instruments, Concord, MA,

USA) based on the intensity of the signals of the different elements. The detector used was the Silicon Drift Detector (SDD, X-MaxN 150 mm^2 detector).

2.5. Preparation of Samples for Light and Scanning Electron Microscopy

Pieces of teeth were embedded in acrylic resin (JET, Campo Limpo Paulista, SP, Brazil) and cut with a diamond disc on a calcified tissue cutting machine (Elsaw, Elquip, São Carlos, SP, Brazil). Sections were carefully sanded with a gradual decrease in the sandpaper granulation, from 600 to 4000, using water. The width of the sections was circa 100 µm for bright field and polarized light microscopy. For SEM, thick sections were used, which were etched with 10% acetic acid (*Purussaurus* sp. enamel) for 15 s or with 37% phosphoric acid for 30 s (enamel from the other 3 species). Immediately after acid etching, the teeth were copiously washed with running water for 10 min to remove traces of the acid. Samples were then dehydrated overnight and covered with a gold pellicle. The SEM photographs were taken at the same microscope described above.

2.6. Light Microscopy

Tooth sections were photographed on an Axiovert 2.1 microscope (Zeiss, Karlstadt, Germany) with and without phase contrast filters, and on an Axioscope 40 (Zeiss, Karlstadt, Germany) using crossed polarizing filters and the $\lambda/1$ red filter.

2.7. Quantification of Major Enamel Biochemical Components

Major enamel biochemical components (mineral, organic, and adsorbed water in volume and weight percentages) were measured in each sample (non-demineralized ground sections with thickness in the range of 100–400 µm), at five discrete regions of interest (ROI; 12 µm × 12 µm) along a longitudinal line running from the enamel surface to the enamel dentine junction, as described previously in detail [43,44]. Unlike thermogravimetric analysis [45], the methods used here did not require the destruction of the samples. Basic physical parameters related to the mineral, organic, and water contents (mineral unit-cell composition, mineral X-ray linear attenuation coefficient, mineral density, and refractive indexes of all major biochemical components), required for quantification, were derived from human enamel. Transverse microradiography [46], the gold-standard technique for quantification of dental enamel mineral volume [47], was used to quantify mineral content. The following assumptions were considered for the enamel mineral composition: unit cell composition of $Ca_{8.856}Mg_{0.088}Na0.292K_{0.010}(PO4)_{5.312}(HPO_4)_{0.280}(CO_3)_{0.407}(OH)_{0.702}Cl_{0.078}(CO_3)_{0.050}$) and density of 2.99 g/cm^3 [48], with 37% calcium weight and 18% phosphorus weight (Ca/P ratio of 2.06). Digital 2D microradiographic images (enamel ground sections and aluminium step-wedge with 17 aluminium foils, with thickness ranging from 20 to 340 µm) were obtained from microtomographic equipment (Skyscan 1172, Bruker, Belgium) operated at 60 kV, emitting peak X-ray energy of 10 kV [43]. After mineral volume quantification, birefringence measurements (mean of five measurements) using underwater immersion (and after immersion in water for 5 days prior to the analysis) were obtained using polarizing microscopy (Axioskop 40, Carl Zeiss, Germany) equipped with a 0–5 orders Berek compensator, and then organic and water volumes were quantified from the mathematical interpretation of birefringence [25,44]. The assumed refractive indexes of organic matter and water were 1.56 and 1.00, respectively. The sum of all measured volumes resulted in 100% enamel volume at each ROI. Volume percentages were converted to weight percentages using the following densities: 2.99 g/cm^3 (mineral), 1.45 g/cm^3 (organic; glutamic acid), and 1.0 g/cm^3 (water).

2.8. Superficial Enamel Sampling with Acid

The outer enamel of all 4 teeth was etched twice. This method is ideal to obtain superficial etches for microelement determination and the study of environmental contaminants in children in vivo [5,49]. The method was used by Brudevold et al. [50], with the biopsy depth and other parameters studied in detail [51]. This method was then adapted to obtain enamel peptides by Porto et al. [2,52], having been also used by Stewart et al. [8,9] and Nogueira et al. [3]. In the *Purussaurus* sp. tooth, 3 different regions were selected for sampling due to the color differences, as described above, and as seen in Figure 1. This procedure was carried out inside a Class 100 hood, using only ultraclean solutions diluted just before use. The biopsy was analyzed in order to perform the determination of 27 chemical elements by Inductively Coupled Plasma-Mass Spectrometry (ICP-MS). In this study, lead-free adhesive tape (Magic Tape, 810 Scotch®, 3M, Sumare, SP, Brazil) with a circular perforation (diameter = 2 mm) was firmly pressed onto the surface of the tooth, delimiting the microbiopsy site (the tape with the circular perforations can be observed in Figure 1A). The sampling site was etched according to the following procedure. The enamel surface was rinsed with water (MilliQ) and etched with 20 L 10% (vol/vol) HCl for approximately 20 s. The microbiopsy solution was then transferred to an ultraclean centrifuge tube (1.5 mL) (Axygen Scientific, Inc., Union City, NJ, USA). The samples were closed with parafilm and sent for microelement determination by ICP-MS (NexIon 5000, Perkin Elmer) in the Laboratory of Toxicology and Essentiality of Metals (University of São Paulo, FCFRP, Ribeirão Preto, SP, Brazil), without opening the tubes before the analyses. The second etch was made in the same way, but the acid was then used to obtain enamel peptides, as described in a separate manuscript.

2.9. Chemical Analysis by ICP-MS

Several macro- and microelements were determined on the superficial enamel samples by Inductively Coupled Plasma-Mass Spectrometry (ICP-MS) in the Laboratory of Metals Toxicology, University of Sao Paulo in Ribeirao Preto (Brazil). The quantification limit (LOQ) for the different elements in the study is presented in Supplementary Table S2. Calcium concentrations determined by SEM-EDS in each sample (Table 1) were used to calculate the mass of dental enamel. Based on this, the amount of each element was expressed in ppm (μg/g) of dental enamel, as a more universal concentration.

Table 1. Mean ± SD (n = 4–5 measurements in each sample) of macroelement composition (weight%) determined in the internal enamel by SEM-EDS. The first two rows are Crocodylia samples (*Purussaurus* sp., and *M. niger*), and the last two are Rodentia samples (*Neoepiblema* sp. and *H. hydrochaeris*). The Ca/P ratio corresponds to the expected for hydroxyapatite composition. Ca: calcium, P: phosphorus, Ca/P: ratio of calcium per phosphorus.

Sample	Ca (wt%)	P (wt%)	Ratio Ca/P
Fossil alligator	36.8 ± 24.4	16.5 ± 0.2	2.2 ± 0.03
Extant alligator	33.3 ± 0.4	16.6 ± 0.1	2.0 ± 0.03
Fossil rodent	38.6 ± 0.3	17.8 ± 0.1	2.2 ± 0.02
Extant rodent	36.8 ± 1.5	17.7 ± 1.3	2.1 ± 0.12

3. Results

Here, we compare 10-Million-year-old fossils and modern enamel from related species of two different vertebrate taxa. For clarity, we first describe the macro- and microscopic features of the teeth and enamel, while the chemical composition is described later.

3.1. Morphological Description

Figure 1 shows images made from the fossil alligator tooth used to obtain superficial enamel acid etch. The tooth measured ≈7 cm in length and ≈4 cm in width (Figure 1A). Figure 1B shows a larger view of the white rectangle that can be observed in Figure 1A. Small cracks on the surface form a fairly regular rectangular lattice. The area etched with acid is the small circle, which has a ca. 2 mm diameter. This image shows that the etching only slightly modified the superficial aspect of the enamel and that such changes can only be seen under magnification. This image also shows the variation in color seen in the enamel, as well as the cracks at right angles. Some gray "squares" can be seen in the lower part of the figure, where the enamel was unintentionally removed together with the tape. The gray observed in the picture is the underlying dentine. Another observation here is that the enamel in this fossil sample is poorly attached to the dentine. The light strength with which the glue on the tape was attached to the enamel was sufficient to detach the enamel from the dentine in this area with many cracks. The many details in shape and color seen in the enamel greatly contrast with the gray appearance of the dentine and are a clear indication that enamel does act as a reflective surface for light, and this indicates the very high crystallinity of this calcified tissue. The same was observed in all other specimens, both fossil and modern. The underlying dentine, on the other hand, displays a gray appearance that is compatible with the incorporation of other ions into its structure, and also a greater ability to absorb light, the opposite of the reflection of light seen in the enamel. Figure 1C–E demonstrates that the acid etching extraction technique causes minimal damage to the fossil surface. The superficial aspect of Figure 1E shows a rough surface on the etched area.

Figure 2 shows images made from the extant alligator tooth. In Figure 2A, the whole tooth is shown, and the dental enamel covers the tooth crown, which is seen to the right side of the arrow and displays a yellowish appearance, except for the area etched with acid, amplified in Figure 2B. The enamel of this species has a porous superficial appearance where the acid was applied.

In comparison wish the alligator fossil enamel, the modern enamel does not show many right-angle cracks. We believe this difference can be the result of weathering before the burial of the fossil or lithostatic compression.

Figure 3 shows a fragment of the fossilized rodent tooth. The superficial enamel is shown in Figure 3A. The enamel has a brown aspect and has some white deposits. On the left side of this image, the underlying dentine is apparent. In Figure 3B, the etched enamel is seen as a circle, showing minimal damage to the specimen (in B, the bar is 1 mm, and the circular area has a diameter of 2 mm). In Figure 3E, the prismatic enamel is seen in the etched enamel.

In sharp contrast to the *Purussaurus* sp. enamel, no cracks were observed. It is possible that alligator's fossils have being exposed to greater taphonomic effects than the rodent's fossils, or the fossils have reacted differently to similar conditions.

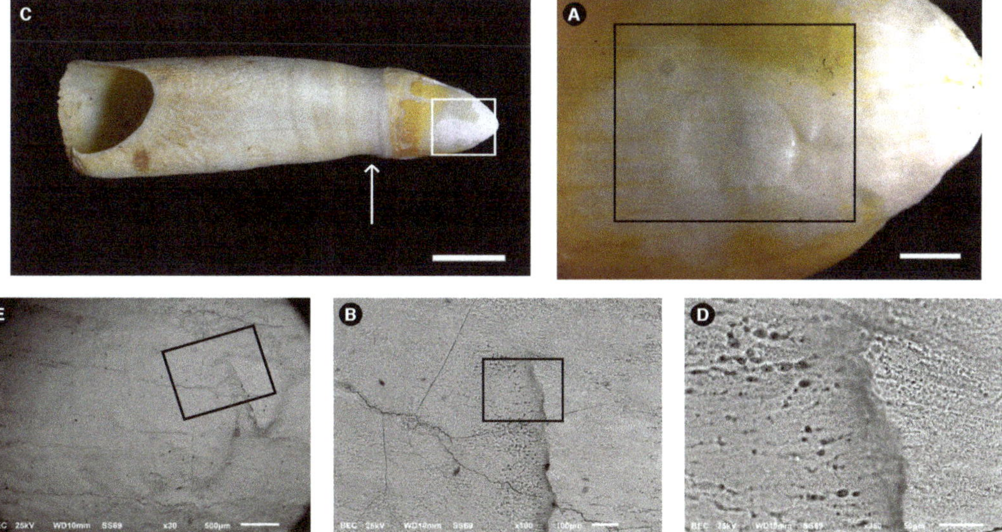

Figure 2. Images of a *Melanosuchus niger* (Crocodylia, Alligatoridae) tooth used to obtain superficial enamel acid etch samples. Images were made with photograph cameras (**A**,**B**) and with a Scanning

Electron Microscope using Back-scattered-electrons (BSE) mode (**C–E**). Rectangles are always amplified in the next figure of the series. (**A**). Picture taken without a magnifying lens; the arrow indicates the limit between the crown (to the right) and the root (to the left). The crown exhibits a yellow appearance on its surface; Bar= 1cm. (**B**). Details of the rectangle are depicted in (**A**); the area etched with the acid is shown as a circle, but some acid has leached to the sides, as seen from the brighter areas on the enamel. Bar = 1 mm. (**C**). Amplification of the rectangle shown in (**B**). Bar = 500 μm. (**D**). Amplification of the rectangle shown in (**C**). Bar = 100 μm. (**E**). Amplification of the rectangle shown in (**D**). The high porosity of the etched enamel is clearly observed. No enamel prisms can be observed. Bar = 50 μm.

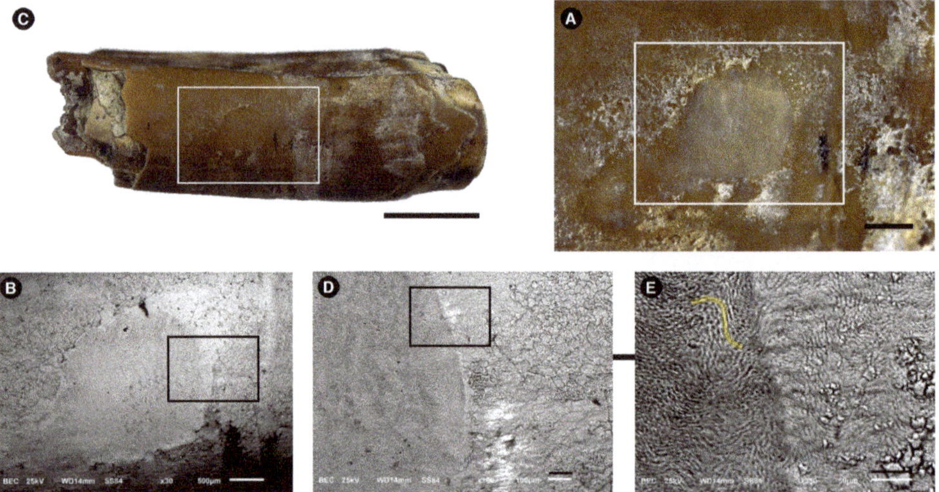

Figure 3. Images of a *Neoepiblema* sp. (Rodentia, Caviomorpha) lower incisor used to obtain superficial enamel acid etch samples. Images were made with photograph cameras (**A,B**) and with a Scanning Electron Microscope using Back-scattered-electrons (BSE) mode (**C–E**). Rectangles are always amplified in the next figure of the series. (**A**), tooth fragment picture showing the superficial enamel (brown aspect) that has some white deposits and is broken on the left side of the tooth, where the underlying dentine is apparent. Bar = 0.5 cm. (**B**). Details of the white rectangle depicted in (**A**). Bar = 1 mm. (**C**), amplification of the white rectangle shown in **B**. The circular area (grey) was etched with the acid. Bar = 500 μm. (**D**), amplification of the black rectangle shown in (**C**). Bar = 100 μm. (**E**). Amplification of the black rectangle shown in (**D**). In the etched area (left), the enamel's rods/prisms (yellow line) are apparent. Bar = 50 μm.

Figure 4 shows the tooth enamel of the extant rodent. The superficial nature of the enamel biopsy can be observed in Figure 4A,C. Circular transparent spots with a diameter of ≈200 μm can be observed. Figure 4C is an amplification of the rectangle shown in B. The circular area was etched with acid. The circular spots are superficial, having disappeared in the etched area, and have a composition with less dense elements (suggesting an organic composition) than the Ca and P that compose the hydroxyapatite of the enamel. Figure 4D is an amplification of the rectangle is shown in Figure 4C, and Figure 4E is an amplification of the rectangle shown in Figure 4D. Prismatic enamel is seen on the left (asterisk), in the etched area.

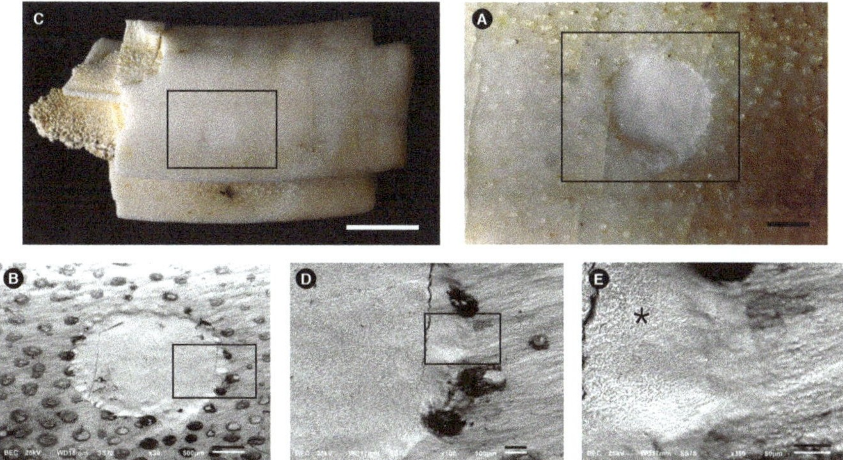

Figure 4. Images of a *Hydrochoerus hydrochaeris* (Rodentia, Caviomorpha) tooth used to obtain superficial enamel acid etch samples. Images were made with photograph cameras (**A**,**B**) and with a Scanning Electron Microscope using Back-scattered-electrons (BSE) mode (**C–E**). Rectangles are always amplified in the next figure of the series. (**A**), tooth picture taken with magnification lens showing the circular etched area inside the black rectangle. Bar = 0.5 cm. (**B**), details of the rectangle depicted in (**A**). Circular transparent spots with a diameter of ≈200 μm can be observed. Bar = 1 mm. (**C**), amplification of the rectangle shown in (**B**). The circular area was etched with acid. The circular spots are superficial, having disappeared in the etched area, and have a composition with less dense elements (suggesting an organic composition) than the Ca and P that compose the hydroxyapatite of the enamel. Bar = 500 μm. (**D**), amplification of the rectangle shown in (**C**). Bar = 100 μm. (**E**), amplification of the rectangle shown in (**D**). Prismatic enamel is seen on the left (asterisk) in the etched area. Bar = 50 μm.

3.2. Microstructure (SEM Analysis)

Figure 5 exhibits SEM (conventional method with gold coat) pictures of cross sections of the the enamel of the four species. The enamel of the fossilized alligator (Figure 5A–C) is formed by rod-like structures that are comparable in shape and width to the prismatic structures of mammalian enamel. These structures run roughly perpendicular to the enamel surface and dentine-enamel junction (DEJ), having a width of approximately 5 μm. In extant alligators (Figure 5D–G), there is a typical aprismatic structure, showing growth lines that are likely to be equivalent to the mammalian Retzius lines. These lines were localized near the DEJ and likely represent periods of physiological stress during enamel formation. The enamel of the fossilized rodent (Figure 5H–K) is formed by prisms that are roughly parallel to each other and perpendicular to the (DEJ) and enamel surface. The DEJ exhibits an irregular surface. The enamel of the extant rodent shows two distinct patterns of the arrangement of the prisms. The prisms run nearly parallel to the DEJ and extend approximately 25–30 μm near the DEJ. A thin layer of parallel prisms is also observed near the enamel surface. In the inner enamel, a different pattern is observed, with a large band (≈80 μm) of prisms exhibiting intense decussation.

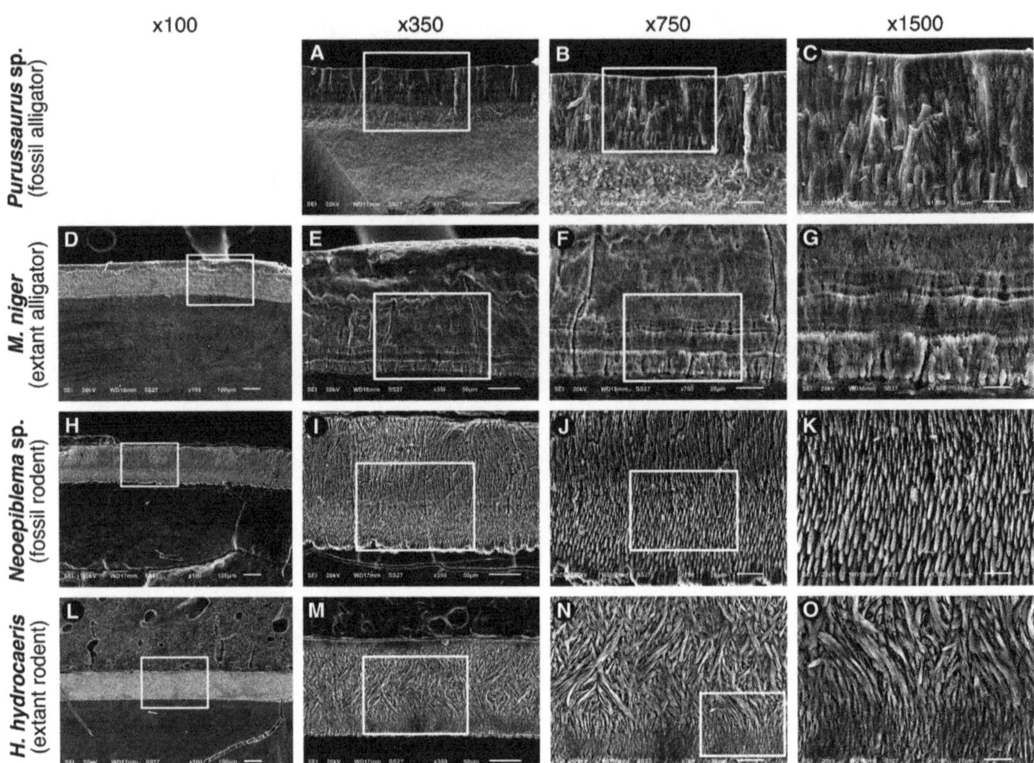

Figure 5. Images obtained by SEM analysis of enamel cross-sections from the four (4) specimens *Purussaurus* sp., *Melanosuchus niger*, *Neoepiblema* sp., and *Hydrochoerus hydrochaeris*. (**A–C**), the enamel of *Purussaurus* sp. is formed by rod-like structures that run roughly perpendicular to the enamel surface and dentine-enamel junction (DEJ), having a width of approximately 5 μm. (**D–G**), *Melanosuchus niger* has a typical aprismatic structure, showing growth lines that are likely to be the equivalent to the mammalian Retzius lines. These lines were localized near the DEJ and are likely to represent periods of physiological stress during enamel synthesis. (**H–K**), the enamel of *Neoepiblema* sp. is formed by prisms that are roughly parallel and perpendicular to the DEJ and enamel surface. The DEJ exhibits an irregular surface. (**L–O**), the enamel of *Hydrochoerus hydrochaeris* shows 2 distinct patterns of prims arrangement. The prisms run nearly parallel from the DEJ and extend approximately 25–30 μm near the DEJ. A thin layer of parallel prisms is also observed near the enamel surface. In most parts of enamel, the prisms exhibit a pattern of intense decussation.

Supplementary Figure S1 shows photomicrographs of ground sections (≈100–400 μm) of the four teeth used in this study. The images show that enamel (e) is translucent in all specimens, while dentine (d) is dark/opaque in all specimens. We show an exception to this, where the mineral has been lost, possibly due to acids from the environment (Supplementary Figure S1A: arrow head; Figure S1B: asterisk). The white arrows in panels A and C show marked apposition lines of the dentine. In D, black arrows in the enamel show appositional lines. In F, the arrow indicates a very pronounced line; the inner enamel is less mineralized, and the outer enamel appears more mineralized. In G, the enamel of the extant rodent is shown; the arrow indicates a less mineralized area in the middle of the enamel throughout the enamel. In H, a larger magnification of the black square is shown under polarized light. Due to the thickness of this section, the prismatic aspect of this enamel is difficult to recognize, but in the outer enamel (asterisk) diagonal lines are enamel rods.

3.3. Determination of Organic, Mineral, and Water Content of Dental Enamel

In this study, major enamel biochemical components (mineral, organic, and adsorbed water in volume and weight percentages) were measured at five discrete regions of interest (ROI; 12 mm × 12 mm) along a longitudinal line running from the enamel surface to the DEJ in the enamel of each tooth.

Figures 6 and 7 show the images made at each step of the initial analysis of the fossil tooth ground sections from alligator and rodent fossils under polarizing light. The sections were submerged in water during sample preparation. Without an interference filter, dental enamel shows high interference colors (Figures 6A,B and 7A,B), indicating a relatively high ground section thickness. With the Red I filter, the additional interference color is shown at the −45° (45° counterclockwise) diagonal position (Figure 6C), and the subtraction interference color is shown at the +45° (45° clockwise) diagonal position (Figure 6D), indicating negative birefringence, based on the Michel-Levy interference color chart [53].

The step-by-step analyses of the modern samples are not shown here, but figures of polarized light microscopy and microradiography are shown for each sample as supplemental material (Supplementary Figures S2, S4, S5, S7, S9 and S10). The enamel of all samples exhibited negative birefringence.

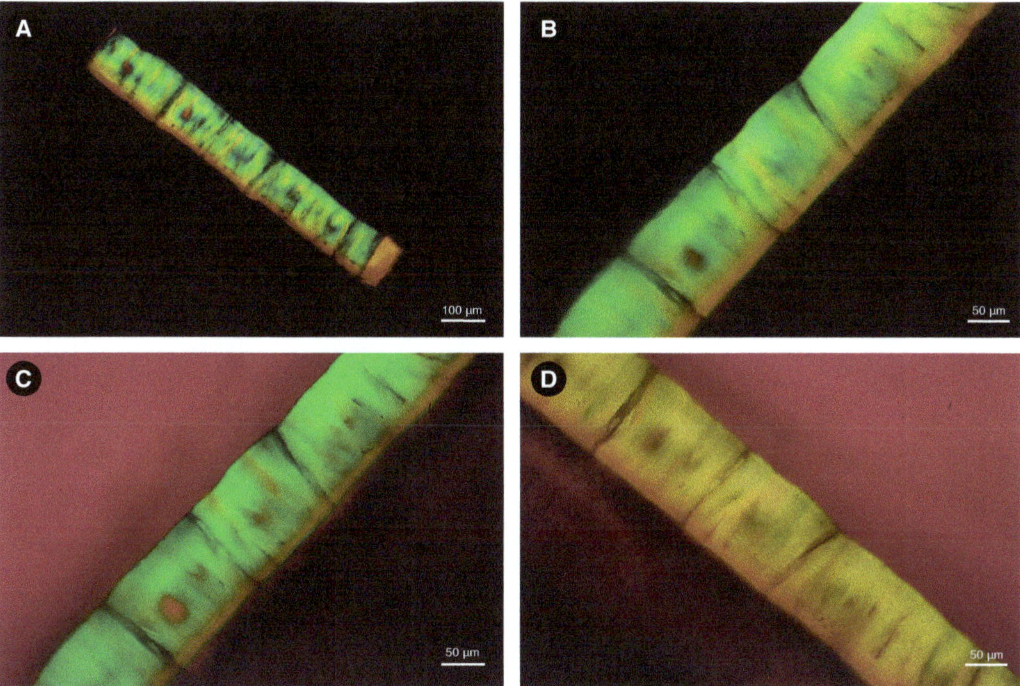

Figure 6. Image of a ground section of dental enamel sample from *Purussaurus* sp. taken under polarizing microscopy with water immersion. Without an interference filter, dental enamel shows high interference colors (**A**,**B**), indicating a relatively high ground section thickness. With the Red I filter, the addition interference color is shown at the −45° diagonal position (**C**), and the subtraction interference color is shown at the +45° diagonal position (**D**), indicating negative birefringence.

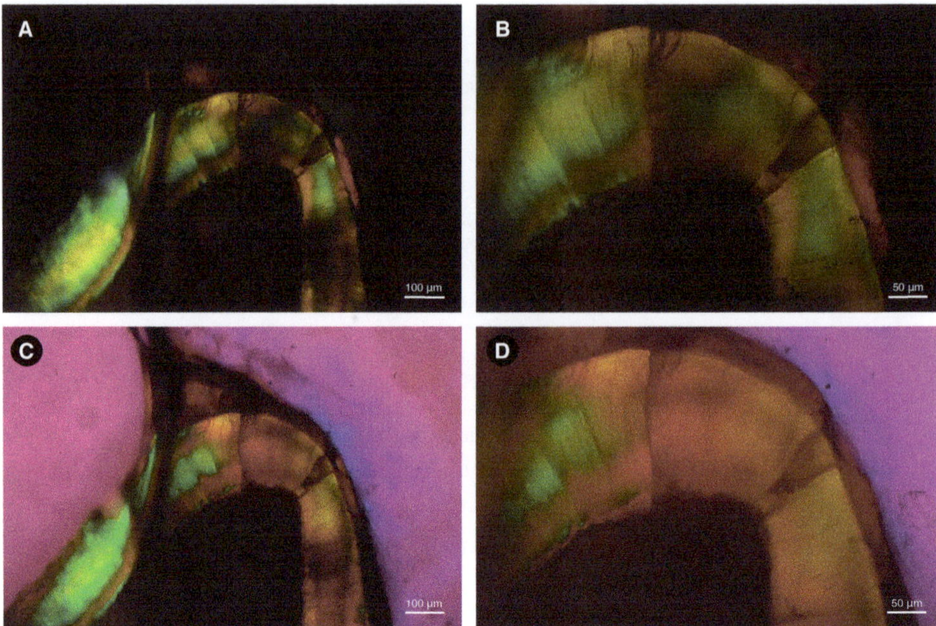

Figure 7. Image of a ground section of dental enamel sample from *Neoepiblema* sp. taken under polarizing microscopy with water immersion. Without an interference filter, dental enamel shows high interference colors (**A**,**B**), indicating a relatively high ground section thickness. With the Red I filter, the addition interference color is shown at the −45° diagonal position (**C**), and the subtraction interference color is shown at the +45° diagonal position (**D**), indicating negative birefringence.

Table 2 summarizes the mineral, organic, and water volume (vol%) and weight (wt%) obtained based on the Equation used for human enamel, which also exhibits negative birefringence. These values refer to five ROI (regions of interest) selected in the enamel layer, taking into account that the enamel thickness (distance from enamel surface to the DEJ varied among the samples. The enamel thickness of the fossilized alligator is 115 μm; the enamel thickness of the extant alligator is 190 μm. The fossilized rodent has an enamel thickness of 205 μm, while the enamel thickness of the extant rodent is 145 μm. The density of the organic component was estimated based on the density of glutamic acid (1.5 g/cm^3). The "water" component corresponds to the adsorbed water located in the enamel pores, outside the mineral crystallites.

The mineral vol (%) found in the enamel of the fossilized alligator varied from 83.4 to 89.2, corresponding to 92 to 95.3 wt%. The organic matter vol (%) was 8.1 in the outer enamel, corresponding to 4.5 wt (%), with an increase in the middle part of the enamel (10.3% at 55 μm, corresponding to 5.7% by weight) and a decrease in the organic content in the deeper enamel (5.8 vol%; and 3.1 wt%). The water content did not vary more than 0.2 from the outer to the inner ROI, starting at 6.6 vol (%) and decreasing to 6.4% in the inner enamel, corresponding to 2.4 and 2.3 (wt%), respectively.

In the sample of the modern alligator enamel, mean mineral content was 59.4 (vol%) and 75.2 (wt%). The values for organic volume ranged from 33.9 to 49% (vol), corresponding to 20.98 to 33.51% (wt). The water content varied from 1.17 to 5.18% (volume), corresponding to a mean water wt (%) of 1.4. The biggest difference in mineral and organic matter was found between the surface of the enamel and the second ROI, with less mineral and higher organic content in the surface, as compared with the 2nd to 5th ROIs.

The measurements of the fossilized rodent enamel showed a mean mineral content of 52.3 (vol%) and 70 (wt%) and an organic content of 39.2 (vol%) and 26.3 (wt%). The largest difference between different depths was found in the outer enamel, with the first ROI indicating an 8 (wt%) difference compared with the 2nd ROI, which was more homogeneous with the deeper layers analyzed. The outer layer displayed less mineral and a higher organic content. The water content varied from 8.04 to 10.13% of total enamel volume, and 4.1 (wt%).

The extant rodent enamel showed a very homogeneous mineral content, ranging from 68.1 to 71.6 vol% (82.9–85.3 wt%), and an organic content of 18.5 ± 1.2 (vol%), corresponding to 11.2 ± 0.8 (wt%). The water content exhibited little variation between the ROIs, with a mean vol% of 11.7 (± 0.4), corresponding to 4.8 ± 0.2 (wt%).

Supplementary Figures S3, S6, S8 and S11 show graphs that visually represent the variation of biochemical content along the enamel depth.

Table 2. Enamel components (volume and weight percentages) measured in samples from *Purussaurus* sp. (fossilized alligator), *Melanosuchus niger* (extant alligator), *Neoepiblema* sp. (fossilized rodent), and *Hydrochoerus hydrochaeris* (extant rodent). The "water" component corresponds to adsorbed water during sample preparation.

	Distance from Enamel Surface (µm)	Mineral		Organic [1]		Water	
		vol (%)	wt (%)	vol (%)	wt (%)	vol (%)	wt (%)
fossil alligator	15	85.3	93.1	8.2	4.5	6.6	2.4
	35	83.8	92.3	9.8	5.4	6.4	2.4
	55	83.3	92.0	10.3	5.7	6.3	2.3
	75	89.2	95.3	4.6	2.5	6.3	2.2
	95	87.7	94.5	5.8	3.1	6.5	2.3
	Mean	85.3	93.1	8.2	4.5	6.4	2.3
	SD	2.5	1.5	2.5	1.4	0.1	0.1
extant alligator	15	47.7	65.0	49.0	33.5	3.3	1.5
	55	62.9	77.7	33.9	21.0	3.3	1.4
	95	59.5	75.3	35.4	22.5	5.2	2.2
	135	59.8	75.2	37.6	23.7	2.7	1.1
	175	59.2	74.5	39.7	25.1	1.2	0.5
	Mean	59.5	75.2	37.6	23.7	3.3	1.4
	SD	5.8	4.9	6.0	4.9	1.4	0.6
fossil rodent	40	42.8	61.3	47.1	33.8	10.1	4.9
	70	53.2	70.8	37.3	24.9	9.5	4.2
	100	52.0	69.7	38.8	26.1	9.2	4.1
	130	52.3	69.9	39.7	26.6	8.0	3.6
	160	52.3	69.9	39.2	26.3	8.5	3.8
	Mean	52.3	69.9	39.2	26.3	9.2	4.1
	SD	4.3	3.9	3.8	3.6	0.8	0.5
extant rodent	20	71.7	85.4	16.9	10.1	11.5	4.6
	45	69.9	84.2	18.5	11.2	11.6	4.7
	70	69.4	83.9	18.4	11.2	12.3	5.0
	95	68.4	83.2	19.2	11.7	12.5	5.1
	120	68.1	82.9	20.2	12.4	11.7	4.8
	Mean	69.4	83.9	18.5	11.2	11.7	4.8
	SD	1.4	1.0	1.2	0.8	0.4	0.2

[1] Density estimated based on the density of glutamic acid ($1.5\,g/cm^3$).

3.4. Inorganic Composition of the Superficial Enamel

3.4.1. Energy-Dispersive Spectroscopy (EDS)

Relative quantification of the major chemical elements by SEM-EDS resulted in a ratio of 2.0–2.2 between Ca and P for both modern and fossil enamel (Table 1), characteristic of hydroxyapatite. These results corroborate the previous assumption of using the unit cell composition proposed by Elliot [48] for the determination of organic, mineral, and water content of dental enamel.

3.4.2. Inductively Coupled Plasma-Mass Spectrometry (ICP-MS)

The ICP-MS analysis results identified 27 elements from superficial enamel samples. Calcium values were used to determine the amount of enamel obtained in each sample. Thus, the results of 27 microelements found in each sample are expressed as μg/g of dental enamel (Figure 8). The most abundant element was Zn (4–23%), followed by Pb (0.05–3.7%), Fe (0.5–1.7%), Mg (0.2–1.5%), Al (0.06–1.4%), a series of elements with values near 0.5% (K, Cd, Cr, Mn, and Co), and other still less abundant elements (Be, V, Ni, Cu, As, Se, Rb, Ag, Ba, Tl, Bi, Pd, Th, La, Ce, Sm, and U) near or lower than 0.01%. Not all elements were detected in all samples. The results are summarized in Figures 8 and 9, and all values are presented in Table S2 of the Supplementary Materials.

An increase in the concentration of Pb, Co, Cd, Ce, Th, As, Cu, Bi, Ag, Tl, and U (U234, U235, and U238) is seen in the fossilized enamel, with different magnitudes of change (Figure 8). No expressive difference between fossil and modern samples was detected for the other microelements.

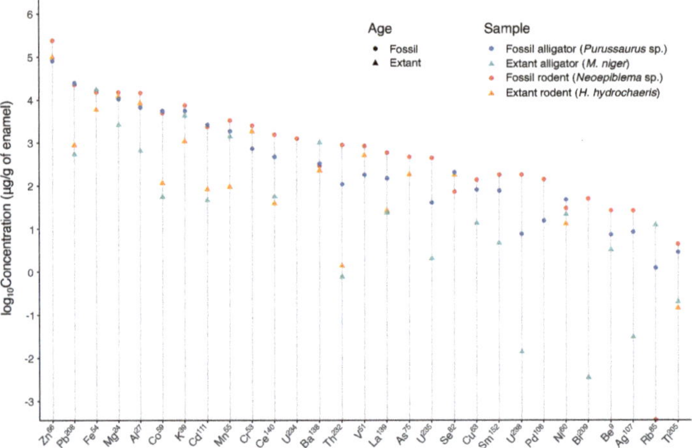

Figure 8. Log$_{10}$ concentration (μg/g of enamel) of 29 microelements detected in the enamel samples. The fossil samples are identified by a circular icon and the modern samples by a triangular icon. Light and dark blue are used for Crocodylia species, while red and orange are used for Rodentia species.

Figure 9 summarizes the results of this study regarding the enamel inorganic element composition. The macroelements (Ca, O, P, and C), determined by SEM-EDS are shown in the pie chart. The microelements, represented by the gray slice of the pie, cannot be estimated by SEM-EDS but might be <2%. A tentative relative amount (%) is drawn based on the ICP-MS results for the microelements. The total of microelements recovered from ICP-MS analysis was transformed to 100%, and based on this, the right-side chart was constructed for microelement distribution. Thus, each element appears with a relative proportion. At first glance, the most abundant microelement is Zn, followed by Pb or Fe, depending on the sample. Note that Fe (shown in orange) has a similar absolute percentage (Supplementary Table S2) in extant alligator (1.7%) and fossil rodent (1.5%), but this similar relative amount is not evident in Figure 9, since no Zn was measured in extant alligator.

Since the enamel of extinct alligator was etched in three different areas (1. Darkest region of enamel; 2. The predominant brown color of enamel; and 3. Yellow region of enamel), these areas are mapped in the upper-left part of Figure 9, so that the results from each area can be separately shown in the graph.

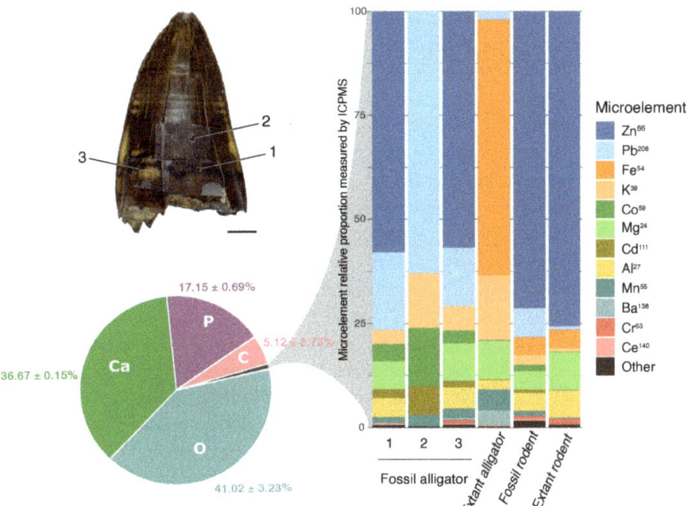

Figure 9. Pie chart displaying the mean (±SD) of the major chemical macroelements (Ca, O, P, and C) found in the enamel (weight percent, wt%) of all samples analyzed by SEM-EDS in this study. On the right, the bar chart shows the proportion of microelements lower than 1% found in each sample (and determined by ICP-MS). In the bars, each element appears as a relative percentage based on the more abundant microelements determined by ICP-MS, except calcium. The picture on the upper left shows the three different enamel regions analyzed in the fossilized alligator (*Purussaurus* sp.) enamel: 1. The darkest enamel area; 2. The predominant brown color enamel; and 3. The yellow enamel area.

4. Discussion

This study highlights several important morphological and chemical aspects of modern and fossil enamel. Since several points regarding the recovery of ancient organic molecules in fossils have been questioned in the last decade, this broader investigation was undertaken with specialists from several areas of the dental enamel research, and a smaller number of samples, to be able to make more precise determinations.

This study brings to light some points that can attract more attention in the future, so that higher quality information can be gained from fossils, both in the interest of better understanding early life in ancient environments, and also, possibly, the metabolic consequences of the evolution of cells and proteins and how they were formed, folded, and worked in environments with different metal concentrations. The discrepancy between the availability of chemical elements (mainly metals) in modern and ancient environments is what makes some chemical elements nowadays toxic to organisms, such as lead, which has a high affinity for hydroxyapatite due to its chemical similarities with calcium. Skeletons were used to prove the anthropogenic contamination of the Earth with lead based on the capacity of bone to harbor lead in its mineral for long times such as decades or centuries [54].

This study started based on the need to have a more solid knowledge of the similarities and differences between modern and fossil samples due to our interest in recovering ancient enamel-specific peptides. Although it is well-known in paleontology that teeth are the best-preserved fossils, and sometimes only enamel is recovered in some fossils, the comparison of modern and ancient enamel has been performed only by a few groups.

To our knowledge, so far such studies were carried out to observe chemical aspects, with a particular interest in the diagenetic processes [55–57] and isotopic analysis [16,58,59] and, as we did before, aiming at recovering protein material from fossils [3,11,12]. Another set of studies did show morphological features of dental enamel [13,14], which had already demonstrated that microscopic aspects were well-preserved. However, those studies did not establish any association between the well-defined morphology of prisms and other structures, with the possibility that such preserved structure might also indicate that fewer (bio)chemical alterations had taken place over time. Thus, this study started from the observation that similar physical aspects were observed (even by the naked eye) in the enamel of modern and fossilized species of correlated taxa: this was a consequence of the high crystallinity and high density of the enamel that seemed to have been preserved.

To the best of our knowledge, this study reports, for the first time, spatially resolved major enamel biochemical component volumes at discrete histological layers of dental enamel. The sum of the organic and water volumes constitutes the pore volume, which, summed with mineral volume, yields the total enamel volume. Corroborating the enamel mineral theoretical composition used for microradiography, with a Ca/P ratio of 2.06 [48], experimental SEM-EDS data indicated Ca/P ratios in the range of 2.0 to 2.2 (Table 2). Two types of water have been reported in dental enamel, lattice water (inside the mineral crystallites) and adsorbed water (located in the pores, outside the mineral crystallites) [60]. The water content measured here represents the adsorbed water, the main pathway for the diffusion of materials in dental enamel. Similar to sound human dental enamel [43,44,46], all samples presented negative birefringence under water immersion, typical of dental enamel.

The intensity of negative birefringence in (mature) enamel is directly proportional to the solid volume (mineral and organic), so negative birefringence is explained by an increased organic volume in sites with low mineral volume [25,44]. In humans, as other mammals who use teeth intensively for food breakdown in preparation for initial digestion by salivary enzymes in the oral cavity, average mature mid-crown dental enamel presents 93% mineral volume (based on a mineral density of 2.99 g/cm^3), 5.5% water volume, and 1.5% organic volume [25,46]. From the occlusal to cervical regions of the tooth crown and from outer to inner enamel, following a decreasing gradient of mastication-related mechanical load, mineral content decreases [61], accompanied by an increase in both water and organic contents [62]. Possibly, variations in major enamel component volumes described might reflect variations in the use of teeth for mastication, the occurrence or not of continuous amelogenesis, and the rate of tooth replacement. Particularly, by contrasting the fossil with the extant alligator, the higher mineral content of the*Purussaurus* sp. (fossil) enamel might be explained by the more intense use of teeth for food breakdown, with a wider range of lower jaw movements, though other specimens (extinct and extant crocodiles) need to be analyzed. Overall, our data suggest that all samples provide a relatively large amount of organic matter and would be suitable sources of enamel peptides. There are, though, several points of uncertainty that can lead to much better results with dental enamel in the future. In particular, the determination of mineral composition, organic and water volumes, and weight percentages, as described in this study, is based on human enamel density, and this is clearly an assumption that is unlikely to reflect the reality in the enamel of other species, and the dental enamel density in fossils also needs to be determined. However, this can be achieved.

Differences in the elements found in modern and fossilized specimens seem to reflect differences in the permineralization of the fossil. The higher mineral content of the alligator fossil enamel, as compared with the rodent fossil enamel, may reflect that they came from different localities, as well as the local differences in the mineral content of the sediments from which the fossils were recovered.

Furthermore, the determination of organic material will certainly benefit from the development of other forms of microscopic techniques to help the determination of organic composition, such as the slow decalcification of samples after the use of alternative fixatives, as already used in the 1960s for microenzimology and successfully applied to recover enzyme function in mineralizing teeth [24,63]. Because the dental enamel is so calcified, techniques normally used for other tissues are difficult to apply to this tissue, but the determination of proteins will be possible with some developments. Interestingly, in the 1960s, alternative fixatives were developed to avoid the use of aldehyde-based fixatives that destroy enzyme activity. This fixative is also incompatible with mass spectrometry determination of peptides and proteins since they cross-link different parts of proteins covalently. So, certainly, there is much to be learned from fossil enamel using alternative fixative methods and mass spectrometry.

The discussion of the organic, inorganic, and water content of the different depths of human and mouse enamel (the ones that have been more studied so far) would be long and outside of the scope of this article. However, there are already several independent lines of evidence (based on different techniques) that suggest that 96 wt% of the mineral in dental enamel might only be found in the very superficial human enamel and possibly also in other mm-thick enamels of mammals. The very thick enamel of mammals, as far as is known, has a very long mineralization process that results in the high mineral content of the outer surface [64]. There is also long-standing evidence that protein remains in the enamel in higher amounts, and careful microscopic technique has been able to show remaining scaffolds of proteins (called "tuft" protein in the past), as discussed by Robinson and Hudson [65].

According to Smith et al. [19], in a study on the comparative proportion of mineral and volatiles in the developing enamel of normal and genetically modified mice, "volatiles were also found in amounts that were often higher than expected, especially in more mature enamel". This study also showed some other interesting aspects that must be kept in mind in the future when techniques "adapted" to studying enamel will (hopefully) be used for many more groups. One such aspect is, for instance, the idea that large amounts of protein will not hinder enamel maturation, as observed in some genetically modified mice. As stated by Smith et al. [19], "the results of this study suggest that maturational growth of enamel crystals can occur in the presence of relatively large amounts of proteins and/or their fragments. The crown problems [...] have nothing to do with protein or mineral but they occur because amelogenesis is shut down early and the enamel organ cells transform into a dysplastic epithelium that secretes a thin calcified material that is not enamel". Supplementary Figure S4 shows a superficial layer in the extant alligator enamel that appears to be enamel but might be an indication that "amelogenesis was shut down early" in this specimen, which is a modern tooth that showed higher content of organic material.

It is the fact that nature has time and a countless number of possibilities that makes the diversity of phenotypes so great. The study of ancient species brings to light unknown structural aspects that resulted from genetic variation. Since genetically modified mouse enamel has been studied in detail for most enamel proteins, the correlation of some findings from such models with structural aspects of past species' enamel might help us better understand past organisms. This is the case in transgenes that overexpressed ameloblastin, an enamel protein that is secreted in the secretion phase of amelogenesis and is part of the calcium-binding phosphoprotein (SCPP) family that evolved from a common ancestral gene [66,67]. Higher levels of expression of ameloblastin in a background lacking amelogenin resulted in short and randomly oriented apatite crystals, and the enamel resembled that of the reptile *iguana* [68].

In addition to many other implications, the variety and different proportions of enamel proteins have an impact on the research looking for ancient peptides and proteins. Structural and ultrastructural features might provide indications for the set of enamel proteins that can be found for paleoproteomics research. The ancient proteins might be preserved longer than we expect, but we might not be able to recover them so efficiently without knowing their sequences and post-translation modifications, as well as how they interact with each other and with minerals.

Bartlett et al. [21] showed that the knockout of Amelx and Mmp20 in mice resulted in the presence of "fan-like" crystal arrangements in the deeper enamel (20 μm of enamel, close to the DEJ), while the middle and outer enamel of such animals show a dysplastic and more mineralized layer [21]. Interestingly, the enamel of the extant alligator exhibited mineral structures that resemble fans in form (Figure 5F–G), with the same orientation (the larger border of the fan towards the outside of the enamel). Additoinal studies of the modern alligators' enamel can provide more information on the enamel proteome, the detailed crystal arrangement, and, possibly, the relationship between both.

The ICP-MS data must be viewed with caution, since only a single measurement was made. Nonetheless, though the exploratory results of this study, we consider these data important for their contribution to a wider view of some aspects of the enamel in modern and fossil samples, and they may help formute new hypotheses and the need to avoid some of the mentioned problems of this study in the future. The lack of Zn in one of the samples is one of the problems. This is an error that cannot be fixed, since only one sample was taken for measurement, and might result from the high levels of microelements found in blank samples and the exclusion of many data based on this. Zinc is a very important mineral for metabolism, is present in many proteins (particularly in enzymes), and is always found in dental enamel. The results of this study are preliminary and need to be repeated in a larger set of samples, with samples collected at least in triplicates.

The time scale seems to be important to explain the presence of lead in higher concentrations on the outer enamel of fossils, in addition to other factors such as the acidity of the soil and the fossilization process. This accumulation on the outer enamel over millions of years is different from the low levels of lead found in the enamel of most humans or animals (that do not live in contaminated areas). It is clear that superficial enamel can lose minerals as it can also gain minerals (a net loss of Ca and P being the cause of caries). During these pH changes, lead can be incorporated into the enamel, even to deeper enamel [69]. However, this does not change the fact that lead on the surface of modern human enamel reflects the contamination of the environment [5,7,70–73], and the fact that in pre-industrial times the composition of superficial enamel is expected to have less lead, as lower lead levels were found in skeletons of prehistorical humans [54]. Therefore, the fact that lead and other minerals show a gradient in surface enamel (F, Cu, and other metals also show this gradient [74]) points out the need to carefully analyze superficial enamel, taking into account time scale and how many micrometers from the surface the ions were determined.

5. Conclusions

Specimens analyzed in this study showed high crystallinity and high density in both modern and fossilized enamel. Few structural changes were detected relative to the age of the sample. The polarizing microscopy indicated the birefringence property for the modern and fossilized samples (negative birefringence). The microradiography determination of major enamel biochemical components showed unexpectedly high organic weight (%) in two specimens (23.72% in extant allitagor and 26.30% in fossilized rodent). Relative quantification of the major chemical elements by SEM-EDS resulted in a ratio of 2.0–2.2 between Ca and P for both modern and fossil enamel, characteristic of hydroxyapatite. The ICP-MS analyses recovered 27 microelements (in addition to Ca) from superficial enamel samples. For the majority of chemical elements, it was not possible to establish differences between modern and fossil enamel. Nevertheless, an increase in the concentration of Pb, Co, Cd, Ce, Th, As, Cu, Bi, Ag, and Tl was seen in the fossilized enamel, most likely deposited during permineraliztion, with different magnitudes of change. We believe that future studies should be conducted in order to test if larger sampling agrees with our preliminary results and improves the inference power regarding the resistance of fossil enamel through deep time. Furthermore, we expect to recover and identify the organic matrix of dental enamel from the Miocene species studied here (work in progress). Results of this study show the superior preservation of the dental enamel over long time windows, a tissue particularly important for the current interest in the knowledge of ancient environments and peptide recovery, and, as such, for paleontology of the 21st century.

Supplementary Materials: The following supporting information can be downloaded at: https://www.mdpi.com/article/10.3390/biology11111636/s1. Figure S1. A photomicrograph panel of ground sections (~100 µm) of the 4 teeth used in this study. The images show that dental enamel (e) is translucent in all specimens, while dentine (d) is dark/opaque in all specimens. There is some superficial mineral loss in dentine of *Purussaurus sp.*, possibly due to acids from the environment (A: arrow head; B: asterisk). The white arrows of panels A and C show marked apposition lines in the dentine. In D, black arrows in the enamel show appositional lines. In F the blue arrow indicates a very pronounced line. In G the *H. hydrochaeris* enamel is shown; the arrow indicates an area in middle enamel that seems different from the inner and the outer enamel. In H a larger magnification of the black square is shown under polarized light. Due to the thickness of this section, the prismatic aspect of this enamel is difficult to recognize, but in the outer enamel (asterisk) diagonal lines are enamel rods. Table S1. ICP-MS method quantification limits (LOQ) of the elements in the study. Table S2. Twenty-seven (27) microelements in the enamel composition determined by ICP-MS. The sample concentration results in ppb were converted to concentration (ppm) of enamel mass based on Calcium percentage values recovered by EDS-SEM. Purussaurus (1): Darkest region of enamel, (2) Predominant brown enamel, (3) Yellow region of enamel. Figure S2. Microradiographic image of a ground section of dental enamel sample from *Purussaurus* sp. The arrow indicates the enamel-dentine junction (EDJ). Points of biochemical components (mineral, water, and organic) measurements (at 15, 35, 55, 75, and 95 µm from the enamel surface) were located along the longitudinal line shown in the enamel layer. The enamel thickness is 115 µm. Figure S3. Mineral (A), organic (B), and total water (B) enamel component volumes of *Purussaurus* sp. at histological points located along the longitudinal line shown in Supplementary Figure S2. Figure S4. Ground section of dental enamel sample from *Melanosuchus niger* under polarizing microscopy. The image was taken with the Red I filter (most of the enamel is shown in blue). The ground section was submitted to water immersion before analysis. On the surface of the enamel there is a layer that is different from the underlying enamel (orange). Figure S5. Microradiographic image of a ground section of dental enamel sample from *M. niger*. Points of biochemical components (mineral, water, and organic) measurements (at 15, 55, 95, 135, and 175 µm from the enamel surface) were located along the longitudinal line shown in the enamel layer. The enamel thickness is 190 µm. Figure S6. Mineral (A), organic (B), and total water (B) enamel component volumes of *M. niger* at histological points located along the longitudinal line shown in Figure S5. Figure S7. Microradiographic image of a ground section of dental enamel sample from *Neoepiblema* sp. Points of biochemical components (mineral, water, and organic) measurements (at 40, 70, 100, 130, and 160 µm from the enamel surface) were located along the longitudinal line shown in the enamel layer. The enamel thickness is 205 µm. Figure S8. Mineral (A), organic (B), and total water (B) enamel component volumes of *Neoepiblema* sp. at histological points located along the longitudinal line shown in Figure S7. Figure S9. Image of a ground section of dental enamel sample from *Hydrochoerus hydrochaeris* under polarizing microscopy. The image was taken with the Red I filter. The ground section was submitted to water immersion before analysis. Figure S10. Microradiographic image of a ground section of dental enamel sample from *H. hydrochaeris*. Points of biochemical components (mineral, water, and organic) measurements (at 20, 45, 70, 95, and 120 µm from the enamel surface) were located along the longitudinal line shown in the enamel layer. The enamel thickness is 145 µm. Figure S11. Mineral (A), organic (B), and total water (B) enamel component volumes of *H. hydrochaeris* at histological points located along the longitudinal line shown in Figure S10.

Author Contributions: Conceptualization, C.P.-L., J.T.-F. and R.F.G.; data curation, A.S.H. and F.P.M.; Formal analysis, C.P.-L., J.T.-F., V.H.F.-S., F.B.d.S. and F.B.J.; funding acquisition, R.F.G. and M.C.L.; investigation, C.P.-L., J.T.-F., N.M.-R., A.S.H., F.P.M., J.A.M., F.B.d.S., F.B.J., S.R.P.L. and R.F.G.; methodology, F.B.d.S. and R.F.G.; project administration, C.P.-L. and R.F.G.; resources, A.S.H., J.A.M., R.F.G. and M.C.L.; supervision, R.F.G. and M.C.L.; visualization, C.P.-L., J.T.-F., N.M.-R. and V.H.F.-S.; writing—original draft, C.P.-L., J.T.-F., N.M.-R., A.S.H., F.B.d.S., S.R.P.L., R.F.G. and M.C.L.; writing—review and editing, C.P.-L., R.F.G., A.S.H., S.R.P.L., V.H.F.-S. and F.B.J. All authors have read and agreed to the published version of the manuscript.

Funding: This research was funded by FAPESP grant number 2021/05059-0, 2018/24069-3 and 20/07997-4; CNPq grant number 313028-2018-4

Institutional Review Board Statement: Not applicable.

Informed Consent Statement: Not applicable.

Data Availability Statement: Not applicable.

Acknowledgments: Special thanks to Professor Edson Guilherme and the Laboratory of Paleontological Studies, Universidade Federal do Acre (Brazil), for valuable contribution and support in making available the paleontological material of the UFAC collection. We also thank the two anonymous reviewers for the comments and suggestions.

Conflicts of Interest: The authors declare no conflict of interest.

References

1. Wiens, J.J.; Chippindale, P.T.; Hillis, D.M. When Are Phylogenetic Analyses Misled by Convergence? A Case Study in Texas Cave Salamanders. *Syst. Biol.* **2003**, *52*, 501–514. [CrossRef]
2. Porto, I.M.; Laure, H.J.; Tykot, R.H.; de Sousa, F.B.; Rosa, J.C.; Gerlach, R.F. Recovery and identification of mature enamel proteins in ancient teeth. *Eur. J. Oral Sci.* **2011**, *119* (Suppl. 1), 83–87. [CrossRef] [PubMed]
3. Nogueira, F.C.S.; Neves, L.X.; Pessoa-Lima, C.; Langer, M.C.; Domont, G.B.; Line, S.R.P.; Paes Leme, A.F.; Gerlach, R.F. Ancient enamel peptides recovered from the South American Pleistocene species Notiomastodon platensis and Myocastor cf. coypus. *J. Proteom.* **2021**, *240*, 104187. [CrossRef] [PubMed]
4. Porto, I.M.; Line, S.R.P.; Laure, H.J.; Gerlach, R.F. Comparison of three methods for enamel protein extraction in different developmental phases of rat lower incisors. *Eur. J. Oral Sci.* **2006**, *114* (Suppl. 1), 272–275; discussion 285–286, 382. [CrossRef] [PubMed]
5. Gomes, V.E.; Rosário de Sousa, M.d.L.; Barbosa, F., Jr.; Krug, F.J.; Pereira Saraiva, M.d.C.; Cury, J.A.; Gerlach, R.F. In vivo studies on lead content of deciduous teeth superficial enamel of preschool children. *Sci. Total Environ.* **2004**, *320*, 25–35. [CrossRef]
6. Barbosa, F.; Tanus, S.J.E.; Gerlach, R.F.; Parsons, P.J. A Critical Review of Biomarkers Used for Monitoring Human Exposure to Lead: Advantages, Limitations, and Future Needs. *Environ. Health Perspect.* **2005**, *113*, 1669–1674. [CrossRef]
7. Costa de Almeida, G.R.; Pereira Saraiva, M.d.C.; Barbosa, F.; Krug, F.J.; Cury, J.A.; Rosário de Sousa, M.d.L.; Rabelo Buzalaf, M.A.; Gerlach, R.F. Lead contents in the surface enamel of deciduous teeth sampled in vivo from children in uncontaminated and in lead-contaminated areas. *Environ. Res.* **2007**, *104*, 337–345. [CrossRef]
8. Andre Stewart, N.; Ferian Molina, G.; Issa, J.P.M.; Andrew Yates, N.; Sosovicka, M.; Rezende Vieira, A.; Peres Line, S.R.; Montgomery, J.; Fernanda Gerlach, R. The identification of peptides by nanoLC-MS/MS from human surface tooth enamel following a simple acid etch extraction. *RSC Adv.* **2016**, *6*, 61673–61679. [CrossRef]
9. Stewart, N.A.; Gerlach, R.F.; Gowland, R.L.; Gron, K.J.; Montgomery, J. Sex determination of human remains from peptides in tooth enamel. *Proc. Natl. Acad. Sci. USA* **2017**, *114*, 13649–13654. [CrossRef]
10. Delgado, S.; Vidal, N.; Veron, G.; Sire, J.Y. Amelogenin, the major protein of tooth enamel: A new phylogenetic marker for ordinal mammal relationships. *Mol. Phylogenetics Evol.* **2008**, *47*, 865–869. [CrossRef]
11. Cappellini, E.; Welker, F.; Pandolfi, L.; Ramos-Madrigal, J.; Samodova, D.; Rüther, P.L.; Fotakis, A.K.; Lyon, D.; Moreno-Mayar, J.V.; Bukhsianidze, M.; et al. Early Pleistocene enamel proteome from Dmanisi resolves Stephanorhinus phylogeny. *Nature* **2019**, *574*, 103–107. [CrossRef] [PubMed]
12. Welker, F.; Ramos-Madrigal, J.; Kuhlwilm, M.; Liao, W.; Gutenbrunner, P.; de Manuel, M.; Samodova, D.; Mackie, M.; Allentoft, M.E.; Bacon, A.M.; et al. Enamel proteome shows that Gigantopithecus was an early diverging pongine. *Nature* **2019**, *576*, 262–265. [CrossRef] [PubMed]
13. Line, S.R.P. Incremental markings of enamel in ectothermal vertebrates. *Arch. Oral Biol.* **2000**, *45*, 363–368. [CrossRef]
14. Line, S.R.P.; Bergqvist, L.P. Enamel structure of paleocene mammals of the São José de Itaboraí basin, Brazil. 'Condylarthra', Litopterna, Notoungulata, Xenungulata, and Astrapotheria. *J. Vertebr. Paleontol.* **2005**, *25*, 924–928. [CrossRef]
15. Vašinová Galiová, M.; Nývltová Fišáková, M.; Kynický, J.; Prokeš, L.; Neff, H.; Mason, A.Z.; Gadas, P.; Košler, J.; Kanický, V. Elemental mapping in fossil tooth root section of Ursus arctos by laser ablation inductively coupled plasma mass spectrometry (LA-ICP-MS). *Talanta* **2013**, *105*, 235–243. [CrossRef]
16. Ivanova, V.; Shchetnikov, A.; Semeney, E.; Filinov, I.; Simon, K. LA-ICP-MS analysis of rare earth elements in tooth enamel of fossil small mammals (Ust-Oda section, Fore-Baikal area, Siberia): Paleoenvironmental interpretation. *J. Quat. Sci.* **2022**, *37*, 1246–1260. [CrossRef]
17. Demarchi, B.; Hall, S.; Roncal-Herrero, T.; Freeman, C.L.; Woolley, J.; Crisp, M.K.; Wilson, J.; Fotakis, A.; Fischer, R.; Kessler, B.M.; et al. Protein sequences bound to mineral surfaces persist into deep time. *eLife* **2016**, *5*, e17092. [CrossRef]
18. Chun, Y.H.P.; Lu, Y.; Hu, Y.; Krebsbach, P.H.; Yamada, Y.; Hu, J.C.C.; Simmer, J.P. Transgenic rescue of enamel phenotype in Ambn null mice. *J. Dent. Res.* **2010**, *89*, 1414–1420. [CrossRef]

19. Smith, C.E.; Hu, Y.; Richardson, A.S.; Bartlett, J.D.; Hu, J.C.C.; Simmer, J.P. Relationships between protein and mineral during enamel development in normal and genetically altered mice. *Eur. J. Oral Sci.* **2011**, *119*, 125–135. [CrossRef]
20. Hatakeyama, J.; Fukumoto, S.; Nakamura, T.; Haruyama, N.; Suzuki, S.; Hatakeyama, Y.; Shum, L.; Gibson, C.W.; Yamada, Y.; Kulkarni, A.B. Synergistic roles of amelogenin and ameloblastin. *J. Dent. Res.* **2009**, *88*, 318–322. [CrossRef]
21. Bartlett, J.D.; Smith, C.E.; Hu, Y.; Ikeda, A.; Strauss, M.; Liang, T.; Hsu, Y.H.; Trout, A.H.; McComb, D.W.; Freeman, R.C.; et al. MMP20-generated amelogenin cleavage products prevent formation of fan-shaped enamel malformations. *Sci. Rep.* **2021**, *11*, 10570. [CrossRef] [PubMed]
22. Saiani, R.A.; Porto, I.M.; Marcantonio Junior, E.; Cury, J.A.; de Sousa, F.B.; Gerlach, R.F. Morphological characterization of rat incisor fluorotic lesions. *Arch. Oral Biol.* **2009**, *54*, 1008–1015. [CrossRef] [PubMed]
23. Porto, I.M.; Merzel, J.; de Sousa, F.B.; Bachmann, L.; Cury, J.A.; Line, S.R.P.; Gerlach, R.F. Enamel mineralization in the absence of maturation stage ameloblasts. *Arch. Oral Biol.* **2009**, *54*, 313–321. [CrossRef] [PubMed]
24. Porto, I.M.; Saiani, R.A.; Chan, K.L.A.; Kazarian, S.G.; Gerlach, R.F.; Bachmann, L. Organic and inorganic content of fluorotic rat incisors measured by FTIR spectroscopy. *Spectrochim. Acta. Part A Mol. Biomol. Spectrosc.* **2010**, *77*, 59–63. [CrossRef] [PubMed]
25. Sousa, F.B.; Vianna, S.S.; Santos-Magalhães, N.S. A new approach for improving the birefringence analysis of dental enamel mineral content using polarizing microscopy. *J. Microsc.* **2006**, *221*, 79–83. [CrossRef]
26. de Souza-Guerra, C.; Barroso, R.C.; de Almeida, A.P.; Peixoto, I.T.A.; Moreira, S.; de Sousa, F.B.; Gerlach, R.F. Anatomical variations in primary teeth microelements with known differences in lead content by micro-Synchrotron Radiation X-Ray Fluorescence (μ-SRXRF)—A preliminary study. *J. Trace Elem. Med. Biol.* **2014**, *28*, 186–193. [CrossRef]
27. Jordan, T.E. Retroarc foreland and related basins. In *Tectonics of Sedimentary Basins*; Blackwell Scientific Publications: Hoboken, NJ, USA, 1995; pp. 331–391.
28. Silva, A.J.P.; Lopes, R.C.L.; Vasconcelos, A.M. Bacias sedimentares paleozoicas e meso-cenozoicas interiores. In *Geologia, Tectonica e Recursos Minerais do Brasil: Texto, Mapas & SIG*; CPRM: Brasilia, Brazil, 2003; p. 55.
29. Cunha, P.d.C. Bacia do Acre. *Bol. De Geociências Da Petrobras* **2007**, *15*, 207–215.
30. Bissaro-Júnior, M.C.; Kerber, L.; Crowley, J.L.; Ribeiro, A.M.; Ghilardi, R.P.; Guilherme, E.; Negri, F.R.; Souza Filho, J.P.; Hsiou, A.S. Detrital zircon U–Pb geochronology constrains the age of Brazilian Neogene deposits from Western Amazonia. *Palaeogeogr. Palaeoclimatol. Palaeoecol.* **2019**, *516*, 64–70. [CrossRef]
31. Latrubesse, E.; Bocquentin, J.; Santos, J.C.R.; Ramonell, C.G. Paleoenvironmental model for the late cenozoic of southwestern Amazonia: Paleontology and geology. *Acta Amaz.* **1997**, *27*, 103–117. [CrossRef]
32. Latrubesse, E.M.; Cozzuol, M.; da Silva-Caminha, S.A.F.; Rigsby, C.A.; Absy, M.L.; Jaramillo, C. The Late Miocene paleogeography of the Amazon Basin and the evolution of the Amazon River system. *Earth-Sci. Rev.* **2010**, *99*, 99–124. [CrossRef]
33. Cozzuol, M.A. The Acre vertebrate fauna: Age, diversity, and geography. *J. South Am. Earth Sci.* **2006**, *21*, 185–203. [CrossRef]
34. Lundberg, J.G.; Sabaj Pérez, M.H.; Dahdul, W.M.; Aguilera, O.A. The Amazonian neogene fish fauna. In *Amazonia: Landscape and Species Evolution: A Look into the Past*; Wiley-Blackwell: Chichester, UK, 2009; pp. 281–301.
35. Riff, D.; Romano, P.S.R.; Oliveira, G.R.; Aguilera, O.A.; Hoorn, C. Neogene crocodile and turtle fauna in northern South America. In *Amazonia. Landscapes and Species Evolution: A Look Into the Past*; Wiley-Blackwell: Chichester, UK, 2010; pp. 259–280.
36. Negri, F.R.; Bocquentin-Villanueva, J.; Ferigolo, J.; Antoine, P.O. A review of Tertiary mammal faunas and birds from western Amazonia. In *Amazonia: Landscape and Species Evolution: A Look into the Past*; Wiley-Blackwell: Chichester, UK, 2009; pp. 243–258.
37. Hsiou, A.S. Lagartos e Serpentes (Lepidosauria, Squamata) do Mioceno Médio-Superior da Região Norte da América do Sul; IGEO/UFRGS; Porto Alegre, RS, Brazil. 2010. Available online: http://hdl.handle.net/10183/23712 (accessed on 6 June 2022).
38. Aureliano, T.; Ghilardi, A.M.; Guilherme, E.; Souza-Filho, J.P.; Cavalcanti, M.; Riff, D. Morphometry, Bite-Force, and Paleobiology of the Late Miocene Caiman Purussaurus brasiliensis. *PLoS ONE* **2015**, *10*, e0117944. [CrossRef] [PubMed]
39. Rio, J.P.; Mannion, P.D. Phylogenetic analysis of a new morphological dataset elucidates the evolutionary history of Crocodylia and resolves the long-standing gharial problem. *PeerJ* **2021**, *9*, e12094. [CrossRef] [PubMed]
40. Kerber, L.; Candela, A.M.; Ferreira, J.D.; Pretto, F.A.; Bubadué, J.; Negri, F.R. Postcranial Morphology of the Extinct Rodent Neoepiblema (Rodentia: Chinchilloidea): Insights Into the Paleobiology of Neoepiblemids. *J. Mamm. Evol.* **2022**, *29*, 207–235. [CrossRef]
41. De Figueiredo, F.A.T.; Ramos, J.; Kawakita, E.R.H.; Bilal, A.S.; de Sousa, F.B.; Swaim, W.D.; Issa, J.P.M.; Gerlach, R.F. Lead line in rodents: An old sign of lead intoxication turned into a new method for environmental surveillance. *Environ. Sci. Pollut. Res.* **2016**, *23*, 21475–21484. [CrossRef]
42. Meisnar, M.; Lozano-Perez, S.; Moody, M.; Holland, J. Low-energy EDX – A novel approach to study stress corrosion cracking in SUS304 stainless steel via scanning electron microscopy. *Micron* **2014**, *66*, 16–22. [CrossRef]
43. De Andrade Dantas, E.L.; de Figueiredo, J.T.; Macedo-Ribeiro, N.; Oliezer, R.S.; Gerlach, R.F.; de Sousa, F.B. Variation in mineral, organic, and water volumes at the neonatal line and in pre- and postnatal enamel. *Arch. Oral Biol.* **2020**, *118*, 104850. [CrossRef]
44. De Medeiros, R.; Soares, J.; De Sousa, F. Natural enamel caries in polarized light microscopy: Differences in histopathological features derived from a qualitative versus a quantitative approach to interpret enamel birefringence. *J. Microsc.* **2012**, *246*, 177–189. [CrossRef]
45. Little, M.F.; Casciani, F.S. The nature of water in sound human enamel: A preliminary study. *Arch. Oral Biol.* **1966**, *11*, 565–571. [CrossRef]
46. Angmar, B.; Carlström, D.; Glas, J.E. Studies on the ultrastructure of dental enamel: IV. The mineralization of normal human enamel. *J. Ultrastruct. Res.* **1963**, *8*, 12–23. [CrossRef]
47. Arends, J.; ten Bosch, J.J. Demineralization and remineralization evaluation techniques. *J. Dent. Res.* **1992**, *71*, 924–928. [CrossRef] [PubMed]

48. Elliott, J.C. Structure, Crystal Chemistry and Density of Enamel Apatites. In *Ciba Foundation Symposium 205-Dental Enamel*; John Wiley & Sons, Ltd.: Hoboken, NJ, USA, 2007; pp. 54–72. [CrossRef]

49. de Almeida, G.R.C.; de Souza Guerra, C.; Tanus-Santos, J.E.; Barbosa, F.; Gerlach, R.F. A plateau detected in lead accumulation in subsurface deciduous enamel from individuals exposed to lead may be useful to identify children and regions exposed to higher levels of lead. *Environ. Res.* **2008**, *107*, 264–270. [CrossRef] [PubMed]

50. Brudevold, F.; Reda, A.; Aasenden, R.; Bakhos, Y. Determination of trace elements in surface enamel of human teeth by a new biopsy procedure. *Arch. Oral Biol.* **1975**, *20*, 667–673. [CrossRef]

51. Costa de Almeida, G.R.; Molina, G.F.; Meschiari, C.A.; Barbosa de Sousa, F.; Gerlach, R.F. Analysis of enamel microbiopsies in shed primary teeth by Scanning Electron Microscopy (SEM) and Polarizing Microscopy (PM). *Sci. Total Environ.* **2009**, *407*, 5169–5175. [CrossRef]

52. Porto, I.M.; Laure, H.J.; de Sousa, F.B.; Rosa, J.C.; Gerlach, R.F. New techniques for the recovery of small amounts of mature enamel proteins. *J. Archaeol. Sci.* **2011**, *38*, 3596–3604. [CrossRef]

53. Wood, E.A. *Crystals and Light: An Introduction to Optical Crystallography*; Courier Corporation: North Chelmsford, MA, USA, 1977.

54. Ericson, J.E.; Shirahata, H.; Patterson, C.C. Skeletal Concentrations of Lead in Ancient Peruvians. *N. Engl. J. Med.* **1979**, *300*, 946–951. [CrossRef]

55. Jacques, L.; Ogle, N.; Moussa, I.; Kalin, R.; Vignaud, P.; Brunet, M.; Bocherens, H. Implications of diagenesis for the isotopic analysis of Upper Miocene large mammalian herbivore tooth enamel from Chad. *Palaeogeogr. Palaeoclimatol. Palaeoecol.* **2008**, *266*, 200–210. [CrossRef]

56. Kocsis, L.; Trueman, C.N.; Palmer, M.R. Protracted diagenetic alteration of REE contents in fossil bioapatites: Direct evidence from Lu–Hf isotope systematics. *Geochim. Et Cosmochim. Acta* **2010**, *74*, 6077–6092. [CrossRef]

57. Kohn, M.J.; Schoeninger, M.J.; Barker, W.W. Altered states: Effects of diagenesis on fossil tooth chemistry. *Geochim. Et Cosmochim. Acta* **1999**, *63*, 2737–2747. [CrossRef]

58. Schoeninger, M.J.; Hallin, K.; Reeser, H.; Valley, J.W.; Fournelle, J. Isotopic alteration of mammalian tooth enamel. *Int. J. Osteoarchaeol.* **2003**, *13*, 11–19. [CrossRef]

59. Domingo, L.; Cuevas-González, J.; Grimes, S.T.; Hernández Fernández, M.; López-Martínez, N. Multiproxy reconstruction of the palaeoclimate and palaeoenvironment of the Middle Miocene Somosaguas site (Madrid, Spain) using herbivore dental enamel. *Palaeogeogr. Palaeoclimatol. Palaeoecol.* **2009**, *272*, 53–68. [CrossRef]

60. LeGeros, R.Z.; Bonel, G.; Legros, R. Types of "H2O" in human enamel and in precipitated apatites. *Calcif. Tissue Res.* **1978**, *26*, 111–118. [CrossRef] [PubMed]

61. Theuns, H.M.; van Dijk, J.W.E.; Jongebloed, W.L.; Groeneveld, A. The mineral content of human enamel studied by polarizing microscopy, microradiography and scanning electron microscopy. *Arch. Oral Biol.* **1983**, *28*, 797–803. [CrossRef]

62. Setally Azevedo Macena, M.; de Alencar e Silva Leite, M.L.; de Lima Gouveia, C.; de Lima, T.A.S.; Athayde, P.A.A.; de Sousa, F.B. A comparative study on component volumes from outer to inner dental enamel in relation to enamel tufts. *Arch. Oral Biol.* **2014**, *59*, 568–577. [CrossRef] [PubMed]

63. Porto, I.M.; Rocha, L.B.; Rossi, M.A.; Gerlach, R.F. In situ zymography and immunolabeling in fixed and decalcified craniofacial tissues. *J. Histochem. Cytochem. Off. J. Histochem. Soc.* **2009**, *57*, 615–622. [CrossRef]

64. Aoba, T. Recent observations on enamel crystal formation during mammalian amelogenesis. *Anat. Rec.* **1996**, *245*, 208–218. [CrossRef]

65. Robinson, C.; Hudson, J. Tuft protein: Protein cross-linking in enamel development. *Eur. J. Oral Sci.* **2011**, *119*, 50–54. [CrossRef] [PubMed]

66. Delgado, S.; Casane, D.; Bonnaud, L.; Laurin, M.; Sire, J.Y.; Girondot, M. Molecular Evidence for Precambrian Origin of Amelogenin, the Major Protein of Vertebrate Enamel. *Mol. Biol. Evol.* **2001**, *18*, 2146–2153. [CrossRef]

67. Kawasaki, K.; Weiss, K.M. Mineralized tissue and vertebrate evolution: The secretory calcium-binding phosphoprotein gene cluster. *Proc. Natl. Acad. Sci.* **2003**, *100*, 4060–4065. [CrossRef] [CrossRef]

68. Lu, X.; Ito, Y.; Kulkarni, A.; Gibson, C.; Luan, X.; Diekwisch, T.G.H. Ameloblastin-rich enamel matrix favors short and randomly oriented apatite crystals. *Eur. J. Oral Sci.* **2011**, *119*, 254–260. [CrossRef]

69. Molina, G.F.; Costa de Almeida, G.R.; de Souza Guerra, C.; Cury, J.A.; de Almeida, A.P.; Barroso, R.C.; Gerlach, R.F. Lead deposition in bovine enamel during a pH-cycling regimen simulating the caries process. *Caries Res.* **2011**, *45*, 469–474. [CrossRef] [PubMed]

70. Brudevold, F.; Aasenden, R.; Srinivasian, B.N.; Bakhos, Y. Lead in enamel and saliva, dental caries and the use of enamel biopsies for measuring past exposure to lead. *J. Dent. Res.* **1977**, *56*, 1165–1171. [CrossRef] [PubMed]

71. Cleymaet, R.; Retief, D.H.; Quartier, E.; Slop, D.; Coomans, D.; Michotte, Y. A comparative study of the lead and cadmium content of surface enamel of Belgian and Kenyan children. *Sci. Total Environ.* **1991**, *104*, 175–189. [CrossRef]

72. Cleymaet, R.; Bottenberg, P.; Slop, D.; Clara, R.; Coomans, D. Study of lead and cadmium content of surface enamel of schoolchildren from an industrial area in Belgium. *Community Dent. Oral Epidemiol.* **1991**, *19*, 107–111. [CrossRef]

73. Costa de Almeida, G.R.; de Sousa Guerra, C.; de Angelo Souza Leite, G.; Antonio, R.C.; Barbosa, F.; Tanus-Santos, J.E.; Gerlach, R.F. Lead contents in the surface enamel of primary and permanent teeth, whole blood, serum, and saliva of 6- to 8-year-old children. *Sci. Total Environ.* **2011**, *409*, 1799–1805. [CrossRef]

74. Weatherell, J.A.; Robinson, C.; Hallsworth, A.S. Variations in the chemical composition of human enamel. *J. Dent. Res.* **1974**, *53*, 180–192. [CrossRef]

Article

Midfacial Morphology and Neandertal–Modern Human Interbreeding

Steven E. Churchill [1,2], Kamryn Keys [3] and Ann H. Ross [3,*]

[1] Department of Evolutionary Anthropology, Duke University, Durham, NC 27708, USA; churchy@duke.edu
[2] Centre for the Exploration of the Deep Human Journey, University of the Witwatersrand, Johannesburg 2050, South Africa
[3] Human Identification & Forensic Analysis Laboratory, Department of Biological Sciences, North Carolina State University, Raleigh, NC 27695, USA; kkeys@ncsu.edu
* Correspondence: ahross@ncsu.edu

Simple Summary: Studies of human fossils, and the DNA extracted from them, reveal a complex history of interbreeding between various human lineages over the last one hundred thousand years. Of particular interest is the nature of the population interactions between the Neandertals of Ice Age Europe and western Asia and the modern humans that eventually replaced them. Here, we used six measurements of the facial skeleton, in samples of Neandertal and early modern human fossils, in an exploratory study aimed at trying to identify geographic regions (from the Near East to western Europe) where interbreeding may have been prevalent enough to have left a signal in the facial morphology of the early modern humans of those regions. Although fossil sample sizes were in some cases very small, the results are consistent with the Near East having played an important role in the introduction of Neandertal genes into the genomes of living humans.

Abstract: Ancient DNA from, Neandertal and modern human fossils, and comparative morphological analyses of them, reveal a complex history of interbreeding between these lineages and the introgression of Neandertal genes into modern human genomes. Despite substantial increases in our knowledge of these events, the timing and geographic location of hybridization events remain unclear. Six measures of facial size and shape, from regional samples of Neandertals and early modern humans, were used in a multivariate exploratory analysis to try to identify regions in which early modern human facial morphology was more similar to that of Neandertals, which might thus represent regions of greater introgression of Neandertal genes. The results of canonical variates analysis and hierarchical cluster analysis suggest important affinities in facial morphology between both Middle and Upper Paleolithic early modern humans of the Near East with Neandertals, highlighting the importance of this region for interbreeding between the two lineages.

Keywords: hybridization; introgression; ancient DNA (aDNA); hominin paleontology; paleoanthropology

Citation: Churchill, S.E.; Keys, K.; Ross, A.H. Midfacial Morphology and Neandertal–Modern Human Interbreeding. *Biology* **2022**, *11*, 1163. https://doi.org/10.3390/biology11081163

Academic Editor: Zhifei Zhang

Received: 4 July 2022
Accepted: 1 August 2022
Published: 3 August 2022

Publisher's Note: MDPI stays neutral with regard to jurisdictional claims in published maps and institutional affiliations.

Copyright: © 2022 by the authors. Licensee MDPI, Basel, Switzerland. This article is an open access article distributed under the terms and conditions of the Creative Commons Attribution (CC BY) license (https://creativecommons.org/licenses/by/4.0/).

1. Introduction

The first two decades of the 21st Century brought a radical transformation to our understanding of the evolutionary history of the genus *Homo* during the Middle and Late Pleistocene. This metamorphosis has been driven by advances in ancient DNA (aDNA) and related studies, the increasing use of sophisticated, computationally-intensive morphometric analyses (such as 3D geometric morphometrics of the enamel–dentine junction of hominin teeth) and virtual imaging techniques, a continued emphasis on fieldwork and the recovery of new fossil specimens, and rigorous radiometric dating of important sites and specimens. In combination, these approaches have illuminated a complex picture of multiple, sympatric lineages of *Homo* persisting through the Pleistocene, with considerable interbreeding occurring between them. Highlights from this picture include: (1) the discovery of a previously unknown lineage (the Denisovans), apparently representing a sister species to the Neandertals, from an isolated fossil from 48–30 Ka-old layers in a Siberian cave [1]; (2) the recognition of persistent yet temporally and geographicallycomplex patterns of interbreeding between *Homo* lineages through the Middle and Late Pleistocene [2]; (3) the discovery of multiple, late-surviving, basal (based on their possession of ancestral morphological traits) species of *Homo* in Africa and Asia [3–5], one of which (*H. naledi*) may have been a source of introgressed archaic human genes in the modern African genome [6–9]; (4) virtual reconstruction, reanalysis, and dating of early *Homo sapiens* specimens that reveal an earlier origins of our species in Africa (at ca. 315 Ka) [10] and possibly an earlier incursion of modern humans into Europe (by ca. 210 Ka) [11,12] than previously known; and (5) the rediscovery of fossil material that has improved our understanding of the skeletal morphology of the Denisovans [13] and clarified the morphological and taxonomic affinities of Middle–Late Pleistocene Asian (possibly Denisovan) specimens [14].

While the paleontological and genomic records show interbreeding between *Homo* lineages to have been common across the Middle and Late Pleistocene Old World [6,15–18], much attention has been focused on the nature of introgression between the Neandertal and modern human lineages in Europe and western Asia. The picture that has emerged from aDNA and morphological studies suggests a divergence of the modern human and Neandertal/Denisovan lineages in the Middle Pleistocene (between 700–500 Ka) [19], separation of the Neandertal and Denisovan lineages shortly thereafter [20], and then repeated episodes of hybridization between the Neandertal and modern human lineages in both western Asia and Europe [2,21,22], as well as between the Denisovan and other lineages [17]. However, the nature of Neandertal–modern human hybridization remains unclear. Initial analysis of the Neandertal genome [23] noted similar amounts of Neandertal ancestry in recent humans from both Europe and Asia, suggesting that interbreeding had been limited to the Near East before 100 Ka (but see Sankararaman et al. [21] who estimate this date to be 65–47 Ka), during early range expansion of modern humans and prior to the divergence of living European and Asian populations. This scenario is also consistent with the nature of introgressed modern human genes in a ca. 50 Ka-old Neandertal from the Altai Mountains of Siberia [24]. Oddly, however, subsequent analyses of the Denisovan, Neandertal, and modern human genomes [25,26] revealed a greater amount of Neandertal ancestry in East Asian modern human populations than in those from Europe, indicating the introgression of Neandertal genes into the lineage leading to modern-day Asians after the separation of European and Asian lineages. The finding of a greater Neandertal contribution to East Asian than European populations is consistent with the complete absence of Neandertal mitochondrial DNA (mtDNA) in both early modern European fossils and living Europeans [27] and with the lack of evidence for gene flow from modern humans in the DNA of four late (<45 Ka) Neandertals from Belgium, France, and western Russia [28]. That modern-day Europeans do not carry a greater percentage of Neandertal nuclear DNA than Asians is surprising, given the potentially longer period of overlap between Neandertals and early modern humans in Europe than elsewhere [29], and thus greater opportunity for interbreeding, and also given morphological evidence (reviewed below) of admixture in the European fossil record. Nonetheless, Neandertal genes do not appear to be apportioned equally in living Europeans and Asians [30]. One geographic region in which Neandertal DNA may have uniquely introgressed into Asian but not European lineages of modern humans is on the eastern edge of the Neandertal range: the genome of a 45 Ka-old modern human from Ust'-Ishim in Siberia documents Neandertal introgression consistent with a hybridization event some 13–7 Ka earlier [31]. Hybridization in western Asia may constitute a "second pulse" of introgression after initial contact in the Near East, as suggested by some analyses of the genomic data [32]. Continued eastward expansion of modern human populations after hybridizing with Neandertals on the easternmost edge of their distribution could account for living Asians carrying, on average, a higher percentage of Neandertal ancestry than do living Europeans.

However, more recent morphometric and aDNA studies, focused primarily on early modern European fossils, have produced compelling evidence of Neandertal–modern human hybridization in Europe (Figure 1). Ancient DNA from Kostenki 14, a 38.7–36.2 Ka-old modern human from the western portion of the Russian Plain, contains longer segments of Neandertal DNA than are found in living Europeans, producing an estimated hybridization date of ca. 54 Ka [33]. Unlike the somewhat older specimen from farther east at Ust'-Ishim [31], the Kostenki individual is genetically closer to present-day Europeans than East Asians [33]. Likewise, a relatively high proportion (6–9%) of the genome of a 42–37 Ka-old modern human from Romania, Oase 1, appears to have derived from Neandertals, consistent with this individual having had a Neandertal ancestor some four-to-six generations earlier [34]. This finding confirmed earlier suggestions, based on the presence of derived Neandertal features in the skull and postcranial skeleton, that the fossils from the Peştera cu Oase, as well as those of nearby Muierii, showed morphological evidence of Neandertal admixture [35–37] (but see [38]). A similarly recent hybridization event (six or seven generations earlier) has been inferred from aDNA of early modern humans at Bacho Kiro, Bulgaria, around 45.9–42.6 Ka [39], while a somewhat more distant event (70–80 generations earlier) has been recognized in the genome of a >45 Ka-old cranium from Zlatý kůň in Czechia [40]. Note, however, that multivariate analysis of cranial metrics of comparably-aged crania from the Czechian site of Mladeč failed to detect a morphological signal of admixture [41]. Despite the apparent lack of such a signal, the aDNA evidence unequivocally indicates hybridization events in Europe. This genetic evidence, however, derives entirely from sites in central and eastern Europe, and thus the extent (if any) of interbreeding that occurred between Neandertals and modern humans in western Europe remains unclear.

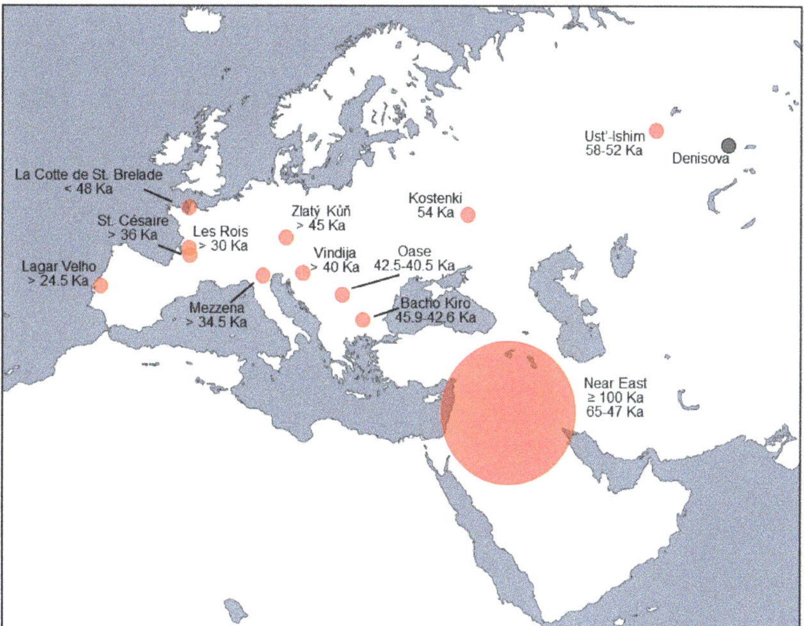

Figure 1. Map of western Eurasia showing areas and estimated dates of possible Neandertal–modern human hybridization (in red) based on fossil samples from indicated sites. Ancient DNA from a Neandertal fossil from Denisova Cave (black dot) has been interpreted as reflecting Neandertal–modern human admixture in the Near East at 100 Ka or earlier. See text for details.

However, morphological indicators of interbreeding have long been recognized in the fossil record of Late Pleistocene Europe (including western Europe), in the form of both the persistence of derived Neandertal features in European early modern humans [35,37,42–44], and in the existence of possible hybrid or near-hybrid individuals [45–47]. For example, traditional and morphometric comparisons of a sample of teeth from Middle Paleolithic deposits on the south coast of Jersey (Channel Islands) may provide morphological evidence of Neandertal–modern human hybridization in the most western reaches of Europe [48]. These teeth, which postdate 48 Ka, evince a mix of Neandertal and modern human traits and tend to fall between Neandertal and recent human samples in morphometric shape space, leading Compton et al. [48] to suggest that they may derive from a hybrid population. These authors [48] also suggest that the morphologically intermediate Neandertal teeth from 45–38 Ka-old sediments at Palomas (Spain) [49] may likewise reflect an admixed population.

Of interest, too, is the presence of Neandertal features in early- and mid-Upper Paleolithic modern human fossils from France (at Les Rois [47]) and Portugal (at Lagar Velho [46]), which, when combined with a similar finding in the fossils from the Channel Islands [48] discussed above, may reflect an appreciable hybridization zone in western Europe. In addition, a number of relatively late-surviving Neandertals in western, southern, and central Europe have been argued to evince a greater degree of modern human morphology than their older conspecifics, suggesting gene flow from invading modern humans into the resident Neandertal populations across most of Europe. These somewhat more modern-looking Neandertals have been identified at St. Cesaire, France [50,51], Vindija, Croatia [52–55], and Riparo Mezzena, Italy [45].

Efforts to interpret the morphological evidence are hampered by unclear expectations of how admixture is expressed in skeletal morphology (see [38]). While much of the literature has focused on the persistence of Neandertal autapomorphies in early modern European fossils, a high frequency of dental and sutural anomalies may also provide a skeletal signal of hybridization in a population [56]. In this regard, Ackermann [56] pointed to a high frequency (36%) of rotated mandibular premolars in the Krapina (Croatia) Neandertal sample as a potential indicator of admixture. At ca. 130 Ka [57], the Krapina Neandertals antedate the widespread incursion of modern humans into Europe (which may have begun as early as 48 Ka [29]). Thus, if they do represent an admixed population, then either some populations of central European Neandertals must have interbred with earlier waves of modern human migration into the Balkans (as possibly represented by the ca. 210 Ka-old modern human from Apidima, Greece [11]; note however that the Apidima 1 cranium is not universally accepted as representing a modern human [58]), or they must represent the introgression of modern human genes via gene flow from a hybrid zone in the Near East (established perhaps as early as 194 Ka [12]), in advance of the actual in-migration of modern humans. This latter mechanism has been invoked to explain the presence of some modern human-like features in the morphology of the ca. 40 Ka-old Neandertals from the G_3 layer at the nearby Croatian site of Vindija [52]. Thus, the morphological evidence may be documenting a richer history of Neandertal–modern human admixture in Europe than seen in the current aDNA record, but one that was undoubtedly complex [32].

To further explore the potential morphological signal of Neandertal–modern human admixture, we here assess the degree of morphometric similarity in the facial skeletons of Neandertals and early modern humans across seven geographic regions: the Near East, and eastern, central, southeastern (Balkans), western, southwestern (Iberian and Apennine peninsulas and Mediterranean France), and northern Europe. We focus on the facial skeleton because it has been shown to be a good indicator of population affinity [59], because it may reflect those affinities more faithfully than the cranial vault in Pleistocene humans [60], because Neandertals are characterized by distinctive facial morphology [61], and because at least one early modern European sample (Muierii [35]) has been argued to express Neandertal-like aspects in its facial form. While we expect, a priori, all modern human groups to be more similar to one another than any are to the Neandertals [62], we might also expect—if there was regional variation in levels of hybridization—that the ways that early modern European groups differ from one another in facial morphology might reflect variation in the Neandertal contributions to their genomes.

2. Materials and Methods

2.1. Samples and Measurements

Facial metric data were obtained from the literature [63–68] for 12 fossil and recent human samples, totaling 316 specimens (see Supplementary Materials). Neandertals (n = 13) were subdivided into Near Eastern (NE NEAN; n = 5), southern European (SE NEAN; n = 5), and western European (WE NEAN; n = 3) samples, reflecting geographic regions known to have some genetic separation [69] (but note that sample size considerations necessitated combining the Neandertals from southeastern and southwestern Europe into a single southern European sample). Data from anatomically modern human fossils (AMH, n = 201), from Upper Paleolithic and Mesolithic contexts across Europe and the Near East, were also subdivided by geographic region: Near Eastern/North African (NE/NA AMH; n = 3), and eastern (EE AMH; n = 24), southeastern (SE AMH; n = 23), central (CE AMH; n = 49), northern (NE AMH; n = 25), southwestern (SW AMH; n = 35), and western (WE AMH; n = 42) European samples. Temporal representation across the samples is not uniform: the NE/NA AMH is an Upper Paleolithic sample, the NE AMH is a Mesolithic sample, and the others are combined Upper Paleolithic/Mesolithic samples (with two of them, SE AMH and EE AMH, being largely Mesolithic) (see Supplementary Materials for details of sample composition). Since Neandertal–modern human hybridization has been hypothesized to have occurred in the Near East at potentially early (>100 Ka) and later (65–47 Ka) dates (see above), we also included a sample of Near Eastern, Middle Paleolithic-associated fossils (NE MP AMH; n = 5), from the sites of Skhūl and Qafzeh (Israel), with radiometric dates $\geq$ 90 Ka. We also included data from a recent human sample from east Africa (EA RECENT; n = 83) as an outgroup (that is, a group not expected to have significant Neandertal ancestry).

The selection of traditional craniometric measurements was based on the consistency and availability of measurements from different sources, which limited the study to six measurements (Table 1, Figure 2). An additional consideration was that these measurements are part of the standard forensic set that has been tested for reliability and repeatability, allowing for the incorporation of data from several sources with minimal inter-observer error [70–72]. Furthermore, midfacial measures have been found to be the most morphologically informative when examining craniofacial variation in contemporary contexts, as they have been linked to climatic adaptations (which may differentiate regional groups), and because they are minimally affected by developmentally plastic alterations of the cranial vault [73,74]. Craniofacial measurements used in this study are listed in Table 1 and the landmarks upon which they are based are illustrated in Figure 2. To include as many specimens as possible, it was necessary to impute missing variables in some individuals. However, individuals were not included in this analysis if they were missing more than two variables ($\leq$33%). Because sample sizes were small within regions, simple mean substitution was used to estimate the missing values. Sex variation is negligible within each group included in population studies and thus, males and females were pooled to incorporate all of the observed biological information and to increase sample sizes [75]. To examine the effect of climate on facial morphology, each sample was scored and numerically coded using the climate map for biodiversity [76]. The recent human samples were scored according to Metzger [76], while fossil samples were scored using paleoclimatic data [77–79]. Radiocarbon dates associated with each fossil specimen (see Supplementary Materials) were taken from the literature [63–68] and were used to evaluate the effect of time on facial morphology across our samples.

Table 1. Traditional craniometric measurements utilized in this study.

Measurement	Abbreviation	Description
Upper Facial Height (n-pr)	NPH	[71]
Bifrontal Breadth (fmo-fmo)	FMB	[71]
Nasal Height (n-ns)	NLH	[71]
Nasal Breadth (al-al)	NLB	[71]
Orbital Height	OBH	[71]
Orbital Breadth (d-ec)	OBB	[71]

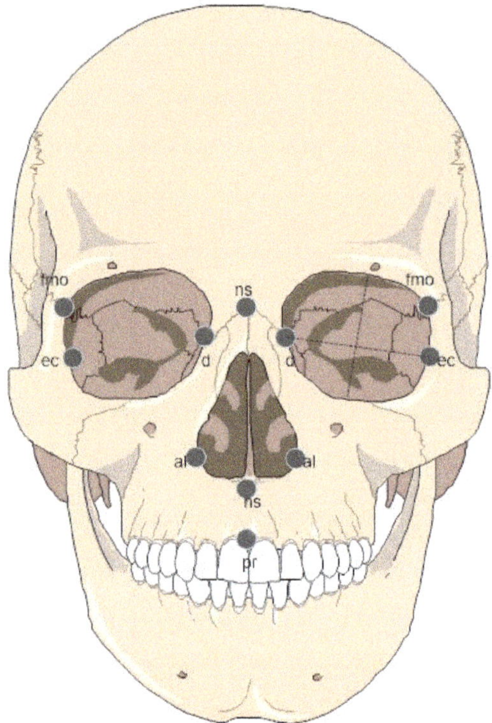

Figure 2. Facial landmark locations for the measurements used.

Because significant facial size differences have been shown between *Homo* groups [62], variables were size adjusted following Darroch and Mosimann [80,81], without log transformation. Size and shape variables were calculated from the raw measurements, where size is defined as the geometric mean (GM) of all six variables. The GM of the six variables was calculated as follows:

$$\text{Size} = \left(\prod_{i=1} X_i \right)^{1/n} \tag{1}$$

and the raw variables were divided by the GM to create new shape variables ($Y = X/SIZE$, where X is the raw measurement). While these newly created shape variables do not remove absolute size (only geometric morphometric approaches truly remove size), they are scale-free and provide a better understanding of the geometric or shape-related similarity among the groups [81].

2.2. Multivariate Statistics

A canonical variates analysis (CVA: a linear combination of predictor variables that summarize among-population variation) was conducted using the newly calculated shape variables [82]. Among-group differentiation was measured using Mahalanobis squared distances, which is a similarity measure and a function of group means and the pooled variance–covariance matrix [82]. A hierarchical (or agglomerative) cluster analysis using average linkage was conducted on the Mahalanobis squared distances to examine group similarity [70]. All statistical analyses were conducted in SAS 9.4 [83] and the hierarchical cluster analysis was conducted in JMP 16 Pro [84].

2.3. Spatial Analyses

A Pearson's product–moment correlation coefficient was performed to assess the relationship between climatic zones, geometric mean (i.e., size), and the shape variables. Further, a Pearson's product–moment correlation was used to assess size and shape changes over time. Dutilleul's [85] estimator was used to correct for spatial autocorrelations and was performed using PASSaGE: Pattern Analysis, Spatial Statistics and Geographic Exegesis Version 2 [86].

3. Results

Table 2 presents the group means for the raw variables.

Table 2. Group means for facial variables (standard deviations in parentheses).

	CE AMH	EA RECENT	EE AMH	NE MP AMH	NE NEAN	NE/NA AMH	NE AMH	SE AMH	SW AMH	SE NEAN	WE AMH	WE NEAN
NPH	67 (5.21)	63 (4.8)	70 (5.28)	75 (2.7)	88 (6.06)	65 (4.16)	72 (4.45)	67 (4.15)	70 (4.6)	88 (14.14)	68 (5.2)	85 (4.9)
FMB	96 (4.87)	97 4.04)	96 (5.32)	103 (5.54)	114 (4.18)	103.5 (3.5)	101 (4.62)	94 (5.05)	99 (5.01)	111 (10.52)	98 (4.56)	113 (1.2)
NLH	50 (4.7)	48 (3.55)	52 (3.5)	54 (1.3)	63 (3.9)	46 (1.76)	52 (3.19)	49 (3.16)	52 (3.71)	59 (4.28)	50 (4.09)	60 (0.6)
NLB	25 (2.21)	28 (1.87)	25 (2.58)	30 (1.41)	34 (2.88)	26 (1.54)	24 (2.06)	24 (2.05)	25 (1.9)	34 (4.3)	25 (2.17)	32 (3.8)
OBB	42 (3.04)	39 (1.76)	43 (2.54)	44 (1.79)	46 (1.24)	42 (3.8)	43 (1.73	42 (.67)	42 (2.56)	44 (1.22)	41 (2.9)	44 (2.1)
OBH	30 (2.41)	33 (1.9)	32 (2.25)	33 (3.39)	36 (0.89)	28 (1.53)	32 (2.46)	30 (2.5)	33 (3.31)	37 (1.52)	30 (2.0)	37 (1.2)

3.1. Multivariate Statistics

The Mahalanobis squared distances are presented in Table 3. No significant differences were detected between SE AMH and WE AMH, or between NE and SW AMH samples. Likewise, the western European Neandertal sample is not significantly different from either the Near Eastern or southern European Neandertal samples. While all the other groups are significantly different from one another, all the AMH samples are fairly similar given the small distance values. The Near Eastern Middle Paleolithic AMH sample is closest to the southwestern European AMH sample. Furthermore, the East African recent sample is similar to various AMH groups. Interestingly, the Near Eastern Neandertal sample is most dissimilar to the Near Eastern North Africa AMH sample.

Table 4 presents the significant canonical roots, with 80 percent of the variation depicted on the first two canonical variates (CAN 1 and CAN 2). The canonical structure (Table 5) shows that the variation exhibited on CAN 1 is related to orbital, nasal breadth, and a moderate degree to NPH. At the same time, that on the second axis (CAN 2) is related to size (geometric mean), and moderately to NPH. The plot of CAN 1 and CAN 2 (Figure 3) shows that Neandertals have narrow orbits, wide nasal breadths, and large faces overall. The European AMH have wider orbits, narrower nasal breadths, and smaller geometric means. The Near Eastern Middle Paleolithic AMH sample has an orbital and nasal breadth and geometric mean values that are intermediate between the Neandertal and European AMH samples. The Near Eastern/North African AMH sample has intermediate orbital and nasal breadth, combined with a small geometric mean. The East African recent sample has narrow orbits, wide nasal breadth, and the smallest geometric mean.

Table 3. Mahalanobis squared distances.

	CEA AMH	EA RECENT	EE AMH	NE MP AMH	NE NEAN	NE/NA AMH	NE AMH	SE AMH	SW AMH	SE NEAN	WE AMH
EA RECENT	7.15	0									
EE AMH	1.22	10.63	0								
NE MP AMH	7.78	9.62	7.96	0							
NE NEAN	35.19	34.54	32.99	11.88	0						
NE/NA AMH	51.76	52.84	57.27	59.71	81.24	0					
NE AMH	2.59	11.59	1.99	7.64	28.19	54.78	0				
SE AMH	0.89	11.13	1.51	11.56	42.00	55.13	2.57	0			
SWAMH	1.70	7.15	1.22	6.08	27.66	54.25	0.82 *	2.60	0		
SE NEAN	32.87	29.27	32.20	11.27	2.99 *	69.72	26.82	38.68	25.92	0	
WE AMH	0.56 *	8.04	1.81	7.81	32.83	52.40	1.15	1.14	1.18	30.58	0
WE NEAN	28.71	26.83	27.15	9.25	1.28 *	74.71	21.29	34.40	20.87	2.03 *	25.76

* not significantly different. All other groups are significantly different from each at >0.03 level.

Table 4. Significant canonical axis for the shape transformed variables.

No.	Eigenvalue	Cumulative %	Proportion	Likelihood Ratio	Approximate F	Num DF	Den DF	Pr > F
1	1.80	0.47	0.47	0.08	12.17	77	1793	<0.0001
2	1.28	0.80	0.33	0.23	8.62	60	1572	<0.0001
3	0.51	0.96	0.03	0.78	4.79	45	1111	<0.0001
4	0.12	0.98	0.03	0.70	2.48	32	1542	<0.0001
5	0.09	0.99	0.02	0.87	2.04	21	868	0.004

Table 5. Total canonical structure for the shape transformed variables.

Variable	CAN 1	CAN 2	CAN 3	CAN 4	CAN 5
GM	−0.27	0.93	−0.08	0.05	−0.07
NPHs	0.50	0.60	0.15	0.04	0.06
FMBs	−0.06	−0.25	0.31	−0.56	0.39
NLHs	0.38	0.25	−0.04	0.12	−0.13
NLBs	−0.81	−0.11	0.12	0.29	0.11
OBBs	0.71	−0.12	0.07	0.07	−0.07
OBHs	−0.37	−0.45	−0.28	−0.31	−0.43

The hierarchical cluster analysis using the Mahalanobis squared distances shows two distinct clusters: (1) anatomically modern humans (fossil and recent), and (2) Neandertals (Figure 4a). All of the European fossil AMH groups cluster together. The Near Eastern Middle Paleolithic sample leaf branches off this European AMH cluster, but is significantly dissimilar. Surprisingly, the East African recent human sample branches off of the stem leading to the European AMHs and the NE MP AMHs (Figure 4a). Among the modern human samples, the most dissimilar is the Near Eastern/North African AMH series (Figure 4a).

The constellation plot, which arranges the groups as endpoints, further illustrates the similarity/dissimilarity of the groups (Figure 4b). The morphological distance between cluster joints is illustrated by the length of the line between them. The constellation plot shows that the most distinct group is the Near Eastern/North African AMH, which is no closer to the other modern human samples than it is to Neandertals. With this one exception, all of the anatomically modern human samples are closer to one another than they are to the Neandertal samples.

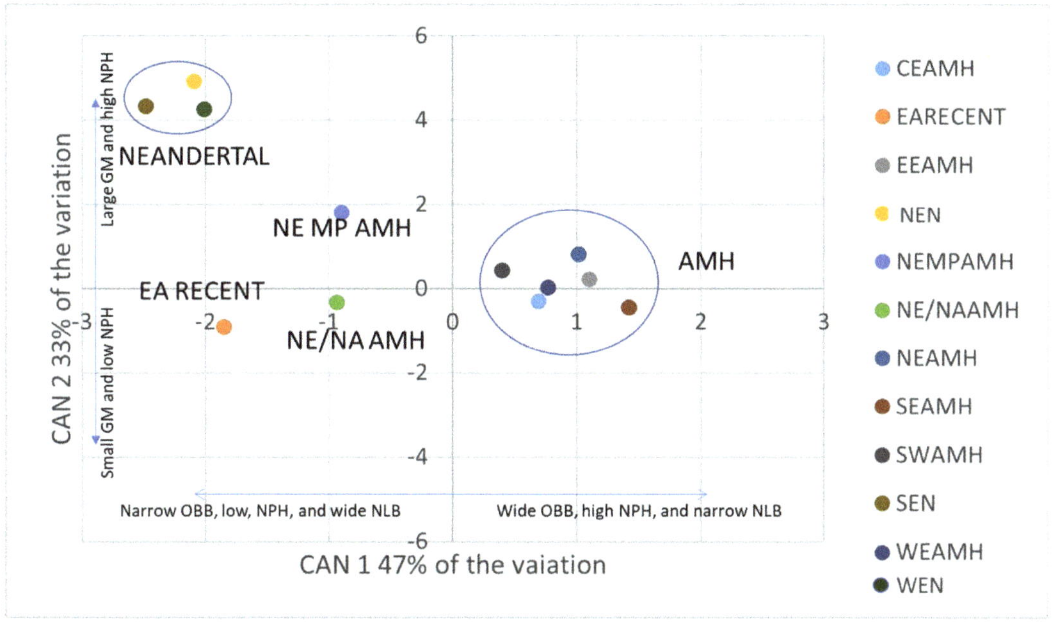

Figure 3. A plot of CVA canonical axis 2 (CAN 2) on axis 1 (CAN 1) represents 80% of the total variation. Neandertals cluster together (narrow orbits, wide nasal breadth, and large geometric mean); Near Eastern Middle Paleolithic AMH fall in between the Neandertal and the other AMH samples. The Near Eastern/North African AMH sample has intermediate orbit and nasal breadths, and a small geometric mean and is closest to the recent East African sample.

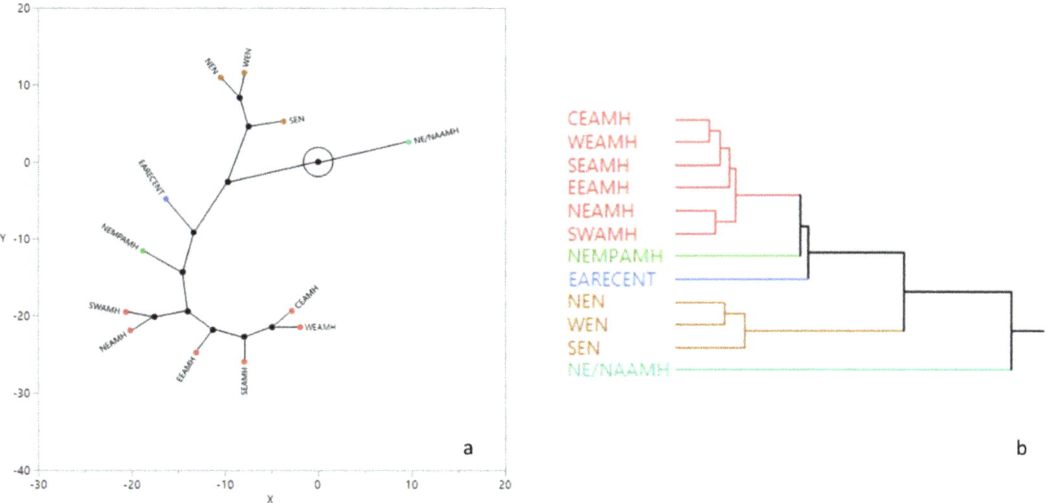

Figure 4. Constellation plot (**a**) and dendrogram (**b**) results from hierarchical cluster analysis showing group relationships.

3.2. Spatial Analysis

The Pearson's product-moment correlation using the Ditulleul [85] method to correct for spatial autocorrelation revealed no significant correlation between the climatic zones and shape variables (NPHs, r = −0.523, *p*-value = 0.234; FMBs, r = 0.132, *p*-value = 0.307; NLHs, r = −0.244, *p*-value = 0.383; NLBs, r = 0.468, *p*-value = 0.303; OBBs, r = −0.365, *p*-value = 0.314; OBHs, r = 0.303, *p*-value = 0.336), nor between climatic zones and geometric means (r = −0.162; *p*-value = 0.340), indicating that climate does not model either craniofacial shape or size (at least in these samples). The only significant variable correlated with temporality was the geometric mean, our measure of size (r = 0.56, *p*-value > 0.001; NPHs, r = 0.23, *p*-value = 0.088; FMBs, r = −0.137, *p*-value = 0.061; NLHs, r = 0.016, *p*-value = 0.882; NLBs, r = 0.184, *p*-value = 0.178; OBBs, r = −0.184, *p*-value = 0.112; OBHs, r = −0.171, *p*-value = 0.106).

4. Discussion

Neandertals possessed distinctive facial morphology, characterized by large, superoinferiorly tall yet mediolaterally narrow faces with pronounced midfacial prognathism, absolutely and relatively tall orbits, flat or convex infraorbital plates (and thus, no canine fossae), wide nasal apertures and strongly projecting external noses, and prominent, double-arched supraorbital tori [87–89]. This constellation of features is, unfortunately, not perfectly represented by the measures of overall facial shape used in this analysis (for example, the inclusion of a measure of facial length might be expected to have high utility in distinguishing Neandertal from modern groups [90]). Nonetheless, the variables used here should be expected to reasonably represent overall facial size and shape (that is, relatively wide versus relatively narrow faces), the shape of the orbits (relative orbital width), and the shape of the nasal aperture (relative nasal width). The large size and distinctive morphology of the Neandertal face does indeed appear to be captured on the first two canonical axes of the CVA (Figure 2), which in combination separate the Neandertal samples from the fossil and recent modern human samples. The strong negative values of the Neandertal samples on CAN 1 reflect faces with wide nasal apertures and narrow orbits (relative to overall facial size), while their strong positive values on CAN 2 denote large faces with elevated upper facial heights (relative to overall facial size). Note that upper facial height (NPH) has moderate yet positive loadings on both CAN 1 and 2 (Table 5), which we interpret as reflecting, on CAN 1, the utility of this variable in differentiating modern human groups, while CAN 2 accounts for residual variation in NPH (that is, variation not explained on CAN 1) that serves (in part) to separate Neandertal from modern human samples. All of the anatomically modern European groups have positive values on canonical axis 1 and largely neutral values on axis 2, reflecting smaller faces overall, with relatively narrower noses and wider orbits.

At the risk of over-interpreting the position of very small fossil samples, we find the intermediate positions (between Neandertal and European AMH samples) of the Near Eastern Middle Paleolithic AMH and the Near Eastern/North African AMH samples to be interesting, as this might reflect a greater contribution of Neandertal genes to facial morphology in the Near East. Relative to European fossil modern humans, the NE MP AMH sample exhibits a more Neandertal-like constellation of facial size and morphological features (Figure 3). Given the antiquity of this sample, it is possible that these fossils represent a population that had not yet fully evolved the derived modern human condition of smaller faces, and which may retain (to some degree) ancestral features such as wide nasal apertures (which in turn may be related to the plesiomorphic retention of facial prognathism [90,91]). However, the modern human pattern is already evident in the facial morphology of two specimens from Jebel Irhoud (Morocco) [10], which considerably antedate the fossils from Skhūl and Qafzeh. Interestingly, the fossils from Irhoud, as well as from Skhūl and Qafzeh, fall within the range of variation of recent modern humans—and distinct from Neandertals—in a principal components analysis (PCA) of facial shape metrics [10]. However, Qafzeh 9 falls on the edge of the recent human distribution, and in the direction of Neandertals in shape space, as does (but to a lesser extent) Skhūl 5 (Figure 4a in [10]). Qafzeh 6, on the other hand, falls well away from the Neandertals. Note, however, that this PCA was conducted on size-scaled 3D geometric morphometric shape variables [10], and thus does not factor in variation within and between groups in facial size. Overall, the intermediate position of the NE MP AMH sample between the Neandertal and other fossil AMH samples (in our analysis) could be interpreted as reflecting sufficient Neandertal–modern human hybridization in the Near East, at a sufficiently early date (>100 Ka), to affect the facial morphology of the early modern human fossils from Skhūl and Qafzeh.

It is also interesting to note that the NE/NA AMH sample appears (based on its position on CAN 1) to be similar to the NE MP AMH sample in facial shape but to have a facial size (based on its position on CAN 2) that falls among most other fossil and recent modern human samples. Again, at the risk of over-interpreting morphology in a very small fossil sample, this may be a signal of Neandertal introgression in the Near East (but perhaps later in time, consistent with the inference of hybridization there between 65–47 Ka [21]). The east African recent human sample falls with Neandertals on CAN 1, and with most of the other modern human groups on CAN 2 (Figure 3). This sample, representing the Teita from Kenya [71], has mean nasal breadths (NLB; Table 2) that fall at the high end of the range of sample means, likely representing climatic adaptation to the hot and humid climate of southeastern Kenya. We thus interpret their position on canonical axis 1 to reflect convergence in nasal morphology, rather than homology.

Hierarchical clusters based on Mahalanobis squared distances produce an interesting picture (Figure 4). Under the (arguably false) assumption that the facial metrics employed in this study are selectively neutral (i.e., following neutral microevolutionary processes [59]) and thus accurately reflecting relationships between populations), and assuming no hybridization whatsoever between groups, we would expect clusters that show two distinct clades (Neandertals versus modern humans), and clear geographic (or possibly temporal plus geographic) structure within each clade. Expectations given hybridization—particularly involving temporally complex and potentially small-scale hybridization events, distributed across geographic space—are unclear. While the cluster analysis did largely produce separate Neandertal and modern human clades, the Neandertal clade is nested within the modern human cluster (with the Near Eastern/North African AMH sample being the sole outgroup). With the exception of this outgroup, the plots show good (but not perfect) geographic structure: on the AMH branch, all European AMHs cluster together, joined by the NE MP AMH sample, which is then joined by the EA RECENT sample. Within the European AMH cluster, however, geographic structure breaks down (for example, northern and southwestern early modern Europeans cluster together). Likewise, the Neandertal clade does not show the expected geographic structure, in which western and southern European Neandertals should be closer to one another than either is to Near Eastern Neandertals. Complexity in the hierarchical clustering of both the AMH and Neandertal samples is no doubt due to small fossil sample sizes, variation in the average geological age of the AMH samples (although we find that only overall facial size significantly varies with time), the effects of natural selection on aspects of facial morphology, and the imperfect way in which facial morphology reflects populational relationships. Still, we find it interesting that, in multivariate shape space, the Near Eastern/North African AMH sample is equidistant from the Neandertal and all other modern human samples. This result is consistent with a significant Neandertal signal in the Upper Paleolithic peoples of the Near East and northeastern Africa, and thus consistent with evidence suggesting that the Near East was a locus of hybridization on the order of 65–47 Ka [21]. While molecular and morphological evidence (reviewed above) clearly indicates some level of interbreeding between Neandertals and modern humans across the entire Neandertal range, such interbreeding is not clearly reflected in the limited analysis performed here.

The spatial analysis failed to detect significant relationships between any aspect of facial size or shape and climatic zones, which is surprising given the large body of literature that documents such relationships across global samples of modern humans (see [73,88]). In particular, we would expect both overall facial size and nasal size to covary inversely with temperature, and nasal breadth to covary positively with precipitation and temperature [73]. It is likely that two factors, in combination, obscured a climatic signal in the data. First, the bulk of the data represents populations from a somewhat limited geographic area (relative to the global samples that are usually employed to assess ecogeographic variation). These data also primarily represents Mesolithic populations, who may not have experienced the colder climatic extremes that are likely to affect facial morphology [73]. Second, Neandertals retained the plesiomorphic condition of wide noses [88], despite inhabiting cold–temperate environments. Thus, our samples included two groups with absolutely and relatively wide noses—the Neandertals and the Kenyan Teita—who lived in the climatic extremes of cold/dry and hot/wet environments, respectively. Given the relative primacy of nasal morphology in human adaptation to climate, the combination of these groups with similar nasal morphology yet very different climatic zones is likely to have weakened the morphological signal of climatic variation but may also reflect the complex and multifactorial mechanisms that shape craniofacial morphology [92].

Again, we caution that these results are based on very small sample sizes, and any interpretation of them should be viewed with caution. While this analysis was both exploratory in nature and limited to facial metrics available from the literature, the results suggest that there may be utility in expanding this approach, both by including metrics that may capture additional aspects of Neandertal facial morphology (such as midfacial prognathism) and by augmenting sample sizes to the extent possible (e.g., incorporating data from Natufian specimens in the NA/NE AMH sample).

5. Conclusions

This exploratory, multivariate analysis incorporated only six variables, which reflect only the size and shape of the face overall, orbital shape, and nasal aperture shape. Furthermore, the analysis was conducted on samples that, when temporally and taxonomically constrained, were woefully small, or, when of adequate sample size, represented populations (namely, Mesolithic peoples) not temporally close to potential Neandertal–modern human hybridization events. Despite these limitations, the separation of Neandertals from all modern humans in the multivariate space created by the first two canonical axes of the CVA, combined with the hierarchical clustering of distinct Neandertal and European AMH clades when Mahalanobis distances are considered, shows the utility of analyzing facial morphology for the information it may contain about population relationships and potential Neandertal–modern human interbreeding. Two samples, one representing Middle Paleolithic-associated early modern humans from the Near East, the other representing Near Eastern and northeast African Upper Paleolithic modern humans, were found to be either intermediate between the Neandertals and all other modern human samples in multivariate space (both samples, but especially the former), or to form a hierarchical branch in the cluster analysis unlike any other modern human sample (the later sample). While caution should be used in interpreting the results of analyses based on small sample sizes, these results could be considered consistent with the Near East being a substantial locus of Neandertal–modern human hybridization. This in no way negates the abundant aDNA and morphological evidence that suggests that such hybridization occurred across all (except perhaps northern) Europe, but simply that interbreeding in Europe was of a nature that it did not leave a clear and interpretable signal (at least given the limitation of the current study) in the facial morphology of most Late Pleistocene, early Holocene European modern humans.

Supplementary Materials: The following supporting information can be downloaded at: https://www.mdpi.com/article/10.3390/biology11081163/s1, Sample provenience, climate zone scores, and raw data.

Author Contributions: Conceptualization, A.H.R. and S.E.C.; methodology, A.H.R.; analysis, A.H.R.; data management, K.K.; writing—original draft preparation, S.E.C., K.K. and A.H.R.; writing—review and editing, S.E.C., K.K. and A.H.R. All authors have read and agreed to the published version of the manuscript.

Funding: This research received no external funding.

Institutional Review Board Statement: IRB approval is exempt as all of the data were gathered from published research.

Data Availability Statement: The data presented in this study are available in the Supplementary Material.

Conflicts of Interest: The authors declare no conflict of interest.

References

1. Krause, J.; Briggs, A.W.; Kircher, M.; Maricic, T.; Zwyns, N.; Derevianko, A.; Pääbo, S. A Complete mtDNA Genome of an Early Modern Human from Kostenki, Russia. *Curr. Biol.* **2010**, *20*, 231–236. [CrossRef] [PubMed]
2. Gokcumen, O. Archaic hominin introgression into modern human genomes. *Am. J. Phys. Anthr.* **2019**, *171*, 60–73. [CrossRef]
3. Brown, P.; Sutikna, T.; Morwood, M.J.; Soejono, R.P.; Jatmiko; Saptomo, E.W.; Due, R.A. A new small-bodied hominin from the Late Pleistocene of Flores, Indonesia. *Nature* **2004**, *431*, 1055–1061. [CrossRef]
4. Berger, L.R.; Hawks, J.; de Ruiter, D.J.; Churchill, S.E.; Schmid, P.; Delezene, L.K.; Kivell, T.L.; Garvin, H.M.; Williams, S.A.; DeSilva, J.M.; et al. Homo naledi, a new species of the genus Homo from the Dinaledi Chamber, South Africa. *eLife* **2015**, *4*, e09560. [CrossRef] [PubMed]
5. Détroit, F.; Mijares, A.S.; Corny, J.; Daver, G.; Zanolli, C.; Dizon, E.; Robles, E.; Grün, R.; Piper, P.J. A new species of Homo from the Late Pleistocene of the Philippines. *Nature* **2019**, *568*, 181–186. [CrossRef] [PubMed]
6. Hammer, M.F.; Woerner, A.E.; Mendez, F.L.; Watkins, J.C.; Wall, J.D. Genetic evidence for archaic admixture in Africa. *Proc. Natl. Acad. Sci. USA* **2011**, *108*, 15123–15128. [CrossRef]
7. Berger, L.R.; Hawks, J.; Dirks, P.H.; Elliott, M.; Roberts, E.M. Homo naledi and Pleistocene hominin evolution in subequatorial Africa. *eLife* **2017**, *6*, e24234. [CrossRef] [PubMed]
8. Ragsdale, A.P.; Gravel, S. Models of archaic admixture and recent history from two-locus statistics. *PLoS Genet.* **2019**, *15*, e1008204. [CrossRef]
9. Hollfelder, N.; Breton, G.; Sjödin, P.; Jakobsson, M. The deep population history in Africa. *Hum. Mol. Genet.* **2021**, *30*, R2–R10. [CrossRef]
10. Hublin, J.-J.; Ben-Ncer, A.; Bailey, S.E.; Freidline, S.E.; Neubauer, S.; Skinner, M.M.; Bergmann, I.; LE Cabec, A.; Benazzi, S.; Harvati-Papatheodorou, K.; et al. New fossils from Jebel Irhoud, Morocco and the pan-African origin of Homo sapiens. *Nature* **2017**, *546*, 289–292. [CrossRef]
11. Harvati, K.; Röding, C.; Bosman, A.; Karakostis, F.A.; Grün, R.; Stringer, C.; Karkanas, P.; Thompson, N.C.; Koutoulidis, V.; Moulopoulos, L.; et al. Apidima Cave fossils provide earliest evidence of Homo sapiens in Eurasia. *Nature* **2019**, *571*, 500–504. [CrossRef] [PubMed]
12. Hershkovitz, I.; Weber, G.W.; Quam, R.; Duval, M.; Grün, R.; Kinsley, L.; Ayalon, A.; Bar-Matthews, M.; Valladas, H.; Mercier, N.; et al. The earliest modern humans outside Africa. *Science* **2018**, *359*, 456–459. [CrossRef] [PubMed]
13. Chen, F.; Welker, F.; Shen, C.-C.; Bailey, S.; Bergmann, I.; Davis, S.; Xia, H.; Wang, H.; Fischer, R.; Freidline, S.E.; et al. A late Middle Pleistocene Denisovan mandible from the Tibetan Plateau. *Nature* **2019**, *569*, 409–412. [CrossRef] [PubMed]
14. Ni, X.; Ji, Q.; Wu, W.; Shao, Q.; Ji, Y.; Zhang, C.; Liang, L.; Ge, J.; Guo, Z.; Li, J.; et al. Massive cranium from Harbin in northeastern China establishes a new Middle Pleistocene human lineage. *Innovation* **2021**, *2*, 100130. [CrossRef]
15. Reich, D.; Patterson, N.; Kircher, M.; Delfin, F.; Nandineni, M.R.; Pugach, I.; Ko, A.M.-S.; Ko, Y.-C.; Jinam, T.A.; Phipps, M.E.; et al. Denisova Admixture and the First Modern Human Dispersals into Southeast Asia and Oceania. *Am. J. Hum. Genet.* **2011**, *89*, 516–528. [CrossRef]
16. Lachance, J.; Vernot, B.; Elbers, C.C.; Ferwerda, B.; Froment, A.; Bodo, J.-M.; Lema, G.; Fu, W.; Nyambo, T.B.; Rebbeck, T.R.; et al. Evolutionary History and Adaptation from High-Coverage Whole-Genome Sequences of Diverse African Hunter-Gatherers. *Cell* **2012**, *150*, 457–469. [CrossRef]
17. Slon, V.; Mafessoni, F.; Vernot, B.; De Filippo, C.; Grote, S.; Viola, B.; Hajdinjak, M.; Peyrégne, S.; Nagel, S.; Brown, S.; et al. The genome of the offspring of a Neanderthal mother and a Denisovan father. *Nature* **2018**, *561*, 113–116. [CrossRef]
18. Bailey, S.E.; Hublin, J.-J.; Antón, S.C. Rare dental trait provides morphological evidence of archaic introgression in Asian fossil record. *Proc. Natl. Acad. Sci. USA* **2019**, *116*, 14806–14807. [CrossRef]
19. Bergström, A.; Stringer, C.; Hajdinjak, M.; Scerri, E.M.L.; Skoglund, P. Origins of modern human ancestry. *Nature* **2021**, *590*, 229–237. [CrossRef]

20. Rogers, A.R.; Bohlender, R.J.; Huff, C.D. Early history of Neanderthals and Denisovans. *Proc. Natl. Acad. Sci. USA* **2017**, *114*, 9859–9863. [CrossRef]

21. Sankararaman, S.; Patterson, N.; Li, H.; Pääbo, S.; Reich, D. The Date of Interbreeding between Neandertals and Modern Humans. *PLoS Genet.* **2012**, *8*, e1002947. [CrossRef] [PubMed]

22. Villanea, F.A.; Schraiber, J.G. Multiple episodes of interbreeding between Neanderthal and modern humans. *Nat. Ecol. Evol.* **2018**, *3*, 39–44. [CrossRef] [PubMed]

23. Green, R.E.; Krause, J.; Briggs, A.W.; Maricic, T.; Stenzel, U.; Kircher, M.; Patterson, N.; Li, H.; Zhai, W.; Fritz, M.H.-Y.; et al. A Draft Sequence of the Neandertal Genome. *Science* **2010**, *328*, 710–722. [CrossRef] [PubMed]

24. Kuhlwilm, M.; Gronau, I.; Hubisz, M.; De Filippo, C.; Prado-Martinez, J.; Kircher, M.; Fu, Q.; Burbano, H.A.; Lalueza-Fox, C.; De La Rasilla, M.; et al. Ancient gene flow from early modern humans into Eastern Neanderthals. *Nature* **2016**, *530*, 429–433. [CrossRef] [PubMed]

25. Meyer, M.; Kircher, M.; Gansauge, M.-T.; Li, H.; Racimo, F.; Mallick, S.; Schraiber, J.G.; Jay, F.; Prüfer, K.; De Filippo, C.; et al. A High-Coverage Genome Sequence from an Archaic Denisovan Individual. *Science* **2012**, *338*, 222–226. [CrossRef]

26. Wall, J.D.; Yang, M.A.; Jay, F.; Kim, S.K.; Durand, E.Y.; Stevison, L.; Gignoux, C.; Woerner, A.; Hammer, M.F.; Slatkin, M. Higher Levels of Neanderthal Ancestry in East Asians than in Europeans. *Genetics* **2013**, *194*, 199–209. [CrossRef]

27. Currat, M.; Excoffier, L. Modern Humans Did Not Admix with Neanderthals during Their Range Expansion into Europe. *PLoS Biol.* **2004**, *2*, e421. [CrossRef]

28. Hajdinjak, M.; Fu, Q.; Hübner, A.; Petr, M.; Mafessoni, F.; Grote, S.; Skoglund, P.; Narasimham, V.; Rougier, H.; Crevecoeur, I.; et al. Reconstructing the genetic history of late Neanderthals. *Nature* **2018**, *555*, 652–656. [CrossRef] [PubMed]

29. Hublin, J.-J. The modern human colonization of western Eurasia: When and where? *Quat. Sci. Rev.* **2015**, *118*, 194–210. [CrossRef]

30. Quilodrán, C.S.; Tsoupas, A.; Currat, M. The Spatial Signature of Introgression After a Biological Invasion with Hybridization. *Front. Ecol. Evol.* **2020**, *8*, 569620. [CrossRef]

31. Fu, Q.; Li, H.; Moorjani, P.; Jay, F.; Slepchenko, S.M.; Bondarev, A.A.; Johnson, P.L.F.; Aximu-Petri, A.; Prüfer, K.; De Filippo, C.; et al. Genome sequence of a 45,000-year-old modern human from western Siberia. *Nature* **2014**, *514*, 445–449. [CrossRef] [PubMed]

32. Vernot, B.; Akey, J.M. Complex History of Admixture between Modern Humans and Neandertals. *Am. J. Hum. Genet.* **2015**, *96*, 448–453. [CrossRef] [PubMed]

33. Seguin-Orlando, A.; Korneliussen, T.S.; Sikora, M.; Malaspinas, A.-S.; Manica, A.; Moltke, I.; Albrechtsen, A.; Ko, A.; Margaryan, A.; Moiseyev, V.; et al. Genomic structure in Europeans dating back at least 36,200 years. *Science* **2014**, *346*, 1113–1118. [CrossRef] [PubMed]

34. Fu, Q.; Hajdinjak, M.; Moldovan, O.T.; Constantin, S.; Mallick, S.; Skoglund, P.; Patterson, N.; Rohland, N.; Lazaridis, I.; Nickel, B.; et al. An early modern human from Romania with a recent Neanderthal ancestor. *Nature* **2015**, *524*, 216–219. [CrossRef] [PubMed]

35. Soficaru, A.; Doboş, A.; Trinkaus, E. Early modern humans from the Peştera Muierii, Baia de Fier, Romania. *Proc. Natl. Acad. Sci. USA* **2006**, *103*, 17196–17201. [CrossRef]

36. Rougier, H.; Milota, S.; Rodrigo, R.; Gherase, M.; Sarcină, L.; Moldovan, O.; Zilhão, J.; Constantin, S.; Franciscus, R.G.; Zollikofer, C.P.E.; et al. Peştera cu Oase 2 and the cranial morphology of early modern Europeans. *Proc. Natl. Acad. Sci. USA* **2007**, *104*, 1165–1170. [CrossRef]

37. Trinkaus, E. European early modern humans and the fate of the Neandertals. *Proc. Natl. Acad. Sci. USA* **2007**, *104*, 7367–7372. [CrossRef]

38. Harvaati, K.; Roksandic, M. The Human Fossil Record from Romania: Early Upper Paleolithic European Mandibles and Neanderthal Admixture. In *Paleoanthropology of the Balkans and Anatolia*; Springer: Berlin/Heidelberg, Germany, 2016; pp. 51–68. [CrossRef]

39. Hajdinjak, M.; Mafessoni, F.; Skov, L.; Vernot, B.; Hübner, A.; Fu, Q.; Essel, E.; Nagel, S.; Nickel, B.; Richter, J.; et al. Initial Upper Palaeolithic humans in Europe had recent Neanderthal ancestry. *Nature* **2021**, *592*, 253–257. [CrossRef]

40. Prüfer, K.; Posth, C.; Yu, H.; Stoessel, A.; Spyrou, M.A.; Deviese, T.; Mattonai, M.; Ribechini, E.; Higham, T.; Veleminský, P.; et al. A genome sequence from a modern human skull over 45,000 years old from Zlatý kůň in Czechia. *Nat. Ecol. Evol.* **2021**, *5*, 820–825. [CrossRef]

41. Bräuer, G.; Broeg, H.; Stringer, C.B. Earliest Upper Paleolithic crania from Mladeč, Czech Republic, and the question of Neanderthal-modern continuity: Metrical evidence from the fronto-facial region. In *Neanderthals Revisited: New Approaches and Perspectives. Vertebrate Paleobiology and Paleoanthropology*; Hublin, J.J., Harvati, K., Harrison, T., Eds.; Springer: Dordrecht, The Netherlands, 2006; pp. 269–279. [CrossRef]

42. Frayer, D.W. The Persistence of Neanderthal Features in post-Neanderthal Europeans. In *Continuity or Replacement: Controversies in Homo Sapiens Evolution*; Bräuer, G., Smith, F., Eds.; A.A. Balkema: Rotterdam, The Netherlnads, 1992; pp. 179–188.

43. Bräuer, G.; Broeg, H. On the Degree of Neandertal-modern Continuity in the Earliest Upper Palaeolithic Crania from the Czech Republic: Evidence from Non-metrical Features. In *The Origins and Past of Modern Humans—Towards Reconciliation*; Omoto, K., Tobias, P.V., Eds.; World Scientific: Singapore, 1998; pp. 106–125.

44. Churchill, S.E.; Smith, F.H. Makers of The Early Aurignacian Of Europe. *Am. J. Phys. Anthropol.* **2000**, *113*, 61–115. [CrossRef]

45. Condemi, S.; Mounier, A.; Giunti, P.; Lari, M.; Caramelli, D.; Longo, L. Possible Interbreeding in Late Italian Neanderthals? New Data from the Mezzena Jaw (Monti Lessini, Verona, Italy). *PLoS ONE* **2013**, *8*, e59781. [CrossRef] [PubMed]

46. Duarte, C.; Maurício, J.; Pettitt, P.B.; Souto, P.; Trinkaus, E.; van der Plicht, H.; Zilhão, J. The early Upper Paleolithic human skeleton from the Abrigo do Lagar Velho (Portugal) and modern human emergence in Iberia. *Proc. Natl. Acad. Sci. USA* **1999**, *96*, 7604–7609. [CrossRef] [PubMed]

47. Rozzi, R.; Fernando, V.; d'Errico, F.; Vanhaeren, M.; Grootes, P.M.; Kerautret, B.; Dujardin, V. Cutmarked Human Remains Bearing Neandertal Features and Modern Human Remains Associated with the Aurignacian at Les Rois. *J. Anthropol. Sci.* **2009**, *87*, 153–185.

48. Compton, T.; Skinner, M.M.; Humphrey, L.; Pope, M.; Bates, M.; Davies, T.W.; Parfitt, S.A.; Plummer, W.P.; Scott, B.; Shaw, A.; et al. The morphology of the Late Pleistocene hominin remains from the site of La Cotte de St Brelade, Jersey (Channel Islands). *J. Hum. Evol.* **2021**, *152*, 102939. [CrossRef]

49. Walker, M.J.; Gibert, J.; López, M.V.; Lombardi, A.V.; Pérez-Pérez, A.; Zapata, J.; Ortega, J.; Higham, T.; Pike, A.; Schwenninger, J.-L.; et al. Late Neandertals in Southeastern Iberia: Sima de las Palomas del Cabezo Gordo, Murcia, Spain. *Proc. Natl. Acad. Sci. USA* **2008**, *105*, 20631–20636. [CrossRef] [PubMed]

50. Wolpoff, M.H. *Paleoanthropology*, 2nd ed.; McGraw-Hill: Boston, MA, USA, 1999.

51. Trinkaus, E.; Churchill, S.E.; Ruff, C.B.; Vandermeersch, B. Long Bone Shaft Robusticity and Body Proportions of the Saint-Césaire 1 Châtelperronian Neanderthal. *J. Archaeol. Sci.* **1999**, *26*, 753–773. [CrossRef]

52. Di Vincenzo, F.; Churchill, S.E.; Manzi, G. The Vindija Neanderthal scapular glenoid fossa: Comparative shape analysis suggests evo-devo changes among Neanderthals. *J. Hum. Evol.* **2012**, *62*, 274–285. [CrossRef]

53. Wolpoff, M.H.; Smith, F.H.; Malez, M.; Radovčić, J.; Rukavina, D. Upper pleistocene human remains from Vindija cave, Croatia, Yugoslavia. *Am. J. Phys. Anthr.* **1981**, *54*, 499–545. [CrossRef]

54. Smith, F.H.; Trinkaus, E. Les origines de l'homme moderne en Europe centrale: Un cas de continuité. In *Aux Origines d'Homo sapiens*; Hublin, J.-J., Tillier, A.-M., Eds.; Presses Universitaires de France: Paris, France, 1991; pp. 251–290.

55. Ahern, J.C.; Karavanić, I.; Paunović, M.; Janković, I.; Smith, F.H. New discoveries and interpretations of hominid fossils and artifacts from Vindija Cave, Croatia. *J. Hum. Evol.* **2003**, *46*, 27–67. [CrossRef] [PubMed]

56. Ackermann, R.R. Phenotypic traits of primate hybrids: Recognizing admixture in the fossil record. *Evol. Anthr. Issues News Rev.* **2010**, *19*, 258–270. [CrossRef]

57. Rink, W.J.; Schwarcz, H.P.; Smith, F.H.; Radovčiĉ, J. ESR ages for Krapina hominids. *Nature* **1995**, *378*, 24. [CrossRef]

58. Rosas, A.; Bastir, M. An assessment of the late Middle Pleistocene occipital from Apidima 1 skull (Greece). *L'Anthropologie* **2020**, *124*, 102745. [CrossRef]

59. Smith, H.F. Which cranial regions reflect molecular distances reliably in humans? Evidence from three-dimensional morphology. *Am. J. Hum. Biol.* **2008**, *21*, 36–47. [CrossRef] [PubMed]

60. Weber, G.W.; Hershkovitz, I.; Gunz, P.; Neubauer, S.; Ayalon, A.; Latimer, B.; Bar-Matthews, M.; Yasur, G.; Barzilai, O.; May, H. Before the massive modern human dispersal into Eurasia: A 55,000-year-old partial cranium from Manot Cave, Israel. *Quat. Int.* **2019**, *551*, 29–39. [CrossRef]

61. Trinkaus, E. Modern Human versus Neandertal Evolutionary Distinctiveness. *Curr. Anthr.* **2006**, *47*, 597–620. [CrossRef]

62. Freidline, S.E.; Gunz, P.; Harvati, K.; Hublin, J.-J. Middle Pleistocene human facial morphology in an evolutionary and developmental context. *J. Hum. Evol.* **2012**, *63*, 723–740. [CrossRef] [PubMed]

63. Arensburg, B. New Upper Palaeolithic Human Remains from Israel. In *Eretz-Israel: Archaeological, Historical and Geographical Studies*; Israel Exploration Society (IES): Jerusalem, Israel, 1977; pp. 208–215. Available online: http://www.jstor.org/stable/23618749 (accessed on 5 July 2022).

64. Brewster, C.; Meiklejohn, C.; Von Cramon-Taubadel, N.; Pinhasi, R. Craniometric analysis of European Upper Palaeolithic and Mesolithic samples supports discontinuity at the Last Glacial Maximum. *Nat. Commun.* **2014**, *5*, 4094. [CrossRef]

65. Lahr, M.M. The question of robusticity and the relationship between cranial size and shape inHomo sapiens. *J. Hum. Evol.* **1996**, *31*, 157–191. [CrossRef]

66. Trinkaus, E.; Milota, S.; Rodrigo, R.; Mircea, G.; Moldovan, O. Early modern human cranial remains from the Peştera cu Oase, Romania. *J. Hum. Evol.* **2003**, *45*, 245–253. [CrossRef] [PubMed]

67. Vandermeersch, B.; Arensburg, B.; Ofer, B.-Y.; Belfer-Cohen, A. Upper Paleolithic Human Remains from Qafzeh Cave, Israel. Mitekufat Haeven. *J. Isr. Prehist. Soc.* **2013**, *43*, 7–21. Available online: http://www.jstor.org/stable/23784046. (accessed on 5 July 2022).

68. Weaver, T.D.; Stringer, C.B. Unconstrained cranial evolution in Neandertals and modern humans compared to common chimpanzees. *Proc. R. Soc. B Boil. Sci.* **2015**, *282*, 20151519. [CrossRef]

69. Fabre, V.; Condemi, S.; Degioanni, A. Genetic Evidence of Geographical Groups among Neanderthals. *PLoS ONE* **2009**, *4*, e5151. [CrossRef] [PubMed]

70. Ross, A.; Williams, S. Ancestry Studies in Forensic Anthropology: Back on the Frontier of Racism. *Biology* **2021**, *10*, 602. [CrossRef] [PubMed]

71. Howells, W.W. *Cranial Variation in Man*; Harvard University Press: Cambridge, MA, USA, 1973.

72. Stone, J.H.; Chew, K.; Ross, A.H.; Verano, J.W. Craniofacial plasticity in ancient Peru. *Anthr. Anz.* **2015**, *72*, 169–183. [CrossRef]

73. Hubbe, M.; Hanihara, T.; Harvati, K. Climate Signatures in the Morphological Differentiation of Worldwide Modern Human Populations. *Anat. Rec.* **2009**, *292*, 1720–1733. [CrossRef]

74. Ross, A.H.; Ubelaker, D.H. Effect of Intentional Cranial Modification on Craniofacial Landmarks. *J. Craniofacial Surg.* **2009**, *20*, 2185–2187. [CrossRef]
75. Sardi, M.L.; Rozzi, F.R.; Dahinten, S.L.; Pucciarelli, H.M. Amerindians: Testing the hypothesis about their homogeneity. *Comptes Rendus Palevol.* **2004**, *3*, 403–409. [CrossRef]
76. Metzger, M.J.; Bunce, R.G.H.; Jongman, R.H.G.; Sayre, R.; Trabucco, A.; Zomer, R. A high-resolution bioclimate map of the world: A unifying framework for global biodiversity research and monitoring. *Glob. Ecol. Biogeogr.* **2012**, *22*, 630–638. [CrossRef]
77. Frumkin, A.; Bar-Yosef, O.; Schwarcz, H.P. Possible paleohydrologic and paleoclimatic effects on hominin migration and occupation of the Levantine Middle Paleolithic. *J. Hum. Evol.* **2011**, *60*, 437–451. [CrossRef]
78. Nicholson, C.M. Eemian paleoclimate zones and Neanderthal landscape-use: A GIS model of settlement patterning during the last interglacial. *Quat. Int.* **2017**, *438*, 144–157. [CrossRef]
79. Pederzani, S.; Aldeias, V.; Dibble, H.L.; Goldberg, P.; Hublin, J.-J.; Madelaine, S.; McPherron, S.P.; Sandgathe, D.; Steele, T.E.; Turq, A.; et al. Reconstructing Late Pleistocene paleoclimate at the scale of human behavior: An example from the Neandertal occupation of La Ferrassie (France). *Sci. Rep.* **2021**, *11*, 1419. [CrossRef] [PubMed]
80. Darroch, J.N.; Mosimann, J.E. Canonical and principal components of shape. *Biometrika* **1985**, *72*, 241–252. [CrossRef]
81. Ross, A.H. Regional isolation in the Balkan region: An analysis of craniofacial variation. *Am. J. Phys. Anthr.* **2003**, *124*, 73–80. [CrossRef] [PubMed]
82. Pietrusewsky, M. Traditional Morphometrics and Biological Distance: Methods and an Example. In *Biological Anthropology of the Human Skeleton*, 3rd ed.; Katzenberg, M.A., Grauer, A.L., Eds.; Wiley Blackwell: Oxford, UK, 2019; pp. 547–591.
83. SAS Institute Inc. *SAS 9.4*; SAS Institute Inc.: Cary, NC, USA, 1989.
84. SAS Institute Inc. *JMP, Version 16 ed.*; SAS Institute Inc.: Cary, NC, USA, 2016.
85. Dutilleul, P.; Clifford, P.; Richardson, S.; Hemon, D. Modifying the t Test for Assessing the Correlation Between Two Spatial Processes. *Biometrics* **1993**, *49*, 305. [CrossRef]
86. Rosenberg, M.S.; Anderson, C.D. PASSaGE: Pattern Analysis, Spatial Statistics and Geographic Exegesis. Version 2. *Methods Ecol. Evol.* **2010**, *2*, 229–232. [CrossRef]
87. Franciscus, R.G. Later Pleistocene Nasofacial Variation in Western Eurasia and Africa and Modern Human Origins. Ph.D. Thesis, University of New Mexico, Albuquerque, NM, USA, 1995.
88. Franciscus, R.G. Internal nasal floor configuration in Homo with special reference to the evolution of Neandertal facial form. *J. Hum. Evol.* **2003**, *44*, 701–729. [CrossRef]
89. Hublin, J.-J. Climatic changes, paleogeography, and the evolution of the Neandertals. In *Neandertals and Modern Humans in Western Asia*; Akazawa, T., Aoki, K., Bar-Yosef, O., Eds.; Plenum: New York, NY, USA, 1998; pp. 295–310.
90. Trinkaus, E. Neandertal faces were not long; modern human faces are short. *Proc. Natl. Acad. Sci. USA* **2003**, *100*, 8142–8145. [CrossRef]
91. Trinkaus, E. The Neandertal face: Evolutionary and functional perspectives on a recent hominid face. *J. Hum. Evol.* **1987**, *16*, 429–443. [CrossRef]
92. Ross, A.H.; Ubelaker, D.H. Complex Nature of Hominin Dispersals: Ecogeographical and Climatic Evidence for Pre-Contact Craniofacial Variation. *Sci. Rep.* **2019**, *9*, 11743. [CrossRef]

Article

Evolution of the Family Equidae, Subfamily Equinae, in North, Central and South America, Eurasia and Africa during the Plio-Pleistocene

Omar Cirilli [1,2,†], Helena Machado [3,†], Joaquin Arroyo-Cabrales [4], Christina I. Barrón-Ortiz [5], Edward Davis [3,6], Christopher N. Jass [5], Advait M. Jukar [7,8,9], Zoe Landry [10], Alejandro H. Marín-Leyva [11], Luca Pandolfi [12], Diana Pushkina [13], Lorenzo Rook [2], Juha Saarinen [13], Eric Scott [14,15], Gina Semprebon [16], Flavia Strani [17,18], Natalia A. Villavicencio [19,20,21], Ferhat Kaya [13,22,*] and Raymond L. Bernor [1,23,*,‡]

[1] Laboratory of Evolutionary Biology, Department of Anatomy, College of Medicine, Howard University, Washington, DC 20059, USA
[2] Earth Science Department, Paleo[Fab]Lab, University of Florence, Via La Pira 4, 50121 Firenze, Italy
[3] Earth Sciences Department, University of Oregon, 100 Cascade Hall, Eugene, OR 97403, USA
[4] Instituto Nacional de Antropología e Historia Laboratorio de Arqueozoología "M. en C. Ticul Álvarez Solórzano", Subdirección de Laboratorios y Apoyo Académico, Ciudad de Mexico 00810, Mexico
[5] Quaternary Palaeontology Program, Royal Alberta Museum, 9810 103a Ave NW, Edmonton, AB T5J 0G2, Canada
[6] Clark Honors College, University of Oregon, Eugene, OR 97401, USA
[7] Department of Geosciences, University of Arizona, 1040 E 4th St., Tucson, AZ 85721, USA
[8] Department of Paleobiology, National Museum of Natural History, Smithsonian Institution, 10th St and Constitution Ave NW, Washington, DC 20013, USA
[9] Division of Vertebrate Paleontology, Yale Peabody Museum of Natural History, 170 Whitney Ave, New Haven, CT 06520, USA
[10] Department of Earth Sciences, University of Ottawa, 25 Templeton Street, Ottawa, ON K1N 6N5, Canada
[11] Laboratorio de Paleontología, Facultad de Biología, Universidad Michoacana de San Nicolás de Hidalgo, Edif. R 2. Piso. Ciudad Universitaria, Morelia 58030, Mexico
[12] Department of Science, University of Basilicata, Via dell'Ateneo Lucano 10, 85100 Potenza, Italy
[13] Department of Geosciences and Geography, University of Helsinki, FI-00014 Helsinki, Finland
[14] Cogstone Resource Management, Inc., 1518 W., Taft Avenue, Orange, CA 92865, USA
[15] Department of Biology, California State University, 5500 University Parkway, San Bernardino, CA 92407, USA
[16] Department of Biology, Bay Path University, 588 Longmeadow Street, Longmeadow, MA 01106, USA
[17] PaleoFactory, Dipartimento di Scienze della Terra, Sapienza Università di Roma, Piazzale Aldo Moro 5, 00185 Rome, Italy
[18] Departamento de Ciencias de La Tierra, Instituto Universitario de Investigación en Ciencias Ambientales de Aragón (IUCA), Universidad de Zaragoza, 50009 Zaragoza, Spain
[19] Corporación Laguna de Taguatagua, Av. Libertador Bernardo O'Higgins 351, Santiago 1030000, Chile
[20] Instituto de Ciencias Sociales, Universidad de O'Higgins, Av., Libertador Bernardo O'Higgins 611, Rancagua 2852046, Chile
[21] Biomolecular Laboratory, Institut Català de Paleoecologia Humana i Evolucio Social, 43007 Tarragona, Spain
[22] Department of Archaeology, University of Oulu, FI-90014 Oulu, Finland
[23] Human Origins Program, Department of Anthropology, National Museum of Natural History, Smithsonian Institution, Washington, DC 20013, USA
* Correspondence: kaya.ferhat@oulu.fi (F.K.); rbernor@howard.edu or rlbernor@gmail.com or BernorR@si.edu (R.L.B.)
† These authors contributed equally to this work.
‡ R.L.B. is the senior author of the present manuscript.

Citation: Cirilli, O.; Machado, H.; Arroyo-Cabrales, J.; Barrón-Ortiz, C.I.; Davis, E.; Jass, C.N.; Jukar, A.M.; Landry, Z.; Marín-Leyva, A.H.; Pandolfi, L.; et al. Evolution of the Family Equidae, Subfamily Equinae, in North, Central and South America, Eurasia and Africa during the Plio-Pleistocene. *Biology* 2022, 11, 1258. https://doi.org/10.3390/biology11091258

Academic Editor: Zhifei Zhang

Received: 18 July 2022
Accepted: 19 August 2022
Published: 24 August 2022

Publisher's Note: MDPI stays neutral with regard to jurisdictional claims in published maps and institutional affiliations.

Copyright: © 2022 by the authors. Licensee MDPI, Basel, Switzerland. This article is an open access article distributed under the terms and conditions of the Creative Commons Attribution (CC BY) license (https://creativecommons.org/licenses/by/4.0/).

Simple Summary: The family Equidae enjoys an iconic evolutionary record, especially the genus *Equus* which is actively investigated by both paleontologists and molecular biologists. Nevertheless, a comprehensive evolutionary framework for *Equus* across its geographic range, including North, Central and South America, Eurasia and Africa, is long overdue. Herein, we provide an updated taxonomic framework so as to develop its biochronologic and biogeographic frameworks that lead to well-resolved paleoecologic, paleoclimatic and phylogenetic interpretations. We present *Equus'* evolutionary framework in direct comparison to more archaic lineages of Equidae that coexisted but progressively declined over time alongside evolving *Equus* species. We show the varying correlations between body size, and we use paleoclimatic map reconstructions to show the environmental changes accompanying taxonomic distribution across *Equus* geographic and chronologic ranges. We present the two most recent phylogenetic hypotheses on the evolution of the genus *Equus* using osteological characters and address parallel molecular studies.

Abstract: Studies of horse evolution arose during the middle of the 19th century, and several hypotheses have been proposed for their taxonomy, paleobiogeography, paleoecology and evolution. The present contribution represents a collaboration of 19 multinational experts with the goal of providing an updated summary of Pliocene and Pleistocene North, Central and South American, Eurasian and African horses. At the present time, we recognize 114 valid species across these continents, plus 4 North African species in need of further investigation. Our biochronology and biogeography sections integrate Equinae taxonomic records with their chronologic and geographic ranges recognizing regional biochronologic frameworks. The paleoecology section provides insights into paleobotany and diet utilizing both the mesowear and light microscopic methods, along with calculation of body masses. We provide a temporal sequence of maps that render paleoclimatic conditions across these continents integrated with Equinae occurrences. These records reveal a succession of extinctions of primitive lineages and the rise and diversification of more modern taxa. Two recent morphological-based cladistic analyses are presented here as competing hypotheses, with reference to molecular-based phylogenies. Our contribution represents a state-of-the art understanding of Plio-Pleistocene *Equus* evolution, their biochronologic and biogeographic background and paleoecological and paleoclimatic contexts.

Keywords: Equidae; Equinae; hipparionini; protohippini; equini; paleoecology; paleoclimatology; biochronology; phylogeny; evolution

1. Introduction

Studies on the evolution of the family Equidae started in the middle of the 19th century following the opening of the western interior of the United States. Marsh [1] produced an early orthogenetic scheme of Cenozoic horse evolution detailing changes in the limb skeleton and cheek teeth. Gidley [2] challenged the purported orthogonal evolution of horses with his own interpretation of equid evolution. Osborn [3] chose not to openly debate the phylogeny of Equidae but rather displayed Cenozoic horse diversity in his 1918 treatise on the unparalleled American Museum of Natural History's collection of fossil equids. Matthew [4], however, did produce a phylogeny detailing 10 stages, actual morphological grades, ascending from *Eohippus* to *Equus*. Stirton [5] provided a widely accepted augmentation of Matthew's earlier work with his revised orthogonal scheme of North American Cenozoic Equidae. Simpson [6] published his book on horses, which was the most authoritative account up to that time. His scheme was vertical rather than horizontal in its view of North American equid evolution, with limited attention paid to the extension of North American taxa into Eurasia and Africa. MacFadden [7] updated in a significant way Simpson's [6] book, depicting the phylogeny, geographic distribution, diet and body sizes for the family Equidae.

This work principally focused on the evolution of *Equus* and its close relatives in North and South America, Eurasia and Africa during the Plio-Pleistocene (5.3 Ma–10 ka). We documented several American lineages that overlap *Equus* in this time range, whereas in Eurasia and Africa, only hipparionine horses co-occurred with *Equus* beginning at ca. 2.6 Ma, which we included in this work for their biogeographical and paleoecological significances. These taxa were reviewed by MacFadden [7] and Bernor et al. [8] including the references therein.

In the present manuscript, we aimed to revise and discuss the most recent knowledge on the Equinae fossil record, with the following goals:

(1) Provide a systematic revision of tridactyl and monodactyl horses from 5.3 Ma to 10 ka. These taxa are discussed in chronological order, from the oldest to the youngest occurrence in each region starting from North and Central America, South America, Eastern Asia (i.e., Central Asia, China, Mongolia and Russia), Indian Subcontinent, Europe and Africa, with their temporal ranges, geographic distributions, time of origins and extinctions. Part of this information is also reported in Table S1;

(2) Integrate the distribution of the fossil record with paleoclimate and paleoecological data in order to provide new insights into the evolution of Equinae and their associated paleoenvironments;

(3) Compare and discuss the latest morphological and genetic-based phylogenetic hypotheses on the emergence of the genus *Equus*. Recently, Barrón–Ortiz et al. [9] and Cirilli et al. [10] provided phylogenies of *Equus*, including fossil and extant species, with different resulting hypotheses. We compare and discuss the results of these two competing hypotheses here;

(4) Provide a summary synthesis of the major patterns in the evolution, adaptation and extinction of Equidae 5.3 Ma–10 ka.

2. Materials and Methods

We provide a revised taxonomy of all Plio-Pleistocene *Equus* across the Americas, Eurasia and Africa summarizing previously published research, as a group of 19 researchers from Europe and North and South America. We provide an updated chronology and geographic distribution for these species. We followed the international guidelines for fossil and extant horse measurements published by Eisenmann et al. [11] and Bernor et al. [12]. Over the last 30 years, these methods have been applied to several case studies in Equinae samples from North and South America, Eurasia and Africa, which led to the identification of new species as well as to the clarification of the taxonomic fossil record. More recently, Cirilli et al. [13] and Bernor et al. [14] provided a combination of analyses to analyze fossil and extant samples including univariate, bivariate and multivariate analyses on cranial and postcranial elements using boxplots, bivariate plots, Log10 ratio diagrams and PCA. We found that robust, overlapping statistical and analytical methods led to a finer resolution of the taxonomy and, ultimately, biogeographical and paleoecological studies. We provide essential information on those species we recognized in the record under consideration.

The taxonomic revision of the 5.3 Ma to 10 ka equid genera and species is given below and includes the authorship, chronological and paleobiogeographic ranges and some historical and evolutionary considerations on the taxon. A reduced emended diagnosis of the species is reported in the Supplementary Materials in order to offer the most relevant anatomical features to identify the taxon. Additional information is reported in Table S1.

We compiled a global Neogene dataset of Equinae body mass estimates and paleodietary information for extensive palaeoecological analyses. These data were collected during several museum visits and complemented with data from publications and databases. Paleodietary information included results from traditional mesowear [15], low-magnification microwear [16] and isotopic analyses of equids from American, Eurasian and African localities (Tables S2 and S3). Net primary productivity (NPP) estimates were calculated for equine localities from the mean hypsodonty and mean longitudinal loph counts of large herbivorous mammal communities using the equation of Oksanen et al. [17]. The equation was as follows: $NPP = 2601 - 144HYP - 935 LOP$, where HYP is the mean ordinated hypsodonty, and LOP is the mean longitudinal loph count.

For the selected localities, body mass estimates based on metapodial measurements of equine paleopopulations [18,19], univariate mesowear scores calculated using the method of Saarinen et al. [20], and NPP estimates [17] were included to test whether the diet and productivity of the paleoenvironments were connected with body size patterns in equine horses. For this purpose, we used ordinary least squares linear models with body mass estimates as the dependent variable and the mean mesowear scores and NPP estimates as the explaining variables. These analyses were based on Eurasian and African *Equus* because of the large amount of data available and high variation in the ecology and environments of that genus during the Pleistocene, particularly in Europe but also, to some extent, in Asia and Africa. Because of slight methodological differences concerning the North American equine data [21] (Section 6.2 in the Supplementary Materials), we discuss them separately from Eurasian and African *Equus* in the context of the patterns revealed by the Eurasian and African *Equus* models. We also discuss the paleoenvironmental and habitat properties of key equine species based on what is known regarding the vegetation type and climate in their environments/paleoenvironments. Furthermore, we compared the patterns in the equine body size evolution, dietary ecologies and paleoenvironments between the continents and discuss how differential changes in biome distribution on the different continents could explain the observed differences in the body size patterns between continents.

We assembled data on the large herbivorous mammals (i.e., Artiodactyla, Perissodactyla, Proboscidea and Primates) from the NOW database [22] and calculated the mean ordinated crown height for each locality (Table S4) following Fortelius et al. [23] for lists with at least two species with a hypsodonty value. All NOW localities between 7 Ma to recent from North and South America, Eurasia and Africa were included in the study and divided into four different age groups: 7–4 Ma; 4–2.5 Ma; 2.5–1.5 Ma; 1.5 to the recent. Mean ordinated crown height is a robust proxy for humidity and productivity at the regional scale [23–26]. We plotted the results onto present-day maps and interpolated between the localities using Quantum GIS 3.14.16 Pi. For the interpolations, thematic mapping and grid interpolation were used with the following settings: 20 km grid size; 800 km search radius; 800 grid borders. The interpolation method employed an inverse distance-weighted algorithm (IDW).

Finally, we discuss the most recent phylogenetic outcomes on the origin of the genus *Equus*. We compared the morphological-based cladistic results of Barrón-Ortiz et al. [9] and Cirilli et al. [10] with the genomic-based analyses of Orlando et al. [27], Jónsson et al. [28] and Heintzman et al. [29] in order to identify similarities between the two different cladistic approaches.

3. Systematics of the Equinae since 5.3 Ma in North, Central and South America

3.1. North and Central America

Horses have been commonly found in numerous terrestrial North and Central American vertebrate faunas. From the Middle Miocene through the Early Pleistocene, the diversity of horses encompassed the genera *Astrohippus, Boreohippidion, Calippus, Cormohipparion, Dinohippus, Nannipus, Neohipparion, Pliohippus, Protohippus, Pseudohipparion* and an Hipparionini of indeterminate genus or species. The genus *Equus* is commonly interpreted to have first appeared during the Blancan North American Land Mammal Age (NALMA), though recent analyses propose an earlier origin of crown-group *Equus* that extends into the Middle to Late Pliocene [29]. Nevertheless, the genus peaked in diversity and widespread geographic distribution during the Pleistocene. In the particular case of North and Central America, we use *Equus* in the broad sense (i.e., sensu lato), as the generic taxonomy of this group of equids has not been resolved and is an area of ongoing study (e.g., [9,10,29]).

What follows is a summary of the species of equids present in North and Central America from the Late Miocene to the Late Pleistocene (Hemphillian to Rancholabrean NALMAs) based on a review of the literature. In the particular case of *Equus* sensu lato, we recognized potentially valid species based on a meta-analysis of relevant studies (published between 1901 and 2021; n = 68) discussing fossil specimens of this group of equids; details of this analysis are provided in the Supplementary Materials.

1. *Calippus elachistus*, Hulbert, 1988 [30] (Right Mandibular Fragment with m2–m3, UF342139). This species seems to be restricted to Florida (USA) from the Late Miocene to the Late Hemphillian. The occlusal dimensions of its cheeck teeth are much smaller than any other species of *Calippus*, except *Ca. regulus*, with slightly smaller occlusal dimensions in the early to middle wear stages and significantly smaller basal crown lengths than *Ca. regulus*.

2. *Calippus hondurensis*, Olson and McGrew, 1941 [31] (Partial Skull Containing Left P2–M3 and Right P2–4, WM 1769). This species has been reported in Puntarenas (Costa Rica), Gracias (Honduras), Mexico (Guanajuato, Hidalgo, Jalisco and Zacatecas) and in the USA (Florida). It may be distinguished by its small size and relatively small protocone.

3. *Dinohippus leardi*, Drescher, 1941 [32] (M1, CIT 2645). This species has been recorded from the Late Miocene in California (USA). The size of the molars is similar to that of *Pliohippus nobilis* or larger in unworn teeth.

4. *Dinohippus spectans*, Cope, 1880 [33] (Left M2 with Associated or Referred P2, AMNH 8183). This species has been recorded in Oregon, Nevada, California, Texas and Idaho (USA) dating to the Late Miocene, with molar teeth of larger size than those of any of the extinct American horses, except *Equus excelsus*, approximately equal to those of *Hippidion principale*.

5. *Astrohippus ansae*, Matthew and Stirton, 1930 [34] (Partial Left Maxilla with P2-M3, UC30225). This species is a Hemphilian–Blancan species recorded in Zacatecas (Mexico), New Mexico, Oklahoma and Texas (USA).

6. *Astrohippus stockii*, Lance, 1950 [35] (Palate with P2-M2, Front Portion of M3 on the Right Maxilla and P2-M2 on the Left One, CIT3576). This species has been recorded in Chihuahua, Guanajuato and Jalisco (Mexico) and Florida, New Mexico and Texas (USA), from the latest Hemphilian to Blancan NALMAs. *Astrohippus stockii* is smaller than *A. ansae*, but it possesses higher-crowned cheek teeth [35]. In recent phylogenetic analyses, *A. stockii* was recovered as the sister group to the clade composed of "*Dinohippus*" *mexicanus* plus *Equus* sensu lato [8,36] or the sister group to the clade composed of successive species of "*Dinohippus*" (i.e., "*Dinohippus*" *leardi*, "*Dinohippus*" *interpolatus*, *Dinohippus leidyanus* and "*Dinohippus*" *mexicanus*) plus *Equus* sensu lato [37].

7. *Dinohippus interpolatus*, Cope, 1893 [38] (First and Second Upper Molars, Plate XII, Figures 3 and 4). This is a late Hemphillian-Blancan species that has been recorded in Hidalgo and Zacatecas (Mexico) and in California, Kansas, New Mexico and Texas (USA).

8. *Dinohippus leiydianus*, Osborn, 1918 [3] (Skull, Jaws, Vertebrae, Fore and Hind Limbs, Considerable Portions of the Ribs and Other Parts of the Skeleton of One Individual, AMNH 17224). This species comes from the late Hemphillian–Blancan with records in Alberta (Canada) and in Arizona, California, Kansas, Nebraska and Oklahoma (USA).

9. *Dinohippus mexicanus*, Lance, 1950 [35] (Partial Left Maxilla with P2-M3 and Part of the Zygomatic Arch, LACM-CIT 3697). This species has been found in Chihuahua, Guanajuato, Jalisco, Hidalgo, Nayarit and Zacatecas (Mexico) and in California, Florida, New Mexico and Texas (USA) from Hemphillian to Blancan NALMAs. It is a medium-sized monodactyl equine horse.

10. *Cormohipparion occidentale*, Leidy, 1856 [39] (Four Left and One Right Upper Cheek Teeth, ANSP 11287). This species is a Hemphillian–Blancan species recorded in California, Florida, Nebraska, New Mexico and Oklahoma (USA). It is a large and hypsodont North American hipparion.

11. *Nannippus aztecus*, Mooser, 1968 [40] (Fragmented Right Maxillary with P3–M3, FO 873). This species has been recorded in Mexico (i.e., Chihuahua, Guanajuato and Jalisco) and in the USA (i.e., Alabama, Florida, Louisiana, Mississippi, Oklahoma and Texas) from the latest Hemphillian to Blancan NALMAs. It is a small-sized horse.

12. *Nannippus lenticularis*, Cope, 1893 [3] (Two Upper Cheek Teeth). This species has been recorded from Hemphillian to Blancan NALMAs in Alberta (Canada) and in Alabama, Kansas, Nebraska, Oklahoma and Texas (USA).

13. *Nannippus peninsulatus*, Cope, 1885 [41] (M2, AMNH8345). This is a Hemphillian–Blancan species with records in Guanajuato, Hidalgo, Jalisco and Michoacan (Mexico) and in Arizona, Florida, Kansas, Nebraska, New Mexico and Texas (USA). *Nannippus peninsulatus* was a highly cursorial equid that appears to have been functionally monodactyl [42,43]. It had an estimated body mass of 59.6 kg [44].

14. *Neohipparion eurystyle*, Cope, 1893 [38] (TMM 40289-1). This species has been recorded in Alberta (Canada); Guanajuato, Hidalgo, Jalisco and Zacatecas (Mexico); Alabama, Cali-

fornia, Florida, Kansas, Nebraska, Oklahoma and Texas (USA) from the Hemphillian to Blancan NALMAs. It is a very hypsodont medium-sized hipparion.

15. *Neohipparion leptode*, Merriam, 1915 [45] (Lower Molar, UCMP 19414). This species is a Hemphillian–Blancan species recorded in California, Kansas, Nebraska, Nevada, Oklahoma and Oregon (USA). It is a large hipparion.

16. *Hipparionini genus* and Species Indeterminate. Hulbert and Harington [46] reported a remarkable specimen of a Hipparionini equid from the Canadian Arctic, which represents the northernmost fossil record of an equid reported to date. It was found in an Early Pliocene deposit (~3.5–4 Ma) from the Strathcona Fiord, Beaver Pond locality, Ellesmere Island, Canada [46]. The specimen consists of associated maxillae and premaxillae with the right dI1 and dP2–dP4 and the left dP1–dP4 of a young foal (approximately 6–10 months of age) [46]. It is a relatively large hipparionine equid (estimated adult tooth row length of 150 mm), with deciduous premolars that have low crowns; complex enamel plications; oval, isolated protocones; a facial region that shows a reduced preorbital fossa located posterior to the infraorbital foramen [46]. This combination of traits is not known in any contemporaneous North American hipparionines, but it is found in some Asiatic hipparionines, particularly *Plesiohipparion*, indicating possible affinities with this group and suggesting a previously unrecognized dispersal event from Asia into North America [8,46]. Alternatively, the Ellesmere Island hipparionine could represent a previously unknown autochthonous lineage of high-latitude North American hipparionines that potentially evolved from the mid-Miocene North American *Cormohipparion* [46].

17. *Neohipparion gidley*, Merriam, 1915 [45] (Left M3, UCMP 21382). This species is a Hemphillian species recorded in California and Oklahoma (USA). It is the largest of the North American hipparions.

18. *Boreohippidion galushai*, MacFadden and Skinner, 1979 [47] (Partial Skull with Well-Preserved Dentition, AMNH 100077). This is a late Hemphillian horse from Arizona (USA).

19. *Cormohipparion emsliei*, Hulbert, 1987 [48] (partial skull with most of the right maxilla including dP1, P2-M3; right and left premaxillae with I1–I3; edentulous fragment of the left maxilla with alveoli for dP1 and P2). UF 94700. All elements possibly belong to the same individual as they present similar stages of tooth wear and preservation. It is a species recorded in the latest Hemphillian to Blancan NALMAs in Alabama, Florida and Louisiana (USA). It is a medium-sized species of *Cormohipparion*.

20. *Pseudohipparion simpsoni*, Webb and Hulbert, 1986 [49] (Associated P3-M1, UF 12943). This is a latest Hemphillian species recorded in Florida, Kansas, Oklahoma and Texas (USA).

21. *Pliohippus coalingensis*, Merriam, 1914 [50] (UCMVP 21341). This species is a Pliocene horse from California (USA).

22. *Nannippus beckensis*, Dalquest and Donovan, 1973 [51] (Partial Skull with Right and Left P2–M3, TMM41452-1). This is a Blancan species from Texas (USA). This species is a medium-sized and moderately hypsodont hipparion.

23. *Equus simplicidens*, Cope, 1892 [52] (Left M1, TMM 40282-6). This species is interpreted to have been a medium- to large-sized equid with primitive dentition [53], recorded in Baja California (Mexico) and Arizona, California, Idaho, Kansas, Nebraska and Texas (USA) from Blancan to Irvingtonian. The species was initially based upon fragmented molars, with sizes comparable to *E. occidentalis* and *E. caballus* [52]. According to Skinner [54], *E. simplicidens* shows great similarities in the skull and dentition with the modern *Equus grevyi*, and the differences in the skull are small and expected in temporal and geographic separation. Gidley [55] described the Hagerman horses based upon characters common to all zebrine horses, with taxonomically significant differences from non-zebrines, but the characters used to distinguish it from other zebrines are of doubtful validity [56]. Comparing *E. simplicidens* with the East African Grevy's zebra, *E. grevyi*, Skinner [54] included both in the subgenus *Dolichohippus* [57]. This proposal was questioned by Forsten and Eisenmann [57], as the cranial similarities found by Skinner [54] might be allometrically related to the large skull size [57]. Furthermore, Skinner [54] did not compare the basicranium, missing the comparison of Franck's Index (i.e., the distance from the staphylioin to the

hormion and from the hormion to the basion). The index was considered phylogenetically important, as the lengthening of the hormion to the basion distance seems to have led to a decrease in the index during *Equus* evolution, with a high index being related to a more primitive character than a lower derived index [57]. In Forsten and Eisenmann's [57] analysis, both *E. simplicidens* and *Pliohippus* (*Dinohippus*), considered the generic ancestor of *Equus*, presented a high index, while *E. grevyi* and the other extant species presented lower indices. Following Matthew [4], Forsten and Eisenmann [57] also suggested *E. simplicidens* should be included in the subgenus *Plesippus* [4,58]. *Equus simplicidens* has long been considered the earliest common ancestor of *Equus* [57], but a recent analysis of the genus *Equus* suggested that *Plesippus* and *Allohippus* should be elevated to a generic rank, indicating *Allohippus stenonis* as the sister taxon to *Equus* and *Plesippus simplicidens* and *P. idahoensis* as the sister taxa to the *Allohippus* plus *Equus* clades [9]. On the other hand, recent cladistic analyses combined with morphological and morphometrical comparisons of skulls suggest *E. simplicidens* as the ancestor of *Equus*, not endorsing *Plesippus* and *Allohippus* at the genus or subgenus level [10].

24. *Equus idahoensis*, Merriam, 1918 [59] (Upper Left Premolar, UCMP 22348). This is a Blancan and early Irvingtonian species recorded in Arizona, California, Idaho and Nevada (USA). The type locality is Locality 3036C, in the beds of the Idaho locality, near Froman Ferry on the Snake River, 8 mi SW Caldwell, Idaho. According to Winans [56], none of the traits from the original description of *E. idahoensis* are unique to this species. However, large samples of specimens (e.g., Grandview, Idaho; 111 Ranch, Arizona) have been referred to as *E. idahoensis*, which have distinctive morphological features [9,60] that indicate that this is a potentially valid species. The cheek teeth are large and heavily cemented.

25. *Equus enormis*, Downs and Miller, 1994 [61] (The holotype, IVCM 32, is a partial skull and right and left mandibles, with the right distal humerus, right radius-ulna, MCIII, unciform, magnum, trapezoid and MCIII, phalanges 1, 2, and 3 of the manus; partial pelvis, right femur, MTIII with MTII and MTIV and phalanx 3 of the pes from Vallecito Creek, Anza-Borrego Desert State Park, San Diego County, California, USA). This species is known primarily from the late Blancan – Irvingtonian of California (USA). *Equus enormis* is a large-sized monodactyl horse with an estimated height at the withers of 1.5 m.

26. *Equus cumminsii*, Cope, 1893 [38] (Fragmentary Upper Molar, TMM 40287-14). This small species has been recorded in Kansas and Texas (USA) from Blancan to early Irvingtonian (NALMAs). Although it is poorly represented by fossils and the type of specimen is a single damaged tooth, some authors consider this species as an early ass based on dental morphology [60,62,63].

27. *Equus calobatus*, Troxell, 1915 [64] (Left MTIII, YPM 13470). This species is a large stilt-legged horse reported from the late Blancan to early Rancholabrean NALMAs, with records in Alberta (Canada); Aguascalientes (Mexico); Colorado, Kansas, Nebraska, New Mexico, Oklahoma and Texas (USA). The original discovery consisted of "unusually long and slender" limb bones [64] from Rock Creek, Texas, but no single holotype specimen was designated. Hibbard [65], therefore, selected YPM 13460 as the lectotype. Because the lectotype and cotypes are limb bones with no distinctive characters other than the large size and relative slenderness of the metapodials, there are few morphological criteria available for evaluating this species. Multiple studies [29,56] have synonymized *E. calobatus* with *Equus* (or *Haringtonhippus*) *francisci*, but other studies consider it a valid species [66,67].

28. *Equus scotti*, Gidley, 1900 [68] (Associated Skeleton with Skull, Mandible, Complete Feet and Forelimb Bones, One Complete Foot and Hindlimb and All the Cervical, Several Dorsal and Lumbar Vertebrae, AMNH 10606). This species is recorded from the late Blancan to Rancholabrean NALMAs in Alberta, Ontario, Saskatchewan and Yukon (Canada) and in California, Florida, Idaho, Kansas, Nebraska, New Mexico, Oklahoma and Texas (USA). Winans [54] also interpreted *E. scotti* to be on average slightly smaller than *E. simplicidens*, but the measurements provided in that study actually indicate the opposite, and subsequent review confirms that *E. scotti* was of a larger form than *E. simplicidens*.

29. *Equus stenonis anguinus*, Azzaroli and Voorhies, 1993 [69] (Complete Skull and Jaw, USNM 23903). This is a late Blancan species recorded in Arizona and Idaho (USA). It is described as similar to *E. stenonis* from the Early Pleistocene of Italy, with skull dimensions falling within the size range of this latter species [69]; however, the limb bones are, on average, more elongated. *Equus stenonis anguinus* possesses a preorbital pit and a deep narial notch as do the European *E. stenonis* samples.

30. *Equus conversidens*, Owen, 1869 [70] (Fragmentary Right and Left Maxilla with All Cheek Teeth, IGM4008, Old Catalog Number MNM-403). This is a widespread species reported to have ranged from the Irvingtonian to Rancholabrean with a geographic distribution encompassing North and Central America: Alberta (Canada); Aguacaliente de Cartago (Costa Rica); Apopa Municipality (El Salvador); Aguascalientes, Chiapas, Hidalgo, Jalisco, Michoacán, Nuevo Leon, Puebla, Oaxaca, San Luis Potosi, Sonora, Estado de Mexico, Tlaxcala, Yucatán and Zacatecas (Mexico); Azuero Peninsula and El Hatillo (Panama); Arizona, California, Florida, Kansas, Nebraska, New Mexico, Oklahoma, Texas and Wyoming (USA). The holotype specimen from the Tepeyac Mountain was described by Owen [70] based upon photos [71]. Owen considered the species to be almost identical to *E. curvidens* (South American *E. neogeus*) but with cheek tooth rows converging towards their anterior ends. Cope [72] interpreted the anterior convergence of the cheek tooth rows to be an artifact of the restoration compounded by the photography and assigned the specimen to *E. tau*, albeit without any stated justification. Gidley [69] interpreted the two sides of the maxilla to be from different individuals, since they were found separately and with missing broken edges. Hibbard [65] provided a reconsideration of the specimen and confirmed that Cope [68] was correct regarding the distortion of the palate and that Gidley [73] was incorrect regarding the two sides deriving from different individuals. Azzaroli [74] described a fragmentary skull (LACM 308/123900) from Barranca del Muerto near Tequixquiac, Mexico, in which the "two tooth rows converge rostrally, giving evidence that the palate of the holotype (of *E. conversidens*) was correctly mounted and that Owen's name is after all appropriate".

31. *Equus lambei*, Hay 1915 [75] (~200 ka–~10 ka) (nearly complete skull from a female, USNM8426, collected from Gold Run Creek in the Klondike Region, Yukon Territory, Canada). This species inhabited the steppe–tundra grasslands of Beringia, with remains having been recovered from Siberia, Alaska, and the Yukon (extending slightly into the adjacent Northwest Territories). Recent genomic evidence suggests that *E. lambei* and *E. ferus* may represent a single species [29,76–78], although further research is required before this phylogeny can be resolved (see Supplementary Materials, Section 3.4 supplementary text for a discussion on *E. ferus* in North America).

32. *Equus* (or *Haringtonhippus*) *francisci*, Hay, 1915 [76] (Complete Cranium, Mandible and MTIII, TMM 34–2518). This species has been recorded from Irvingtonian to Rancholabrean localities in the Yukon (Canada); Aguascalientes, Estado de Mexico, Jalisco, Puebla, Sonora, and Zacatecas (Mexico); Alaska, Arizona, Florida, Kansas, Nebraska, New Mexico, Texas and Wyoming (USA). It is the oldest name assigned to the stilt-legged group and was first described as being similar to *E. tau* but with different P3-M1 proportions, which is expected in teeth at different stages of wear and, therefore, probably not a significant taxonomical difference (56). Eisenmann et al. [67] reassigned *E. francisci* to the genus *Amerhippus*, as *Amerhippus francisci*. Heintzman et al. [29] assigned stilt-legged, non-caballine specimens from Gypsum Cave, Natural Trap Cave, the Yukon and elsewhere to their new genus *Haringtonhippus* under the species *Ha. francisci*, based upon complete mitochondrial and partial nuclear genomes as well as morphological data and a crown group definition of the genus *Equus*. Barrón-Ortiz et al. [9] considered *Haringtonhippus* to be a synonym of *Equus* and regarded both *E. francisci* and *E. conversidens* as distinct taxa based upon morphological criteria.

33. *Equus fraternus*, Leidy, 1860 [79] (Upper Left P2, AMNH 9200). This species has been recorded in Alberta (Canada) and in Florida, Illinois, Mississippi, Nebraska, Pennsylvania, South Carolina and Texas (USA) from Irvintonian to Rancholabrean. Winans [56] considered the dental characteristics used to identify the species to vary with wear and that the

specimens used in its diagnosis represented more than one individual, making it uncertain whether to attribute it to one species. Azzaroli [74,80] referred several complete skulls, mandibles and other bones from the southeastern USA to this species.

34. *Equus pseudaltidens*, Hulbert, 1995 [67] (right maxillary with worn DP2, DP3, DP4, M1 and M2 and unerupted P2, P3 and P4 (BEG 31186-35); right and left mandibles with worn i1, di2, dp2, dp3, dp4, m1 and m2 and unerupted p2, p3 and p4 (BEG 31186-36); cranium lacking occiput (BEG 31186-37); right third metacarpal (BEG 31186-3); right and left femora (BEG 31186-2, 34); right and left tibiae (BEG 31186-1, 10); right and left third metatarsals (BEG 31186-4, 7); first phalanx (BEG 31186-24), all thought to belong to the same individual and estimated to have been approximately 3 years old [81]). This species was originally described as *Onager altidens* by Quinn [81]. The use of *Onager* instead of *Hemionus* by Quinn [81] was invalid [67]. Referral of this species to the genus *Equus* makes it a homonym of *Equus altidens*, von Reichenau, 1915 [53,67,82]. Therefore, Hulbert [67] proposed the replacement name *E. pseudaltidens* for *E. altidens* (Quinn). Also referred to this species are a pair of maxillae (BEG 31186-23) and a right mandible (BEG 31186-22) of an animal approximately 1 year old, and 24 deciduous and permanent upper and lower teeth recovered from the type locality [81]. *Equus pseudaltidens* is known from the Irvingtonian–Rancholabrean and it has been reported from the Gulf Coastal Plain of Texas [67,81] and possibly from Coleman, Florida [67]. It is a stilt-legged equid with metapodial dimensions that are similar to extant hemionines [67,76]. Compared to other stilt-legged equids discussed here, *Equus pseudaltidens* is smaller than *E. calobatus* but larger than both *E. francisci* and *E. cedralensis* [67,76,83]. Kurtén and Anderson [84] synonymized *E.* (*Hemionus*) *pseudaltidens* with *Equus* (*Hemionus*) *hemionus*. Winans [85] assigned it to her *E. francisci* species group. Hulbert [67] considered that *E. pseudaltidens* and *E. francisci* were distinct species and hypothesized that they are sister taxa. Azzaroli [74] synonymized *E. pseudaltidens* (as *Onager altidens*) with *E. semiplicatus*. Eisenmann et al. [67] considered *E. pseudaltidens* distinct from *E. semiplicatus* and *E. francisci* and assigned it along with the latter species to *Amerhippus*. Heintzman et al. [29] considered *E. altidens* (=*E. pseudaltidens*) a junior synonym of *Haringtonhippus francisci*.

35. *Equus verae*, Sher, 1971 [86] (Holotype Mandible with a Full Row of Teeth, GIN 835-123/21, River Bolshaja Chukochya Exp. 21, Kolyma Lowland, Northeast Yakutia). This species is a large-bodied, stout-legged Early Pleistocene (Olyorian)–Rancholabrean species recorded in Northeastern Siberia (Russia) and the Yukon (Canada). *E. verae* is much larger than the *E. stenonis* species and similarities in the teeth and the size of limb bones with *E. suessenbornensis*, suggesting a subspecies position for *E. verae* as well as for *E. coliemensis* (see below). Eisenmann [87,88] suggested that *E. verae* may belong to the subgenus *Sussemionus*, but this has not been substantiated by other authors.

36. *Equus occidentalis*, Leidy, 1865 [89] (Lectotype Left P3, VPM 9129). It is a Rancholabrean NALMA horse with records in Mexico (i.e., Baja California and Sonora) and the USA (i.e., Arizona, California, Nevada, New Mexico and Oregon). Leidy [89] named *E. occidentalis* from two upper premolars and one lower molar from two widely separated geographic localities but did not designate a holotype. Gidley [73] selected a left P3 from Tuolumne County, California, as the lectotype. Merriam [90] referred to *E. occidentalis* thousands of bones of a large and stout-limbed equid recovered from Rancho La Brea. Savage [91] and Miller [92] believed that the equid from Rancho La Brea, identified as *E. occidentalis*, did not conform to the lectotype designated by Gidley [73], but neither of these authors proposed a new name. Azzaroli [74] decided to retain the name *E. occidentalis sensu* Merriam [90] and selected the skull figured by Merriam [90] as his lectotype. However, since Gidley [73] had already designated a lectotype for the species, according to ICZN Article 74, no subsequent lectotype designations can be made. Brown et al. [93] concluded that some of Leidy's original fossils of *E. occidentalis* (exclusive of the Tuolumne County tooth) most likely come from the McKittrick asphalt deposits; this locality was not named by Leidy [89], because the town of McKittrick, California, was not named until 1900. These authors also confirmed that many specimens from the original type series of *E. occidentalis* closely resemble the large Pleistocene horses

from McKittrick and Rancho La Brea [93]. Barrón-Ortiz et al. [94] recognized the presence of this species outside of the North American Western Interior during the Late Pleistocene. Barrón-Ortiz et al. [9] recognized it as a valid species closely related to *E. neogeus*.

37. *Equus cedralensis*, Alberdi et al., 2014 [83] (fragment of a mandibular ramus formed by two specimens: one p2-m3 right row (DP-2675 I-2 15) and a second fragment of the symphysis with the anterior dentition (DP-2674 I-2 8), articulated together from Rancho La Amapola, Cedral, San Luis Potosí, Mexico, and stored at the Paleontological Collection (DP-INAH) of the Laboratorio de Arqueozoología "M. en C. Ticul Álvarez Solórzano" Subdirección de Laboratorios y Apoyo Académico, INAH in Mexico City). This species is primarily known from the Rancholabrean of Mexico (i.e., Aguascalientes, Chihuahua, Estado de Mexico, Michoacán, Puebla and San Luis Potosí). *Equus cedralensis* is an equid with a small body mass (estimated mean mass of 138 kg) [83,95]. *Equus cedralensis* was diagnosed as stout-legged, but a recent analysis placed it within stilt-legged horses and with its dental morphology being similar to *Ha. francisci*. Jimenez-Hidalgo and Diaz-Sibaja [96] considered it a junior synonym of *Ha. francisci*. *Equus cedralensis* differs from the holotype of *Ha. francisci*, as it is smaller in size and the lower first and second incisors possess enamel cups.

38. *Equus mexicanus*, Hibbard, 1955 [65] (cranium lacking the LM3 (No. 48 (HV-3)) from Tajo de Tequixquiac, Estado de México, Mexico, and stored at the Museo Nacional de Historia Natural; the specimen is cataloged as IGM4009). This species is known from the Rancholabrean of Mexico (i.e., Aguascalientes, Chiapas, Estado de Mexico, Jalisco, Michoacán, Oaxaca, Puebla, San Luis Potosi and Zacatecas) and the USA (i.e., California, Oregon and Texas). *Equus mexicanus* is a large body sized species (estimated mean mass of 458 kg) [83,95]. Winans [85] placed *E. mexicanus* in her *Equus laurentius* species group, but as with other species groups in this study, this was not a strict synonymy. Azzaroli [74] recognized *E. mexicanus* as a valid taxon, noting that previous investigations had proposed synonymy with and tentatively identified as *E. pacificus* but rejecting this, since the latter species was initially based upon a single tooth. Barrón-Ortiz [97] assigned specimens identified as *E. mexicanus* to *E. ferus scotti*. Barrón-Ortiz et al. [94] assigned specimens identified as *E. mexicanus* to *E. ferus*. Barrón-Ortiz et al. [9] recognized *E. mexicanus* as a valid taxon distinct from *E. ferus*.

3.2. South America

Two genera, *Equus* and *Hippidion*, inhabited the South American continent with records from the Late Pliocene to the Late Pleistocene [98–100]. *Hippidion* is an endemic genus of South American horses characterized mostly by the retraction of the nasal notch, a particular tooth morphology (considered more primitive than Equus and comparable with Pliohippus) and the robustness and shortness of its limb bones [98]. The genus is, at present, represented by three species: *Hippidion saldiasi*, *Hippidion devillei* and *Hippidion principale* [98]. On the other hand, the South American *Equus* is represented by a single species, *E. neogeus*, with caballine affinities and metapodial variation corresponding to an intraspecific characteristic representing a smooth cline [9,99,100].

1. *Equus neogeus*, Lund, 1840 [101] (MTIII, 866 Zoologisk Museum). This is a Middle–Late Pleistocene species (Ensenandan and Lujanian SALMA) and the only representative of the genus in the South American continent [9,98–100]. Most records are from the Late Pleistocene, but its earliest appearance is recorded in the Middle Pleistocene in Tarija, Bolivia, dated at approximately 1.0–0.8 Ma [102,103]. It has a wide geographic range distribution, encompassing all of South America, except for the Amazon basin and latitudes below 40° [100]. The species probably became extinct sometime during the Late Pleistocene–Holocene transition as suggested by the youngest direct radiocarbon date of 11,700 BP (Río Quequén Salado, Argentina [104]).

2. *Hippidion saldiasi*, Roth, 1899 [105] (p2, Museo Nacional de La Plata). This is a Late Pleistocene species, dated between 12,000 and 10,000 years BP, mostly known from Argentinian and Chilean Patagonia, with records in Central Chile and the Atacama Desert [98,106]. The

last records for the species were radiocarbon dated between 12,110 and 9870 BP in Southern Patagonia (Cerro Bombero, Argentina [107]) and Cueva Lago Sofía, Chile [108].

3. *Hippidion principale*, Lund, 1846 [109] (M2, Peter W. Lund Collection, ZMK). This is a Late Pleistocene species (Lujanian SALMA), with records in Argentina, Bolivia, Brazil and Uruguay [98]. This species represents the largest *Hippidion*. There are few radiocarbon-dated records for this species, with the youngest situated at approximately 16,130 BP (Arroyo La Carolina, Argentina) [104]; however, remains found in archaeological contexts dating close to 13,200 BP in Tagua, Central Chile [110], suggest a later presence for this taxon.

4. *Hippidion devillei*, Gervais, 1855 [111] (P2–P3 Row and Fragmented Astragalus, IPMNHN). This species has been reported in Uquia (Argentina, Late Pliocene–Early Pleistocene), Tarija (Bolivia) and Buenos Aires (Argentina) from the Middle Pleistocene (Ensenadan SALMA) to the Late Pleistocene (Lujanian SALMA) and in Brazil [98]. This taxon has been directly radiocarbon dated only from cave contexts in the high Andes of Peru, with the youngest record of 12,860 BP [112,113].

4. Systematics of the Equinae since 5.3 Ma in Eurasia and Africa

4.1. Eastern and Central Asia (China, Mongolia, Russia, Uzbekistan, Kazakhstan and Tajikistan)

The fossil record of the three-toed horses from Eastern Asia includes four different genera (i.e., *Plesiohipparion, Cremohipparion, Proboscidipparion* and *Baryhipparion*) with seven identified species. As for the Indian Subcontinent, Europe and Africa (see below), the Miocene–Pliocene boundary marks the extinction of the genera *Hippotherium, Hipparion* s.s., *Sivalhippus* and *Shanxihippus* [8]. On the other hand, Sun and Deng [114] argued that the *Equus* Datum in China is represented by the simultaneous appearance of five stenonine *Equus* species: *E. eisenmannae, E. sanmeniensis, E. huanghoensis, E. qyingingensis* and *E. yunnanensis*. Subsequently, Sun et al. [115] indicated the *E. qyingingensis* FAD at 2.1 Ma.

1. *Plesiohipparion houfenense*, Teilhard de Chardin and Young, 1931 [116] (MN13–MN15; 6–3.55Ma). The lectotype RV 31031 includes the right p3–m3 from Jingle, Shanxi. The earliest *P. houfenense* first occurs in the Late Miocene Khunuk Formation, Kholobolchi Nor, Mongolia [8,117–119] and the Late Miocene/Early Pliocene Goazhuan Formation of the Yushe Basin (5.8–4.2 Ma) [8,120]. It also occurs into the Pliocene of China.

2. *Proboscidipparion pater*, Matsumoto, 1927 [121] (MN14–MN15; 5–3.5 Ma). The lectotype THP14321 is a skull with a mandible estimated to be 4–3.55 Ma [122]. This species is reported from the Yushe Basin (China), and it may be the original source for the evolution of the Pliocene European species *Proboscidipparion crassum* and *Proboscidipparion heintzi*.

3. *Plesiohipparion huangheense*, Qiu et al., 1987 [122] (MN15; 5.0–3.55). The lectotype THP 10097 is a lower jaw fragment, including the cheek teeth, from the Yushe Basin [122]. It is a Chinese species reported from Inner Mongolia at 3.9 Ma [7,119,122] and more broadly from the MN15 of China and India [8]. Ultimately, *Pl.* aff. *huangheense* has been reported from the Early Pleistocene in Gulyazi, Turkey [123].

4. *Cremohipparion licenti*, Qiu et al., 1987 [122] (MN15, circa 4.0 Ma). The holotype is THP20764, an incomplete cranium from the Yushe Basin [122]. This distinctly Chinese species is the latest occurring member of the genus *Cremohipparion* and is reported from the Yushe Basin [8].

5. *Baryhipparion insperatum*, Qiu et al., 1987 [122] (MN16–MNQ17; 3.55–1.8 Ma). The holotype is THP19009, an incomplete cranium with the mandible from the Yushe Basin [122]. It is reported that this species is from the Pliocene in China.

6. *Plesiohipparion shanxiense*, Bernor et al., 2015 [124] (MNQ17; 2.5–1.8 Ma). The holotype is F:AM111820, a complete skull with the mandible [124]. This species, previously recognized as *Plesiohipparion* cf. *P. houfenense* [117], is the largest and, at the same, the time youngest member of the genus in Eastern Eurasia. It is believed to be 2.0 Ma in age [124]. It may represent the last evolutionary stage of the genus *Plesiohipparion* in China. The absence of the POF suggest an evolutionary relationship with *Pl. houfenense*.

7. *Proboscidipparion sinense*, Sefve, 1927 [125] (MN17-MQ1; 2.5–1.0 Ma). The holotype is PMU M3925, a complete cranium from Henan Province, China. *Proboscidipparion sinense* occurs later in the record and is approximately one-seventh larger than *P. pater*. *Proboscidipparion sinense* is the latest occurring hipparion in China extending its range up to 1.0 Ma [126].

8. *Equus eisenmannae*, Qiu et al., 2004 [127] (2.55–1.86 Ma). The holotype is IVPP V13552, a complete cranium with and the mandible from Longdan [127]. It is a large-sized horse, mostly known from the Early Pleistocene locality of Longdan (China), similar in size to *E. livenzovensis* (see below), and the primitive features of the cranial morphology suggest a close evolutionary relationship with *E. simplicidens* [10,13]. At the present time, its evolutionary linkage with other Chinese *Equus* is not known.

9. *Equus sanmeniensis*, Teilhard de Chardin and Piveteau, 1930 [128] (2.5–0.8 Ma). The lectotype is NIH 002 (Paris), a complete cranium with the mandible from Nihewan, Hebei Province. It is a large-sized, Early Pleistocene species from North and northwest China, Siberia (Aldan River, Bajakal Lake area), Kazakhstan and Tajikistan [114,129]. Sun and Deng [114] suggested a morphological similarity between *E. sanmeniensis* and *E. simplicidens*, although diversified from *E. stenonis*. This evolutionary hypothesis has also been supported by Cirilli et al. [10,13] through morphometric studies on crania. *Equus sanmeniensis* has been reported from the Early to Middle Pleistocene [114,130].

10. *Equus huanghoensis*, Chow and Liu, 1959 [131] (2.5–1.7 Ma). The holotype is IVPP V2385–2389, with three upper premolars and two molars from Huanghe, Shanxi (Sun and Deng, 2019). It is a large-sized, Early Pleistocene species from the localities of Nihewan (Hebei), Linyi (Shanxi), Sanmenxia Pinglu (Shanxi), Xunyi (Shaanxi) and Nanjing (Jiangsu). Sun and Deng [114] supported the hypothesis provided by Deng and Xue [130] that *E. huanghoensis* is a stenonid horse, considering the Nihewan sample as one of the *Equus* species with the largest palatal length with *E. eisenmannae*. The morphometric analyses of Cirilli et al. [10,13] show a primitive morphology of the cranium, similar to *E. simplicidens* and distinct from *E. stenonis*. Ao et al. [132] indicated the age of 1.7 Ma as the youngest record of this species in China.

11. *Equus yunnanensis*, Colbert, 1940 [133] (2.5–0.01 Ma). The lectotype is IVPP V 4250.1, an almost complete but deformed cranium. It is a medium-sized, Early Pleistocene species from the Chinese localities of Yuanmou (Yunnan), Liucheng (Guangxi), Jianshi and Enshi (Hubei), Hanzhong (Shaanxi) and Huili (Sichuan) and from Irrawaddy in Myanmar. The species was initially described by Colbert [133] on isolated cheek teeth, while better knowledge of this species came with the new discoveries from the Yuanmou locality. Deng and Xue [130] proposed a close evolutionary relationship with *E. wangi*, whereas Sun and Deng [114] suggested a close evolutionary relationship with *E. teilhardi*, suggesting that these species are distinct from all other Chinese stenonid horses [114].

12. *Equus teilhardi*, Eisenmann, 1975 [134] (2.0–1.0 Ma). The holotype is NIH001, an incomplete mandible. It is a medium-sized, Early Pleistocene species from northwestern and North China. Sun et al. [135] proposed a close evolutionary relationship between *E. teilhardi* and *E. yunnanensis*, later supported by the cladistic analysis of Sun and Deng [114].

13. *Equus qyingyangensis*, Deng and Xue, 1999 [130] (2.1–1.2 Ma). The holotype is NWUV 1128, an incomplete cranium. It is a medium-sized, Early Pleistocene species from northwestern and North China. Eisenmann and Deng [136] recognized some close anatomical features between *E. simplicidens* and *E. qyingyangensis*, suggesting a close evolutionary relationship between these two species. The latest phylogenetic results of Sun and Deng [114] and Cirilli et al. [10] support this last hypothesis. A new *E. qyingyangensis* sample has recently been described from Jinyuan Cave [115], with an FAD of 2.1 Ma.

14. *Equus wangi*, Deng and Xue, 1999 [130] (2.0–ca. 1.0 Ma). The holotype is NWUV 1170, a complete upper and lower cheek teeth rows from Gansu Province, Early Pleistocene [130]. Sun and Deng [114] reported a large size for *E. wangi*, similar to *E. eisenmanne*, *E. sanmeniensis* and *E. huanghoensis*. The phylogenetic position of *E. wangi* is not well defined, although Sun and Deng [114] highlighted a possible closer relationship with *E. eisenmannae* than any other stenonine *Equus*.

15. *Equus pamirensis*, Sharapov, 1986 [137] (Early Pleistocene). The holotype is IZIP 1-438 (Institute of Zoology and Parasitology, Uzbekistan), a complete upper tooth row from Kuruksai [137], approximately 2 Ma (MN17), possibly earlier. It is a large species of stenonine horse described from the Kuruksai, 18 km NE of Baldzuan, Tajikistan, in the Kuruksai River valley in the Afghan–Tajik Depression [138]. The site has been correlated to the middle Villafranchian. The taxonomy of this horse is contentious [138], with it being referred to variously as *Equus* (*Allohippus*) aff. *sivalensis* [137] and *E. stenonis bactrianus* [139]. We currently regard this as a distinct species.

16. Central Asian Small *Equus* sp. from Kuruksai [138] (Early Pleistocene). This is a small species of horse that co-occurs with the larger *E. pamirensis* at Kuruksai. Metrically, the metapodials fall within the range of variation of *E. stehlini* [138]. Similar small horse remains have also been found in the Early–Middle Pleistocene Lakhuti 1 locality in the Afghan–Tajik Depression [138].

17. *Equus* (*Hemionus*) *nalaikhaensis*, Kuznetsova and Zhegallo, 1996 [140]. This species was found in the late Early Pleistocene and early Middle Pleistocene (approximately 1 my, Jaramillo paleomagnetic episode) in Mongolia. The lectotype PIN 3747/500 is represented by the incomplete skull of an old male from Nalajkha [141].

18. *Equus coliemensis*, Lazarev, 1980 [142]. The holotype is a skull with very worn teeth (col. IA 1741). The type locality is the river Bolshaja Chukochya, Kolyma lowland, northeast Yakutia, Siberia, Russia. It is reported from the late Early Pleistocene in northeastern Siberia (Russia). Recently, Eisenmann [87] included *E. coliemensis* in the subgenus *Sussemonius*.

19. *Equus lenensis*, Rusanov, 1968 [143]. The holotype skull comes from the Lena River delta (GIN Yakutia col. 33). The type locality is the river Bolshaja Chukochya, Kolyma lowland, northeast Yakutia, Siberia (Russia). It is reported from the Middle Pleistocene from northeastern Siberia (Russia). Lazarev [144] considered *E. lenensis* to be close to the North American *E. lambei*, even larger and more heavily built than the latter. It is also known from the Middle Pleistocene in Yakutia *Equus orientalis* (*Equus caballus/ferus orientalis*) and *Equus nordostensis* (*Equus ferus/caballus nordostensis*) [143–145]. *Equus nordostensis* is characterized by a large size based on the skull, with low plication of the "marks" of the upper teeth and a long protocone [144]. According to Kuzmina [129], it is junior synonym of *E. mosbachensis*. *Equus orientalis* has a large skull and a long snout with an elongated teeth row, flat protocone and rare plication of the upper teeth [144].

20. *Equus beijingensis*, Liu, 1963 [146] (late Middle Pleistocene). The holotype is V2573-2574, a palate and jaw from Zhoukoudian, China [146,147]. It is a Chinese caballine horse recovered mostly form locality 21 of Zhoukoudian. Unfortunately, the species is not well represented, although Forsten [147] indicated a similar size with *E. sanmeniensis*. Liu [146] also indicated *E. sanmeniensis* as the possible ancestor for *E. beijingensis*; nevertheless, this hypothesis was discarded by Fortsen [147] and Deng and Xue [130], who identified *E. beijingensis* as a caballine horse, characterized by a U-shaped linguaflexid. Its evolutionary position is not well defined, although Forsten [147] and Deng and Xue [130] proposed that *E. beijingensis* is a relative of European *E. ferus* (their *E. mosbachensis*) or of a North Americam caballine horse [130].

21. *Equus valeriani*, Gromova, 1946 [148] (Late Middle–Late Pleistocene). The hypodigm includes upper and lower cheek teeth, figured in Eisenmann et al. [149]. It is an enigmatic taxon, described by Gromova [148], from Samarkand, Uzbekistan. According to Gromova [148] and Eisenmann et al. [149], it shows a stenonine metaconid-metastylid in the lower cheek teeth but with a long protocone in the upper cheek teeth. Its possible occurrence has also been proposed in Syrie (Kéberien Géométrique d'Umm el Tlel), although this identification still remains uncertain [149].

22. *Equus dalianensis*, Zhow et al., 1985 [150] (Late Pleistocene). The holotype is V821966, an incomplete mandible preserving two lower cheek teeth rows. It is a Chinese caballine horse described from Gulongshan Cave, Liaoning. Forsten [147] demonstrated close morphological and morphometrical similarities with *E. ferus gemanicus*, *E. ferus orientalis* and *E. ferus chosaricus*, suggesting that they all represent individual populations of a

single widespread species, *E. ferus*. Deng and Xue [130] suggested a common origin for *E. dalianensis* and *E. przewalskii*, with no ancestor-descendant relationships between them. Nevertheless, a recent genomic analysis by Yuan et al. [151] revealed that *E. dalianensis* is a separate clade of caballine horses, distinct from *E. przewalskii*.

23. *Equus ovodovi*, Eisenmann and Vasiliev, 2011 [152] (0.04–0.01 Ma). The holotype is IAES 21, a fragmentary palate from Proskuriakova Cave [152]. It was described in the Late Pleistocene site of Proskuriakova Cave (Khakassia, southwestern Siberia, Russia). It was first considered a species related to *E. hydruntinus* and modern hemiones, although the genomic analyses by Orlando et al. [153] suggest a relationship with wild asses, representing a new separated fossil clade with no extant relatives [150]. For this reason, Orlando et al. [153] and Eisenmann and Vasiliev [152] included this species in the subgenus *Sussemionus*. Molecular studies suggest that *E. ovodovi* is a sister to extant zebras and is nested within the clade that includes both extant zebra and asses [154]. *Equus ovodovi* has been recognized in the Late Pleistocene of southern and eastern Russia and more recently in China [152,154,155].

24. *Equus hemionus*, Pallas, 1774 [156] (0.0 Ma). Pallas [156] did not refer to any holotype or lectotype but gave a detailed description of the anatomical features of the species, associated with an illustration of an animal located near Lake Torej-Nur, Transbaikal area [156] (V.19, pp. 394–417, Pl. VII). *Equus hemionus* is known as the Asiatic wild ass, distributed in China, India, Iran, Mongolia and Turkmenistan. Historically, it has also been reported in Afghanistan, Armenia, Azerbaijan, Georgia, Iraq, Jordan, Kuwait, Kyrgyzstan, Russia, Saudi Arabia, Syria, Tajikistan, Turkey and Ukraine. Four different subspecies are identified, mostly describing the present areal distribution: *E. hemionus hemionus* (Mongolia), *E. hemionus khur* (India), *E. hemionus kulan* (Turkmenistan) and *E. hemionus onager* (Iran). Another extinct subspecies was recognized in Syria, *E. hemionus hemippus*. The fossil record is not well studied, but crania and mandibles of the species have been reported from Narmada Valley in Central India [157], the Indian state of Gujarat [158], and the Son Valley in Northern India [159]. This species is distinguished from *E. namadicus* by its smaller size and smaller protocones on the premolars. Radiocarbon dates from these deposits suggest that this species entered South Asia during the last glacial period, most likely from West Asia [160,161]. The most recent genetic analyses suggest that the onager and kiang populations diverged evolutionarily ca. 0.4–0.2 Ma [28].

25. *Equus kiang*, Moorcroft, 1841 [162] (0.0 Ma). Moorcroft [162] did not designate any holotype or lectotype but provided a general description of the species [162]. *Equus kiang* is known as the Tibetan ass, with a distribution in China, Pakistan, India, Nepal and, possibly, Bhutan. Three subspecies have been identified, *E. kiang kiang*, *E. kiang holdereri* and *E. kiang polyodont*, with different authors pointing out their subspecific statuses [162–165]. At the present time, no paleontological information is available for *E. kiang*. Jonsson et al. [28] distinguished *E. kiang* from *E. hemionus* as a valid taxon. The most recent morphological cladistic analysis found *E. kiang* and *E. hemionus* to be stenonine horses [10].

26. *Equus przewalskii*, Poliakov, 1881 [166] (~0.1–0.0 Ma). The holotype, a skull (number 512) and skin (number 1523), are found in the collection of the Laboratory of Evolutionary Morphology, Moscow (museum exposition), originally obtained by N. M. Polyakov in Central Asia, southern Dzhungaria, in 1878 [129]. For a complete description of the species see Groves [167] and Grubb and Groves [168]. It is an extant species of caballine horse that is found in small geographic areas of Central Asia, although, historically, it once ranged from Eastern Europe to eastern Russia [169]. In China, *E. przewalskii* is common in the Late Pleistocene (~0.1–0.012 Ma) sites in the northern and central regions of the country, but it is absent in the Holocene, except in northwestern China [170]. It shows close morphological similarities with *Equus ferus* (see below), and both are members of caballine horses [10,28,153]. While genomic analyses have shown that the Przewalski's horses are the descendants of the first domesticated horses from the Botai culture in Central Asia (Kazakhstan) around 5.5 ka [171,172], subsequent morphological studies have shown that Botai horses are not domestic horses but harvested wild Prezwalski's horses [173].

4.2. Indian Subcontinent

The Siwalik Group and co-eval sediments of the Himalayan Foreland Basin preserve an exceptional record of hipparionine and equine horses. The earliest lineage, *Cormohipparion*, appears in the record at approximately 10.8 Ma. The diverse indigenous *Sivalhippus* lineage ranges from 10.4 to 6.8 Ma, "*Hipparion*" from 10 to 9.6 Ma and *Cremohipparion* from 8.8 to 7.2 Ma [174]. Across the Mio–Pliocene boundary, a turnover in hipparion taxa seems to have taken place, with the older lineages being replaced in the Late Pliocene by *Plesiohipparion* and *Eurygnathohippus* along with a distinct but poorly known species "*Hippotherium*" *antelopinum* [119,175]. These Late Pliocene hipparionines are replaced by stenonine equids represented by *Equus sivalensis* and a small species of ass-like *Equus* in the Early Pleistocene [176]. By the Middle Pleistocene, a third species of large stenonine horse, *Equus namadicus*, is common in peninsular India deposits; this species went extinct in the Late Pleistocene [177]. *Equus hemionus* doesn't appear in the record until ~0.03 Ma [177], and *Equus caballus* is found in Holocene archaeological sites [178].

1. "*Hippotherium*" *antelopinum*, Falconer and Cautley, 1849 [179] (3.6–2.6 Ma). The lectotype is NHMUK PV M.2647, a subadult right maxilla fragment with P2-M3. It is a species of hipparionine horse from the late Pliocene age deposits between the rivers Yamuna and Sutlej in India. This taxon has been the subject of much nomenclatural confusion. Lydekker named the lectotype and along with the hypodigm, placed it within the genus *Hipparion*. Later authors [180–182] have referred Miocene hipparionine material collected on the Potwar Plateau in Pakistan to this táxon and reassigned the species to the genus *Cremohipparion*. However, given that the hypodigm of "*Hippotherium*" *antelopinum* comes from the Late Pliocene and does not preserve any apomorphies of *Cremohipparion*; we refer to the late Pliocene specimens as "*Hippotherium*" *antelopinum*, separate from the Potwar Plateau specimens from the Dhok Pathan Formation, which can still be taxonomically referred to as *Cremohipparion*, but a formal description of the species with a new type of specimen is required. A more comprehensive study currently in preparation will attempt to resolve this issue

2. *Plesiohipparion huangheense*, Qiu et al., 1987 [122] (3.6–2.6 Ma). Jukar, et al. [119] identified NHMUK PV OR 15790, a mandibular fragment with p4-m1, originally classified as "*Hippotherium*" *antelopinum*, as *P. huangheense* from the Late Pliocene of the Siwalik Hills.

3. *Eurygnathohippus* sp. (3.6–2.6 Ma). Four mandibular cheek teeth from the Late Pliocene of the Siwaliks from the Potwar Plateau in Pakistan and the Siwalik Hills in India were identified by Jukar et al. [175] as *Eurygnathohippus* sp. These specimens all bear the characteristic single pli-caballinid and ectostylid.

4. *Equus sivalensis*, Falconer and Cautley, 1849 [179] (2.6–0.6 Ma). The lectotype is NHMUK PV M.16160, an incomplete cranium from the Siwaliks [183]. It is a large species of stenonine horse found in the Siwaliks of the Indian Subcontinent, ranging from the Potwar Plateau in the west to the Nepal Siwaliks in the east. The exact temporal distribution is unknown; however, based on paleomagnetic dating of the Pinjor Formation where the species was found, it likely ranges in age from 2.6 to 0.6 Ma [176,184]. However, some potentially older occurrences from just below the Gauss–Matuyama boundary (>2.6 Ma) have also been reported [185,186].

5. *Equus* sp. (~2.2–1.2 Ma). This is small species of *Equus* with smaller slender metapodials has been reported from the Pinjor Formation of the Upper Siwaliks. This species has been referred to as *Equus sivalensis minor* [187], *Equus* sp. A [188] or *Equus* cf. *E. sivalensis* [189]. A set of postcranial remains, including metapodials, astragali and phalanges, which were formerly tentatively referred to as "*Hippotherium*" *antelopinum*, are now believed to belong to this small species of *Equus* [138,184]. Based on specimens collected from the Mangla–Samwal Anticline and the Pabbi Hills in Pakistan, this species likely ranges in age from ~2.2 to 1.2 Ma [176]. Geographically, it ranges from the Pabbi Hills to the river Yamuna in the east.

6. *Equus namadicus*, Falconer and Cautley, 1849 [179] (~0.5–0.015 Ma). The lectotype is NHMUK PV M.2683, an incomplete cranium from the Siwaliks [183]. It is a large-

sized stenonine horse from the Middle and Late Pleistocene of the Indian Subcontinent. The stratigraphic range includes the Middle Pleistocene Surajkund Formation in Central India [190] and Late Pleistocene deposits throughout peninsular India [177]. However, Lydekker [191] reported some specimens from the uppermost Upper Siwaliks, which might suggest that the species extends back to the early Middle Pleistocene.

4.3. Europe

As in Eastern and Central Asia, the Mio–Pliocene boundary represents a relevant turnover of three-toed horses in Europe, with the extinction of the genera *Hippotherium* and *Hipparion* s.s. The Pliocene and Pleistocene are characterized by the persistence of *Cremohipparion*, and the dispersion of *Proboscidipparion* and *Plesiohipparion* [7], represented by five species. The *Equus* Datum is represented by the oldest species, *Equus livenzovensis*, at ca. 2.6 Ma in Russia, Italy, France and Spain [176], which led to the *Equus stenonis'* evolution and to the radiation of the African fossil species [10,13,184,192]. At the present time, we recognize 13 species in the genus *Equus* during the Pleistocene.

1. *Plesiohipparion longipes*, Gromova, 1952 [193] (7–3.0 Ma). The holotype is PIN2413/5030, a complete mt3 from Pavlodar [193]. It has been identified from Pavlodar (Kazhakstan), Akkasdagi and Calta (Turkey) [194] and Baynunah (UAE) [8,195]. In all cases, *Pl. longipes* was recognized by its extreme length dimensions of mc3s and mt3s. None of these attributions of *Plesiohipparion* display the characteristics of extremely angled and pointed metaconids and metastylids of *Pl. houfenense*, *Pl. huangheense*, *Pl. rocinantis* or *Pl. shanxiense*. If all these taxa are referable to *Plesiohipparion*, the chronologic range would be 7 Ma to the Early Pleistocene and the geographic range being from China to Spain.
2. *"Cremohipparion" fissurae*, Crusafont and Sondaar, 1971 [196] (MN14-15). The holotype is an mt3 from Layna, Spain, figured in Crusafont and Sondaar [196] (Pl. 1). It was originally described as *"Hipparion" fissurae*. The species is recorded from the MN15 Pliocene localities in Spain [197] and more recently from the MN14 of Puerto de la Cadena [198]. The most recent analyses provided by Cirilli et al. [192] suggest an attribution to the genus *Cremohipparion*, yielding this species as the possible last representative of the genus in Europe. Its evolutionary framework is not yet defined.
3. *Proboscidipparion crassum*, Gervais, 1859 [199] (ca. 4.0–2.7 Ma; MN14-16). Deperet [200] described and figured the sample from Roussillon, France, without assigning a holotype or lectotype. The sample figured by Deperet [200] (Pl. V, Figures 6–10; Pl. VI) represents the hypodigm for the species. The sample is a small-sized equid with remarkable similarities with *Pr. heintzi* [192,194]. No complete crania are known from this species, whereas it is well documented by isolated upper and lower cheek teeth and postcranial elements. Bernor and Sen [194] showed that *Pr. crassum* also has a very short mc3, whereas Cirilli et al. [192] gave substantial indications in cranial and postcranial elements for its attribution to the genus *Proboscidipparion*. This species is mostly known from the Pliocene of France (Perpignan, Montpellier) but also from Dorkovo (Bulgaria) and Reg Crag (England) [197,201,202].
4. *Proboscidipparion heintzi*, Eisenmann and Sondaar, 1998 [203] (MN15). The holotype is MNHN.F.ACA49A, a complete mc3 from Calta, Turkey [203]. It was originally identified and described *"Hipparion" heintzi* [203], whereas Bernor and Sen [194] restudied and allocated this taxon to *Pr. heintzi*, recognizing the similarity of the Calta juvenile skull MNHN.F.ACA336 to Chinese *Pr. pater* in the retracted and anteriorly broadly open nasal aperture accompanied by very elongate anterostyle of dP2 [190]. It has only been reported from the locality of Calta.
5. *Plesiohipparion rocinantis*, Hernández-Pacheco, 1921 [204] (3.0–2.58 Ma). The lectotype is a p3/4 figured in Alberdi [205] (Pl. 6, Figure 4). It is the largest three-toed horse from Europe. Qiu et al. [122], followed by Bernor et al. [124,206] and Bernor and Sun [207], recognized this species as being a member of the *Plesiohipparion* clade by cranial and postcranial morphological features. *Plesiohipparion rocinantis* is reported between 3.0 and 2.6 Ma [208,209] from La Puebla de Almoradier, Las Higuerelas and Villaroya (Spain); Roca-Neyra (France); Red Crag (England); Kvabebi (Georgia). This species may include

"Hipparion" moriturum from Ercsi (Hungary) and the sample from Sèsklo (Greece) previously ascribed to *Plesiohipparion* cf. *Pl. shanxiense* [210]. It represents the last occurrence of *Plesiohipparion* in Europe at the Plio–Pleistocene transition [8,192].

Rook et al. [211] reported the occurrence of *"Hipparion"* sp. from the Early Pleistocene locality of Montopoli, Italy (2.6 Ma). The *"Hipparion"* sp. from Montopoli is represented by a single incomplete upper cheek tooth that, however, shows some morphological features distinct from the first *Equus, E. livenzovensis* occurring in the same locality [211]. Together with Villarroya, Roca-Neyra, Sèsklo, Guliazy and Kvabebi, the Montopoli specimen represents one of the last occurrences of three-toed horses in Europe and may, in fact, be a small species of *Cremohipparion*.

6. *Equus livenzovensis*, Bajgusheva, 1978 [212] (2.6–2.0 Ma). The holotype is POMK L-4, a fragmentary skull from Liventsovka [212] in the Early Pleistocene. The species represents the *Equus* Datum in Western Eurasia, Early Pleistocene localities dated at the Plio–Pleistocene boundary (2.58 Ma) such as Liventsovka (Russia), Montopoli (Italy) and El Rincón–1 (Spain) [184,212–219]. Recent research [214,216,220] suggests the occurrence of *E. livenzovensis* in Eastern European localities, dated between 2.58 and 2.0 Ma. The species shows the typical stenonine morphology, even though it is a large-sized horse.

7. *Equus major*, Delafond and Depéret, 1893 [221] (?2.6–1.9 Ma). The hypodigm is represented by a P2-M1 and a 1ph3 figured by Delafond and Depéret [221] from Chagny, France. It is poorly represented in the Early Pleistocene of Europe, being represented by few remains and localities. It was described by an incomplete upper cheek tooth row and postcranial elements from the Early Pleistocene locality of Chagny (Central France). Following the ICZN guidelines, Alberdi et al. [216] established that *E. major* has priority over *Equus robustus* Pomel, 1853 [222]; *Equus stenonis* race *major* Boule, 1891 [223]; *Equus bressanus* Viret, 1954 [224]; *Equus major euxinicus* Samson, 1975 [225]. *Equus major* has been reported from the European sites of Senèze, Chagny, Pardines and Le Coupet (France); Tegelen (the Netherlands); East Reunion and Norfolk (England) [216], and, as reported by Forsten [214], it may also be present at Liventsovka (Russia). It is a large-sized monodactyl horse, the largest species of the European Early Pleistocene.

8. *Equus stenonis*, Cocchi, 1867 [226] (2.45–ca. 1.6 Ma). The holotype is IGF560, a complete cranium from the Upper Valdarno Basin, Italy [13]. It was the most widespread *Equus* species during the Early Pleistocene (MNQ17 and MNQ18). In the last century, *E. stenonis* samples were identified under several subspecies as *E. stenonis vireti, E. stenonis senezensis, E. stenonis stenonis, E. stenonis granatensis, E. stenonis guthi, E. stenonis mygdoniensis, E. stenonis anguinus, E. stenonis pueblensis* and *E. stenonis olivolanus*. Recently, Cirilli et al. [13] reevaluated these subspecies, considering most of them to be ecomorphotypes of the same species, resulting in recognizing that *E. stenonis* is a monotypic, polymorphic species. At the present time, the species is reported as having occurred from Georgia to Spain in the circum-Mediterranean area as well as the Levant.

9. *Equus senezensis*, Prat, 1964 [227] (2.2–2.0 Ma). The lectotype includes a left P2-M3 and two mc3s, figured in Prat [227] (Pl.1 Figure C; Pl.2 Figure D,E). It is a medium-sized horse, distributed in France and Italy. Earlier referred to as *E. stenonis senezensis* [227], it has been recognized as a different species by Alberdi et al. [216] and Cirilli et al. [13]. It is morphologically similar to the European *E. stenonis* but with a reduced size. It is reported from the type locality of Senèze and possibly from Italy between 2.2 and 2.1 [228,229].

10. *Equus stehlini*, Azzaroli, 1964 [230] (1.9–1.78 Ma). The holotype is IGF563, an incomplete cranium from the Upper Valdarno Basin. It is the smallest Early Pleistocene *Equus* species. Over the last decades, it was considered both either a species or a subspecies of *E. senezensis* Azzaroli [216,230]. Cirilli [229] found that it can be considered a different Early Pleistocene species that probably evolved from *E. senezensis*. At the present time, *E. stehlini* is known only from Italy [228,229].

11. *Equus altidens*, von Reichenau, 1915 [82] (1.8–0.78 Ma). The lectotype is a right p2 figured in von Reichenau [82] (Pl. 6, Figure 17) from Sussenborn, Germany. It is a medium-sized horse, intermediate in size between *E. stehlini* and *E. stenonis*, occurring in Western

Eurasia in the late Early to early Middle Pleistocene. Originally described from the Middle Pleistocene in Süssenborn [82], over the last decades its chronologic range was extended to the late Early Pleistocene, representing the most widespread species after 1.8 Ma until the Middle Pleistocene [231,232]. Recently, Bernor et al. [14] reported its first occurrence in Dmanisi (Georgia, 1.85–1.76 Ma), supporting the hypothesis of a dispersion of this species from east to west, being part of a faunal turnover that included several other mammalian species during this time frame [14,233,234]. It was also identified in Moldova (Tiraspol, layer 5) [129]. Several populations of *E. altidens* have been identified as different subspecies (i.e., *E. altidens altidens* and *E. altidens granatensis*). Nevertheless, within the nomen, *E. altidens* may also be included in other European taxa such as *Equus marxi* von Reichenau, 1915 [82]; *Equus hipparionoides* Vekua, 1962 [235]; *E. stenonis mygdoniensis* Koufos, 1992 [236]; *Equus granatensis* Eisenmann, 1995 [13,216,232,233,237,238]. Its origin remains controversial. Guerrero Alba and Palmqvist [239] proposed a possible African origin, claiming it to be part of the *E. numidicus–E. tabeti* evolutionary lineage. This latter hypothesis was also reported by Belmaker [240]. Eisenmann [87] included *E. altidens* in the new subgenus *Sussemionus*, with other Early and Middle Pleistocene species. More recently, Bernor et al. [14] proposed a new evolutionary hypothesis, considering that *E. altidens* originated in Western Asia and is potentially related to living *E. hemionus* and *E. grevyi*.

12. *Equus suessenbornensis*, Wüst, 1901 [241] (1.5–0.6 Ma). The lectotype is IQW1964/1177, a P2-M3 from Süssenborn, Germany. It is a large horse, larger than *E. stenonis* and *E. livenzovensis* but smaller than *E. major*. As for *E. altidens*, the species has been described from the Middle Pleistocene in Süssenborn [241], even if its best-known sample comes from the Georgian locality of Akhalkalaki (ca. 1.0 Ma) [242]. In addition to Akhalkalaki and Süssenborn, it has been reported in Central European localities such as Stránská Skála (Czech Republic) and Ceyssaguet and Solilhac (France) [216,243,244]. Over the last decades, its biochronologic range was extended to the late Early Pleistocene, with its earliest occurrence in the Italian localities of Farneta and Pirro Nord [216,231] and in the Spanish sites of Barranco León 5 and Fuente Nueva 3 [245].

13. *Equus apolloniensis*, Koufos et al., 1997 [246] (1.2–0.9 Ma). The holotype is LGPUT-APL-148, a nearly complete cranium from Apollonia, Greece [246]. It is a peculiar species of the late Early Pleistocene, mostly recorded from the locality of Apollonia–1 (Mygdonia Basin, Greece) and, possibly, from other localities of the Balkans and Anatolia [244,246,247]. As reported by Gkeme et al. [247], this species differs from *E. stenonis* and other European Early Pleistocene *Equus*, with a distinct cranial morphology, and its size is intermediate between *E. stenonis* and *E. suessenbornensis*. Koufos et al. [246] interpreted *E. apolloniensis* as an intermediate species between *E. stenonis* and *E. suessenbornensis*, whereas Eisenmann and Boulbes [248] considered *E. apolloniensis* as "a step within the lineage of asses".

14. *Equus wuesti*, Musil, 2001 [249] (1.1–0.9 Ma). The holotypes are IQW1980/17067 and IQW1981/17619, two fragmentary mandibles with p2-m3 from Untermassfeld, Germany [249]. It has been established from the Epivillafranchian locality of Untermassfeld (Germany). Musil [249] reported isolated teeth, mandibles and long bones with primitive and derivate characteristics, and a larger size when compared with the widespread *E. altidens*. This evidence was also reported by Palombo and Alberdi [232], highlighting a more robust morphology of the postcranial elements. However, scholars disagree on its possible origin. Forsten [250] considered *E. wuesti* close to and derived from *E. altidens*, whereas others [249,251] recognized *E. wuesti* as the possible source for *E. altidens*. Nevertheless, considering the latest *E. altidens* discoveries [14,231,232,238], this last hypothesis seems to not be well supported. Palombo and Alberdi [232] suggest also that it can represent an ecomorphotype of *E. altidens*.

15. *Equus petralonensis*, Tsoukala, 1989 [252] (ca. 0.4 Ma). The holotype is PEC-500, an mc3 from Petralona Cave, Greece [252]. It is a slender and gracile horse from Greece (Petralona Cave). However, the taxonomic status of this species has been actively debated. Forsten [250] considered *E. petralonensis* as being a member of the *E. altidens* group, together with the Early Pleistocene equids from Libakos, Krimini and Gerakaou.

Eisenmann et al. [67] synonymized *E. petralonensis* with *Equus hydruntinus*, within the subspecies *E. hydruntinus petralonensis*. Despite the taxonomic controversy surrounding this species, *E. petralonensis* may be considered a stenonine horse because of its mandibular cheek tooth morphology [246].

16. *Equus graziosii*, Azzaroli, 1969 [253] (MIS 6). No access number is available for the holotype, which is figured in Azzaroli [254] (Pl. XLV, Figure 1a,b) as a complete cranium from Val di Chiana, Arezzo. It is an enigmatic species from the late Middle–Late Pleistocene. It was described as a different species by Azzaroli [254] based on a partial cranium, mandibles and postcranial elements. According to Azzaroli [254], the cranium and maxillary and mandibular cheek teeth have typical asinine features, although some other authors have highlighted the morphological similarities with *E. hydruntinus* [244,255,256]. The evolutionary and phylogenetic position of *E. graziosii* is still questionable.

17. *Equus hydruntinus*, Regalia, 1907 [257] (0.6–0.01 Ma). There is no holotype. The hypodigm includes isolated upper and lower molars and the fragments of radius and tibia from Grotta Castello (Sicily, Italy). It is known also as the European wild ass, which is a small-sized and slender horse from the Middle and Late Pleistocene in Europe. Its geographic range spans across Europe and is documented in numerous localities [244]. Apparently, the first specimens of this species were also found in Central Asia in several Uzbekistan Late Weichselian localities associated with the Late Paleolithic [129,258]. Its evolutionary history has been debated by many authors, who have proposed different scenarios such as a direct origin from *E. altidens* [150,259,260], *Equus tabeti* [237,239] or, more recently, as a new taxon arrival from Asia [228,238]. Nevertheless, the most recent DNA analyses relate *E. hydruntinus* as a morphotype of the modern *E. hemionus*, proposing the subspecies *E. hemionus hydruntinus* [261]. However, several morphological features distinguish *E. hydruntinus* from *E. hemionus* (for a detailed discussion, see [244]), allowing for the consideration of *E. hydruntinus* as a still valid name for fossil identification. The oldest remains of *E. hydruntinus* were reported from the Middle Pleistocene levels of Vallparadìs, ca. 0.6 Ma [262,263], although the species has been documented in Western Eurasia until the Holocene [244]. Different subspecies have been suggested including *E. hydruntinus minor* (Lunel Viel, France), *E. hydruntinus danubiensis* (Romania), *E. hydruntinus petralonensis* (Petralona Cave, Greece) and *E. hydruntinus davidi* (Saint-Agneau, France). As reported by Boulbed and Van Asperen [244], *E. hydruntinus* adapted to semiarid, steppe conditions with a preference for temperate climates, although it could tolerate limited cold conditions.

18. *Equus ferus*, Boddaert, 1758 [264] (ca. 0.7–0.6 Ma). Boddaert [264] did not refer to any holotype or lectotype but gave a detailed description of the anatomical features of the species [264] (p. 159). It first appears in Western Eurasia in the early Middle Pleistocene, although a precise age is not available at the present time. The species has been questioned mostly from a taxonomic viewpoint, wherein many subspecies or different species have been erected to identify the Middle and Late Pleistocene fossil samples of the caballine horses as *Equus mosbachensis*, *E. mosbachensis tautavelensis*, *E. mosbachensis campdepeyri*, *E. mosbachensis micoquii*, *E. mosbachensis palustris*, *Equus steinheimensis*, *Equus torralbae*, *Equus achenheimensis*, *E. ferus taubachensis*, *E. ferus piveteaui*, *E. ferus germanicus*, *E. ferus antunesi*, *E. ferus gallicus*, *E. ferus latipes*, *E. ferus arcelini* and *Equus caballus*. These taxa are based on the size or morphological differences among the different fossil samples, representing an interesting case of morphological variability within the same lineage. These different subspecies have been considered to be chrono species by Eisenmann and Kuznetsova [265]. Nevertheless, van Asperen [266] noted that differences in the size and morphology of Middle and Late Pleistocene caballine horses can be observed, although they are not more variable than modern ponies or highly homogeneous groups such as Arabian horses or *E. przewalskii* [244]. Moreover, no unidirectional or evolutionary trend in size and shape can be identified, whereas morphology and size fluctuate over time [19,266]. For this reason, the proposal of Boulbes and van Asperen [244] to consider these species/subspecies as ecomorphological variants of the same species, *E. ferus*, seems the most parsimonious position. Moreover, as reported by van Asperen [267] following the ICZN, the correct

species to indicate the wild caballine horses should be *E. ferus* and not *E. caballus*, which refers to domesticated forms. In addition, the genomic studies of Weinstock et al. [77] and Orlando et al. [153] support the genetic variation of the Middle and Late Pleistocene caballine horses.

4.4. Africa

The 5.3 Ma to 10 ka record of Equidae in Africa include two groups of Equinae: Hipparionini and Equini. Churcher and Richardson [268] provided a comprehensive review of African Equidae that was updated by Bernor et al. [269] for its taxonomic content, biogeography and paleoecology with some consideration of the molecular evolution. Churcher and Richardson's review [268] of the literature that led to their revision was extensive, and the reader is referred to their article for a complete rendering of the record. We documented 10 (+4 not well defined) species of hipparions with three genera *Cremohipparion*, *Sivalhippus* and *Eurygnathohippus*, and 13 species of *Equus* in the African record, but there certainly could be more or less, although significant synonymies were cited by Bernor et al. [269] and pending new studies, especially on the *Equus* samples. The *Equus* Datum in Africa has been a matter of debate over the last years. Bernor et al. [269] and Rook et al. [176] reported the first known occurrence of the genus *Equus* in in East Africa at 2.33 Ma in the Omo Shungura Formation, member G (*Equus* sp.). Materials from these earliest occurring *Equus* are not well represented across the skull, mandible, dentition and postcranial elements. Nevertheless, recent research in the North African sequence of Oued Boucherit (Algeria) have recalibrated the localities of Aïn Boucherit, El Kherba and Aïn Hanech. The lowermost stratigraphic level of Ain Boucherit has been dated at 2.44 Ma [270], where *Equus numidicus*, *Equus* cf. *E. numidicus* and *Equus tabeti* have been reported [270,271]. Therefore, this new North African age for the *Equus* Datum anticipates the earliest occurrence of *Equus* in the north rather than in East Africa.

1. *Cremohipparion periafricanum*, Villalta and Crusafont, 1957 [272] (6.8–4.0 Ma). The lectotype is a P2-M3 figured in Alberdi [205] (Pl.3, Figure 3) from Vadecebro II, Spain. It is a small (dwarf) hipparion that is a close relative (or senior synonym) of *Cr. nikosi* from the Quarry 5 levels of Samos, Greece, dated 6.8 Ma [206]. Fragmentary remains of *Cr.* aff. *periafricanum* have been reported from Tizi N'Tadderth, Morocco [273], and Sahabi, Libya [274].
2. *Eurygnathohippus feibeli*, Bernor and Harris, 2003 [275] (6.8–4.0 Ma). The holotype is KNM-LT139, a partial right forelimb including a fragmentary radius, mt3, a1ph3, a2ph3, partial mc2, a1ph2, a2ph2, a2ph3 and a partial mc4 [275]. There were additional dental and postcranial elements from Lower and Upper Nawata that were referred to this species. Bernor and Harris [275] suggested that the Ekora 4 cranium, ca. 4.0 Ma, was a late surviving member of *Eu. feibeli*. Whereas Churcher and Richardson [268] recognized "*Hipparion*" *sitifense* in the North African Late Miocene–Pliocene horizons, Bernor and Harris [275] and Bernor and Scott [276] noted that the type material described by Pomel [277] could not be located and could potentially be confused either with *Cr. periafricanum* or *Eu. feibeli*.
3. *Sivalhippus turkanensis*, Hooijer and Maglio, 1973 [278] (6.5–4.0 Ma). The holotype is KNM-LT136, an adult female cranium. Bernor and Harris [275] assigned this as being a species of *Eurygnathohippus*, *Eu. turkanense*. Subsequent studies of the *Sivalhippus* clade [174] and by Sun et al. [279] demonstrate the close identity of cranial, dental and, in particular, postcranial anatomy of *Si. turkanensis* and *Si. perimensis* and the extension of this genus into China and Africa in the Late Miocene.
4. *Eurygnathohippus hooijeri*, Bernor and Kaiser, 2006 [280] (5.0 Ma). The holotype is SAMPQ-L22187, a complete adult female skull with associated dentition, mandible and dentition and postcranial characteristics from the earliest Pliocene Langebaanweg E Quarry, South Africa [280]. Hooijer [281] originally described the specimen under the nomen "*Hipparion*" cf. *H. baardi*.

5. *Eurygnathohippus woldegabrieli*, Bernor et al., 2013 [282] (4.4–4.2 Ma). The holotype is ARA-VP-3/21, an incomplete mandible [282]. The hypodigm include also 156 dental and postcranial specimens from 14 localities at Aramis (Middle Awash), Ethiopia. The type specimen is ARA-VP-3/21 a mandible including symphysis, right partial ramus with p2 and p3, left ramus with p2 and 3 preserved and pr-m3 poorly preserved. Mandibular incisor teeth and canines, if originally present are lacking.

6. *Eurygnathohippus afarense* Eisenmann, 1976 [283] (KH3, Hadar, ca. 3.0 Ma). The holotype is AL363-18, a partial cranium from the Kada Hadar Member [283]. *"Hipparion" afarense* was nominated for skeletal material originating from the Kada Hadar 3 horizon, Hadar, Ethiopia. Eisenmann [283] also referred a mandible, AL177-2, to *E. afarense*.

7. *Eurygnathohippus hasumense*, Eisenmann, 1983 [284] (3.8–3.2 Ma). The holotype is KNM-ER 2776, a p4-m2 from zones B and C of the Kubi Algi Formation [284]. She included cheek teeth of common morphology from the Chemeron Formation, Kenya and the Denen Dora Member of the Hadar Formation, Ethiopia. A cranium with associated mandible was included in this hypodigm, AL340-8 [269] (Figure 13 in Bernor et al. [269]), and a partial skeleton including cheek teeth and complete postcranial elements, AL155-6 from DD2, ca. 3.2 Ma. Bernor et al. [281,285] analyzed a series of Ethiopian *Eurygnathohippus* establishing a phyletic relationship that currently includes *Eu. feibeli*, *Eu. woldegabrieli*, *Eu. "afarense"*, *Eu. hasumense* and *Eu. cornelianus*.

8. *Eurygnathohippus pomeli* Eisenmann and Geraads, 2006, [286] (ca. 2.5 Ma). The holotype is AaO-3647, an almost complete, but transversely crushed skull [286]. It was originally described as *"Hipparion" pomeli*. Reference [286] reported a well-preserved assemblage of hipparionini Ahl al Oughlam near Casablanca. Eisenmann and Geraads [286] argued that the sample is homogeneous and biochronologically correlative with eastern African faunas that are ca. 2.5 Ma, roughly contemporaneous with Omo Shungura D. Bernor and Harris [275] recognized this as a species of *Eurygnathohippus*. The northward extension of Plio–Pleistocene *Eurygnathohippus* into North Africa is remarkable as was its extension into India at this time [175].

9. *Eurygnathohippus cornelianus*, van Hoepen, 1930 [287] (ca. 2.6–1.0 Ma). The hypodigm includes a mandibular dentition from Cornelia, Orange Free State, with hypertrophied i1s and i2s and atrophied i3s placed immediately posterior to the i2s [287] (plates 20–22). Leakey [288] (pl. 20, 4 figures) reported the occurrence of *"Stylohipparion albertense"* (=*Eu. cornelianus*) from Bed II, Olduvai Gorge, Tanzania based on premaxillae and mandibular symphyses with identical incisor morphology. Hooijer [281] reported an adult skull from Olduvai BKII which he referred to *Hipparion* cf. *H. ethiopicum* which is likely a member of the *Eu. cornelianus* lineage. Eisenmann [284] did not recognize the existence of *Eu. cornelianus* at Olduvai, but referred an immature cranium, KNM-ER3539 to *Eu. cornelianus*. Armour-Chelu et al. [289] and Armour-Chelu and Bernor [290] argued that the first evidence of this clade may be from the Upper Ndolanya Beds, Tanzania, circa 2.6 Ma. It is also likely present in the Omo Shungura F, dated 2.36. Bernor et al. [8,269] advanced the hypothesis that *Eu. cornelianus* is a member of an evolving lineage that occurred in East and South Africa between ca. 2.4 and less than 1 Ma. The specimen/locality content of *Eurygnathohippus* is currently under investigation.

10. *Hipparion (Eurygnathohippus) steytleri*, van Hoepen, 1930 [287]. Van der Made et al. [291] argued that the author named *Hipparion steytleri* based on a right M1/1, left M3 and left m1-2 that formed a type series. Van Hoepen [287] also named *Eu. cornelianus* on the basis of a mandibular symphysis. Van der Made et al. [291] considered the nomen *H. stytleri* to have priority over *Eu. cornelianus*, but the cheek teeth of *H. stytleri* are of insufficient diagnostic value in themselves to define a valid species as stipulated by Article 75.5 of the ICZN. On the other hand, Van Hoepen [287] exercised considerable foresight in recognizing the highly derived state of the type specimen mandible of *Eu. cornelianus* because of its very wide mandibular symphysis with hugely hypertrophied i1 and i2 with very reduced, peg-like i3 situated immediately posterior along the mid-line of i2. Moreover, Leakey [288] illustrated a series of *Eu. cornelianus* mandibular symphyses and premaxillae from Olduvai Bed 1. Bernor et al. [292] described a juvenile skull of *Eu. cornelianus* (RMNH67) from

Olduvai Gorge [293] which has a long preorbital bar, faint preorbital fossa and a dP2 with an extended anterostyle. *Eurygnathohippus cornelianus* sensu strictu has a known chronologic range of 2–1 Ma, but the lineage apparent extends lower in time.

11. *Hipparion (Eurygnathohippus) libycum* Pomel, 1897 [273]. The hypodigm includes a left p3/4 figured by Pomel ([273], pl. 1, Figures 5–7), two lower cheek teeth from "carriers des gres ouvertes a la campagne Brunie" in Oran (pl. 1, Figures 1–7) and a distal epiphysis of a third metatarsal from "carriers de grès du quartier", St-Pierre, Oran. Hopwood [289] assigned the left p3/4 figured by Pomel ([273], pl. 1, Figures 5–7) as a lectotype [287]. It shows gracile metapodials. Van der Made et al. [287] reported that the original type specimens from Oran are in the Central Faculty of Algiers (MGFCA) but have provided no accession numbers for the holotype.

12. *Hipparion (Eurygnathohippus) ambiguum*, Pomel, 1897 [277]. The hypodigm includes a right P2 from Beni Fouda (Ain Boucherit) [277] (pl. 2, Figures 2–4). Previously the repository was unknown. Van der Made et al. [291] (Figures 2a,b and 3), redrafted an image by Pomel [277] and reported that the specimen is currently maintained in the Central Faculty of Algiers (MGFCA) but offered no photographic images or measurements of this specimen. Van der Made et al. [291] have provided no accession number for the holotype. The type locality of Aïn Boucherit has a magnetostratigraphic date of 2.44 [291] whereas the East African localities referred to by Van der Made et al. [291] have an age range of 3.8–1.2 Ma [269].

13. *Hipparion (Eurygnatohippus) massoesylium*, Pomel [277]. The hypodigm includes five teeth from "puits Carouby" and "aux portes d'Oran" (also Puits Karoubi), left P4, M1-3 and right M3 (pl. 1, Figures 8–10). Van der Made et al. [291] considered these specimens to be the holotype of the species. Van der Made et al. [291] have reported that the holotype is kept in the Central Faculty of Algiers (MGFCA) but have provided no accession numbers for the associated cheek teeth.

14. *"Hipparion" sitifense*, Pomel, 1897 [277]. The hypodigm includes four specimens including two teeth from St. Arnaud, a calcaneum from a nearby locality and a tooth figured by Thomas [294]. Pomel figured an M1/2 (Pl. 1, Figures 13–16), a right P4 (Pl. 1, Figures 11 and 12) and a calcaneum (Pl.2, Figures 9 and 10). The M1/2 figured by Pomel represents the lectotype [291]. The latter authors re-figured the original illustrations from Pomel [277], but this figure does not have a scale bar, and the specimens do not have formal institutional accession numbers. Eisenmann [295] stated that it is not known where the "type specimens" from Saint Arnaud et al. are, whereas Van der Made et al. [291] report that the original material is in the Central Faculty of Algiers (MGFCA; p. 44). Van der Made et al. [291] nominated a lectotype citing a specimen figured by Pomel [277] (pl. 1, Figures 13–15) without specifically designating a specimen accession number, institution, element and providing a redrafted original figure of the specimen without the benefit of a scale. It cannot be known if *"H. sitifense"* is referable to the genus *Hipparion* or other small hipparionins *Cremohipparion* or *Eurygnathohippus* (for which mandibular cheek teeth are needed). Arambourg [296,297] reported a specimen "from the type locality (for which we cannot be certain)" described and figured a mandibular specimen "with rounded metaconid-metastylid and no ectostylid". It should be noted that this material is not a legitimate sample of "the original type series of Pomel, 1897". Arambourg [296,297] assigned material from Ain el Hadj Baba, Mascara, Saint Donat and Aïn el Bey to *"H. sitifense"*. These are all small sized.

The *Equus* Datum in Africa has been a matter of debate over the last years. Bernor et al. [269] and Rook et al. [176] reported the first known occurrence of the genus *Equus* in East Africa at 2.33 Ma in the Omo Shungura Formation, member G (*Equus* sp.). Material of these earliest occurring *Equus* are not well represented across the skull, mandible, dentition and postcranial elements. During this interval of time, the most common European species is *E. stenonis* and possibly some late surviving populations of *E. livenzovensis* [13], *E. eisenmannae*, and *E. sanmeniensis* in China. Nevertheless, recent research in the North African sequence of Oued Boucherit (Algeria) have recalibrated the localities of Aïn Boucherit, El Kherba and Aïn Hanech. The lowermost stratigraphic level of Aïn Boucherit has been dated at 2.44 Ma [270], where *Equus numidicus*, *Equus* cf. *E. numidicus* and *Equus tabeti* have been reported [270,271]. Therefore, this new North African age for the *Equus* Datum anticipates the earliest occurrence of Equus in the North rather than in East Africa.

15. *Equus numidicus*, Pomel, 1897 [277] (2.44–1.2 Ma). The hypodigm includes a right P2 [273] (Pl. 2, Figure 2). Arambourg reported cranial and postcranial elements from Aïn Boucherit and Aïn Jourdel (Pl. 18, Figures 6 and 7; Pl. 19–20; Figures 58 and 62). It is a medium-sized horse approximately the size of a large zebra that originated from Aïn Boucherit and Aïn Hanech, Algeria, ca. 2.44–2.0 Ma [270,271,295]. No complete crania are known, although incomplete cranial remains associated with postcranial elements have been reported [296]. The evolutionary relationships of *E. numidicus* are not well defined. Azzaroli [298] pointed out a possible relationship to *E. stenonis*, although it was no longer investigated. However, some anatomical features of the upper and lower cheek teeth and postcranial elements resemble those of *E. stenonis* [297,298].

16. *Equus tabeti*, Arambourg, 1970 [297] (2.44–1.2 Ma). The holotype is a partial palate (1949.2:773) figured in Arambourg [297] (Pl. 21, Figure 3). It is a medium-small sized species of *Equus* with "asinine" maxillary cheek teeth, stenonine mandibular cheek teeth and slender third metapodials and phalanges [284,297]. No complete crania are known, even if some incomplete crania and mandibles have been reported by Arambourg [297]. The type material originates from Aïn Hanech, Algeria [297]. Geraads [299] estimated the age of Aïn Hanech to be 1.2 Ma. Eisenmann [284] reported the possible presence of *Equus* cf. *tabeti* at Koobi Fora, Kenya believing that it is a primitive ass and may have been derived from *E. numidicus*. The recent analyses of Duval et al. [270] suggest an earlier occurrence of *E. tabeti* and *E. numidicus* at 2.44 Ma in the lowermost levels of Aïn Boucherit. Beside North Africa, *E. tabeti* is reported from Early Pleistocene sites of "Ubeidiya", Bizat Ruhama and Qafzed (Levantine corridor [300–302]) and from East Africa [183].

17. *Equus koobiforensis*, Eisenmann, 1983 [284] (2.1–1.0 Ma). The holotype specimen is KNM-ER 1484, a complete skull recovered from the *Notochoerus scotti* Zone, Area 130, just below the KBS Tuff, ca. 1.9 Ma. Eisenmann [284] reported a number of close dental similarities shared by *E. koobiforensis* and European *E. stenonis* but did not suggest a direct phylogenetic relationship between these taxa. Azzaroli [258] stated that *E. koobiforensis* was essentially a Grevy's zebra. Bernor et al. [183] cited the likely evolutionary relationship between North American Pliocene *E. simplicidens*, European *E. stenonis* and *E. koobiforensis*. Cirilli et al. [10] demonstrated the explicit cladistic relationships between the *E. simplicidens*-*E. stenonis*-*E. koobiforensis*-*E. grevyi* clade, whereas Cirilli et al. [13] reinforced this result on robust statistical grounds. No precise age is available for its last occurrence in the fossil record.

18. *Equus oldowayensis*, Hopwood, 1937 [293] (1.9–1.0 Ma). The holotype is a lower jaw from an animal approximately 2 years old [293] (Figures 1 and 2; Catalogue Number VIII, 353, in the Bayerische Paläontologische Staatssammlung, Munich). Hopwood [293] also designated a lower incisive region with the left incisors and right first incisor (BMNH M14199) as the paratype. The original Olduvai collection deposited in Munich, which included the type of *E. oldowayensis*, was destroyed together with its catalogue, during WW II (K. Heissig personal communication with Churcher and Hooijer; [269]). *Equus oldowayensis* is usually reported from the type locality of Olduvai (1.8 Ma, Tanzania), but no precise ages are available for its chronologic range. Recently, Bernor et al. [183] reported an incomplete cranium from Olorgesailie (ca. 1.0 Ma, Kenya).

19. *Equus capensis*, Broom, 1909 [303] (ca. 2.0?–? Ma). It was a large-bodied horse estimated to be 150 cm at the withers, with a body mass of approximately 400 kg [304]. It originated from South Africa. Churcher [305] synonymized *E. helmei*, *E. cawoodi*, *E. kubmi*, *E. zietsmani* and specimens of *E. harrisi* and *E. plicatus* into *E. capensis*. *Equus capensis* was widely distributed in the Plio–Pleistocene of South Africa, although no information is known about its first and last occurrence in the fossil record.

20. *Equus mauritanicus*, Pomel, 1897 [277] (1.0 Ma). The hypodigm include isolated teeth and postcranial elements, figured in Pomel [277] (Pl.3–8). It is reported from a large sample from Tighenif (Algeria). Churcher and Richardson [268] referred *E. mauritanicus* to a subspecies of *E. burchellii* (=*E. quagga mauritanicus*). Eisenmann [284,295] recognized *E. mauritanicus* as a distinct species of *Hippotigris* and claimed similarities to *E. stenonis* in the dentition. Eisenmann [284] further introduced the notion of cross-breeding between *E. mauritanicus* and Quaggas. Churcher and Richardson [268] reported an extensive distribution of *E. (Hippotigris) "burchelli"* (=*quagga*) from North, East and South Africa. The possible relationships of *E. mauritanicus* with plain zebras have been suggested by Eisenmann [295] and more recently from Bernor et al. [14] by morphological analyses. Beside the type locality of Tighenif, *E. mauritanicus* is reported also in Oumm Qatafa (Juden Desert, Egypt, Middle Pleistocene [306]).

21. *Equus melkiensis* Bagtache et al. 1984 [307] (Late Pleistocene). The holotype is a short mc3 (I.P.H. Allo. 61–1314) recovered from Allobroges, Algeria and of latest Pleistocene age. Eisenmann [236] reported *E. melkiensis* also from Morocco. This species has been identified also at Gesher Benot Ya'akov and Nahal Hesi (Israel; [87,301]) and Oumm Qatafa (Egypt, Middle Pleistocene, [306]).

22. *Equus algericus*, Bagtache et al. 1984 [307] (Late Pleistocene). The holotype is IPH61-103, a m2 from Allobroges, Algeria [307] (Figure 1). It is reported to be a caballine species with a withers height of approximately 1.44 m. *Equus algericus* was also reported from Morocco [308], which are purported to have the characteristic caballine metaconid-metastylid (=double knot) morphology.

23. *Equus grevyi*, Oustalet, 1882 [309] (0.5–0.0 Ma). Oustalet [309] (v.10, pp. 12–14) described the anatomical features of the species, with an associated illustration (Figures 1 and 2). He also reported that the living animal was donated to the Museum National d'Histoire Naturelle in Paris, where the holotype should be kept. *Equus grevyi* is the largest living wild equid with a withers height of 140–160 cm. Azzaroli [259], Bernor et al. [14,183] and Cirilli et al. [10,13] all have cited the close relationship between European *E. stenonis*, *E. koobiforensis* and *E. grevyi*. *Equus grevyi* is currently distributed in the arid regions of Ethiopia and Northern Kenya and has recently vanished from Somalia, Djibouti and Eritrea [269]. Eisenmann [284] recognized *Equus* cf. *E. grevyi* from the *Metridiochoerus compactus* zone, the Guomde Formation and Galana Boi beds of Kenya based on both cheek tooth and postcranial remains. Bernor et al. [183] and Cirilli et al. [10,13] have recognized *E. grevyi* as a terminal member of the *E. simplicidens–E. stenonis–E. koobiforensis–E. grevyi* clade. Recently, O'Brien et al. [310] reported an incomplete cranium ascribed to *E. grevyi* from the Kapthurin Formation (Kenya) dated between 547 and 392.6 ka. This age represents the best dated *E. grevyi* FAD, at the present time.

24. *Equus quagga*, Boddaert, 1785 [264] (?1.0–0.0 Ma). Boddaert [264] did not identify a holotype or lectotype for *E. quagga* but gave a description of the anatomical features of the species (p. 160). It has a shoulder height ranging from a mean of 128 cm in males and 123 cm in females. *Equus quagga* is one of the most widely distributed African ungulates ranging from southern Sudan and southern Ethiopia to northern Nambia and northern South Africa. Several subspecies have been recognized including *E. quagga crawshaii*, *E. quagga borensis*, *E. quagga boehmi*, *E. quagga chapmani*, *E. quagga burchellii* and *E. quagga quagga*. Fossil remains have been reported from North to South Africa. Leonard et al. [311] suggest that the various subspecies of *E. quagga* differentiated between 120 and 290 ka. Fossil remains of quagga have been reported from South African Plio–Pleistocene karst deposits, but their certain identity in North and East Africa are somewhat elusive [269]. Pedersen et al. [312] have identified a South African region as the likely source for the origin of the plain zebras from which all extant populations expanded from at approximately 370 ka. Moreover, the genetic analyses of Pedersen et al. [312] have reported a remarkable gene flow in the extant *E. quagga* subspecies, highlighting the challenge of identifying the subspecific designation only by morphology, identifying at least four genetic clusters for *E. quagga boehmi* and *E. quagga crawshayi*.

25. *Equus zebra*, Linnaeus, 1758 [313] (?0.5–0.0 Ma). Linnaeus [313] does not refer any holotype or lectotype but gives a description of the anatomical features of the species (p. 101). It is a medium-sized, long-legged zebra with a mean shoulder height ranging from 124–127 cm. There are two recognized sub-species, *E. zebra zebra* (Cape Mountain Zebra) and *E. zebra hartmannae* (Hartmann's Mountain Zebra). Churcher and Richardson [264] report a relatively small sample from the Middle Pleistocene to recent fossil remains of *E. zebra* in South Africa.

26. *Equus (Asinus) africanus*, Heuglin and Fitzinger, 1866 [314]. The lectotype is designated as a skull of an adult female collected by von Heuglin near Atbarah River, Sudan, and in Stuttgart (SMNS32026). *Equus (Asinus) africanus* are equines of small size and stocky build. Churcher [315] reported the earliest occurrence of this taxon from the middle of Bed II, Olduvai Gorge (>1.2 Ma). This identification was based on a single third metatarsal, which was short (231 mm) and slender, although not to the extent of *E. tabeti*. Two subspecies are recognized, *E. (A.) africanus africanus* von Heuglin and Fitzinger, 1866 [314] (Nubian wild ass) and *E. (A.) africanus somaliensis* Noak, 1884 [316] (Somali wild ass) [317–319]. The African wild ass *E. africanus* is widely believed to have been the ancestor of the domestic donkey [269] and, more recently, believed to be the descendant of the Early Pleistocene species *E. tabeti* [320]. Nevertheless, this last evidence has only been suggested and not yet proven.

5. Biochronology and Biogeography

Rook et al. [176] recently provided a comprehensive biochronology and biogeography for latest Neogene–Pleistocene *Equus*-bearing horizons of Eurasia, Africa and North and South America. Following this recent summary, we report here a synthesis of major Equinae evolutionary events for the last 5.3 million years across these continents. Berggren and Van Couvering [321] suggest the application of the term biochron for units of geologic time that are based on paleontological data without reference to lithostratigraphy or rock units. Mammal biochronologic scales have been developed for Europe (ELMA), Asia (ALMA), North America (NALMA) and South America (SALMA) and, most recently, for Africa (AFLMA). These timescales are variously expressed in terms of conventional mammal biostratigraphic zones or as land mammal ages (LMAs). Each timescale based on land mammals in different continental landmasses has its own history of development reflecting the uniqueness of the records and the extent to which faunal succession has been resolved.

5.1. NALMA Timescale and Equinae Evolution

The Hemphillian NALMA ranges from approximately 8 Ma to about 4.9 Ma, with four faunal stages (Hh1-Hh4), and correlates with the Late Miocene to Early Pliocene in age, the most recent age (Hh4) extending over the Mio–Pliocene boundary (5.3–4.9 to 4.6 Ma). Equids recorded from this interval include the genera *Dinohippus* (i.e., *D. leidyanus, D. interpolatus, D. leardi, D. spectans* and *D. mexicanus,*), the hipparions *Cormohipparion* (i.e., *Co. occidentale* and *Co. emsliei*), *Nannippus* (i.e., *Na. aztecus, Na. beckensis, Na. lenticularis* and *Na. peninsulatus*), *Neohipparion* (i.e., *Ne. eurystyle, Ne. gidley* and *Ne. leptode*), *Calippus* (i.e., *Ca. elachistus,* Ca. *hondurensis*), *Astrohippus* (i.e., *A. stocki* and *A. ansae*) and *Boreohippidion* (i.e., *B. galushai*). These last are taxa that hold-over from the Late Miocene.

Hemphillian Hh4 ranges into the Early Pliocene until 4.6 Ma, and a range of 4.9–4.6 Ma for the earliest Blancan has been proposed [322]. In fact, the succeeding Blancan NALMA has been defined by the first appearance in North America of arvicoline rodents circa 4.8 Ma. The Blancan has recently been subdivided into five intervals: Blancan I (4.9–4.62 Ma.), Blancan II (4.62–4.1 Ma.), Blancan III (4.1–3.0 Ma.), Blancan IV (3.0–2.5 Ma.) and Blancan V (2.5–1.9 Ma.) [322]. The equid *Dinohippus* is known to persist into Blancan I and II intervals, while *E. simplicidens, E. idahoensis* and *E. cumminsii* are known from the Blancan III and later Blancan assemblages. The diminutive hipparionine horse *Na. peninsulatus* is reported from the Blancan V interval but does not survive into the Irvingtonian. The Blancan IV/V boundary corresponds closely to the base of the Pleistocene.

The Irvingtonian NALMA is subdivided into three units: Irvingtonian I (~1.9–0.85 Ma), Irvingtonian II (0.85–0.4 Ma) and Irvingtonian III (0.4–0.195 Ma). Early Irvingtonian assemblages includes the equids *E. scotti*, *E. conversidens* [60], and *Equus* (or *Haringtonhippus*) *francisci* (possibly including *E. calobatus*).

Finally, the Rancholabrean NALMA extends from 0.195 to about 0.11 Ma with the onset of the Holocene. Common Rancholabrean equid species include *E. scotti*, *E. conversidens* and *E.* (or *Haringtonhippus*) *francisci*. *Equus occidentalis* is also abundant in the American southwest during this period. Fossils resembling *E. ferus* (e.g., *E. lambei*) have also been documented from Rancholabrean faunas.

5.2. SALMA Timescale and Equinae Evolution

Two lineages of Equidae occurred in South America during the Pleistocene, *Hippidion* and *Equus*. Although there are no records of *Hippidion* in Central or North America, most evidence suggests that both lineages originated in Holarctica and then migrated independently to South America during the important biogeographical event known as the Great American Biotic Interchange (GABI) [103,323,324]. However, there is no record for Equidae in South America until the beginning of Pleistocene, following the formation of the Isthmus of Panama from the early Pliocene onward (approximately 3 Ma) [103,324]. The earliest record of Equidae in South America is *Hippidion principale* from Early Pleistocene deposits (Uquian) of Argentina [323,324]. However, the age of the first record of *Equus* in South America is controversial. Traditionally, its earliest record is from middle Pleistocene (Ensenadan SALMA) of Tarija outcrops in Southern Bolivia, based on a biostratigraphic sequence at Tolomosa Formation and independently calibrated to occur between ~0.99 and <0.76 Ma [103]. Nevertheless, there is no consensus regarding the age of these deposits and some researchers consider the deposition in Tarija to have occurred only during the Late Pleistocene [325] (and references therein). Recently, it was proposed that only one species of *Equus* lived in South America during the Pleistocene, *E. neogeus* [99]. This species is considered a fossil-index for deposits of Lujanian SALMA (late Pleistocene-earliest Holocene; 0.8 to 0.011 Ma). Although *E. neogeus* was widely distributed in South America, only few localities are calibrated by independent chronostratigraphic data, indicating a Lujanian SALMA [98]. Therefore, the dispersal of *Equus* into South America occurred during the GABI, but if it is considered that *Equus'* earliest record is in the Late Pleistocene, it thus followed the fourth and latest phase of the GABI or *Equus* migrated to South America during GABI 3, considering its early record to be in the Middle Pleistocene [103]. All equids that occurred in South America during the Pleistocene (*Hippidion* and *Equus*) became extinct in the early Holocene [99,324].

5.3. ALMA Timescale and Equinae evolution

Equids first appear in Asia in the Miocene, between 11.4 and 11.1 Ma with the dispersal of the three-toed horse *Cormohipparion* [8,274,326]. Thereafter, a diverse assemblage of hipparionines is seen between ~10.2 and 6.0 Ma. Hipparionines are found across the Mio-Pliocene boundary which lies in the Dhok Pathan Formation (ca. 5.3 Ma) on the Potwar Plateau in Pakistan [327]. Hipparionines are rare in South Asia between 6.0 and 2.6 Ma, but recently, three taxa have been reported between 3.6 Ma and 2.6 Ma: *"Hippotherium"*, *Plesiohipparion*, and *Eurygnathohippus* [119,174]. The youngest indeterminate hipparionine records are dated paleomagnetically ~2.6–2.5 Ma, around the same time the first *Equus* occurs in South Asia [328,329]. *Equus* makes its appearance just above the Gauss-Matuyama boundary, which coincides with the Plio-Pleistocene transition.

In China *E. eisenmannae*, a large yet primitive stenonine horse from the Longdan loessic section in Linxia Basin, Gansu Province, was magnetically dated to 2.55 Ma for its lower fossil-producing horizon, which is the earliest record of *Equus* in China. Although represented by poorly preserved fossils, *Equus* sp. from Zanda Basin in southern Tibet [118] was dated to 2.48 Ma, suggesting fast dispersion of *Equus* even in higher elevations.

In the Siwaliks, remains of *Equus* have been found in sediments ranging from 2.6 to 0.6 Ma, a period termed the Pinjor Faunal Zone [328]. As noted in Bernor et al. [182], two morphotypes of *Equus* have been recorded—a large taxon called *Equus sivalensis* and a smaller taxon sometimes called *Equus sivalensis minor* for specimens from the the Upper Pinjor Formation near the town of Mirzapur [330], *Equus* cf. *E. sivalensis* from the Pabbi Hills [192] and *Equus* sp. (small) for specimens from the Mangla–Samwal anticline [190]. *Equus sivalensis* has been recorded from the entire temporal range of the Pinjor Faunal Zone [329]; however, the temporal range of the smaller horse appears to be restricted to ~2.2–1.2 Ma.

Equus sanmeniensis is magnetically dated 1.7–1.6 Ma from the Shangshazui stone artifact site in the classical Nihewan Basin and to 1.66 Ma from the nearby Majuangou III hominin tool site. This species was also recorded from the *Homo erectus* site at Gongwangling, Lantian, Shaanxi Province, and magnetically dated to a slightly younger ~1.54–1.65 Ma. *Equus yunnanensis* from the *Homo erectus* site at Niujianbao in Yuanmou Basin was magnetically dated to 1.7 Ma. [331] In the Late Pleistocene, *Equus namadicus* and *Equus hemionus* are known from the Indian peninsula [332].

5.4. ELMA Timescale and Equinae Evolution

The Miocene fossil record (Vallesian and Turiolian Land Mammal Ages) as well as the Pliocene one (Ruscinian and early Villafranchian Land Mammal Ages) in Europe do not have monodactyl horses. These times are characterized by different hipparionine horses' evolutionary lines and, during the Pliocene, three lineages of hipparions persisted in Europe: *Cremohipparion* (*C. fissurae*), *Plesiohipparion* (*Pl. longipes, Pl. rocinantis*) and *Proboscidipparion* (*Pr. heintzi, Pr. crassum*) [8,206,270]. It is during the early to middle Villafranchian transition, that *E. livenzovensis* first occurs in Southwest Russia and Italy at around 2.6 Ma (beginning of middle Villafranchian; Early Pleistocene) and constitutes the regional *Equus* First Appearance Datum [10,13,184,192]. In Europe, the earliest representatives of the genus *Equus* co-existed with the last hipparionin horses (the genera *Plesiohipparion, Proboscidipparion* and *Cremohipparion*) in the Early Pleistocene [10], although at the present time the effective co-existance of *Equus* and hipparion is found in the localities of Montopoli (Italy) and Roca-Neyra (France).

Equus livenzovensis appears to be at the base of the radiation of the later lineage of fossils horses, the European Pleistocene *Equus stenonis* group (=stenonine horses). The European stenonine horses have been recently revised [10,13]. In addition to *E. livenzovensis*, the species (and their chronological ranges) included in this group are *E. stenonis* (end of middle Villafranchian to early late Villafranchian; Early Pleistocene, 2.4–1.7 Ma; [13]), *E. stehlini* (late Villafranchian; Early Pleistocene, 1.8–1.6 Ma; [229]), *E. altidens*, and *E. suessenbornensis* (end of the late Villafranchian to Early Galerian, Early Pleistocene to Early Middle Pleistocene, 1.6–0.6 Ma; [184]). The most relevant turnover for the *Equus* species occurs in the Middle Pleistocene (ca 0.6 Ma) with the first occurrence of *E. ferus* (or *E. ferus mosbachensis*) in Mosbach (Germany) and *E. hydruntinus* from Vallparadìs (Spain) [240]. The arrival of these two species marks the extinction of the Early and early Middle Pleistocene *Equus* species.

5.5. AFLMA Timescale and Equinae Evolution

The Miocene to Pleistocene mammal record of Africa is overall less complete than the fossil record on other continents with no established land mammal age scheme for Africa at a continental scale was established until very recent times [333].

The completeness of the mammal fossil record across the continent is extremely variable with regions in which the Neogene record is totally missing and others (such as Kenya or Ethiopia) with a relatively continuous and well documented record [334].

Without an established continental-scale biochronology, Africa's "biochronology" is based on stratigraphic ordering in different sedimentary basins and is largely dependent on radiochronology with limited use of magnetostratigraphy. As an example, Pickford [335] subdivided Miocene faunas from Kenyan sites into Faunal Sets I to VII; suggesting age spans for late Miocene sets are 12.0–10.5 Ma (V), 10.5–7.5 (VI), and 7.5–5.5 (VII).

The Late Miocene–Early Pliocene boundary (Sugutan and Baringian Land Mammal Ages) is poorly represented in Africa. Hipparionine horses are first found in North and East Africa circa 10.5 Ma [8] the richest locality being the Algerian site of Bou Hanifia and the Ethiopian site of Chorora [269]. At the end of Late Miocene (Baringian), diversification of the hipparionine genus *Eurygnathohippus* (exhibiting evolutionary relationships to Siwalik hipparionines; [175]) was well underway, as was significant the branching by endemic elephants and marked successes by new bovid tribes and suines arriving from Eurasia.

The Plio–Pleistocene time period has been rigorously studied biochronologically in Africa by temporal distributions of elephants and suids, often in conjunction with the dating of hominin finds. *Equus* first occurs during the Early Pleistocene, even if this occurrence event in Africa is delayed relative to Eurasia, where it is at ca 2.6 Ma. Indeed, the most recent results identify the presence of the genus *Equus* in Nord Africa at ca. 2.44 Ma (*E. numidicus*, [269]) and in East Africa in lower Member G of the Omo Shungura Formation ca. 2.33 Ma (Shunguran Land Mammal Age; [176]). The first occurring African *Equus* is apparently related to European *E. stenonis* [10,13]. Representatives of the genus *Equus* are two times as abundant as *Eurygnathohippus* during the Early Pleistocene (Natronian Land Mammal Age) like in the Ethiopian locality of Daka [336], with *Eurygnathohippus* sharply declining in its numbers in East and South Africa after 1 Ma (Naivashan Land Mammal Age). Unfortunately, little is known of the first occurrence of the living species. Recently, a report by O'Brien et al. [310] of an incomplete *E. grevyi* cranium from the Kapthurin Formation (Kenya, 547–392.6 ka) represents the best dated *E. grevyi* FAD.

5.6. General Remarks about Equinae Biochronology

Unlike high-resolution biostratigraphic tools available in the marine realm, mammalian biochronology is not permissive of recognizing strictly synchronous events at global scale. Nevertheless, a review of the currently available evidence of the Land Mammal Ages, defined and calibrated across different continents (either in North and South America, Eurasia and Africa), allows us to recognize major faunal change (corresponding to the limit between subsequent Land Mammal Ages) always correlatable within the magnetochronostratigraphic scale and to place in this framework the main evolutionary events occurring around the world along the Equinae evolutionary history.

6. Paleoecology

In this chapter, we present some new paleoecological insights in diet and body size on the subfamily equinae, with special remarks for the genus *Equus*.

6.1. Relationships of Diet, Habitats and Body Size in Equines, with Pleistocene Equus from Eurasia and Africa as a Particular Example

In general, equines of the genus *Equus* tend to have mesowear values indicating grazing diets, but there is considerable variation in diets and body size (Table S2), which likely reflect differences in habitat preference and social strategies, as well as the effect of available vegetation in different paleoenvironments. Our new analyses based on the most extensive compilation of body mass and palaeodiet data of particularly in Eurasian equines largely confirm previous hypotheses of the relationship between body size, diet, behavior and environments of equines during the Neogene, and during the Quaternary in particular. A few main hypotheses have been presented for the main factors that affected body size evolution and body size variation (also within species) of equine horses during the Neogene and the Quaternary. First, resource availability and quality (mainly regulated by annual primary productivity as well as nutritional properties and chemical defenses of available plants) have been suggested to be a limiting factor for some of the small-sized equine species or populations [13,19,244,267]. Conversely, the positive effect of seasonally high productivity and high resource quality due to low plant defense mechanisms has been suggested to explain the particularly large body size of Pleistocene herbivorous mammals in general [337–340]. Third, the effects of differential habitat heterogeneity, social structures and population density (intraspecific competition) on limiting the body size of some *Equus* species/paleopopulations, especially purely grazing ones that were abundant in open environments, has been discussed [19,20,340]. Observations of modern and Pleistocene *Equus* populations indicate that large-sized species tend to have smaller group sizes and population densities and more mixed or browse-dominated diets than small-sized species which are graze dedicated [19].

Ordinary least squares multiple regression models indicate that both diet and productivity of environments (estimated NPP values) are related to the variation in body size, both between and within the species of *Equus* during the Pleistocene-present (Figure 1). The "All *Equus*" model includes a wide range of extant and Pleistocene *Equus* populations from Eurasia and Africa, and it shows a significant negative effect of mesowear score and a significant positive effect of NPP on *Equus* body mass (although there is remarkable scatter especially in the residuals of the NPP estimate effect on body mass). The pattern is similar in the most abundant Pleistocene species/lineages of European *Equus*, including *E. stenonis*, *E. altidens* and the Middle–Late Pleistocene caballines (*E. ferus* + *E. mosbachensis*), although the patterns were statistically less robust (but note the small sample size especially in the case of *E. stenonis* and *E. altidens*). The connection between diet and body size was particularly robust, especially for the all-*Equus* model, indicating that very large species of *Equus* had more browse-dominated diets than small and medium-sized species/populations (Figure 1). Furthermore, the association of large size and more browse-dominated diets is shown for Africa as well as Eurasia (Figure 2).

These results indicate that both primary production and differences in the dietary niche had an effect on the body size of *Equus*. Large size in *Equus* is associated with more browse-dominated diets and more productive paleoenvironments, while small and medium-sized horses typically occupied less productive environments and had more purely grazing diets. The association of diet and body size appears stronger than the association of estimated NPP and diet, except perhaps in *E. ferus/mosbachensis*. A possible explanation for this is that the diet includes a signal of niche partitioning between sympatric species of *Equus*. As Saarinen et al. [19] discuss, when a small and a large species of *Equus* occurred sympatrically, the larger species was typically less abundant in the fossil assemblage and (in all such cases, including the new ones added in the present study) had more browse-dominated diet. Thus, the effect of larger group sizes of grazing, gregarious populations of *Equus* on limiting their body size via the effect of larger population densities and more intense intraspecific competition appears to have been a significant mechanism limiting their body size, although there was also a more general positive effect of primary production on body size.

This model of the relationship of body size with diet (affected by available vegetation and dietary niche partitioning), population density (associated with dietary preferences and social strategies of different *Equus* species) and habitat (vegetation openness, heterogeneity and resource availability) based on data from Eurasian and African *Equus* is by and large supported by more general observations from earlier equines and from *Equus* from North America. In Eurasian and African hipparionines, large body size is typically associated with browse-dominated diets, such as in *Hippotherium* from the early late Miocene of Europe, while smallest body sizes tend to occur in grazing taxa, such as the small species of *Cremohipparion* from the circum-Mediterranean environments during the late late Miocene (Table S2) [8].

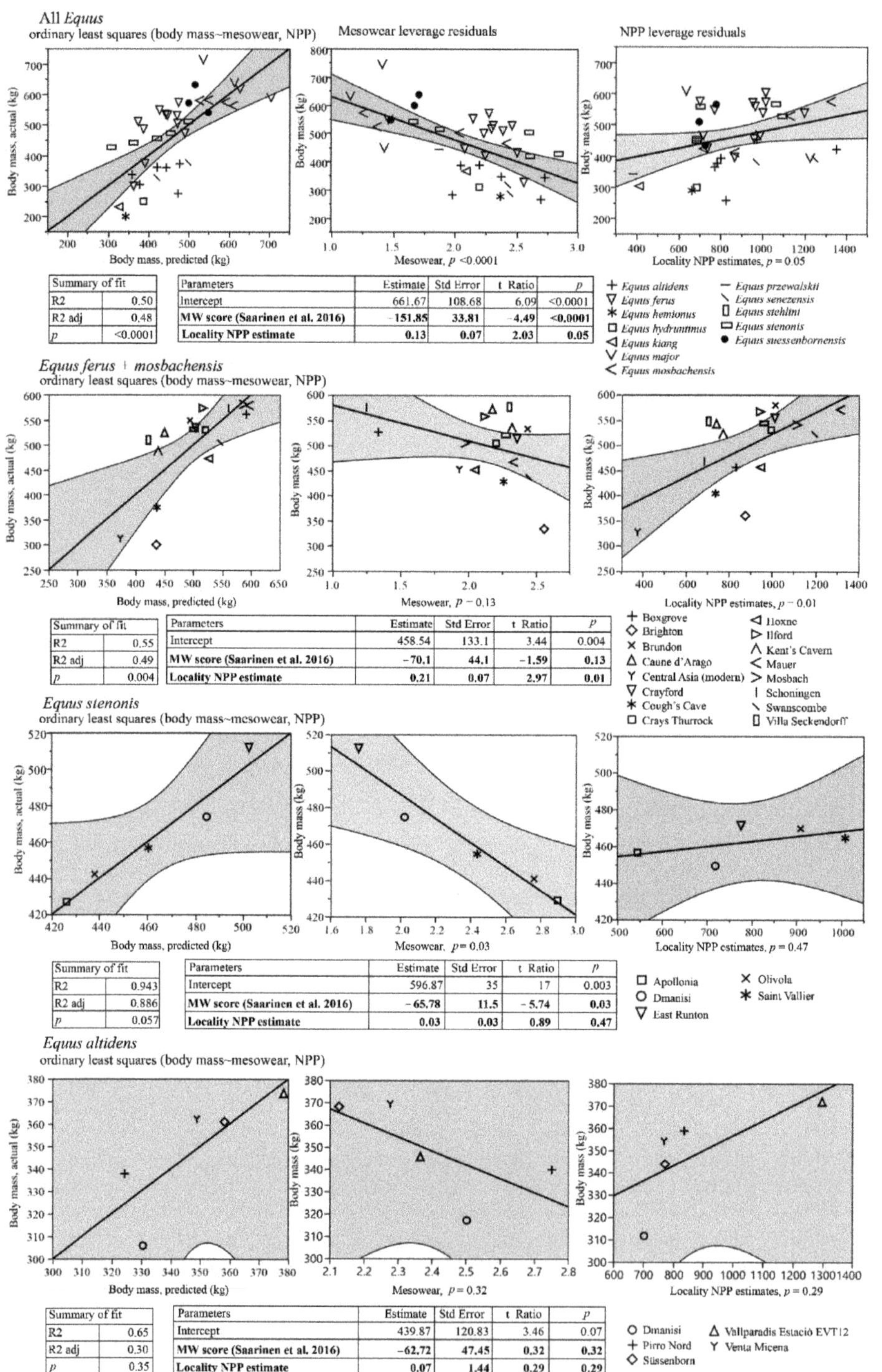

Figure 1. Ordinary least squares models of the effect of the mesowear score and the estimated NPP of the localities on the body mass estimates of the genus *Equus* from the Pleistocene of Eurasia.

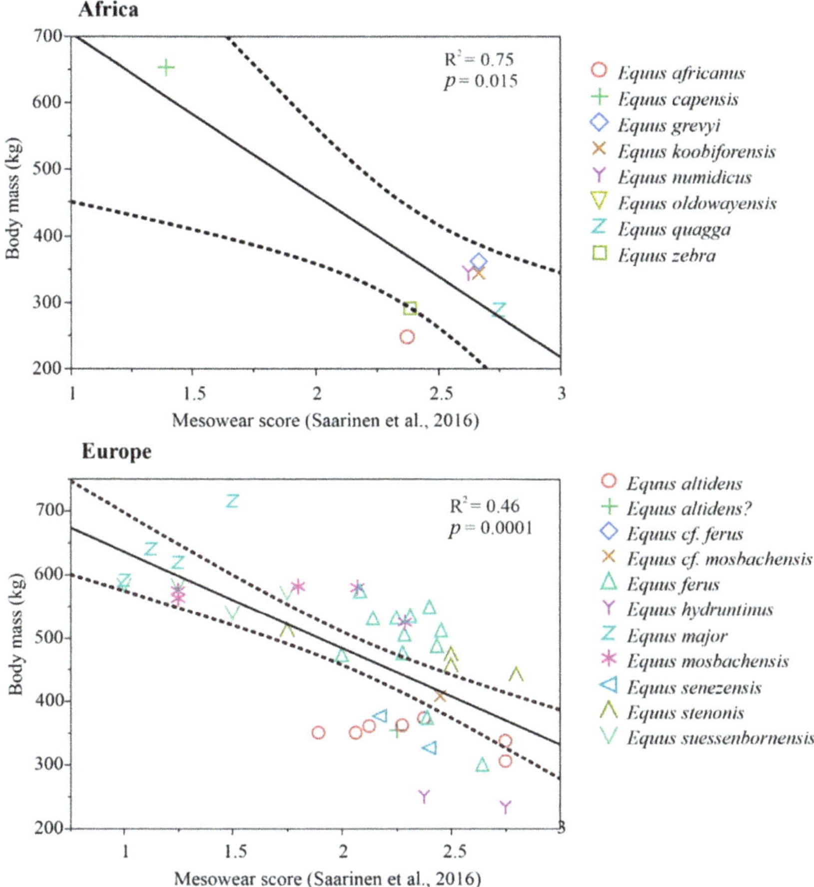

Figure 2. Linear regressions between body mass and the mesowear of *Equus* in Africa and Europe during the Quaternary. On both continents, the body mass of *Equus* was significantly negatively related to the mesowear score, although in Africa, this pattern was entirely driven by one very large-sized species, *E. capensis*. Adapted from [20].

Tracking the abrasion incurred on molars in deep time (hypsodonty index), the level of abrasion incurred by individuals cumulatively in their lifetimes (mesowear) and the short-term acquisition of microwear scars from exogenous grit and/or food items has shed light on dietary and environmental shifts through time for North American Equini. By using three dietary proxies with different temporal resolution capabilities, the amounts of different levels of dietary abrasion as well as the possible causes of this abrasion was elucidated. Highly hypsodont members of the subfamily Equinae appear in North America about 17.5 Ma (late early Miocene-Hemingfordian) [341]. It is at this time that high degrees of large pitting are found in their dental enamel, and they begin to show scratch textures regardless of dietary classification indicating heavy exposure to exogenous grit. Also, highly hypsodont equines first appeared in the late Middle Miocene (ca. 14 million years ago), well after the projected availability of pervasive open grasslands (earliest Miocene) [342,343] or even the latest Oligocene [344]. Thus, the appearance of hypsodonty in Miocene equines was not synchronous with the appearance of grasslands in North America, and exogenous grit appears to have been a contributing factor to increased exposure to abrasion and subsequent increases in crown height in fossil horses rather than grazing alone [21,345,346].

Mesowear patterns closely mirror hypsodonty trends (i.e., higher mesowear scores when hypsodonty increases.) The species of *Equus* shown in Table S3 have abrasive mesowear consistent with grazing. Thus, grass was an important dietary item as it is today. Also, as these taxa, while grazing, were feeding low to the ground, they would have encountered grit encroaching on their food items which would have contributed further to their abrasive mesowear. Even so, microwear has shown that roughly 80% of these taxa (Table S3) seasonally or regionally engaged in mixed feeding, or in one case, browsing. This demonstrates that hypsodonty, although advantageous for consuming grass, does not preclude consuming leaves or other browse at times. The other and earlier Equini shown in Table S3 (i.e., *Cormohipparion*, *Pseudhipparion*, *Calippus*, *Hipparion*, *Dinohippus*, *Neohipparion* and *Nannipus*) exhibited a less grazing type mesowear and/or microwear indicative of a tendency to either consume grass or a mixture of grass and browse. These results are consistent with the fact that even after open grasslands became pervasive in North America, forest vegetation was apparently also available until the late Miocene [21,342,347].

Analyses of body mass and mesowear of Pleistocene *Equus* from Mexico and the USA (Alaska) show that the body sizes of most of the species were relatively small compared to the larger species from the Pleistocene of Europe, and all of these species had grass-dominated diets, although some variation occurred at a smaller scale than in the Pleistocene of Europe (Table S2). The relatively small body size and lesser variation in size and diet in North American *Equus* paleopopulations compared to Europe could reflect less productive paleoenvironments in general, with the estimated NPP values for nearly all of the North American and Mexican localities being comparatively low (between ca. 299 and 730 $g(C)/m^2/a$) (Table S2). The highest estimated NPP (of the sites analyzed here) is at Rancho la Brea (ca. 839 $g(C)/m^2/a$), which is also the only one of those localities that has a very large-sized species, *E. occidentalis*. Pérez-Crespo et al. [348] noted that the sympatric *E. conversidens* and *Equus* (or *Haringtonhippus*) *francisci* from Valsequillo, Mexico, had differences in diet and body size that corresponded with the "Eurasian and African *Equus* model", with the larger species *E. conversidens* having more browse-based and the smaller *Equus* (or *Haringtonhippus*) *francisci* having more grazing dietary signal. The observation that equines in general tend to have heavily grazing diets in North America since the late Miocene makes sense (and more so than in the Eurasia and Africa, where more browsing and mixed-feeding forms occurred, even among Pleistocene *Equus*), as it explains the evolution of the prominent grazing adaptations of this group in North America. Interestingly, it seems that body sizes of North American equines were on average smaller than European equines, mostly lacking the very largest body size category (above ca. 550 kg in *Equus*), with *E. occidentalis* being the only exception. Species of this largest size category such as *E. major*, *E. suessenbornensis* and *E. mosbachensis* in Eurasia and *E. capensis* in Africa, were the ones with most browse-dominated diets during the Pleistocene. It is possible that the evolution of these giant species of *Equus* is associated with changes in the dietary niche and population densities in the more wooded, comparatively high-productivity paleoenvironments of Pleistocene Europe in particular, while the more open-adapted, grazing horses in North America and much of Africa and Central Asia attained more modest sizes due either to less productive environments or grazing, gregarious ecological strategies that limited individual body size.

6.2. North America

The major evolution and diversification of equids occurred in North America even though a number of successive dispersals took place to Eurasia. Equidae apparently evolved in isolation from Eurasia in North America from the middle Eocene to the late Oligocene [44]. During the Tertiary, equids were very widespread in North America. In fact, at most fossil localities, they are the most common medium- to large-sized mammals recovered [44]. Equids achieved their maximum diversity in the late Miocene [44,349] resulting in the evolution of the subfamily Equinae [350].

The Miocene was also a time of craniodental reorganization of the equid skull. [351,352]. These trends began in parahippine and merychippine equids and eventually led to the genus *Equus*, which is thought to have evolved from *Dinohippus* in the Pliocene [346,353]. The dramatic changes in equid skulls, teeth and limbs have long been thought to reflect evolutionary adaptations to a changing environment, often thought of as evidence that grasslands expanded during this time [354–362]. Gross changes in dental morphology reveal a true shift toward more abrasive diets including grass in the derived Equinae from the middle Miocene onward reflecting their adaptation to grazing in more open environments [346,363]. While grazing was clearly an important long-term dietary strategy for derived Equinae once they appear in North America, some also engaged periodically in short-term mixed feeding regionally or seasonally.

Interestingly, it appears that there was less variation in the dietary ecology of North American derived equines than the species that encountered a wide range of environments from wooded to open in Eurasia. Middle Miocene equines such as *Co. quinni* and *Co. goorisi* were small sized and had mixed to grazing diets (Table S2). The species of *Equus* from the Pleistocene of Mexico and North America mostly have relatively small body sizes and mostly grazing mesowear (Table S2). The relatively small body sizes could be related to the grazing, gregarious lifestyle or the relatively low-productivity, open paleoenvironments of the North American paleopopulations summarized in this study. In Mexico and Southern USA, the small species *E. cedralensis* and *E. conversidens* had grass-dominated mesowear signals and occupied low-productivity, sometimes nearly desert-like environments as suggested by low estimated NPP values of their localities (Table S2). *Equus mexicanus* and *E. scotti* were large (mean body mass between 450 and 480 kg), but not as large as the "very large", predominantly mixed-feeding or browse-dominated "woodland" species in the Eurasia, such as *E. major* and *E. suessenbornensis*, both with average body masses around 550–600 kg (Table S2). The only species in the "very large" size category from the North American sites is *E. occidentalis*. This species occupied a relatively high-productivity paleoenvironment in Rancho la Brea (Table S2), where its mesowear signal indicates predominantly grazing diet [363]. However, the proportion of sharp cusps is also relatively high in the Rancho la Brea population of *E. occidentalis*, and microwear texture analyses indicate that it consumed a significant proportion of woody browse during the pre-LGM cool stages at Rancho la Brea, so at least periodically significant inclusion of browse is indicated for the diet of also this very large equine [364]. In Alaska, the small to medium-sized *E. lambei* occupied a cold mammoth steppe paleoenvironment with relatively modest estimated NPP, and its mesowear signal indicates a grazing or at least heavily grass-dominated diet (Table S2).

6.3. Eurasia

The earliest hipparionines that dispersed from North America to Eurasia at the beginning of the late Miocene were medium-sized (around 160 kg in body mass) species of the genus *Cormohipparion*, such as *Co. sinapensis* from Sinap, Turkey. These were relatively slender and modest-sized and probably occupied relatively open environments from East Asia to Turkey [8]. Early on, however, the larger, more robust hipparionines of the genus *Hippotherium* emerged and were widespread across Eurasia. *Hippotherium primigenium* was a relatively large species (body mass 200–250 kg) with predominantly browsing diets that occupied primarily forest and woodland environments in Central Europe during the early late Miocene [8]. Considerable dietary variation occurred in *Hi. primigenium*, being purely browsing in the forested paleoenvironment of Höwenegg and more grass-dominated in the locally more open floodplain environment in Eppelsheim, in Germany [365,366]. Two species of *Hippotherium*, *H. primigenium* and *H. kammerschmittae*, from the later late Miocene had browse-dominated diets in Dorn Dürkheim, Germany [366]. During the later late Miocene (Turolian), hipparionines diversified in Eurasia, and included several species and lineages of different body size and dietary ecology. In the Mediterranean realm and Western Asia for example, medium-sized (ca. 100–200 kg) species of *Hipparion* and *Cremohipparion* species were mixed-feeders, whereas larger species of the genus *Hippotherium* (with body masses more than 200 kg) retained browse-dominated or mixed diets, while the small species of *Cremohipparion* (body mass less than 100 kg) had grazing diets (Tables S2 and S3) [8]. Hipparionines thus seem to by and large reflect the model of larger sizes being related to more browse-dominated diets and smaller sizes to grazing diets, as in the example of *Equus* during the Pleistocene in Eurasia. Similarly high diversity of hipparionines occurred in East Asia during the latest Miocene.

During the Pliocene, the diversity of hipparionines in Eurasia dropped drastically and there was a turnover in the species composition, with a few, in general large-sized (200–350 kg) species in the genera *Plesiohipparion*, *Baryhipparion* and *Proboscidipparion* surviving [8]. All these genera had mostly mixed-feeding diets, with *Plesiohipparion* and *Proboscidipparion* having wide geographic ranges from East Asia to Europe [8]. Considerable ecological flexibility seems typical, especially for *Proboscidipparion*. While *Pr. sinense* occupied relatively open environments in East Asia and had a mixed but relatively abrasion-rich mesowear signal [8], *Proboscidipparion* sp. from Red Crag, England (latest Pliocene, ca. 2.7 Ma) had a browse-dominated diet [8,367] and lived in a warm-temperate forest environment [368] (Tables S2 and S3).

The earliest species of *Equus* to disperse from North America to Eurasia were relatively large sized but ecologically quite generalized, grazing, open-adapted species such as *E. eisenmannae* in East Asia and *E. livenzovensis in* Western Asia and Europe. These taxa had average body masses around 500 kg and at least *E. eisenmannae* had mesowear values indicating typical grazing diet for the genus (Tables S2 and S3). At the beginning of the Pleistocene, the first of the specialized, very large-sized and robust woodland horses with mixed and even browse-dominated diets, *E. major*, emerged in Western Europe. This species typically occurs in Early Pleistocene sites in Europe where palaeoenvironmental proxies such as pollen records and large mammal ecometrics indicate relatively wooded and productive paleoenvironments such as in Red Crag (UK) and Tegelen (Netherlands) [19,369]. *Equus major* was one of the largest species of equid, with mean body mass around 600 kg, and maximum body mass of ca. 800 kg.

The common Early Pleistocene species of European *Equus*, *E. stenonis*, occurred in a wide range of localities suggesting broad tolerance of environmental conditions. It was a medium-large species of *Equus*, with mean body mass between 400 and 500 kg (Table S2). Available paleodietary evidence indicates mostly grazing diets for *E. stenonis* [370,371] (Tables S2 and S3), but the small sample from East Runton, UK, had a more mixed dietary signal (Table S2). Analysis of body mass, mesowear and the NPP of *E. stenonis* paleopopulations indicates a strong inverse relationship of the amount of grass in diet and body size (Figure 1). There was also a geographic pattern of body size in *E. stenonis*, with the Western European populations associated with more high-productivity environments having on average larger body sizes than Eastern European populations, which occurred in less productive environments [13]. *Equus senezensis* was a smaller species with mean body mass around 350 kg and a grazing diet and occupied mostly open landscape [372,373] (Tables S2 and S3).

Later during the Early Pleistocene, the large-sized *E. suessenbornensis* and the small-sized *E. altidens* became prominent species in Eurasia, being the dominant species there during the late Early and early Middle Pleistocene. *Equus suessenbornensis* was a very large-sized and robust species, comparable in size to the earlier Pleistocene Western European *Equus major* (with mean body mass within paleopopulations ranging from over 500 kg to slightly over 600 kg). Similar to other very large species of *Equus*, *E. suessenbornensis* typically had mixed to even browsing diets [19,374] (Tables S2 and S3), and although being widespread in Europe, it was typically less abundant than the small *E. altidens*, also where these two co-occurred. *Equus altidens* was the earliest identified hemionine and it shares interesting similarities in mesowear signal to extant hemionines. As in modern hemionines, most paleopopulations of *E. altidens* show heavily grazing mesowear signals [375] (Table S3), but also relatively abundant association of low occlusal relief and sharp cusps in some localities [375]. This kind of mesowear signal suggests diet based mostly on grasses but also including a significant component of dry, open environment browse such as aridity-resistant shrubs [15]. Some populations also display microwear patterns compatible with a mixed diet suggesting a certain degree of dietary plasticity for this species [369,375] (Table S3). Mean body mass estimates of *E. altidens* vary around 350 kg (Table S2). In general, *E. altidens* tends to be associated with paleoenvironments where dental ecometrics of large herbivorous mammal communities indicate relatively modest primary production estimates (between ca. 700 and 900). In Guadix-Baza Basin, Andalucia, Spain, the paleoenvironments of *E. altidens* have been suggested to have been similar to present Mediterranean woodlands and forest-steppes [374,376,377]. The diet of *E. altidens* reflects differences in paleoenvironments, being purely grazing in Venta Micena and Vallparadìs (EVT12 layer; MIS 31) but more mixed in Barranco León and Fuente Nueva 3 in Guadix-Baza Basin, Andalucia, where the paleoenvironment was more Mediterranean forest or woodland type and in layer EVT7 (MIS 21) of Vallparadìs where environmental conditions became more humid, and seasonality might have increased following the "0.9 Ma event" [374,375]. In Süssenborn, Germany, this species occurred in a paleoenvironment which has been interpreted periodically cool and relatively open, but not periglacial, based on the faunal association [378].

The Middle Pleistocene marks the arrival of caballine horses in Eurasia, a significant turnover event. *Equus mosbachensis* (=*E. ferus mosbachensis*/*E. ferus*), the typical caballine during the early Middle Pleistocene in Europe, was a very large and robust form (mean body mass from over 500 kg to nearly 600 kg), and it displayed more diverse dietary adaptations including grazing, mixed or even browse-dominated diets [19,373,379] (Tables S2 and S3). Large, browse-dominated forms of this taxon are associated with relatively wooded paleoenvironments such as Boxgrove in the UK and Schöningen in Germany [19,380,381]. Even grazing populations can be found in habitats dominated by wooded landscapes (e.g., open woodlands), feeding also in closed environments such as in Fontana Ranuccio (0.4 Ma) [374,382] (Table S3). The cold-stage paleopopulation of *E. mosbachensis* from Caune d'Arago, France, had somewhat smaller average body size and more grazing diet (Tables S2 and S3). The wild horse (*E. ferus*) was abundant and widespread in Eurasia during the late Middle and Late Pleistocene, with small-sized forms having more grazing dietary signals and occurring in sites with smaller estimated NPP than larger forms of the species (Figure 1, Table S2). The smallest forms of *E. ferus* with most grazing dietary signals come from sites where associated paleobotanical evidence indicates very open and grass-dominated "mammoth steppe" environments, such as Brighton (MIS 6 glacial) and Gough's Cave (MIS2 glacial) in UK (Table S2) [383,384]. Conversely, large forms of *E. ferus* typically occurred in more wooded paleoenvironments, such as Grays Thurrock (MIS 9 interglacial) and Brundon and Ilford (MIS 7 interglacial) in the UK, and Taubach (last interglacial, MIS 5) in Germany, and had less purely grazing diets (Table S2) [385,386]. Further east, a medium-large sized form of *E. ferus* ("*E. ferus latipes*") had a grass-dominated diet at the locality of Kostenki 14 in western Russia (Table S2), where it occupied a cool and arid steppe environment [387]. The northernmost populations from Taimyr and Yakutia, Northern Siberia, "*E. lenensis*", mostly lived in cold, low-productivity steppe-tundra environments, although the northern edge of boreal forest advanced in these areas during warmer stages [388]. They are characterized by small average body size and grass-dominated mesowear signal, although some individuals show sharper and more high-relief cusps indicating inclusion of browse or non-grass herbaceous vegetation in their diet (Table S2). *Equus hydruntinus* had a more limited range in Eurasia during the late Pleistocene, and similarly to other hemionines, it seems to have been associated with relatively open habitats and it consistently had grass-dominated diets (Tables S2 and S3) [389].

The extant equids in Eurasia are currently limited to the Central Asian hemionines (*E. hemionus* and *E. kiang*) and the Przewalski horse (*E. przewalski*). These are all relatively small-sized members of the genus, they all have grazing diets, and they occupy the steppe environments of Central Asia (Tables S2 and S3). Similar to *E. altidens*, the hemionines today have a comparatively high proportion of low and sharp mesowear among *Equus*, indicating some inclusion of "dry browse" or non-grass herbs in diet, in addition to grass (Table S2).

6.4. Africa

The earliest hipparionine with palaeodietary evidence from Africa is *Cormohipparion* sp. from the late Miocene of Chorora, Ethiopia (ca. 8.5 Ma), which has a browse-dominated mesowear signal and medium body size (ca. 160 kg) (Tables S2 and S3) [8,274,390–394]. Since this earliest record, most of the equines in Africa show mesowear and other paleodietary evidence suggesting grass-dominated to grazing diets. The hipparionines of the genus *Eurygnathohippus* were relatively large in size (over 200 kg in mean body mass) and yet they had remarkably grass-dominated diets (Tables S1 and S2), unlike the large-sized hipparionines in Eurasia, which tend to have more mixed or browse-dominated diets. This could reflect adaptation of the African derived hipparionines of the genus *Eurygnathohippus* to graze in relatively productive, but grass-dominated savanna environments.

After the arrival of *Equus* in Africa in the Pleistocene, most of the African equine species had grazing diets and were of small body size compared to a much wider range of sizes and diets in Eurasia (Figure 2), probably reflecting similarity in their adaptations to grazing in grass-dominated African savanna environments. The only clear exception to this pattern is the very large-sized South African species *E. capensis*, which had a more mixed or even browse-dominated dietary signal, paralleling the relationship between diet and body size observed for the Pleistocene of Europe (Figure 2; Table S1).

The extant African zebras (*E. quagga*, *E. grevyi* and *E. zebra*) all have relatively small body size compared with the large Pleistocene species of *Equus* (particularly in Eurasia) and they typically have some of the most purely grazing diets among the equids (Table S2). The Grevy's zebra (*E. grevyi*) is the largest of these species, and the largest extant species of wild *Equus*, but it has a relatively tall and slender morphology, with elongate metapodials compared to the Quagga, and mean body mass estimates are relatively modest (around 360 kg on average) compared to many of the Pleistocene taxa, resembling however those of *E. koobiforensis* from the Pleistocene of East Africa. Our mesowear data suggest that the proportion of sharp but low-relief cusps is higher in *E. grevyi* than the rest of extant zebras, indicating perhaps a somewhat higher proportion of dry browse such as dry-adapted shrubs in its diet (Table S2). The Africa wild ass (*E. africanus*) also has a relatively high proportion of low and sharp mesowear, despite mostly grazing dietary signal, which could also reflect inclusion of dry browse in the arid environments of this species (Table S2) [395].

7. Climate and Evolution

Figure 3 provides the distribution of Equinae in North America, 7–4 Ma. As with the succeeding climate and evolution maps, the numbers on these maps are tied to Table S4. During this time frame, 10 genera are recognized from North America (*Calippus*, *Dinohippus*, *Cormohipparion*, *Nannippus*, *Neohipparion*, *Astrohippus*, *Boreohippidion*, *Pliohippus*, *Pseudohipparion* and *Equus*), 3 from Eurasia (*Plesiohipparion*, *Proboscidipparion* and *Cremohipparion*) and 3 from Africa (*Cremohipparion*, *Eurygnathohippus* and *Sivalhippus*). The Equinae in North America include Ca. *elaschistus*, Ca. *hondurensis*, D. *interpolatus*, D. *leardi*, D. *leydianus*, D. *spectans*, Co. *occidentale*, Na. *aztecus*, Na. *lenticularis*, Na. *peninsulatus*, Ne. *leptode*, A. *ansae*, A. *lenticularis*, Bo. *galushai*, D. *mexicanus*, Co. *emsliei*, Plio. *coalingensis*, Ne. *eurystyle*, Ne. *gidleyi*, Ps. *simpsoni*, Na. *beckensis*, *Equus/Plesippus simplicidens*, E. *cumminsi*, E. *enormis* and *Equus/Plesippus idahoensis*. The Eurasian record includes Pl. *longipes*, Pl. *houfenense*, Pr. *pater*, Pr. *crassum*, Cr. *fissurae* and Pl. *huangheense*, wheres Africa has Cr. *periafricanum*, Eu. *feibeli*, S. *turkanensis*, Eu. *hooijeri* and Eu. *woldegabrieli* (taxa 1–36, Table S4). *Calippus*, *Dinohippus*, *Cormohipparion*, *Nannippus*, *Neohipparion* and *Astrohippus* have records extending back to 10 Ma and amongst these *Cormohipparion*, *Nannippus*, and *Neohipparion* are hipparionine horses. *Dinohippus* has a chronology beginning at 10.3 Ma and D. *mexicanus* (5.3–4.6 Ma) is demonstrably related to *Equus*. Four species of *Equus*, E. *cumminsi*, E. *enormis*, E. *idahoensis* and E. *simplicidens* first occur in the Blancan (since 4.9 Ma) and in particular, *E. simplicidens* would appear to be related to first occurring Eurasian *Equus* [10,190]. Large mammal Mean HYP in North America ranges from 2.0–2.5, whereas Europe has the lowest mean HYP ranging from 1.5–2, with higher values (2.0–2.5) in Turkey, Greece and Spain. Africa has mostly 2.0 values with slightly lower values in the horn of Africa, whereas Asia shows mostly 2.0 values with localized areas ranging between 2.0 and slightly above 2.5 in China, Mongolia, Kazakhstan and Iran.

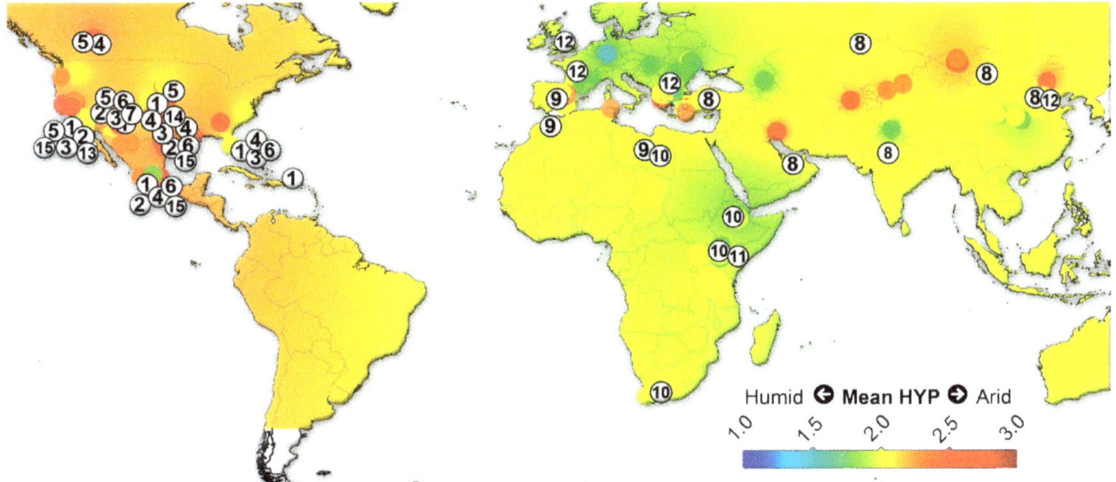

Figure 3. Spatial distribution of the large herbivorous genera mean ordinated crown height through time ranges 7 to 4 Ma in North, Central and South America, Eurasia and Africa. The mean ordinated hypsodonty map represents the paleoclimatological conditions grading from most humid (blue) to most arid (red). Numbers in white circles show coded number of each taxa given in Table S4. The mean ordinated hypsodonty values are represented by the color-coded circles indicate the spatial position of the localities that mean hypsodonty scores calculated (Table S5). IDW interpolation algorithm hypothetically interpolates no data (no locality) area based on the actual data. These areas should be ignored.

Figure 4 presents the distribution of taxa between 4 and 2.6 Ma. During this time frame, 4 genera are recognized from North America (*Pliohippus, Nannippus, Equus* and *Plesiohipparion*), 6 from Eurasia (*Plesiohipparion, Proboscidipparion, Cremohipparion,* "*Hippotherium*", *Eurygnathohippus, Baryhipparion*) and 2 from Africa (*Eurygnathohippus, Cremohipparion*). North American taxa carrying over into this interval include *Plio. coalingensis, Na. beckensis, Equus/Plesippus simplicidens, E. cumminsi* and *Equus/Plesippus idahoensis*. The persisting Eurasia taxa include *Pl. longipes, Pl. houfenense, Pr. pater, Pr. crassum, Cr. fissurae* and *Pl. huangheense*, whereas most African species disappeared, with the only survival of the hipparion genera *Eurygnathohippus* and *Cremohipparion*. First occurring taxa include *Plesiohipparion* sp. in Ellesmere Island (N. America), *Ba. insperatum, Cr. licenti,* "*Hippotherium*" *antelopinum, Eurygnathohippus* sp, *Pr. heintzi, Pl. rocinantis, E. afarensis, Eu. hasumense* and *Eu. cornelianus* (Africa). The African clade *Eurygnathohippus* is found there during this interval and is represented by a lower cheek tooth from India at the end of this temporal interval. "*Hippotherium*" *antelopinum* is a medium sized hipparionine whose type locality is in India. North America records the immigration of *Plesiohipparion* into Greenland. North America and South America have mostly Mean HYP between 2.0 and 2.5, with some areas in the west recording values of around 2.5 and others between 1.5–2.0.

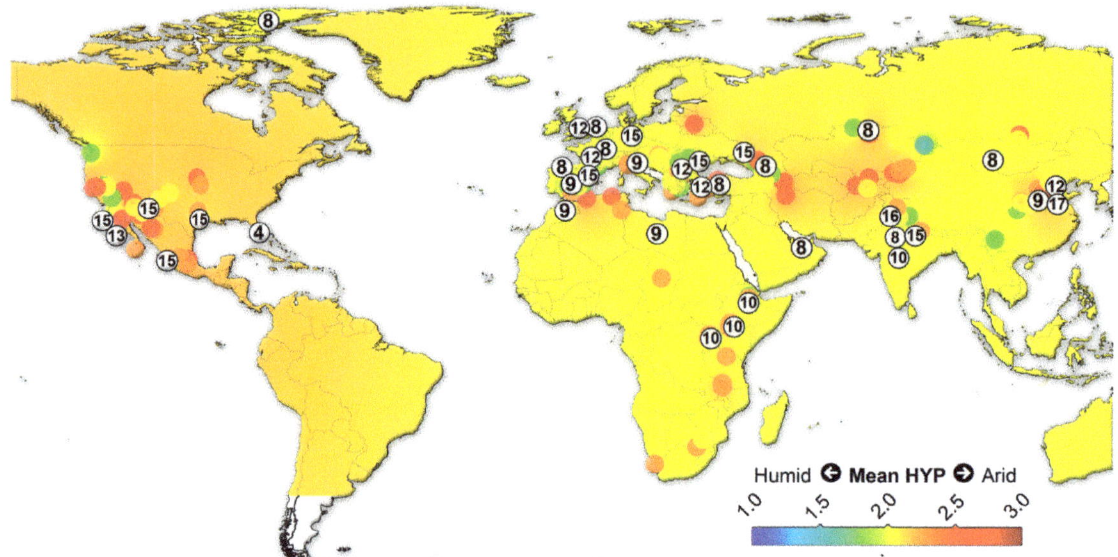

Figure 4. Spatial distribution of the large herbivorous genera mean ordinated crown height through time ranges 4 to 2.6 Ma in North, Central and South America, Eurasia and Africa. The mean ordinated hypsodonty map represents the paleoclimatological conditions grading from most humid (blue) to most arid (red). Numbers in white circles show coded number of each taxa given in Table S4. The mean ordinated hypsodonty values are represented by the color-coded circles indicate the spatial position of the localities that mean hypsodonty scores calculated (Table S5). IDW interpolation algorithm hypothetically interpolate no data (no locality) area based on the actual data. These areas should be ignored.

Figure 5 presents the distribution of taxa between 2.58 and 1.5 Ma, including the *Equus* Datum in Eurasia at the beginning of the Pleistocene. During this time frame, 2 genera are recognized from North America (*Nannippus* and *Equus*), 4 from Eurasia (*Plesiohipparion, Proboscidipparion, Baryhipparion* and *Equus*) and 2 from Africa (*Eurygnathohippus* and *Equus*). North America records *E. calobatus, E. scotti, E. stenonis anguinus, E. conversidens, E. ferus/lambei, E. francisci, E. fraternus* and *E. pseudaltidens* during this interval, in addiction to *Na. beckensis, Equus/Plesippus simplicidens, E. cumminsi* and *Equus/Plesippus idahoensis*. Europe records several species of *Equus* in this interval, including: *E. livenzovensis, E. major, E. stenonis, E. senezensis, E. stehlini* and *E. altidens*. The Indian subcontinent has *E. sivalensis* and *Equus* sp. in India. China records several *Equus* species during this interval including *E. eisenmannae, E. sanmeniensis E. huanghoensis, E. yunnanensis, E. qingyangensis, E. teihardi* and *E. wangi*, with *Ba. insperatum, Pl. shanxiense* and *Pr. sinense*. Central Asia includes *E. pamirensis*, whereas Africa includes the *Equus* species in North Africa (*E. numidicus*) and *Equus* sp. in East Africa. Overall, the African record includes *E. tabeti* in North Africa with *Eu. pomeli, E. koobiforensis, E. oldowayensis* in East Africa and *E. capensis* and *E. zebra* in South Africa. The earliest species of *Equus* are found in some localities at ca. 2.6 Ma in Europe, Siwalik Hills and China, which shows values between 2.0–2.5. These values are more diffused in Eurasia compared with Figure 4, although are still present values between 1.5–2.0 in China, Russia, Caucasus and Europe. Africa overall has values between 2.0 and 2.5 through most of the continent, with isolated areas between 2.5–3.0 in North and East Africa. During this interval, most continental mammal records are dry with mean hypsodonty mostly being 2.0 or higher with several higher incidences of 2.5 or higher.

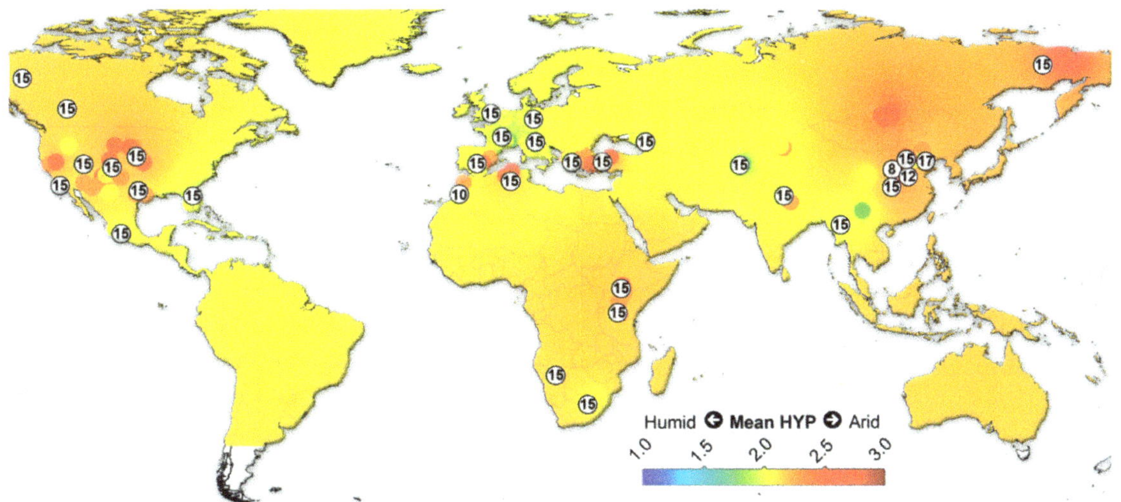

Figure 5. Spatial distribution of the large herbivorous genera mean ordinated crown height through time ranges 2.58 to 1.5 Ma in North, Central and South America, Eurasia and Africa. The mean ordinated hypsodonty map represents the paleoclimatological conditions grading from most humid (blue) to most arid (red). Numbers in white circles show coded number of each taxa given in Table S4. The mean ordinated hypsodonty values are represented by the color-coded circles indicate the spatial position of the localities that mean hypsodonty scores calculated (Table S5). IDW interpolation algorithm hypothetically interpolate no data (no locality) area based on the actual data. These areas should be ignored.

Figure 6 presents the distribution of taxa between 1.5 Ma to recent. During this time frame, 2 genera are potentially recognized from North America (*Equus* and *Haringtonhippus*), 2 from South America (*Equus* and *Hippidion*), 2 from Eurasia (*Equus* and *Proboscidipparion*) 2 from Africa (*Equus* and *Eurygnathohippus*). This time frame includes a number of taxa that carry over from the 2.58–1.5 Ma interval including *Equus/Plesippus simplicidens*, *E. cumminsi*, *Equus/Plesippus idahoensis*, *E. calobatus*, *E. scotti*, *E. conversidens*, *E. ferus/lambei*, *E. francisci*, *E. fraternus* and *E. pseudaltidens* (North America); *E. sameniensis*, *E. yunnanensis*, *E. qingyangensis*, *E. teilhardi*, *E. wangi* (China); *E. sivalensis* (India); *E. numidicus*, *E. tabeti* (North Africa); *E. koobiforensis*, *E. oldowayensis* (East Africa); *E. capensis*, *E. zebra* (South Africa). Taxa first occurring between 1.5 Ma to recent are *E. verae*, *E. cedralensis*, *E. mexicanus* and *E. occidentalis* for North America; *E. neogeus*, *Hippidion devillei*, *Hippidion saldiasi*, *Hippidion principale* for South America; *E. beijingensis*, *E. dalianensis*, *E. hemionus* (also India), *E. kiang* (also Nepal) and *E. przewalskii* in China; *E. nalaikensis* and *E. colimensis*, *E. lenensis* and *E. ovodovi* in Central and North Asia; *E. suessenbornensis*, *E. apolloniensis*, *E. wuesti*, *E. hipparionoides*, *E. marxi*, *E. ferus*, *E. hydruntinus*, *E. petraolnensis* and *E. graziosi* in Europe; *E. mauritanicus*, *E. melkiensis* and *E. algericus* in North Africa and *E. africanus*, *E. grevyi* and *E. quagga* in East and South Africa. This time frame record also records the last occurrence of the hipparionini horses in Asia with *Proboscidipparion* and Africa with *Eurygnathohippus* [8]. Mean HYP shows North and South America having more moderate climates around values of 2.0 and 2.5, with values lower than 2.0 in local areas in the East and the West. Europe likewise has Mean HYP values around 2.0 with values lower than 2.0 in Central and Eastern Europe, whereas Asia, the Indian Subcontinent and Africa have higher values hovering around Mean HYP of 2.5. Mean HYP values between 1.5–2.0 are also found in China.

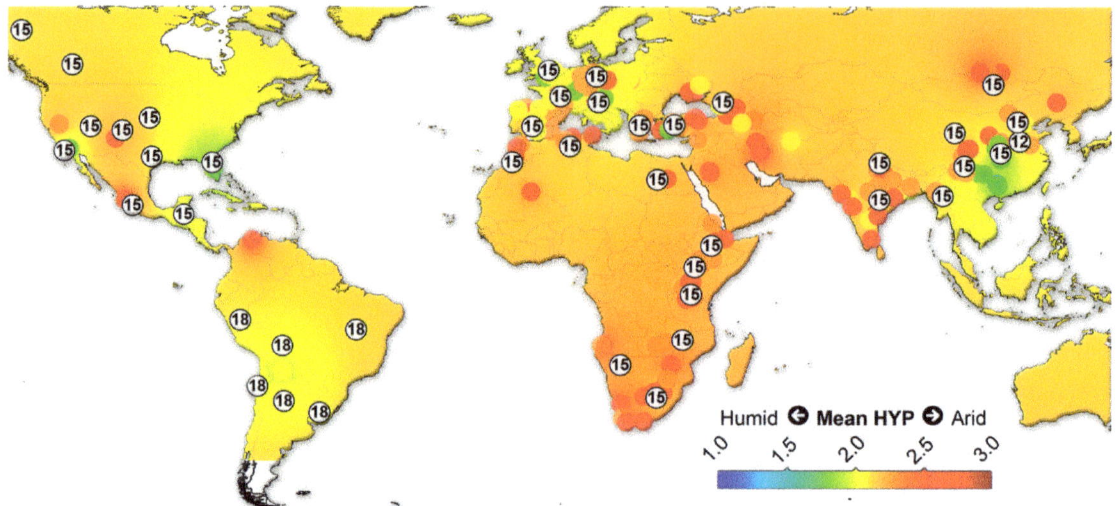

Figure 6. Spatial distribution of the large herbivorous genera mean ordinated crown height through time ranges 1.5 Ma to recent in North, Central and South America, Eurasia and Africa. The mean ordinated hypsodonty map represents the paleoclimatological conditions grading from most humid (blue) to most arid (red). Numbers in white circles show the coded numbers of each taxon given in Table S4. The mean ordinated hypsodonty values are represented by the color-coded circles indicate the spatial position of the localities that mean hypsodonty scores calculated (Table S5). IDW interpolation algorithm hypothetically interpolate no data (no locality) area based on the actual data. These areas should be ignored.

The mean hypsodonty map patterns indicate that while most of Eurasia and Africa occupied by mild or humid values, arid environmental conditions were prominent in North America between 7 and 4 Ma (Figure 3). At the late Miocene–Early Pliocene transition, moist conditions occurred in Europe with an arid belt extending eastward into Asia and Africa. By the end of the Pliocene arid conditions remained in North America and aridity began to increase on the mid-latitudes of Eurasia and along the Rift Valley of East Africa and the western corners of North Africa (Figure 4). Mean ordinated crown height patterns of the large herbivorous mammal communities indicate that while Southeast Asia, Central and Western Europe, and Florida and California in North America occupied by semi-humid or humid values, arid environmental conditions persisted or increased drastically rest of the World and in particular in East and South Africa, 1.5 Ma to recent (Figure 6).

Overall, these maps exhibit a general trend of increased drying over time. Most occurrences of *Equus* are with Mean HYP values of 2.0–2.5. Very few archaic Equinae taxa continue across the Pliocene–Quaternary boundary. North America retained the more primitive lineages of *Pliohippus* and *Nannippus* up into the 2.58–1.5 Ma interval. Hipparionines persisted up into the Pleistocene of Europe, as late as 1.0 Ma in China and slightly later than 1.0 in Africa. The extinction of older North American and Eurasian-African lineages would appear to be associated with the expansion of more open country dry conditions.

8. Phylogeny

As reported in the introduction, recently two morphological-based cladistic analyses have re-evaluated the origin of the genus *Equus*. Herein, we present the state of art of these cladistic hypotheses, with a separate section on the contribution from the molecular phylogenies.

Barrón-Ortiz et al. [9] undertook a phylogenetic analysis using a matrix of 32 characters (22 cranial, 6 mandibular, 3 autopodial, and estimated body size). The authors included in the matrix 21 Equini species, two of which were considered as outgroups, *Acritohippus stylodontus*, and *Pliohippus pernix*. Barrón-Ortiz et al. [9] undertook the analysis using TNT 1.1 [396] with the implicit enumeration option (exhaustive search), using equal weighting for the characters, and without a collapsing rule. They treated all characters as unordered.

Cirilli et al. [10] have combined a new matrix including 30 Operational Taxonomic Units (OTUs) and 129 characters (72 cranial, 40 mandibular and 17 on autopodia), 68 of which were new and the other extrapolated from the recent published matrices on perissodactyl phylogeny [9,114,397–400]. The characters were mainly coded by direct observations. The ingroup included a comprehensive sample of 26 equid species and the outgroup was represented by the Brazilian tapir *Tapirus terrestris*, the rhinocerotoid *Hyrachyus eximius*, the Rhinocerotidae *Trigonias osborni* and the early-diverging equid *Merychippus insignis*. The analysis was performed in PAUP 4.0b10, with Heuristic search, TBR and 1000 replications with additional random sequence, and gaps treated as missing. In this analysis, 24 characters have been ordered and 105 characters unordered. All characters were equally weighted.

8.1. What Is Equus? Paradigms, Phylogeny, and Taxonomy

The primary objectives of the study conducted by Barrón-Ortiz et al. [9] were to review and discuss different paradigms for understanding generic-level taxonomy, particularly in regards to the delimitation of mammalian genera, and to evaluate how those different paradigms impact the concept and contents of *Equus* in a given phylogenetic tree. Barrón-Ortiz et al. [9] established a new phylogenetic tree of derived Equini for that analysis. The tree served as a model for the evaluation of how distinct paradigms impact our placement of generic names on any given tree.

8.1.1. What Is a Genus?

Although several studies have discussed limitations of and provided alternatives to the Linnaean taxonomic system [401–406], Linnaean taxonomy continues to be widely used to study and communicate about past and present biodiversity [407]. This is especially true when it comes to the binomial nomen (genus and species). Because of the widespread use of binomial nomenclature within and outside of the life sciences, the genus is perhaps the most important higher-level taxonomic rank. Therefore, the question about how we define and delimit genera is not a trivial one as it affects how we view, study, and communicate about biological organisms. *Equus* is a model taxon for such discussions because of its complex generic and species-level history. At the core of the delimitation of *Equus* within any phylogenetic tree lie the philosophical and practical issues regarding the definition of genera and how best to reconcile taxonomy with evolutionary history.

Different paradigms exist for understanding and delimiting genera. In the case of mammals, Barrón-Ortiz et al. [9] identified four that are commonly used in combination with monophyly to delimit genera: phylogenetic gaps, uniqueness of adaptive zone, crown group definition, and divergence time [407–411]. One of the primary distinctions between the paradigms is the way genera are conceived. At one extreme, the uniqueness of adaptive zone paradigm conceives genera as having some level of biological reality beyond monophyly (i.e., a genus occupies a unique adaptive zone). An adaptive zone corresponds to a particular mode of life or a unique ecological situation [6,409,412,413]. At the other extreme, under the divergence time and crown group paradigms, genera are arbitrarily defined [168,410,411,414] and are not conceived as having biological reality other than monophyly. The phylogenetic gaps paradigm occupies an intermediate position. Under this paradigm, genera are not necessarily conceived as having some level of biological reality, but the gaps between monophyletic groups of species used to delimit genera arise from biological processes such as speciation, extinction, evolutionary and adaptive radiations, and unequal rates of evolution [409]. Because genera may be conceived under different paradigms, it is important for researchers to explicitly state their operational paradigm when considering questions of generic-level taxonomy.

8.1.2. Morphological Phylogenetic Analysis of Derived Equini

For the second study objective, Barrón-Ortiz et al. [9] conducted a morphological phylogenetic analysis of derived Equini. The phylogenetic analysis produced three equally most parsimonious trees of 85 steps with consistency (CI) and retention (RI) indices of 0.57 and 0.80, respectively [9] Figure 1 for the strict consensus tree.

Using the strict consensus tree obtained in their analysis, Barrón-Ortiz et al. [9] evaluated how the four paradigms commonly used to delimit mammalian genera impacted the concept and contents of *Equus*, although we emphasize that the same could be applied to any phylogenetic hypothesis. The results of this evaluation and taxonomic implications are summarized below.

8.1.3. Phylogentic Gaps and the Concept of *Equus*

Under the phylogenetic gaps paradigm, a genus is comprised of a single species or a monophyletic group of species, separated from other single species or monophyletic groups of species of the same rank by a decided gap [409]. In the context of a phylogenetic analysis, the gaps between single species or monophyletic groups of species can be measured by the number of synapomorphic traits. Application of this paradigm to the strict consensus tree of Barrón-Ortiz et al. [9], suggested that *Equus* should be delimited to clade 6, as this clade shows the most synapomorphic traits, including the species *E. neogeus*, *E. occidentalis*, *E. ferus*, *E. mexicanus*, *E. hemionus*, *E. quagga*, *E. conversidens* and *E. francisci*. The Early Pleistocene *E. stenonis* and the North American *E. simplicidens* and *E. idahoensis* are not included in this clade.

8.1.4. Uniqueness of Adaptive Zone and the Concept of *Equus*

In the uniqueness of adaptive zone paradigm, a genus is comprised of a single species or a monophyletic group of species that occupies a different adaptive zone (i.e., a unique mode of life) from the one occupied by species of another genus [409,413]. Application of this criterion in the context of a phylogenetic analysis requires: (1) the identification of traits (i.e., character states) that allow or potentially allow a single species or a monophyletic group of species to occupy a unique adaptive zone and (2) identifying where those traits occur in the tree.

The unique mode of life of *Equus* could potentially be defined as "ungulate mammals that are adapted to live in generally open, arid habitats and that can thrive on low-quality, high-fiber foods such as grasses and other coarse and tough vegetation" [9,19]. Potential morphological adaptations for this mode of life include modifications to the locomotory [415] and digestive systems, particularly the dentition [9]. Based on the position of the majority of purported, adaptive zone-related traits, *Equus* is assigned to clade 6, or possibly clade 7, in the strict consensus tree of Barrón-Ortiz et al. [9], under the uniqueness of adaptive zone paradigm. The identification to the clade 7 would include *E. stenonis* in the genus *Equus*, but not *E. simplicidens* and *E. idahoensis*.

8.1.5. Crown Group and the Concept of *Equus*

This paradigm follows a nominalist perspective to the definition of taxa. The nominalist perspective assumes that the limits of named taxa are arbitrary conventions, and then proceeds to spell out those conventions [414]. Under the crown group paradigm, a genus is defined as the clade that includes the most recent common ancestor of all extant species assigned to that genus, and all descendants of that ancestor. Therefore, under this paradigm, *Equus* is defined as the clade that includes the most recent common ancestor of all extant species assigned to *Equus*, and all descendants of that ancestor. *Equus* is constrained to clade 6 in the strict consensus tree of Barrón-Ortiz et al. [9] based on the crown group paradigm, which include the same species obtained under the phylogenetic gaps paradigm.

8.1.6. Divergence Time and the Concept of *Equus*

The divergence time paradigm states that a species or a monophyletic group of species should be regarded as a distinct genus if it diverged well-before the Miocene-Pliocene boundary (4–7 Ma) [168,410,411]. Application of this paradigm in the context of a phylogenetic analysis requires the creation of a time-calibrated phylogeny. Based on the time-calibrated phylogeny of Equini of Barrón-Ortiz et al. [9], *Equus* is delimited to clade 9 under the divergence time paradigm. Here, *E. stenonis* and the North American *E. simplicidens* and *E. idahoensis* should be included in the *Equus* clade.

8.1.7. Taxonomic Implications

Barrón-Ortiz et al. [9] concluded that *Equus* should be delimited to clade 6 in their phylogenetic analysis, based on the fact that three out of the four paradigms used to define mammalian genera identified clade 6 as the most suitable position for *Equus*. This taxonomic arrangement excludes *E. stenonis*, *E. idahoensis*, *E. simplicidens*, and *"Dinohippus" mexicanus* from the genus *Equus* and it implies that *Haringtonhippus* is a junior synonym of *Equus*. Some researchers have assigned *E. simplicidens* and *E. idahoensis* to *Plesippus* at the generic or subgeneric rank [4,57,59,416,417], with *Plesippus simplicidens* selected as the type species [4,416]. Likewise, *E. stenonis* has been referred to *Allohippus* at the generic or subgeneric rank [416]. Based on the results of their analysis, Barrón-Ortiz et al. [9] suggested that *Plesippus* and *Allohippus* should be elevated to generic rank, *"Dinohippus" mexicanus* should be assigned to a new genus, and *Haringtonhippus* should be synonymized with *Equus*.

8.2. Cirilli et al. [10] Phylogeny: Equus Modeled as a Sigle Monophyletic Clade

The results from Cirilli et al. [10] differ from the previous phylogeny of Barròn-Ortiz et al. [9]. Cirilli et al. [10] obtained a single most parsimonious tree from the matrix used, and the genus *Equus* is modeled as a single clade with node 52 being supported by 18 unambiguous synapomorphies, and 13 of these have a CI $\geq$ 0.500 [10], (Figure 2), not allowing the endorsement of *Plesippus* or *Allohippus* at generic or subgeneic level. In particular, the genus is defined by a linear lateral outline of the skull, the absence or reduction of the buccinator fossa, the presence of a shallow depression on the lingual margin of the protocone, the squared shape of the protocone on P2, the presence of an elongated pli caballine on P3 and P4, the squared shape of the protocone on P3 and P4, a V-shaped morphology of the linguaflexid, part of the metaconid-metastylid complex, a squared morphology of the lingual side of the metastylid, a strong and broad 3rd phalanx, a reduced lateral second and fourth metapodials. Moreover, additional analyses as the bootstrap tree supports the *Equus* clade with 99/100 replications [10], (Supplemental Materials). According to Brochu and Sumrall [418], clades within a cladogram are named if two criteria are met: the clade is stable and unlikely to collapse, and there is a need to discuss the group. In addition, Bryant [419], Cantino et al. [420], and Schulte et al. [421] provide some guidelines for the establishment of clade names, including the application of methods for measuring nodal support, careful consideration of those taxa that are likely to move around in different analyses, and use of multiple basal taxa as specifiers for node-based groups. A recently proposed phylogenetic nomenclatural system [422–424] specified that all supraspecific taxonomic nomina be explicitly defined on the basis of common ancestry. In the work by Cirilli et al. [10], *E. simplicidens* is considered as the common ancestor of all the *Equus* species, and place at the base of their radiation, separated from the genus *Dinohippus*. This view is supported by recent molecular analyses of the group, where all the extant equid taxa are grouped into a single genus, *Equus* [27–29,425]. A large *Equus* clade, including some fossil taxa, is also identified by Heintzman et al. [29], where a new clade composed by representatives of *Haringtonhippus* is supposed to diverge from *Equus* during the early Pliocene. It would appear that *Haringtonhippus* is convergent in cranial and postcranial features with Asian *E. hemionus* and perhaps Pleistocene *E. altidens*.

8.2.1. Phylogenetic Gaps, Crown Group, Adaptive Zone, and Divergence Time Applied to the Phylogeny of Cirilli et al. [10]

The phylogenetic gaps criterion identify a genus as a taxonomic category containing a single species, or a monophyletic group of species, which is separated from other taxa of the same rank by a relevant gap. The results from Cirilli et al. [10] support the definition of *Equus* as being a single monophyletic clade since *E. simplicidens*, grouped separately from the species included in the genus *Dinohippus*.

As reported above, the concept of the adaptive zone implies that ecological factors contribute to the speciation process. In this regard, for the genus *Equus* may be taken in consideration the progressive shift to a diet mostly based on C4 grasses. The palaeoecological studies based on the North American record provide some insights between the last representative of the genus *Dinohippus, D. mexicanus*, and the first forms of *Equus*. As reported by MacFadden et al. [426] and Semprebon et al. [427], fossil and extant species of *Equus* have been almost grazers or mixed feeders (except for some large species) [19], whereas some populations of the late Hemphillian *D. mexicanus* from Florida show a browsing signal. However, other late Hemphillian *Dinohippus* samples were identified as mostly grazer, suggesting that the dietary transition from browser to mixed feeders and grazer may already have occurred in North America. Nevertheless, the presence of some individual with δ13C values of 24.7 and 21.5 per mil in the *D. mexicanus* sample studied by MacFadden et al. [426] suggests that some individuals of this sample were feeding on C4 grasses. This evidence indicates that some populations of *Dinohippus* shifted to a more grazing diet, which may have led to speciation process to new forms adapted to new environments. This would have affected not only the diet, but also the increase of the body mass, from ca. 300 kg in *D. mexicanus* to 300 and 400 kg in *E. simplicidens* [10,426]. Moreover, MacFadden et al. [426] reported that the dietary shift from browser to widespread grazing in *Equus* may have occurred during the early Pliocene in North America between, 4.8 and 4.5 Ma, a time frame coherent with the first occurrences of *E. simplicidens*.

The crown group as defined as being a collection of species composed of the living representatives of the collection, the most recent common ancestor of the collection, and all descendants of the most recent common ancestor. In this regard, the results from Cirilli et al. [10] include zebras, asses and the caballine horses in a single clade with the most recent common ancestor identified as being *E. simplicidens*.

Moreover, the concept of the most recent common ancestor is directly linked also to divergence time, and to the estimations based on the genomic analyses from Orlando et al. [27], which estimated a time frame of 4.5–4.0 Ma for the origin of the most recent common ancestor for *Equus*. This estimated ages confirm some one of the oldest discoveries of *E. simplicidens* in North America [428–430] and therefore supporting the hypothesis of *E. simplicidens* as the first representative of the genus *Equus*.

To summarize, phylogenetic gaps, crown group, adaptive zone and divergence time are congruent for identify of the *Equus* clade at node 52 of the phylogeny from Cirilli et al. [9], in agreement with the MPT, Bootstrap and UPGMA tree.

8.2.2. Living and Fossil Equids

The results by Cirilli et al. [10] support the taxonomic division of caballines (domestic horse and Przewalski's wild horse) and noncaballines (zebras and African and Asiatic asses) proposed by morphological data [62] and other molecular and combined studies [431–435] ([8] UPGMA analyses, supplemental information). Similarly to previous studies, Cirilli et al. [10] reported a paraphyletic origin of the extant zebra species as proposed in the literature using cranial morphology [397], palaeogenetics [27,436,437], and nuclear data [438–440], but with a low levels of support of the nodes. Other molecular analyses instead suggested a monophyly of the zebra species with the mountain zebra placed as the sister taxon of Burchell's and Grevy's zebras [432,440–442]; a result affected, anyway, by the absence of fossil representatives of this group in the analyses.

8.3. The Contribution from the Molecular Phylogeny

In the last two decades, new perspectives on the evolution of the genus *Equus* have been reported with the contribution from the molecular phylogeny. Orlando et al. [27] coded the genome of a fossil horse dated ca. 780–560 ka, identifying that the most recent common ancestor for the genus *Equus* emerged at ca. 4.5–4.0 Ma in North America, which is now in agreement with oldest findings and occurrences of the North American *E. simplicidens*. Analogous results were obtained also by Vilstrup et al. [425], which however highlighted the distinction of the North American stilt legged horses from the living asses, supporting a different evolution which led to a similar morphology. Moreover, Vilstrup et al. [425] identified zebras and asses as distinct clades, proposing an estimated age for origin of the plain zebras at ca. 0.7 ± 0.1 Ma, and divergence from the Grevy's zebas at ca. 1.5 Ma.

More inputs came from Jónsson et al. [28], which identified all the living equids belonging to a single genus, *Equus*. Moreover, Jónsson et al. [28] estimated that living zebras and asses cluster into a single monophyletic clade originating at ca. 2 Ma, that the African and Asiatic asses diverged slightly later, at ca. 1.8 Ma, and that the living zebras already diverged at ca. 1 Ma. These estimated ages are in agreement with the first occurrence of fossil species related to living zebras (*E. koobiforensis*, *E. mauritanicus*) or asses (*E. altidens*, *E. tabeti*). *Equus hemionus* and *E. kiang* diverged later, between 356–233 ka. Jónsson et al. [28] estimated also that the gene flow between caballine and stenonine horses ceased between 3.4 and 2.1 Ma, which is in agreement with the dispersal of the stenonine horses in Eurasia at the beginning of the Pleistocene.

Heintzmann et al. [29] used the crown group definition for the genus *Equus* and focused their study on the North American species, especially on the stilt legged species, which were previously identified close to Asian asses [56,62,69]. Nevertheless, the genetic analyses of Orlando et al. [153] and Vilstrup et al. [425] separated these species from the Asian asses and placed them close to the caballine horses. The new phylogeny from Heintzmann et al. [29] has identified the North American stilt horses as a distinct branch from the living and fossil *Equus*, diverging between 5.7–4.1 Ma, during the late Hemphillian or early Blancan. This separation anticipates the origin of the most recent common ancestor identified by Orlando et al. [27]. Heintzmann et al. [29] proposed a new genus, *Harringtonhippus*, for the stilt legged horses from North America, represented by the species *Ha. Francisci*. However, this taxonomy is not accepted by other authors on philosophical grounds [9].

Vershinina et al. [78] identified two dispersal event for the caballine horses. The first occurred between 0.95–0.45 Ma, in east to west direction, consistent with the oldest findings of the caballine horses in Eurasia. The second occurred at 0.2–0.05 Ma, bidirectional but predominantly west to east, due the identification of metapopulations of Eurasian Late Pleistocene horses in Alaska and Northern Yukon, which provided the opportunity for a gene flow between the North American and Eurasia horses during the Late Pleistocene.

Another interesting perspective comes from the subgenus *Sussemionus*. This subgenus was proposed by Eisenmann [256] as an informal group of species from the Early and Middle Pleistocene of Eurasia. Later, Eisenmann [87] formalized the subgenus, characterized by a combination of some dental features [87]. Eisenmann [87] included in this subgenus the species *E. coliemensis*, *E. suessenbornensis*, *E. verae*, *E. granatensis*, *E. hipparionoides* and *E. altidens*. Following the description of the author, the genus includes some anatomical features observed in the Süssenborn sample and in the modern Asian asses. Nevertheless, recent molecular studies [29,151,153,427,443] identified this subgenus separated from the living species even included in the genus *Equus*, surviving until the late Holocene with the species *E. ovodovi*, a Late Pleistocene species from Siberia. However, it should be noted that this subgenus has never been tested with a morphological based cladistic analysis, which is needed to address its taxonomic status.

9. Conclusions

Nineteen collaborating international scientists provide herein a detailed review and synthesis of fossil Equinae occurrences from the Plio-Plesitocene and recent of North,

Central and South America, Eurasia and Africa including fossil and living species. At the present time, our review has identified valid 114 (+4) species of Equinae from 5.3 Ma to recent including 38 from North America, 4 from South America, 26 from East Asia, 6 from the Indian Subcontinent, 18 from Europe and 26 from Africa. In all continents other than South America, more primitive equine clades persisted after *Equus* appeared, and extinction of these more archaic clades were diachronous at the continental to inter-continental scale. While actively researched over the last several decades, Equinae taxonomy is not wholly settled and there are challenges to unifying them at the genus and higher taxonomic levels. That being said, the taxonomy of Equinae reviewed herein has allowed us to provide well resolved biochronology and biogeography, paleoecology and paleoclimatic context of the 5.3 Ma–recent Equinae records.

The paleoclimatic maps from 7 Ma to recent have shown a more suitable environment for the evolution of the modern Equini in North America rather Eurasia and Africa during the Pliocene, with more arid conditions which favored speciation of *Equus* and its dispersal into Eurasia and Africa at the beginning of the Pleistocene. This result is congruent with the hypothesis of several previous morphological, paleoecological and molecular studies cited herein which support the origin of *Equus* during the Pliocene in North America.

Finally, we presented the most recent cladistic morphological based hypotheses on the origin of the genus *Equus*, combined with the results from the molecular phylogenies. Phylogenetic evidence suggests that the genus *Equus* is closely related with *Dinohippus*, from which evolved. The North American *E. simplicidens* represents the ancestral species for the origin of the stenonine horses in Eurasia and Africa, culminated with the evolution of modern zebras and asses. This last point is supported also by the molecular analyses, which have hypothesized that North American stilt legged horses diverged from the living asses and are not phylogenetically linked. However, more studies are needed to shed light on the evolution of the caballine horses, which at the present time remains unresolved. Lastly, we acknowledge the different interpretations of morphological and molecular based cladistic analyses and the need to better integrate these studies going forward.

Supplementary Materials: The following supporting information can be downloaded at: https://www.mdpi.com/article/10.3390/biology11091258/s1, Table S1: taxonomy; Table S2: body mass and diet; Table S3: Equini paleoecology_summary table; Table S4: Taxa coded on the MeanHYP maps; Table S5: Mean Hyposodonty data; Table S6: North and Central American Equus sensu lato.

Author Contributions: Conceptualization, R.L.B. and O.C.; methodology, O.C., H.M., C.I.B.-O., J.S., G.S., F.K. and R.L.B.; North and Central America, H.M., C.I.B.-O., Z.L., A.H.M.-L., J.A.-C., E.D. and E.S.; South America, H.M. and N.A.V.; eastern Asia, O.C., R.L.B., A.M.J. and D.P.; Indian Subcontinent, A.M.J., R.L.B., O.C.; Europe, O.C., R.L.B. and D.P.; Africa, R.L.B. and O.C.; Biochronology and Biogeography, L.R., L.P., R.L.B., O.C. and F.K.; Paleoecology, J.S., F.S., G.S. (with the contribution of C.I.B.-O., A.H.M.-L., D.P., Z.L., O.C and R.L.B.); Climate and Evolution, F.K., O.C. and R.L.B.; Phylogeny: 8.1, C.I.B.-O., C.N.J. and H.M.; 8.2, O.C., L.P. and R.L.B.; 8.3, O.C.; Conclusions, R.L.B. and O.C. Simple Summary, R.L.B. and O.C. All authors have read and agreed to the published version of the manuscript.

Funding: This research was funded by the National Science Foundation (DBI:ABI 1759882 and 1759821) to R.L. Bernor and E. Davis, respectively, under the aegis of the FuTRES Equid working group for which they, O. Cirilli and H. Machado have been funded. F.K. has been funded by Finnish Cultural Foundation (project nr. 00220063). Z.L. acknowledge the Admission Scholarship from University of Ottawa, and a Canadian Graduate Scholarship-Doctoral Program from Natural Sciences and Engineering Research Council of Canada. A.M.J. thanks the Yale Institute for Biospheric Studies for funding. L.P. thanks the European Commission's Research Infrastructure Action, EU-SYNTHESYS projects AT-TAF-2550, DE-TAF-3049, GB-TAF-2825, HU-TAF-3593, HU-TAF-5477, ES-TAF-2997, the SYNTHESYS Project http://www.synthesys.info/ (access on 5 August 2022) which is financed by European Community Research Infrastructure Action under the FP7 "Capacities" Program, and the research project "Ecomorphology of fossil and extant Hippopotamids and Rhinocerotids" granted to L.P. by the University of Florence ("Progetto Giovani Ricercatori Protagonisti" initiative). J.S. wishes to acknowledge the Academy of Finland (AoF. project nr. 340775/346292, "NEPA-Non-analogue ecosystems in the past"). F.S. is supported by Sapienza "5 per mille" funds (ref. SPC: 2021-0070-1350-175998).

Institutional Review Board Statement: Not applicable.

Informed Consent Statement: Not applicable.

Data Availability Statement: All data generated by this study are available in this manuscript and the accompanying Supplementary Materials (Supplementary Materials Tables S1–S6).

Acknowledgments: We acknowledge the curators of the museums that allowed us to study the collections of fossil and extant species cited herein, the scientists who have made their original data available to the scientific community. We also acknowledge Vera Eisenmann for her generosity publicly sharing her data on fossil horses on her website (http://vera-eisenman.com; access on 5 August 2022). C.I.B.O. would like to thank Marisol Montellano-Ballesteros, Christopher N. Jass, Leonardo S. Avila, William T. Taylor, Alwynne Beaudoin and Duncan Ross Parliament for discussions on horse taxonomy. We thank three anonymous reviewers for their comments which have improved the quality of this manuscript. This is the FuTRES publication no. 32.

Conflicts of Interest: The authors declare no conflict of interest.

References

1. Marsh, C.O. Polydactyle Horses, Recent and Extinct. *Am. J. Sci.* **1879**, *17*, 499–505. [CrossRef]
2. Gidley, J.W. Revision of the Miocene and Pliocene Equidae in North America. *Bull. Am. Mus. Nat. Hist.* **1907**, *23*, 865–934.
3. Osborn, H.F. Equidae of the Oligocene, Miocene, and Pliocene of North America. Iconographic type revision. *Mem. Am. Mus. Nat. Hist. New Ser.* **1918**, *2*, 1–326.
4. Matthew, W.D. A new link in the ancestry of the horse. *Am. Mus. Novit.* **1924**, *131*, 130–131.
5. Stirton, R.A. Phylogeny of North American Equidae. *Univ. Calif. Publ. Geol. Sci.* **1940**, *25*, 165–198.
6. Simpson, G.G. *Horses: The Story of the Horse Family in the Modern World and through Sixty Million Years of History*; Oxford University Press: New York, NY, USA, 1951; pp. 1–247.
7. MacFadden, B.J. Fossil Horses—Evidence for Evolution. *Science* **2005**, *307*, 1728–1730. [CrossRef] [PubMed]
8. Bernor, R.L.; Kaya, F.; Kaakinen, A.; Saarinen, J.; Fortelius, M. Old World hipparion evolution, biogeography, climatology and ecology. *Earth Sci. Rev.* **2021**, *221*, 103748. [CrossRef]
9. Barrón-Ortiz, C.I.; Avilla, L.S.; Jass, C.N.; Bravo-Cuevas, V.M.; Machado, H.; Mothé, D. What is *Equus*? Reconciling taxonomy and phylogenetic analyses. *Front. Ecol. Evol.* **2019**, *7*, 343. [CrossRef]
10. Cirilli, O.; Pandolfi, L.; Rook, L.; Bernor, R.L. Evolution of Old World *Equus* and origin of the zebra-ass clade. *Sci. Rep.* **2021**, *11*, 10156. [CrossRef]
11. Eisenmann, V.; Alberdi, M.T.; De Giuli, C.; Staesche, U. Methodology. In *Studying Fossil Horses*; Woodburne, M., Sondaar, P.Y., Eds.; E.J. Brill Press: Leiden, The Netherlands, 1988; pp. 1–71.
12. Bernor, R.L.; Tobien, H.; Hayek, L.A.C.; Mittmann, H.W. *Hippotherium primigenium* (Equidae, Mammalia) from the late Miocene of Höwenegg (Hegau, Germany). *Andrias* **1997**, *10*, 1–230.
13. Cirilli, O.; Saarinen, J.; Pandolfi, L.; Rook, L.; Bernor, R.L. An updated review on *Equus stenonis* (Mammalia, Equidae): New implications for the European Early Pleistocene *Equus* taxonomy and paleoecology, and remarks on the Old World *Equus* evolution. *Quat. Sci. Rev.* **2021**, *269*, 107155. [CrossRef]
14. Bernor, R.L.; Cirilli, O.; Buskianidze, M.; Lordkipanidze, D.; Rook, L. The Dmanisi *Equus*: Systematics, biogeography and paleoecology. *J. Hum. Evol.* **2021**, *158*, 103051. [CrossRef] [PubMed]
15. Fortelius, M.; Solounias, N. Functional characterization of ungulate molars using the abrasion-attrition wear gradient: A new method for reconstructing paleodiets. *Am. Mus. Novit.* **2000**, *3301*, 1–36. [CrossRef]
16. Solounias, N.; Semprebon, G. Advances in the reconstruction of ungulate ecomorphology with application to early fossil equids. *Am. Mus. Novit.* **2002**, *3366*, 1–49. [CrossRef]
17. Oksanen, O.; Žliobaitė, I.; Saarinen, J.; Lawing, A.M.; Fortelius, M. A Humboldtian approach to life and climate of the geological past: Estimating palaeotemperature from dental traits of mammalian communities. *J. Biogeogr.* **2019**, *46*, 1760–1776. [CrossRef]
18. Scott, K.M. Postcranial dimensions of ungulates as predictors of body mass. In *Body Size in Mammalian Palaeobiology*; Damuth, J., MacFadden, B.J., Eds.; Cambridge University Press: New York, NY, USA, 1990; pp. 301–355.
19. Saarinen, J.; Cirilli, O.; Strani, F.; Meshida, K.; Bernor, R.L. Testing equid body mass estimate equations on modern zebras—with implications to understanding the relationship of body size, diet and habitats of *Equus* in the Pleistocene of Europe. *Front. Ecol. Evol.* **2021**, *9*, 622412. [CrossRef]
20. Saarinen, J.; Eronen, J.; Fortelius, M.; Seppä, H.; Lister, A.M. Patterns of diet and body mass of large ungulates from the Pleistocene of Western Europe, and their relation to vegetation. *Palaeontol. Electron.* **2016**, *19*, 1–58. [CrossRef]
21. Mihlbachler, M.C.; Rivals, F.; Solounias, N.; Semprebon, G.M. Dietary change and evolution of horses in North America. *Science* **2011**, *331*, 1178–1181. [CrossRef]

22. The NOW Community. New and Old World Database of Fossil Mammals (NOW). Licensed under CC BY 4.0. 2022. Available online: https://nowdatabase.org/now/database/ (accessed on 4 April 2022).

23. Fortelius, M.; Eronen, J.T.; Jernvall, J.; Liu, L.; Pushkina, D.; Rinne, J.; Tesakov, A.; Vislobokova, I.A.; Zhang, Z.; Zhou, L. Fossil mammals resolve regional patterns ofEurasian climate change during 20 million years. *Evol. Ecol. Res.* **2002**, *4*, 1005–1016.

24. Eronen, J.T.; Evans, A.S.; Fortelius, M.; Jernvall, J. The impact of regional climate on the evolution of mammals: A case study using fossil horses. *Evolution* **2010**, *64*, 398–408. [CrossRef]

25. Eronen, J.T.; Polly, P.D.; Fred, M.; Damuth, J.; Frank, D.C.; Mosbrugger, V.; Scheidegger, C.; Stenseth, N.C.; Fortelius, M. Ecometrics: The grains that bind the past and present together. *Integr. Zool.* **2010**, *5*, 88–101. [CrossRef] [PubMed]

26. Liu, L.; Puolamäki, K.; Eronen, J.T.; Mirzaie Ataabadi, M.; Hernesniemi, E.; Fortelius, M. Dentalfunctional traits of mammals resolve productivity in terrestrial ecosystemspast and present. *Proc. R. Soc. Lond. Biol. Sci.* **2012**, *279*, 2793–2799.

27. Orlando, L.; Vilstrup, L.; Ginolhac, A.; Zhang, G.; Froese, D.; Albrechtsen, A.; Stiller, M.; Schubert, M.; Cappellini, E.; Petersen, B.; et al. Recalibrating *Equus* evolution using the genome sequence of an early Middle Pleistocene horse. *Nature* **2013**, *499*, 74–80. [CrossRef]

28. Jónsson, H.; Schubert, M.; Seguin-Orlando, A.; Ginolhac, A.; Petersen, L.; Fumagalli, M.; Albrechtsen, A.; Petersen, B.; Korneliussen, T.S.; Vilstrup, J.T.; et al. Speciation with gene flow in equids despite extensive chromosomal plasticity. *Proc. Natl. Acad. Sci. USA* **2014**, *111*, 18655–18660. [CrossRef] [PubMed]

29. Heintzman, P.D.; Zazula, G.D.; MacPhee, R.D.; Scott, E.; Cahill, J.A.; McHorse, B.K.; Kapp, J.D.; Stiller, M.; Wooller, M.J.; Orlando, L.; et al. A new genus of horse from Pleistocene North America. *eLife* **2017**, *6*, e29944. [CrossRef]

30. Hulbert, R.C. *Calippus* and *Protohippus* (Mammalia, Perissodactyla, Equidae) from the Miocene (Barstovian-early Hemphillian) of the Gulf Coastal Plain. *Bull. Fla. State Mus. Biol. Sci.* **1988**, *32*, 221–340.

31. Olson, E.C.; McGrew, P.O. Mammalian fauna from the Pliocene of Honduras. *Bullettin Geol. Soc. Am.* **1941**, *52*, 1219–1244. [CrossRef]

32. Drescher, A.B. Later Tertiary Equidae from the Tejon Hills, California. *Carneg. Inst. Wash. Pub.* **1941**, *530*, 1–23.

33. Cope, E.D. A new *Hippidium*. *Am. Nat.* **1880**, *14*, 223.

34. Matthew, W.D.; Stirton, R.A. Equidae from the Pliocene of Texas. *Bullettin* **1930**, *19*, 349–396.

35. Lance, J.F. Palaeontología y estratigrafía del Plioceno de Yepómera, estado de Chihuahua 1a Parte: Equidos, excepto Neohipparion. *Inst. Geol. Univ. Nac. Autónoma México* **1950**, *54*, 83–94.

36. Prado, J.E.; Alberdi, M.T. A cladistic analysis of the horses of the tribe Equini. *J. Paleontol.* **1996**, *39*, 663–680.

37. Kelly, T.S. New Middle Miocene equid crania from California and their implications for the phylogeny of the Equini. *Nat. Hist. Mus. Los Angel. Cty. Contrib. Sci.* **1998**, *473*, 1–43. [CrossRef]

38. Cope, E.D. *A Preliminary Report on the Vertebrate Paleontology of the Llano Estacado*; Geological Survey of Texas: Austin, TX, USA, 1893; pp. 1–136.

39. Leidy, J. Notices of some remains of extinct Mammalia, recently discovered by Dr. F.V. Hayden, in the badlands of Nebraska. *Proc. Acad. Nat. Sci. Phila.* **1856**, *8*, 59.

40. Mooser, O. Fossil Equidae from the middle Pliocene of the Central Plateau of Mexico. *Southwest. Nat.* **1968**, *13*, 1–12. [CrossRef]

41. Cope, E.D. Pliocene horses of southwestern Texas. *Am. Nat.* **1885**, *19*, 1208–1209.

42. Sondaar, P.Y. The osteology ofthe manus of fossil and Recent Equidae. *Ned. Akad. Wettenschappen* **1968**, *25*, 1–76.

43. MacFadden, B.J. Systematics and phylogeny of *Hipparion, Neohipparion, Nannippus,* and *Cormohipparion* (Mammalia, Equidae), from the Miocene and Pliocene of the New World. *Bull. Am. Mus. Nat. Hist.* **1984**, *179*, 196.

44. MacFadden, B.J. *Fossil Horses: Systematics, Paleobiology, and Evolution of the Family Equidae*; Cambridge University Press: Cambridge, UK, 1992.

45. Merriam, J.C. *New Species of the Hipparion Group from the Pacific Coast and Great Basin Provinces of North America*; University of California Press: Oakland, CA, USA, 1915; Volume Voume 9, pp. 1–8.

46. Hulbert, R.C.; Harrington, R. An early Pliocene hipparionine horse from the Canadian Arctic. *Paleontology* **1999**, *42*, 1017–1025. [CrossRef]

47. MacFadden, B.J.; Skinner, M. Diversification and biogeography of the one-toed horses Onohippidium and Hippidion. *Yale Peabody Mus. Nat. Hist. Postilla* **1979**, *175*, 1–9.

48. Hulbert, R.C. Late Neogene Neohipparion (Mammalia, Equidae) from the Gulf Coastal Plain of Florida and Texas. *J. Paleontol.* **1987**, *61*, 809–830. [CrossRef]

49. Webb, S.D.; Hulbert, R.C. Systematics and evolution of *Pseudhipparion* (Mammalia, Equidae) from the late Neogene of the Gulf Coastal Plain and the Great Plains. In *Vertebrates, Phylogeny, and Philosophy*; Flanagan, K.M., Lillegraven, J.A., Eds.; University of Wyoming: Laramie, WY, USA, 1986; pp. 237–272.

50. Merriam, J.C. Correlation between the Tertiary of the Great Basin and That of the Marginal Marine Province in California. *Science* **1914**, *40*, 643–645. [CrossRef] [PubMed]

51. Dalquest, W.W.; Donovan, T.J. A new three-toed horse (Nannippus) from the late Pliocene of Scurry County, Texas. *J. Paleont.* **1973**, *47*, 34–45.

52. Cope, E.D. A contribution to the vertebrate paleontology of Texas. *Proc. Am. Philos. Soc.* **1892**, *30*, 123–131.

53. Dalquest, W.W.; Schultz, G.E. *Ice Age Mammals of Northwestern Texas*; Midwestern State University Press: Wichita Falls, TX, USA, 1992.

54. Orlando, M.F. Order Perissodactyla. In *Early Pleistocene Preglacial and Glacial Rocks and Faunas of North-Central Nebraska*; Skinner, M.F., Hibbard, C.W., Eds.; American Museum of Natural History: New York, NY, USA, 1972; Volume 148, pp. 117–125.

55. Gidley, J.W. A new Pliocene horse from Idaho. *J. Mammal.* **1930**, *16*, 52–60. [CrossRef]

56. Winans, M.C. Revision of North American fossil species of the genus *Equus* (Mammalia: Perissodactyla: Equidae). Ph.D. Thesis, University of Texas, Austin, TX, USA, 1985; p. 264.

57. Forsten, A.; Eisenmann, V. *Equus* (*Plesippus*) *simplicidens* (Cope), not *Dolichohippus*. *Mammalia* **1995**, *59*, 85–89. [CrossRef]

58. Gazin, C.L. A study of the fossil horses remains from the Upper Pliocene of Idaho. *Proc. U.S. Natl. Mus.* **1936**, *83*, 281–320. [CrossRef]

59. Merriam, J.C. New Mammalia from the Idaho Formation. *Bull. Dep. Geol.* **1918**, *10*, 523–530.

60. Scott, E. *Equus idahoensis* from the Pliocene of Arizona, and Its Role in Plesippine Evolution in the American Southwest. 2005. Available online: http://www.sbcounty.gov/museum/discover/divisions/geo/pdf/SVP_2005b_EquusIdahoensis.pdf (accessed on 10 March 2022).

61. Downs, T.; Miller, G.J. Late Cenozoic equids from the Anza-Borrego Desert of California. *Nat. Hist. Mus. Los Angel. Co. Contrib. Sci.* **1994**, *440*, 1–90. [CrossRef]

62. Forsten, A. Mitochondrial-DNA time-table and ethe evolution of *Equus*: Comparison of molecular and paleontological evidence. *Ann. Zool. Fenn.* **1992**, *28*, 301–309.

63. Azzaroli, A. The genus *Equus* in Europe. In *European Neogene Mammal Chronology*; Lindsay, E.H., Fahlbusch, V., Mein, P., Eds.; Plenum Press: New York, NY, USA, 1990; pp. 339–355.

64. Troxell, E.L. The Vertebrate Fossils of Rock Creek, Texas. *Am. J. Sci.* **1915**, *39*, 613–638. [CrossRef]

65. Hibbard, C.W. Pleistocene vertebrates from the Upper Becerra (Becerra Superior) formation, Valley of Tequixquiac, Mexico, with notes on other Pleistocene forms. *Contrib. Mus. Paleontol. Univ. Mich.* **1955**, *12*, 47–96.

66. Eisenmann, V.; Howe, J.; Pichardo, M. Old World Hemiones and New World slender species (Mammalia, Equidae). *Palaeovertebrata* **2008**, *36*, 159–233. [CrossRef]

67. Hulbert, R.C. *Equus* from Leisey Shell Pit 1A and other Irvingtonian localities from Florida. *Bull. Fla. Mus. Nat. Hist.* **1995**, *37*, 553–602.

68. Gidley, J.W. A new species of Pleistocene horse from the Staked Plains of Texas. *Bull. Am. Mus. Nat. Hist.* **1900**, *13*, 111–116.

69. Azzaroli, A.; Voorhies, M. The genus *Equus* in North America: The Blancan species. *Palaeontogr. Ital.* **1993**, *80*, 175–198.

70. Owen, R. On Fossil Remains of Equines from Centrai and South America. *Philos. Trans. Lond.* **1869**, *159*, 559–573.

71. Mooser, O.; Dalquest, W.W. Pleistocene mammals from Aguascalientes, Central Mexico. *J. Mammal.* **1975**, *56*, 781–820. [CrossRef]

72. Cope, E.D. Extinct Mammalia of the Valley of Mexico. *Proc. Am. Philos. Soc.* **1884**, *22*, 1–15.

73. Gidley, J.W. Tooth characters and revision of the North American species of the genus. *Bull. Am. Mus. Nat. Hist.* **1901**, *14*, 91–142.

74. Azzaroli, A. The genus *Equus* in North America—The Pleistocene species. *Paleontogr. Ital.* **1998**, *85*, 1–60.

75. Hay, O.P. Description of a new species of Pleistocene horse, *Equus lambei*, from the Pleistocene of Yukon Territory. *Proc. U. S. Natl. Mus.* **1917**, *53*, 435–443. [CrossRef]

76. Hay, O.P. Contributions to the knowledge of the mammals of the Pleistocene of North America. *Proc. U. S. Natl. Mus.* **1915**, *48*, 515–575. [CrossRef]

77. Weinstock, J.; Willerslev, E.; Sher, A.; Tong, W.; Ho, S.; Rubenstein, D.; Storer, J.; Burns, J.; Martin, L.; Bravi, C.; et al. Evolution, systematics, and phylogeography of Pleistocene horses in the New World: A molecular perspective. *PLoS Biol.* **2005**, *3*, e241. [CrossRef]

78. Vershinina, A.O.; Heintzman, P.D.; Froese, D.G.; Zazula, G.; Cassatt-Johnstone, M.; Dalen, L.; Der Sarkissian, C.; Dunn, S.G.; Ermini, L.; Gamba, C.; et al. Ancient horse genomes reveal the timing and extent of dispersals across the Bering Land Bridge. *Mol. Ecol.* **2021**, *23*, 6144–6161. [CrossRef]

79. Leidy, J. Description of Vertebrate Fossils. In *Post-Pleiocene Fossils of South-Carolina*; Holmes, F.S., Ed.; Russel and Jones: Charleston, SC, USA, 1860; pp. 99–122.

80. Azzaroli, A. A Synopsis of the Quaternary species of *Equus* in North America. *Boll. Soc. Paleont. Ital.* **1995**, *34*, 205–221.

81. Quinn, J.H. Pleistocene Equidae of Texas. *Econ. Geol. Rep. Investig.* **1957**, *33*, 1–51.

82. von Reichenau, W. Beitrage zur naheren Kenntnis fossiler Pferde aus deutschem Pleistozan, insbesondere über die Entwicklung und die Abkaustadiem del Gebisses von Hochterrassenpferde (*Equus mosbachensis* v. R.). *Abh. Gross. Hess. Geolog. Darmstadt* **1915**, *7*, 1–55.

83. Alberdi, M.T.; Arroyo-Cabrales, J.; Marin-Leyva, A.H.; Polaco, O.J. Study of Cedral Horses and their place in the Mexican Quaternary. *Rev. Mex. Cienc. Geológicas* **2014**, *31*, 221–237.

84. Kurten, B.; Anderson, E. *Pleistocene Mammals of North America*; Columbia University Press: New York, NY, USA, 1980; pp. 1–442.

85. Winans, M.C. A quantitative study of the North American fossil species of the genus *Equus*. *Evol. Perissodactyls* **1989**, *72*, 262–297.

86. Sher, A.V. *Mlekopitaiushchie i Stratigrafia Pleistotsena Krainego Severo-Vostoka SSSR i Severnoi Ameriki*; Nauka: Moskva, Russia, 1971.

87. Eisenmann, V. *Sussemionus*, a new subgenus of *Equus* (Perissodactyla, Mammalia). *Comp. Rendus. Biol.* **2010**, *333*, 235–240. [CrossRef] [PubMed]

88. Eisenmann, V. The equids from Liventsovka and other localities of the Khaprovskii Faunal Complex, Russia: A revision. *Geobios* **2022**, *70*, 17–33. [CrossRef]

89. Leidy, J. Bones and teeth of horses from California and Oregon. *Proc. Acad. Nat. Sci. Phila.* **1865**, *17*, 94.

90. Merriam, J.C. Preliminary report on the horses of Rancho La Brea. *Bull. Dep. Geol. Sci.* **1913**, *7*, 397–418.

91. Savage, D.E. Late Cenozoic vertebrates of the San Francisco Bay Region. *Bull. Dep. Geol. Sci.* **1951**, *28*, 215–314.

92. Miller, W.E. Pleistocene vertebrates of the Los Angeles Basin and vicinity (exclusive of Rancho La Brea). *Bull. Los Angel. Cty. Mus.* **1971**, *10*, 1–124.

93. Brown, K.E.; Akersten, W.A.; Scott, E. Equus occidentalis Leidy from "Asphalto", Kern County, California. In *La Brea and Beyond: The Paleontology of Asphalt-Preserved Biotas*; Harris, J.M., Ed.; Natural History Museum of Los Angeles County: Los Angeles, CA, USA, 2015; pp. 81–89.

94. Barrón-Ortiz, C.I.; Rodrigues, A.T.; Theodor, J.M.; Kooyman, B.P.; Yang, D.Y.; Speller, C.F. Cheek tooth morphology and ancient mitochondrial DNA of late pleistocene horses from the western interior of North America: Implications for the taxonomy of North American Late Pleistocene. *PLoS ONE* **2017**, *12*, e0183045. [CrossRef]

95. Marín-Leyva, A.H.; Arroyo-Cabrales, J.; García-Zepeda, M.L.; Ponce-Saavedra, J.; Schaaf, P.; Pérez-Crespo, V.P.; Pedro Morales-Puente, P.; Cienfuegos-Alvarado, E.; Alberdi, M.T. Feeding ecology and habitat of Late Pleistocene *Equus* horses fromwest-central Mexico using carbon and oxygen isotopes variation. *Rev. Mex. Cienc. Geol.* **2016**, *33*, 157–169.

96. Jimenez-Hidalgo, E.; Diaz-Sibaja, R. Was *Equus cedralensis* a non-stilt legged horse? Taxonomical implication for the Mexican Pleistocene horses. *Ameghiniana* **2020**, *57*, 284–288. [CrossRef]

97. Barrón-Ortiz, C. The Late Pleistocene Extinction in North America: An Investigation of Horse and Bison Fossil Material and Its Implication for Nutritional Extinction Models. Ph.D. Thesis, University of Calgary, Calgary, AB, Canada, 2016. [CrossRef]

98. Prado, J.L.; Alberdi, M.T. *Fossil Horses of South America*; Springer: Berlin/Heidelberg, Germany, 2017; pp. 1–150.

99. Machado, H.; Grillo, O.; Scott, E.; Avilla, L. Following the footsteps of the South American *Equus*: Are autopodia taxonomically informative? *J. Mammal. Evol.* **2018**, *25*, 397–405. [CrossRef]

100. Machado, H.; Avilla, L. The Diversity of South American *Equus*. Did Size Really Matter? *Front. Ecol. Evol.* **2019**, *7*, 235. [CrossRef]

101. Lund, P.W. Nouvelles Recherches sur la Faune fossile du Brésil. *Ann. Sci. Nat.* **1840**, *13*, 310–319.

102. MacFadden, B.J.; Siles, O.; Zeitler, P.; Johnson, N.M.; Campbell, K.E. Magnetic polarity stratigraphy of the middle Pleistocene (Ensenadan) Tarija Formation of southern Bolivia. *Quat. Res.* **1983**, *19*, 172–187. [CrossRef]

103. MacFadden, B.J.; Zeitler, P.K.; Anaya, F.; Cottle, J.M. Confirmation of the middle Pleistocene age of the sedimentary sequence from Tarija, Bolivia. *Quat. Res.* **2013**, *79*, 268–273. [CrossRef]

104. Prado, J.L.; Martinez-Maza, C.; Alberdi, M.T. Megafauna extinction in South America: A new chronology for theArgentine Pampas. *Palaeogeogr. Palaeoclim. Palaeoecol.* **2015**, *425*, 41–49. [CrossRef]

105. Roth, S. El mamífero misterioso de la Patagonia *Grypotherium domesticum*. II. Descripción de los restos encontrados en la caverna de Última Esperanza. *Rev. Mus. Plata* **1899**, *9*, 421–453.

106. Labarca, R.; Caro, F.J.; Villavicencio, N.A.; Capriles, J.M.; Briones, E.; Latorre, C.; Santoro, C.M. A Partially Complete Skeleton of *Hippidion saldiasi* Roth, 1899 (Mammalia: Perissodactyla) from the Late Pleistocene of the High Andes in Northern Chile. *J. Vertebr. Paleontol.* **2021**, *40*, e1862132. [CrossRef]

107. Paunero, R.S.; Rosales, G.; Prado, J.L.; Alberdi, M.T. Cerro bombero: Registro de hippidion saldiasi roth, 1899 (equidae, perissodactyla) en el holoceno temprano de Patagonia (Santa Cruz, Argentina). *Estud. Geol.* **2008**, *64*, 89–98. [CrossRef]

108. Massone, M.; Prieto, A. Evaluacion de la modalidad cultural Fell 1 en Magallanes. *Chungara* **2004**, *1*, 303–315. [CrossRef]

109. Lund, P.W. Meddelelse af det ubbytte de i 1844 undersögte Knoglehuler have avgivet til Kundskaben om Brasiliens Dyreverden för sidste Jordomvaeltning. *Ebenda* **1846**, *12*, 1–94.

110. Montané, J. Paleo-Indian remains from Lagunade Tagua Tagua, Central Chile. *Science* **1968**, *161*, 1137–1138. [CrossRef] [PubMed]

111. Gervais, P. *Recherches sur les Mammiferes Fossiles de l'Amerique Meridionale*; Bertrand, P., Ed.; Librarie-Editeur: Paris, France, 1855; pp. 1–63.

112. Shockey, B.J.; Salas-Gismondi, R.; Baby, P.; Guyot, J.-L.; Baltazar, M.C.; Huamán, L.; Clack, A.; Stucchi, M.; Pujos, F.; Emerson, J.M.; et al. New Pleistocene Cave Faunas of the Andes of Central Perú: Radiocarbon Ages and the Survival of Low Latitude, Pleistocene DNA. *Palaeontol. Electron.* **2009**, *12*, 15.

113. Villavicencio, N.A.; Werdelin, L. The casa del diablo cave (puno, peru) and the late pleistocene demise of megafauna in the andean altiplano. *Quat. Sci. Rev.* **2018**, *195*, 21–31. [CrossRef]

114. Sun, B.; Deng, T. The *Equus* Datum and the Early Radiation of *Equus* in China. *Front. Ecol. Evol.* **2019**, *7*, 429. [CrossRef]

115. Sun, B.; Liu, W.; Liu, J.; Liu, L.; Jin, C. *Equus qiyngyangensis* in Jinyan Cave and its paleozoographic significance. *Quat. Int.* **2020**, *591*, 35–46. [CrossRef]

116. Teilhard de Chardin, P.; Young, C.C. Fossil mammals from the late Cenozoic of northern China. *Palaeontol. Sin.* **1931**, *9*, 1–67.

117. Flynn, L.J.; Bernor, R.L. Late Tertiary mammals from the Mongolian People's Republic. *Am. Mus. Novit.* **1987**, *2872*, 1–16.

118. Wang, X.; Li, Q.; Xie, G.; Saylor, J.E.; Tseng, Z.J.; Takeuchi, G.T.; Deng, T.; Wang, Y.; Hou, S.; Liu, J.; et al. Mio-Pleistocene Zanda Basin biostratigraphy and geochronology, pre-Ice Age fauna, and mammalian evolution in western Himalaya. *Palaeogeogr. Palaeoclimatol. Palaeoecol.* **2013**, *374*, 81–95. [CrossRef]

119. Jukar, A.; Sun, B.; Bernor, R.L. The first occurrence of *Plesiohipparion huangheense* (Qiu, Huang & Guo, 1987) (Equidae, Hipparionini) from the late Pliocene of India. *Boll. Soc. Paleo. Ital.* **2018**, *57*, 125–132.

120. Qiu, Z.X.; Qiu, Z.D.; Deng, T.; Li, C.K.; Zhang, Z.Q.; Wang, B.Y.; Wang, X. Neogene land mammal stages/ages of China. In *Fossil Mammals of Asia: Neogene Biostratigraphy and Chronology*; Wang, X., Flynn, L.J., Fortelius, M., Eds.; Columbia University Press: New York, NY, USA, 2013; pp. 29–90.

121. Matsumoto, H. On Hipparion richthofeni Koken. *Sci. Rep. Tohoku Univ.* **1927**, *10*, 59–75.
122. Qiu, Z.; Huang, W.; Guo, Z. The Chinese hipparionine fossils. *Palaeontol. Sin. New Ser.* **1987**, *175*, 1–250.
123. Bernor, R.L.; Lipscomb, D. The systematic position of "*Plesiohipparion*" aff. *huangheense* (Equidae, Hipparionini) from Gülyazi, Turkey. *Mitt. Bayer. Staatssamml. Paläontol. Hist. Geol.* **1991**, *31*, 107–123.
124. Bernor, R.L.; Sun, B.; Chen, Y. *Plesiohipparion shanxiense* n. sp. from the Early Pleistocene (Nihowanian) of E Shanxi, China. *Boll. Soc. Paleontol. Ital.* **2015**, *54*, 197–210.
125. Sefve, I. Die Hipparionen Nord-Chinas. *Palaeo. Sin. Ser.* **1927**, *4*, 1–54.
126. Liu, P.; Lovlie, R. Magnetostratigraphic age of Pleistocene loess/paleosol sections at Kehe, Shanxi. *J. Stratigr.* **2007**, *31*, 240–246.
127. Qiu, Z.; Deng, T.; Banyue, W. Early Pleistocene mammalian fauna from Longdan, Dongxiang, Gansu, China. *Palaeontol. Sin. New Ser.* **2004**, *191*, 1–245.
128. Teilhard de Chardin, P.; Piveteau, J. Les mammifères fossiles de Nihowan (Chine). *Ann. Paléont.* **1930**, *19*, l–134.
129. Kuzmina, I.E. *Horses of North Eurasia from the Pliocene till the Present Time*; Vereschagin, N.K., Ed.; Russian academy of Sciences: Saint-Petersburg, Russia, 1997; Volume 273, pp. 1–224.
130. Deng, T.; Xue, X. *Chinese Fossil Horses of Equus and Their Environment*; China Ocean Press: Beijing, China, 1999.
131. Chow, M.; Liu, H. Fossil equine teeth from Shansi. *Vert. PalAsiat.* **1959**, *1*, 133–136.
132. Ao, H.; An, Z.; Dekkers, M.J.; Li, Y.; Xiao, G.; Zhao, H.; Qiang, X. Pleistocene magnetochronology of the fauna and Paleolithic sites in the Nihewan Basin: Significance for environmental and hominin evolution in North China. *Quat. Geochronol.* **2013**, *18*, 78–92. [CrossRef]
133. Colbert, E. Pleistocene mammals from the Ma Kai Valley of northern Yunnan, China. *Am. Mus. Novit.* **1940**, *1099*, 1–10.
134. Eisenmann, V. Nouvelles interpretations des restes d'équidés (Mammalia, Perissodactyla) de Nihowan (Pléistocène inférieur de la Chine du Nord): *Equus teilhardi* nov. sp. *Geobios* **1975**, *8*, 125–134. [CrossRef]
135. Sun, B.; Deng, T.; Liu, Y. Early Pleistocene *Equus* (Equidae, Perissodactyla) from Andersson Loc. 32 in Qixian, Shanxi, China. *Hist. Biol.* **2017**, *31*, 211–222. [CrossRef]
136. Eisenmann, V.; Deng, T. *Equus qingyangensis* (Equidae, Perissodactyla) of the Upper Pliocene of Bajiazui, China: Evidence for the North American origin of an Old World lineage distinct from *E. stenonis*. *Quaternaire* **2005**, *2*, 113–122.
137. Sharapov, S. *Kuruksaiskii Kompleks Pozdnepliotsenovykh Mlekopitajushchikh Afgano-Adzhikistanskoi Depressii*; Izdatelstvo Donish: Dushanbe, Tajikistan, 1986; p. 270.
138. Forsten, A.; Sharapov, S. Fossil equids (Mammalia, Equidae) from the Neogene and Pleistocene of Tadzhikistan. *Geodiversitas* **2000**, *22*, 293–314.
139. Vangengeim, E.A.; Sotnikova, M.V.; Alekseeva, L.I.; Vislobokova, I.A.; Zhegallo, V.I.; Zazhigin, V.S.; Shevyreva, N.S. *Biostratigrafiya Pozdnego Pliotsena—Rannego Pleistotsena Adzhikistana*; Nuaka: Moscow, Russia, 1988; p. 125.
140. Kuznetsova, T.V.; Zhegallo, V.I. Taxonomic Diversity of Equids from the Nalaikha Locality (Mongolia). In Proceedings of the International Conference on the Conditions of the Theriofauna in Russia and Adjacent Countries, Moscow, Russia, 1–3 February 1995; pp. 174–178.
141. Kuznetsova, T.V.; Zhegallo, V.I. *Equus* (*Hemionus*) *nalaikhaensis* from the Pleistocene of Mongolia, the Earliest Kulan of Central Asia. *Paleontol. J.* **2009**, *43*, 574–583. [CrossRef]
142. Lazarev, P.A. *Anthropogenic Horses of Yakutia*; Nauka: Moscow, Russia, 1980; p. 190. (In Russian)
143. Rusanov, B.S. *Biostratigraphy of Cenoizoic Deposits of Southern Yakutia*; Nauka: Moscow, Russia, 1968; p. 459.
144. Lazarev, P.A. *Large mammals of Anthropogene in Yakutia*; Nauka: Moscow, Russia, 2008; p. 160. (In Russian)
145. Spasskaya, N.; Pavlinov, I.; Sharko, F.; Boulygina, E.; Tsygankova, S.; Nedoluzhko, A.; Mashchenko, E. Morphometric and genetic analyses of diversity of the Lena horse (*Equus lenensis* Russanov, 1968; Mammalia: Equidae). *Russ. J. Theriol.* **2021**, *20*, 82–95. [CrossRef]
146. Liu, H.Y. A new species of *Equus* from Locality 21 of Zhoukoudian. *Vert PalAsiat* **1963**, *7*, 318–322.
147. Forsten, A. Chinese fossil horses of the genus *Equus*. *Acta Zool. Fenn.* **1986**, *181*, 1–40.
148. Gromova, V.I. O novoj iskopaiemoj loshadi iz srednej Azii. *Dokl. Akad. Nauk SSSR* **1946**, *54*, 349–351.
149. Eisenmann, V.; Helmer, D.; Segui, M.S. The big *Equus* from Geometric Kebaran of Umm el Tlel, Syria: Equus valeriani, Equus capensis or Equus caballus? *Proc. Fifth Int. Symp. Archaeozoology Southwest. Asia Adjac. Areas* **2002**, *1*, 62–73.
150. Zhou, X.; Sun, Y.; Xu, Q.; Li, Y. Note on a new Late Pleistocene *Equus* from Dalian. *Vertebr. Palasiat.* **1985**, *23*, 69–76.
151. Yuan, J.; Hou, X.; Barlow, A.; Preick, M.; Taron, U.H.; Alberti, F.; Basler, N.; Deng, T.; Lai, X.; Hofreiter, M.; et al. Molecular identification of Late and terminal Pleistocene *Equus ovodovi* from northeastern China. *PLoS ONE* **2019**, *14*, e0216883. [CrossRef] [PubMed]
152. Eisenmann, V.; Vasiliev, S. Unexpected finding of a new *Equus* species (Mammalia, Perissodactyla) belonging to a supposedly extinct subgenus in Late Pleistocene deposits of Khakassia (southwestern Siberia). *Geodiversitas* **2011**, *33*, 519–530. [CrossRef]
153. Orlando, L.; Metcalf, J.L.; Alberdi, M.T.; Telles-Antunes, M.; Bonjean, D.; Otte, M.; Martin, F.; Eisenmann, V.; Mashkour, M.; Morello, F.; et al. Revising the recent evolutionary history of equids using ancient DNA. *Proc. Natl. Acad. Sci. USA* **2009**, *106*, 21754–21759. [CrossRef] [PubMed]
154. Yuan, J.; Sheng, G.; Preick, M.; Sun, B.; Hou, X.; Chen, S.; Taron, U.H.; Barlow, A.; Wang, L.; Hu, J.; et al. Mitochondrial genomes of Late Pleistocene caballines horses from China belong to a separate clade. *Quat. Sci. Rev.* **2020**, *250*, 106691. [CrossRef]

155. Plasteeva, N.; Vasiliev, S.; Kosintsev, P. *Equus* (*Sussemionus*) *ovodovi* Eisenmann et Vasiliev, 2011 from the Laye Pleistocene of Western Siberia. *Russ. J. Theriol.* **2015**, *14*, 187–200. [CrossRef]
156. Pallas, P.S. *Equus* hemionus. *Novi. Comment. Acad. Sci. Imp. Petropol.* **1774**, *19*, 394–417.
157. Biswas, S.K. Regional Tectonic Framework, Structure and Evolution of the Western Marginal Basins of India. *Tectonophysics* **1987**, *135*, 302–327.
158. Costa, A.G. A new Late Pleistocene fauna from arid coastal India: Implications for inundated coastal refugia and human dispersals. *Quat. Int.* **2017**, *436*, 253–269. [CrossRef]
159. Dassarma, D.C.; Biswas, S. Newly discovered Late Quaternary vertebrates from the terraced alluvial fills of the Son Valley. *Indian J. Earth Sci.* **1977**, *4*, 122–136.
160. Turvey, S.T.; Sathe, V.; Crees, J.J.; Jukar, A.M.; Chakraborty, P.; Lister, A.M. Late Quaternary megafaunal extinctions in India: How much do we know? *Quat. Sci. Rev.* **2021**, *252*, 106740. [CrossRef]
161. Stewart, M.J.; Louys, G.J.; Price, N.A.; Drake, H.S.; Petraglia, M.D. Middle and Late Pleistocene mammal fossils of Arabia and surrounding regions: Implications for biogeography and hominin dispersals. *Quat. Int.* **2019**, *515*, 12–29. [CrossRef]
162. Moorcroft, W. In Ladakh and Kashmir; in Peshawar, Kabul, Kunduz, and Bokhara. In *Travels in the Provinces of Hindustan and the Panjab*; John Murray: London, UK, 1841; p. 312.
163. Shah, N. Status and action plan for the kiang (*Equus kiang*). In *Equids: Zebras, Asses and Horses. Status Survey and Conservation Action Plan*; Moehlman, P.D., Ed.; IUCN: Gland, Switzerland, 2002; pp. 72–81.
164. Wang, Y.X. *A Complete Check-List of Mammal Species and Subspecies in China: A Taxonomic and Geographic Reference*; China Forestry Publishing House: Beijing, China, 2002.
165. Schaller, G.B. *Wildlife of the Tibetan Steppe*; University of Chicago Press: Chicago, IL, USA, 1998.
166. Poliakov, I.S. Przewalski horse (*Equus Przevalskii* n. sp.). In *Izvestija Russkogo Geograficheskogo Obshchestva*; Russian Academy of Sciences: Saint Petersburg, Russia, 1881; Volume 17, pp. 1–19. (In Russian)
167. Groves, C. Morphology, habitat, and taxonomy. In *Przewalski's Horse: The History and Biology of an Endangered Species*; Boyd, L., Houpt, K.A., Eds.; State University of New York Press: Albany, NY, USA, 1994; pp. 39–59.
168. Groves, C.; Grubb, P. *Ungulate Taxonomy*; Johns Hopkins University Press: Baltimore, MD, USA, 2011.
169. Heptner, V.; Nasimovich, A.A.; Bannikov, A.G. Mammals of the Soviet Union. In *Artiodactyla and Perissodactyla*; Vysshaya Shkola: Moscow, Russia, 1961; Volume 1.
170. Deng, T. Ages of some Late Pleistocene faunas based on the presence of *Equus przewalskii* (Perissodactyla, Equidae). *Vertebr. Palasiat.* **1999**, *23*, 51–56.
171. Gaunitz, C.; Fages, A.; Hanghøj, K.; Albrechtsen, A.; Khan, N.; Schubert, M.; Seguin-Orlando, A.; Owens, I.J.; Felkel, S.; Bignon-Lau, O.; et al. Ancient genomes revisit the ancestry of domestic and Przewalski's horses. *Science* **2018**, *360*, 111–114. [CrossRef]
172. Librado, P.; Orlando, L. Genomics and Evolutionary History of Equids. *Annu. Rev. Anim. Biosci.* **2020**, *7*, 16–41. [CrossRef]
173. Taylor, W.T.T.; Barrón-Ortiz, C.I. Rethinking the evidence for early horse domestication at Botai. *Sci. Rep.* **2021**, *11*, 7440. [CrossRef]
174. Wolf, D.; Bernor, R.L.; Hussain, S.T. A systematic, biostratigraphic, and paleobiogeographic reevaluation of the Siwalik hipparion-ine horse assemblage from the Potwar Plateau, Northern Pakistan. *Palaeontogr. Abt.* **2013**, *300*, 1–115. [CrossRef]
175. Jukar, A.M.; Sun, B.; Nanda, A.C.; Bernor, R.L. The first occurrence of *Eurygnathohippus* (Mammalia, Perissodatyla, Equidae) outside Africa and its biogeographic significance. *Boll. Soc. Paleo. Ital.* **2019**, *58*, 171–179.
176. Rook, L.; Bernor, R.L.; Avilla, L.D.; Cirilli, O.; Flynn, L.; Jukar, A.M.; Sanders, W.; Scott, E.; Wang, X. Mammal Biochronology (Land Mammal Ages) around the World from the Late Miocene to Middle Pleistocene and Major Events in Horse Evolutionary History. *Front. Ecol. Evol.* **2019**, *7*, 278. [CrossRef]
177. Jukar, A.M.; Lyons, S.K.; Wagner, P.J.; Uhen, M.D. Late Quaternary extinctions in the Indian Subcontinent. *Paleogeol. Paleoclim. Paleoecol.* **2021**, *562*, 110137. [CrossRef]
178. Thomas, P.K.; Joglekar, P. Holocene Faunal Studies. *Man Environ.* **1994**, *19*, 179–203. [CrossRef]
179. Falconer, H.; Cautley, P.T. Illustrations-Part IX Equidae, Camelidae, Sivatherium. In *Fauna Antiqua Sivalensis, Being the Fossil Zoology of the Sewalik Hills in the North of India*; Smith, Elder and Co.: London, UK, 1849.
180. Hussain, S.T. Revision of Hipparion (Equidae, Mammalia) from the Siwalik Hills of Pakistan and India. *Geol. Mijnb.* **1971**, *147*, 80–81.
181. MacFadden, B.J.; Woodburne, M.O. Systematics of the Neogene Siwalik Hipparions (Mammalia, Equidae) based on cranial and dental morphology. *J. Vert. Paleo* **1982**, *2*, 185–218. [CrossRef]
182. Bernor, R.L.; Hussain, S.T. An Assessment of the Systematic, Phylogenetic and Biogeographic Relationships of Siwalik Hipparion-ine horses. *J. Vert. Paleo* **1985**, *5*, 32–87. [CrossRef]
183. Lydekker, R. Containing the order Ungulata, suborders Perissodactyla, Toxodontia, Condylartha, and Ambylopoda. In *Catalogue of the Fossil Mammalia in the British Museum (Natural History) Part III*; Taylor Francis: London, UK, 1886.
184. Bernor, R.L.; Cirilli, O.; Jukar, A.M.; Potts, R.; Buskianidze, M.; Rook, L. Evolution of Early *Equus* in Italy, Georgia, the Indian Subcontinent, East Africa, and the Origins of African Zebras. *Front. Ecol. Evol.* **2019**, *7*, 166. [CrossRef]
185. Patel, R.; Nainwal, H.C.; Sharma, T.; Rana, R.S. National Workshop on Indian Siwalik: Recent Advances and Future Research. In *Equus cf. sivalensis from the Tatrot Formation (Upper Pliocene) of Jhil-Bankabara Area Sirmaur District, Himachal Pradesh, India*; Geological Survey of India: Lucknow, India, 2017; pp. 51–53.

186. Moigne, A.M.; Dambricourt Malassé, A.; Singh, M.; Kaur, A.; Gaillard, C.; Karir, B.; Pal, S.; Bhardwaj, V.; Abdessadok, S.; Chapon Sao, C.; et al. The faunal assemblage of the paleonto-archeological localities of the Late Pliocene Quranwala Zone, Mosol Formation, Siwalik Range, NW India. *Palevol* **2016**, *15*, 359–378. [CrossRef]

187. Gaur, R.; Chopra, S.R.K. Taphonomy, fauna, environment and ecology of Upper Sivaliks (Plio-Pleistocene) near Chandigarh, India. *Nature* **1984**, *308*, 353–355. [CrossRef]

188. Hussain, S.T.; van den Bergh, G.; Steensma, K.J.; de Visser, J.A.; de Vos, J.; Mohammad, A.; van Dam, J.; Sondaar, P.; Malik, S.B. Biostratigraphy of the Plio-Pleistocene continental sediments (Upper Siwaliks) of the Mangla-Samwal anticline, Azad Kashmir, Pakistan. *Proc. K. Ned. Akad. Wet.* **1992**, *95*, 65–80.

189. Dennell, R.; Coard, R.; Turner, A. The biostratigraphy and magnetic polarity zonation of the Pabbi Hills, northern Pakistan: An Upper Siwalik (Pinjor Stage) Upper Pliocene–Lower Pleistocene fluvial sequence. *Palaeogeogr. Palaeoclimatol. Palaeoecol.* **2006**, *234*, 168–185. [CrossRef]

190. Sonakia, A.; Biswas, S. Fossil mammals including early man from the Quaternary deposits of the Narmada and Son basins of Madhya Pradesh, India. *Palaeontol. Indica* **2011**, *53*, 1–84.

191. Lydekker, R. Siwalik and Narbada Equidae. *Mem. Geol. Soc. India Palaeontol. Indica Ser.* **1882**, *2*, 67–98.

192. Cirilli, O.; Bernor, R.L.; Rook, L. New insights on the Early Pleistocene equids from Roca-Neyra (France, central Europe): Implications for the Hipparion LAD and the *Equus* FAD in Europe. *J. Paleontol.* **2021**, *95*, 406–425. [CrossRef]

193. Gromova, V. *Le Genre Hipparion*; Bureau de Recherches Geologiques et Minieres: Orléans, France, 1952; Volume 12, pp. 1–288.

194. Bernor, R.L.; Sen, S. The Early Pliocene *Plesiohipparion* and *Proboscidipparion* (Equidae, Hipparionini) from Çalta, Turkey (Ruscinian Age, c. 4.0 Ma). *Geodiversitas* **2017**, *39*, 285–314. [CrossRef]

195. Bernor, R.L.; Beech, M.; Bibi, F. Equidae from the Baynunah Formation. In *Sands of Time*; Springer: Berlin/Heidelberg, Germany, 2022; pp. 259–280.

196. Crusafont, P.M.; Sondaar, P. Una nouvelle espece d'Hipparion du Pliocéne terminal d'Espagne. *Palaeovertebrata* **1971**, *4*, 59–66. [CrossRef]

197. Alberdi, M.T.; Alcalà, L. A study of the new samples of the Pliocene Hipparion (Equidae, Mammalia) from Spain and Bulgaria. *Trans. R. Soc. Edinb. Earth Sci.* **1999**, *89*, 167–186. [CrossRef]

198. Piñero, P.; Agustí, J.; Oms, O.; Fierro, I.; Montoya, P.; Mansino, S.; Ruiz-Sánchez, F.; Alba, D.M.; Alberdi, M.T.; Blain, H.A.; et al. Early Pliocene continental vertebrate Fauna at Puerto de la Cadena (SE Spain) and its bearing on the marine-continental correlation of the Late Neogene of Eastern Betics. *Paleogeogr. Palaeoclim. Palaeoecol.* **2017**, *479*, 102–114. [CrossRef]

199. Gervais, P. *Zoologie et Paléontologie Française (Animaux Vertébrés): Deuxième Edition*; Arthus Bertrand: Paris, France, 1859; pp. 1–544.

200. Depéret, C. Les animaux pliocènes du Roussilon. *Mem. Soc. Geol. Fr.* **1890**, *3*, 1–198.

201. Forsten, A. The hipparions (Mammalia, Equidae) of Suffolk, England. *Trans. R. Soc. Edinb. Earth Sci.* **2001**, *92*, 115–120. [CrossRef]

202. Forsten, A. Latest Hipparion Cristol, 1832 in Europe. A review of the Pliocene *Hipparion crassum* Gervais Group and other finds (Mammalia, Equidae). *Geodiversitas* **2002**, *24*, 465–486.

203. Eisenmann, V.; Sondaar, P. Pliocene vertebrate locality of Calta, Ankara, Turkey. Hipparion. *Geodiversitas* **1998**, *20*, 409–439.

204. Hernández-Pacheco, E. La llanura manchega y sus mamíferos fósiles: Yacimiento de La Puebla de Almoradier. *Com. Investig. Pleontol. Mem.* **1921**, *28*, 1–41.

205. Alberdi, M.T. El género Hipparion en España, revision e historia evolutive. *Trab. Sobre Neógeno Cuatern.* **1974**, *1*, 1–139.

206. Bernor, R.L.; Koufos, G.D.; Woodburne, M.O.; Fortelius, M. The evolutionary history and biochronology of European and southwestern Asian late Miocene and Pliocene hipparionine horses. In *The Evolution of Western Eurasian Neogene Mammal Faunas*; Bernor, R.L., Fahlbusch, V., Mittman, H.W., Eds.; Columbia University Press: New York, NY, USA, 1996; pp. 307–338.

207. Bernor, R.L.; Sun, B.Y. Morphology through ontogeny of Chinese *Proboscidipparion* and *Plesiohipparion* and observations on their Eurasian and African relatives. *Vertebr. Palasiat.* **2015**, *53*, 73–92.

208. Agustì, J.; Oms, O. On the age of the last hipparionine faunas in western Europe. *Comptes Rendus Acad. Sci. Paris Sci.* **2001**, *332*, 291–297. [CrossRef]

209. Azanza, B.; Morales, M.; Andrés, M.; Cerdeño, E.; Alberdi, M.T. Los macromamíferos y la datación del yacimiento clásico de Villarroya. In *Villarroya, Yacimiento Clave de la Paleontologìa Riojana*; Alberdi, M.T., Azanza, B., Cervante, E., Eds.; Instituto de Estudios Riojanos: La Rioja, Spain, 2016; pp. 267–280.

210. Athanassiou, A. A Villafranchian Hipparion-bearing mammal fauna from Sésklo (E. Thessaly, Greece): Implications for the question of Hipparion-*Equus* sympatry in Europe. *Quaternary* **2018**, *12*, 12. [CrossRef]

211. Rook, L.; Cirilli, O.; Bernor, R.L. A late occurring "Hipparion" from the middle Villafranchian of Montopoli, Italy (Early Pleistocene; MN16b; ca. 2.5 Ma). *Boll. Soc. Paleontol. Ital.* **2017**, *56*, 333–339.

212. Bajgusheava, V.S. Krupnaja Loshad Khaprovskogo Kompleksa is alluvija Severo Vostochnoto Priazovija. *Invesija Servo Kavk. Nauchnogo Zentra Vychei Shkoly* **1978**, *1*, 98–102.

213. Gromova, V.I. Istorija loshadej (roda *Equus*) v Starom Svete. Chast' 1. Obzor i opisanie form. *Tr. Paleontol. Inst.* **1949**, *17*, 1–373.

214. Forsten, A. The fossil horses (Equidae, Mammalia) from the Plio–Pleistocene of Liventsovka near Rostov–Don, Russia. *Geobios* **1997**, *31*, 645–657. [CrossRef]

215. Alberdi, M.T.; Cerdeño, E.; López–Martínez, N.; Morales, J.; Soria, M.D. La Fauna Villafranquiense de El Rincon–1 (Albacete, Castilla—La Mancha). *Estud. Geol.* **1997**, *53*, 69–93. [CrossRef]

216. Alberdi, M.T.; Ortiz–Jaureguizar, E.; Prado, J.L. A quantitative review of European stenonoid horses. *J. Paleontol.* **1998**, *72*, 371–387. [CrossRef]
217. Azzaroli, A. On *Equus livenzovensis* Baigusheva 1978 and the "stenonid" lineage of Equids. *Paleontogr. Ital.* **2000**, *87*, 1–17.
218. Bernor, R.L.; Cirilli, O.; Wang, S.Q.; Rook, L. *Equus* cf. *livenzovenzis* from Montopoli, Italy (early Pleistocene; MN16b; ca. 2.6 Ma). *Boll. Soc. Paleontol. Ital.* **2018**, *57*, 203–216.
219. Cherin, M.; Cirilli, O.; Azzarà, B.; Bernor, R.L. *Equus stenonis* (Equidae, Mammalia) from the Early Pleistocene of Pantalla (Italy) and the dispersion of stenonines horses in Europe. *Boll. Soc. Paleontol. Ital.* **2021**, *60*, 1–18.
220. Terhune, C.; Curran, S.; Croitor, R.; Drăguşin, V.; Gaudin, T.; Petrulescu, A.; Robinson, C.; Robu, M.; Werdelin, L. Early Pleistocene fauna of the Olteţ Valley of Romania: Biochronological and biogeographical implications. *Quat. Int.* **2020**, *553*, 14–33. [CrossRef]
221. Delafond, F.; Depéret, C. Etude des Gites Minéraux de la France. In *Les Terrains Tertiaires de la Bresse et Leurs Gites de Lignites et de Minerais de Fer*; Imprimerie National: Paris, France, 1893; pp. 1–332.
222. Pomel, A. *Catalogue Méthodique et Descriptif des Vertebres Fossiles Découverts dans le Bassin Hydrographique Supérieur de la Loire et Surtout Dans la Vallée de Son Affluent Principal*; L'Allirée Chez, J., Baillière, B., Eds.; Hachette: Paris, France, 1853; pp. 1–123.
223. Boule, M. *Bulletin de la Société Géologique de France*; BSGF—Earth Sciences Bulletin: Vienne, France, 1891; Volume 18, p. 801.
224. Viret, J. Le loess à bancs durcis de Saint Vallier (Drome) et sa faune de mammifères villagranchiens. *Nouv. Arch. Mus. D'histoire Nat. Lyon* **1954**, *4*, 1–200.
225. Samson, P. Les équidés fossiles de Roumanie (Pliocène moyen—Pléistocène supérieur). *Geol. Romana* **1975**, *14*, 165–354.
226. Cocchi, I. L'uomo fossile dell'Italia Centrale. *Mem. Della Soc. Ital. Sci. Nat.* **1867**, *2*, 1–180.
227. Prat, F. Les equidés villafranchiens en France, genre *Equus*. *Cah. Quat.* **1980**, *2*, 1–290.
228. Palombo, M.R.; Alberdi, M.T.; Bellucci, L.; Sardella, R. An intriguing middle–sized horse from Coste San Giacomo (Anagni Basin, central Italy). *Quat. Res.* **2017**, *87*, 347–362. [CrossRef]
229. Cirilli, O. *Equus stehlini* Azzaroli, 1964 (Perissodactyla, Equidae). A revision of the most enigmatic horse from the Early Pleistocene of Europe, with new insights on the evolutionary history of European medium- and small-sized horses. *Riv. Paleont. Strat.* **2022**, *128*, 241–265. [CrossRef]
230. Azzaroli, A. The two Villafranchian Horses of the Upper Valdarno. *Palaeontogr. Ital.* **1964**, *59*, 1–12.
231. Alberdi, M.T.; Palombo, M.R. The late Early to early Middle Pleistocene stenonoid horses from Italy. *Quat. Int.* **2013**, *288*, 25–44. [CrossRef]
232. Palombo, M.R.; Alberdi, M.T. Light and shadows in the evolution of South European stenonoid horses. *Foss. Impr.* **2017**, *73*, 115–140. [CrossRef]
233. Rook, L.; Martínez–Navarro, B. Villafranchian: The long story of a Plio–Pleistocene European large mammal biochronologic unit. *Quat. Int.* **2010**, *219*, 134–144. [CrossRef]
234. Bartolini-Lucenti, S.; Cirilli, O.; Pandolfi, L.; Bernor, R.L.; Bukhsianidze, M.; Carotenuto, F.; Lordkipanidze, D.; Tsikaridze, N.; Rook, L. Zoogeographic significance of Dmanisi large mammal assemblage. *J. Hum. Evol.* **2022**, *163*, 103125. [CrossRef]
235. Vekua, A.K. *The Lower Pleistocene Mammalian Fauna of Akhalkalaki*. Tbilisi. 1962, pp. 1–191. (In Georgian). Available online: http://www.rhinoresourcecenter.com/index.php?s=1&act=pdfviewer&id=1435388432&folder=143 (accessed on 16 July 2022).
236. Koufos, G.D. Early Pleistocene equids from Mygdonia basin (Macedonia, Greece). *Palaeontogr. Ital.* **1992**, *79*, 167–199.
237. Eisenmann, V. Equus granatensis of Venta Micena and evidence for primitive non–stenonid horses in the Lower Pleistocene. In *The Hominids and Their Environment during the Lower and Middle Pleistocene of Eurasia*; Gilbert, J., Sanchez, F., Gilbert, L., Ribot, F., Eds.; Museo de Prehistoria y Paleontología: Orce, Spain, 1995; pp. 175–189.
238. Koufos, G.D.; Vlachou, T.; Gkeme, A. The Fossil Record of Equids (Mammalia: Perissodactyla: Equidae) in Greece. In *Fossil Vertebrates of Greece*; Vlachos, E., Ed.; Springer: Berlin/Heidelberg, Germany, 2021; Volume 2, pp. 351–401.
239. Guerrero-Alba, S.; Palmqvist, P. Estudio morfometrico del caballo de Venta Micena (Orce, Granada) y su comparacion con los equidos modernos y del Plio- Pleistoceno en el viejo mundo. *Paleontol. Evol.* **1997**, *30*, 93–148.
240. Belmaker, M. Early Pleistocene faunal connections between Africa and Eurasia: An ecological perspective. In *Out of Africa I: The First Hominin Colonization of Eurasia*; Fleagle, J.G., Shea, J.J., Grine, F.E., Baden, A.L., Leakey, R.E., Eds.; Springer: Berlin/Heidelberg, Germany, 2010; pp. 183–205.
241. Wüst, E. Untersuchungen über das Pliozan und das Alteste Pleistozan Thüringens, nordlich vom Thüringer Walde und westlich von der Saale. *Abh. Nat. Ges. Halle* **1900**, *23*, 1–352.
242. Vekua, A.K. The Lower Pleistocene mammalian fauna of Akhalkalaki (Southern Georgia, USSR). *Palaeontogr. Ital.* **1986**, *74*, 63–96.
243. Musil, R. Die Pferdefunde der lokalität Stránská Skála. *Anthropos* **1972**, *20*, 185–192.
244. Boulbes, N.; Van Asperen, E.N. Biostratigraphy and palaeoecology of European *Equus*. *Front. Ecol. Evol.* **2019**, *7*, 301. [CrossRef]
245. Alberdi, M.T. Estudio de los caballos de los yacimientos de Fuente Nueva–3 y Barranco León–5 (Granada). In *Ocupaciones Humanas en el Pleistoceno Inferior Y Medio de la Cuenca de Guadix–Baza*; Toro, I., Martínez–Navarro, B., Agustí, J., Eds.; Arqueología Monografías, Junta de Andalucía, Consejería de Cultura: Andalucía, Spain, 2010; pp. 291–306.
246. Koufos, G.D.; Kostopoulos, D.S.; Sylvestrou, I.A. *Equus apolloniensis* n. sp. (Mammalia, Equidae) from the latest Villafranchian locality of Apollonia, macedonia, Greece. *Palaeontol. Evol.* **1997**, *30*, 49–76.
247. Gkeme, A.G.; Koufos, G.D.; Kostopoulos, D.S. Reconsidering the Equids from the Early Pleistocene Fauna of Apollonia 1 (Mygdonia Basin, Greece). *Quaternary* **2021**, *4*, 12. [CrossRef]

248. Eisenmann, V.; Boulbes, N. New results on equids from the Early Pleistocene site of Untermassfeld. In *The Pleistocene of Untermassfeld Near Meiningen (Thütingen, Germany)*; Kahlke, R.D., Ed.; Romisch Germanischen Zentralmuseums: Weimar, Germany, 2020; pp. 1295–1321.

249. Musil, R. Die Equiden–Reste aus dem Unterpleistozän von Untermassfeld. In *Das Pleistozän von Untermassfeld bei Meiningen (Thüringen)*; Kahlke, R.D., Ed.; Monographien, Römisch–Germanisches Zentralmuseum, Forschungsinstitut für vor und Frühgeschichte: Liebniz, Germany, 2001; pp. 557–587.

250. Forsten, A. A review of *Equus stenonis* Cocchi (Perissodactyla, Equidae) and related forms. *Quat. Sci. Rev.* **2000**, *18*, 1373–1408. [CrossRef]

251. Lister, A.M.; Parfitt, S.A.; Owen, F.J.; Collinge, S.E.; Breda, M. Metric analysis of ungulate mammals in the early Middle Pleistocene of Britain, in relation to taxonomy and biostratigraphy: II: Cervidae, Equidae and Suidae. *Quat. Int.* **2010**, *228*, 157–179. [CrossRef]

252. Tsoukala, E. Contribution to the Study of the Pleistocene Fauna of Large Mammals (Carnivora, Perissodactyla, Artiodactyla) from Petralona Cave Chalkidiki (N. Greece). Ph.D. Thesis, Aristotle University of Thessaloniki, Thessaloniki, Serres, 1989.

253. Azzaroli, A. On a Late Pleistocene ass from Tuscany with notes on thehistory of ass. *Palaeontol. Ital.* **1979**, *71*, 27–47.

254. Azzaroli, A. Pleistocene and living Horses of the old World. *Palaeontogr. Ital.* **1966**, *61*, 1–15.

255. Caloi, L. Il genere *Equus* nell'Italia centrale. *Studi Geol. Camerti* **1995**, 469–486.

256. Eisenmann, V. Pliocene and Pleistocene Equids: Palaeontology versus molecular biology. *Cour. Forsch. Senckenberg* **2006**, *256*, 71–89.

257. Regalia, E. Sull'*Equus (Asinus) hydruntinus* Regalia della grotta di Romanelli (Castro, Lecce). *Arch. Antropol. Etnol.* **1907**, *37*, 375–390.

258. Vishnyatsky, L.B. The Palaeolithic of Central Asia. *J. World Prehistory* **1999**, *13*, 69–122. [CrossRef]

259. Musil, R. Die Equiden–Reste aus dem Pleistozän von Süssenborn bei Weimar. *Paläontologische Abh. Paläozool.* **1969**, *3*, 617–666.

260. Azzaroli, A. Ascent and decline of monodactyl equids: A case for prehistoric overkill. *Ann. Zool. Fenn.* **1992**, *28*, 151–163.

261. Bennett, E.A.; Champlot, S.; Peters, J.; Arbuckle, B.S.; Guimaraes, S.; Pruvost, M.; Bar-David, S.; Davis, S.J.M.; Gautier, M.; Kaczensky, P.; et al. Taming the late Quaternary phylogeography of the Eurasiatic wild ass through ancient and modern DNA. *PLoS ONE* **2017**, *12*, e0174216.

262. Aurell-Garrido, J.; Madurell-Malapeira, J.; Alba, D.M.; Moyà-Solà, S. The stenonian and caballoid equids from the Pleistocene section of Vallparadís (Terrassa, Catalonia, Spain). *Cidaris* **2010**, *30*, 61–66.

263. Madurell-Malapeira, J.; Minwer-Barakat, R.; Alba, D.M.; Garcés, M.; Gómez, M.; Aurell-Garrido, J.; Ros-Montoya, S.; Moya-Sola, S.; Berastegui, X. The Vallparadís section (Terrassa, Iberian peninsula) and the latest Villafranchian faunas of Europe. *Quat. Sci. Rev.* **2010**, *29*, 2972–2982. [CrossRef]

264. Boddaert, P. *Elenchus Animalium, Volumen 1: Sistens Quadrupedia Huc Usque Nota, Erorumque Varietates*; C. R. Hake: Rotterdam, The Netherlands, 1785; p. 174.

265. Eisenmann, V.; Kuznetsova, T.V. Early Pleistocene equids (Mammalia, Perissodactyla) of Nalaikha (Mongolia) and the emergence of modern *Equus* Linnaeus, 1758. *Geodiversitas* **2004**, *26*, 535–561.

266. van Asperen, E.N. Implications of age variation and sexual dimorphism in modern equids forMiddle Pleistocene equid taxonomy. *Int. J. Osteoarchaeol.* **2013**, *23*, 1–12. [CrossRef]

267. van Asperen, E.N. Ecomorphological adaptations to climate and substrate in late middle Pleistocene caballoid horses. *Palaeogeogr. Palaeoclimatol. Palaeoecol.* **2010**, *297*, 584–596. [CrossRef]

268. Churcher, C.S.; Richardson, M.L. Equidae. In *Evolution of African Mammals*; Maglio, V.J., Cooke, H.S.B., Eds.; Harvard University Press: Cambridge, MA, USA, 1978; pp. 379–422.

269. Bernor, R.L.; Armour-Chelu, M.; Gilbert, H.; Kaiser, T.; Schulz, E. Equidae. In *Cenozoic Mammals of Africa*; Werdelin, L., Sanders, W., Eds.; University of California Press: Berkeley, CA, USA, 2010; pp. 685–721.

270. Duval, M.; Sahnouni, M.; Parés, J.P.; van der Made, J.; Abdessadok, S.; Harichane, Z.; Cheheb, R.C.; Boulaghraif, K.; Pérez-Gonzalez, A. The Plio-Pleistocene sequesnce of Oued Boucherti (Algeria): A unique chronologically-constrained archeological and paleontological record in North Africa. *Quat. Sci. Rev.* **2021**, *271*, 107116. [CrossRef]

271. Sahnouni, M.; Pares, J.M.; Duval, M.; Caceres, I.; Harichane, Z.; Van der Made, J.; Perez-Gonzalez, A.; Abdessadok, S.; Kandi, N.; Derradji, A.; et al. 1.9-million- and 2.4 million-year-old artifacts and stone toolcutmarked bones from Aïn Boucherit, Algeria. *Science* **2017**, *362*, 1297–1307. [CrossRef] [PubMed]

272. Villalta, J.F.; Crusafont, M. Les gisements de mammifères du Néogène espagnol. *Compt. Rendus Soc. Geol. Fr.* **1957**, *1*, 9–10.

273. Cirilli, O.; Zouhri, S.; Boughabi, S.; Benvenuti, M.; Bernor, R.L.; Papini, M.; Rook, L. The hipparionine horses (Perissodactyla: Mammalia) from the Late Miocene of Tizi N'Tadderht (southern Ouarzazate basin; Central High Atlas; Morocco). *Riv. Paleontol. Strat.* **2020**, *126*, 1–12.

274. Bernor, R.L.; Boaz, N.T.; Cirilli, O.; El-Shawaihdi, M.; Rook, L. Sahabi *Eurygnathohippus feibeli*; its systematic, stratigraphic, chronologic and biogeographic contexts. *Riv. Paleont. Strat.* **2020**, *126*, 561–581.

275. Bernor, R.L.; Harris, J. Systematics and Evolutionary Biology of the Late Miocene and Early Pliocene Hipparionine Horses from Lothagam, Kenya. In *Lothagam: The Dawn of Humanity in Eastern Africa*; Leakey, M., Harris, J., Eds.; Columbia University Press: New York, NY, USA, 2003; pp. 387–438.

276. Bernor, R.L.; Scott, R.S. New Interpretations of the Systematics, Biogeography and Paleoecology of the Sahabi Hipparions (latest Miocene), Libya. *Geodiversitas* **2003**, *25*, 297–319.

277. Pomel, A. Les equidés: Carte géologie de l'Algérie. *Monogr. Paléontol.* **1897**, *12*, 1–44.

278. Hooijer, D.A.; Maglio, V.J. The earliest Hipparion south of the Sahara, in the late Miocene of Kenya. *Proc. K. Ned. Akad. Wet.* **1973**, *76*, 311–315.

279. Sun, B.; Zhang, X.; Liu, Y.; Bernor, R.L. *Sivalhippus ptychodus* and *Sivalhippus platyodus* (Perissodactyla, Mammalia) from the late Miocene of China. *Riv. Ital. Paleo. Strat.* **2018**, *124*, 1–22.

280. Bernor, R.L.; Kaiser, T. Systematics and Paleoecology of the Earliest Pliocene Equid, *Eurygnathohippus hooijeri* n. sp. from Langebaanweg, South Africa. *Mitt. Aus Hamburischen Zool. Mus. Inst.* **2006**, *103*, 147–183.

281. Hooijer, D.A. The late Pliocene Equidae of Langebaanweg, Cape Province, South Africa. *Zool. Verh.* **1976**, *148*, 3–39.

282. Bernor, R.L.; Gilbert, H.; Semprebon, G.; Simpson, S.; Semaw, S. *Eurygnathohippus woldegabrieli* sp. nov. (Perissodactyla: Mammalia) from the Middle Pliocene of Aramis, Ethiopia (4.4 Ma.). *J. Vert. Paleo.* **2013**, *33*, 1472–1485. [CrossRef]

283. Eisenmann, V. Le protostylide: Valeur systematique et signification phylétique chez les espèces actuelles et fossiles du genre *Equus* (Perissodactyla, Mammalia). *Z. Saugetierkd.* **1976**, *41*, 349–365.

284. Eisenmann, V. Family Equidae. The Fossil Ungulates: Proboscidea, Perissodactyla, and Suidae. In *Koobi Fora Research Project*; Harris, J.M., Ed.; Clarendon Press: London, UK, 1983; Volume 2, pp. 156–214.

285. Bernor, R.L.; Scott, R.S.; Haile-Selassie, Y. A Contribution to the Evolutionary History of Ethiopian Hipparionine Horses: Morphometric Evidence from the Postcranial Skeleton. *Geodiversitas* **2005**, *27*, 133–158.

286. Eisenmann, V.; Geraads, D. *Hipparion pomeli* sp. nov. from the late Pliocene of Ahl al Oughlam, Morocco, and a revision of the relationships of Pliocene and Pleistocene African hipparions. *Palaeo Afr.* **2006**, *42*, 51–98.

287. van Hoepen, E.C.N. Fossiele Pferde van Cornelia. *Bloemfontein* **1930**, *2*, 13–24.

288. Leakey, L.S.B. A Preliminary Report on the Geology and Fauna. In *Olduvai Gorge 1951–1961*; Cambridge University Press: Cambridge, UK, 1965; Volume 1.

289. Armour-Chelu, M.; Bernor, R.L.; Mittmann, H.-W. Hooijer's hypodigm for "*Hipparion*" cf. *ethiopicum* (Equidae, Hipparioninae) from Olduvai, Tanzania and comparative material from the East African Plio-Pleistocene. *Abh. Bayer. Akad. Wiss.* **2006**, *30*, 15–24.

290. Armour-Chelu, M.; Bernor, R.L. Equidae. In *Geology and Paleontology of Laetoli*; Harrison, T., Ed.; Springer: Berlin/Heidelberg, Germany, 2011; pp. 295–326.

291. van der Made, J.; Boulaghraif, K.; Chelli Cheheb, R.; Caceres, I.; Harichane, Z.; Sahnouni, M. The last North African hipparions —Hipparion decline and extinction follows a common pattern. *N. Jb. Geol. Palaont. Abh.* **2022**, *303*, 39–97. [CrossRef]

292. Bernor, R.L.; Coillot, T.; Wolf, D. Phylogenetic Signatures in the Juvenile Skull and Dentition of Olduvai Gorge *Eurygnathohippus cornelianus* (Mammalia: Equidae). *Riv. Ital. Paleontol. Stratigr.* **2014**, *120*, 243–252.

293. Hopwood, A.T. Die fossilen Pferde von Oldoway. *Wiss. Ergeb. Oldoway Exped.* **1937**, *4*, 112–136.

294. Thomas, P. Recherches stratigraphiques et paleontologiques sur quelques formations d'eau douce de l'Algerie. *Mem. Soc. Geol. Fr.* **1884**, *3*, 1–50.

295. Eisenmann, V. *Les Chevaux (Equus Sensu Lato) Fossiles et Actuels: Crânes et Dents Jugales Supérieures*; Editions du Centre National de la Recherche Scientifique: Paris, France, 1980.

296. Arambourg, C. Vertébrés continentaux du Miocène supérieur de l'Afrique du Nord. *Publ. Serv. Cart. Geol. L'algerie* **1959**, *4*, 1–161.

297. Arambourg, C. Les Vertébrés du Pléistocène de l'Afrique du Nord. *Arch. Mus. Natl. D'histoiry Nat.* **1970**, *10*, 1–127.

298. Azzaroli, A. On Villafranchian Palaearctic Equids and their allies. *Palaeontogr. Ital.* **1982**, *72*, 74–97.

299. Geraads, D.; Raynal, J.-P.; Eisenmann, V. The earliest human occupation of North Africa: A reply to Sahnouni et al. (2002). *J. Hum. Evol.* **2004**, *46*, 751–761. [CrossRef] [PubMed]

300. Tchernov, E. Memoires et Travaux du Centre de Recherche Francais de Jerusalem 5. In *The Lower Pleistocene Mammals of 'Ubeidiya (Jordan Valley)*; Association Paleorient: Paris, France, 1986.

301. Yeshurun, R.; Zaidner, Y.; Eisenmann, V.; Martinez-Navarro, B.; Bar-Oz, G. Lower Paleolithic hominin ecology at the fringe of the desert: Faunal remains from Bizat Ruhama and Nahal Hesi, Northern Negev, Israel. *J. Hum. Evol.* **2011**, *60*, 492–507. [CrossRef]

302. Tchernov, E. The Faunal Sequence of the Southwest Asian Middle Paleolithic in Relation to Hominid Dispersal Events. In *Neandertals and Modern Humans in Western Asia*; Akazawa, T., Aoki, K., Bar-Yosef, O., Eds.; Springer: Berlin/Heidelberg, Germany, 2002; pp. 77–94.

303. Broom, R. On evidence of a large horse recently extinct in South Africa. *Ann. S. Afr. Mus.* **1909**, *7*, 281–282.

304. Eisenmann, V. *Equus capensis* (Mammalia, Perissodactyla) from Elandsfontein. *Palaeontol. Afr.* **2000**, *36*, 91–96.

305. Churcher, C.S. The fossil equidae from the Krugersdorp Caves. *Ann. Transvaal Mus.* **1970**, *26*, 145–168.

306. Morom, N.; Lazagabaster, I.A.; Shafir, R.; Natalio, F.; Eisenmann, V.; Kolsa Horwitz, L. The Middle Pleistocene mammalian fauna of Oumm Qatafa Cave, Judean Desert: Taxonomy, taphonomy and paleoenvironment. *J. Quat. Sci.* **2022**, *37*, 612–638. [CrossRef]

307. Bagtache, B.; Hadjouis, D.; Eisenmann, V. Presence d'un *Equus* caballine (*E. algericus* n. sp.) et d'une autre espece nouvelle d'*Equus* (*E. melkiensis* n. sp.) dans l'Atérien des Allbroges, Algerie. *Comptes Rendus Séances L'académie Sci. Paris* **1984**, *298*, 609–612.

308. Aoüraghe, H.; Debenath, A. Les Equides du Pleistocene Superior de la grotte Zourah a el Harhoura, Maroc. *Quaternarie* **1999**, *10*, 283–292. [CrossRef]

309. Oustalet, E. Une nouvelle espèce de zèbre, le zèbre de Grévy (*Equus grevyi*). *Nature* **1882**, *10*, 12–14.

310. O'Brien, K.; Tryon, C.A.; Blegen, N.; Kimeu, B.; Rowan, J.; Faith, J.T. First appearance of Grévy's zebra (*Equus grevyi*), from the Middle Pleistocene of the Kapthurin Formation, Kenya, sheds lights on the evolution and paleoecology of large zebras. *Quat. Sci. Rev.* **2021**, *256*, 106835. [CrossRef]

311. Leonard, J.A.; Rohland, N.; Glaberman, S.; Fleischer, R.C.; Caccone, A.; Hofreiter, M. A rapid loss of stripes: The evolutionary history of the extinct quagga. *Biol. Lett.* **2005**, *1*, 291–295. [CrossRef]

312. Pedersen, C.E.T.; Albrechtsen, A.; Etter, P.D.; Johnson, E.A.; Orlando, L.; Chikhi, L.; Siegismund, H.R.; Heller, R. A southern African origin and cryptic structure in highly mobile plains zebras. *Nat. Ecol. Evol.* **2018**, *2*, 491–498. [CrossRef]

313. Linnaeus, C. *Systema Naturae per Regna Tria Naturae, Secundum Classes, Ordines, Genera, Species cum Characteribus, Differentiis, Synonimis, Locis: Editio Decima, Reformata*; Laurentii Salvii: Stockholm, Sweden, 1758; pp. 1–824.

314. von Heuglin, T. *Systematische Übersicht der Vögel Nord-Ost-Afrika's: Mit Einschluss der Arabischen Küste des Rothen Meeres und der Nil-Quellen-Länder Südwärts bis zum IV. Grade Nördl. Breite*; K. Akademie der Wissenschaften: Berlin, Germany, 1866.

315. Churcher, C.S. Oldest ass recovered from Olduvai Gorge, Tanzania, and the origin of asses. *J. Paleontol.* **1982**, *56*, 1124–1132.

316. Noack, N. Neues aus der Tierhandlung von Karl Hagenbeck, sowie aus dem Zoologischen Garten in Hamburg. *Zool. Gart.* **1884**, *25*, 100–115.

317. Groves, C.P. The taxonomy, distribution, and adaptations of recent equids. In *Equids in the Ancient World*; Meadow, R.H., Uerpmann, H.-P., Eds.; Reichert: Wiesbaden, Germany, 1986; pp. 11–65.

318. Denzau, G.; Denzau, H. (Eds.) *Wildesel*; Thorbecke: Stuttgart, Germany, 1999; pp. 1–221.

319. Groves, C.P.; Smeenk, C. The nomenclature of the African wild ass. *Zool. Meded.* **2007**, *81*, 121–135.

320. Sam, Y. African origins of modern asses as seen from paleontology and DNA: What about the Atlas wild ass? *Geobios* **2020**, *58*, 73–84. [CrossRef]

321. Berggren, W.A.; Van Couvering, J.A. The late Neogene: Biostratigraphy, geochronology and paleoclimatology of the last 15 million years in marine and continental sequences. *Palaeogeogr. Palaeoclimatol. Palaeoecol.* **1974**, *16*, 1–216.

322. Bell, C.J.; Lundelius, E.L., Jr.; Barnosky, A.D.; Graham, R.W.; Lindsay, E.H.; Ruez, D.R., Jr.; Semken, H.A.; Webb, S.D.; Zakrzwski, R.J. The Blancan, Irvingtonian, and Rancholabrean Mammal Ages. In *Late Cretaceous and Cenozoic Mammals of North America: Biostratigraphy and Geochronology*; Woodburne, M.O., Ed.; Columbia University Press: New York, NY, USA, 2004; pp. 232–314.

323. Woodburne, M.O. The great American biotic interchange: Dispersals, tectonics, climate, sea level and holding pens. *J. Mammal. Evol.* **2010**, *17*, 245–264. [CrossRef] [PubMed]

324. Avilla, L.S.; Bernardes, C.; Mothé, D. A new genus for *Onohippidium galushai* Macfadden and Skinner, 1979 (Mammalia, Equidae), from the late Hemphillian of North America. *J. Vertebr. Paleontol.* **2015**, *35*, e925909. [CrossRef]

325. Coltorti, M.; Abbazzi, L.; Ferretti, M.P.; Iacumin, P.; Paredes Rios, F.; Pellegrini, M.; Pieruccini, P.; Rustioni, M.; Rook, L. Last Glacial mammals in South America: A new scenario from the Tarija Basin (Bolivia). *Naturwissenschaften* **2007**, *94*, 288–299. [CrossRef] [PubMed]

326. Bernor, R.L.; Golich, U.B.; Harzahauser, M.; Semprebon, G.M. The Pannonian C hipparions from the Vienna Basin. *Palaeogeogr. Palaeoclimatol. Palaeoecol.* **2017**, *476*, 28–41. [CrossRef]

327. Barry, J.C.; Behrensmeyer, A.K.; Badgley, C.E.; Flynn, L.J.; Peltonen, H.; Cheema, I.U.; Pilbeam, D.; Lindsay, E.H.; Raza, S.M.; Rajpar, A.R.; et al. The Neogene Siwaliks of the Potwar Plateau, Pakistan. In *Fossil Mammals of Asia*; Wang, X., Flynn, L.J., Fortelius, M., Eds.; Columbia University Press: New York, NY, USA, 2013; pp. 373–399.

328. Opdyke, N.D.; Lindsay, E.; Johnson, G.D.; Johnson, N.; Tahirkheli, R.A.K.; Mirza, M.A. Magnetic polarity stratigraphy and vertebrate paleontology of the upper siwalik subgroup of northern Pakistan. *Palaeogeogr. Palaeoclimatol. Palaeoecol.* **1979**, *27*, 1–34. [CrossRef]

329. Patnaik, R. Indian Neogene Siwalik mammalian biostratigraphy: An overview. In *Fossil Mammals of Asia: Neogene Biostratigraphy and Chronology*; Wang, X., Flynn, L.J., Fortelius, M., Eds.; Columbia University Press: New York, NY, USA, 2013; pp. 423–444.

330. Gaur, R.; Chopra, S.R.K. On a new subspecies of *Equus* from Pinjor Formation of Upper Sivaliks—With remarks on Sivalik *Equus*. *J. Palaeontol. Soc. India* **1984**, *29*, 19–25.

331. Zhu, R.X.; Potts, R.; Pan, Y.X.; Yao, H.T.; Lü, L.Q.; Zhao, X. Early evidence of the genus *Homo* in East Asia. *J. Hum. Evol.* **2008**, *55*, 1075–1085. [CrossRef]

332. Chauhan, P.R. Large mammal fossil occurrences and associated archaeological evidence in Pleistocene contexts of peninsular India and Sri Lanka. *Quater. Int.* **2008**, *192*, 20–42. [CrossRef]

333. Van Couvering, J.; Delson, E. African Land Mammal Ages. *J. Vert. Pal.* **2020**, *40*, 5. [CrossRef]

334. Werdelin, L. Chronology of Neogene Mammal localities. In *Cenozoic Mammals of Africa*; Werdelin, L., Sanders, W.L., Eds.; University of California Press: Berkeley, CA, USA, 2010; pp. 27–43.

335. Pickford, M. Preliminary Miocene mammalian biostratigraphy for Western Kenya. *J. Hum. Evol.* **1981**, *10*, 73–97. [CrossRef]

336. Gilbert, H.; Bernor, R.L. Equidae. In *Homo Erectus—Pleistocene Evidence from the Middle Awash, Ethiopia*; Gilbert, H., Asfaw, B., Eds.; University of California Press: Berkeley, CA, USA, 2008; pp. 133–166.

337. Geist, V. The relation of social evolution and dispersal in ungulates during the Pleistocene, with emphasis on Old World deer and the genus Bison. *Quat. Res.* **1971**, *1*, 285–315. [CrossRef]

338. Guthrie, D.R. *Frozen Fauna of the Mammoth Steppe—The Story of Blue Babe*; University of Chicago Press: Chicago, IL, USA, 1990.

339. McNab, B.K. Geographic and temporal correlations of mammalian size reconsidered: A resource rule. *Oecologia* **2010**, *164*, 13–23. [CrossRef] [PubMed]

340. Saarinen, J. Ecometrics of Large Herbivorous Land Mammals in Relation to Climatic and Environmental Changes during the Pleistocene. Ph.D. Thesis, University of Helsinki, Helsinki, Finland, 2014.

341. Damuth, J.; Janis, C. On the relationships between hypsodonty and feeding ecology in ungulate mammals, and its utility in paleoecology. *Biol. Rev. Camb. Philos. Soc.* **2011**, *86*, 733–758. [CrossRef]
342. Strömberg, C. Using phytolith assemblages to reconstruct the origin and spread of grass-dominated habitats in the great plains of North America during the late Eocene to early Miocene. *Palaeogeo. Palaeoclim. Palaeoecol.* **2004**, *207*, 239–275. [CrossRef]
343. Strömberg, C. Decoupled taxonomic radiation and ecological expansion of open-habitat grasses in the Cenozoic of North America. *Proc. Natl. Acad. Sci. USA* **2005**, *102*, 11980–11984. [CrossRef]
344. Strömberg, C. Evolution of Grasses and Grassland Ecosystems. *Annu. Rev. Earth Planet. Sci.* **2011**, *39*, 517–544. [CrossRef]
345. Janis, C.; Sahney, S.; Benton, M. Grit not grass: Concordant patterns of early origin of hypsodonty in Great Plains ungulates and Glires. *Palaeogeo. Palaeoclim. Palaeoecol.* **2012**, *366*, 1–10.
346. Semprebon, G.M.; Rivals, F.; Janis, C.M. The role of grass vs. exogenous abrasives in the paleodietary patterns of North American ungulates. *Front. Ecol. Evol.* **2019**, *7*, 65. [CrossRef]
347. Strömberg, C.; McInerney, F. The Neogene transition from C3 to C4 grasslands in North America: Assemblage analysis of fossil phytoliths. *Paleobiology* **2011**, *37*, 50–71. [CrossRef]
348. Pérez-Crespo, V.A.; Bravo-Cuevas, V.M.; Arroyo-Cabrales, J. Feeding habits of *Equus conversidens* and *Haringtonhippus francisci* from Valsequillo, Puebla, México. *Hist. Biol.* **2021**, *34*, 1252–1259. [CrossRef]
349. Hulbert, R.C. Taxonomic evolution in North American Neogene horses (Subfamily Equinae): The rise and fall of an adaptive radiation. *Paleobiology* **1993**, *19*, 216–234. [CrossRef]
350. Hulbert, R.C.; MacFadden, B.J. Morphological transformation and cladogenesis at the base of the adaptive radiation of Miocene hypsodont horses. *Am. Mus. Novit.* **1991**, *3000*, 1–61.
351. Radinsky, L. Allometry and reorganization of horse skull proportions. *Science* **1983**, *221*, 1189–1191. [CrossRef] [PubMed]
352. Radinsky, L. Ontogeny and phylogeny in horse skull evolution. *Int. J. Org. Evol.* **1984**, *38*, 1–15. [CrossRef]
353. Hulbert, R.C. Phylogenetic interrelationships and evolution of North American late Neogene Equidae. In *The Evolution of Perissodactyls*; Prothero, D.R., Schoch, R.M., Eds.; Clarendon Press: Oxford, UK, 1989; pp. 176–196.
354. Osborn, H.F. *The Age of Mammals*; MacMillan Company: New York, NY, USA, 1910.
355. Scott, W.B. *A History of Land Mammals in the Western Hemisphere*; MacMillian Company: New York, NY, USA, 1937.
356. Simpson, G.G. *Tempo and Mode in Evolution*; Columbia University Press: New York, NY, USA, 1944.
357. Stirton, R.A. Observation of evolutionary rates in hypsodonty. *Evolution* **1947**, *1*, 32–41. [CrossRef]
358. Webb, S.D. A history of savanna vertebrates in the New World. Part I: North America. *Annu. Rev. Ecol. Syst.* **1977**, *8*, 355–380. [CrossRef]
359. Webb, S.D. The rise and fall of the late Miocene ungulate fauna in North America. In *Coevolution*; Nitecki, M.H., Ed.; University of Chicago Press: Chicago, IL, USA, 1983; pp. 267–306.
360. Janis, C.M. The significance of fossil ungulate communities as indicators of vegetation structure and climate. In *Fossils and Climate*; Brenchley, P.J., Ed.; John Wiley & Sons: New York, NY, USA, 1984; pp. 85–104.
361. Janis, C.M. Tertiary mammal evolution in the context of changing climates, vegetation, and tectonic events. *Annu. Rev. Ecol. Syst.* **1993**, *24*, 467–500. [CrossRef]
362. Webb, S.D.; Opdyke, N.D. Global climatic influence on Cenozoic land mammal faunas. National Research Council Panel on Effects of Past Global Change on Life. In *Effects of Past Global Change on Life*; National Academies Press: Washington, DC, USA, 1995.
363. Cohen, J.; DeSantis, L.R.G.; Lindsay, E.L.; Meachen, J.A.; O'Keefe, F.R.; Southon, J.R.; Binder, W.J. Dietary stability inferred from dental mesowear analysis in large ungulates from Racho La Brea and opportunistic feeding during the late Pleistocene. *Palaeogeo. Palaeoclim. Palaeoecol.* **2021**, *570*, 110360. [CrossRef]
364. Jones, D.B.; DeSantis, L.R.G. Dietary ecology od ungulates from La Brea tar pits in southern California: A multi-proxy approach. *Palaeogeo. Palaeoclim. Palaeoecol.* **2017**, *466*, 110–117. [CrossRef]
365. Kaiser, T.M.; Solounias, N.; Fortelius, M.; Bernor, R.L.; Schrenk, F. Tooth mesowear analysis on *Hippotherium primigenium* from the Vallesian Dinotheriensande (Germany). A blind test study. *Carolinea* **2000**, *58*, 103–114.
366. Kaiser, T.M.; Bernor, R.L.; Scott, R.S.; Franzen, J.L.; Solounias, N. New Interpretations of the Systematics and Palaeoecology of the Dorn Dürkheim 1 Hipparions (Late Miocene, Turolian Age [MN11]) Rheinhessen, German. *Senckenberg Lethea* **2003**, *83*, 103. [CrossRef]
367. Rivals, F.; Lister, A.M. Dietary flexibility and niche partitioning of large herbivores through the Pleistocene of Britain. *Quat. Sci. Rev.* **2016**, *146*, 116–133. [CrossRef]
368. Head, M.J. Pollen and dinoflagellates from the Red Crag at Walton-on-the-Naze, Essex: Evidence for a mild climatic phase during the early Late Pliocene of eastern England. *Geol. Mag.* **1998**, *135*, 803–817. [CrossRef]
369. Zalasiewicz, J.A.; Mathers, S.J.; Hughes, M.J.; Gibbard, P.L.; Peglar, S.M.; Harland, R.; Nicholson, R.A.; Boulton, G.S.; Cambridge, P.; Wealthall, G.P. Stratigraphy and Palaeoenvironments of the Red Crag and Norwich Crag Formations between Aldeburgh and Sizewell, Suffolk, England. *Philos. Trans. R. Soc. Lond.* **1988**, *322*, 221–272.
370. Rivals, F.; Athanassiou, A. Dietary adaptations in an ungulate community from the late pliocene of greece. *Palaeogeogr. Palaeoclimatol. Palaeoecol.* **2008**, *265*, 134–139. [CrossRef]

371. Strani, F.; DeMiguel, D.; Bellucci, L.; Sardella, R. Dietary response of early pleistocene ungulate communities to the climate oscillations of the Gelasian/Calabrian transition in central Italy. *Palaeogeogr. Palaeoclimatol. Palaeoecol.* **2018**, *499*, 102–111. [CrossRef]

372. Strani, F.; DeMiguel, D.; Sardella, R.; Bellucci, L. Paleoenvironments and climatic changes in the italian peninsula during the early pleistocene: Evidence from dental wear patterns of the ungulate community of Coste San Giacomo. *Quat. Sci. Rev.* **2015**, *121*, 28–35. [CrossRef]

373. Strani, F.; Pushkina, D.; Bocherens, H.; Bellucci, L.; Sardella, R.; DeMiguel, D. Dietary adaptations of early and middle pleistocene equids from the anagni basin (Frosinone, Central Italy). *Front. Ecol. Evol.* **2019**, *7*, 176. [CrossRef]

374. Saarinen, J.; Oksanen, O.; Zliobaite, I.; Fortelius, M.; DeMiguel, D.; Azanza, B.; Bocherens, H.; Luzón, C.; Solano-Garcia, J.; Yravedra, J.; et al. Pliocene to Middle Pleistocene climate history in the Guadix-Baza Basin, and the environmental conditions of early *Homo* in Europe. *Quat. Sci. Rev.* **2021**, *268*, 107132. [CrossRef]

375. Strani, F.; DeMiguel, D.; Alba, D.; Moyà-Solà, S.; Bellucci, L.; Sardella, R.; Madurell-Malapeira, J. The effects of the "0.9 Ma event" on the Mediterranean ecosystems during the Early-Middle Pleistocene transition as revealed by dental wear patterns of fossil ungulates. *Quat. Sci. Rev.* **2019**, *210*, 80–89. [CrossRef]

376. Blain, H.-A.; Lozano-Fernández, I.; Agustí, J.; Bailon, S.; Menéndez, L.; Patrocinio, M.; Ros-Montoya, S.; Manuel, J.; Toro-Moyano, I.; Martínez-Navarro, B.; et al. Refining upon the climatic background of the early Pleistocene hominid settlement in western Europe: Barranco León and Fuente Nueva-3 (Guadix- Baza Basin, SE Spain). *Quat. Sci. Rev.* **2016**, *144*, 132–144. [CrossRef]

377. Sánchez-Bandera, C.; Oms, O.; Blain, H.-A.; Lozano-Fernández, I.; Bispal-Chinesta, J.F.; Agustí, J.; Saarinen, J.; Fortelius, M.; Titton, S.; Serrano-Ramos, A.; et al. New stratigraphically constrained palaeoenvironmental reconstructions for the first human settlement in Western Europe: The Early Pleistocene herpetofaunal assemblages from Barranco León and Fuente Nueva 3 (Granada, SE Spain). *Quat. Sci. Rev.* **2020**, *243*, 106466. [CrossRef]

378. Kahlke, R.D.; García, N.; Kostopoulos, D.S.; Lacombat, F.; Lister, A.M.; Mazza, P.P.; Spassov, N.; Titov, V.V. Western Palaearctic palaeoenvironmental conditions during the Early and early Middle Pleistocene inferred from large mammal communities, and implications for hominin dispersal in Europe. *Quat. Sci. Rev.* **2011**, *30*, 1368–1395. [CrossRef]

379. Rivals, F.; Julien, M.-A.; Kuitems, M.; Van Kolfschoten, T.; Serangeli, J.; Drucker, D.G.; Bocherens, H.; Conard, N.J. Investigation of equid paleodiet from Schöningen 13 II-4 through dental wear and isotopic analyses: Archaeological implications. *J. Hum. Evol.* **2015**, *89*, 129–137. [CrossRef] [PubMed]

380. Roberts, M.B. Excavation of the Lower Paleolithic site at Amey's Eartham Pit, Boxgrove, West Sussex: A preliminary report. *Proc. Prehist. Soc.* **1986**, *52*, 215–245. [CrossRef]

381. Urban, B.; Bigga, G. Environmental reconstruction and biostratigraphy of late Middle Pleistocene lakeshore deposits at Schöningen. *J. Hum. Evol.* **2015**, *89*, 57–70. [CrossRef]

382. Strani, F.; DeMiguel, D.; Bona, F.; Sardella, R.; Biddittu, I.; Bruni, L.; De Castro, A.; Guadagnoli, F.; Bellucci, L. Ungulate dietary adaptations and palaeoecology of the Middle Pleistocene site of Fontana Ranuccio (anagni, central italy). *Palaeogeogr. Palaeoclimatol. Palaeoecol.* **2018**, *496*, 238–247. [CrossRef]

383. Leroi-Gourhan, A. Pollen analysis of sediment samples from Gough's Cave, Cheddar. *Proc. Univ. Bristol. Spelaeol. Soc.* **1986**, *17*, 141–144.

384. Bates, M.R.; Bates, C.R.; Gibbard, P.L.; Macphail, R.I.; Owen, F.J.; Parfitt, S.A.; Preece, R.C.; Roberts, M.B.; Robinson, J.E.; Whittaker, J.E.; et al. Late Middle Pleistocene deposits at Norton Farm on the West Sussex coastal plain, southern England. *J. Quat. Sci.* **2000**, *15*, 61–89. [CrossRef]

385. West, R.G.; Lambert, C.A.; Sparks, B.W.; Dickson, J.H. Interglacial deposits at Ilford, Essex. *Philos. Trans. R. Soc.* **1964**, *247*, 185–212.

386. Gibbard, P.L. *Pleistocene History of the Lower Thames Valley*; Cambridge University Press: Cambridge, UK, 1994.

387. Sedov, S.N.; Khokhlova, O.S.; Sinitsyn, A.A.; Korkka, A.V.; Ortega, B.; Solleiro, E.; Rozanova, M.S.; Kuznetsova, A.M. Kazdym Late Pleistocene paleosol sequences as an instrument for the local paleographic reconstruction of the Kostenki 14 key section (Voronezh oblast) as an example. *Eurasian Soil Sc.* **2010**, *43*, 876–892. [CrossRef]

388. Ukraintseva, V.V. *Mammoths and the Environment*; Cambridge University Press: Cambridge, UK, 2013.

389. van Asperen, E.N.; Stefaniak, K.; Proskurnyak, I.; Ridush, B. Equids from Emine-Bair-Khosar Cave (Crimea, Ukraine): Co-occurrence of the stenonid *Equus hydruntinus* and the caballoid *E. ferus latipes* based on skull and postcranial remains. *Palaeontol. Electron.* **2012**, *15*, 28. [CrossRef]

390. Bernor, R.L.; Kaiser, T.M.; Nelson, S.V. The Oldest Ethiopian Hipparion (Equinae, Perissodactyla) from Chorora: Systematics, Paleodiet and Paleoclimate. *Cour. Forsch. Inst. Senckenberg* **2004**, *246*, 213–226.

391. Bernor, R.L. The Latest Miocene Hipparionine (Equidae) from Lemundong'o, Kenya. *Kirtlandia* **2007**, *56*, 148–151.

392. Bernor, R.L.; Kaiser, T.M.; Wolf, D. Revisiting Sahabi Equid Species Diversity, Biogeographic Patterns and Diet Preferences. *Gharyounis Bull.* **2008**, *5*, 159–167.

393. Bernor, R.L.; Haile Selassie, Y. Equidae. In *Ardipithecus Kadabba: Late Miocene Evidence from the Middle Awash, Ethiopia*; Haile-Selassie, Y., Woldegabriel, G., Eds.; University of California Press: Berkeley, CA, USA, 2009; pp. 397–428.

394. Bernor, R.L.; Boaz, N.T.; Rook, L. *Eurygnathohippus feibeli* (Perissodactyla: Mammalia) from the late Miocene of Sahabi (Libya) and its evolutionary and Biogeographic Significance. *Bolletino Soc. Paleontol. Ital.* **2012**, *51*, 39–48.

395. Schulz, E.; Kaiser, T.M. Historical distribution, habitat requirements and feeding ecology of the genus *Equus* (Perissodactyla). *Mammal. Rev.* **2013**, *43*, 111–123. [CrossRef]

396. Goloboff, P.; Farris, J.; Nixon, K. TNT, a free program for phylogenetic analysis. *Cladistics* **2008**, *24*, 774–786. [CrossRef]

397. Bennett, D.K. Stripes do not a zebra make, Part I: A cladistic analysis of *Equus*. *Syst. Zool.* **1980**, *29*, 272–287. [CrossRef]

398. Antoine, P.O. Phylogénie et évolution des Elasmotheriina (Mammalia, Rhinocerotidae). *Mémoir. Mus. Natl. Hist.* **2002**, *188*, 1–359.

399. Antoine, P.O.; Downing, K.F.; Crochet, J.Y.; Duranthon, F.; Flynn, L.J.; Marivaux, L.; Métais, G.; Rajpar, A.R.; Roohi, G. A revision of *Aceratherium blanfordi* Lydekker, 1884 (Mammalia: Rhinocerotidae) from the Early Miocene of Pakistan: Postcranials as a key. *Zool. J. Linn. Soc. Lond.* **2010**, *160*, 139–194. [CrossRef]

400. Pandolfi, L. *Persiatherium rodleri*, gen. et sp. nov. (Mammalia, Rhinocerotidae), from the Late Miocene of Maragheh (Northwestern Iran). *J. Vert Paleontol.* **2015**, *36*, e1040118. [CrossRef]

401. de Queiroz, K.; Gauthier, J. Phylogenetic taxonomy. *Annu. Rev. Ecol. Syst.* **1992**, *23*, 449–480. [CrossRef]

402. Ereshefsky, M. *The Poverty of the Linnaean hierarchy. A Philosophical Study of Biological Taxonomy*; Cambridge University Press: Cambridge, UK, 2001.

403. Papavero, N.; Llorente-Bousquets, J.; Minoro Abel, J. Proposal of a new system of nomenclature for phylogenetic systematics. *Arq. Zool.* **2001**, *36*, 1–145. [CrossRef]

404. Béthoux, O. Propositions for a character-state-based biological taxonomy. *Zool. Scr.* **2007**, *36*, 409–416. [CrossRef]

405. Cantino, P.D.; de Queiroz, K. PhyloCode: International Code of Phylogenetic Nomenclature (Version 4c). 2010. Available online: http://www.ohio.edu/phylocode (accessed on 5 June 2022).

406. Zachos, F.E. Linnaean ranks, temporal banding, and time-clipping: Why not slaughter the sacred cow? *Biol. J. Linn. Soc.* **2011**, *103*, 732–734. [CrossRef]

407. Vences, M.; Guayasamin, J.M.; Miralles, A.; de la Riva, I. To name or not to name: Criteria to promote economy of change in Linnaean classification schemes. *Zootaxa* **2013**, *3636*, 201–244. [CrossRef]

408. Hennig, W. *Phylogenetic Systematics*; University Illinois Press: Champaign, IL, USA, 1966.

409. Mayr, E. *Principles of Systematic Zoology*; McGraw-Hill: New York, NY, USA, 1969.

410. Groves, C.P. *Primate Taxonomy*; Smithsonian Institution Press: Washington, DC, USA, 2001.

411. Groves, C.P. The what, why, and how of primate taxonomy. *Int. J. Primatol.* **2004**, *25*, 1105–1126. [CrossRef]

412. Simpson, G.G. *The Major Features of Evolution*; Columbia University Press: New York, NY, USA, 1953.

413. Mayr, E. Taxonomic categories in fossil hominids. *Cold Spring Harb. Symp. Quant. Biol.* **1950**, *15*, 109–118. [CrossRef]

414. de Queiroz, K. Replacement of an essentialistic perspective on taxonomic definitions as exemplified by the definition of "Mammalia". *Syst. Biol.* **1994**, *43*, 497–510. [CrossRef]

415. Janis, C.M.; Bernor, R.L. The Evolution of Equid Monodactyly: A Review Including a New Hypothesis. *Front. Ecol. Evol* **2019**, *7*, 119. [CrossRef]

416. Schultz, J.R. *Plesippus francescana* (Frick) from the late Pliocene, Coso Mountains, California, with a review of the genus *Plesippus*. In *Contributions to Palæontology. Publications of the Carnegie Institution of Washington*; Carnegie institution of Washington: Washington, DC, USA, 1936; Volume 1, pp. 1–13.

417. Eisenmann, V.; Baylac, M. Extant and fossil *Equus* (Mammalia, Perissodactyla) skulls: A morphometric definition of the subgenus *Equus*. *Zool. Scr.* **2000**, *29*, 89–100. [CrossRef]

418. Brochu, A.B.; Sumrall, C.D. Phylogenetic Nomenclature and Paleontology. *J. Paleontol.* **2001**, *75*, 754–757. [CrossRef]

419. Bryant, H.N. Explicitness, Stability, and Universality in the Phylogenetic Definition and Usage of Taxon Names: A Case Study of the Phylogenetic Taxonomy of the Carnivora (Mammalia). *Syst. Biol.* **1996**, *45*, 174–189. [CrossRef]

420. Cantino, P.D.; Olmstead, R.G.; Wagstaff, S.J. A Comparison of Phylogenetic Nomenclature with the Current System: A Botanical Case Study. *Syst. Biol.* **1997**, *46*, 313–331. [CrossRef]

421. Schulte, J.S.; Macey, J.R.; Larson, A.; Papenfuss, T.J. Molecular Tests of Phylogenetic Taxonomies: A General Procedure and Example Using Four Subfamilies of the Lizard Family Iguanidae. *Mol. Phylogenetics Evol.* **1998**, *10*, 367–376. [CrossRef]

422. de Queiroz, K.; Gauthier, J. Phylogeny as a Central Principle in Taxonomy: Phylogenetic Definitions of Taxon Names. *Syst. Biol.* **1990**, *39*, 307–322. [CrossRef]

423. de Queiroz, K.; Gauthier, J. Toward a phylogenetic system of biological nomenclature. *Trends Ecol. Evol.* **1994**, *9*, 27–31. [CrossRef]

424. de Queiroz, K.; Cantino, P.D. Taxon Names, Not Taxa, Are Defined. *Taxon* **2001**, *50*, 821–826. [CrossRef]

425. Vilstrup, J.T.; Seguin-Orlando, A.; Stiller, M.; Ginolhac, A.; Raghavan, M.; Nielsen, S.C.A.; Weinstock, J.; Froese, D.; Vasiliev, S.K.; Ovodov, N.D. Mitochondrial Phylogenomics of Modern and Ancient Equids. *PLoS ONE* **2013**, *8*, e55950.

426. MacFadden, B.; Solounias, N.; Cerling, T.E. Ancient Diets, Ecology, and Extinction of 5-Million-Year-Old Horses from Florida. *Science* **1999**, *283*, 824–827. [CrossRef]

427. Semprebon, G.M.; Rivals, F.; Solounias, N.; Hulbert, R.C. Paleodietary reconstruction of fossil horses from the Eocene through Pleistocene of North America. *Palaeogeogr. Palaeoclim. Palaeoecol.* **2016**, *442*, 110–127. [CrossRef]

428. Miller, W.E. The Late Pliocene Las Tunas Local Fauna from Southernmost Baja California, Mexico. *J. Paleontol.* **1980**, *54*, 762–805.

429. MacFadden, B.J.; Carranza-Castañeda, O. Cranium of *Dinohippus mexicanus* (Mammalia: Equidae) from the Early Pliocene (latest Hemphillian) of Central Mexico, and the origin of *Equus*. *Bull. Fla. Mus. Nat. Hist.* **2002**, *43*, 163–185.

430. Carranza-Castañeda, O. *Dinohippus mexicanus* (Early—Late, Late, and Latest Hemphillian) and the transition to genus *Equus*, in Central Mexico faunas. *Front. Earth Sci.* **2019**, *7*, 89. [CrossRef]

431. Lowenstein, J.M.; Rayder, O.A. Immunological systematics of the extinct quagga (Equidae). *Experientia* **1985**, *41*, 1192–1193. [CrossRef]
432. George, M.; Rayder, O.A. Mitochondrial DNA evolution in the genus *Equus*. *Mol. Biol. Evol.* **1986**, *3*, 535–546.
433. Oakefull, E.A.; Clegg, J.B. Phylogenetic Relationships within the Genus *Equus* and the Evolution of α and θ Globin Genes. *J. Mol. Evol.* **1998**, *47*, 772–783. [CrossRef]
434. Kruger, K.; Gaillard, C.; Stranzinger, G.; Rieder, S. Phylogenetic analysis and species allocation of individual equids using microsatellite data. *J. Anim. Breed. Genet.* **2005**, *122*, 78–86. [CrossRef] [PubMed]
435. Steiner, C.C.; Ryder, O.A. Molecular phylogeny and evolution of the Perissodactyla. *Zool. J. Linn. Soc.* **2011**, *163*, 1289–1303. [CrossRef]
436. Orlando, L.; Mashkour, M.; Burke, A.; Douady, C.J.; Eisenmann, V.; Hanni, C. Geographic distribution of an extinct equid (*Equus hydruntinus*: Mammalia, Equidae) revealed by morphological and genetical analyses of fossils. *Mol. Ecol.* **2006**, *15*, 2083–2093. [CrossRef] [PubMed]
437. Orlando, L.; Male, D.; Alberdi, M.T.; Prado, J.L.; Prieto, A.; Cooper, A.; Hanni, C. Ancient DNA Clarifies the Evolutionary History of American Late Pleistocene Equids. *J. Mol. Evol.* **2008**, *66*, 533–538. [CrossRef]
438. Flint, J.; Ryder, O.A.; Clegg, J.B. Comparison of theα-globin gene cluster structure in Perissodactyla. *J. Mol. Evol.* **1990**, *30*, 36–42. [CrossRef]
439. Wallner, B.; Brem, G.; Muller, M.; Achmann, R. Fixed nucleotide differences on the *Y* chromosome indicate clear divergence between *Equus przewalskii* and *Equus caballus*. *Anim. Genet.* **2003**, *34*, 453–456. [CrossRef]
440. Kaminski, M. The biochemical evolution of the horse. *Comp. Biochem.* **1979**, *63*, 175–178. [CrossRef]
441. Ishida, N.; Oyunsuren, T.; Mashima, S.; Mukoyama, H.; Naruya, S. Mitochondrial DNA sequences of various species of the genus *Equus* with special reference to the phylogenetic relationship between Przewalskii's wild horse and domestic horse. *J. Mol. Evol.* **1995**, *41*, 180–188. [CrossRef]
442. Oakenfull, E.A.; Lim, H.N.; Ryder, O.A. A survey of equid mitochondrial DNA: Implications for the evolution, genetic diversity and conservation of *Equus*. *Conserv. Genet.* **2000**, *1*, 341–355. [CrossRef]
443. Cai, D.; Zhu, S.; Gong, M.; Zhang, N.; Wen, J.; Qiyao, L.; Sun, W.; Shao, X.; Guo, Y.; Cai, Y.; et al. Radiocarbon and genomic evidence for the survival of *Equus Sussemionus* until the late Holocene. *eLife* **2022**, *11*, e73346. [CrossRef]

Review

Twentieth-Century Paleoproteomics: Lessons from Venta Micena Fossils

Jesús M. Torres [1], Concepción Borja [1], Luis Gibert [2], Francesc Ribot [3] and Enrique G. Olivares [1,4,5,*]

[1] Departamento de Bioquímica y Biología Molecular III e Inmunología, Universidad de Granada, 18016 Granada, Spain
[2] Departament de Geoquímica, Petrologia i Prospecció Geològica, Universitat de Barcelona, 08028 Barcelona, Spain
[3] Museo de Prehistoria y Paleontología Josep Gibert, 18858 Orce, Spain
[4] Instituto de Biopatología y Medicina Regenerativa, Centro de Investigación Biomédica, Universidad de Granada, 18100 Armilla, Spain
[5] Unidad de Gestión Clínica Laboratorios, Hospital Clínico Universitario San Cecilio, 18016 Granada, Spain
* Correspondence: engarcia@ugr.es; Tel.: +34-958248809

Simple Summary: Two independent research groups led by Olivares (Spain) and Lowenstein (USA) investigated the immunological reactions of proteins extracted from the controversial Orce skull (VM-0), a 1.3-million-year-old fossil found at the Venta Micena site in Orce, Granada (Spain) and initially believed to come from an unidentified hominin. Work by both groups with polyclonal and monoclonal antibodies showed that proteins from this fossil reacted most strongly to antibodies against modern human proteins. Other hominin and mammal fossils from Venta Micena were also studied.

Abstract: Proteomics methods can identify amino acid sequences in fossil proteins, thus making it possible to determine the ascription or proximity of a fossil to other species. Before mass spectrometry was used to study fossil proteins, earlier studies used antibodies to recognize their sequences. Lowenstein and colleagues, at the University of San Francisco, pioneered the identification of fossil proteins with immunological methods. His group, together with Olivares's group at the University of Granada, studied the immunological reactions of proteins from the controversial Orce skull fragment (VM-0), a 1.3-million-year-old fossil found at the Venta Micena site in Orce (Granada province, southern Spain) and initially assigned to a hominin. However, discrepancies regarding the morphological features of the internal face of the fossil raised doubts about this ascription. In this article, we review the immunological analysis of the proteins extracted from VM-0 and other Venta Micena fossils assigned to hominins and to other mammals, and explain how these methods helped to determine the species specificity of these fossils and resolve paleontological controversies.

Keywords: fossil proteins; ELISA; paleoproteomics; RIA; Venta Micena site; VM-0; VM-1960

Citation: Torres, J.M.; Borja, C.; Gibert, L.; Ribot, F.; Olivares, E.G. Twentieth-Century Paleoproteomics: Lessons from Venta Micena Fossils. *Biology* 2022, *11*, 1184. https://doi.org/10.3390/biology11081184

Academic Editors: Mary H. Schweitzer and Ferhat Kaya

Received: 1 July 2022
Accepted: 3 August 2022
Published: 6 August 2022
Corrected: 17 October 2022

Publisher's Note: MDPI stays neutral with regard to jurisdictional claims in published maps and institutional affiliations.

Copyright: © 2022 by the authors. Licensee MDPI, Basel, Switzerland. This article is an open access article distributed under the terms and conditions of the Creative Commons Attribution (CC BY) license (https://creativecommons.org/licenses/by/4.0/).

1. Introduction

All living organisms carry their own evolutionary history in their cells, and this history can be read in analyses of nucleic acid sequences or protein amino acid sequences. Thus, phylogenetic trees constructed from DNA or proteins have helped to clarify evolutionary relationships among species. Although DNA and proteins are also determinants of morphology, the genetic information that morphology provides is indirect and difficult to interpret, since numerous genes and complex interrelations are involved in configuring the structures of a living organism. In addition, convergence or parallel evolution phenomena can lead to similarities between unrelated species in one or more morphological characteristics. Nevertheless, in classical paleontology, species identification and classification are based exclusively on morphological features of the fossil record. As Wilson and Cann remarked, "The fossil record, on the other hand, is infamously spotty because a handful of surviving bones may not represent the majority of organisms that left no trace of themselves. Fossils cannot, in principle, be interpreted objectively: the physical characteristics by which they are classified necessarily reflect the models the paleontologists wish to test" [1]. Alongside morphological features, the analysis of biomolecules that survive in fossils can be of great help in identifying and classifying these remains, especially when they are fragmented and their morphological classification is controversial. Molecular paleontology methods have been developed to deal with this challenge.

2. Short Survival of Fossil DNA, Longer Survival of Fossil Proteins

Although recent decades have seen spectacular developments in molecular paleontology, this branch of science is not as recent as has been suggested. In the 1950s, Abelson first demonstrated the presence of amino acids and peptides in fossils [2]. In their initial work the detection of amino acids present only in certain proteins (e.g., hydroxyproline in collagen) made it possible to infer which type of protein these amino acids came from, but no additional genetic information could be obtained regarding the species to which the rest of the amino acids belonged. In 1963, Wykoff published electron microscopy images of collagen fibrils in dinosaur bones more than 200 million years old—another example of the early stages of molecular paleontology [3]. The molecule that has most often been investigated in the tissues of extinct species, ancient bones, or fossil remains is DNA, as the direct carrier of genetic information. The first ancient DNA sequence was obtained by Wilson's group, who studied a museum specimen of tissue from a quagga—a species from the horse family that became extinct in the late 19th century. To study the DNA remnants in the sample it was necessary to amplify them with a technique first developed in 1984: molecular cloning [4].

The main drawback that limits the scope of molecular paleontology is that any biomolecules that may survive in fossil remains must necessarily be altered and present at very low concentrations. When an organism dies, most of its biomolecules, as well as the organism itself, quickly disappear. However, under special circumstances in which rapid dehydration or rapid burial in an anaerobic environment occurs, hard (bone, shell, etc.) and even soft tissues (skin, muscle, etc.) can survive, and may thus contain biomolecules—albeit not in an intact form [5]. Proteins found in fossils are thus usually denatured and fractionated into peptides. In addition, after death, a process of racemization takes place: amino acids with the *L*-form spatial configuration are converted to the isomeric *D*-form [6]. DNA, an even more fragile molecule than proteins, is usually fractionated into sequences of only a few hundred base pairs containing abundant lesions such as baseless sites, oxidized pyrimidines, and chain cross-linkings [7]. Accordingly, Lindahl noted that it would be unlikely that any useful DNA could ever be extracted from very ancient fossils [8]. In fact, although the entire Neanderthal genome has been sequenced [9], most studies that focused on DNA found that it is unlikely to survive for more than 100,000 years. Nonetheless, notable exceptions to date are the sequencing of this biomolecule in a 400,000 year old *Homo heidelbergensis* fossil [10], and the genomic data obtained from a 560–780-thousand-year-old equid specimen [11] and from two mammoth specimens more than 1 million years old [12]. In addition, Woodward et al. published nine DNA sequences of the gene encoding cytochrome b, which were extracted from an 80-million-year-old dinosaur bone [13]. A drawback of these results was that the sequences did not show a significant degree of similarity to equivalent sequences from birds and reptiles, i.e., dinosaurs' closest extant relatives. However, later analyses of the sequences obtained by Woodward and colleagues revealed a greater similarity to human DNA than to that of other animals. This similarity, therefore, ruled out the possibility of dinosaur DNA and showed that the results were probably due to the inadvertent contamination of the sample during processing [14].

Although the ideal outcome is to read genetic information directly from the DNA nucleotide sequence, proteins also provide useful information, albeit indirectly, about amino acid sequences. In contrast to DNA, which appears to survive for only a thousand years, some proteins, under certain conditions, can persist in fossils for millions of years. Proteins bind to the mineral phase (hydroxyapatite) of bone, and this binding provides considerable protection from degradation by exogenous agents. Moreover, the amount of hydroxyapatite crystals increases after death, and this may favor protein encapsulation [15,16]. Compared to DNA, however, proteins present a technical obstacle in that they are not amplifiable, so their concentration cannot be increased—as can be attempted for DNA with polymerase chain reaction techniques. Although initial studies conducted between the 1950s and 1970s identified amino acids and peptides in fossils up to millions of years old, they did not provide information on the species specificity of these biomolecules, that is, on their ascription to or kinship with other species [2,3].

3. Detection of Fossil Proteins with Immunological Methods: Applications in Paleontological Controversies

Proteins undergo profound changes over time; nonetheless, these molecules, although fragmented or altered, can in some cases retain intact amino acid sequences. The protein fragments may be detectable with antibodies, which identify sequences comprising between 4 and 12 amino acids (epitopes) [17]. Mass spectrometry (MS) is also able to detect amino acid sequences in peptides [18]; however, this technology had not yet been implemented for fossil proteins in the twentieth century. In this period, most studies that aimed to identify fossil proteins were carried out with antibodies. Jerold M. Lowenstein was the first to identify genetic information contained in fossil proteins by applying radioimmunoassay (RIA) [19], an immunological technique able to specifically detect proteins in quantities as low as 10^{-13} M. Lowenstein and colleagues found human collagen, the most abundant protein in bone, in fossil samples of 20,000-year-old *Homo sapiens*, 50,000-year-old *Homo neanderthalensis*, 0.5-million-year-old *Homo erectus*, and 1.9-million-year-old *Australopithecus robustus* [20,21]. Collagen was also detected with dot-blotting in a 10-million-year-old fossil bone [22]. Osteocalcin, another abundant protein in bone, was identified by Ulrich et al. with antibodies in 13-million-year-old fossil bovine bones and 30-million-year-old rodent teeth. These researchers observed that osteocalcin in bovine fossil material still retained its functional ability to bind calcium [23]. Osteocalcin was also detected in a sample of 75-million-year-old dinosaur bone [24].

Particularly interesting is the detection of proteins in *Ramapithecus* fossils. In the 1960s, some paleoanthropologists considered this species, which lived 8 to 20 million years ago, to be a hominid, and it was thus suggested that the human lineage had diverged from that of apes about 20 million years ago. Molecular data, however, contradicted this hypothesis. Sarich and Cronin used immunological techniques to study modern chimpanzee, gorilla and human albumin, and concluded that these three species diverged from a common ancestor only 5 million years ago [25]. If this hypothesis is correct, it rules out hominin ancestry for *Ramapithecus*. Lowenstein produced antibodies by injecting an extract prepared from this fossil into a rabbit, and found that these antibodies reacted more strongly with gorilla, orangutan and gibbon sera than with chimpanzee or human sera. According to these results, *Ramapithecus* was genetically as closely related to Asian monkeys as to African monkeys, and more distantly related to humans [26]. Currently, paleontologists do not include *Ramapithecus* in the human lineage and consider it more closely related to orangutans.

At the turn of the twentieth century, a skull of modern human appearance was discovered in Sussex, England, which appeared in association with a jaw displaying ape-like morphological characteristics. Because the morphology of these bones was consistent with then-current theories of human evolution, the so-called Piltdown Man (*Eoanthropus dawsoni*) was accepted in 1912 by English anthropological authorities as the missing link between apes and humans, and was considered the first English human. It was not until 1953 when it became evident, based on an analysis of fluoride content, that the purported fossil was a fraud: a 500-year-old human skull to which the artificially aged jaw of a monkey had been added and the teeth modified to make them look human [27]. It remained to be determined whether the jaw was from a chimpanzee or orangutan. Lowenstein et al. studied a sample from the jaw and observed that antibodies to orangutan collagen reacted more strongly with an extract from this sample than did antibodies to human or chimpanzee collagen [28].

4. The Case of Orce Man

4.1. The Orce fossils

The Orce fossils assigned to hominins by Josep Gibert and colleagues include a skull fragment (the so-called Orce skull, VM-0), a humeral diaphysis (VM-1960), and a distal fragment of humerus (VM-3691) [29]. These remains were found at the Venta Micena site near the town of Orce in the province of Granada, southern Spain. The age of this site has been estimated magnetostratigraphically as 1.3 million years; if accurate, this would make the fossils the oldest evidence of a hominin presence in Europe. It has been hypothesized that Orce Man colonized Europe from the south by crossing the Strait of Gibraltar [30]. There is currently general agreement that early humans occupied the Orce area between 1.3 and 1.2 million years ago, based on a limited number of stone tools and on evidence of anthropic actions on bones detected at Venta Micena quarry 3 [31]. Additional evidence has come from nearby sites: two human deciduous molars, probably belonging to the same individual, were discovered at the Barranco Leon site [32–34], and stone artefacts were found at the Fuentenueva-3a and Barranco León-5 sites [35,36]. However, as often occurs in paleoanthropology, the Orce fossils became the subject of intense controversy [37]. In contrast to the position of Gibert and colleagues [38], some Spanish paleontologists claimed that the Orce skull belonged to an equid or a ruminant, and that the humeri were too incomplete to be identified with certainty [39,40]. Nevertheless, reputable paleoanthropologists such as Phillip V. Tobias, after close examination of the fossils assigned to hominins, supported Gibert's conclusions [41]. Given the uncertainties surrounding the morphological features, immunological studies of the fossil molecules were undertaken.

4.2. Methods to Investigate Proteins in Fossils

The Venta Micena fossils were dated to 1.3 million years, an age that far exceeded the limit of detectability of DNA in bones, so the likelihood of finding this biomolecule was slim. Attempts to amplify autochthonous DNA from equid fossils found at Venta Micena equivalent in age to the hominin materials from this site were unsuccessful. An alternative approach based on the prolonged survival of proteins in fossil bones (as noted above) was to analyze these biomolecules. Venta Micena fossil proteins were investigated by two independent groups: Lowenstein et al. at the University of California, San Francisco, and Olivares et al. at the University of Granada, Spain. The San Francisco team used RIA for the immunological detection of fossil proteins, while the Granada group used enzyme-linked immunosorbent assay (ELISA), a technique equivalent in sensitivity and detectability to RIA [42].

4.2.1. Preparation of Fossil Extracts

Aseptic conditions were rigorously maintained, and disposable materials were used to avoid external contamination during extraction and testing processes. The fossil sample was ground to a fine powder and treated successively with phosphate-buffered saline (PBS), EDTA, and acetic acid. The supernatant was collected after each treatment (Figure 1).

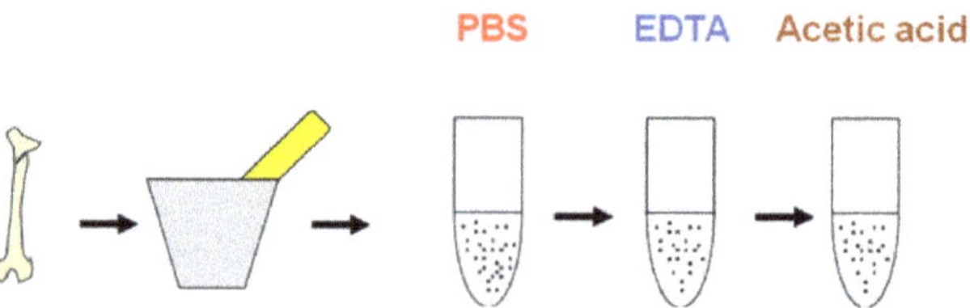

Figure 1. Preparation of fossil extracts.

4.2.2. Immunoassays for the Detection of Albumin

Fossil extracts were placed in the wells of a plastic microtiter plate. During this process, some proteins became irreversibly attached to the plastic. In the ELISA, appropriate mouse antiserum or monoclonal antibody to specific albumin was added, and the second antibody used was a biotinylated goat antimouse immunoglobulin, followed by extravidin-peroxidase. Extravidin binds biotin, and peroxidase catalyzes a color reaction that can be quantified in a microplate autoreader. In the RIA, rabbit antisera were added to various species of albumins, followed by the second antibody, ^{125}I-labeled goat antirabbit immunoglobulin. Radioactivity was quantified in a scintillation counter. Absorbance (ELISA) or radioactivity (RIA) obtained with fossil extracts was compared to the calibration curves for native albumins from different species in order to quantify the albumin detected.

4.2.3. Detection of IgG by Quantitative Dot-Blotting

Fossil extracts were placed on a nitrocellulose membrane, which has a higher capacity than plastic to attach proteins. The appropriate anti-IgG antibodies labeled with peroxidase were then added. We then cut out the nitrocellulose circles within which the reaction took place, and transferred each circle to a microtiter plate well. Peroxidase catalyzes a color reaction that can be quantified in a microplate autoreader.

4.3. Detection of Albumin in Fossil Bones

Both the Granada and the San Francisco groups studied Venta Micena fossils VM-0 and VM-1960 attributed to hominins, along with fossil CV-1, a fragment of humerus found at the Cueva Victoria site in the province of Murcia, Spain, and dated to an estimated 0.9 million years [42,43]. For comparison, fossils of different mammals from Venta Micena and Cueva Victoria were also analyzed by the two groups. An extract from fossil VM-0 was tested with antisera against albumin from different species, and both groups found greater reactivity with antisera against human albumin, whereas reactivity with other antibodies, especially antihorse albumin, was much lower (Figure 2). The conclusion, therefore, was that the albumin detected in VM-0 was closer to that of humans than other species. Lowenstein and colleagues also detected collagen and transferrin with immunological reactions similar to human proteins in fossil VM-0. Samples from the VM-1960 humerus, also attributed to a hominin, yielded results similar to those for VM-0 at both laboratories. However, neither group detected albumin in CV-1, a hominin fossil from the Cueva Victoria site. In their studies of other mammals, both groups observed reactions similar to horse albumin in equid fossils, and the San Francisco group detected reactions similar to bison albumin in two bovine fossils. Together, these results confirm the presence of albumin and other proteins in fossils believed to be approximately 1.3 million years old, and demonstrated that it is possible to identify species-specific characteristics in these fossil proteins with immunological techniques [42].

4.4. Fossil Proteins or Contamination

One of the most important considerations in molecular paleontology is to determine whether the biomolecules detected are an integral part of the fossil (endogenous) or originate from exogenous contamination. Special protocols for fossil sample preparation are required to verify the endogenous or exogenous origin of the biomolecules (Figure 1).

Although the possibility of contamination was unlikely (for example) in the case of horse albumin detected in the equid fossils, the albumin found in VM-0 and VM-1960 could have originated from contact with sweat or saliva during handling by paleontologists. It has also been suggested that seepage from recent human remains may have contaminated the fossils [44]. This latter possibility was easily ruled out, since soil collected from the site where the hominin fossils were found was analyzed and no albumin was detected (Figure 3).

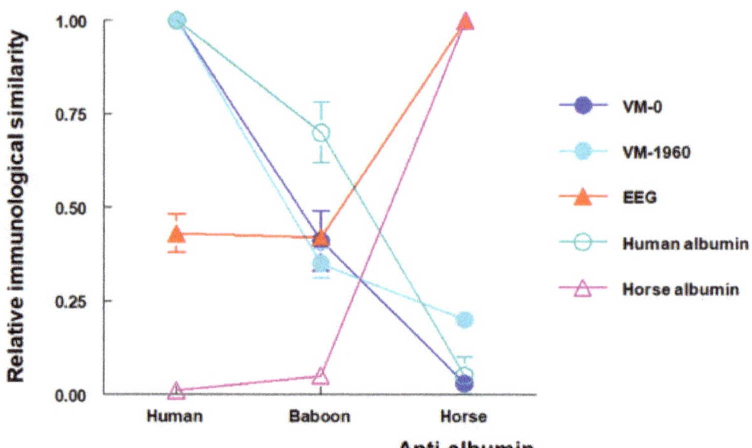

Figure 2. Reactivity of EDTA extracts of fossils from the Venta Micena site with three polyspecific mouse sera against human, baboon, and horse albumin in an enzyme-linked immunosorbent assay. VM-0 (skull fragment) and VM-1960 (humeral diaphysis) were assigned to hominins; EEG is a skull fragment of an equid fossil. All three specimens were dated to 1.3 million years old. The results are expressed as relative immunological similarity, defined as the ratio of each reaction to the homologous (most specific) albumin determination. VM-0 and VM-1960 produced a pattern of reactions similar to modern human albumin, whereas EEG produced a pattern of reactions similar to modern horse albumin. From Borja et al. [42], with permission.

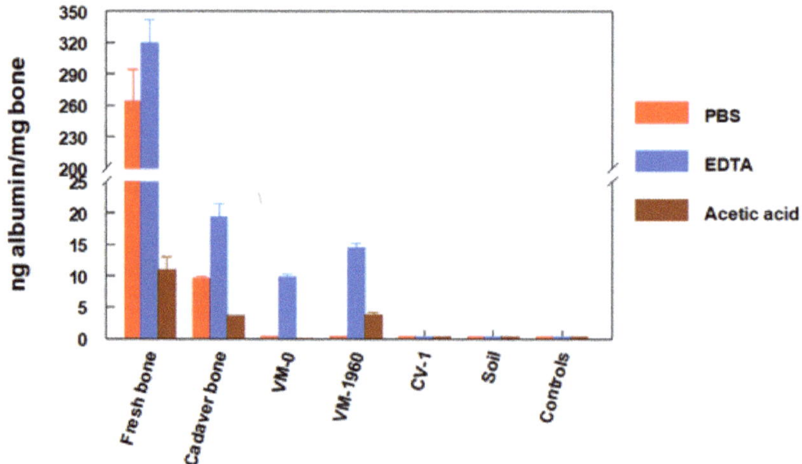

Figure 3. Enzyme-linked immunosorbent assay quantification of albumin in the PBS, EDTA and acetic acid extracts of human fresh bone (a fragment of femoral diaphysis), cadaver bone (a fragment of human occipital buried for approximately 10 years), fossils assigned to hominins (VM-0, VM-1960, CV-1), and soil collected around VM-0. Also shown are the buffers used for extractions, tested in the absence of bone. A mouse anti-human albumin polyclonal antiserum was used in all assays. From Lowenstein et al. [45], with permission.

The fossil proteins were bound to the mineral phase of the bone, from which they were released by treating the samples with a decalcifying EDTA solution [42]. Proteins from exogenous contamination are not bound to the mineral phase and can therefore be extracted without dissolving the bone, i.e., by simply washing the sample in phosphate-buffered saline solution (PBS). Albumin not bound to the mineral phase can be detected in fresh, surgically removed human bone that contains albumin from retained blood, which is also easily eluted with PBS. Albumin not bound to the mineral phase has been detected in human bones that were buried for as long as 10 years (Figure 3) [45]. Fossils VM-0 and VM-1960 did not contain unbound albumin, and this protein was detected only when the mineral phase was dissolved with EDTA. Therefore, the albumin in VM-0 and VM-1960 was integrated into these fossils and did not arise from exogenous contamination. Given the endogenous origin of this albumin and its immunological reactivity similar to modern human albumin, these results support the ascription of both fossils to a hominin [38,42,45].

Subsequent criticism of these data was based on the work of Cattaneo et al. [46]. These authors, in an effort to reevaluate the results obtained by different groups that reported the detection of proteins in fossil bones and archaeological artefacts, buried recent human and bovine bones in garden soil, and reported that under these conditions, no albumin was detectable after 1 month in human bone and after 3 months in bovine bone. Palmqvist, based on these results, inferred that it is not possible to detect proteins in bones beyond a few months [47]. In connection with efforts to detect ancient DNA, Poinar pointed out that "There are many types of fossilization processes, and to assume that the breakdown of DNA is similar in all of them, or is equivalent to that in non-fossilized material, is not scientific" [48]; this reasoning may also apply to fossil proteins. A bone becomes a fossil only after being subjected to particular conditions over a prolonged period—conditions that are not replicated by burying fresh bones in garden soil for several months. The Venta Micena site, located near the shore of an ephemeral alkaline lake, displays very specific preservation conditions in which mammal bones were covered and buried in an impermeable deposit of carbonate mud shortly after the animals' death. This mud consisted of calcite crystals that formed a film around the bone surfaces and protected them from alteration until complete burial and later excavation [49].

To shed further light on the fate of proteins in fresh buried and fossilized bone, Lowenstein and colleagues compared the amount of albumin detected in fresh human bone, human bone buried in a cemetery for 10 years, and in fossils VM-0 and VM-1960. Their findings showed that while the amounts of albumin detected in fossils were logically much lower than those found in fresh bone, the amounts of albumin detected in bone from the 10-year-old cemetery burial were not much higher than those observed in the fossils [45] (Figure 3). This apparent contradiction can be explained considering that after death, under normal conditions all proteins in most bones tend to disappear within a relatively brief period. However, under special conditions that lead to fossilization, such as those at the Venta Micena site, the proteins are "frozen" in the mineral phase of the bone and can be preserved for millions of years [24,38].

4.5. Monoclonal Antibodies to Study the Integrity of Fossil Proteins

Despite their persistence, fossil proteins are inevitably fractionated or denatured, although some amino acid sequences detectable with antibodies may survive. In the studies discussed above, antibodies obtained from the blood serum of an animal that was previously immunized against a protein (antiserum) were used. This antiserum contained a mixture of different polyclonal antibodies, each of which recognized an independent part (epitope) of the protein used for immunization. For more fine-grained molecular studies, monoclonal antibody technology can be used to produce a single type of antibody that reacts with a single epitope of the protein [50]. By independently testing different monoclonal antibodies against human albumin, it is possible to analyze different epitopes of this molecule individually, and to determine which of them have survived in the fossil albumin. In studies of the reactivity of monoclonal antibodies against human albumin with extracts of the hominin fossils VM-0 and VM-1960, it was found that each of the monoclonal antibodies showed a different degree of reactivity (Figure 4). Higher or lower reactivity indicated a better or worse degree of preservation of different epitopes recognized in the fossil albumin. Monoclonal antibodies, therefore, not only made it possible to confirm the data obtained with polyclonal antibodies, but also provided an opportunity to analyze the integrity of different epitopes of the protein [42,45].

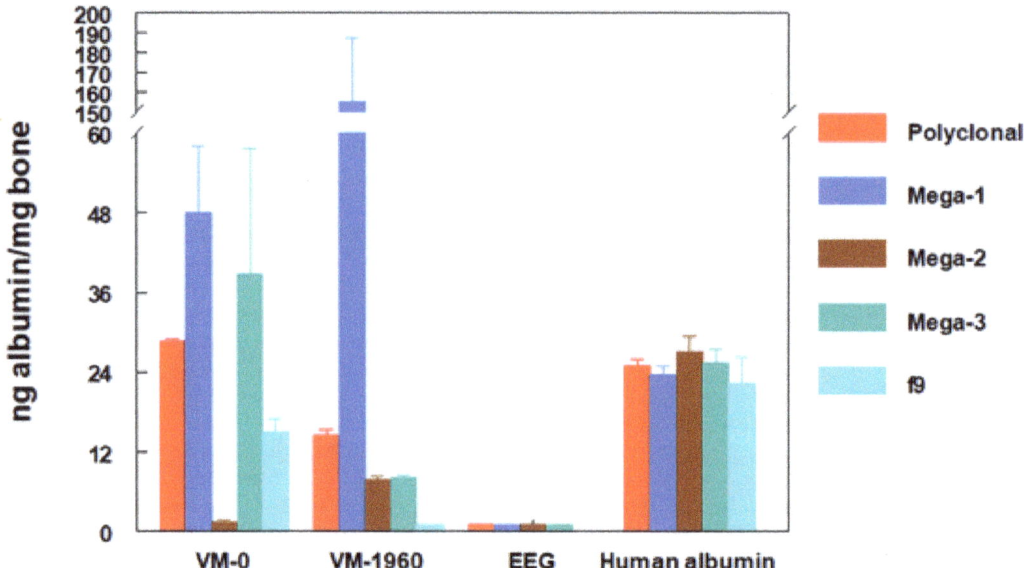

Figure 4. Enzyme-linked immunosorbent quantification of albumin in EDTA extracts of fossils VM-0 and VM-1960, assigned to hominins, and fossil EEG, assigned to an equid, with monoclonal antibodies. Four monoclonal anti-human albumin antibodies (MEGA-1, MEGA-2, MEGA-3 and 8F6F9) were used with the fossils. Polyclonal mouse anti-human albumin serum was used for comparison. Human albumin (25 ng) was tested for comparison. From Lowenstein et al. [45], with permission.

If a polyclonal antiserum is used, the fractionated, degraded and denatured fossil protein, which has lost part of its epitopes, will capture fewer antibodies than the native protein used to generate an ELISA or RIA calibration curve. Therefore, for an equivalent number of molecules, the signal produced by the antiserum in reaction with the fossil protein will be weaker than with the native protein. However, if the epitope recognized by a monoclonal antibody is preserved in the fossil protein, the antibody will produce a signal with the fractionated fossil protein (lower molecular weight) that is quantitatively equivalent to that produced with the native protein (higher molecular weight) used for the ELISA or RIA calibration curve (Figure 5). Consequently, when the signals are situated on the calibration curve in order to quantify the amount of fossil protein, higher amounts than expected may be detected (Figure 5). For this reason, the term ng-equivalent was coined to indicate that the amounts of fossil protein detected in assays with monoclonal antibodies reflect the reactivity of each monoclonal antibody rather than the actual amount of fossil protein present in the sample [42,45,51]. In fact, when different monoclonal antibodies recognizing different epitopes of human albumin were used, the quantification of fossil albumin varied depending on the reactivity of each monoclonal antibody, thus reflecting the presence or absence of each recognized epitope (Figure 4). Thus, the use of monoclonal antibodies can evidence the greater or lesser preservation of different epitopes recognized in fossil proteins [45].

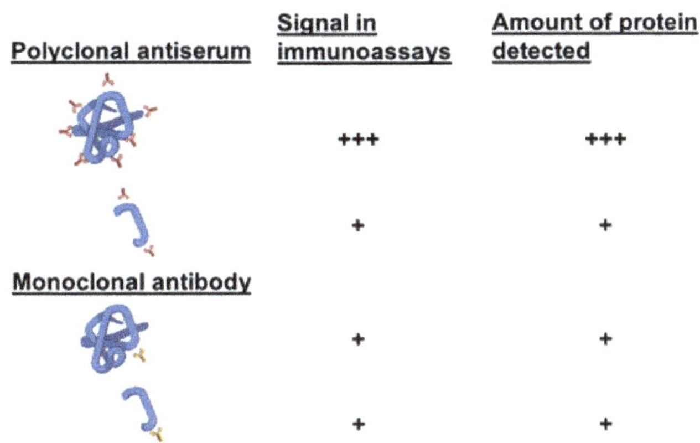

Figure 5. Differences in reactivity and quantification of fossil proteins (short peptide) in comparison to the native protein in assays with a polyclonal antiserum or a monoclonal antibody. Figure created in BioRender.com.

4.6. Proteins Other than Collagen in Fossils

A novel aspect of research with the Venta Micena specimens was the detection of albumin, a protein detected for the first time in such ancient fossils. Although collagen and osteocalcin, the most abundant proteins in bone, had previously been identified in fossil bones that were millions of years old [22–24], in principle the detection of albumin seemed unlikely since this protein is much less abundant in bone than collagen or osteocalcin. Moreover, because this protein is highly soluble, it was assumed that it would be rapidly washed out of bones during the degradation process. However, as Tuross and colleagues pointed out, the key phenomenon of protein preservation, i.e., the encapsulation of these biomolecules in hydroxyapatite crystals, appears to especially affect plasma proteins [52,53]; this could explain why albumin and other plasma proteins remained detectable in the Venta Micena fossils. Albumin, in some cases, becomes so highly concentrated in bone after death that it can reach levels higher than those found in living animals [52]. However, collagen—although abundant in bone—is an uninformative molecule in genetic terms because its amino acid sequence is highly repetitive and similar across different species. In contrast, albumin provides more evolutionary information. The fact that this protein has evolved more rapidly than collagen makes albumin a better discriminator between species [54]—clearly an advantage in paleoproteomics. It is thus not surprising that this protein has been studied with immunological methods in several extinct species such as the mammoth, the Steller elephant seal, the Tasmanian sea lion, and the quagga, in which the results have helped to resolve controversies regarding their phylogenetic affinities [55–58]. The DNA of some of these species was later sequenced, yielding phylogenetic trees very similar to those derived from the immunological study of albumin [59].

Another plasma protein detected in Venta Micena fossils was immunoglobulin G (IgG). Extracts from fossil equid bones and from fossil bones ascribed to hominins were tested with anti-human IgG and anti-horse IgG polyclonal antisera (Figure 6). The equid fossils showed higher reactions with anti-horse IgG than with anti-human IgG, while the hominin fossils reacted more strongly with anti-human IgG than with anti-horse IgG. These results demonstrate the feasibility of detecting species-specific IgG in different types of fossils. Samples from VM-1960, which had shown a stronger reaction to anti-human albumin, also showed a more marked reaction to anti-human IgG, whereas Cueva Victoria fossils, in which no albumin was detected [42] (Figure 3), showed no reaction to anti-IgG sera [51]. In parallel with the results of assays with albumin, no IgG was found when fossil samples were washed with PBS. The presence of IgG was seen only after decalcification with EDTA— a result interpreted to show that the proteins detected were embedded in the mineral phase

(Figure 6) [42,51]. Although some paleontologists have denied the presence of hominins at the Venta Micena site based on morphological and other criteria (but have thus far not used molecular methods themselves) [35], the immunological findings in Venta Micena fossils strongly support the assignment to hominins of cranial fragment VM-0 and the two humeral fragments VM-1960 and CV-3691. Phillip V. Tobias, who described *Homo habilis*, opined: "The convergence of the two laboratories, working independently of one another, by two different methods, provides very strong evidence in support of the conclusion that all three of the bones from Venta Micena are of hominid origin. When the molecular and the skeletal data are considered together, the picture afforded by the bio-anthropological evidence is that the three bones are of human origin" [41].

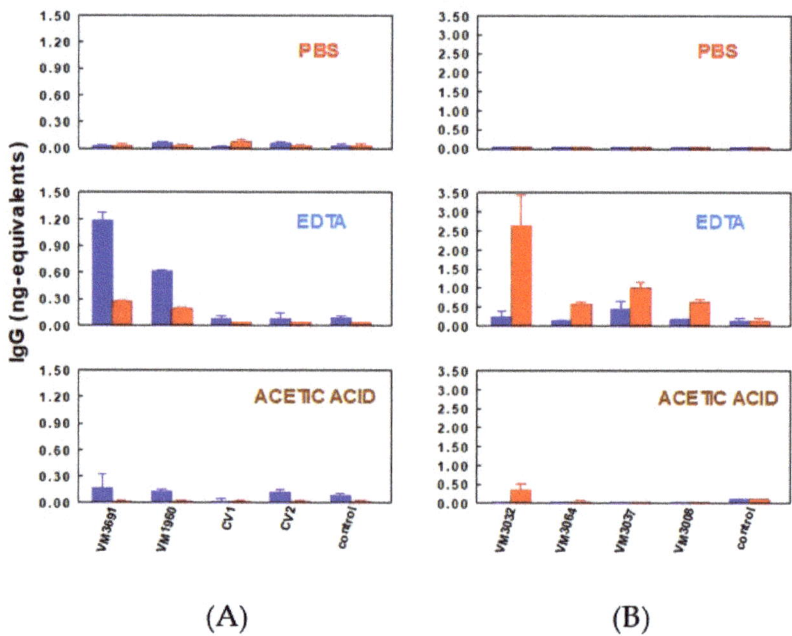

Figure 6. Quantities of human IgG (blue bars) and horse IgG (red bars) detected in PBS, EDTA and acetic acid extracts of (**A**) hominin and (**B**) equid fossils from Venta Micena, determined with quantitative dot-blotting. The extraction solutions were used as negative controls. From Torres et al. [51], with permission.

The presence of proteins other than collagen in fossils is consistent with microscopic observations of cells and tissues in dinosaur bones [60,61]. Mongelard detected albumin with immunological methods in 1700-year-old rodent fossils [62], and this protein was recently identified with MS methods in different hominin fossils [63] and in a mammoth femur [64]. Furthermore, extracts of these fossils can be used to immunize animals as a way to obtain anti-fossil protein sera that can be tested with native proteins to determine protein and species specificities. For example, rabbit anti-Ramapithecus bone extract reacted with sera from different modern primates [56], rat anti-dinosaur bone extract reacted with avian and mammalian hemoglobin [65], and rabbit anti-sauropod eggshell extract reacted with chicken ovalbumin [66]. Similarly, antisera were obtained in mice against VM-0 and against a fragment of Neanderthal humeral diaphysis (fossil CU-1) found at the Cueva Umbría in Orce, Granada (50,000–70,000 years old). Anti-VM-0 serum showed higher reactivity with human albumin than did non-immune serum, and weaker albeit positive reactivity with human IgG. In related work, anti-Neanderthal serum reacted more strongly than anti-VM-0 serum with human albumin, IgG, hemoglobin and transferrin [67].

Although twentieth-century paleoproteomics research used less powerful technologies and less stringent protocols compared to currently available tools to assess contamination, confirm endogeneity, and authenticate species assignment, the body of knowledge provided then has paved the way for the extraordinary development of twenty-first-century paleoproteomics. However, peptides extracted from fossils may be altered, and this limits the applicability of MS methods [68]. In these cases, immunological methods hold potential to aid in efforts to increase the accuracy and reliability of fossil protein identification.

4.7. Twenty-First-Century Paleoproteomics and the Venta Micena Fossils

Paleoproteomics is a relatively young yet rapidly growing field of molecular science in which proteomics-based sequencing technology is used to identify species and propose evolutionary relationships among extinct taxa. As a complementary approach to paleogenomics, the study of ancient proteins has the potential to disclose older, more complete phylogenies due to the relative stability of amino acids in proteins compared to nucleic acids in DNA [69]. Mass spectrometry provides unprecedented information on modern and ancient proteomes, and can yield protein sequence data from extinct organisms as well as historical and prehistorical artifacts [70]. Since the seminal application of MS in paleoproteomics [71], extraction and analysis protocols, software for data processing, protein databases, and high-mass-accuracy instrumentation have all seen significant progress. These advances are potentially applicable to the study of Venta Micena fossils. Given the ages of these fossils—older than 1 million years—this approach to research can be considered deep time paleoproteomics [72]. Another promising approach to extend research on plasma proteins such as albumin or IgG is affinity purification coupled with MS, which has been used to selectively identify proteins [73,74]. In addition, antibodies to specific protein or peptide components can be used to detect and separate proteins from a heterogeneous mixture, making them ideal tools to study degraded or fragmented ancient proteins (targeted paleoproteomics) [72]. Furthermore, since collagen has been detected by immunological methods in VM-0, this fossil, along with the other Venta Micena fossils assigned to hominins, could also be analyzed with zooarchaeology by mass spectrometry (ZooMS), an efficient proteomics-based method of species identification by collagen peptide mass fingerprinting. This method is especially useful to determine the affiliation of fossils that are morphologically controversial, such as VM-0 [75]. Combining previous knowledge in the immunodetection of fossil proteins with recent advances in MS approaches will make it possible to access the great store of potential information locked in the Venta Micena fossil record, and will contribute to the growth of paleoproteomics.

5. Conclusions

In paleoproteomics research, immunological techniques predominated during the twentieth century, whereas the twenty-first century has seen significant developments and improvements in the application of more informative MS methods. Immunological methods have been used to obtain molecular data that shed light on the species ascriptions of the Venta Micena fossils. These methods showed that the fossils contained well-preserved serum proteins, i.e., albumin, IgG, and transferrin. This research also demonstrated the species specificity of fossil proteins, thus helping to resolve scientific controversies that arose regarding the morphological data used to ascribe the Orce skull and other Venta Micena fossils to hominins. The use of monoclonal antibodies that target human albumin made it possible to determine the degree of preservation of different epitopes recognized in the fossil albumin. The combination of immunological methods with antibodies together with MS (targeted paleoproteomics) could represent an important advance in the study of these fossils by providing data that can help determine the phylogenetic relationships between these fossils and other known species of hominins.

Author Contributions: All authors contributed to the writing and editing, and approved the final version. All authors have read and agreed to the published version of the manuscript.

Funding: Plan Andaluz de Investigación, Desarrollo e Innovación, Groups CTS-564 and CTS-202.

Institutional Review Board Statement: Not applicable.

Informed Consent Statement: Not applicable.

Data Availability Statement: Not applicable.

Acknowledgments: We thank K. Shashok for editing the use of English in this manuscript.

Conflicts of Interest: The authors declare no conflict of interest.

Abbreviations

ELISA	Enzyme-linked immunosorbent assay
IgG	Immunoglobulin G
MS	Mass spectrometry
RIA	Radioimmunoassay

References

1. Wilson, A.C.; Cann, R.L. The recent African genesis of humans. *Sci. Am.* **1992**, *266*, 68–73. [CrossRef] [PubMed]
2. Abelson, P. *Organic Constituents of Fossils*; Carnegie Institute of Washington Yearbook: Washington, DC, USA, 1954; Volume 53, p. 5.
3. Wykoff, R.W.G. *The Biochemistry of Animal Fossils*; Scientechnica: Bristol, UK, 1972.
4. Higuchi, R.; Bowman, B.; Freiberger, M.; Ryder, O.A.; Wilson, A.C. DNA sequences from the quagga, an extinct member of the horse family. *Nature* **1984**, *312*, 282–284. [CrossRef] [PubMed]
5. Curry, G.B.; Eglinton, G. Molecules through time—Fossil molecules and biochemical systematics—Proceedings of a Royal-Society discussion meeting on Biomolecular Paleontology held on 20 and 21 march, 1991—Preface. *Philos. Trans. R. Soc. Lond. B Biol. Sci.* **1991**, *333*, 311–312. [CrossRef]
6. Hare, P.E.; Hoering, T.C.; King, K. (Eds.) *Biogeochemistry of Amino Acids*; John Wiley & Sons Inc: New York, NY, USA, 1980.
7. Lindahl, T. Instability and decay of the primary structure of DNA. *Nature* **1993**, *362*, 709–715. [CrossRef]
8. Lindahl, T. Facts and artifacts of ancient DNA. *Cell* **1997**, *90*, 1–3. [CrossRef]
9. Prufer, K.; Racimo, F.; Patterson, N.; Jay, F.; Sankararaman, S.; Sawyer, S.; Heinze, A.; Renaud, G.; Sudmant, P.H.; de Filippo, C.; et al. The complete genome sequence of a Neanderthal from the Altai Mountains. *Nature* **2014**, *505*, 43–49. [CrossRef]
10. Meyer, M.; Fu, Q.; Aximu-Petri, A.; Glocke, I.; Nickel, B.; Arsuaga, J.L.; Martinez, I.; Gracia, A.; de Castro, J.M.; Carbonell, E.; et al. A mitochondrial genome sequence of a hominin from Sima de los Huesos. *Nature* **2014**, *505*, 403–406. [CrossRef]
11. Orlando, L.; Ginolhac, A.; Zhang, G.; Froese, D.; Albrechtsen, A.; Stiller, M.; Schubert, M.; Cappellini, E.; Petersen, B.; Moltke, I.; et al. Recalibrating Equus evolution using the genome sequence of an early Middle Pleistocene horse. *Nature* **2013**, *499*, 74–78. [CrossRef]
12. van der Valk, T.; Pecnerova, P.; Diez-del-Molino, D.; Bergstrom, A.; Oppenheimer, J.; Hartmann, S.; Xenikoudakis, G.; Thomas, J.A.; Dehasque, M.; Saglican, E.; et al. Million-year-old DNA sheds light on the genomic history of mammoths. *Nature* **2021**, *591*, 265–269. [CrossRef]
13. Woodward, S.R.; Weyand, N.J.; Bunnell, M. DNA sequence from Cretaceous period bone fragments. *Science* **1994**, *266*, 1229–1232. [CrossRef] [PubMed]
14. Zischler, H.; Hoss, M.; Handt, O.; von Haeseler, A.; van der Kuyl, A.C.; Goudsmit, J.; Pääbo, S. Detecting dinosaur DNA. *Science* **1995**, *268*, 1192–1193. [CrossRef] [PubMed]
15. Herman, A.; Addadi, L.; Weiner, S. Interactions of sea-urchin skeleton macromolecules with growing calcite crystals—A study of intracrystalline proteins. *Nature* **1988**, *331*, 546–548. [CrossRef]
16. Smith, A.J.; Matthews, J.B.; Wilson, C.; Sewell, H.F. Plasma proteins in human cortical bone: In vitro binding studies. *Calcif. Tissue Int.* **1985**, *37*, 208–210. [CrossRef]
17. Buus, S.; Rockberg, J.; Forsstrom, B.; Nilsson, P.; Uhlen, M.; Schafer-Nielsen, C. High-resolution mapping of linear antibody epitopes using ultrahigh-density peptide microarrays. *Mol. Cell. Proteom.* **2012**, *11*, 1790–1800. [CrossRef]
18. Graves, P.R.; Haystead, T.A. Molecular biologist's guide to proteomics. *Microbiol. Mol. Biol. Rev.* **2002**, *66*, 39–63. [CrossRef]
19. Olivares, E.G. Not a first: Identifying hominin fossils from their proteins. *Nature* **2019**, *573*, 196. [CrossRef]
20. Lowenstein, J.M. Immunological reactions from fossil material. *Philos. Trans. R. Soc. Lond. B Biol. Sci.* **1981**, *292*, 143–149. [CrossRef] [PubMed]
21. Lowenstein, J.M.; Scheuenstuhl, G. Immunological methods in molecular palaeontology. *Philos. Trans. R. Soc. Lond. B Biol. Sci.* **1991**, *333*, 375–380. [CrossRef]

22. Rowley, M.J.; Rich, P.V.; Rich, T.H.; Mackay, I.R. Immunoreactive collagen in avian and mammalian fossils. *Naturwissenschaften* **1986**, *73*, 620–623. [CrossRef] [PubMed]
23. Ulrich, M.M.; Perizonius, W.R.; Spoor, C.F.; Sandberg, P.; Vermeer, C. Extraction of osteocalcin from fossil bones and teeth. *Biochem. Biophys. Res. Commun.* **1987**, *149*, 712–719. [CrossRef]
24. Muyzer, G.; Sandberg, P.; Knapen, M.H.J.; Vermeer, C.; Collins, M.; Westbroek, P. Preservation of the bone protein osteocalcin in dinosaurs. *Geology* **1992**, *20*, 871–874. [CrossRef]
25. Sarich, V.M.; Wilson, A.C. Immunological time scale for hominid evolution. *Science* **1967**, *158*, 1200–1203. [CrossRef]
26. Lowenstein, J.M. Fossil proteins and evolutionary time. *Pontif. Acad. Sci. Scr. Var.* **1983**, *50*, 151–162.
27. Lewin, R. *Bones of Contention: Controversies in the Search for Human Origins*, 2nd ed.; The University of Chicago Press: Chicago, IL, USA, 1987.
28. Lowenstein, J.M.; Molleson, T.; Washburn, S.L. Piltdown jaw confirmed as orang. *Nature* **1982**, *299*, 294. [CrossRef]
29. Gibert, J.; Ribot, F.; Ferrandez, C.; Martinez, B.; Caporicci, R.; Campillo, D. Anatomical study: Comparison of the cranial fragment from Venta Micena, (Orce; Spain) with fossil and extant mammals. *Hum. Evol.* **1989**, *4*, 283–305. [CrossRef]
30. Scott, G.R.; Gibert, L.; Gibert, J. Magnetostratigraphy of the Orce Region (Baza Basin), SE Spain: New chronologies for Early Pleistocene faunas and hominid occupation sites. *Quat. Sci. Rev.* **2007**, *26*, 415–435. [CrossRef]
31. Gibert, J.; Gibert Beotas, L.; Ferràndez-Cañadell, C.; Iglesias; González, F. Venta Micena, Barranco León-5 and Fuentenueva-3: Three archaeological sites in the Early Pleistocene deposits of Orce, south-east Spain. In *The Human Evolution Source Book*; Ciochon, R.L., Fleagle, J.G., Eds.; Pearson Prentice Hall: Englewood Cliffs, NJ, USA, 2006; pp. 327–335.
32. Gibert, J.; Gibert, L.; Albadalejo, S.; Ribot, F.; Sánchez, F.; Gibert, P. Molar tooth fragment BL5-0: The oldest human remain found in the Plio-Pleistocene of Orce (Granada province, Spain). *Hum. Evol.* **1999**, *14*, 3–19. [CrossRef]
33. Ribot, F.; Gibert, L.; Ferrandez-Canadell, C.; Garcia Olivares, E.; Sanchez, F.; Leria, M. Two Deciduous Human Molars from the Early Pleistocene Deposits of Barranco Leon (Orce, Spain). *Curr. Anthropol.* **2015**, *56*, 134–142. [CrossRef]
34. Toro-Moyano, I.; Martinez-Navarro, B.; Agusti, J.; Souday, C.; Bermudez de Castro, J.M.; Martinon-Torres, M.; Fajardo, B.; Duval, M.; Falgueres, C.; Oms, O.; et al. The oldest human fossil in Europe, from Orce (Spain). *J. Hum. Evol.* **2013**, *65*, 1–9. [CrossRef] [PubMed]
35. Barsky, D.; Titton, S.; Sala-Ramos, R.; Bargalló, A.; Grégoire, S.; Saos, T.; Serrano-Ramos, A.; Oms, O.; Solano García, J.-A.; Toro-Moyano, I.; et al. The Significance of Subtlety: Contrasting Lithic Raw Materials Procurement and Use Patterns at the Oldowan Sites of Barranco León and Fuente Nueva 3 (Orce, Andalusia, Spain). *Front. Earth Sci.* **2022**, *10*, 893776. [CrossRef]
36. Gibert, J.; Gibert, L.; Iglesias, A.; Maestro, E. Two 'Oldowan' assemblages in the Plio-Pleistocene deposits of the Orce region, southeast Spain. *Antiquity* **1998**, *72*, 17–25. [CrossRef]
37. Alba, D.M. A fistful of fossils: The rise and fall of the Orce Man and the politics of paleoanthropological science. *J. Hum. Evol.* **2022**, *165*, 103166. [CrossRef]
38. Gibert, J.; Campillo, D.; Arques, J.M.; Garcia-Olivares, E.; Borja, C.; Lowenstein, J.M. Hominid status of the Orce cranial fragment reasserted. *J. Hum. Evol.* **1998**, *34*, 203–217. [CrossRef]
39. Agusti, J.; Moya-Sola, S. On the identity of the cranial fragment attributed to homo-sp in Venta Micena, Granada, Spain. *Estud. Geol.-Madr.* **1987**, *43*, 535–538.
40. Martinez-Navarro, B. The skull of Orce: Parietal bones or frontal bones? *J. Hum. Evol.* **2002**, *43*, 265–270. [CrossRef]
41. Tobias, P.V. Some comments on the case for Early Pleistocene hominids in South-Eastern Spain. *Hum. Evol.* **2006**, *13*, 91–96. [CrossRef]
42. Borja, C.; Garcia-Pacheco, M.; Olivares, E.G.; Scheuenstuhl, G.; Lowenstein, J.M. Immunospecificity of albumin detected in 1.6 million-year-old fossils from Venta Micena in Orce, Granada, Spain. *Am. J. Phys. Anthropol.* **1997**, *103*, 433–441. [CrossRef]
43. Gibert, L.; Scott, G.R.; Scholz, D.; Budsky, A.; Ferrandez, C.; Ribot, F.; Martin, R.A.; Leria, M. Chronology for the Cueva Victoria fossil site (SE Spain): Evidence for Early Pleistocene Afro-Iberian dispersals. *J. Hum. Evol.* **2016**, *90*, 183–197. [CrossRef]
44. Kooyman, B.; Newman, M.E.; Ceri, H. Verifying the reliability of blood residue analysis on archaeological tools. *J. Archaeol. Sci.* **1992**, *19*, 265–269. [CrossRef]
45. Lowenstein, J.M.; Borja, C.; García-Olivares, E. Species-specific albumin in fossil bones from Orce, Granada, Spain. *Hum. Evol.* **1999**, *14*, 21–28. [CrossRef]
46. Cattaneo, C.; Gelsthorpe, K.; Phillips, P.; Sokol, R.J. Blood residues on stone tools: Indoor and outdoor experiments. *World Archaeol.* **1993**, *25*, 29–43. [CrossRef]
47. Palmqvist, P. A critical re-evaluation of the evidence for the presence of hominids in Lower Pleistocene times at Venta Micena, Southern Spain. *J. Hum. Evol.* **1997**, *33*, 83–89. [CrossRef]
48. Poinar, G.O. Recovery of antediluvian DNA. *Nature* **1993**, *365*, 700. [CrossRef]
49. Gibert, L.; Orti, F.; Rosell, L. Plio-Pleistocene lacustrine evaporites of the Baza Basin (Betic Chain, SE Spain). *Sediment. Geol.* **2007**, *200*, 89–116. [CrossRef]
50. Singh, S.; Kumar, N.K.; Dwiwedi, P.; Charan, J.; Kaur, R.; Sidhu, P.; Chugh, V.K. Monoclonal Antibodies: A Review. *Curr. Clin. Pharmacol.* **2018**, *13*, 85–99. [CrossRef] [PubMed]
51. Torres, J.M.; Borja, C.; Olivares, E.G. Immunoglobulin G in 1.6 million-year-old fossil bones from Venta Micena (Granada, Spain). *J. Archaeol. Sci.* **2002**, *29*, 167–175. [CrossRef]
52. Tuross, N. Albumin preservation in the Taima-taima mastodon skeleton. *Appl. Geochem.* **1989**, *4*, 255–259. [CrossRef]

53. Tuross, N.; Behrensmeyer, A.K.; Eanes, E.D.; Fisher, L.W.; Hare, P.E. Molecular preservation and crystallographic alterations in a weathering sequence of wildebeest bones. *Appl. Geochem.* **1989**, *4*, 261–270. [CrossRef]
54. Wilson, A.C.; Carlson, S.S.; White, T.J. Biochemical evolution. *Ann. Rev. Biochem.* **1977**, *46*, 573–639. [CrossRef]
55. Lowenstein, J.M.; Ryder, O.A. Immunological systematics of the extinct quagga (equidae). *Experientia* **1985**, *41*, 1192–1193. [CrossRef]
56. Lowenstein, J.M.; Sarich, V.M.; Richardson, B.J. Albumin systematics of the extinct mammoth and tasmanian wolf. *Nature* **1981**, *291*, 409–411. [CrossRef] [PubMed]
57. Rainey, W.E.; Lowenstein, J.M.; Sarich, V.M.; Magor, D.M. Sirenian molecular systematics including the extinct stellers sea cow (Hydrodamalis-gigas). *Naturwissenschaften* **1984**, *71*, 586–588. [CrossRef] [PubMed]
58. Shoshani, J.; Lowenstein, J.M.; Walz, D.A.; Goodman, M. Proboscidean origins of mastodon and woolly mammoth demonstrated immunologically. *Paleobiology* **1985**, *11*, 429–437. [CrossRef]
59. Thomas, R.H.; Schaffner, W.; Wilson, A.C.; Paabo, S. DNA phylogeny of the extinct marsupial wolf. *Nature* **1989**, *340*, 465–467. [CrossRef] [PubMed]
60. Pawlicki, R.; Nowogrodzka-Zagorska, M. Blood vessels and red blood cells preserved in dinosaur bones. *Ann. Anat. -Anat. Anz.* **1998**, *180*, 73–77. [CrossRef]
61. Schweitzer, M.H.; Johnson, C.; Zocco, T.G.; Horner, J.R.; Starkey, J.R. Preservation of biomolecules in cancellous bone of Tyrannosaurus rex. *J. Vertebr. Paleontol.* **1997**, *17*, 349–359. [CrossRef]
62. Montgelard, C. Albumin preservation in fossil bones and systematics of Malpaisomys insularis (Muridae, Rodentia), an extinct rodent of the Canary Islands. *Hist. Biol.* **1992**, *6*, 293–302. [CrossRef]
63. Welker, F.; Ramos-Madrigal, J.; Gutenbrunner, P.; Mackie, M.; Tiwary, S.; Rakownikow Jersie-Christensen, R.; Chiva, C.; Dickinson, M.R.; Kuhlwilm, M.; de Manuel, M.; et al. The dental proteome of Homo antecessor. *Nature* **2020**, *580*, 235–238. [CrossRef]
64. Cappellini, E.; Jensen, L.J.; Szklarczyk, D.; Ginolhac, A.; da Fonseca, R.A.; Stafford, T.W.; Holen, S.R.; Collins, M.J.; Orlando, L.; Willerslev, E.; et al. Proteomic analysis of a pleistocene mammoth femur reveals more than one hundred ancient bone proteins. *J. Proteome Res.* **2012**, *11*, 917–926. [CrossRef]
65. Schweitzer, M.H.; Marshall, M.; Carron, K.; Bohle, D.S.; Arnold, E.V.; Barnard, D.; Horner, J.R.; Starkey, J.R. Heme compounds in dinosaur trabecular bone. *Proc. Natl. Acad. Sci. USA* **1997**, *94*, 6291–6296. [CrossRef]
66. Schweitzer, M.H.; Chiappe, L.; Garrido, A.C.; Lowenstein, J.M.; Pincus, S.H. Molecular preservation in Late Cretaceous sauropod dinosaur eggshells. *Proc. Biol. Sci.* **2005**, *272*, 775–784. [CrossRef] [PubMed]
67. Borja, C. Detección y Caracterización de Proteínas Fósiles Mediante Técnicas Inmunes. Ph.D. Thesis, Universidad de Granada, Granada, Spain, 1995.
68. Poinar, H.N.; Hofreiter, M.; Spaulding, W.G.; Martin, P.S.; Stankiewicz, B.A.; Bland, H.; Evershed, R.P.; Possnert, G.; Paabo, S. Molecular coproscopy: Dung and diet of the extinct ground sloth Nothrotheriops shastensis. *Science* **1998**, *281*, 402–406. [CrossRef] [PubMed]
69. Buckley, M. Paleoproteomics: An Introduction to the Analysis of Ancient Proteins by Soft Ionisation Mass Spectrometry. In *Paleogenomics: Genome-Scale Analysis of Ancient DNA*; Lindqvist, C., Rajora, O.P., Eds.; Springer International Publishing: Cham, Switzerland, 2019; pp. 31–52.
70. Cleland, T.P.; Schroeter, E.R. A Comparison of Common Mass Spectrometry Approaches for Paleoproteomics. *J. Proteome Res.* **2018**, *17*, 936–945. [CrossRef] [PubMed]
71. Asara, J.M.; Schweitzer, M.H.; Freimark, L.M.; Phillips, M.; Cantley, L.C. Protein sequences from mastodon and Tyrannosaurus rex revealed by mass spectrometry. *Science* **2007**, *316*, 280–285. [CrossRef] [PubMed]
72. Schroeter, E.R.; Cleland, T.P.; Schweitzer, M.H. Deep Time Paleoproteomics: Looking Forward. *J. Proteome Res.* **2022**, *21*, 9–19. [CrossRef]
73. Low, T.Y.; Syafruddin, S.E.; Mohtar, M.A.; Vellaichamy, A.; ARahman, N.S.; Pung, Y.F.; Tan, C.S.H. Recent progress in mass spectrometry-based strategies for elucidating protein-protein interactions. *Cell. Mol. Life Sci.* **2021**, *78*, 5325–5339. [CrossRef]
74. ten Have, S.; Boulon, S.; Ahmad, Y.; Lamond, A.I. Mass spectrometry-based immuno-precipitation proteomics—the user's guide. *Proteomics* **2011**, *11*, 1153–1159. [CrossRef] [PubMed]
75. Brown, S.; Higham, T.; Slon, V.; Pääbo, S.; Meyer, M.; Douka, K.; Brock, F.; Comeskey, D.; Procopio, N.; Shunkov, M.; et al. Identification of a new hominin bone from Denisova Cave, Siberia using collagen fingerprinting and mitochondrial DNA analysis. *Sci. Rep.* **2016**, *6*, 23559. [CrossRef]

biology

Article

Using Macro- and Microscale Preservation in Vertebrate Fossils as Predictors for Molecular Preservation in Fluvial Environments

Caitlin Colleary [1,2,*], Shane O'Reilly [3], Andrei Dolocan [4], Jason G. Toyoda [5], Rosalie K. Chu [5], Malak M. Tfaily [5,6], Michael F. Hochella, Jr. [1,7] and Sterling J. Nesbitt [1]

[1] Department of Geosciences, Virginia Tech, Blacksburg, VA 24061, USA
[2] Cleveland Museum of Natural History, Cleveland, OH 44106, USA
[3] Atlantic Technological University, ATU Sligo, Ash Lane, F91 YW50 Sligo, Ireland
[4] Texas Materials Institute, University of Texas at Austin, Austin, TX 78712, USA
[5] Environmental Molecular Sciences Laboratory, Pacific Northwest National Laboratory, Richland, WA 99354, USA
[6] Department of Environmental Science, University of Arizona, Tucson, AZ 87519, USA
[7] Earth Systems Science Division, Pacific Northwest National Laboratory, Richland, WA 99354, USA
[*] Correspondence: ccolleary@cmnh.org

Simple Summary: Fossils are the only direct evidence of life throughout Earth's history. We examine the biology of ancient animals to learn about evolution and past ecosystems. Biomolecules are a relatively new source of information from fossil records because new technology is now being used in paleontology that makes it possible to detect molecular remains in fossils. However, molecules extracted from fossils are complex mixtures with environmental and other sources of organic compounds. Additionally, macroscale preservation is well-known to vary greatly across fossil localities. Therefore, a goal in molecular paleontology is to develop ways to predict where molecules may be preserved and differentiate between endogenous and exogenous sources. Here, we use a powerful combination of methods that focus on high-resolution mass spectrometry to evaluate the molecular-scale preservation of a dinosaur quarry from the Triassic Period. We found that despite very good overall preservation at this locality, there is no evidence of endogenous molecules, demonstrating that molecular preservation is variable and that good macro- and microscale preservation cannot necessarily be used as predictors for biomolecule preservation in the fossil record.

Citation: Colleary, C.; O'Reilly, S.; Dolocan, A.; Toyoda, J.G.; Chu, R.K.; Tfaily, M.M.; Hochella, M.F., Jr.; Nesbitt, S.J. Using Macro- and Microscale Preservation in Vertebrate Fossils as Predictors for Molecular Preservation in Fluvial Environments. *Biology* **2022**, *11*, 1304. https://doi.org/10.3390/biology11091304

Academic Editor: Mary H. Schweitzer

Received: 30 June 2022
Accepted: 26 August 2022
Published: 2 September 2022

Publisher's Note: MDPI stays neutral with regard to jurisdictional claims in published maps and institutional affiliations.

Copyright: © 2022 by the authors. Licensee MDPI, Basel, Switzerland. This article is an open access article distributed under the terms and conditions of the Creative Commons Attribution (CC BY) license (https://creativecommons.org/licenses/by/4.0/).

Abstract: Exceptionally preserved fossils retain soft tissues and often the biomolecules that were present in an animal during its life. The majority of terrestrial vertebrate fossils are not traditionally considered exceptionally preserved, with fossils falling on a spectrum ranging from very well-preserved to poorly preserved when considering completeness, morphology and the presence of microstructures. Within this variability of anatomical preservation, high-quality macro-scale preservation (e.g., articulated skeletons) may not be reflected in molecular-scale preservation (i.e., biomolecules). Excavation of the Hayden Quarry (HQ; Chinle Formation, Ghost Ranch, NM, USA) has resulted in the recovery of thousands of fossilized vertebrate specimens. This has contributed greatly to our knowledge of early dinosaur evolution and paleoenvironmental conditions during the Late Triassic Period (~212 Ma). The number of specimens, completeness of skeletons and fidelity of osteohistological microstructures preserved in the bone all demonstrate the remarkable quality of the fossils preserved at this locality. Because the Hayden Quarry is an excellent example of good preservation in a fluvial environment, we have tested different fossil types (i.e., bone, tooth, coprolite) to examine the molecular preservation and overall taphonomy of the HQ to determine how different scales of preservation vary within a single locality. We used multiple high-resolution mass spectrometry techniques (TOF-SIMS, GC-MS, FT-ICR MS) to compare the fossils to unaltered bone from extant vertebrates, experimentally matured bone, and younger dinosaurian skeletal material from other fluvial environments. FT-ICR MS provides detailed molecular information about complex mixtures, and TOF-SIMS has high elemental spatial sensitivity. Using these techniques, we did not find convincing evidence of a molecular signal that can be confidently interpreted as endogenous, indicating that very good macro- and microscale preservation are not necessarily good predictors of molecular preservation.

Keywords: molecular taphonomy; fossils; preservation; mass spectrometry; dinosaurs

1. Introduction

If original biological compounds (biomolecules) are preserved on long timescales (>10 million years), then more of the biological remains of extinct organisms can be uncovered from the fossil record than previously considered possible, expanding what is known about ancient biology and taphonomic processes. Studies of the preservation of biomolecules often focus on exceptionally preserved fossils that retain soft tissues ([1] and references therein); however, exceptional preservation (e.g., hair, feathers, skin) requires specific conditions to occur (including sediment chemistry and microbial activity) and is rare [2]. Therefore, most fossils, particularly those of terrestrial vertebrates which are often entirely skeletal remains, are not often considered exceptionally preserved. The quality of preservation of bone varies from intact well-preserved, articulated skeletons to weathered bone fragments [3–5]. The characterization of good preservation is dependent on scale: (1) good macro-level preservation is the presence of articulated skeletons or features on the bones (e.g., muscle scars); (2) good micro-level preservation is the retention of the microstructures in bone that are often examined in histological studies (e.g., external fundamental systems, three-dimensionally preserved canaliculi); and (3) good molecular-level preservation is the retention of original biomolecules (e.g., nucleic acids, proteins, lipids). With the increased use of high-resolution mass spectrometry, more studies on terrestrial vertebrate fossils have begun to demonstrate that biomolecules preserve more readily than previously considered and in a greater range of depositional settings [5–10], opening much of the vertebrate terrestrial record to recovering more data for ancient animals [11], even from weathered bone fragments, which have traditionally been considered poorly preserved [12].

The terrestrial fossil record is heavily biased toward fossils preserved in fluvial sedimentary settings and the majority of studies that have investigated the preservation of biomolecules in dinosaurs and other terrestrial vertebrates have done so in fossils that are preserved in stream channel, flood plain, delta, and coastal paleoenvironments [5,6,13–15]. Course-grained sandstones and conglomerates are common lithologies in these paleoenvironments and are not normally considered to be conducive to exceptional or good preservation [16], although some have suggested that the porosity of sandstones may improve molecular preservation [6]. Additionally, the influence of water on molecular-scale preservation in fluvial environments has not been experimentally tested, despite being hypothesized to not be conducive to the preservation of biomolecules [17] and recent studies have reported molecular preservation in marine depositional environments [7,8].

Therefore, to explore the molecular preservation of a fluvial terrestrial fossil assemblage, we examined the Hayden Quarry (HQ) (Chinle Formation, Petrified Forest Member, Ghost Ranch, NM, USA), a locality that preserves an extraordinary Late Triassic (~212 Ma) record of dinosaur evolution and paleoenvironmental change [18–20]. The Petrified Forest Member is a series of paleo-channels, with alternating mudstones and siltstones and poorly sorted sandstones and conglomerates [19] and the HQ is divided into four active quarries (H1–H4). The depositional environment is interpreted as episodes of transient flooding, along with periods of standing water and preserves terrestrial and semiaquatic animals [18], (Supplementary Materials). The HQ has a diverse assemblage of terrestrial vertebrates, with high-fidelity preservation at both the macro-scale (i.e., complete skeletons and very small vertebrates, with vertebrae discovered as small as ~1 mm) and microscale, with histological analyses showing high-fidelity preservation of bone microstructures [21].

Incorporating taphonomy into fossil studies is essential for understanding the geological history of the samples to make predictions about what may be preserved [22]. Here, we examine whether high-quality macro- and micro-level preservation are good indicators of high-quality molecular-level preservation. To examine the organic preservation of fossils preserved in a fluvial environment, we used mass spectrometry to examine animal biomolecules (lipids and amino acids) and environmental compounds (e.g., phenol). We analyzed three different types of fossilized tissues from one paleo-channel in the HQ (H4): the femur of the early theropod dinosaur *Tawa hallae* (GR1065) [23], a phytosaur tooth (GR1064), and a coprolite with digested bone from an indeterminate vertebrate (GR1063). We also analyzed an additional bone from H4 (GR1066) to test for any differences between analyzing thin section and non-thin sectioned bone using these techniques. Additionally, we compared the Triassic fossil bone to dinosaur bones from the Cretaceous and Jurassic Periods from similar depositional environments, bone from extant vertebrates, experimentally matured bone (i.e., bone from extant vertebrates that was subjected to a range of temperatures to accelerate the degradation process), and a matrix sample as a control.

Institutional Abbreviations: Ghost Ranch, Hayden Quarry (GR), Los Angeles Museum of Natural History (LACM), Mammoth Site of Hot Springs, South Dakota (MS).

2. Materials and Methods

2.1. Specimens

Thin sections from the Hayden Quarry (HQ) were analyzed to compare different types of fossils and to test the differences in analyzing thin sections and whole bone fragments using surface mass spectrometric techniques (e.g., time of flight secondary ion mass spectrometry). The fossils were embedded in CastoliteAP, a clear polyester resin (vacuumed to remove bubbles). After curing, a Buehler (IsoMet 4000 Lake Bluff, IL, USA) saw with a diamond wafering blade was used to create thin sections that were subsequently glued to glass slides with Aron Alpha (Type 201) cyanoacrylate. The excess material was ground down using a Hillquist thin-section machine and hand polishing on an Ecomet 2 speed grinder-polisher to a thickness that the microanatomy could be viewed using a light microscope. The HQ specimens thin sectioned and analyzed include: a tooth (phytosaur; GR1064), a bone (femur from the early theropod *Tawa hallae*; GR1065) and a coprolite from an unidentified vertebrate (GR1063) collected in Hayden Quarry 2 (H2). Additionally, dinosaur rib fragments used in analysis include a theropod from the Hell Creek Formation in Montana (Hell Creek), USA (LACM 23844, Late Cretaceous Period), a sauropod from the Morrison Formation in Utah (Morrison), USA (LACM 154089, Late Jurassic Period), and two dinosaur rib fragments from Hayden Quarry 4 (H4) (GR1065, GR1066, Triassic Period). All of these fossils were weathered out of or excavated from fluvial depositional environments with sandstone or mudstone lithologies [13,18,24]. Specimens were excavated in situ and areas sampled showed no signs of surface weathering. A matrix sample collected in conjunction with the rib fragment (GR1066) from H4 was included to compare to the compounds found in the fossils. Recently deceased alligator (*Alligator mississippiensis*, TMM M-12613) and elephant (*Loxodonta*, MS-E01) rib bone samples were used to compare unaltered bone chemistry using the same techniques (Table 1).

2.2. Maturation Experiments

Experimentally matured bone samples (modern elephant, MS-E01) were reanalyzed from a previous study [25]. A diamond saw (Dremel®) was sterilized to prevent contamination and used to cut a fresh elephant rib bone (deceased zoo animal, ME-E-01) into three 2 mm^2 fragments. The fragments were sealed in 3 mm × 15 mm platinum capsules; however, this does not prevent water from evaporating. They were loaded into cold sealed pressure vessels in the Hydrothermal Laboratory at Virginia Tech. These short-term experiments accelerate the degradation of the bone and were conducted for 24 h at 100 °C, 200 °C and 250 °C at atmospheric pressure (based on the protocol in [26]).

Table 1. Specimen information and analyses done.

Specimen Number	Specimen	Locality	Time Period	Analyses
GR1063	Coprolite	Hayden Quarry (H2)	Triassic	TOF-SIMS
GR1064	Phytosaur tooth	Hayden Quarry (H2)	Triassic	TOF-SIMS
GR1065	*Tawa hallae* femur	Hayden Quarry (H2)	Triassic	TOF-SIMS
GR1066	Rib fragment	Hayden Quarry (H4)	Triassic	TOF-SIMS, GC-MS, FT-ICR MS
GR1067	Rib fragment	Hayden Quarry (H2)	Triassic	GC-MS, FT-ICR MS
	Matrix sample	Hayden Quarry (H4)	Triassic	TOF-SIMS, GC-MS
LACM 154089	Sauropod rib	Morrison	Jurassic	TOF-SIMS, FT-ICR MS
LACM 23844	Theropod rib	Hell Creek	Cretaceous	TOF-SIMS, FT-ICR MS
	Mammoth rib			FT-ICR MS
TMMM-12613	Alligator rib		Modern	TOF-SIMS
MS-E01	Elephant rib		Modern	TOF-SIMS, FTIR-MS
MS-E01	Elephant rib (experimentally matured)		Modern	TOF-SIMS

2.3. Time-of-Flight Secondary Ion Mass Spectrometry (TOF-SIMS)

TOF-SIMS analysis was performed using an ION-TOF GmbH, Germany TOF.SIMS 5 at The University of Texas at Austin, Texas Materials Institute. A pulsed (20 ns, 10 kHz) analysis ion beam of Bi_3^+ clusters at 30-kV ion energy was raster-scanned over 500×500 μm^2 areas. Bi_3^+ polyatomic sputtering was used to reduce the fragmentation of large organic molecules. A constant flux, 21 eV electron beam was used during data acquisition to reduce sample charging. Secondary ions had positive polarity and an average mass resolution of 1–3000 (m/δm). The base pressure during acquisition was $<1 \times 10^{-8}$ mbar. Mass calibration was performed by identifying the peak positions of CH_2^+, CH_3^+, $C_2H_3^+$, and $C_3H_3^+$ secondary ions. Regions of interest were chosen to reduce the effects of topography (which even at micron-scales, can influence the time it takes certain molecules to reach the analyzer, which degrades the mass resolution, thereby impeding correct molecular assignment).

The benefit of surface mass spectrometry is it is minimally destructive and can be used to examine the spatial distribution of fossils. One of the drawbacks of using this technique when evaluating protein preservation is that it does not provide certain types of data (e.g., peptide sequences). Therefore, we determined a chemical fingerprint of 86 peaks for protein degradation products (amino acids and amino acid fragments) and the inorganic components of bone. We mapped the distribution of positive spectra of ionized molecules on the bone surface using TOF-SIMS in fresh bone and dinosaur fossils (Table 1, Figure 1). Amino acids, amino acid fragments and mineral elements were chosen to characterize the preservation of each sample. The fingerprint was developed by combining relevant peaks from previous analyses [25,27] and choosing additional peaks that are present in the samples (Supplementary Materials, Table S1). Matrix (rock samples not containing fossils) from the same horizon as the fossils in the HQ and fresh bone were used to compare the degradation of bone that occurs during fossilization.

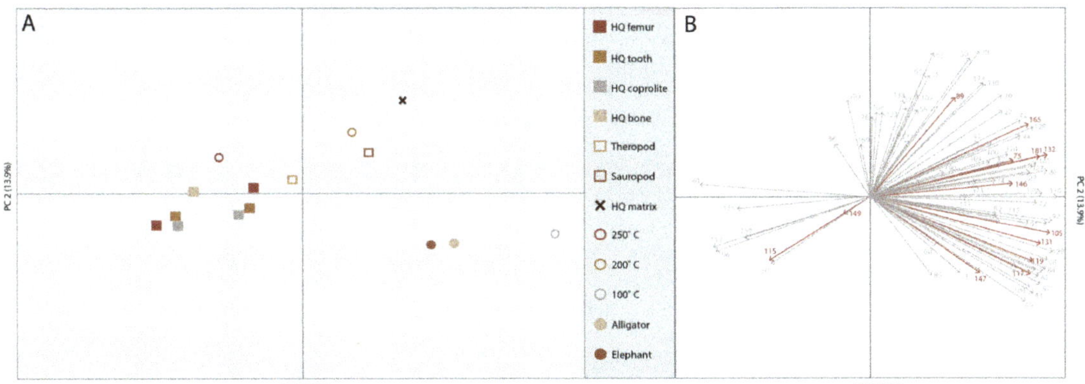

Figure 1. Principal component analysis. (**A**) PCA of chemical signature of 86 organic and inorganic peaks. HQ femur, HQ tooth and HQ coprolite are all thin sections from H2, and HQ bone is an untreated bone fragment from H4. The HQ matrix is from H4. The modern bones (alligator and elephant) and the 100 °C are more similar chemically to one another than the other samples. The HQ fossils show little variation between one another, but there is a greater amount of variation with the associated matrix. Therefore, there are amino acids present in the fossils that are not present in the associated matrix. (**B**) PCA Loadings show how each one of the 86 peaks influences the specimen placement in the PCA.

2.4. Lipid Analyses

Two Triassic Period rib fossils (GR1066, GR1067) were separated from matrix sediment manually and the fossil surfaces were cleaned using a dental drill and solvent-cleaned steel drill- bits. Powders were drilled from cleaned fossils. The powder from the cleaning procedure for H2 was also retained. Sediment matrix was powdered using a solvent-cleaned mortar and pestle. Between 250 and 600 mg of powdered sample was each weighed into 12 mL glass tubes. Samples were extracted for 30 min (ratio of solvent: sample was 5:1) with 9:1 (*v/v*) dichloromethane/methanol [28] using sonication in an ultrasonic bath at room temperature (~21 °C). The extract was separated from solid residue by centrifugation. Extracted residues were re-extracted with fresh solvent for a total of three extraction and supernatants from each step were combined to give a total lipid extract (TLE). TLEs were concentrated to minimal volume under a gentle stream of high-purity N_2 gas. A portion each TLE was silylated with N,O- Bis(trimethylsilyl)trifluoroacetamide /trimethylchlorosilane mixed with pyridine (9:1 *v/v*) at 70 °C for 2 h. A portion each TLE was reacted with N,O-Bis(trimethylsilyl)trifluoroacetamide/trimethylchlorosilane mixed with pyridine (9:1 *v/v*) at 70 °C for 2 h. This reaction replaces active hydrogens on polar functional groups (e.g., hydroxy groups) with a trimethylsilyl moiety to increase analyte volatility and thermal stability, thereby ensuring functionalized lipids are amenable to GC analysis. Aliquots of the derivatized samples were analyzed by gas chromatography/mass spectrometry (Agilent 5890 GC hyphenated to an Agilent 5975C Mass Selective Detector). The GC was equipped an Agilent J&W HP-5MS non-polar capillary column (30 m length, 0.25 mm inner diameter, 250 µm film thickness). The GC temperature program was: 70 °C for 2 min, ramp at 10 °C min^{-1} to 130 °C, followed by a ramp to 300 °C at 4 °C min^1 and a final hold time of 20 min. The mass spectrometer was operated in electron impact ionization mode (70 eV), with a mass scan range from m/z 50 to 600. All glassware was fired (550 °C overnight) and all solvents used were high-purity (OmniSolv). Procedural blanks were run to monitor background contamination.

2.5. Fourier-Transform Ion Cyclotron Resonance Mass Spectrometry (FT-ICR MS)

Mortar and pestles were washed, then wrapped in aluminum foil, rinsed in nano pure water, rinsed in ethanol and then were combusted at 400 °C for 8 h. The aluminum foil remained on the pestle and lined the mortar throughout grinding to prevent contamination. Once they cooled down, they were used to powder the fossil samples. Blanks were created by rinsing and collecting 2 mL of nano pure water, 2 mL of MeOH and ~2 mL of $CHCl_3$ from each mortar and pestle. We analyzed bone and sediment from HQ, two dinosaur bones (LACM 154089 and LACM 154089) and a modern elephant bone (MS-E-01). We combined two rib fragments from Ghost Ranch (GR1066, GR1067) as one sample. Each sample was powdered using the mortar and pestle until we had ~1.5 g for each sample. The elephant sample was prepared in liquid nitrogen prior to powdering (Table 1).

Three sets of extractions were conducted on each bone sample: water, methanol (MeOH,) and chloroform ($CHCl_3$) because each solvent is elective for a specific type of organic compound based on its polarity [29]. Water extractions were done first to extract water-soluble small molecules such as sugars, amino sugars and amino acids. 5 mL of nano pure water was added to the powdered samples in leach-free falcon tubes and vortexed at 1000 rpm for 2 h. Samples were then centrifuged for 5 min at 4500 rpm to pellet the samples. These steps were repeated using methanol (MeOH) to sequentially extract other semi polar organics such as lignin-like compounds with both polar and non-polar sides due to its hyperbranched structure and chloroform (CHCl3) to extract non-polar lipids [30]. The samples were then stored overnight at 4 °C. Prior to infusion into the mass spectrometer the water extracted samples were acidified to a pH of 2, concentrated and desalted using Bond Elut PPL cartridges and following procedures from [31]. The methanol extracted samples were run without clean up and the chloroform samples had methanol added in a 1:1 (v:y) to aide in ionization.

The mass spectrometry analysis was performed using a 12T Fourier transform ion cyclotron resonance mass spectrometer (FT-ICR MS) (Bruker solariX, Billerica, MA, USA) outfitted with a standard electrospray ionization (ESI) interface. Samples were directly infused into the mass spectrometer using a 250 µL Hamilton syringe at a flow rate of 3 µL/min. The coated glass capillary temperature was set to 180 °C and data were acquired in positive and negative mode for better overall coverage of detected molecules. The needle voltage was set to +4.2 kV negative mode and −4.4 kV in positive mode. The data were collected by co-adding 200 scans with a mass range of 100–900 *m/z*, at 4 M with a resolution of 240 K at 400 *m/z*. Formulae were assigned by first converting the raw spectra into a list of peaks using Bruker Data Analysis (version 4.2) and applying an FTMS peak picker with a signal-to-noise ratio set to 7 and absolute intensity set to 100. Once the raw spectra were converted to a list of peaks and their resulting mass-to-charge (*m/z*) ratio data was internally calibrated, formula assigned and peaks aligned using Formularity [32] and following the Compound Identification Algorithm [33,34]. Predicted chemical formulae were assigned with C, H, O, N and S and excluding all other elements with the following rules: O > 0 AND N <= 4 AND S < 2 AND 3 * P <= 0. Alignment tolerance was set to 0.5 ppm and calibration tolerance was set to 0.1 ppm. The molecular formulae in each sample were evaluated on van Krevelen diagrams [35], based on their H:C ratios (*y*-axis) and O:C ratios (*x*-axis) assigning them to the major biochemical classes (e.g., lipid, protein, lignin, carbohydrate, etc.) according to [36]. The H:C and O:C ranges for biochemical classification are provided in Supplementary Materials Table S2.

3. Results

In this study, we analyzed different fossil types from a single fossil locality (HQ) using various analytical methods for evaluating the preservation of different molecular classes, as well as inorganic signals to understand the overall molecular taphonomy of the Hayden Quarry. To test if different types of fossils (i.e., bone, tooth, coprolite) can be chemically distinguished from one another, we used surface mass spectrometry (TOF-SIMS) and compared the samples using multivariate statistics. The principal component analysis (PCA) compared the three different fossil types (i.e., bone, tooth, coprolite), the additional dinosaurian fossils, the unaltered bones of extant vertebrates and the experimentally matured bone to determine the variance between each of the samples (Figure 1). Additionally, we included both thin sections and one whole bone sample from the HQ to see if embedding them in resin altered the molecular signal. All the HQ samples including both thin sections and the bone sample plot together, demonstrating no change in the molecular signal from embedding them. When comparing the samples, there is little variation between all of the fossils from the Hayden Quarry (thin sections and polished bone fragment), which all plot together, along with the 250 °C matured bone. Although, the 200 °C matured bone and the Morrison bone do show slight variation from the others and plot more closely to the sediment matrix sample from the HQ. The HQ matrix sample does show a greater amount of variance from the HQ fossil samples. The two unaltered bones from extant vertebrates (elephant and alligator) and the 100 °C matured bone have similar molecular signatures to one another but differ from the rest of the samples. TOF-SIMS molecular maps (Figure 2) of the HQ polished bone fragment (GR1066) show that the presence of specific elements and molecules varies between certain bone features. Calcium (Ca) and iron (Fe) are ubiquitous across the bone surface but show a decreased abundance in certain areas (around microstructures in the bone) that have a higher abundance of strontium (Sr). Bone microstructures do have macroscopic evidence of mineral infilling. There is one area of the bone that shows a low abundance of Ca, Fe and Sr and a high abundance of the amino acids glycine (Gly) and alanine (Ala). Otherwise, these amino acids are in extremely low abundance, if present at all, across the bone surface.

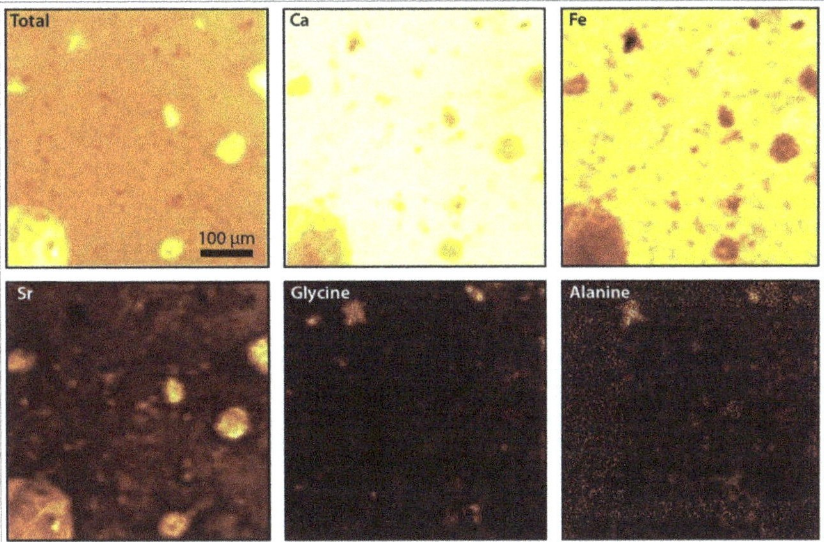

Figure 2. TOF-SIMS images of HQ bone fragment (GR1066). Each image is the same 500 µm × 500 µm area of the bone. The lighter areas represent a greater concentration of a given element or molecule, while the darker areas represent a lower concentration. The total map shows all 86 peaks in the analysis, showing a higher concentration in pore spaces. The rest of the maps show the distribution of specific elements and molecules. Specific features of the bone, like pore spaces, have a higher concentration of strontium (Sr) and lower concentrations of iron (Fe) while some elements, like calcium (Ca) are ubiquitous across the sample surface. Amino acids like glycine and alanine are both present in a single area of the bone.

To evaluate the presence of additional biomolecules, we examined lipids in fossil bone and the surrounding matrix (Figure 3). Lipids were restricted to fatty acids, including unsaturated fatty acids and the lipid profiles were very similar between the fossils and the matrix. The associated matrix had more than eight times the extractable lipids than the fossil bone. Additionally, we compared the fossil, bone and matrix samples by evaluating the presence of additional organic molecules present in the HQ using high-resolution FT-ICR MS. In the PCA analysis (Figure 4), the placement of the elephant and mammoth bones were most heavily influenced by lipid- and protein-like compounds, the Hell Creek and Morrison bones were most heavily influenced by oxygenated, phenol- and amino sugar-like compounds and both the HQ bone and HQ matrix showed little variation from one another and placement was most heavily influenced by unsaturated- and condensed hydrocarbon-like compounds. FT-ICR MS at high magnetic fields provides detailed molecular information about complex mixtures, like those found in fossils, due to its high resolution and mass accuracy, a requirement to assign unique, unambiguous molecular formulae to each peak across an entire molecular weight distribution ($200 < m/z < 1500$) Therefore, the absence of endogenous material in the samples tested in this study suggests that these compounds are indeed not present.

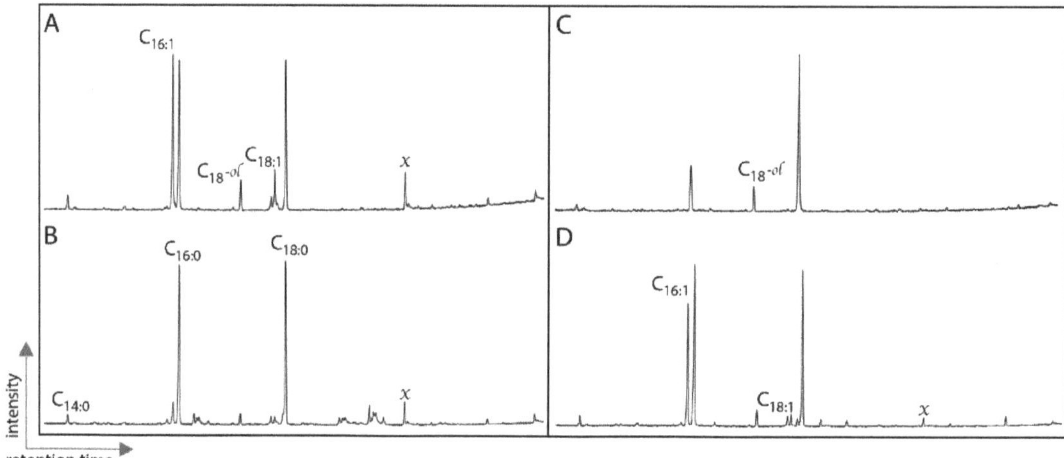

Figure 3. Partial total ion chromatograms of silyated total lipid extracts from Ghost Ranch Hayden Quarry 4: (**A**) fossil bone and (**B**) associated matrix. Hayden Quarry 2 (**C**) fossil bone and (**D**) associated matrix. Tetradecanoic acid (C14:0), hexadecanoic acid (C16:0, C16:1), octadecanoic acid (C18:0, C18:1, octadecan-1-ol (C18:0^{ol}) and contaminants (x).

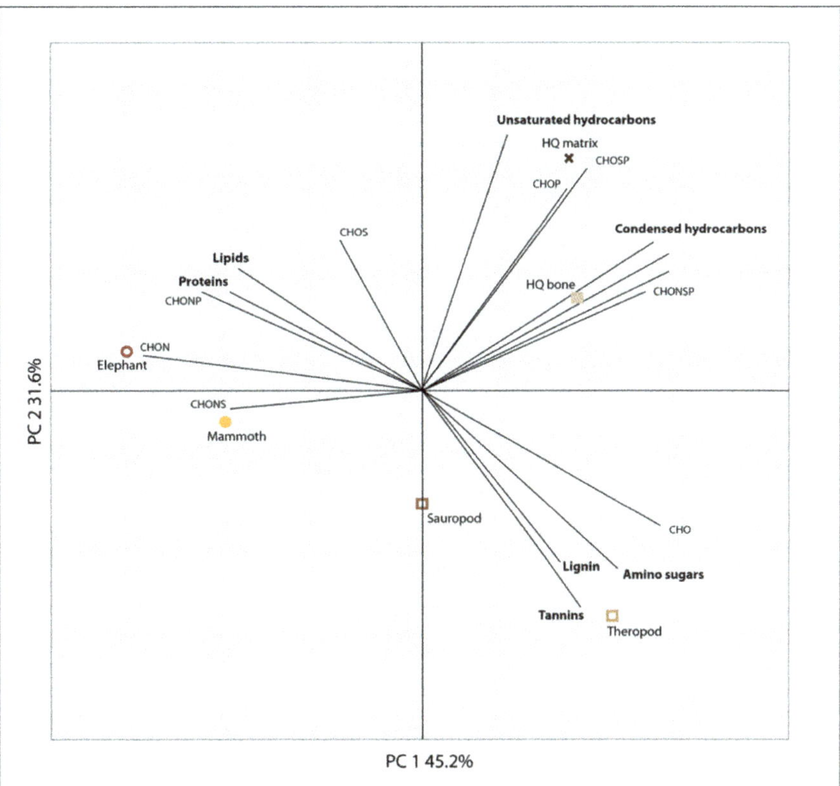

Figure 4. Principal component analysis of FT-ICR MS data. Examination of fossil bones by grinding them up and not demineralizing them reveals additional compounds in greater abundance than proteins and lipids. Proteins and lipids are in the greatest abundance in the elephant and mammoth bones, whereas the Morrison and Hell Creek bones have higher abundances of lignin, tannins and amino sugars and the HQ bones have a higher abundance of unsaturated and condensed hydrocarbons. Amino acids were found in the dinosaur bones and matrix but may be masked by the environmental compounds that are more abundant. Tannins could be any oxygenated compound and lignin could be phenols of plant or microbial origin.

4. Discussion

Depositional environment plays a large role in fossil preservation [25,37,38]. Fluvial formations are common sources of fossil material, particularly of terrestrial vertebrates. The variation in preservation in fluvial burial environments is indicative of the variation in the environments themselves and leads to extremes in preservation, ranging from complete skeletons to fragmentary bone [39]. Therefore, molecular preservation in fluvial environments likely varies as much as macro-scale vertebrate preservation. Amino acids and amino acid fragments were detected in dinosaurian fossils dating back to the Late Triassic (~212 Ma) and the presence of amino acids, specifically glycine, has been included as supporting evidence of ancient collagens [5–7]. We detected an amino acid signature unique to the fossil bones, teeth and coprolites and distinct from the surrounding matrix. However, the absence of lipids and the main signal of condensed hydrocarbon in the HQ bone and matrix, which are both very similar, cast doubt that the amino acids detected in the HQ fossils have an animal origin.

Molecular analytical techniques being applied to fossil studies often fall into two general categories: (1) those that extract targeted compounds and inject them into a mass spectrometer (e.g., DI, LC-MS) to examine what is preserved in the fossil (e.g., paleoproteomics) [40] and (2) those that use surface analytical techniques (e.g., Raman, TOF-SIMS) to analyze the entire fossil and examine taphonomy (e.g., degradation products, mechanisms of preservation) [41]. To date, no single technique excels at accomplishing both of these goals. The benefit of using surface-sampling techniques includes the ability to examine the spatial distribution of molecules that may vary in different parts of a fossil; however the trade-off is a lack of resolution regarding the compounds themselves. TOF-SIMS specifically, while of great utility in semi-non-destructive fossil analyses on heme and pigments, is not the best mass spectrometric technique for evaluating proteins in fossils ([42] and citations therein). However, the ability to evaluate the structure of the bone and make comparisons between the fossil and the matrix are strong additions to molecular taphonomy studies interested in exploring where molecular data may be best preserved for targeting future analyses and for examining the distinction or interplay between fossil and matrix. TOF-SIMS is a highly element sensitive and selective technique, both chemically and spatially; however, in this application, the amino acids we were examining are relatively small molecules with limited fragmentation patterns, which likely produce secondary ion fragments that generally match the fragments generated by many other organic materials in the environment. This highlights a continuing challenge in paleomolecular studies, which is determining the source of organic material in extremely organic-rich, complex environments.

The methods used in this study were chosen to compare the overall molecular taphonomy of the HQ. When examining the variation between the specimens, we used TOF-SIMS to develop a chemical fingerprint of 86 amino acid and amino acid fragments and found this was a good way of distinguishing between fossil and extant bone and between the fossils and associated matrix, but not between the different types of fossils. This demonstrates that there is a similar organic signal in all of the HQ fossil material that is distinct from the associated matrix and is also distinct from the dinosaur bones from the Cretaceous and Jurassic sites. Therefore, because of the difference between the fossils, the molecular signal is not indicative of generalized contamination. However, because there is no distinction between the bone, tooth and coprolite in the HQ, this signal is likely not evidence of ancient animal remains but may instead be taphonomic or environmental. Recent work has suggested that fossil bone may be a good host to modern microbial growth [43] but we did not find evidence of microbial lipids that would suggest that is the case here. The ability to confidently assign the remnants of biomolecules to ancient animals, ancient microbes or modern microbes will remain one of the major goals of ancient molecular studies, particularly on such long timescales. The experimentally matured bone (all MS-E01, heated to increasing temperatures) depict an interesting alteration in the chemical signature. When compared to the fossils, the molecular signal in the heated bones is altered in a predictable way, with the increase in temperature correlating to the age of the fossils. Heating the bone to 100 °C caused some alteration, but it is still most comparable to the modern bone. Heating the bone to 250 °C led to a molecular signal similar to that found in the HQ bone, which we are interpreting as the absence of a molecular signal from the bone. It is possible that the alteration seen in the modern bone is related to the evaporation of water during the heating process and is worthy of further exploration.

Fossils and sedimentary rocks of this age would typically contain the end products of lipid diagenesis—hydrocarbon skeletons such as steranes or *n*-alkanes—as major extractable compounds [44] and are much more likely to be from modern (or very young) sources. Previous work has shown that lipids can be transported from fossils to surrounding matrices [45] or the opposite, from matrices to fossil bone [43]. Thus, comparison between fossils and matrices must include assessment of the relative abundances, occurrences (presence or absence) and distributions of individual lipids (or classes of lipids). In contrast to Liebenau et al. (2015) results, extractable lipids in our specimens were much higher in the matrix sample than fossils. Given that the distributions of lipids in fossil specimens were very similar to the matrices and the observed concentration gradient, it is more likely that detected lipids in the fossils are exogenous. Given, the presence of exceptionally labile lipids (monounsaturated fatty acids) as major lipids in the fossils analyzed, the similar lipid profiles between fossils and matrices and the much higher concentration of lipids in the matrices, it is likely that the lipids detected in the fossil bones were sourced from the matrix and transported in and that they are from modern/recent biological sources. It is possible that the fatty acids detected are from microbial communities that have colonized the fossil bone matrices, as proposed by Saitta et al. (2018). Lipid data casts additional doubt on the source of the amino acid signature being detected.

We also examined the inorganic components of the bone and additional sources of organics, including environmental contaminants (Figure 4). High levels of phenols were found in the Morrison and Hell Creek fossils, which could be a signal of lignin, representing contamination from vascular plants. Lignin is a biopolymer that preserves on very long timescales (~315 Ma) [17]; however, it is not possible to determine with these methods if it is ancient or recent, although a more recent source of the lignin may be from plant roots seeking out sources of phosphorous from buried bone. The source of phenols may also be microbial because microbes produce phenols when under stressed conditions [46]. The HQ samples had hydrocarbons present in high abundance and the bone and matrix samples were similar in composition. Additionally, iron is present in high intensity across the surface of the HQ bone fragment, while other elements like strontium are isolated to pore spaces in the bone (Figure 2). The high intensity of iron is consistent with proposed conditions in burial environments that may favor soft tissue preservation ([12] and references therin) and strontium levels have been shown to increase with time and during diagenesis [47]. The molecular composition of all of the fossils is similar despite being from different sources (e.g., bone, tooth, coprolite), time periods and burial environments, which may be evidence for a taphonomic source of organics being introduced to fossils that are found in similar depositional environments. Additional analyses that target specific compounds may be able to determine the source, specifically extraction-based proteomic techniques.

5. Conclusions

Despite very good macro and micro-level preservation at the Triassic Hayden Quarry, there is no direct evidence for original biomolecules preserved in the fossils. Amino acid evidence shows variation between the fossils and the matrix. This alone is not definitive evidence of original biomolecular preservation; however, we were unable to determine the source of these amino acids with the methods used in this study. Lipid data have no indication of animal lipids and show no distinction in lipid content between the fossil bone and matrix, and the FT-IRC MS data show that the HQ bone and matrix are similar to one another and high in hydrocarbons. Therefore, we have concluded that we detected no original organic preservation at this site. The Hayden Quarry is an example of a fluvial burial environment where we find remarkable macro- and microscale preservation, but no evidence of molecular-scale preservation using these methods. Therefore, as molecular fossil studies continue with the goal of finding trends in preservation to target specimens with probable molecular-scale preservation, we have to continue to consider that variability across fossil localities, which is well-documented at the macro- and microscales, makes it very difficult to make overarching rules about molecular preservation based on burial environment.

Supplementary Materials: The following supporting information can be downloaded at: https://www.mdpi.com/article/10.3390/biology11091304/s1, Table S1: Amino acid peak assignments (TOF-SIMS). Peaks were assigned based on mass and previous studies; Table S2: H:C and O:C ranges for biochemical classification (FTICR-MS).

Author Contributions: C.C. conceived of the study, designed the study, performed TOF-SIMS analysis, analyzed the data and wrote the manuscript. S.O. carried out lipid analysis, assisted with data analysis and helped with the manuscript. A.D. carried out TOF-SIMS analysis, assisted with data analysis and helped with the manuscript. J.G.T. carried out sample preparation and organic extractions for FT-ICR MS analyses and helped with data analysis. R.K.C. performed FT-ICR MS analyses, helped with data analysis and helped with the manuscript. M.M.T. helped with interpreting and analyzing the FT-ICR MS data. M.F.H.J. conceived of and facilitated FT-ICR MS analyses and helped with manuscript. S.J.N. conceived the study, coordinated the study, made the HQ thin sections and helped with the manuscript. All authors gave final approval of the manuscript. All authors have read and agreed to the published version of the manuscript.

Funding: A member of the National Nanotechnology Coordinated Infrastructure (NNCI), supported by NSF grants ECCS 1542100, ECCS 2025151, and NSF EAR 1349667 (to SJN). NanoEarth is also supported by Virginia Tech's Institute for Critical and Applied Sciences (ICTAS). A portion of this research was performed on a project award (50015) from the Environmental Molecular Sciences Laboratory (EMSL), a DOE Office of Science User Facility sponsored by the Biological and Environmental Research program under Contract No. DE-AC05-76RL01830. TOF-SIMS analyses in this study were funded by a Virginia Space Grant Graduate STEM Research Fellowship awarded to CC. Travel and advisory support for this study were also provided by the Virginia Tech National Center for Earth and Environmental Nanotechnology Infrastructure (NanoEarth).

Institutional Review Board Statement: Not applicable.

Informed Consent Statement: Not applicable.

Data Availability Statement: Not applicable.

Acknowledgments: We would like to acknowledge Ruth Elsey and colleagues at the Rockefeller Wildlife Refuge, Louisiana, the curators and collections staff at the Ruth Hall Museum of Paleontology at Ghost Ranch, the Natural History Museum at Los Angeles County and the Mammoth Site of Hot Springs, South Dakota for the specimens used in this study, everyone excavating in the Hayden Quarry in 2017, Robert Bodnar (Virginia Tech) for the use of his lab, supplies and for his thoughts on the maturation experiments, Hector Lamadrid for assistance with the maturation experiments and Roger Summons (Massachusetts Institute of Technology) for the use of his laboratory facilities for lipid analysis. We also acknowledge S. Augusta Maccracken and Chris Griffin for insightful comments on the manuscript and the editors of this Special Issue, particularly Mary Schweitzer for her continued leadership in this field. Figure color palette is from Wes Anderson's *Fantastic Mr. Fox*.

Conflicts of Interest: The authors declare no conflict of interest.

References

1. Briggs, D.E.; Summons, R.E. Ancient biomolecules: Their origins, fossilization, and role in revealing the history of life. *Bioessays* **2014**, *36*, 482–490. [CrossRef] [PubMed]
2. Allison, P.A.; Briggs, D.E. Exceptional fossil record: Distribution of soft-tissue preservation through the Phanerozoic. *Geology* **1993**, *21*, 527–530. [CrossRef]
3. Behrensmeyer, A.K. Taphonomic and ecologic information from bone weathering. *Paleobiology* **1978**, *4*, 150–162. [CrossRef]
4. Behrensmeyer, A.K.; Kidwell, S.M.; Gastaldo, R.A. Taphonomy and paleobiology. *Paleobiology* **2000**, *26*, 103–147. [CrossRef]
5. Bertazzo, S.; Maidment, S.C.R.; Kallepitis, C.; Fearn, S.; Stevens, M.M.; Xie, H. Fibres and cellular structures preserved in 75-million–year-old dinosaur specimens. *Nat. Commun.* **2015**, *6*, 7352. [CrossRef] [PubMed]
6. Schweitzer, M.H.; Suo, Z.; Avci, R.; Asara, J.M.; Allen, M.A.; Arce, F.T.; Horner, J.R. Analyses of soft tissue from *Tyrannosaurus rex* suggest the presence of protein. *Science* **2007**, *316*, 277–280. [CrossRef]
7. Surmik, D.; Boczarowski, A.; Balin, K.; Dulski, M.; Szade, J.; Kremer, B.; Pawlicki, R. Spectroscopic studies on organic matter from Triassic reptile bones, Upper Silesia, Poland. *PLoS ONE* **2016**, *11*, e0151143. [CrossRef]
8. Lindgren, J.; Sjövall, P.; Thiel, V.; Zheng, W.; Ito, S.; Wakamatsu, K.; Hauff, R.; Kear, B.P.; Engdahl, A.; Alwmark, C.; et al. Soft-tissue evidence for homeothermy and crypsis in a Jurassic ichthyosaur. *Nature* **2018**, *564*, 359–365. [CrossRef]
9. Wiemann, J.; Fabbri, M.; Yang, T.-R.; Stein, K.; Sander, P.M.; Norell, M.A.; Briggs, D.E.G. Fossilization transforms vertebrate hard tissue proteins into N-heterocyclic polymers. *Nat. Commun.* **2018**, *9*, 1–9. [CrossRef]
10. Fabbri, M.; Wiemann, J.; Manucci, F.; Briggs, D.E. Three-dimensional soft tissue preservation revealed in the skin of a non-avian dinosaur. *Palaeontology* **2020**, *63*, 185–193. [CrossRef]
11. Wiemann, J.; Menéndez, I.; Crawford, J.M.; Fabbri, M.; Gauthier, J.A.; Hull, P.M.; Norell, M.A.; Briggs, D.E.G. Fossil biomolecules reveal an avian metabolism in the ancestral dinosaur. *Nature* **2022**, *606*, 522–526. [CrossRef] [PubMed]

12. Ulmann, P.V.; Pandya, S.H.; Nellermoe, R. Patterns of soft tissue and cellular preservation in relation to fossil bone tissue structure and overburden depth at the Standing Rock Hadrosaur Site, Maastrichtian Hell Creek Formation, South Dakota, USA. *Cretac. Res.* **2019**, *99*, 1–13. [CrossRef]

13. White, P.D.; Fastovsky, D.E.; Sheehan, P.M. Taphonomy and suggested structure of the dinosaurian assemblage of the Hell Creek Formation (Maastrichtian), eastern Montana and western North Dakota. *Palaios* **1998**, *13*, 41–51. [CrossRef]

14. Lee, Y.C.; Chiang, C.-C.; Huang, P.-Y.; Chung, C.-Y.; Huang, T.D.; Wang, C.-C.; Chen, C.-I.; Chang, R.-S.; Liao, C.-H.; Reisz, R.R. Evidence of preserved collagen in an Early Jurassic sauropodomorph dinosaur revealed by synchrotron FTIR microspectroscopy. *Nat. Commun.* **2017**, *8*, 14220. [CrossRef]

15. Schroeter, E.R.; DeHart, C.J.; Cleland, T.P.; Zheng, W.; Thomas, P.M.; Kelleher, N.L.; Bern, M.; Schweitzer, M.H. Expansion for the *Brachylophosaurus canadensis* collagen I sequence and additional evidence of the preservation of Cretaceous protein. *J. Proteome Res.* **2017**, *16*, 920–932. [CrossRef]

16. Allison, P.A. Konservat-Lagerstätten: Cause and classification. *Paleobiology* **1988**, *14*, 331–344. [CrossRef]

17. Eglinton, G.; Logan, G.A. Molecular preservation. *Philos. Trans. R. Soc. Lond. B Biol. Sci.* **1991**, *333*, 315–328. [CrossRef] [PubMed]

18. Irmis, R.B.; Nesbitt, S.J.; Padian, K.; Smith, N.D.; Turner, A.H.; Woody, D.; Downs, A. A Late Triassic dinosauromorph assemblage from New Mexico and the rise of dinosaurs. *Science* **2007**, *317*, 358–361. [CrossRef]

19. Irmis, R.B.; Mundil, R.; Martz, J.W.; Parker, W.G. High-resolution U–Pb ages from the Upper Triassic Chinle Formation (New Mexico, USA) support a diachronous rise of dinosaurs. *Earth Planet. Sci. Lett.* **2011**, *309*, 258–267. [CrossRef]

20. Whiteside, J.H.; Lindström, S.; Irmis, R.B.; Glasspool, I.J.; Schaller, M.F.; Dunlavey, M.; Nesbitt, S.J.; Smith, N.D.; Turner, A.H. Extreme ecosystem instability suppressed tropical dinosaur dominance for 30 million years. *Proc. Natl. Acad. Sci. USA* **2015**, *112*, 7909–7913. [CrossRef]

21. Werning, S.A. Evolution of Bone Histological Characters in Amniotes, and the Implications for the Evolution of Growth and Metabolism. Ph.D. Dissertation, University of California, Berkeley, CA, USA, 2013.

22. Behrensmeyer, A.K.; Kidwell, S.M. Taphonomy's contributions to paleobiology. *Paleobiology* **1985**, *11*, 105–119. [CrossRef]

23. Nesbitt, S.J.; Smith, N.D.; Irmis, R.B.; Turner, A.H.; Downs, A.; Norell, M.A. A complete skeleton of a Late Triassic saurischian and the early evolution of dinosaurs. *Science* **2009**, *326*, 1530–1533. [CrossRef] [PubMed]

24. Dodson, P.; Behrensmeyer, A.K.; Bakker, R.T.; McIntosh, J.S. Taphonomy and paleoecology of the dinosaur beds of the Jurassic Morrison Formation. *Paleobiology* **1980**, *6*, 208–232. [CrossRef]

25. Colleary, C.; Lamadrid, H.M.; O'Reilly, S.S.; Dolocan, A.; Nesbitt, S.J. Molecular preservation in mammoth bone and variation based on burial environment. *Sci. Rep.* **2021**, *11*, 1–9. [CrossRef]

26. Colleary, C.; Dolocan, A.; Gardner, J.; Singh, S.; Wuttke, M.; Rabenstein, R.; Habersetzer, J.; Schaal, S.; Feseha, M.; Clemens, M.; et al. Chemical, experimental, and morphological evidence for diagenetically altered melanin in exceptionally preserved fossils. *Proc. Natl. Acad. Sci. USA* **2015**, *112*, 12592–12597. [CrossRef]

27. Orlando, L.; Ginolhac, A.; Zhang, G.; Froese, D.; Albrechtsen, A.; Stiller, M.; Schubert, M.; Cappellini, E.; Petersen, B.; Moltke, I.; et al. Recalibrating *Equus* evolution using the genome sequence of an early Middle Pleistocene horse. *Nature* **2013**, *499*, 74–78. [CrossRef]

28. O'Reilly, S.S.; Mariotti, G.; Winter, A.R.; Newman, S.A.; Matys, E.D.; McDermott, F.; Pruss, S.B.; Bosak, T.; Summons, R.E.; Klepac-Ceraj, V. Molecular biosignatures reveal common benthic microbial sources of organic matter in ooids and grapestones from Pigeon Cay, The Bahamas. *Geobiology* **2016**, *15*, 112–130. [CrossRef]

29. Tfaily, M.M.; Chu, R.K.; Tolić, N.; Roscioli, K.M.; Anderton, C.R.; Paša-Tolić, L.; Robinson, E.W.; Hess, N.J. Advanced solvent based methods for molecular characterization of soil organic matter by high-resolution mass spectrometry. *Anal. Chem.* **2015**, *87*, 5206–5215. [CrossRef]

30. Tfaily, M.M.; Chu, R.K.; Toyoda, J.; Tolić, N.; Robinson, E.W.; Paša-Tolić, L.; Hess, N.J. Sequential extraction protocol for organic matter from soils and sediments using high resolution mass spectrometry. *Anal. Chim. Acta.* **2017**, *972*, 54–61. [CrossRef]

31. Dittmar, T.; Koch, B.; Hertkorn, N.; Kattner, G. A simple and efficient method for the solid-phase extraction of dissolved organic matter (SPE-DOM) from seawater. *Limnol. Oceanogr. Methods.* **2008**, *6*, 230–235. [CrossRef]

32. Tolić, N.; Liu, Y.; Liyu, A.; Shen, Y.; Tfaily, M.M.; Kujawinski, E.B.; Longnecker, K.; Kuo, L.; Robinson, E.W.; Paša-Tolić, L.; et al. Formularity: Software for automated formula assignment of natural and other organic matter from ultrahigh-resolution mass spectra. *Anal. Chem.* **2017**, *89*, 12659–12665. [CrossRef] [PubMed]

33. Minor, E.C.; Steinbring, C.J.; Longnecker, K.; Kujawinski, E.B. Characterization of dissolved organic matter in Lake Superior and its watershed using ultrahigh resolution mass spectrometry. *Org. Geochem.* **2012**, *43*, 1–11. [CrossRef]

34. Kujawinski, E.B.; Behn, M.D. Automated analysis of electrospray ionization Fourier transform ion cyclotron resonance mass spectra of natural organic matter. *Anal. Chem.* **2006**, *78*, 4363–4373. [CrossRef] [PubMed]

35. Van Krevelen, D.W. Graphical-statistical method for the study of structure and reaction processes of coal. *Fuel* **1950**, *29*, 269–284.

36. AminiTabrizi, R.; Wilson, R.M.; Fudyma, J.D.; Hodgkons, S.B.; Heyman, H.M.; Rich, V.I.; Saleaska, S.R.; Chanton, J.P.; Tfaily, M.M. Controls on soil organic matter degradation and subsequent greenhouse gas emissions across a permafrost thaw gradient in Northern Sweden. *Front. Earth Sci.* **2020**, *8*, 557961. [CrossRef]

37. Collins, M.J.; Nielsen–Marsh, C.M.; Hiller, J.; Smith, C.I.; Roberts, J.P.; Prigodich, R.V.; Wess, T.J.; Csapò, J.; Millard, A.R.; Turner–Walker, G. The survival of organic matter in bone: A review. *Archaeometry* **2002**, *44*, 383–394. [CrossRef]

38. Keenan, S.W. From bone to fossil: A review of the diagenesis of bioapatite. *Am. Mineral.* **2016**, *101*, 1943–1951. [CrossRef]

39. Behrensmeyer, A.K. Vertebrate preservation in fluvial channels. *Palaeogeogr. Palaeoclimatol. Palaeoecol.* **1988**, *63*, 183–199. [CrossRef]
40. Cleland, T.P.; Schroeter, E.R.; Feranec, R.S.; Vashishth, D. Peptide sequences from the first *Castoroides ohioensis* skull and the utility of old museum collections for palaeoproteomics. *Proc. R. Soc. B.* **2016**, *283*, 20160593. [CrossRef]
41. Wiemann, J.; Briggs, D.E. Validation of biosignatures confirms the informative nature of fossil organic Raman spectra. *bioRxiv* **2021**. [CrossRef]
42. Schweitzer, M.H.; Schroeter, E.R.; Cleland, T.P.; Zheng, W. Paleoproteomics of Mesozoic dinosaurs and other Mesozoic fossils. *Proteomics* **2019**, *19*, 1800251. [CrossRef] [PubMed]
43. Saitta, E.T.; Liang, R.; Lau, M.C.Y.; Brown, C.M.; Longrich, N.R.; Kaye, T.G.; Novak, B.J.; Salzberg, S.L.; Norell, M.A.; Abbott, G.D.; et al. Cretaceous dinosaur bone contains recent organic material and provides an environment conducive to microbial communities. *eLife* **2018**, *8*, e46205. [CrossRef] [PubMed]
44. Gold, D.A.; O'Reilly, S.S.; Luo, G.; Briggs, D.E.G.; Summons, R.E. Prospects for sterane preservation in sponge fossils from museum collections and the utility of sponge biomarkers for molecular clocks. *Bull. Peabody Mus. Nat. Hist.* **2016**, *57*, 181–189. [CrossRef]
45. Liebenau, K.; Kiel, S.; Vardeh, D.; Treude, T.; Thiel, V. A quantitative study of the degradation of whale bone lipids: Implications for the preservation of fatty acids in marine sediments. *Org. Geochem.* **2015**, *89*, 23–30. [CrossRef]
46. Sun, X.; Li, X.; Shen, X.; Wang, J.; Yuan, Q. Recent advances in microbial production of phenolic compounds. *Chin. J. Chem. Eng.* **2021**, *30*, 54–61. [CrossRef]
47. Tuross, N.; Behrensmeyer, A.K.; Eanes, E.D. Strontium increases and crystallinity changes in taphonomic and archaeological bone. *J. Archaeol. Sci.* **1989**, *16*, 661–672. [CrossRef]

Review

Was There a Cambrian Explosion on Land? The Case of Arthropod Terrestrialization

Erik Tihelka [1], Richard J. Howard [2], Chenyang Cai [1,3] and Jesus Lozano-Fernandez [1,4,*]

[1] School of Earth and Biological Sciences, University of Bristol, Bristol BS8 1TQ, UK
[2] Department of Earth Sciences, The Natural History Museum, London SW7 5BD, UK
[3] State Key Laboratory of Palaeobiology and Stratigraphy, Nanjing Institute of Geology and Palaeontology, and Center for Excellence in Life and Paleoenvironment, Chinese Academy of Sciences, Nanjing 210008, China
[4] Department of Genetics, Microbiology and Statistics & Biodiversity Research Institute (IRBio), University of Barcelona, 08028 Barcelona, Spain
* Correspondence: jesus.lozano@ub.edu

Simple Summary: The transition of life from the aquatic realm onto land represented one of the fundamental episodes in the evolution of the Earth that laid down the foundations for modern ecosystems as we know them today. This key transition in the history of life is poorly known, owing to the scarcity of ancient terrestrial fossil deposits; complex terrestrial ecosystems with plants and animals appear in the fossil record during the Silurian and Devonian. However, recent molecular clock studies and new lines of palaeontological evidence point to a possibly much earlier origin of life on land, dating back as far as the Cambrian. Here, we review this controversy, using the arthropods as a case study of the possible cryptic Cambrian explosion on land. In particular, we highlight approaches for reconciling the disagreement between molecular clock estimates and the fossil record for the arthropod colonization of land.

Abstract: Arthropods, the most diverse form of macroscopic life in the history of the Earth, originated in the sea. Since the early Cambrian, at least ~518 million years ago, these animals have dominated the oceans of the world. By the Silurian–Devonian, the fossil record attests to arthropods becoming the first animals to colonize land, However, a growing body of molecular dating and palaeontological evidence suggests that the three major terrestrial arthropod groups (myriapods, hexapods, and arachnids), as well as vascular plants, may have invaded land as early as the Cambrian–Ordovician. These dates precede the oldest fossil evidence of those groups and suggest an unrecorded continental "Cambrian explosion" a hundred million years prior to the formation of early complex terrestrial ecosystems in the Silurian–Devonian. We review the palaeontological, phylogenomic, and molecular clock evidence pertaining to the proposed Cambrian terrestrialization of the arthropods. We argue that despite the challenges posed by incomplete preservation and the scarcity of early Palaeozoic terrestrial deposits, the discrepancy between molecular clock estimates and the fossil record is narrower than is often claimed. We discuss strategies for closing the gap between molecular clock estimates and fossil data in the evolution of early ecosystems on land

Keywords: terrestrialization; artrhopods; Cambrian explosion; molecular clocks; palaeontology; phylogenomics

Citation: Tihelka, E.; Howard, R.J.; Cai, C.; Lozano-Fernandez, J. Was There a Cambrian Explosion on Land? The Case of Arthropod Terrestrialization. *Biology* **2022**, *11*, 1516. https://doi.org/10.3390/biology11101516

Academic Editors: Mary H. Schweitzer and Ferhat Kaya

Received: 5 September 2022
Accepted: 14 October 2022
Published: 17 October 2022

Publisher's Note: MDPI stays neutral with regard to jurisdictional claims in published maps and institutional affiliations.

Copyright: © 2022 by the authors. Licensee MDPI, Basel, Switzerland. This article is an open access article distributed under the terms and conditions of the Creative Commons Attribution (CC BY) license (https://creativecommons.org/licenses/by/4.0/).

1. Introduction

Molecular clocks estimate that life on Earth originated over 4 billion of years ago (Ga), perhaps shortly after the formation of our planet [1], with direct evidence provided by the remains of putative unicellular organisms at around 3.5 Ga (e.g., [2–4]). However, the emergence of complex multicellular organisms, such as animals, plants and fungi, only occurred during the last 1000 million years [5] (but see [6] for older estimates). The origin of animals gave rise to an enormous diversity of multicellular body plans, all with a complex embryonic development. This diversity of body plans is already seen in the exceptional early fossil record of animals, during the "Cambrian explosion", beginning around 540 million years ago (Ma) and concluding perhaps as quickly as 521 Ma [7]. During this interval, most major animal phyla appeared almost simultaneously, from a geological perspective, and persisted throughout the Phanerozoic [8,9]. The often unfamiliar body plans of Cambrian marine animals have been preserved on a number of sites with exceptional preservation, known as the Burgess Shale-type (BST) *Konservat-Lagerstätten*, which provide a unique snapshot of the soft-bodied Cambrian biota in the sea [10]. A diverse and abundant marine arthropod fauna is evidenced by the fossil record from at least ~518 Ma, corresponding to the minimum age of the Chengjiang Biota of Yunnan Province, southwestern China; the oldest reliably dated BST [11].

Animals, plants, and life in general, have marine origins [12]. Only a handful of animal phyla contain lineages that can complete each phase of their life cycle outside of moisture-rich environments and can therefore be considered fully terrestrial. This is because land represents a new and hostile environment for marine organisms, with obstacles to overcome ranging from respiration, reproduction, feeding style, and mechanical support [13]. Among these, the most well-known examples are in vertebrates (reptiles, birds and mammals) and of course arthropods, invertebrates with jointed legs and exoskeletons such as spiders and insects. Additionally, soft-bodied groups with generally poor fossil records [14,15], such as molluscs (including the land snails and slugs [16]), onychophorans (velvet worms [17]), annelids (including earthworms [18]), nematoids (roundworms and horsehair worms, including many parasitic groups that have followed their hosts on land [18–21]), tardigrades (water bears [22]), and platyhelminthes (flatworms [23]) contain land-living lineages, but these are mostly dependent on moisture-rich terrestrial environments for survival. Life on land requires a series of adaptations that may be paralleled across different groups—we can refer to this as terrestrialization: the process by which aquatic organisms adapt to terrestrial life. Terrestrialization is a fascinating field of study in evolutionary biology. Much literature has addressed terrestrialization at the physiological level in arthropods (see review in [24,25]). However, most studies have been conducted on isolated lineages and have not taken full advantage of the comparative approach between diverse terrestrial groups [26]. Multiple and independent terrestrialization events allow comparisons of alternative solutions taken up by different groups to the same adaptive challenge, and represent a powerful tool to understand adaptation in an evolutionary framework. This information is, at the same time, necessary to be able to carry out comparative analyses and estimate the timing and rate of emergence of terrestrial adaptations. Although animal phylogenetic diversity (understood as the diversity of body plans) may be higher in the marine realm, terrestrial biodiversity is clearly higher in terms of the number of species—particularly due to the unparalleled species richness of insects [27]. Understanding animal terrestrialization is thus crucial to understanding the origins of biodiversity on Earth and the mechanisms underpinning evolutionary adaptation [28].

There is fossil evidence of simple terrestrial ecosystems formed by single-cell organisms dating back 1000 Ma [29]). The earliest complex terrestrial ecosystems record a fascinating transition in the history of life. Before the Palaeozoic, the only terrestrial life was unicellular, which, until recently, could only be deduced from indirect evidence [30]. It was during the Palaeozoic that plants and animals began to colonize the Earth's landmasess [31], with plants appearing in the fossil record in the form of microfossils called cryptospores in the Middle Ordovician, around 470 Ma, with potential vascular land plants appearing shortly at ~458 Ma [32]. In the case of arthropods, with certain terrestrial myriapods and arachnids from the Silurian–Devonian [33,34]. Hence, the conventional view of the evolution of terrestrial ecosystems posits that during the Silurian–Devonian, animals and plants diversified on land, which was presumably void of complex organisms, bathed in lethal UV rays, and with low atmospheric oxygen (e.g., [35,36]). This model has however recently been challenged by molecular clock dating studies [25,37,38] and new discoveries of Palaeozoic stem groups of terrestrial lineage [39,40], which imply a substantially earlier, Cambrian to Ordovician, origin of complex terrestrial ecosystems, comparable to a "Cambrian explosion on land". Secondly, updated reconstructions of Devonian-Carboniferous atmospheric oxygen suggest that this period did not suffer from substantially low atmospheric oxygen as stipulated earlier [41,42]. Meanwhile, terrestrial sedimentary rock units older than the Early Devonian are rare worldwide (e.g., [43–47]). For example, Western Europe, one of the best explored regions of the world from a palaeontological point of view, has virtually no terrestrial sedimentary rock outcrops older than the latest Silurian [15,48,49]. The scarcity of preserved rock units imposes an important constraint on the preservation potential of the earliest terrestrial ecosystems. It has been argued that the scarcity of terrestrial organisms from this period may be due to limited surviving fossiliferous sediments rather than because they did not exist in the first place [42,50]. This emerging paradigm may imply up to 100 million years of discordance between when diverse terrestrial ecosystems become represented in the fossil record and their putative origin.

In this brief review, we introduce the timescale of arthropod terrestrialization. Arthropods are represented among the oldest fossil records of animals (Figure 1), and represent the bulk of animal diversity on land today, with more than a million described species [51]. The oldest arthropod fossils are undoubtedly marine. They include the trilobites, with representatives dating back to the early Cambrian, ~521 Ma [52], and trace fossils indicating the presence of arthropod locomotion from at least ~528 Ma [53]. In arthropods, there have been a minimum of three to four major terrestrial invasions during the Palaeozoic: that of hexapods (which includes insects and kin), isopods (a group of crustaceans), myriapods, and that of arachnids—assuming that the latter forms a monophyletic group. The multiple and independent terrestrializations in arthropods provide a unique macroevolutionary case study into adaptative solutions embraced by different groups in response to the same challenge. More broadly, the topic of animal and plant terrestrialization provides an exciting opportunity to study a crucial ecosystem-wide transition that shaped the world we find so familiar today, during an elusive epoch of Earth's history that left little direct physical evidence. However, to carry out these studies it is necessary to: (i) clarify how many land settlements have occurred independently in different arthropod lineages, (ii) estimate when these terrestrialization processes occurred and how long they lasted, and (iii) establish robustly which is the aquatic sister group of each terrestrial lineage. We provide an overview of recent progress in these questions and evaluate the support for the argument of a Cambrian explosion on land.

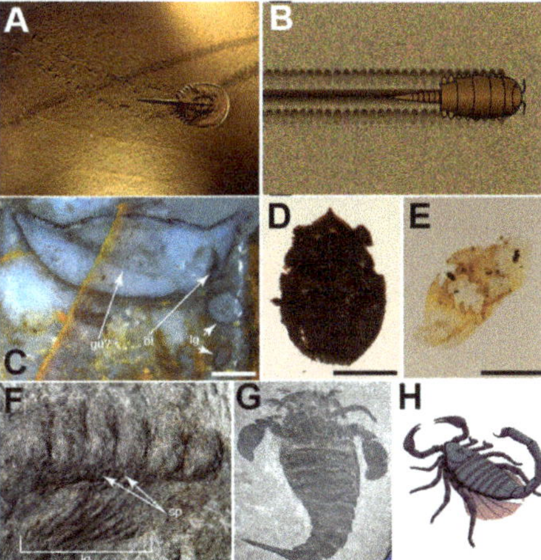

Figure 1. Fossil evidence of arthropod terrestrialization. (**A**) Traces and the body fossil of the horseshoe crab that made it, *Mesolimulus walchi*, morphologically resembling modern forms; (**B**) reconstruction

of a terrestrial Cambrian ichnofossil, possibly made by the euthycarcinoid *Mosineia*, a group in kinship with myriapods; (**C**) Section through the abdomen of a trigonotarbid arachnid preserved in the Early Devonian Rhynie chert, revealing book lungs (bl), a possible trace of the gut (gu?), and sections through the legs (lg); (**D**) Carbonised body fossil of a trigonotarbid arachnid *Palaeotarbus jerami* from the Silurian Ludford Lane; (**E**) Putative myriapod mandibles from the Silurian Ludford Lane; (**F**) Millipede *Pneumodesmus newmani* from the Lower Devonian of Cowie Harbour (Scotland), presenting spiracles (sp) and legs (lg); (**G**) Eurypterid *Eurypterus remipes* from the Silurian; (**H**) Palaeoreconstruction of the Devonian scorpion *Waeringoscorpio westerwaldensis*, with filamentous gills that suggest a potential aquatic adaptation. Image sources: Wikimedia Commons Illustration authors: (**B**) Haug; (**C–E**), Erik Tihelka; (**H**) Junnn11 (@ni075). Institutional repositories: (**C–F**) National Museum of Scotland, Edinburgh: R.08.14 & G.2001.109.1; (**D,E**) Ulster Museum, Belfast: K25850 & LL1.6-23; (**G**) Generaldirektion Kulturelles Erbe, Direktion Archiologie/Erdgeschichte, Mainz, Germany, based on PWL2007/5000-LS. Scale bars: (**C,D**) 500 μm, (**E**) 250 μm, (**G**) ~10 mm.

2. Origin and Terrestrialization of Arthropods

2.1. Arthropod Origins

It is difficult to precisely estimate terrestrial arthropod biodiversity in deep time due to the caveats of the fossil record; terrestrial arthropod fossils are usually limited to sites of exceptional preservation known as *Konservat-Lagerstätten*, and therefore their stratigraphic and environmental distribution is discontinuous. However, we can suppose that, as in the modern biosphere, arthropods were probably the largest component of the diversity and abundance of Palaeozoic land animals, given the lack of initial competition and the phylogenetic diversity of those that are present in the terrestrial Palaeozoic fossil record. Indeed, arthropods are likely to have been the dominant animal group in terms of biodiversity in perpetuity for the past 520 million years [54]. Arthropods are characterised by presenting internal and external body segmentation with regional specialisations (tagmosis: in the case of insects, for example, they possess a thorax where legs and wings are inserted while there are no extremities in the abdomen); an external skeleton composed of articulated sclerotized parts; body segments that originally had associated articulated limbs; growth through successive moults (ecdysis); and an open circulatory system with a dorsal heart with lateral valves [55]. This set of unique characteristics suggests that they are a monophyletic group (descendants of a common ancestor who possessed the diagnostic characteristics of the lineage). Arthropods are represented by chelicerates (with arachnids such as spiders and scorpions, and marine groups such as pycnogonids and horseshoe crabs); myriapods (such as millipedes and centipedes); hexapods (containing insects) and predominantly aquatic 'crustaceans' (for example crabs and prawns), which are collectively known as pancrustaceans; and include important extinct groups, such as the trilobites (Figure 2). Their abundance makes arthropods ecologically essential; for example, myriapods are important processors of detritus in forests, and termites consume such large amounts of cellulose that they are significant for the carbon cycle and atmospheric gas composition [56]. Without arthropods, life and ecosystems on Earth would be radically different. Their surprising diversity (which exceeds 75% of all living species described [57]) can help to elucidate the patterns and processes of macroevolution.

The earliest animals we know as land-dwelling were arthropods [58]. Evaluating the earliest fossil evidence of arthropod life on land can rely on two approaches—phylogenetic bracketing and direct anatomical evidence. Under the former approach, the discovery of a fossil representative belonging to an entirely terrestrial clade can be deemed to provide evidence of life on land, even when the state of preservation of the individual fossils is not particularly impressive. The second, more direct approach, relies on identifying unambiguous terrestrial adaptations in fossil specimens to conclude that these indeed lived on land.

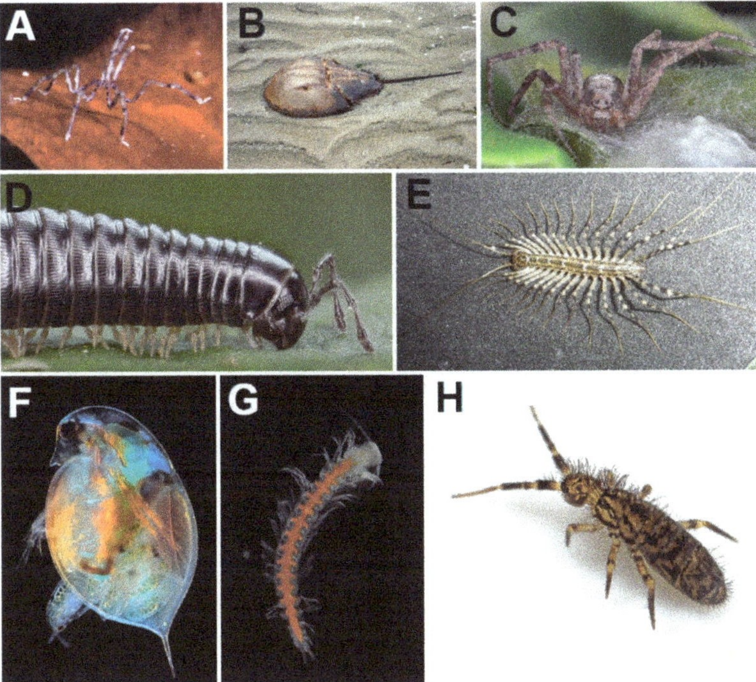

Figure 2. Present diversity of arthropods (**A**) pycnogonid *Endeis flaccida* (chelicerate); (**B**) xiphosuran *Limulus polyphemus* (chelicerate); (**C**) spider *Philodromus aureolus* (arachnid: chelicerate); (**D**) millipede *Cylindroiulus caeruleocinctus* (myriapod); (**E**) centipede *Scutigera coleoptrata* (myriapod); (**F**) branchiopod *Daphnia* sp. (pancrustacean); (**G**) remipede *Morlockia williamsi* (pancrustacean); (**H**) hexapod *Orchesella villosa* (pancrustacean). Image sources: Wikimedia Commons; (**G**) Jørgen Olesen.

The earliest fossil assemblage preserving arthropods belonging to terrestrial clades is the Přídolí-aged Ludlow bone bed Member exposed at Ludford Lane, near Ludlow in Shropshire, western England [34,59–61]. This site contains a range of myriapods (Figure 1E), including scutigeromorph centipedes in the genus *Crussolume* [61], the arthropleurid *Eoarthropleura* [61], and a singular specimen of the trigonotarbid arachnid *Eotarbus jerami* Dunlop 1996 (= *Palaeotarbus jerami*, junior synonymy resolved by Dunlop [62]; Figure 1D). Any of these can be confidently considered to be the oldest terrestrial arthropod body fossils, albeit the fidelity of their preservation does not permit the observation of anatomical adaptations for life on land—most are represented by small shreds of cuticle or, in the case of *Eotarbus*, a dark carbonised specimen. U-Pb zircon dating of the Ludlow bone bed at Ludford Lane in Shropshire constrained the age of the deposit to ~420 Ma [63].

The earliest animal possessing unambiguous terrestrial adaptations is the millipede *Pneumodesmus newmani* from the Lower Devonian Cowie Harbour near Stonehaven in Aberdeenshire, Scotland [33], which is preserved with more fidelity. The terrestrial character of this organism is indisputable since it possesses spiracles, openings on the cuticle that allow air to enter the tracheal system (Figure 1F). Two other diplopod species were reported from the locality, all described by Wilson and Anderson [33]. The *Dictyocaris* Member of the Cowie Formation at Cowie Harbour was initially considered to be Silurian based on palynological evidence (~426.9 Ma [64–66]), but isotopic dating confidently constrained its age to the lowermost Devonian (Lochkovian; ~414 Ma [67]), making it some 6 Mya younger than the Ludford Lane assemblage. Recently, the scorpion *Palaeoscorpius devonicus* [68,69] from the Lower Devonian Hunsrück Slate Lagerstätte in Germany (~405 Ma) was interpreted as possessing adaptations for life on land, namely probable book lungs, indicating that it was likely terrestrial [70].

2.2. Arthropod Phylogeny

The evolutionary relationships among the major arthropod groups have always been a subject of debate, such that by the start of the 21st century virtually all conceivable topologies for the group had been proposed [71]. Identifying the closest relatives of each terrestrial lineage is crucial, not only for comparative studies dealing with adaptation strategies for life on land, but also to understand the potential terrestrialization routes and constrain their timing. To infer these phylogenies, the anatomical structures of living and fossil species provide a treasure trove of comparative data that has been expanded even further during the last few decades by vast quantities of molecular data [72]. In their adaptation to land, arthropods have undergone convergent evolution (independent origins of similar biological systems in different lineages), which has often complicated efforts to assess kinship relationships between them [54]. For example, trachea (respiratory structures adapted to terrestrial environments) are found in several lineages that have conquered the land independently during the Palaeozoic: in a few arachnids, myriapods, isopods, and hexapods. The introduction of genome-scale phylogenetic analyses-phylogenomics—has greatly narrowed down the number of hypotheses on hexapod phylogeny, but crucially, some nodes of the arthropod tree remain difficult to resolve. Such challenging nodes often represent ancient and rapid radiations that are complex to address with any dataset, molecular or morphological, and represent the major lasting controversies in reconstructing the process of the arthropod invasion of land [18,66,73–75].

2.3. Myriapods

According to a classical phylogenetic hypothesis, the exclusively terrestrial myriapods, have been regarded as the sister group of the hexapods. This hypothetical clade, called Tracheata (or Atelocerata), is supported mainly by the presence of tracheae in both groups to carry out gas exchange (reviewed in [76]). Current studies based on molecular data, and also a re-examination of more subtle morphological characters of the nervous system and ommatidia [73,77], discard this hypothesis, and attribute this coincidental morphological convergence to independent convergence [78]. A second hypothesis recovered by early analyses of molecular data implicated myriapods as a sister group to the chelicerates (Myriochelata or Paradoxopoda). However, these results are now considered as caused by a phylogenetic reconstruction bias due to the rapid evolutionary rates of pancrustaceans attracting to the outgroup and pushing myriapods and chelicerates into an artefactual clade when using simpler models of molecular evolution [79]. Today, there is a certain consensus on the main relationships between arthropods, supported by phylogenomic data [78]. The myriapods, the first of the three large terrestrial lineages, are generally accepted as a sister group to the pancrustaceans (hexapods and all crustacean lineages), and the chelicerates as the closest relative of this clade (Figure 3). Thus, the basic division between arthropods consists of those that have mandibles (myriapods and pancrustaceans) and chelicerae. The internal phylogeny of myriapods, though, is currently more contentious. Several recent phylotranscriptomic analyses disagree on the exact relationship between their main lineages [78,80–83] but they do not have an impact on the single terrestrialization event inferred for the group.

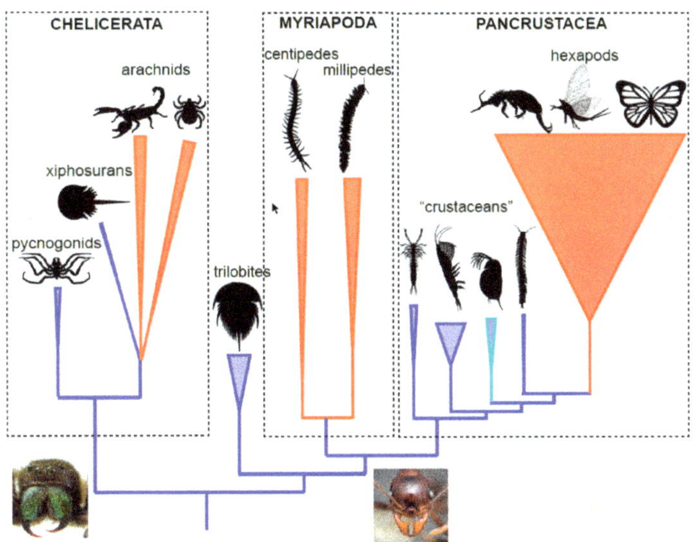

Figure 3. Cladogram with the current consensus on the phylogenetic relationships between the main groups of arthropods. The terrestrial groups are represented in orange colours while the marine clades in blue and turquoise for Branchiopoda (fresh water). The thickness of the terminal branches corresponds to a proportional approximation of the number of described species. At the base of the cladogram, image with detail of chelicerae and mandibles, the defining structures of the two groups. Some of the silhouettes come from Phylopic (phylopic.org/; accessed on 5 November 2022).

2.4. Pancrustacea (Hexapoda)

There is strong molecular and morphological evidence that favours the position of hexapods as nested within the 'crustaceans' (as the clade Pancrustacea, or Tetraconata), and myriapods as the sister group of pancrusteans forming the Mandibulata group, characterized by the presence of this distinctive oral structure [84–86]. In contrast, the exact relationships of hexapods within the Pancrustacea are still unclear, and it is not obvious which is their aquatic sister group. Phylogenomic datasets have variously lent support to the mostly freshwater-dwelling branchiopods [25], or the species-poor and enigmatic remipedes [68,70]. Establishing which 'crustacean' group is the most closely related to hexapods has a great impact on whether the latter group presumably colonised terrestrial environments directly from the sea, or whether they first colonized freshwater environments and later moved to land. Most recent phylogenomic studies, though, using hundreds of molecular markers, have shifted the balance in favour of Remipedia [84,86]. Remipedia are a class of blind and predatory crustaceans that live in coastal aquifers that contain saline groundwater. They were discovered less than 40 years ago [87], and have a very restricted distribution, with fewer than 30 known species described from the anchialine caves in the Caribbean Sea, two species from the Canary Islands and one from Western Australia. Very little is known about the biology of these organisms, which makes it difficult to understand their significance for hexapod terrestrialization.

2.5. Pancrustacea (Isopods)

The suborder Oniscidea (woodlice) represents the most diverse isopod crustacean group, with over 3700 described species [88]. It is the only pancrustacean group besides the hexapods composed almost entirely of terrestrial species; its members are found in almost all terrestrial habitats, ranging from nearshore settings to forests [89]. In particular, the intertidal genus *Ligia* inhabiting shorelines is often regarded as a transitory group [90]. Given their varying degrees of adaptations for life in semi-aquatic and terrestrial environments, woodlice provide a rewarding model group for understanding the transition from marine to terrestrial habitats, which hinges on an understanding of their phylogeny [90,91]. Morphological studies implicate Ligiidae as the basalmost woodlouse clade, implying a single invasion of land directly from the marine realm [92], although some molecular studies have challenged the monophyly of the group (e.g., [93,94]). Overall, isopods remain probably the least-studied terrestrialization event among arthropods. Their fossil record is fragmentary and scarce, with their oldest occurrence from the Cretaceous (summarised in [90]). If terrestrial isopods originated in the late Palaeozoic, potentially the Carboniferous [90], they would represent the most recent arthropod terrestrialization event.

In some sense, other pancrustacean clades such as amphipods and the decapods, also invaded semi-terrestrial habitats (e.g., supralittoral zone of beaches, most soil and leaf litter, edges of freshwater habitats) and these have been considered as terrestrialization events by some (e.g., [13]). Here, we refrain from treating these groups as fully terrestrial, since their adaptation to life is not as developed as in the case of the woodlice. Nonetheless, these taxa represent important study groups for future research in arthropod adaptation to semi-terrestrial habitats.

2.6. Arachnids

Among terrestrial arthropods, only insects outnumber arachnids in terms of the number of described species (1 million versus 112,000, respectively; [51]). The clade Arachnida includes all terrestrial chelicerates, composed mainly of predatory groups such as spiders and scorpions, and parasites such as ticks. However, chelicerates also include marine taxa such as the pycnogonids (sea spiders) and xiphosurans (horseshoe crabs). Neither the currently available morphological nor molecular data have unequivocally resolved the internal kinship relationships between chelicerates [66]. Arachnids have traditionally been regarded as a monophyletic group, implying that a single and irreversible ancestral colonization of land paved the way to this group's evolutionary success. Some recent studies including genome-scale and morphological phylogenies, however, do not support this relationship, instead placing the marine Xiphosura within terrestrial arachnids, and not as a sister group to it [74,95]. The focus of this debate is whether there has been a single common ancestor for all terrestrial arachnids, a single terrestrialization event within a common ancestor of terrestrial arachnids + xiphosurans (with the later transitioning again into aquatic environments soon after), or whether arachnid terrestrialization occurred on two or more separate occasions. Resolving this puzzle is enormously significant, as it rewrites our perception of the evolution of terrestrial adaptations (e.g., the respiratory system, sensory and reproductive systems, and the locomotor appendages). The physiological demands of life on land require a significant modification of these anatomical features, which is probably best illustrated by the respiratory organs, a great variety of which are present in extant chelicerates (book lungs and tracheae in terrestrial groups, and book gills in marine forms) [96,97]. If xiphosurans were a group of marine arachnids, this may suggest that the remaining lineages colonized land independently. A second option would be that xiphosurans recolonized the marine environment from a terrestrial ancestor. Of these two options, the first would be considered more plausible, since the fossil record of Xiphosura extends back more than 400 Ma with exclusively aquatic forms, without traces of a potential terrestrial or amphibious ancestors [98] (Figure 1D). Furthermore, no widespread losses of terrestrial respiratory organs in arthropods are known, once acquired, in line with the predictions of Dollo's law [99].

In addition, even though horseshoe crabs can make momentary incursions into the coasts to spawn eggs, they do not have distinctly terrestrial morphological adaptations and their body structures present great similarity, and probably homology, with that of other aquatic fossil chelicerates [100,101]. Other recent studies using genome-scale datasets, as well as morphological and fossil evidence suggest that marine chelicerates (pycnogonids and Xiphosura) are successive sister groups of a monophyletic lineage of terrestrial arachnids. These results are compatible with a single colonization of land within chelicerates and the absence of wholly marine arachnid orders [66,102].

3. Pre-Devonian Fossil Record of Terrestrial Arthropods

3.1. Trace Fossil Evidence

The oldest traces of activity in the terrestrial environment made by arthropods (ichnofossils) date from the middle Cambrian to the Early Ordovician. The oldest of these include trackways on land from the late Cambrian (~500 Ma) of Ontario, Canada produced by arthropods with at least 11 pairs of similar walking legs and a long tail-spine, presumably made by the extinct euthycarcinoids [103] (Figure 1E). Another site famous for its Cambrian traces of life on land is the middle to late Cambrian Blackberry Hill in central Wisconsin, which preserves diversity of arthropod trackways in a tidal flat and nearshore environment, along with the remains of the oldest euthycarcinoid, *Mosineia* [104]. Massive trackways from the intertidal zone left by euthycarcinoids with walking legs during the Cambrian and Ordovician indicate longer stays on land where these amphibious animals may have come in pursuit of shallow lagoons and freshwater pools [105], albeit their excursions on land may have been short-lived [106]. Other early arthropods to make temporary excursions to near-shore habitats were the trilobites, whose trace fossils in tidal-flat deposits are known since the Cambrian, albeit their traces were likely made subaqueally [107,108]. Trilobites possessed gill lamellae for respiration [109,110], which are unlikely to have provided them with the ability to survive on land for prolonged periods of time. Myriapods have been implicated in producing Ordovician backfilled burrows from Pennsylvania (445 Ma [111]), although the terrestrial nature of this deposit has been later disputed [36]. A slightly younger record of trackways and trails (*Diplichnites* and *Diplopodichnus*) from Cumbria, England (>450 Ma) records myriapods moving alongside the edges of ponds, but these were likely made under water [112,113]. Overall, locomotive traces documented throughout the Cambrian and Ordovician reinforce the view that aerial activities of arthropods (if not terrestrial arthropods) were common on the coasts and along the edges of freshwater bodies during this time.

3.2. Body Fossil Evidence

The availability of land-dwelling arthropod body fossils is fundamentally constrained by the limited number of terrestrial formations before the Devonian and the limited interest these geological units have attracted in the past [42]. Fossils generally require a steady rate of sedimentation to preserve, which is not an easily achievable condition for minute soft-bodied arthropods inhabiting the soil or decaying vegetation matter. As such, palaeontologists have to rely on a restricted set of fossil localities that provide unusual preservation windows for their time.

The earliest relatives of myriapods in the fossil record are the Cambrian to Triassic euthycarcinoids, mentioned earlier for the terrestrial trace fossils. The affinities of this group have been traditionally difficult to pinpoint, but recent findings of exceptionally preserved Devonian specimens establish the group as the stem-group to myriapods [39]. These aquatic arthropods were amphibious, ranging from marine and brackish to freshwater deposits [114]. Their ventures on land have been variously interpreted as short migrations between ephemeral freshwater pools, grazing on microbial mats and detritus, or migrations to fertilise eggs on land like in modern horseshoe crabs [103,115]. Recent synchrotron studies revealed probable respiratory organs in a Devonian euthycarcinoid, consistent with an amphibious lifestyle [116]. Other early myriapod remains are known from the Silurian Kerrera (425 Ma) and Ludlow (420 Ma) deposits in the United Kingdom, albeit it is difficult to determine if they were truly terrestrial [59,117]. The earliest undoubtedly terrestrial fossil myriapod is the millipede *Pneumodesmus newmani* from Cowie in Scotland, originally regarded as Late Silurian [33], but more recently as Lower Devonian (414 Ma [67]). Its terrestrial ecology is indicated by the presence of spiracles.

The earliest hexapod fossils are the Early Devonian (~405 Ma) springtail *Rhyniella praecursor* [118,119] and the enigmatic *Leverhulmia mariae* [120], from the coeval Rhynie and Windyfield chert deposits in Scotland, which became preserved with extraordinary fidelity when silica-rich water from volcanic springs inundated hot springs and the surrounding land. While various systematic positions of the peculiar *Leverhulmia* have been proposed, *Rhyniella* is a crown-group springtail, not that different from species that inhabit soil and leaf litter today [121], suggesting that this clade of hexapods radiated well before the Early Devonian. Nonetheless, insect fossils before the Carboniferous are few; the Rhynie chert is followed by a window of 80 Ma (referred to as the 'hexapod gap') during which no insects are known [122]. The existence of pre-Devonian hexapods is a reasonable assumption, proposed already by early cladistic studies predating the molecular clock methodology [123]. Although a decade-old bounty of 1000 dollars has been put on an undisputable insect fossil from the pre-Devonian [124], this sum remains to be claimed. Instead, the hunt for early hexapods yielded a number of dubious records, like fossils only seen once and never again [125], suspected modern contaminants [106], and miss-identifications, such as purported Devonian insect wings that turned out to be malacostracan tail fans [126,127]. A recent review is provided by [128]. Others represent genuinely difficult fossils to interpret, such as the purported Devonian hexapod *Strudiella devonica* [129], which may however represent a decayed non-insect arthropod [130], or the Devonian *Wingertshellicus/Devonohexapodus* at once interpreted as an aquatic stem-hexapod [131], but not unequivocally accepted [132]. It is interesting to note that even in deposits such as the Rhynie chert where arthropod cuticles are not rare in some facies, the vast majority belong to arachnids, not hexapods as may be expected from modern ecosystems, where insects predominate. Winged insects only came to dominate terrestrial ecosystems by the Carboniferous, leading many to postulate that hexapods may have been species poor until the origin of with wings [42] that appear unequivocally in the fossil record in the latest Mississippian (~322 Ma [133]).

Among arachnids, we find the oldest fossil evidence of arthropod life on land, represented by scorpion remains from the Silurian (~437 Ma [134]). However, their terrestriality is not unambiguous due to the absence of bona fide terrestrial characters, such as book lungs, and have been found in aquatic or semi-aquatic deposits [135,136] (Figure 1H). Putative book lungs have been reported from a fossil scorpion from the Devonian Hunsrück Slate in Germany (~405 Ma [70]). Current molecular, phylogenomic and morphological evidence suggests, however, that scorpions are arachnids related to spiders [102,137,138], in a clade of mostly lung-bearing arachnids known as Arachnopulmonata. Within this clade, the latest phylogenomic results suggest that pseudoscorpions are the closest relatives of scorpions [66,74,139]. This phylogenetic position is hardly reconcilable with a marine origin of scorpions, suggesting that some of these ancestral scorpions may have secondarily returned to the aquatic environment, although without obvious marine adaptations. The earliest member of Trigonotarbida, a group of extinct terrestrial arachnids known to possess book lungs [140,141], is known from the Silurian (~420 Ma [63]) Ludford Lane in England [59]. Trigonotarbids persisted until the Permian and are known in stunning anatomic detail, in part thanks to their preservation in Rhynie chert [142]).

4. Reconciling Rocks and Clocks

4.1. Methodologies to Build Chronologies

The abundant arthropod fossil record is informative on the diversity of the group, the historical evolution of morphological characters, and provides temporal guidelines for molecular dating. Solving the relative times of evolutionary divergences between species and clades in the geological past provides crucial information for dating the origin of terrestrial ecosystems. The reconstruction of these "timetrees", or chronograms, is increasingly methodologically sophisticated and has become the backbone for comparative studies of evolutionary biology and palaeontology. Molecular data inform us both on the understanding of the tree's branching pattern (the phylogeny) and, once calibrated with fossils, on the timing at which these branching events occurred (the timeline). The dates are inferred using the molecular clock technique [143], where the time elapsed since the divergence of different organisms or species is deduced from the differences between their DNA or amino acid sequences. To carry out these analyses, calibration points are routinely used where minimum ages are defined based on the oldest fossil evidence that can be unequivocally assigned to that node, that is, the origin of that group cannot be younger than its oldest fossil [144]. Node dating is the most widely used method [145], and it has developed a lot in recent years, with the implementation of Bayesian methods that allow assigning probabilities to age ranges and to other various parameters based on previous knowledge about the group in question [146]. While the chronologies constrain the real age of the lineages, the fossils inform us of when those organisms became numerically and ecologically abundant. Furthermore, including fossils in phylogenetic analyses helps arrive at more accurate trees and divergence time estimates [147–149]. Therefore, chronologies provide an essential conceptual framework for investigating the evolution of the first terrestrial ecosystems and the interactions over time between organisms and their environment.

4.2. Dating the Arthropod Terrestrialization

Most recent chronologies of arthropod radiation (or subgroups of them) using molecular clocks are generally compatible with paleontological evidence, proposing an origin of the group between the end of the Ediacaran period and the beginning of the Cambrian (with credibility intervals falling with 95% of probability between 551–536 Ma) [37]. These studies also suggest the origin of arachnids and hexapods are in some consensus with the fossil evidence, preceding the oldest fossils by a few tens of thousands of years (Figure 4). In the case of chelicerates, the origin of terrestrial arachnids and of their main diversifications have been inferred to fall between the Cambrian and Ordovician (494–475 Ma) [66]. Molecular evolution rates were likely high during its origin, coinciding with a rapid cladogenesis [37]. When xiphosurans are nested within arachnids, the origin of this clade is inferred in ages comprising mostly Ediacaran [95] to Cambrian period [66]. For hexapods, the estimated ages vary in different studies between 520–450 Ma (summarised in [150]). Likewise, a Cambrian–Ordovician origin has been proposed for myriapods [25,78,83,151–153].

Consequently, there are certain differences when ages inferred from molecular dating studies are compared with the oldest fossil record, where arachnids first appeared in the Silurian (427 Ma) and hexapods in the Devonian (411 Ma). In the case of arachnids, it has been suggested that these differences may be due to the fact that the closest relative of arachnids is an extinct group. Eurypterids (also called 'sea scorpions') have been proposed as a possible sister group (Figure 1G). These aquatic organisms emerged during the Ordovician (~467 Ma) and represented an important component of marine fauna until they disappeared from the fossil record during the end-Permian mass extinction (~252 Ma) [10]. It seems that they could make inroads into the terrestrial environment, as suggested by ichnofossils, and recent studies show that they had respiratory structures adapted to breathing air, possibly since the Cambrian–Ordovician [40]. The latter study suggests that their ancestor may have been semi-terrestrial, similar to eurypterids. Regarding the origin of myriapods, the divergence times inferred are substantially older (524–505 Ma) than their oldest fossil evidence, and they firmly place the earliest members of this group in the Cambrian [37,78,151–153], despite the fact that its oldest fossil is 414 Ma (Figure 4). The reinterpretation of Euthycarcinoidea as the closest relative of myriapods based on the similarity of mouth and eye structures bridges this gap between the fossil record and molecular clocks [39].

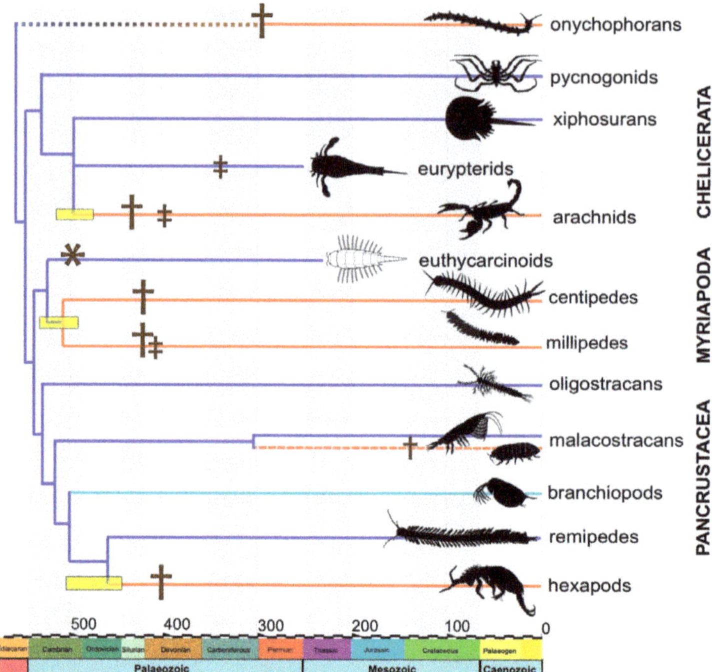

Figure 4. Schematic chronogram with divergence times between the most representative arthropod clades. The internal nodes of the tree fall into the mean estimated divergences taken from recent studies cited in the text. On the x-axis, time runs from most recent (right) to the past (left), and is expressed in millions of years. The yellow rectangles mark the credibility intervals for the different terrestrialization phenomena. The dagger symbol represents the oldest known fossil in that terrestrial group, the double dagger represents the oldest direct evidence of terrestrial breathing structures, and the asterisk the oldest trace evidence of terrestrial behaviour. The terrestrial groups are represented in orange colours while the marine clades in blue and turquoise for Branchiopoda (fresh water). Some of the silhouettes are from Phylopic (phylopic.org/; accessed on 5 November 2022).

4.3. Reconciling the Fossil and Molecular Evidence

The discrepancies between the results derived from molecular clocks and the oldest fossil evidence may be related to the nature of the rock record, especially to the rarity of terrestrial sediments from the Cambrian to the Silurian. It has been suggested that Euramerica, the region from which much of the data on the first terrestrial arthropods and plant megafossils are derived, is almost absent from terrestrial sediments before the upper Silurian and that these are not more widespread until the Early Devonian [50]. This temporal bias in the rock record possibly affects the fossil record of terrestrial organisms and may explain part of the mismatch between molecular and fossil dates. The discrepancy may also be explained by failures in the molecular clock methodology, particularly with the node dating strategy. A recent method has been developed to estimate divergence time in a total-evidence framework, where fossils are directly integrated into the combined analysis of molecular data from living species with morphological data from fossils and living groups [154]. In the process of reconstructing kinship relationships and dating them, fossils are incorporated without having to determine their phylogenetic position a priori, and therefore this phylogenetic uncertainty can be directly integrated in the analysis. Some studies suggest that this approach improves divergence time estimates [147]. Computational limitations currently limit the application of this methodology to determine deep divergences. However, the field is advancing rapidly, and it is predicted that soon these methodologies will help to establish the affinity of fossils, and more carefully assign the age of the lineages and the different terrestrialization processes [155].

5. Conclusions

Ephemeral terrestrial habitats have existed for at least 1 billion years. However, animal terrestrialization and the consequent formation of more complex habitats has been a much more recent process. How recent remains a point of contention. The fossil record provides the only direct source of data to understand the temporal acquisition of characters, while phylogenies and molecular clocks complement this record to constrain the timing of the origin of these groups. The most recent molecular dating suggests that land plants were already present in the middle Cambrian to Early Ordovician [38], although other recent molecular clock estimates push this date back even further, into the Precambrian [156]. Similarly, recent molecular dating studies also suggest a concomitant colonization of the land by arthropods. If myriapods and arachnids really colonized the terrestrial environment so early, it would be possible that millipedes, a group of detritivore myriapods, fed on bacterial mats on the shoreline. Arachnids are a predominantly predatory group, suggesting that they must have originated from a diverse ecosystem. In this scenario, arachnids could have myriapods as potential prey. These ecologies represent habitats highly unfavourable to fossilization, such as high-energy environments characterized by erosion rather than deposition [157]. It is not surprising, then, that direct palaeontological insights may be limited in these cases, and molecular inference can step in to fill the gap.

Author Contributions: Conceptualization, J.L.-F.; writing—original draft preparation, J.L.-F., E.T., R.J.H. and C.C.; writing—review and editing, J.L.-F., E.T., R.J.H. and C.C. All authors have read and agreed to the published version of the manuscript.

Funding: This research was funded by the Strategic Priority Research Program of the Chinese Academy of Sciences (XDB26000000), the National Natural Science Foundation of China (42288201, 42222201), and the Second Tibetan Plateau Scientific Expedition and Research project (2019QZKK0706).

Institutional Review Board Statement: Not applicable.

Informed Consent Statement: Not applicable.

Data Availability Statement: Not applicable.

Acknowledgments: We are grateful to Greg Edgecombe, who provided constructive comments on a previous version of this manuscript, as well as three anonymous reviewers. We would like to thank the authors of images deposited in open access that have been used to generate some of the figures.

Conflicts of Interest: The authors declare no conflict of interest.

References

1. Betts, H.C.; Puttick, M.N.; Clark, J.W.; Williams, T.A.; Donoghue, P.C.J.; Pisani, D. Integrated Genomic and Fossil Evidence Illuminates Life's Early Evolution and Eukaryote Origin. *Nat. Ecol. Evol.* **2018**, *2*, 1556–1562. [CrossRef]
2. Schopf, T.J.M. Rates of Evolution and the Notion of "Living Fossils". *Annu. Rev. Earth Planet. Sci.* **1984**, *12*, 245–292. [CrossRef]
3. Allwood, A.C.; Rosing, M.T.; Flannery, D.T.; Hurowitz, J.A.; Heirwegh, C.M. Reassessing Evidence of Life in 3700-Million-Year-Old Rocks of Greenland. *Nature* **2018**, *563*, 241–244. [CrossRef]
4. Cavalazzi, B.; Lemelle, L.; Simionovici, A.; Cady, S.L.; Russell, M.J.; Bailo, E.; Canteri, R.; Enrico, E.; Manceau, A.; Maris, A.; et al. Cellular Remains in a ~3.42-Billion-Year-Old Subseafloor Hydrothermal Environment. *Sci. Adv.* **2021**, *7*, eabf3963. [CrossRef]
5. Sogin, M.L. The origin of eukaryotes and evolution into Major Kingdoms. In *Early Life on Earth*; Nobel Symposium No. 84; Columbia University Press: New York, NY, USA, 1994; pp. 181–192.
6. Zhu, S.; Zhu, M.; Knoll, A.H.; Yin, Z.; Zhao, F.; Sun, S.; Qu, Y.; Shi, M.; Liu, H. Decimetre-Scale Multicellular Eukaryotes from the 1.56-Billion-Year-Old Gaoyuzhuang Formation in North China. *Nat. Commun.* **2016**, *7*, 11500. [CrossRef]
7. Paterson, S.; Vogwill, T.; Buckling, A.; Benmayor, R.; Spiers, A.J.; Thomson, N.R.; Quail, M.; Smith, F.; Walker, D.; Libberton, B.; et al. Antagonistic Coevolution Accelerates Molecular Evolution. *Nature* **2010**, *464*, 275–278. [CrossRef]
8. Erwin, D.H. Evolutionary Uniformitarianism. *Dev. Biol.* **2011**, *357*, 27–34. [CrossRef]
9. Erwin, D.H.; Valentine, J.W. *The Cambrian Explosion: The Construction of Animal Biodiversity*; Macmillan Learning: Bedford, UK, 2013.

10. Gaines, R.R. Burgess Shale-Type Preservation and Its Distribution in Space and Time. *Paleontol. Soc. Pap.* **2014**, *20*, 123–146. [CrossRef]

11. Yang, C.; Li, X.-H.; Zhu, M.; Condon, D.J.; Chen, J. Geochronological Constraint on the Cambrian Chengjiang Biota, South China. *J. Geol. Soc.* **2018**, *175*, 659–666. [CrossRef]

12. Littlewood, D.T.J. Marine parasites and the tree of life. In *Marine Parasitology*; Rohde, K.K., Ed.; CSIRO Publishing: Melbourne, Australia, 2005; pp. 6–10.

13. Selden, P.A.; Jeram, A.J. Palaeophysiology of Terrestrialisation in the Chelicerata. *Earth Environ. Sci. Trans. R. Soc. Edinb.* **1989**, *80*, 303–310. [CrossRef]

14. De Baets, K.; Dentzien-Dias, P.; Harrison, G.W.M.; Littlewood, D.T.J.; Parry, L.A. Fossil Constraints on the Timescale of Parasitic Helminth Evolution. In *The Evolution and Fossil Record of Parasitism: Identification and Macroevolution of Parasites*; Topics in Geobiology; De Baets, K., Huntley, J.W., Eds.; Springer International Publishing: Cham, Switzerland, 2021; pp. 231–271. ISBN 978-3-030-42484-8.

15. Muscente, A.D.; Schiffbauer, J.D.; Broce, J.; Laflamme, M.; O'Donnell, K.; Boag, T.H.; Meyer, M.; Hawkins, A.D.; Huntley, J.W.; McNamara, M.; et al. Exceptionally Preserved Fossil Assemblages through Geologic Time and Space. *Gondwana Res.* **2017**, *48*, 164–188. [CrossRef]

16. Vermeij, G.J.; Watson-Zink, V.M. Terrestrialization in Gastropods: Lineages, Ecological Constraints and Comparisons with Other Animals. *Biol. J. Linn. Soc.* **2022**, *136*, 393–404. [CrossRef]

17. Garwood, R.J.; Edgecombe, G.D.; Charbonnier, S.; Chabard, D.; Sotty, D.; Giribet, G. Carboniferous Onychophora from Montceau-Les-Mines, France, and Onychophoran Terrestrialization. *Invert. Biol.* **2016**, *135*, 179–190. [CrossRef]

18. van Straalen, N.M. Evolutionary Terrestrialization Scenarios for Soil Invertebrates. *Pedobiologia* **2021**, *87–88*, 150753. [CrossRef]

19. Holterman, M.; Schratzberger, M.; Helder, J. Nematodes as Evolutionary Commuters between Marine, Freshwater and Terrestrial Habitats. *Biol. J. Linn. Soc.* **2019**, *128*, 756–767. [CrossRef]

20. Okamura, B.; Gruhl, A.; De Baets, K. Evolutionary Transitions of Parasites between Freshwater and Marine Environments. *Integr. Comp. Biol.* **2022**, *62*, 345–356. [CrossRef]

21. Tchesunov, A.V.; Ivanenko, V.N. What Is the Difference between Marine and Limnetic-Terrestrial Associations of Nematodes with Invertebrates? *Integr. Zool.* **2022**, *17*, 481–510. [CrossRef]

22. Guidetti, R.; Bertolani, R. Paleontology and Molecular Dating. In *Water Bears: The Biology of Tardigrades*; Zoological Monographs; Schill, R.O., Ed.; Springer International Publishing: Cham, Switzerland, 2018; pp. 131–143. ISBN 978-3-319-95702-9.

23. Sluys, R. The Evolutionary Terrestrialization of Planarian Flatworms (Platyhelminthes, Tricladida, Geoplanidae): A Review and Research Programme. *Zoosyst. Evol.* **2019**, *95*, 543–556. [CrossRef]

24. Dunlop, J.A.; Scholtz, G.; Selden, P.A. Water-to-Land Transitions. In *Arthropod Biology and Evolution: Molecules, Development, Morphology*; Minelli, A., Boxshall, G., Fusco, G., Eds.; Springer: Berlin/Heidelberg, Germany, 2013; pp. 417–439. ISBN 978-3-642-36160-9.

25. Lozano-Fernandez, J.; Carton, R.; Tanner, A.R.; Puttick, M.N.; Blaxter, M.; Vinther, J.; Olesen, J.; Giribet, G.; Edgecombe, G.D.; Pisani, D. A Molecular Palaeobiological Exploration of Arthropod Terrestrialization. *Phil. Trans. Roy. Soc. B.* **2016**, *371*, 20150133. [CrossRef]

26. Felsenstein, J. Phylogenies and the Comparative Method. *Am. Nat.* **1985**, *125*, 1–15. [CrossRef]

27. Benton, M.J. The Origins of Modern Biodiversity on Land. *Phil. Trans. R. Soc. B* **2010**, *365*, 3667–3679. [CrossRef] [PubMed]

28. Román-Palacios, C.; Moraga-López, D.; Wiens, J.J. The Origins of Global Biodiversity on Land, Sea and Freshwater. *Ecol. Lett.* **2022**, *25*, 1376–1386. [CrossRef] [PubMed]

29. Clarke, J.T.; Warnock, R.C.M.; Donoghue, P.C.J. Establishing a Time-Scale for Plant Evolution. *New Phytol.* **2011**, *192*, 266–301. [CrossRef]

30. Prave, A.R. Life on Land in the Proterozoic: Evidence from the Torridonian Rocks of Northwest Scotland. *Geology* **2002**, *30*, 811–814. [CrossRef]

31. Buatois, L.A.; Davies, N.S.; Gibling, M.R.; Krapovickas, V.; Labandeira, C.C.; MacNaughton, R.B.; Mángano, M.G.; Minter, N.J.; Shillito, A.P. The Invasion of the Land in Deep Time: Integrating Paleozoic Records of Paleobiology, Ichnology, Sedimentology, and Geomorphology. *Integr. Comp. Biol.* **2022**, *62*, 297–331. [CrossRef]

32. Xu, H.; Wang, K.; Huang, Z.; Tang, P.; Wang, Y.; Liu, B.; Yan, W. The Earliest Vascular Land Plants from the Upper Ordovician of China 2022. *Biol. Sci.* **2022**, preprint.

33. Wilson, H.M.; Anderson, L.I. Morphology and Taxonomy of Paleozoic Millipedes (Diplopoda: Chilognatha: Archipolypoda) from Scotland. *J. Paleontol.* **2004**, *78*, 169–184. [CrossRef]

34. Dunlop, J. A Trigonotarbid Arachnid from the Upper Silurian of Shropshire. *Palaeontology* **1996**, *39*, 605–614.

35. Ward, P.; Labandeira, C.; Laurin, M.; Berner, R.A. Confirmation of Romer's Gap as a Low Oxygen Interval Constraining the Timing of Initial Arthropod and Vertebrate Terrestrialization. *Proc. Natl. Acad. Sci. USA* **2006**, *103*, 16818–16822. [CrossRef]

36. Davies, N.S.; Rygel, M.C.; Gibling, M.R. Marine Influence in the Upper Ordovician Juniata Formation (Potters Mills, Pennsylvania): Implications for the History of Life on Land. *PALAIOS* **2010**, *25*, 527–539. [CrossRef]

37. Lozano-Fernandez, J.; Tanner, A.R.; Puttick, M.N.; Vinther, J.; Edgecombe, G.D.; Pisani, D. A Cambrian–Ordovician Terrestrialization of Arachnids. *Front. Genet.* **2020**, *11*, 182. [CrossRef] [PubMed]

38. Morris, J.L.; Puttick, M.N.; Clark, J.W.; Edwards, D.; Kenrick, P.; Pressel, S.; Wellman, C.H.; Yang, Z.; Schneider, H.; Donoghue, P.C. The Timescale of Early Land Plant Evolution. *Proc. Natl. Acad. Sci. USA* **2018**, *115*, E2274–E2283. [CrossRef] [PubMed]

39. Edgecombe, G.D.; Strullu-Derrien, C.; Góral, T.; Hetherington, A.J.; Thompson, C.; Koch, M. Aquatic Stem Group Myriapods Close a Gap between Molecular Divergence Dates and the Terrestrial Fossil Record. *Proc. Natl. Acad. Sci. USA* **2020**, *117*, 8966–8972. [CrossRef] [PubMed]

40. Lamsdell, J.C.; McCoy, V.E.; Perron-Feller, O.A.; Hopkins, M.J. Air Breathing in an Exceptionally Preserved 340-Million-Year-Old Sea Scorpion. *Curr. Biol.* **2020**, *30*, 4316–4321.e2. [CrossRef]

41. Royer, D.L.; Donnadieu, Y.; Park, J.; Kowalczyk, J.; Goddéris, Y. Error Analysis of CO_2 and O_2 Estimates from the Long-Term Geochemical Model GEOCARBSULF. *Am. J. Sci.* **2014**, *314*, 1259–1283. [CrossRef]

42. Schachat, S.R.; Labandeira, C.C.; Saltzman, M.R.; Cramer, B.D.; Payne, J.L.; Boyce, C.K. Phanerozoic p O_2 and the Early Evolution of Terrestrial Animals. *Proc. R. Soc. Lond. B* **2018**, *285*, 20172631.

43. Gregor, B. Denudation of the Continents. *Nature* **1970**, *228*, 273–275. [CrossRef]

44. Blatt, H.; Jones, R.L. Proportions of Exposed Igneous, Metamorphic, and Sedimentary Rocks. *GSA Bull.* **1975**, *86*, 1085–1088. [CrossRef]

45. Khain, V.Y.; Ronov, A.B.; Seslavinskiy, K.B. Silurian Lithologic Associations of the World. *Int. Geol. Rev.* **1978**, *20*, 249–268. [CrossRef]

46. Ronov, A.B. The Earth's Sedimentary Shell (Quantitative Patterns of Its Structure, Compositions, and Evolution). *Int. Geol. Rev.* **1982**, *24*, 1313–1363. [CrossRef]

47. Ronov, A.B.; Khain, V.E.; Balukhovsky, A.N.; Seslavinsky, K.B. Quantitative Analysis of Phanerozoic Sedimentation. *Sed. Geol.* **1980**, *25*, 311–325. [CrossRef]

48. Smith, A.B.; McGowan, A.J. The Shape of the Phanerozoic Marine Palaeodiversity Curve: How Much Can Be Predicted from the Sedimentary Rock Record of Western Europe? *Palaeontology* **2007**, *50*, 765–774. [CrossRef]

49. Robardet, M.; Blaise, J.; Bouyx, E.; Gourvennec, R.; Lardeux, H.; Le Hérissé, A.; Le Menn, J.; Melou, M.; Paris, F.; Plusquellec, Y.; et al. Palaeogeography of Western Europe from the Ordovician to the Devonian. *Bull. Soc. Geol. Fr.* **1993**, *164*, 683–695.

50. Kenrick, P.; Wellman, C.H.; Schneider, H.; Edgecombe, G.D. A Timeline for Terrestrialization: Consequences for the Carbon Cycle in the Palaeozoic. *Phil. Trans. R. Soc. B* **2012**, *367*, 519–536. [CrossRef]

51. Zhang, Z.-Q. Animal Biodiversity: An Outline of Higher-Level Classification and Survey of Taxonomic Richness (Addenda 2013). *Zootaxa* **2013**, *3703*, 1–82. [CrossRef] [PubMed]

52. Maloof, A.C.; Rose, C.V.; Beach, R.; Samuels, B.M.; Calmet, C.C.; Erwin, D.H.; Poirier, G.R.; Yao, N.; Simons, F.J. Possible Animal-Body Fossils in Pre-Marinoan Limestones from South Australia. *Nat. Geosci.* **2010**, *3*, 653–659. [CrossRef]

53. Wolfe, J.M.; Daley, A.C.; Legg, D.A.; Edgecombe, G.D. Fossil Calibrations for the Arthropod Tree of Life. *Earth Sci. Rev.* **2016**, *160*, 43–110. [CrossRef]

54. Edgecombe, G.D. Arthropod Phylogeny: An Overview from the Perspectives of Morphology, Molecular Data and the Fossil Record. *Arthropod Struct. Dev.* **2010**, *39*, 74–87. [CrossRef]

55. Grimaldi, D.; Engel, M.S. *Evolution of the Insects*, 1st ed.; Cambridge University Press: Cambridge, UK, 2005.

56. Govorushko, S. Economic and Ecological Importance of Termites: A Global Review. *Entomol. Sci.* **2019**, *22*, 21–35. [CrossRef]

57. Brusca, R.C.; Brusca, G.J. *Invertebrates*, 2nd ed.; Sinauer Associates, Incorporated: Sunderland, MA, USA, 2003; ISBN 978-0-87893-099-9.

58. Little, C. *The Colonisation of Land: Origins and Adaptations of Terrestrial Animals*; Cambridge University Press: Cambridge, UK, 1983; ISBN 978-0-521-25218-8.

59. Jeram, A.J.; Selden, P.A.; Edwards, D. Land Animals in the Silurian: Arachnids and Myriapods from Shropshire, England. *Science* **1990**, *250*, 658–661. [CrossRef]

60. Shear, W.; Selden, P. Eoarthropleura (Arthropoda, Arthropleurida) from the Silurian of Britain and the Devonian of North America. *Neues Jahrb. Geol. Palaontol. Abh.* **1995**, *196*, 347–375. [CrossRef]

61. Shear, W.A.; Jeram, A.J.; Selden, P. Centiped Legs (Arthropoda, Chilopoda, Scutigeromorpha) from the Silurian and Devonian of Britain and the Devonian of North America. *Am. Mus. Novit.* **1998**, *3231*, 1–16.

62. Dunlop, J.A. A Replacement Name for the Trigonotarbid Arachnid *Eotarbus* Dunlop. *Palaeontology* **1999**, *42*, 191. [CrossRef]

63. Brookfield, M.E.; Catlos, E.J.; Suarez, S.E. Myriapod Divergence Times Differ between Molecular Clock and Fossil Evidence: U/Pb Zircon Ages of the Earliest Fossil Millipede-Bearing Sediments and Their Significance. *Hist. Biol.* **2020**, *10*, 2014–2018. [CrossRef]

64. Marshall, J.E.A. Palynology of the Stonehaven Group, Scotland: Evidence for a Mid Silurian Age and Its Geological Implications. *Geol. Mag.* **1991**, *128*, 283–286. [CrossRef]

65. Wellman, C.H. A Land Plant Microfossil Assemblage of Mid Silurian Age from the Stonehaven Group, Scotland. *J. Micropalaeontol.* **1993**, *12*, 47–66. [CrossRef]

66. Howard, R.J.; Puttick, M.N.; Edgecombe, G.D.; Lozano-Fernandez, J. Arachnid Monophyly: Morphological, Palaeontological and Molecular Support for a Single Terrestrialization within Chelicerata. *Arthropod Struct. Devel.* **2020**, *59*, 100997. [CrossRef]

67. Suarez, S.E.; Brookfield, M.E.; Catlos, E.J.; Stöckli, D.F. A U-Pb Zircon Age Constraint on the Oldest-Recorded Air-Breathing Land Animal. *PLoS ONE* **2017**, *12*, e0179262. [CrossRef]

68. Kutscher, F. Friedrich Beiträge Zur Sedimentation Und Fossilführung Des Hunsrückschiefers 32. *Palaeoscorpius devonicus*, Ein Devonischer Skorpion. *Jahrb. Nassau. Ver. Naturkd.* **1971**, *101*, 191.

69. Lehmann, W.M. *Palaeoscorpius devonicus* Ng, n. Sp., Ein Skorpion Aus Dem Rheinischen Unterdevon. *N. Jahrb. Geol. Palaontol. Monat.* **1944**, *7*, 177–185.

70. Kühl, G.; Bergmann, A.; Dunlop, J.; Garwood, R.J.; Rust, J. Redescription and Palaeobiology of *Palaeoscorpius devonicus* Lehmann, 1944 from the Lower Devonian Hunsrück Slate of Germany. *Palaeontology* **2012**, *55*, 775–787. [CrossRef]

71. Zrzavý, J.; Hypša, V.; Vlášková, M. Arthropod Phylogeny: Taxonomic Congruence, Total Evidence and Conditional Combination Approaches to Morphological and Molecular Data Sets. In *Arthropod Relationships*; Fortey, R.A., Thomas, R.H., Eds.; The Systematics Association Special Volume Series; Springer: Dordrecht, The Netherlands, 1998; pp. 97–107. ISBN 978-94-011-4904-4.

72. Giribet, G.; Edgecombe, G.D. The Phylogeny and Evolutionary History of Arthropods. *Curr. Biol.* **2019**, *29*, R592–R602. [CrossRef] [PubMed]

73. Legg, D.A.; Sutton, M.D.; Edgecombe, G.D. Arthropod Fossil Data Increase Congruence of Morphological and Molecular Phylogenies. *Nat. Commun.* **2013**, *4*, 2485. [CrossRef] [PubMed]

74. Ballesteros, J.A.; Santibáñez-López, C.E.; Baker, C.M.; Benavides, L.R.; Cunha, T.J.; Gainett, G.; Ontano, A.Z.; Setton, E.V.W.; Arango, C.P.; Gavish-Regev, E.; et al. Comprehensive Species Sampling and Sophisticated Algorithmic Approaches Refute the Monophyly of Arachnida. *Mol. Biol. Evol.* **2022**, *39*, msac021. [CrossRef] [PubMed]

75. Tihelka, E.; Cai, C.; Giacomelli, M.; Lozano-Fernandez, J.; Rota-Stabelli, O.; Huang, D.; Engel, M.S.; Donoghue, P.C.J.; Pisani, D. The Evolution of Insect Biodiversity. *Curr. Biol.* **2021**, *31*, R1299–R1311. [CrossRef] [PubMed]

76. Bäcker, H.; Fanenbruck, M.; Wägele, J.W. A Forgotten Homology Supporting the Monophyly of Tracheata: The Subcoxa of Insects and Myriapods Re-Visited. *Zool. Anz. J. Comp. Zool.* **2008**, *247*, 185–207. [CrossRef]

77. Giribet, G.; Edgecombe, G.D. Reevaluating the Arthropod Tree of Life. *Annu. Rev. Entomol.* **2012**, *57*, 167–186. [CrossRef]

78. Fernández, R.; Edgecombe, G.D.; Giribet, G.; Edgecombe, G.D.; Giribet, G. Phylogenomics Illuminates the Backbone of the Myriapoda Tree of Life and Reconciles Morphological and Molecular Phylogenies. *Sci. Rep.* **2018**, *8*, 83. [CrossRef]

79. Rota-Stabelli, O.; Campbell, L.; Brinkmann, H.; Edgecombe, G.D.; Longhorn, S.J.; Peterson, K.J.; Pisani, D.; Philippe, H.; Telford, M.J. A Congruent Solution to Arthropod Phylogeny: Phylogenomics, MicroRNAs and Morphology Support Monophyletic Mandibulata. *Proc. R. Soc. B* **2011**, *278*, 298–306. [CrossRef]

80. Fernández, R.; Edgecombe, G.D.; Giribet, G. Exploring Phylogenetic Relationships within Myriapoda and the Effects of Matrix Composition and Occupancy on Phylogenomic Reconstruction. *Syst. Biol.* **2016**, *65*, 871–889. [CrossRef]

81. Szucsich, N.U.; Bartel, D.; Blanke, A.; Böhm, A.; Donath, A.; Fukui, M.; Grove, S.; Liu, S.; Macek, O.; Machida, R.; et al. Four Myriapod Relatives—But Who Are Sisters? No End to Debates on Relationships among the Four Major Myriapod Subgroups. *BMC Evol. Biol.* **2020**, *20*, 144. [CrossRef] [PubMed]

82. Wang, J.; Bai, Y.; Zhao, H.; Mu, R.; Dong, Y. Reinvestigating the Phylogeny of Myriapoda with More Extensive Taxon Sampling and Novel Genetic Perspective. *PeerJ* **2021**, *9*, e12691. [CrossRef] [PubMed]

83. Benavides, L.R.; Edgecombe, G.D.; Giribet, G. Re-Evaluating and Dating Myriapod Diversification with Phylotranscriptomics under a Regime of Dense Taxon Sampling. *Mol. Phylogenetics Evol.* **2022**, *178*, 107621. [CrossRef] [PubMed]

84. Schwentner, M.; Combosch, D.J.; Nelson, J.P.; Giribet, G. A Phylogenomic Solution to the Origin of Insects by Resolving Crustacean-Hexapod Relationships. *Curr. Biol.* **2017**, *27*, 1818–1824. [CrossRef]

85. Regier, J.C.; Shultz, J.W.; Zwick, A.; Hussey, A.; Ball, B.; Wetzer, R.; Martin, J.W.; Cunningham, C.W.; Shultz, J.W.; Zwick, A.; et al. Arthropod Relationships Revealed by Phylogenomic Analysis of Nuclear Protein-Coding Sequences. *Nature* **2010**, *463*, 1079–1083. [CrossRef]

86. Lozano-Fernandez, J.; Giacomelli, M.; Fleming, J.F.; Chen, A.; Vinther, J.; Thomsen, P.F.; Glenner, H.; Palero, F.; Legg, D.A.; Iliffe, T.M.; et al. Pancrustacean Evolution Illuminated by Taxon-Rich Genomic-Scale Data Sets with an Expanded Remipede Sampling. *Genome Biol. Evol.* **2019**, *11*, 2055–2070. [CrossRef]

87. Yager, J. Remipedia, a New Class of Crustacea from a Marine Cave in the Bahamas. *J. Crustacean Biol.* **1981**, *1*, 328–333. [CrossRef]

88. Schmalfuss, H. World Catalog of Terrestrial Isopods (Isopoda: Oniscidea). *Stuttg. Beitr. Naturkd. A* **2003**, *654*, 341.

89. Richardson, A.; Araujo, P.B. Lifestyles of Terrestrial Crustacean. In *The Natural History of the Crustacea. Lifestyles and Feeding Biology*; Oxford University Press: Oxford, UK, 2015; pp. 299–336.

90. Broly, P.; Deville, P.; Maillet, S. The Origin of Terrestrial Isopods (Crustacea: Isopoda: Oniscidea). *Evol. Ecol.* **2013**, *27*, 461–476. [CrossRef]

91. Elisabeth, H. Evolutionary Adaptation of Oniscidean Isopods to Terrestrial Life: Structure, Physiology and Behavior. *Terr. Arthropod Rev.* **2011**, *4*, 95–130. [CrossRef]

92. Schmidt, C. Phylogeny of the Terrestrial Isopoda (Oniscidea): A Review. *Arthr. Syst. Phyl.* **2008**, *66*, 191–226.

93. Dimitriou, A.C.; Taiti, S.; Sfenthourakis, S. Genetic Evidence against Monophyly of Oniscidea Implies a Need to Revise Scenarios for the Origin of Terrestrial Isopods. *Sci. Rep.* **2019**, *9*, 18508. [CrossRef] [PubMed]

94. Tabacaru, I.; Giurginca, A. The Monophyly and the Classification of the Terrestrial Isopods (Crustacea, Isopoda, Oniscidea). *Trav. Inst. Speol. Emile Racovitza* **2021**, *59*, 3–23.

95. Ballesteros, J.A.; Sharma, P.P. A Critical Appraisal of the Placement of Xiphosura (Chelicerata) with Account of Known Sources of Phylogenetic Error. *Syst. Biol.* **2019**, *68*, 896–917. [CrossRef]

96. Shultz, J.W. A Phylogenetic Analysis of the Arachnid Orders Based on Morphological Characters. *Zool. J. Linn. Soc.* **2007**, *150*, 221–265. [CrossRef]

97. Shultz, J.W. Evolutionary Morphology and Phylogeny of Arachnida. *Cladistics* **1990**, *6*, 1–38. [CrossRef] [PubMed]

98. Bicknell, R.D.C.; Pates, S. Pictorial Atlas of Fossil and Extant Horseshoe Crabs, with Focus on Xiphosurida. *Front. Earth Sci.* **2020**, *8*, 98. [CrossRef]

99. Gould, S.J. Dollo on Dollo's Law: Irreversibility and the Status of Evolutionary Laws. *J. Hist. Biol.* **1970**, *3*, 189–212. [CrossRef]

100. Lamsdell, J.C. Evolutionary History of the Dynamic Horseshoe Crab. *Int. Wader Stud.* **2019**, *21*, 1–15.

101. Bicknell, R.D.C.; Kimmig, J.; Budd, G.E.; Legg, D.A.; Bader, K.S.; Haug, C.; Kaiser, D.; Laibl, L.; Tashman, J.N.; Campione, N.E. Habitat and Developmental Constraints Drove 330 Million Years of Horseshoe Crab Evolution. *Biol. J. Linn. Soc.* **2022**, *136*, 155–172. [CrossRef]

102. Lozano-Fernandez, J.; Tanner, A.R.; Giacomelli, M.; Carton, R.; Vinther, J.; Edgecombe, G.D.; Pisani, D. Increasing Species Sampling in Chelicerate Genomic-Scale Datasets Provides Support for Monophyly of Acari and Arachnida. *Nat. Commun.* **2019**, *10*, 2295. [CrossRef] [PubMed]

103. MacNaughton, R.B.; Cole, J.M.; Dalrymple, R.W.; Braddy, S.J.; Briggs, D.E.G.; Lukie, T.D. First Steps on Land: Arthropod Trackways in Cambrian-Ordovician Eolian Sandstone, Southeastern Ontario, Canada. *Geology* **2002**, *30*, 391–394. [CrossRef]

104. Collette, J.H.; Hagadorn, J.W. Three-Dimensionally Preserved Arthropods from Cambrian Lagerstätten of Quebec and Wisconsin. *J. Paleontol.* **2010**, *84*, 646–667. [CrossRef]

105. Vaccari, N.E.; Edgecombe, G.D.; Escudero, C.; Edgecombe, G.D.; Escudero, C. Cambrian Origins and Affinities of an Enigmatic Fossil Group of Arthropods. *Nature* **2004**, *430*, 554–557. [CrossRef] [PubMed]

106. Braddy, S.J.; Gass, K.C.; Gass, T.C. Fossils of Blackberry Hill, Wisconsin, USA: The First Animals on Land, 500 Million Years Ago. *Geol. Today* **2022**, *38*, 25–31. [CrossRef]

107. Mángano, M.G.; Buatois, L.A.; Astini, R.; Rindsberg, A.K. Trilobites in Early Cambrian Tidal Flats and the Landward Expansion of the Cambrian Explosion. *Geology* **2014**, *42*, 143–146. [CrossRef]

108. Mángano, M.G.; Buatois, L.A.; Waisfeld, B.G.; Muñoz, D.F.; Vaccari, N.E.; Astini, R.A. Were All Trilobites Fully Marine? Trilobite Expansion into Brackish Water during the Early Palaeozoic. *Proc. R. Soc. B* **2021**, *288*, 20202263. [CrossRef]

109. Suzuki, Y.; Bergström, J. Respiration in Trilobites: A Reevaluation. *GFF* **2008**, *130*, 211–229. [CrossRef]

110. Hou, J.; Hughes, N.C.; Hopkins, M.J. The Trilobite Upper Limb Branch Is a Well-Developed Gill. *Sci. Adv.* **2021**, *7*, eabe7377. [CrossRef]

111. Retallack, G.J. *Scoyenia* Burrows from Ordovician Palaeosols of the Juniata Formation in Pennsylvania. *Palaeontology* **2001**, *44*, 209–235. [CrossRef]

112. Shillito, A.P.; Davies, N.S. Death near the Shoreline, Not Life on Land: Ordovician Arthropod Trackways in the Borrowdale Volcanic Group, UK. *Geology* **2018**, *47*, 55–58. [CrossRef]

113. Johnson, E.W.; Briggs, D.E.G.; Suthren, R.J.; Wright, J.L.; Tunnicliff, S.P. Non-Marine Arthropod Traces from the Subaerial Ordovician Borrowdale Volcanic Group, English Lake District. *Geol. Mag.* **1994**, *131*, 395–406. [CrossRef]

114. Ortega-Hernández, J.; Legg, D.A.; Tremewan, J.; Braddy, S.J. Euthycarcinoids. *Geol. Today* **2010**, *26*, 195–198. [CrossRef]

115. Collette, J.H.; Gass, K.G.; Hagadorn, J.W. *Protichnites eremita* Unshelled? Experimental Model-Based Neoichnology and New Evidence for a Euthycarcinoid Affinity for This Ichnospecies. *J. Paleontol.* **2012**, *86*, 442–454. [CrossRef]

116. Gueriau, P.; Lamsdell, J.C.; Wogelius, R.A.; Manning, P.L.; Egerton, V.M.; Bergmann, U.; Bertrand, L.; Denayer, J. A New Devonian Euthycarcinoid Reveals the Use of Different Respiratory Strategies during the Marine-to-Terrestrial Transition in the Myriapod Lineage. *R. Soc. Open Sci.* **2020**, *7*, 201037. [CrossRef]

117. Trewin, N.H.; Gurr, P.R.; Jones, R.B.; Gavin, P. The Biota, Depositional Environment and Age of the Old Red Sandstone of the Island of Kerrera, Scotland. *Scott. J. Geol.* **2012**, *48*, 77–90. [CrossRef]

118. Scourfield, D.J. The Oldest Known Fossil Insect. *Nature* **1940**, *145*, 799–801. [CrossRef]

119. Whalley, P.; Jarzembowski, E.A. A New Assessment of *Rhyniella*, the Earliest Known Insect, from the Devonian of Rhynie, Scotland. *Nature* **1981**, *291*, 317. [CrossRef]

120. Fayers, S.R.; Trewin, N.H. A Hexapod from the Early Devonian Windyfield Chert, Rhynie, Scotland. *Palaeontology* **2005**, *48*, 1117–1130. [CrossRef]

121. Greenslade, P.; Whalley, P.E.S. The Systematic Position of *Rhyniella praecursor* (Hirst and Malik, the Earliest Known Hexapod. In Proceedings of the 2nd International Seminar on Apterygota, Siena, Italy, 1 January 1986; Volume 1986, pp. 319–323.

122. Shear, W.A. An Insect to Fill the Gap. *Nature* **2012**, *488*, 34–35. [CrossRef]

123. Kukalová-Peck, J. New Carboniferous Diplura, Monura, and Thysanura, the Hexapod Ground Plan, and the Role of Thoracic Side Lobes in the Origin of Wings (Insecta). *Can. J. Zool.* **1987**, *65*, 2327–2345. [CrossRef]

124. Hošek, P. Fossil Insect Diversity. *Vesmír* **1994**, *73*, 196–200.

125. Crowson, R.A. Comments on Insecta of the Rhynie Chert. *Entomol. Gener.* **1985**, 97–98. [CrossRef]

126. Rohdendorf, B.B. Devonskie Eopteridy-Ne Nasekomye a Rakoobraznye Eumalacostraca. *Entomol. Oboz.* **1972**, *51*, 96–97.

127. Schram, F.R. Miscellaneous Late Paleozoic Malacostraca of the Soviet Union. *J. Paleontol.* **1980**, *54*, 542–547.

128. Haug, C.; Haug, J.T. The Presumed Oldest Flying Insect: More Likely a Myriapod? *PeerJ* **2017**, *5*, e3402. [CrossRef]

129. Garrouste, R.; Clément, G.; Nel, P.; Engel, M.S.; Grandcolas, P.; D'Haese, C.; Lagebro, L.; Denayer, J.; Gueriau, P.; Lafaite, P.; et al. A Complete Insect from the Late Devonian Period. *Nature* **2012**, *488*, 82–85. [CrossRef]

130. Hörnschemeyer, T.; Haug, J.T.; Bethoux, O.; Beutel, R.G.; Charbonnier, S.; Hegna, T.A.; Koch, M.; Rust, J.; Wedmann, S.; Bradler, S.; et al. Is *Strudiella* a Devonian Insect? *Nature* **2013**, *494*, E3–E4. [CrossRef]

131. Haas, F.; Waloszek, D.; Hartenberger, R. *Devonohexapodus bocksbergensis*, a New Marine Hexapod from the Lower Devonian Hunsrück Slates, and the Origin of Atelocerata and Hexapoda. *Org. Divers. Evol.* **2003**, *3*, 39–54. [CrossRef]
132. Kühl, G.; Rust, J. *Devonohexapodus Bocksbergensis* Is a Synonym of *Wingertshellicus backesi* (Euarthropoda)—No Evidence for Marine Hexapods Living in the Devonian Hunsrück Sea. *Org. Divers. Evol.* **2009**, *9*, 215–231. [CrossRef]
133. Brauckmann, C.; Brauckmann, B.; Groning, E. The Stratigraphical Position of the Oldest Known Pterygota (Insecta. Carboniferous, Namurian). *Ann. Soc. Géol. Belg.* **1994**, *117*, 47–56.
134. Dunlop, J.A.; Erik Tetlie, O.; Prendini, L. Reinterpretation of the Silurian Scorpion *Proscorpius osborni* (Whitfield): Integrating Data from Palaeozoic and Recent Scorpioans. *Palaeontology* **2008**, *51*, 303–320. [CrossRef]
135. Waddington, J.; Rudkin, D.M.; Dunlop, J.A. A New Mid-Silurian Aquatic Scorpion—One Step Closer to Land? *Biol. Lett.* **2015**, *11*, 20140815. [CrossRef] [PubMed]
136. Wendruff, A.J.; Babcock, L.E.; Wirkner, C.S.; Kluessendorf, J.; Mikulic, D.G. A Silurian Ancestral Scorpion with Fossilised Internal Anatomy Illustrating a Pathway to Arachnid Terrestrialisation. *Sci. Rep.* **2020**, *10*, 14. [CrossRef]
137. Sharma, P.P.; Kaluziak, S.T.; Pérez-Porro, A.R.; González, V.L.; Hormiga, G.; Wheeler, W.C.; Giribet, G. Phylogenomic Interrogation of Arachnida Reveals Systemic Conflicts in Phylogenetic Signal. *Mol. Biol. Evol.* **2014**, *31*, 2963–2984. [CrossRef]
138. Leite, D.J.; Baudouin-Gonzalez, L.; Iwasaki-Yokozawa, S.; Lozano-Fernandez, J.; Turetzek, N.; Akiyama-Oda, Y.; Prpic, N.-M.; Pisani, D.; Oda, H.; Sharma, P.P.; et al. Homeobox Gene Duplication and Divergence in Arachnids. *Mol. Biol. Evol.* **2018**, *35*, 2240–2253. [CrossRef]
139. Ontano, A.Z.; Gainett, G.; Aharon, S.; Ballesteros, J.A.; Benavides, L.R.; Corbett, K.F.; Gavish-Regev, E.; Harvey, M.S.; Monsma, S.; Santibáñez-López, C.E.; et al. Taxonomic Sampling and Rare Genomic Changes Overcome Long-Branch Attraction in the Phylogenetic Placement of Pseudoscorpions. *Mol. Biol. Evol.* **2021**, *38*, 2446–2467. [CrossRef]
140. Claridge, M.F.; Lyon, A.G. Lung-Books in the Devonian Palæocharinidae (Arachnida). *Nature* **1961**, *191*, 1190–1191. [CrossRef]
141. Kamenz, C.; Dunlop, J.A.; Scholtz, G.; Kerp, H.; Hass, H. Microanatomy of Early Devonian Book Lungs. *Biol. Lett.* **2008**, *4*, 212–215. [CrossRef]
142. Dunlop, J.A.; Garwood, R.J. Terrestrial Invertebrates in the Rhynie Chert Ecosystem. *Phil. Trans. R. Soc. B* **2018**, *373*, 20160493. [CrossRef]
143. Zuckerkandl, E.; Pauling, L. Evolutionary Divergence and Convergence in Proteins. In *Evolving Genes and Proteins*; Bryson, V., Vogel, H.J., Eds.; Academic Press: Cambridge, MA, USA, 1965; pp. 97–166. ISBN 978-1-4832-2734-4.
144. Parham, J.F.; Donoghue, P.C.; Bell, C.J.; Calway, T.D.; Head, J.J.; Holroyd, P.A.; Inoue, J.G.; Irmis, R.B.; Joyce, W.G.; Ksepka, D.T. Best Practices for Justifying Fossil Calibrations. *Syst. Biol.* **2011**, *61*, 346–359. [CrossRef] [PubMed]
145. Yang, Z.; Rannala, B. Bayesian Estimation of Species Divergence Times under a Molecular Clock Using Multiple Fossil Calibrations with Soft Bounds. *Mol. Biol. Evol.* **2006**, *23*, 212–226. [CrossRef] [PubMed]
146. dos Reis, M.; Thawornwattana, Y.; Angelis, K.; Telford, M.J.; Donoghue, P.C.J.; Yang, Z. Uncertainty in the Timing of Origin of Animals and the Limits of Precision in Molecular Timescales. *Curr. Biol.* **2015**, *25*, 2939–2950. [CrossRef] [PubMed]
147. Mongiardino Koch, N.; Parry, L.A. Death Is on Our Side: Paleontological Data Drastically Modify Phylogenetic Hypotheses. *Syst. Biol.* **2020**, *69*, 1052–1067. [CrossRef] [PubMed]
148. Mongiardino Koch, N.; Garwood, R.J.; Parry, L.A. Inaccurate Fossil Placement Does Not Compromise Tip-Dated Divergence Times. *bioRxiv* **2022**. *preprint*. [CrossRef]
149. Wright, A.M.; Bapst, D.W.; Barido-Sottani, J.; Warnock, R.C.M. Integrating Fossil Observations into Phylogenetics Using the Fossilized Birth–Death Model. *Ann. Rev. Earth Planet. Sci.* **2022**, *53*, 12. [CrossRef]
150. Klopfstein, S. The Age of Insects and the Revival of the Minimum Age Tree. *Austral. Entomol.* **2021**, *60*, 138–146. [CrossRef]
151. Rota-Stabelli, O.; Lartillot, N.; Philippe, H.; Pisani, D. Serine Codon-Usage Bias in Deep Phylogenomics: Pancrustacean Relationships as a Case Study. *Syst. Biol.* **2013**, *62*, 121–133. [CrossRef]
152. Rehm, P.; Meusemann, K.; Borner, J.; Misof, B.; Burmester, T. Phylogenetic Position of Myriapoda Revealed by 454 Transcriptome Sequencing. *Mol. Phylogenet. Evol.* **2014**, *77*, 25–33. [CrossRef]
153. Miyazawa, H.; Ueda, C.; Yahata, K.; Su, Z.-H.; Ueda, C.; Yahata, K.; Su, Z.-H. Molecular Phylogeny of Myriapoda Provides Insights into Evolutionary Patterns of the Mode in Post-Embryonic Development. *Sci. Rep.* **2014**, *4*, 4127. [CrossRef]
154. Ronquist, F.; Klopfstein, S.; Vilhelmsen, L.; Schulmeister, S.; Murray, D.L.; Rasnitsyn, A.P. A Total-Evidence Approach to Dating with Fossils, Applied to the Early Radiation of the Hymenoptera. *Syst. Biol.* **2012**, *61*, 973–999. [CrossRef]
155. Howard, R.J.; Edgecombe, G.D.; Legg, D.A.; Pisani, D.; Lozano-Fernandez, J. Exploring the Evolution and Terrestrialization of Scorpions (Arachnida: Scorpiones) with Rocks and Clocks. *Org. Divers. Evol.* **2019**, *19*, 71–86. [CrossRef]
156. Su, D.; Yang, L.; Shi, X.; Ma, X.; Zhou, X.; Hedges, S.B.; Zhong, B. Large-Scale Phylogenomic Analyses Reveal the Monophyly of Bryophytes and Neoproterozoic Origin of Land Plants. *Mol. Biol. Evol.* **2021**, *38*, 3332–3344. [CrossRef] [PubMed]
157. Parry, L.A.; Smithwick, F.; Nordén, K.K.; Saitta, E.T.; Lozano-Fernandez, J.; Tanner, A.R.; Caron, J.-B.; Edgecombe, G.D.; Briggs, D.E.G.; Vinther, J. Soft-Bodied Fossils Are Not Simply Rotten Carcasses—Toward a Holistic Understanding of Exceptional Fossil Preservation. *BioEssays* **2018**, *40*, 1700167. [CrossRef] [PubMed]

Article

Preservation and Taphonomy of Fossil Insects from the Earliest Eocene of Denmark

Miriam Heingård [1,*], Peter Sjövall [2], Bo P. Schultz [3], René L. Sylvestersen [3] and Johan Lindgren [1]

[1] Department of Geology, Lund University, SE-223 62 Lund, Sweden; johan.lindgren@geol.lu.se
[2] Materials and Production, RISE Research Institutes of Sweden, SE-501 15 Borås, Sweden; peter.sjovall@ri.se
[3] Fur Museum, Museum Salling, DK-7884 Nederby, Denmark; bosc@museumsalling.dk (B.P.S.); rlsy@museumsalling.dk (R.L.S.)
* Correspondence: miriam.heingard@geol.lu.se

Simple Summary: Insect fossils dating 55 million-years-old from the Stolleklint Clay and Fur Formation of Denmark are known to preserve both fine morphological details and color patterns. To enhance our understanding on how such fragile animals are retained in the fossil record, we examined a pair of beetle elytra, a wasp and a damselfly using sensitive analytical techniques. In our paper, we demonstrate that all three insect fossils are composed of cuticular remains (that is, traces of the exoskeleton) that, in turn, are dominated by the natural pigment eumelanin. In addition, the beetle elytra show evidence of a delicate lamellar structure comparable to multilayered reflectors that produce metallic hues in modern insects. Our results contribute to improved knowledge on the process of fossilization of insect body fossils in marine environments.

Abstract: Marine sediments of the lowermost Eocene Stolleklint Clay and Fur Formation of northwestern Denmark have yielded abundant well-preserved insects. However, despite a long history of research, in-depth information pertaining to preservational modes and taphonomic pathways of these exceptional animal fossils remains scarce. In this paper, we use a combination of scanning electron microscopy coupled with energy-dispersive X-ray spectroscopy (SEM-EDX), transmission electron microscopy (TEM) and time-of-flight secondary ion mass spectrometry (ToF-SIMS) to assess the ultrastructural and molecular composition of three insect fossils: a wasp (Hymenoptera), a damselfly (Odonata) and a pair of beetle elytra (Coleoptera). Our analyses show that all specimens are preserved as organic remnants that originate from the exoskeleton, with the elytra displaying a greater level of morphological fidelity than the other fossils. TEM analysis of the elytra revealed minute features, including a multilayered epicuticle comparable to those nanostructures that generate metallic colors in modern insects. Additionally, ToF-SIMS analyses provided spectral evidence for chemical residues of the pigment eumelanin as part of the cuticular remains. To the best of our knowledge, this is the first occasion where both structural colors and chemical traces of an endogenous pigment have been documented in a single fossil specimen. Overall, our results provide novel insights into the nature of insect body fossils and additionally shed light on exceptionally preserved terrestrial insect faunas found in marine paleoenvironments.

Keywords: cuticle; Eocene; Fur Formation; insects; melanin; mo-clay; pigment; Stolleklint Clay; structural coloration; Ølst Formation

Citation: Heingård, M.; Sjövall, P.; Schultz, B.P.; Sylvestersen, R.L.; Lindgren, J. Preservation and Taphonomy of Fossil Insects from the Earliest Eocene of Denmark. *Biology* 2022, 11, 395. https://doi.org/10.3390/biology11030395

Academic Editor: Klaus H. Hoffmann

Received: 3 February 2022
Accepted: 24 February 2022
Published: 3 March 2022

Publisher's Note: MDPI stays neutral with regard to jurisdictional claims in published maps and institutional affiliations.

Copyright: © 2022 by the authors. Licensee MDPI, Basel, Switzerland. This article is an open access article distributed under the terms and conditions of the Creative Commons Attribution (CC BY) license (https://creativecommons.org/licenses/by/4.0/).

1. Introduction

Lowermost Eocene deposits of the Limfjord Region, Northwestern Jutland, Denmark, have yielded diverse biotas of exceptionally preserved plant and animal body fossils that frequently retain soft parts, such as feathers and skin [1,2]. The local stratigraphic succession comprises the Fur Formation, a *Konservat-Lagerstätte*, and the underlying, less well known Stolleklint Clay of the Ølst Formation, which together constitute the so-called "mo-clay deposits" [3]. Despite being interpreted as representing a relatively deep marine offshore setting [4], the fine-grained sediments house a wealth of terrestrially derived organisms, of which insects are particularly conspicuous with more than 200 species described to date [1,2]. The insect fossils often preserve fine anatomical details, including segmentation, appendages, wings with well-defined venation, traces of original color patterns and sometimes even residual endogenous biomolecules [1,5].

Insects are currently one of the most ubiquitous and numerically abundant groups of animals on Earth, and they have a fossil record that dates back to the Early Devonian [6,7]. Exceptionally preserved biotas, such as those from the Limfjord Region, act as important "windows" into the evolutionary history of this clade, but are also crucial for understanding taphonomic pathways that may contribute to the retention of delicate anatomical features in the rock record. Recent research has made considerable progress with respect to insect fossilization processes (e.g., [8–19]). However, although previous work on insects from the Stolleklint Clay and Fur Formation has touched upon taphonomic and biostratinomic processes [20–23], most studies have focused on other aspects of the assemblage (e.g., [24–28]). Moreover, in-depth chemical and ultrastructural analyses that can provide valuable information on fossil preservation patterns have so far been conducted almost exclusively on vertebrate remains [29–34], while insects, despite their great abundance, merely have been the subject of a single investigation [5].

In the present contribution, we expand current knowledge on organic preservation by employing an integrated experimental approach to a selection of insect fossils from the Eocene of Denmark. We investigate and illustrate these specimens with the aim of achieving a better understanding of the biostratinomic, taphonomic and diagenetic processes that result in exceptional preservation.

2. Geological Setting

The fossils analysed in this study originate from the Ølst and Fur formations in the Limfjord Region, Northwestern Jutland, Denmark. In this area, the Ølst Formation is represented solely by the Stolleklint Clay—the lowermost unit of the formation [35,36]—which is directly overlain by the Fur Formation. The Stolleklint Clay consists of laminated, clays, whereas the Fur Formation comprises an approximately 60-m-thick sequence of clayey diatomite [37]. The clay sequence formed in a semi-restricted marine basin, well below the wave base under anoxic to dysoxic bottom conditions [35]. The diatomite facies of the Fur Formation have been interpreted as deriving from periodic diatom blooms associated with local upwelling [4,37–39]. Oxygen-depleted bottom conditions are indicated by the generally undisturbed bedding planes, an absence of a benthic biota, and well-preserved, often fully articulated fossils [37]. Both formations contain volcanic ash that originates from eruptions associated with the opening of the North Atlantic Ocean [40]. These layers are numbered in relation to an easily recognizable ash bed (denominated "+1") that occurs in the middle part of the Fur Formation [41]. As of today, almost 200 volcanic ash layers have been recognized in the strata [37]. The processes of dating two of these layers (−17 and +19) have yielded ages of ~55.6 and ~55.4 Ma, respectively [42–44], placing both rock units in the earliest Eocene (Ypresian), during and immediately after the Paleocene-Eocene Thermal Maximum [45,46]. Calcareous concretions are common in the Fur Formation within certain horizons [37]. X-ray diffraction has shown that these consist of low Mg-calcite [47]. Carbon and oxygen isotope compositions further indicate that most of the carbonate has a bacterial origin, being formed as a result of metabolization of organic matter [47].

3. Material and Methods

3.1. Fossil Material

Three insect fossils from the Stolleklint Clay and Fur Formation showing various states of preservation were selected for this study: (1) a pair of isolated but three-dimensional beetle elytra (Coleoptera; FUM-N-17627) collected from a calcium carbonate concretion near ash layer +15 in the Fur Formation; (2) a flattened yet fully articulated wasp (Hymenoptera, Ichneumonidae; FUM-N-11263) preserved in a calcium carbonate concretion, collected from the Fur Formation on the Island of Mors; and (3) a compressed but largely articulated damselfly (Odonata, Zygoptera; FUM-N-10904) found in hardened clays of the Stolleklint Clay. All specimens are housed in collections at Museum Salling, Fur Museum, Fur, Denmark, and were photographed in 96% ethanol using an Olympus SZX16 stereo microscope equipped with an Olympus SC30 digital camera prior to ultrastructural and molecular analyses.

3.2. Scanning Electron Microscopy and Elemental Analysis

All fossils were examined in a Zeiss Supra 40VP FEG-SEM using either an Everhart-Thornley type secondary electron detector (SE2) at an electron energy of 2 keV or a variable pressure secondary electron detector (VPSE) at 15 keV. Elemental analyses and mappings used a X-Max 50 mm^2 silicon drift detector from Oxford Instruments at an electron energy of 15 keV. Complementary imaging and elemental analyses were conducted in a Tescan Mira3 High Resolution Schottky FEG-SEM linked to an energy-dispersive spectrometer (X-MaxN 80, 124 eV, 80 mm^2) from Oxford Instruments using an electron energy of 15 keV. Samples obtained from the beetle elytra (FUM-N-17627) were coated with a 15nanometer-thick gold-palladium film prior to analysis, whereas FUM-N-11263 and FUM-N-10904 were examined uncoated.

3.3. Transmission Electron Microscopy

TEM analyses were performed only on samples from the beetle elytra (FUM-N-17627) because the relict cuticle in FUM-N-11263 and FUM-N-10904 was too thin and spatially incoherent to allow meaningful sampling. Small pieces (~1 mm^2) of elytra were removed from FUM-N-17627 using a sterile scalpel and then dehydrated in a graded ethanol series. Following this procedure, the samples were embedded in epoxy resin (Agar 100) via treatment with acetone. Ultra-thin sections (70 nm) were cut using a Leica EM UC7 ultramicrotome equipped with a diamond knife. The sections were then mounted on copper grids without additional treatment or staining and examined in a JEOL JEM-1400 Plus transmission electron microscope at 100 kV.

3.4. Time-of-Flight Secondary Ion Mass Spectrometry

ToF-SIMS was used for molecular characterization of all fossils. In ToF-SIMS, the sample surface is bombarded by a focused beam of high energy ions, and molecular information is obtained from mass spectra acquired by the secondary ions emitted during this collision process [48]. By scanning the primary ion beam and acquiring mass spectra from each pixel in a selected analysis area, spatially resolved mass spectrometric data can be acquired, which in turn can be presented either as ion images (showing the signal intensity of selected secondary ions across the analysis area) or as mass spectra from selected regions of interest (ROIs) within the analysis area.

ToF-SIMS analyses were carried out in a TOFSIMSIV instrument (IONTOF GmbH, Münster, Germany) using 25 keV Bi$_3^+$ primary ions and low-energy electron flooding for charge compensation. Positive and negative ion data were acquired with the instrument optimized for either high mass resolution (bunched mode, m/Δm $\approx$ 3000, lateral resolution 3–5 µm) or high lateral resolution (m/Δm $\approx$ 300, lateral resolution 0.5–1 µm). Spectra and images were generated using the SurfaceLab software (version 7.1, IONTOF GmbH).

Principal components analysis (PCA) of mass spectral data was conducted using the Solo software (version 7.9.5, Eigenvector Research, Inc., Manson, WA, USA), employing Poisson scaling and prior normalization of the peak intensities to the sum intensity of all included peaks. The analysis included all major "eumelanin" peaks (43 in total) in the mass range *m/z* 48–146 (see [49]). Reference spectra were acquired from pure calcium carbonate (Sigma-Aldrich Sweden AB, Stockholm, Sweden), eumelanin (from *Sepia officinalis*; Sigma-Aldrich Sweden AB) and synthetic eumelanin (Fisher Scientific GTF AB, Göteborg, Sweden).

4. Results

4.1. FUM-N-17627

FUM-N-17627 comprises a pair of three-dimensionally preserved beetle elytra with a faint metallic shine when visualized under conventional light (Figure 1A). At higher magnification, the elytra consist of fragmented, dark-colored matter that is regularly perforated by distinct pits (Figure 1B). SEM analysis further showed that the external surface of the fragments has a hexagonal patterning (Figure 1C) similar to what can be observed in the cuticle of extant beetles, whereas the internal surface is comparatively smooth (Figure 1D,E). In the cross section, the cuticle appears largely amorphous, although the outermost portion displays distinct layering (Figure 1F). When visualized under TEM, the broadly homogenous internal texture is readily apparent (Figure 1G). Notably though, the elytra exhibit an outer lamellar structure with a total thickness of about 500 nm that consists of at least four electron-dense layers separated by thinner, more electron-lucent bands (Figure 1H). In addition, two different electron-lucent features were observed under TEM: long, vertical rifts (Figure 1G,H) and submicron fibrillary structures present predominantly in the ventral (lower) part of the section (Figure 1I–K). Setae- and/or sensilla-like bristles (Figure 1L), and features that may be related to the locking system that attaches the elytra to the thorax (see, e.g., [50,51]) were also evident under SEM (Figure 1M,N).

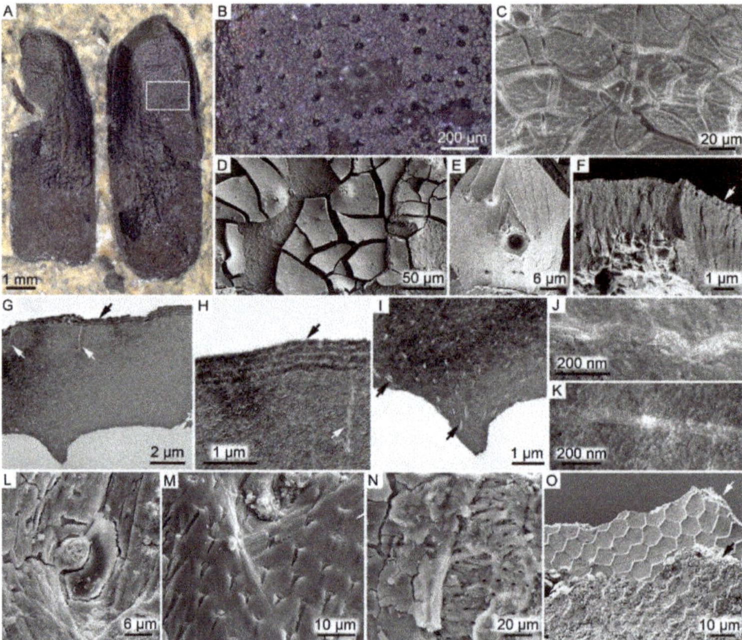

Figure 1. FUM-N-17627 (Coleoptera). (**A**) The fossil elytra in dorsal view prior to sampling, displaying a faint metallic shine under conventional light. (**B**) Higher magnification of the dark matter from the area demarcated in (**A**). (**C**) SEM micrograph of the external (outer) surface of the elytra. Note hexagonal patterning and desiccation cracks. (**D,E**) SEM micrographs of the smooth internal surface of the elytra. Note hexagonally arranged imprints in the sedimentary matrix neighboring the fossil matter. (**F**) SEM micrograph showing a section through the predominantly amorphous cuticular matrix. Note thin layers along the external margin of the elytra (indicated by an arrow). (**G**) TEM micrograph depicting a vertical section through the elytra. Note largely amorphous interior save for thin subvertical rifts (white arrows) and four electron-dense layers that alternate with thinner, electron-translucent bands (black arrow). (**H**) Higher magnification of the thin structures (white arrow)

and epicuticular layering (black arrow). (**I**) TEM micrograph highlighting undulating, electron-lucent fibrillar nanostructures (arrows) in the otherwise largely homogenous cuticular matrix. (**J,K**) Higher magnification of the structures depicted in (**I**) showing individual filaments. (**L–N**) SEM micrographs depicting remains of putative seta, sensilla or microtrichia-like features. (**O**) SEM micrograph showing the two mineral phases identified in direct association with the elytra. The fine texturing with distinct polygonal imprints of the calcium-rich phase (white arrow) is clearly distinguishable from the more coarse-grained iron-rich and silicon-rich phase (black arrow).

Associated with the inferred cuticular remains are two different sedimentary micro-fabrics that display impressions from the fossil. One fabric appears dense with a regular patterning, whereas the other one has less distinct hexagons, appears more granulate (occasionally only as rounded crystals) and is significantly less coherent (Figure 1O).

EDX analysis revealed enrichment of carbon and sulfur in the beetle remains relative to the surrounding sediment (Figure 2). Other investigated elements (e.g., Si, Fe and Ca) are preferentially concentrated to the sediment. The two types of microfabrics that were observed in close association with the elytral fragments differ in composition (Figure 2). The dense mineral is enriched in calcium and oxygen (likely representing the enclosing calcitic concretion), whereas the granulate phase is dominated by silicon, iron and oxygen.

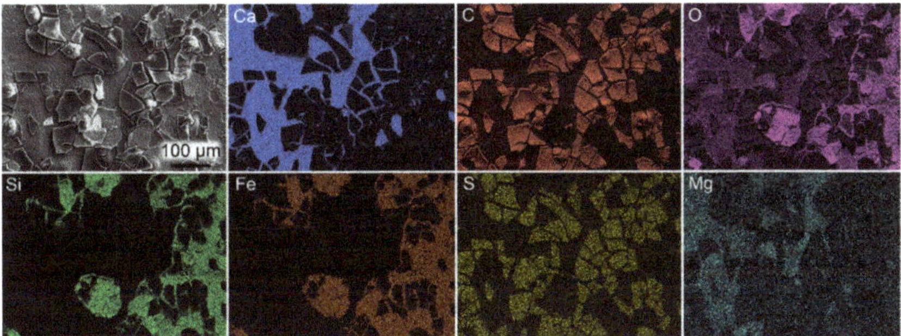

Figure 2. SEM-EDX elemental maps obtained from the fossil beetle elytra (FUM-N-17627). C, carbon (**red**); Ca, calcium (**blue**); Fe, iron (**orange**); Mg, magnesium (**cyan**); O, oxygen (**purple**); S, sulfur (**yellow**); Si, silicon (**green**).

ToF-SIMS data obtained from the surface of the dark matter provided molecular evidence for the presence of the pigment eumelanin (Figure 3). Melanin identification was conducted by comparisons of negative ion spectra acquired specifically from dark-matter areas with reference spectra obtained from synthetic and natural variants of eumelanin. This procedure demonstrated a detailed spectral agreement both with regards to exact m/z values and the relative intensity distribution of all major peaks associated with the eumelanin molecular structure [29,34,52]. Furthermore, spectral comparisons of the fossils and eumelanin standards by PCA that included all eumelanin-related ions indicated peak intensity distributions of the beetle that are consistent with eumelanin standards (Figure S1). Sulfur-containing organic ions were also associated with the dark matter of the fossil (Figure 3C). Negative ion images further showed the presence of a material that displayed high signal intensity from silica-related ions. Mass spectra extracted from theseareas revealed a series of peaks corresponding to mixed $(FeO)_m(SiO_2)_n$ cluster ions, indicating a mixed silicon/iron mineral phase (Figure 3E). Positive ion ToF-SIMS data confirmed the presence of a mixed silicon/iron mineral phase and, furthermore, identified calcite as the calcium-rich mineral (through detailed spectral agreement with a calcium carbonate standard; Figure 3J). Positive ion data further showed the presence of calcium in the form of Ca-containing organic fragment ions on the dark matter surface of the fossil (Figure 3G), although calcium was not observed during our (less sensitive) EDX analyses.

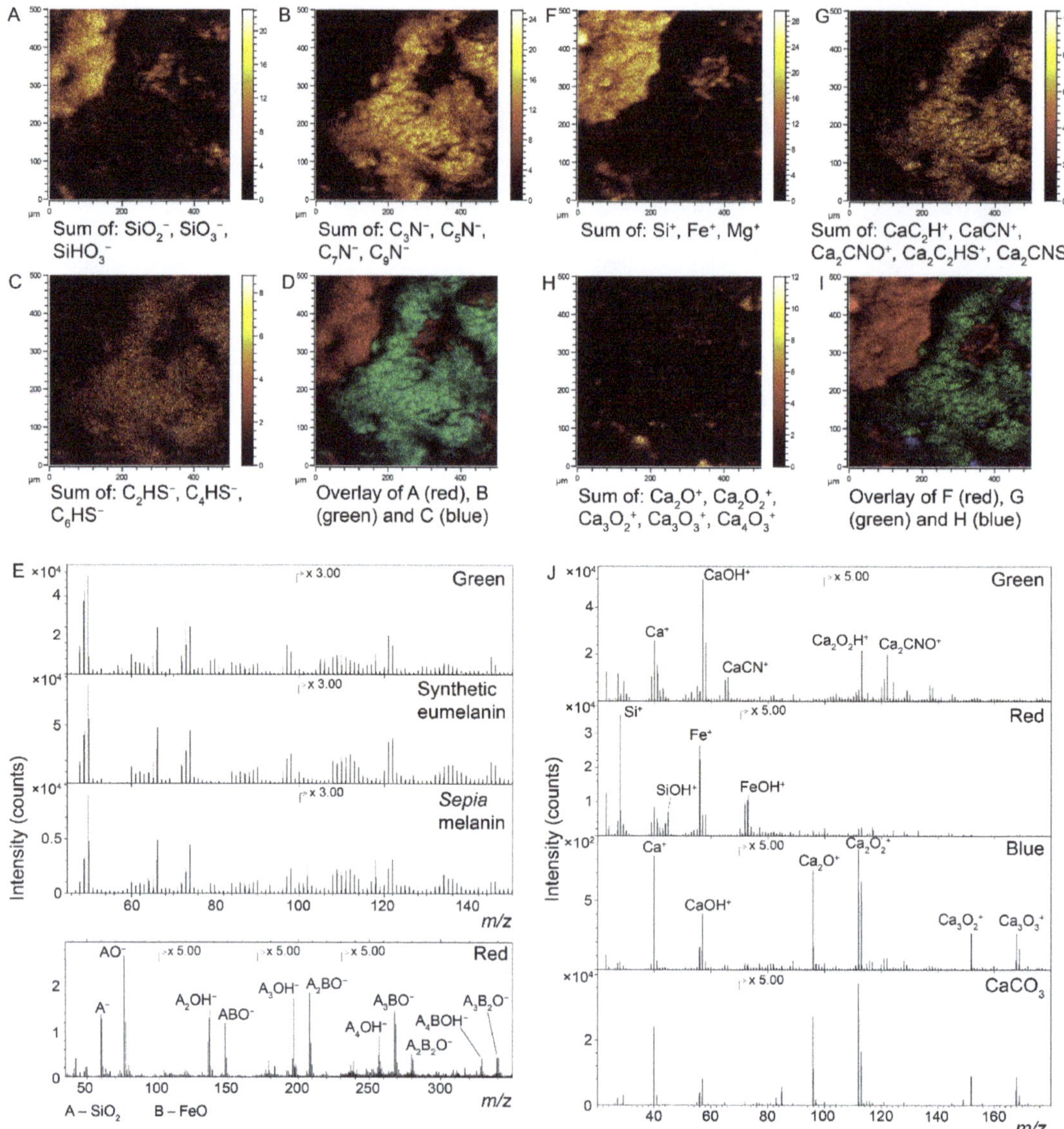

Figure 3. ToF-SIMS characterization of fossil beetle elytra (FUM-N-17627). (**A–D**) Negative ion images representing (**A**) silica, (**B**) eumelanin and (**C**) organic sulfur, together with (**D**) an overlay image in which silica is depicted in red, eumelanin in green and organic sulfur in blue. (**E**) Negative ion spectra obtained from selected ROIs that correspond to the green (eumelanin) and red (silica) areas in (**D**), respectively, where the spectrum from the green area is compared against reference spectra of synthetic and natural eumelanin standards. The spectrum from the red area demonstrates the generation of mixed cluster ions of silica and iron oxide. (**F–I**) Positive ion images representing (**F**) the mixed silica/iron oxide phase, (**G**) Ca-containing organic complexes and (**H**) calcite, together with (**I**) an overlay image in which silica/iron oxide is depicted in red and the Ca-containing organics are green and calcite is in blue. (**J**) Positive ion spectra from the red, green and blue areas in (**I**). The spectrum from the blue area is compared against a reference spectrum of calcite.

4.2. FUM-N-11263 and FUM-N-10904

Both FUM-N-11263 (Hymenoptera; Figure 4) and FUM-N-10904 (Odonata; Figure 5) are preserved as flattened yet fully articulated specimens. Under light microscopy, no obvious internal structures were evident. Instead, the remains visible on the bedding planes are interpreted as being solely cuticular in origin. In the wasp, cuticle fragments are present only in parts of the fossil that are visibly dark-colored (i.e., the head, thorax, wing veins and parts of abdomen). In addition, microtrichia—minute hair-like cuticular protuberances with various functions [53]—can be seen covering nearly the entire wing surface (Figure 4B). In the damselfly, fossilized the remains display color patterns and variations in hue, ranging from black to brown (Figure 5A,B).

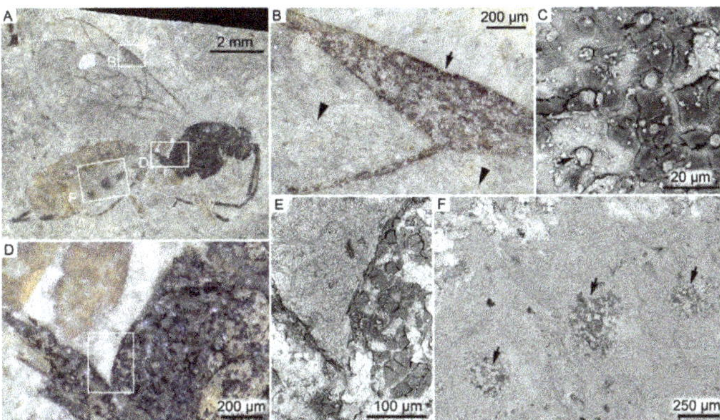

Figure 4. FUM-N-11263 (Hymenoptera). (**A**) Optical microscopy image of the fossil wasp with a dark thorax and head, and yellow-tinted abdomen and legs displayed in lateral view. (**B**) Higher magnification image showing brown pterostigma (arrow), a wing vein and the wing surface with abundant microtrichia (arrowheads). (**C**) SEM micrograph of inferred cuticular fragments in the pterostigma together with broken bristles (arrows). (**D**) Optical microscopy image of the narrow section joining the thorax and abdomen, displaying differences in preservation between the dark-colored matter and yellow-colored legs. (**E**) SEM micrograph of the area demarcated in (**D**) showing inferred cuticular remains. (**F**) SEM micrograph of the ventral side of the abdomen with carbonaceous spots (arrows) presumably representing relict cuticle.

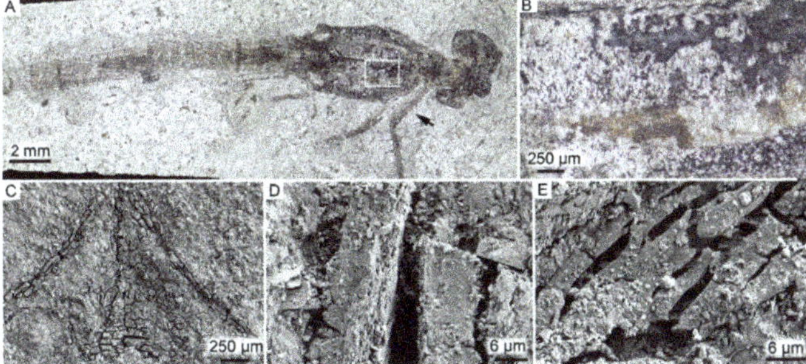

Figure 5. FUM-N-10904 (Odonata). (**A**) The fossil damselfly displayed in oblique dorsal view under conventional light. Note folded wings that rest on the abdomen, and apparent color patterns (e.g., pale shoulder stripes) on the legs and thorax. Arrow indicates location of Figure S3. (**B**) Close-up view of the area demarcated in (**A**) (upper part of the thorax) displaying black- to brown-colored matter. The latter is localized to the shoulder stripe. (**C**) SEM micrograph showing those microstructures that produce the outline of the fossil. (**D**) Higher magnification SEM image of block-like cuticular vestiges. (**E**) SEM micrograph of inferred cuticular residues in one of the legs.

Elemental data from the two specimens (Figure 6) showed that only carbon and sulfur were concentrated to the fossil remains, although this enrichment was rather weak in the wasp. Trace elements, such as iron, occurred in low concentrations and were primarily associated with the sediment.

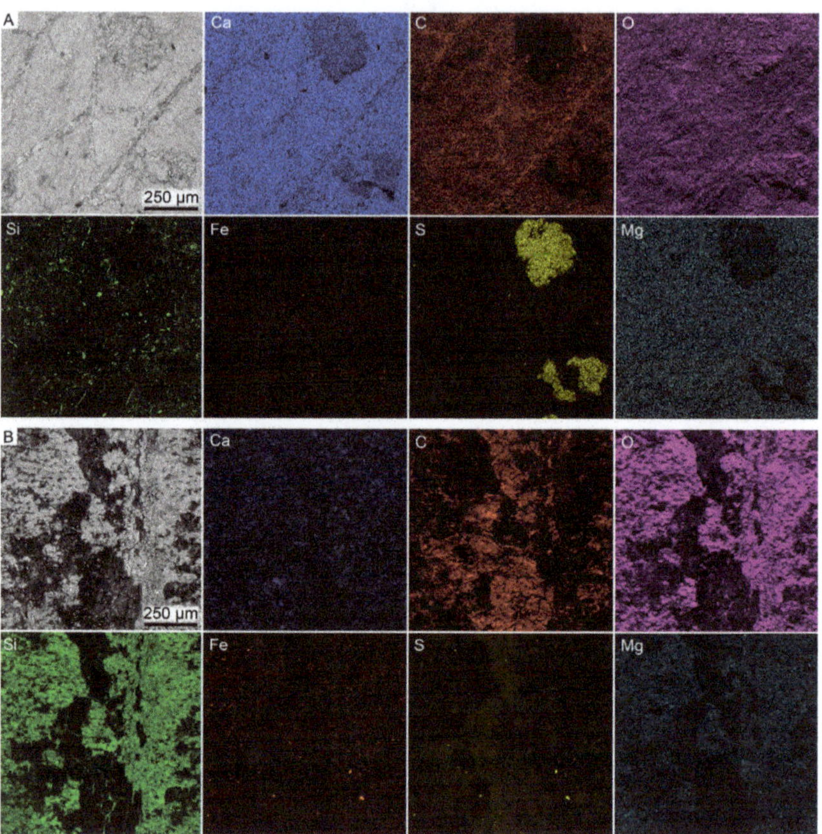

Figure 6. SEM-EDX elemental maps obtained from (**A**) wing veins in the wasp, FUM-N-11263, and (**B**) thorax of the damselfly, FUM-N-10904. C, carbon (red); Ca, calcium (blue); Fe, iron (orange); Mg, magnesium (cyan); O, oxygen (purple); S, sulfur (yellow); Si, silicon (green).

Similarly to the beetle described above, the surfaces of the damselfly residues showed strong spectral agreement with reference spectra of both synthetic and natural eumelanin standards (Figure 7). In the wasp, however, a number of key nitrogen-bearing ions (at *m/z* 50, 74, 98, 122 and 146) showed considerably weaker signal intensities compared to the eumelanin standards, rendering confident molecular determination difficult. Spectral comparisons of the fossils and eumelanin standards by PCA showed peak intensity distributions in the damselfly that were consistent with eumelanin standards, whereas they clearly deviated in the wasp (Figure S1). Additionally, there were only minor differences between the dark and yellow areas of the wasp and adjacent (inorganic) matrix (Figure S2). In the damselfly, the sediment is composed mainly of silicate minerals with additional particulate structures of iron oxide/sulfate (Figures 6B and 7), possibly representing former pyrite framboids. The association between eumelanin and cuticular residues, as well as iron oxide/sulfate and remnant framboids, was further demonstrated by superimposing SEM and ToF-SIMS images of seta on the front leg of the damselfly (Figure S3; location indicated in Figure 5A).

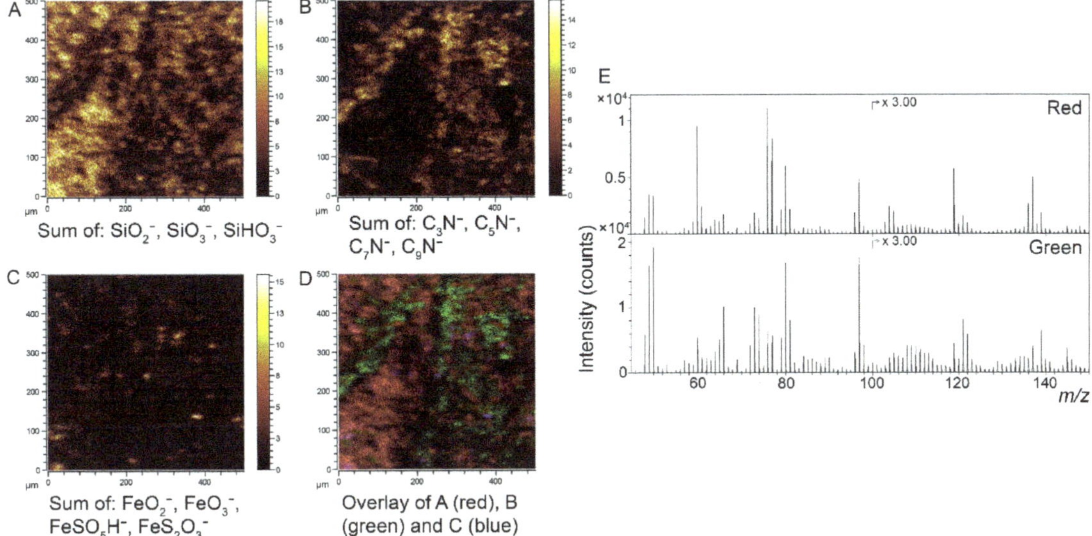

Figure 7. ToF-SIMS characterization of the fossil damselfly (FUM-N-10904). (**A–D**) Negative ion images representing (**A**) silica, (**B**) eumelanin and (**C**) iron oxide/sulfate, together with (**D**) an overlay image in which silica is presented in red, eumelanin in green and iron oxide/sulfate in blue. (**E**) Negative ion spectra from ROIs representing areas with high signal intensity from sediment-related ions (red; top) and eumelanin-associated ions (green; bottom), respectively.

5. Discussion

Insects are generally considered as "soft" organisms that lack naturally biomineralized body parts [54]. However, exceptional preservational conditions occasionally ensure the long-term survival of these otherwise labile animal remains. Decay-prone tissues, e.g., musculature and internal organs, generally require early diagenetic mineral formation to be incorporated in the fossil record [55,56]. Such mineral replacements can result in a high degree of morphological fidelity that include retained three-dimensionality (e.g., [14]). A variety of authigenic minerals are known to be involved in the fossilization of insect carcasses, including pyrite [19,57], calcium carbonate [9], calcium phosphate [58,59] and silica [60]. Alternatively, when authigenic mineralization does not occur, labile tissues can be preserved as organic remains (e.g., [11,15,61,62]).

Although not biomineralized, insect cuticle is a relatively rigid material, something that is attributed to sclerotization—a process in which the exoskeleton is hardened by means of covalent crosslinking between protein molecules [54]. Our microscopic investigation revealed that the insect residues examined herein are exclusively cuticular in origin, indicating that decay progressed until only such comparatively degradation-resistant body parts remained. Our elemental data further showed a concentration of carbon and, to a lesser extent, sulfur, associated with the fossil remnants, indicating that they are predominantly organically preserved. The elevated sulfur levels may indicate diagenetic incorporation of environmental sulfur into the eumelanin molecular structure, a process that has been suggested to enhance the recalcitrance and preservation potential of organic materials ([63] and references therein). As indicated by previously published carbon isotope compositions [47], the sea floor environment facilitated anaerobic microbial decomposition, probably via sulfate reduction [47]. The sulfur produced during this process could have contributed to the precipitation of pyrite [47]. However, although bacterial activity is commonly inferred to be associated with pyritization, fossils preserved in pyrite have yet to be reported from the Stolleklint Clay and Fur Formation to suggest that the conditions for widespread pyritization were not fulfilled in these deposits. Nonetheless, bacterial biofilms have been postulated to have aided the preservation of Fur Formation insects via protection from disintegration while the carcasses were sinking to the bottom [21,22]. Notably though, we did not find any evidence for microbial biofilms in our material.

The bulk of the organic matter in insect exoskeletons comprises a cross-linked chitin-protein complex [54]. Traces of this complex have been reported mainly from comparatively young fossils ([64–67] but see also [68]). In geologically older samples, original components are thought to have transformed into more stable (poly)aromatic and aliphatic compounds [61,69–72]. However, such geopolymers could not be reliably identified in our fossil samples. Instead, our Eocene insects appear to consist predominantly of eumelanin (or breakdown derivatives thereof). ToF-SIMS analyses of the beetle and damselfly provided strong evidence for the presence of preserved eumelanin and, in addition, indicate that this pigment constitutes a major fraction of the organic material in the cuticular residues. In contrast, eumelanin could not be confidently identified in the wasp due the divergent intensity distribution of nitrogen-bearing ions, which suggests further breakdown of the eumelanin biomacromolecule. Melanins in the cuticle of insects not only contribute to visual effects (color patterns) but also have other roles, including immunological defence and UV-protection [73]. Furthermore, cuticle melanogenesis is also intimately linked to the sclerotization process [74,75]. A tyrosine-mediated pathway is responsible for the production of melanins in the cuticle, but when certain dopamine-derived intermediates undergo crosslinking reactions with proteins, the tissue instead hardens (i.e., becomes sclerotized [76]). In most cases, pigments are incorporated into the exoskeleton through this process [75,76]. Despite a growing scientific interest in biochromes and colors of ancient organisms, few studies have hitherto been directed towards pigment residues in fossil insects (we are aware of only three publications in which an original biomolecule has been chemically identified [5,77,78]).

It cannot be completely excluded that remains of other organic components are also present in the fossil. For example, whereas eumelanin is preserved in a state that it is identifiable by ToF-SIMS, other components (chitin, protein and other pigments) could have been broken down into a heterogenous mixture of degradation products (e.g., aliphatic and aromatic hydrocarbons), each with a low concentration that could make it difficult to identify. Furthermore, cuticular components such as chitin would most likely overlap spatially with eumelanin in the sample, thereby preventing the possibility to extract spectra specifically from these other components and, consequently, aggravating their identification.

Interestingly, in all three insects, body parts that likely were originally heavily melanized appear to better preserve cuticular remains. For instance, in the wasp, cuticle fragments were detected only in dark areas of the fossil (i.e., head, thorax and wing veins, as well as in smaller spots on the abdomen). Conversely, the abdomen and legs (where such fragments are not observed) were likely dominated by yellow-colored pigments and, thus, may have lacked substantial eumelanin deposits [54,79]. Similarly, organic structures were not observed to the same extent in the pale shoulder stripes of the damselfly, and the elytra additionally seem to constitute the thick, heavily pigmented dorsal portion only (our imaging analyses did not recover any evidence for the thinner ventral membranous part, hemolymph cavities or trabeculae). These observations not only suggest that eumelanin readily preserves, but also that it may constitute the bulk of the fossil remains or, alternatively, that it provides properties to the fossilized tissues that facilitate their preservation. Accordingly, we hypothesize that the lack of preserved cuticular remains in some areas of the fossils was due to limited initial eumelanin deposits.

In recent years, chemical evidence of eumelanin has been documented in a broad range of animal fossils from the laminated clay of the Stolleklint Clay and in calcareous concretions of the Fur Formation [5,29–31,33,34]. The precise processes responsible for such a widespread presence are, however, not yet fully understood. Still, the high preservation potential of the biochrome is often attributed to its unique molecular structure. Experimental data indicate that it resists enzymatic and chemical degradation mechanisms [80–83], and enhances the strength of tissues and resistance to bacterial decay [84–86]. This may be a consequence of its molecular bond arrangement that has the ability to absorb optical and chemical energy and dissipate it as heat throughout the entire molecular structure, which grants protection against UV-light and suppresses free radicals [29,87,88]. Such factors may contribute to the retention of eumelanin in the fossil record by stabilizing it against decay and providing an inherent resistance to diagenetic alteration [89]. Moreover, the relatively mild geothermal conditions of the Stolleklint Clay and Fur Formation [90,91] presumably limited breakdown (previous studies have indicated that elevated burial temperature is a major factor controlling the preservation of eumelanin [92] as well as other molecular components and structures in arthropods [11,61,66,69]).

The fossil beetle displays a higher degree of structural fidelity than the two other fossils. The rigid cuticle of extant beetles contains internal lamination and multiple pore canals [74]. In our specimen, however, the relict cuticle largely lacks internal laminae and instead appears to be more-or-less amorphous. A similar condition has previously been documented in both experimentally matured cuticles [69] and some fossil beetles [11] and probably reflects degradation and alteration of the chitin–protein complex during diagenesis [69]. Nevertheless, some conspicuous ultrastructural features were still observed in the cuticular remains. The electron-lucent, subvertical structures closely resemble pore canals (see [11], Figure 7C), and the submicron-scale fibril-like features (which, to our knowledge, are previously undocumented in fossil insects) may represent remnant pore canal filaments and/or chitin microfibrils (see [93], Figure 4E). Most notably, however, the cuticle displayed a number of distinct epicuticular layers that correspond to multilayer reflectors that create structural colors in extant insects. Structural colors have a long evolutionary history [10,94,95] and are known to have roles in, e.g., camouflage, mating and visual communication [96–98]. The photonic structures that generate these colors vary extensively in morphology, but the most extensively studied mechanism is the multilayer reflector. These reflectors consist of alternating layers with high and low refractive indices that collectively interact with light [99] and are the most common features that produce structural colors in modern beetles (often in the form of metallic shine or iridescence [97]. The multilayered structures can, for instance, be generated by stratified deposition of pigments, such as melanins or pteridines, and chitin [95,100,101]. In rare cases, such delicate structures are preserved also in fossil insects [10,11,102–104] and potentially can reveal aspects of the original colors and their functions in these ancient animals [10,104]. Notably, all previous reports of fossilized structural coloration in insects are from lacustrine deposits or amber [10,11,102–104], making this the first occurrence of preserved reflectors in a fossil insect preserved in a marine setting. In addition to structural colors, the elytra also provided molecular evidence of eumelanin. To the best of our knowledge, FUM-N-17627 represents the first fossil in which both structural colors and chemical evidence of a pigment have been documented.

Associated with the beetle remains were calcium carbonate-dominated inorganics, representing the entombing concretion, as well as a silicon- and iron-rich mineral. We interpret the former as having formed early during diagenesis and been in close contact with the fossil based on the pristine appearance of the hexagonal cuticular impressions. This interpretation is further supported by our ToF-SIMS data, which show Ca-containing organic fragment ions localized to the dark-colored fossil matter, indicating a calcium-rich coating. Notably, calcium was not detected during our EDX analyses, suggesting that this element is present only at the surface of the dark matter (considering the high surface sensitivity of ToF-SIMS relative to EDX). Calcareous concretions are often considered to have sheltered newly formed fossils from both dissolution and compressional effects [105,106]. Indeed, insect fossils of mo-clay deposits have been previously noted to be especially well-preserved in concretions [1]; however, currently, further research is needed to better understand what role rapid calcium carbonate encapsulation played in the fossilization process. The formation of a silicon- and iron-rich mineral phase is evidently not associated with all insects and, thus, might represent a temporary (and local) event of diatom breakdown and/or an increased concentration of dissolved iron compounds, potentially from nearby volcanic eruptions [56].

6. Conclusions

To gain a better understanding of the retention of insect body fossils in the rock record, we investigated three exceptionally preserved specimens from the lowermost Eocene Stolleklint Clay and Fur Formation of Denmark. Our analyses show that these fossils are all preserved as organic but largely compressed remains of the exoskeleton. Specifically, ToF-SIMS data obtained directly from the cuticle revealed clear evidence of the natural pigment eumelanin, which seemingly dominates the dark-colored residues. Moreover, the beetle elytra exhibit high morphological fidelity with several unique nanostructures. Notably, TEM revealed remnants of an epicuticular multilayer reflector, a biophotonic structure that produces structural colors in modern insects. Our fossils further indicate a potential preservational bias: the preservation potential seems to be greatly diminished in regions of the cuticle that lacked substantial eumelanin deposits. Eumelanin has an inherent resistance to decay and, thus, may remain when most other components have degraded or been lost during diagenesis. The results of our study provide novel insights into the taphonomy of insect assemblages preserved in marine paleoenvironments.

Supplementary Materials: The following supporting information can be downloaded at: https://www.mdpi.com/article/10.3390/biology11030395/s1. Figure S1: PCA analysis; Figure S2: ToF-SIMS spectra of FUM-M-11263; Figure S3: SEM and ToF-SIMS images from FUM-N-10904.

Author Contributions: Conceptualization, M.H., P.S. and J.L.; data curation, M.H. and P.S.; formal analysis, M.H. and P.S.; funding acquisition, P.S. and J.L.; investigation, M.H., P.S. and J.L.; methodology, M.H. and P.S.; project administration, M.H.; resources, M.H., P.S., B.P.S., R.L.S. and J.L.; supervision, J.L.; validation, M.H., P.S., B.P.S., R.L.S. and J.L.; visualization, M.H. and P.S.; writing—original draft, M.H. and P.S.; writing—review and editing, M.H., P.S., B.P.S., R.L.S. and J.L. All authors have read and agreed to the published version of the manuscript.

Funding: Financial support for this project was provided by a Distinguished Young Researcher Grant (Grant number 642-2014-3773; Swedish Research Council) to Johan Lindgren and a Project Grant (Grant number 2019-03731; Swedish Research Council) to Peter Sjövall.

Institutional Review Board Statement: Not applicable.

Informed Consent Statement: Not applicable.

Data Availability Statement: All data generated by this study are available in this manuscript and the accompanying Supplementary Materials.

Acknowledgments: We thank Ola Gustafsson for preparing samples and for the assistance during the transmission electron microscopy analyses and Randolph De La Garza for constructive comments and discussions.

Conflicts of Interest: The authors declare no conflict of interest.

References

1. Bonde, N.; Andersen, S.; Hald, N.; Jakobsen, S.L. *Danekræ—Danmarks Bedste Fossiler*; Gyldendal A/S: Copenhagen, Denmark, 2008; p. 225.
2. Pedersen, G.K.; Pedersen, S.A.S.; Bonde, N.; Heilmann-Clausen, C.; Larsen, L.M.; Lindow, B.; Madsen, H.; Pedersen, A.K.; Rust, J.; Schultz, B.P.; et al. *Moleområdets Geologi—Sedimenter, Fossiler, Askelag og Glacialtektonik*; Dansk Geologisk Forening: Copenhagen, Denmark, 2011; p. 135.
3. Rasmussen, J.A.; Madsen, H.; Schultz, B.P.; Sylvestersen, R.L.; Bonde, N. The lowermost Eocene deposits and biota of the western Limfjord region, Denmark—Field Trip Guidebook. In Proceedings of the 2nd International Mo-Clay Meeting, Nykøbing Mors, Denmark, 2–4 November 2016; p. 35.
4. Bonde, N. Palaeoenvironment in the 'North Sea' as indicated by the fish bearing Mo clay deposit (Paleocene/Eocene), Denmark. *Meded. Werkgr. Tert. Kwart. Geol.* **1979**, *16*, 3–16.
5. Lindgren, J.; Nilsson, D.E.; Sjövall, P.; Jarenmark, M.; Ito, S.; Wakamatsu, K.; Kear, B.P.; Schultz, B.P.; Sylvestersen, R.L.; Madsen, H.; et al. Fossil insect eyes shed light on trilobite optics and the arthropod pigment screen. *Nature* **2019**, *573*, 122–125. [CrossRef]

6.   Grimaldi, D.; Engel, M.S. *Evolution of the Insects*; Cambridge University Press: Cambridge, UK, 2005; p. 755.

7.   Labandeira, C.C.; Beall, B.S.; Hueber, F.M. Early insect diversification: Evidence from a Lower Devonian bristletail from Québec. *Science* **1988**, *242*, 913–916. [CrossRef]

8.   Martínez-Delclòs, X.; Briggs, D.E.G.; Peñalver, E. Taphonomy of insects in carbonates and amber. *Palaeogeogr. Palaeoclimatol. Palaeoecol.* **2004**, *203*, 19–64. [CrossRef]

9.   McCobb, L.M.E.; Duncan, I.J.; Jarzembowski, E.A.; Stankiewicz, B.A.; Wills, M.A.; Briggs, D.E.G. Taphonomy of the insects from the Insect Bed (Bembridge Marls), late Eocene, Isle of Wight, England. *Geol. Mag.* **1998**, *135*, 553–563. [CrossRef]

10. McNamara, M.E.; Briggs, D.E.G.; Orr, P.J.; Noh, H.; Cao, H. The original colours of fossil beetles. *Proc. R. Soc. B* **2012**, *279*, 1114–1121. [CrossRef] [PubMed]

11. McNamara, M.E.; Briggs, D.E.G.; Orr, P.J. The controls on the preservation of structural color in fossil insects. *Palaios* **2012**, *27*, 443–454. [CrossRef]

12. McNamara, M.E.; Briggs, D.E.G.; Orr, P.J.; Gupta, N.S.; Locatelli, E.R.; Qiu, L.; Yang, H.; Wang, Z.; Noh, H.; Cao, H. The fossil record of insect color illuminated by maturation experiments. *Geology* **2013**, *41*, 487–490. [CrossRef]

13. McNamara, M.E. The taphonomy of colour in fossil insects and feathers. *Palaeontology* **2013**, *56*, 557–575. [CrossRef]

14. Barling, N.; Martill, D.M.; Heads, S.W.; Gallien, F. High fidelity preservation of fossil insects from the Crato Formation (Lower Cretaceous) of Brazil. *Cretac. Res.* **2015**, *52*, 605–622. [CrossRef]

15. Pan, Y.; Sha, J.; Fürsich, F.T. A model for organic fossilization of the Early Cretaceous Jehol Lagerstätte based on the taphonomy of "*Ephemeropsis trisetalis*". *Palaios* **2014**, *29*, 363–377. [CrossRef]

16. Greenwalt, D.E.; Rose, T.R.; Siljestrom, S.M.; Goreva, Y.S.; Constenius, K.N.; Wingerath, G. Taphonomy of the fossil insects of the middle Eocene Kishenehn Formation. *Acta Palaeontol. Pol.* **2015**, *60*, 931–947.

17. Bezerra, F.I.; da Silva, J.H.; Miguel, E.D.C.; Paschoal, A.R.; Nascimento, D.R., Jr.; Freire, P.T.C.; Viana, B.C.; Mendes, M. Chemical and mineral comparison of fossil insect cuticles from Crato *Konservat Lagerstätte*, Lower Cretaceous of Brazil. *J. Iber. Geol.* **2020**, *46*, 61–76. [CrossRef]

18. Dias, J.J.; Carvalho, I.S. Remarkable fossil crickets preservation from Crato Formation (Aptian, Araripe Basin), a *Lagerstätten* from Brazil. *J. South Am. Earth Sci.* **2020**, *98*, 102443. [CrossRef]

19. Osés, G.L.; Petri, S.; Becker-Kerber, B.; Romero, G.R.; Rizzutto, M.A.; Rodrigues, F.; Galante, D.; Silva, T.F.; Curado, J.F.; Rangel, E.C.; et al. Deciphering the preservation of fossil insects: A case study from the Crato Member, Early Cretaceous of Brazil. *PeerJ* **2016**, *4*, e2756. [CrossRef]

20. Larsson, S.G. Palaeobiology and mode of burial of the insects of the Lower Eocene Mo-clay of Denmark. *Bull. Geol. Soc. Denmark* **1975**, *24*, 193–209.

21. Rust, J. Biostratinomie von Insekten aus der Fur-Formation von Dänemark (Moler, oberes Paleozän/unteres Eozän). *PalZ* **1998**, *72*, 41–58. [CrossRef]

22. Rust, J.; Andersen, N.M. Giant ants from the Paleogene of Denmark with a discussion of the fossil history and early evolution of ants (Hymenoptera: Formicidae). *Zool. J. Linn. Soc.* **1999**, *125*, 331–348. [CrossRef]

23. Archibald, S.B.; Makarkin, V.N. Tertiary giant lacewings (Neuroptera: Polystoechotidae): Revision and description of new taxa from western north America and Denmark. *J. Syst. Palaeontol.* **2006**, *4*, 119–155. [CrossRef]

24. Andersen, N.M. A fossil water measurer (Insects, Hemiptera, Hydrometridae) from the Paleocene/Eocene of Denmark and its phylogenetic relationships. *Bull. Geol. Soc. Denmark* **1982**, *30*, 91–96. [CrossRef]

25. Andersen, N.M. *Water Striders from the Paleogene of Denmark with a Review of the Fossil Record and Evolution of Semiaquatic Bugs (Hemiptera, Gerromorpha)*; Biologiske Skrifter: Copenhagen, Denmark, 1998; 157p.

26. Rust, J. Fossil record of mass moth migration. *Nature* **2000**, *405*, 530–531. [CrossRef] [PubMed]

27. Bechly, G. A new fossil dragonfly (Anisoptera: Corduliidae) from the Paleocene Fur Formation (Mo clay) of Denmark. *Stutt. Beitr. Naturkd. B* **2005**, *358*, 1–7.

28. Engel, M.S.; Kinzelbach, R.K. A primitive moth from the earliest Eocene Fur Formation ("Mo-clay") of Denmark (*Lepidoptera: Micropterigidae*). *Linzer Biol. Beitr.* **2008**, *40*, 1443–1448.

29. Lindgren, J.; Uvdal, P.; Sjövall, P.; Nilsson, D.E.; Engdahl, A.; Schultz, B.P.; Thiel, V. Molecular preservation of the pigment melanin in fossil melanosomes. *Nat. Commun.* **2012**, *3*, 824–831. [CrossRef]

30. Lindgren, J.; Moyer, A.; Schweitzer, M.H.; Sjövall, P.; Uvdal, P.; Nilsson, D.E.; Heimdal, J.; Engdahl, A.; Gren, J.A.; Schultz, B.P.; et al. Interpreting melanin-based coloration through deep time: A critical review. *Proc. R. Soc. B* **2015**, *282*, 20150614. [CrossRef]

31. Lindgren, J.; Kuriyama, T.; Madsen, H.; Sjövall, P.; Zheng, W.; Uvdal, P.; Engdahl, A.; Moyer, A.E.; Gren, J.A.; Kamezaki, N.; et al. Biochemistry and adaptive colouration of an exceptionally preserved juvenile fossil sea turtle. *Sci. Rep.* **2017**, *7*, 13324. [CrossRef]

32. Vinther, J.; Briggs, D.E.G.; Prum, R.O.; Saranathan, V. The colour of fossil feathers. *Biol. Lett.* **2008**, *4*, 522–525. [CrossRef]

33. Gren, J.A.; Sjövall, P.; Eriksson, M.E.; Sylvestersen, R.L.; Marone, F.; Sigfridsson Clauss, K.G.V.; Taylor, G.J.; Carlson, S.; Uvdal, P.; Lindgren, J. Molecular and microstructural inventory of an isolated fossil bird feather from the Eocene Fur Formation of Denmark. *Palaeotology* **2017**, *60*, 73–90. [CrossRef]

34. Heingård, M.; Sjövall, P.; Sylvestersen, R.L.; Schultz, B.P.; Lindgren, J. Crypsis in the pelagic realm: Evidence from exceptionally preserved fossil fish larvae from the Eocene Stolleklint Clay of Denmark. *Palaeontology* **2021**, *64*, 805–815. [CrossRef]

35. Heilmann-Clausen, C.; Nielsen, O.B.; Gersner, F. 1985. Lithostratigraphy and depositional environment in the upper Paleocene and Eocene of Denmark. *Bull. Geol. Soc. Denmark* **1985**, *33*, 287–323. [CrossRef]

36. Heilmann-Clausen, C. Palæogene aflejringer over danskekalken. In *Danmarks Geologi fra Kridt til i Dag*; Nielsen, O.B., Ed.; Aarhus Geokompendier, 1. Geologisk Institut, Aarhus Universitet: Aarhaus, Denmark, 1995; pp. 70–114.

37. Pedersen, G.K.; Surlyk, F. The Fur Formation, a late Paleocene ash-bearing diatomite from northern Denmark. *Bull. Geol. Soc. Denmark* **1983**, *32*, 43–65. [CrossRef]

38. Bonde, N. Palaeoenvironment as indicated by the "mo-clay formation" (Lowermost Eocene of Denmark). *Tert. Times* **1974**, *2*, 29–36.

39. Pedersen, G.K. Anoxic events during sedimentation of a Palaeogene diatomite in Denmark. *Sedimentology* **1981**, *28*, 487–504. [CrossRef]

40. Larsen, L.M.; Fitton, J.G.; Pedersen, A.K. Paleogene volcanic ash layers in the Danish Basin: Compositions and source areas in the North Atlantic Igneous Province. *Lithos* **2003**, *71*, 47–80. [CrossRef]

41. Bøggild, O.B. Den vulkanske aske i Moleret. *Dan. Geol. Undersøgelse* **1918**, *33*, 84.

42. Westerhold, T.; Röhl, U.; McCarren, H.K.; Zachos, J.C. Latest on the absolute age of the Paleocene-Eocene Thermal Maximum (PETM): New insights from exact stratigraphic position of key ash layers +19 and -17. *Earth Planet. Sci. Lett.* **2009**, *287*, 412–419. [CrossRef]

43. Storey, M.; Duncan, R.A.; Swisher III, C.L. Paleocene-Eocene Thermal Maximum and the opening of the Northeast Atlantic. *Science* **2007**, *316*, 587–589. [CrossRef]

44. Jones, M.T.; Percival, L.M.; Stokke, E.W.; Frieling, J.; Mather, T.A.; Riber, L.; Svensen, H.H. Mercury anomalies across the Palaeocene–Eocene thermal maximum. *Clim. Past* **2019**, *15*, 217–236. [CrossRef]

45. Stokke, E.W.; Liu, E.J.; Jones, M.T. Evidence of explosive hydromagmatic eruptions during the emplacement of the North Atlantic Igneous Province. *Volcanica* **2020**, *3*, 227–250. [CrossRef]

46. Willumsen, P.S. 2004. Palynology of the lower Eocene deposits of northwest Jutland, Denmark. *Bull. Geol. Soc. Den.* **2004**, *52*, 141–157.

47. Pedersen, G.K.; Buchardt, B. The calcareous concretions (cementsten) in the Fur Formation (Paleogene, Denmark): Isotopic evidence of early diagenetic growth. *Bull. Geol. Soc. Den.* **1996**, *43*, 78–86. [CrossRef]

48. Thiel, V.; Sjövall, P. Time-of-flight secondary ion mass spectrometry (TOF-SIMS): Principles and practice in the biogeosciences. In *Principles and Practice of Analytical Techniques in Geosciences*; Grice, K., Ed.; Royal Society of Chemistry: Cambridge, UK, 2015; pp. 122–170.

49. Jarenmark, M.; Sjövall, P.; Ito, S.; Wakamatsu, K.; Lindgren, J. Chemical Evaluation of eumelanin maturation by ToF-SIMS and alkaline peroxide oxidation HPLC analysis. *Int. J. Mol. Sci.* **2021**, *22*, 161. [CrossRef] [PubMed]

50. Sun, J.; Liu, C.; Bhushan, B.; Wu, W.; Tong, J. Effect of microtrichia on the interlocking mechanism in the Asian ladybeetle, *Harmonia axyridis* (Coleoptera: Coccinellidae). *Beilstein J. Nanotechnol.* **2018**, *9*, 812–823. [CrossRef] [PubMed]

51. Gorb, S. Frictional surfaces of the elytra-to-body arresting mechanism in tenebrionid beetles (Coleoptera: Tenebrionidae): Design of co-opted fields of microtrichia and cuticle ultrastructure. *Int. J. Insect Morphol. Embryol.* **1998**, *27*, 205–225. [CrossRef]

52. Lindgren, J.; Sjövall, P.; Carney, R.M.; Uvdal, P.; Gren, J.A.; Dyke, G.; Schultz, B.P.; Shawkey, M.D.; Barnes, K.R.; Polcyn, M.J. Skin pigmentation provides evidence of convergent melanism in extinct marine reptiles. *Nature* **2014**, *506*, 484–488. [CrossRef]

53. Polet, D.T.; Flynn, M.R.; Sperling, F.A.H. A mathematical model to capture complex microstructure orientation on insect wings. *PLoS ONE* **2015**, *10*, e0138282. [CrossRef]

54. Chapman, R.F. *The Insects—Structure and Function*; Cambridge University Press: New York, NY, USA, 2013; 929p.

55. Allison, P.A. The role of anoxia in the decay and mineralization of proteinaceous macro-fossils. *Paleobiology* **1988**, *14*, 139–154. [CrossRef]

56. Briggs, D.E.G. The role of decay and mineralization in the preservation of soft-bodied fossils. *Annu. Rev. Earth Planet. Sci.* **2003**, *31*, 275–301. [CrossRef]

57. Wang, B.; Zhao, F.; Zhang, H.; Fang, Y.; Zheng, D. Widespread pyritization of insects in the early Cretaceous Jehol Biota. *Palaios* **2012**, *27*, 707–711. [CrossRef]

58. Duncan, I.J.; Briggs, D.E.G. Three-dimensionally preserved insects. *Nature* **1996**, *381*, 30–31. [CrossRef]

59. Schwermann, A.H.; Dos Santos Rolo, T.; Caterino, M.S.; Bechly, G.; Schmied, H.; Baumbach, T.; Van Der Kamp, T. Preservation of three-dimensional anatomy in phosphatized fossil arthropods enriches evolutionary inference. *eLife* **2016**, *5*, e12129. [CrossRef] [PubMed]

60. Pierce, W.D. Fossil arthropods of California: No 23. Silicified insects in Miocene nodules from the Calico Mountains. *Bull. South Calif. Acad. Sci.* **1960**, *59*, 40–49.

61. Briggs, D.E.G. Molecular taphonomy of animal and plant cuticles: Selective preservation and diagenesis. *Philos. Trans. R. Soc. B* **1999**, *354*, 7–17. [CrossRef]

62. McNamara, M.E.; Orr, P.J.; Kearns, S.L.; Alcalá, L.; Anadón, P.; Peñalver-Mollá, E. Organic preservation of fossil musculature with ultracellular detail. *Proc. R. Soc. B* **2010**, *277*, 423–427. [CrossRef]

63. McNamara, M.E.; Van Dongen, B.E.; Lockyer, N.P.; Bull, I.D.; Orr, P.J. Fossilization of melanosomes via sulfurization. *Palaeontology* **2016**, *59*, 337–350. [CrossRef]

64. Tanaka, G.; Taniguchi, H.; Maeda, H.; Nomura, S.-I. Original structural color preserved in an ancient leaf beetle. *Geology* **2010**, *38*, 127–130. [CrossRef]

65. Stankiewicz, B.A.; Briggs, D.E.G.; Evershed, R.P.; Flannery, M.B.; Wuttke, M. Preservation of chitin in 25-million-year-old fossils. *Science* **1997**, *276*, 1541–1543. [CrossRef]
66. Stankiewicz, B.A.; Briggs, D.E.G.; Evershed, R.P.; Miller, R.F.; Bierstedt, A. The Fate of chitin in Quaternary and Tertiary strata. In *Nitrogen-Containing Macromolecules in the Bio- and Geosphere*; Stankiewicz, B.A., Van Bergen, P.F., Eds.; American Chemical Society Symposium Series: Washington, DC, USA, 1998; Volume 707, pp. 211–224. [CrossRef]
67. Flannery, M.B.; Stott, A.W.; Briggs, D.E.G.; Evershed, R.P. Chitin in the fossil record: Identification and quantification of D-glucosamine. *Org. Geochem.* **2001**, *32*, 745–754. [CrossRef]
68. Cody, G.D.; Gupta, N.S.; Briggs, D.E.G.; Kilcoyne, A.L.D.; Summons, R.E.; Kenig, F.; Plotnick, R.E.; Scott, A.C. Molecular signature of chitin-protein complex in Paleozoic arthropods. *Geology* **2011**, *39*, 255–258. [CrossRef]
69. Stankiewicz, B.A.; Briggs, D.E.G.; Michels, R.; Collinson, M.E.; Flannery, M.B.; Evershed, R.P. Alternative origin of aliphatic polymer in kerogen. *Geology* **2000**, *6*, 559–562. [CrossRef]
70. Wiemann, J.; Crawford, J.M.; Briggs, D.E.G. Phylogenetic and physiological signals in metazoan fossil biomolecules. *Sci. Adv.* **2020**, *6*, eaba6883. [CrossRef] [PubMed]
71. Gupta, N.S.; Michels, R.; Briggs, D.E.G.; Evershed, R.P.; Pancost, R.D. The organic preservation of fossil arthropods: An experimental study. *Proc. R. Soc. B* **2006**, *273*, 2777–2783. [CrossRef] [PubMed]
72. Stankiewicz, B.A.; Briggs, D.E.G.; Evershed, R.P. Chemical composition of Paleozoic and Mesozoic fossil invertebrate cuticles as revealed by pyrolysis-gas chromatography/mass spectrometry. *Energy Fuels* **1997**, *11*, 515–521. [CrossRef]
73. Shamim, G.; Ranjan, S.K.; Pandey, D.M.; Ramani, R. Biochemistry and biosynthesis of insect pigments. *Eur. J. Entomol.* **2014**, *111*, 149–164. [CrossRef]
74. Noh, M.Y.; Muthukrishnan, S.; Kramer, K.J.; Arakane, Y. Cuticle formation and pigmentation in beetles. *Curr. Opin. Insect Sci.* **2016**, *17*, 1–9. [CrossRef]
75. Sugumaran, M.; Barek, H. Critical analysis of the melanogenic pathway in insects and higher animals. *Int. J. Mol. Sci.* **2016**, *17*, 1753. [CrossRef]
76. Hopkins, T.L.; Kramer, K.J. Insect cuticle sclerotization. *Annu. Rev. Entomol.* **1992**, *37*, 273–302. [CrossRef]
77. Greenwalt, D.E.; Goreva, Y.S.; Siljeström, S.M.; Rose, T.; Harbach, R.E. Hemoglobin-derived porpyrins preserved in a Middle Eocene blood-engorged mosquito. *Proc. Nat. Acad. Sci. USA* **2013**, *110*, 18496–18500. [CrossRef]
78. Labandeira, C.C.; Yang, Q.; Santiago-Blay, J.A.; Hotton, C.L.; Monteiro, A.; Wang, Y.-J.; Goreva, Y.; Shih, C.; Siljeström, S.; Rose, T.R.; et al. The evolutionary converge of mid-Mesozoic lacewings and Cenozoic butterflies. *Proc. R. Soc. B* **2016**, *283*, 20152893. [CrossRef]
79. Badejo, O.; Skaldina, O.; Gilev, A.; Sorvari, J. Benefits of insect colours: A review from social insect studies. *Oecologia* **2020**, *194*, 27–40. [CrossRef]
80. Liu, Y.; Simon, J.D. Isolation and biophysical studies of natural eumelanins: Applications of imaging technologies and ultrafast spectroscopy. *Pigment Cell Res.* **2003**, *16*, 606–618. [CrossRef] [PubMed]
81. Borovanský, J.; Hach, P.; Duchon, J. Melanosome: An unusually resistant subcellular particle. *Cell Biol. Int. Rep.* **1977**, *1*, 549–554. [CrossRef]
82. Ohtaki, N.; Seiji, M. Degradation of melanosomes by lysosomes. *J. Investig. Dermatol.* **1971**, *57*, 1–5. [CrossRef]
83. Riley, P.A. Molecules in focus: Melanin. *Int. J. Biochem. Cell Biol.* **1997**, *29*, 1235–1239. [CrossRef]
84. Bonser, R.H.C. Melanin and the abrasion resistance of feathers. *Condor* **1995**, *97*, 590–591.
85. Burtt, E.H., Jr. An analysis of physical, physiological and optical aspects of avian coloration with emphasis on wood-warblers. *Ornithol. Monogr.* **1986**, *38*, 1–126. [CrossRef]
86. Goldstein, G.; Flory, K.R.; Browne, B.A.; Majid, S.; Ichida, J.M.; Burtt, E.H., Jr. Bacterial degradation of black and white feathers. *Auk* **2004**, *121*, 656–659. [CrossRef]
87. Meredith, P.; Riesz, J. Radiative relaxation quantum yields for synthetic eumelanin. *Photochem. Photobiol.* **2004**, *79*, 211–216. [CrossRef]
88. McGraw, K.J. The antioxidant function of many animal pigments: Are there consistent health benefits of sexually selected colourants? *Anim. Behav.* **2005**, *69*, 757–764. [CrossRef]
89. Glass, K.; Ito, S.; Wilby, P.R.; Sota, T.; Nakamura, A.; Bowers, R.; Vinther, J.; Dutta, S.; Summons, R.; Briggs, D.E.G.; et al. Direct chemical evidence for eumelanin pigment from the Jurassic period. *Proc. Natl. Acad. Sci. USA* **2012**, *109*, 10218–10223. [CrossRef]
90. McNamara, M.E.; Briggs, D.E.G.; Orr, P.J.; Field, D.J.; Wang, Z. Experimental maturation of feathers: Implications for reconstructions of fossil feather colour. *Biol. Lett.* **2013**, *9*, 20130184. [CrossRef] [PubMed]
91. Vickers, M.L.; Lengger, S.K.; Bernasconi, S.M.; Thibault, N.; Schultz, B.P.; Fernandez, A.; Ullman, C.V.; Mccormack, P.; Bjerrum, C.J.; Rasmussen, J.A.; et al. Cold spells in the Nordic Seas during the early Eocene Greenhouse. *Nat. Commun.* **2020**, *11*, 4713. [CrossRef] [PubMed]
92. Glass, K.; Ito, S.; Wilby, P.R.; Sota, T.; Nakamura, A.; Bowers, R.; Miller, K.E.; Dutta, S.; Summons, R.E.; Briggs, D.E.G.; et al. Impact of diagenesis and maturation on the survival of eumelanin in the fossil record. *Org. Geochem.* **2013**, *64*, 29–37. [CrossRef]
93. Van Der Kamp, T.; Riedel, A.; Greven, H. Micromorphology of the elytral cuticle of beetles, with an emphasis on weevils (Coleoptera: Curculionoidea). *Arthropod Struct. Dev.* **2016**, *45*, 14–22. [CrossRef]
94. Vinther, J.; Briggs, D.E.G.; Clarke, J.; Mayr, G.; Prum, R.O. Structural coloration in a fossil feather. *Biol. Lett.* **2009**, *6*, 128–131. [CrossRef]

95. Parker, A.R. 515 million years of structural colour. *J. Opt. A Pure Appl. Opt.* **2000**, *2*, R15–R28. [CrossRef]
96. Parker, A.R. The diversity and implications of animal structural colours. *J. Exp. Biol.* **1998**, *201*, 2343–2347. [CrossRef]
97. Seago, A.E.; Brady, P.; Vigneron, J.-P.; Schultz, T.D. Gold bugs and beyond: A review of iridescence and structural colour mechanisms in beetles (Coleoptera). *J. R. Soc. Interface* **2009**, *6*, S165–S184. [CrossRef]
98. Wilts, B.D.; Michielsen, K.; Kuipers, J.; De Raedt, H.; Stavenga, D.G. Brilliant camouflage: Photonic crystals in the diamond weevil, *Entimus imperialis. Proc. R. Soc. B* **2012**, *279*, 2524–2530. [CrossRef]
99. Kinoshita, S.; Yoshioka, S.; Miyazaki, J. Physics of structural colors. *Rep. Prog. Phys.* **2008**, *71*, 76401–76431. [CrossRef]
100. Onelli, O.; van de Kamp, T.; Skepper, J.N.; Powell, J.; dos Santos Rolo, T.; Baumbach, T.; Vignolini, S. Development of structural colour in leaf beetles. *Sci. Rep.* **2017**, *7*, 1373. [CrossRef] [PubMed]
101. Stavenga, D.G.; Leertouwer, H.L.; Hariyama, T.; De Raedt, H.A.; Wilts, B.D. Sexual dichromatism of the damselfly *Calopteryx japonica* caused by a melanin-chitin multilayer in the male wing veins. *PLoS ONE* **2012**, *7*, e49743. [CrossRef] [PubMed]
102. Parker, A.R.; Mckenzie, D. The cause of 50 million-year-old colour. *Proc. R. Soc. B* **2003**, *270*, S151–S153. [CrossRef] [PubMed]
103. Parker, A.R.; Hegedus, Z.; Watts, R.A. Solar-absorber antireflector on the eye of an Eocene fly (45 Ma). *Proc. R. Soc. B* **1998**, *265*, 811–815. [CrossRef]
104. McNamara, M.E.; Briggs, D.E.G.; Orr, P.J.; Wedmann, S.; Noh, H.; Cao, H. Fossilized biophotonic nanostructures reveal the original colors of 47-million-year-old moths. *PLoS Biol.* **2011**, *9*, e1001200. [CrossRef]
105. Dyke, G.; Lindow, B. Taphonomy and abundance of birds from the Lower Eocene Fur Formation of Denmark. *Geol. J.* **2009**, *44*, 365–373. [CrossRef]
106. De La Garza, R.G.; Madsen, H.; Eriksson, M.E.; Lindgren, J. A fossil seaturtle (Reptilia, Pan-Cheloniidae) with preserved soft tissues from the Eocene Fur Formation of Denmark. *J. Vertebr. Paleontol.* **2021**, *41*, e1938590. [CrossRef]

Article

Crystallographic Texture of the Mineral Matter in the Bivalve Shells of *Gryphaea dilatata* Sowerby, 1816

Alexey Pakhnevich [1,2], **Dmitry Nikolayev** [2] **and Tatiana Lychagina** [2,*]

1 Borissiak Paleontological Institute, Russian Academy of Sciences, 117647 Moscow, Russia
2 Frank Laboratory of Neutron Physics, Joint Institute for Nuclear Research, 141980 Dubna, Russia
* Correspondence: lychagina@jinr.ru

Simple Summary: A new for paleontology method has been applied to study the orientation distribution of the crystals that compose the fossils of mollusk shells. The method is based on the use of neutrons with high penetrating power and makes it possible to study bulk shells without destroying them. In this work, we studied how the habitat conditions and the process of fossilization influenced the distribution of shell crystallite orientations. It was possible to establish a relationship between the distribution of orientations and the shape of the shells.

Abstract: It is assumed that the crystallographic texture of minerals in the shells of recent and fossil mollusks is very stable. To check this, it is necessary to examine the shells of animals that had lain in sediments for millions of years and lived in different conditions. It is revealed that the crystallographic texture of calcite in the shells of *Gryphaea dilatata* from deposits from the Middle Callovian–Lower Oxfordian (Jurassic), which lived in different water areas, is not affected by habitat conditions and the fossilization process. The crystallographic texture was studied using pole figures measured by neutron diffraction. The neutron diffraction method makes it possible to study the crystallographic texture in large samples—up to 100 cm^3 in volume without destroying them. The recrystallization features of the *G. dilatata* valve, which affect the crystallographic texture, were discovered for the first time. This is determined from the isolines appearance of pole figures. The crystallographic texture of the *G. dilatata* mollusks' different valves vary depending on their shape. The pole figures of calcite in the thick-walled valves of *G. dilatata*, *Pycnodonte mirabilis*, and *Ostrea edulis* are close to axial and display weak crystallographic texture.

Keywords: *Gryphaea dilatata*; crystallographic texture; pole figures; neutron diffraction; recrystallization; thick-walled shells

Citation: Pakhnevich, A.; Nikolayev, D.; Lychagina, T. Crystallographic Texture of the Mineral Matter in the Bivalve Shells of *Gryphaea dilatata* Sowerby, 1816. *Biology* **2022**, *11*, 1300. https://doi.org/10.3390/biology11091300

Academic Editors: Zhifei Zhang and Etsuro Ito

Received: 24 June 2022
Accepted: 25 August 2022
Published: 31 August 2022

Publisher's Note: MDPI stays neutral with regard to jurisdictional claims in published maps and institutional affiliations.

Copyright: © 2022 by the authors. Licensee MDPI, Basel, Switzerland. This article is an open access article distributed under the terms and conditions of the Creative Commons Attribution (CC BY) license (https://creativecommons.org/licenses/by/4.0/).

1. Introduction

Bivalve mollusks from the family Gryphaeidae were a frequently occurring fauna element in the seas of the Middle and Late Jurassic of the East European Platform. The most common and numerous were the species of the genus *Gryphaea* Lamarck, 1801. The shells of these mollusks were large and thick-walled. At the same time, the left valve is strongly convex with a curved umbo, while the right one is flattened. Such a shell was necessary as a passive defense against predators (on 169 valves of *Gryphaea dilatata* Sowerby, 1816 from the Callovian deposits near the village of Sukhochevo (Oryol region, Russia), where only one trace of a predator attack was found); it well withstood abrasion in shallow water due to wave activity, and thickened walls coalesced better when jars were formed. The microstructure of gryphaea valves is shown through numerous layers of plates. Such a shell organization turned out to be very evolutionarily successful; therefore, the genus existed from the Late Triassic to the Paleogene [1]. A similar adaptation was also characteristic of the Cenozoic Gryphaeidae, and it also manifested itself within the closely-related bivalve family, Ostreidae.

The shells of recent and fossil *Ostrea edulis* Linnaeus, 1758 (Ostreidae family), consist of calcite. The microstructures of the family Ostreidae representatives are well studied [2–4]. The crystallographic texture of calcite is also known for *O. edulis* valves [5]. Earlier, it was shown [6] that there is no direct connection between the type of microstructure and crystallographic texture, which is why we do not pay much attention to microstructure in the present study.

Crystallographic texture is a collection of orientations of a polycrystalline sample. The shells consist of calcium carbonate polycrystals and are characterized by the anisotropy of physical and mechanical properties, which is closely related to the preferred orientation (texture) of their grains. Texture is formed during shell growth and can be influenced by various environmental factors. Recently, an attempt has been made to study the effects of the elemental content of the shells of the bivalve mollusks, *Mytilus galloprovincialis* Lamarck, 1819, on their crystallographic texture [7].

Quantitative information about crystallographic texture is contained in measured pole figures, which are two-dimensional distributions of relative volumes for specific crystallographic directions on a unit sphere [8]. Values on a pole figure are given in mrd (multiple random distribution) units. The value 1 corresponds to the uniform isotropic distribution of crystalline orientations. Most pole figure measurements of shells are carried out by means of X-rays or electron backscatter diffraction (EBSD) [9–15]. However, these techniques are very local because of the low penetrating power of these radiations into matter. The much greater penetration depth of neutrons in the sample material allows for the investigation of centimeter size samples in transmission geometry, which yields non-destructive texture measurements. Moreover, X-ray or EBSD techniques allow us to study a relatively small part of a valve, while neutron diffraction allows us to measure almost the whole bulk shell [16]. Complete pole figures can be obtained without intensity corrections by means of neutron diffraction; the grain statistics are better, which is of particular interest for coarse-grained samples [17].

The arrangement of crystals in recent shells of *O. edulis* is not highly ordered, i.e., the texture is not too sharp—up to 2.53 mrd—compared to calcite from the shells of recent mussels, such as *M. galloprovincialis*—which are up to 12.53 mrd [18]. It should be emphasized that these numbers were obtained for the whole bulk shells. It can be suspected that the thickened shell of fossil mollusks may have a more ordered texture, which is necessary to preserve its integrity. In this regard, gryphaea shells were selected from three localities in order to solve several problems of the crystallographic texture evaluation of the mineral matter of these mollusks' valves.

Previously, we compared the textures of calcite and aragonite in the shells of recent mussels and oysters from different parts of the areal and geologic ages, but only with a difference of 30 thousand years [18].

The purpose of this work is to compare the crystallographic texture of the valve's mineral matter of the bivalve mollusks, *G. dilatata*, from three remote localities that have been formed under different diagenetic conditions.

The objectives of the study included:

1. Determination of the crystallographic texture sharpness of the *G. dilatata* mineral substance;
2. Assessment of the influence of various habitat conditions and diagenetic processes on the texture;
3. Reveal the crystallographic texture dependence on the valve shape;
4. Reveal the texture similarity in two representatives of the Gryphaeidae family;
5. A comparison of recent and fossil valves of *O. edulis*;
6. Reveal the common features of the crystallographic texture in thick-walled shells.

2. Materials and Methods

The bivalve mollusk shells of *G. dilatata* were chosen as objects to study. This is a widespread species whose shells are found in large amounts in Jurassic deposits. Large samples of adult mollusks have been studied. They were measured one at a time since they contained sufficient substance for analysis. Both convex-left and flattened-right valves were studied.

Figure 1 displays the locations of the collected samples. Valves from the Mikhailovsky quarry near Zheleznogorsk Town (Kursk region, Russia) from the deposits of the Callovian stage of the Middle Jurassic were studied. The valves collected from the quarry near the village of Sukhochevo (Oryol region, Russia) are also from Callovian deposits. Only left valves of this type were found in the urban quarry of Roshal Town (Moscow region, Russia) (Figure 2a–f). All valves were easily extracted, so they did not need to be cleaned. The complex of accompanying gryphaea fauna was analyzed to determine the geologic age.

Figure 1. Collection locations of *Gryphaea dilatata* Sowerby, 1816: 1—Kursk region, near the town of Zheleznogorsk, Mikhailovsky quarry; 2—Oryol region, Kromy district, sand quarry near the village of Sukhochevo; 3—Moscow region, Shatura district, sand quarry on the outskirts of the town of Roshal.

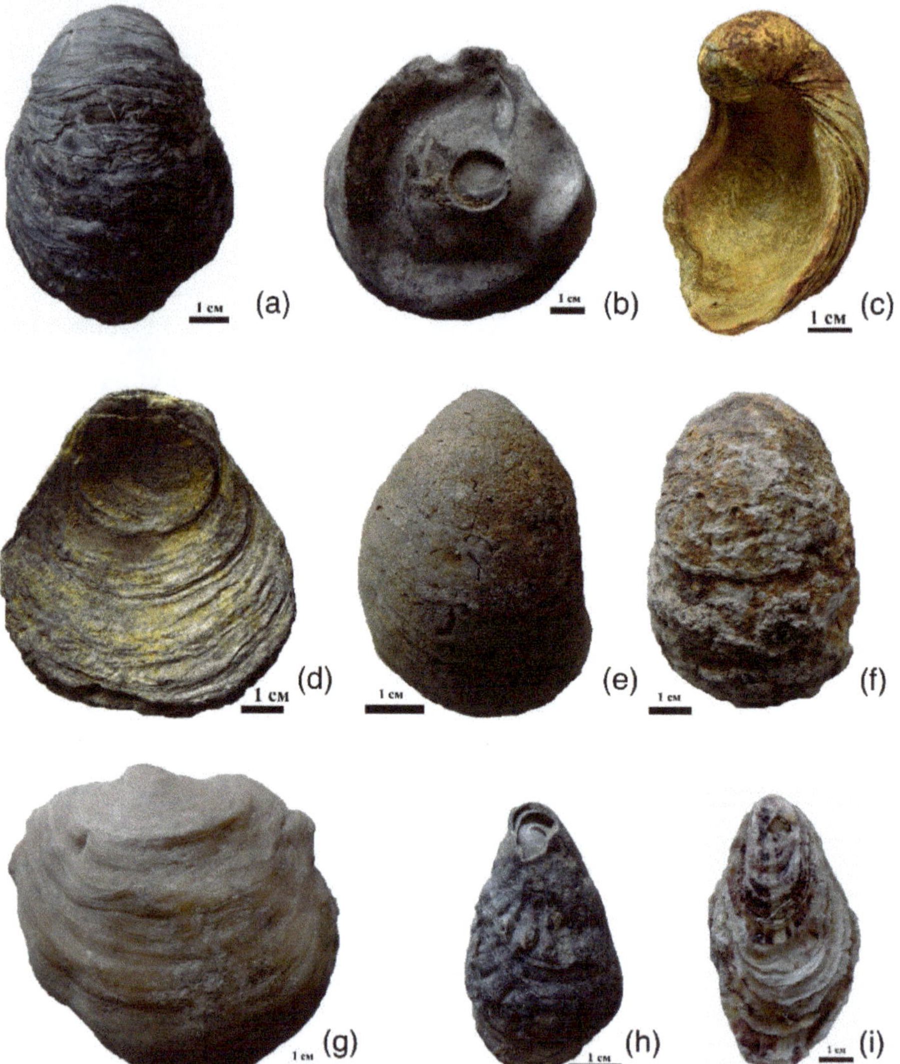

Figure 2. Valves of studied *Gryphaea dilatata* Sowerby, 1816: (**a**) left valve, (**b**) right valve, Kursk region, near Zheleznogorsk, Mikhailovsky quarry, Middle Jurassic, Middle–Upper Callovian; (**c**) left valve, (**d**) right valve Oryol region, Kromy district, sand quarry near the village of Sukhochevo, Middle Jurassic, Middle Callovian; (**e**) left valve, Moscow region, Shatura district, sand quarry on the outskirts of the town of Roshal, Middle Jurassic, Callovian–Upper Jurassic, Lower Oxfordian; (**f**) left valve with a white friable layer on the surface, Oryol region, Kromy district, sand quarry near the village of Sukhochevo, Middle Jurassic, Middle Callovian; (**g**) left valve of *Pycnodonte mirabilis* Rousseau, 1842, Crimea Peninsula, Upper Cretaceous, Maastrichtian; (**h**) right valve of *Ostrea edulis* Linnaeus, 1758, the coast of the Arabatsky Gulf of the Azov Sea in the town of Shchelkino (Crimea Peninsula), Pleistocene, Karangat deposits; (**i**) right valve of *Ostrea edulis* Linnaeus, 1758, coast near the village of Maly Utrish, Black Sea, recent.

To compare with gryphaeas, the left valve of *Pycnodonte mirabilis* Rousseau, 1842, was found (Figure 2g). The genus, *Pycnodonte* Fischer von Waldheim, 1835, also belongs to the family, Gryphaeidae [1]. The valve comes from the Upper Cretaceous, the Maastrichtian deposits of the Crimea Peninsula. A comparison with recent (coast near the village of Maly Utrish, Black Sea; Portugal, port of Lagos) and fossil oysters (coast of the Arabat Bay of the Azov Sea near the town of Shchelkino, Crimea Peninsula; Chushka Spit, coast of the Taman Peninsula—Pleistocene, Karangatian layers) of *O. edulis* was carried out according to a previously published paper [18] (Figure 2h,i).

The valves were glued to glass pins with a special two-component adhesive. The pole figures presented in this paper were measured at the Frank Laboratory of Neutron Physics of the Joint Institute for Nuclear Research (Dubna, Russia). The SKAT texture diffractometer located on channel 7-A of the IBR-2 pulsed nuclear reactor was used [5]. The seventh channel has a long flight base (more than 100 m long), which leads to the good spectral resolution of SKAT. Due to the pulse nature of the neutron flux, the diffractometer implements a time-of-flight measurement technique. SKAT consists of a detector ring (diameter of 2 m) on which 19 detector-collimator complexes are located at the same scattering angle of $2\theta = 90°$.

A sample is rotated 360° with a step of 5° about the horizontal axis with an angle of 45° with respect to the incident neutron beam. Rotation is carried out using a goniometer. Thus, 1368 diffraction patterns are recorded during each sample measurement. It should be noted that due to the time-of-flight technique, the pole figures of all minerals (phases) present in the sample are measured simultaneously, i.e., extracted from the same patterns. A neutron beam cross section of 50×90 mm makes it possible to measure large samples of up to 100 cm^3. Such experimental conditions provide the following advantages: measurements at the same scattering angles lead to the same position of the same diffraction reflections for all detectors, which allows us to avoid corrections. Moreover, since the detector modules are located on the ring at angles of 0° to 180°, one rotation about the horizontal axis of the goniometer is sufficient to measure complete pole figures. One more advantage of neutron texture measurements is that the sample surface does not need to be prepared in a special way. This is due to the results being influenced the whole volume of a sample— and not just the surface—because of the high penetration power of neutrons. The most intense, non-overlapped diffraction reflections from each of the recorded patterns were analyzed using the Pole Figure Extractor program [19] to determine the distribution of the corresponding crystallographic planes of the valves. The intensity of one reflection, corresponding to a crystallographic plane with certain Miller indices, gives one point on the pole figure, which is indicated by these indices. To extract the pole figures, we used an approach based on the approximation of diffraction reflections by a bell-shaped distribution, since the signal-to-background ratio was quite high [20]. All pole figures were normalized and smoothed with the same parameter [21]. The most intense diffraction reflections that correspond to crystallographic planes with Miller indices (0006) and (10–14) for calcite were analyzed. The more ordered the mineral crystals were, the higher the intensity was on the pole figure (pole density), which is expressed in units of isotropic orientation distribution (multiple random distribution, mrd). An increase in pole figure intensity is interpreted as texture intensification. When the value of the pole density is equal to 1, it means that the corresponding crystallographic planes are uniformly distributed in the sample in all directions. The analysis was carried out according to the interpretation of the maximum sharpness and the isolines pattern of the pole figures with the Miller indices, (0006) and (10–14). The pole figures are presented by stereographic projections. The stereographic projection is obtained from the spherical one by means of projecting the sphere point onto the equatorial plane. The position of a given pole on the sphere is commonly characterized in terms of two angles. The angle χ describes the inclination of the pole, where $\chi = 0°$ is the north pole of the unit sphere and the angle φ characterizes the rotation of the pole, as shown in Figure 3.

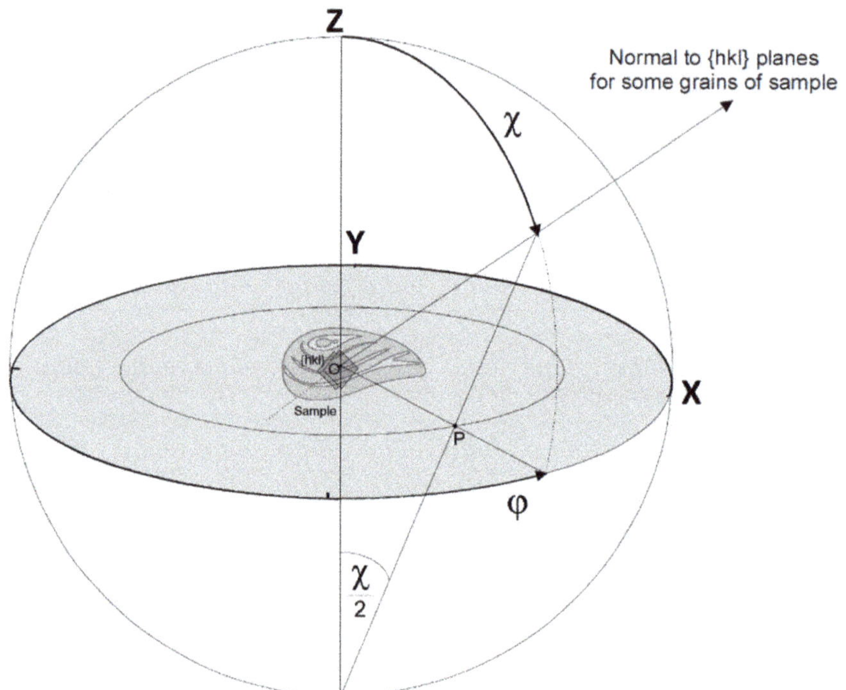

Figure 3. Pole figure construction using stereographic projection. XYZ is a Cartesian coordinate system. χ is the polar angle of a direction in the space, whereas φ is the azimuth angle. Point P is the stereographic projection of a point from the northern hemisphere.

To project a point from the northern hemisphere, it should be connected by a ray with the south pole. The intersection of the projecting ray with the projection plane gives the stereographic projection of this point (point P in Figure 3). The pole figures are usually collected by varying the angle of rotation φ and the tilt angle χ, where φ and χ are varied from 0–360° and 0–90°, respectively. Thus, the intensity of a particular Bragg reflection measured under varying sample orientations yields an intensity distribution I (hkl) as a function of the χ and φ angles over a sphere, which defines the crystallographic orientation distribution of the grains in the shell. This way, the pole figure represents a variation of the pole density for a selected set of crystallographic planes.

The shell microstructure was recorded using Tescan Vega 2 (Czech Republic, Brno) and CamScan-4 (Cambridge, UK) scanning electron microscopes (Borissiak Paleontological Institute, Russian Academy of Science, PIN RAS). Belemnite rostra was scanned with gold spraying, whereas the bivalve mollusks were scanned without spraying in a low vacuum. We used the results of the X-ray microtomography of ferruginous belemnite rosters from the Upper Jurassic deposits of the Volgian stage (Kuntsevo-Filyevsky Park, Moscow, Russia), using a Skyscan 1172 microtomograph (Belgium, Kontich) (PIN RAS), to demonstrate the effect of iron minerals on calcite. The measurements were carried out with a resolution of 3–30 µm, current of 100 mA, voltage of 103–104 kV, rotation angle of 0.7°, rotation by 180°, frame averaging of 8, random movement of 10, and filter Al (1 mm). The NRecon program, version 1.6.4.1 (Bruker, Belgium, Kontich) was used for visualization.

The photographs for Figure 2c,d were taken by S.V. Bagirov (PIN RAS), and the rest were by A.V. Pakhnevich.

3. Results

3.1. Determination of the Shells' Geologic Age

The species, *G. dilatata,* were found in the Middle and Upper Callovian and Lower Oxfordian deposits of the Central part of European Russia [22]. Information about the geologic age of the valves was obtained from fauna assemblages collected together with *G. dilatata,* and from the literature [22,23]. The associated fauna was not extracted along with the valves collected in the Mikhailovsky quarry (Kursk region), so their age was determined according to the data from the book [22]. In this locality, *G. dilatata* are found in clay deposits above the Lower Callovian, that is, they may belong to either the Middle or Upper Callovian layers. Together with mollusk shells, the following fauna complex was found in the locality near the village of Sukhochevo, namely bivalves such as *G. lucerna* Trautschold, 1862; *G. russiensis* Gerasimov, 1984; *Nanogyra nana* Sowerby, 1822; *Pholadomya murchisoni* Sowerby, 1827; *Deltoideum hemideltoideum* Lahusen, 1883, *Goniomya* sp.; traces of vital activity from drilling bivalve, *Lithophaga antiquissima* Eichwald, 1860; rare fragmentary small rostra of belemnites *Cylindroteuthis okensis* Nikitin, 1885; and fragments of ammonite shells from *Erymnoceras* sp.; *Indosphinctes mutatus* Trautschold, 1862; *Hecticoceras* sp. (determined from a gryphaea imprint); and *Kosmoceras* sp. (one of the specimens was identified by a gryphaea imprint). There are numerous drillings on the gryphaea shells. There were also fragments of charred, silicified, and pyritized wood. Due to the presence of species such as *P. murchisoni, I. mutatus, Erymnoceras* sp., *D. hemideltoideum* and *C. okensis*, the Middle Callovian age (Middle Jurassic) was determined.

The complex of fauna and flora from the location on the outskirts of the town of Roshal is different. It contains trace fossils (burrow filling); fragments of shells and internal casts; imprints of ammonites, of which only *Amoeboceras* sp. has been identified; rare tubes of annelids, such as Serpulidae; single teeth from sharks, such as *Sphenodus stschurowskii* Kiprijanoff, 1880; numerous shells from scaphopod mollusks, *Laevidentalium gladiolus* Eichwald, 1846; silicified fragments from crinoid stems (probably redeposited from Carboniferous); numerous drillings of gryphaea shells, among which there are barnacles' drillings; bivalves such as *Astarte cordata* Trautschold, 1861; *Cosmetodon* sp.; *Astarte* sp. cf. *A. duboisiana* d'Orbigny, 1845; *A.* cf. *panderi* Rouillier, 1847 (as well as numerous unidentifiable shell fragments belonging to other species); gastropods such as *Actaeon frearsiana* d'Orbigny, 1845; *Bourgetia reticulata* Gerasimov, 1992; *Oonia calypso* d'Orbigny, 1850; belemnites *C. okensis; C. oweni* Pratt, 1844; *C. subextensa* Nikitin, 1884; *Hibolites hastatus* Blainville, 1827; *Hibolites* sp.; and wood fragments. Most likely, the fossil remains of the Callovian (Middle Jurassic) and Oxfordian (Upper Jurassic) are mixed in the deposit, possibly even with Volgian forms (Upper Jurassic), but this is unconfirmed data since the finding of the Volgian *Astarte* sp. cf. *A. duboisiana* and *A.* cf. *panderi* does not inspire complete confidence. Species such as *A. cordata, Ac. frearsiana,* and *Oo. calypso* are Callovian–Oxfordian [22,23]. However, there are also pure Callovian species, for example, *B. reticulata, C. okensis, C. oweni, C. subextensa,* and *H. hastatus*, as well as pure Oxfordian ones, for example, *S. stschurowskii* and *La. gladiolus*. It is highly likely that the gryphaeas are of early Oxfordian age, since the fauna found in the Oxfordian deposits predominates the exclusively Callovian deposits. Therefore, the geologic age of *G. dilatata* from the Mikhailovsky quarry, which is near the town of Roshal, and the ones near the village of Sukhochevo may not coincide.

3.2. Analysis of the Diagenesis Features

The thicknesses of sediments located above the layers with gryphaea shells differ. It is about 10 m in the Roshal quarry, about 20 m in the Sukhochevo quarry, and at least 25 m in the Mikhailovsky quarry. The rocks where the shells of the mollusks were deposited also differ. They were found in bluish-gray clays in the Mikhailovsky quarry, whereas in the Sukhochevo quarry, they were found in dense layers of ferruginous sand—less often in gray clays—and in loose sands in the Roshalsky quarry.

Given this, in different localities, the pressure of the overlying sediments, the characteristics of the host rock, and the degree of shell ferrugination differ, which, as will be shown below using the example of belemnites, can be of great importance for the preservation of the shell material.

The valves of *G. dilatata* were fossilized under various conditions. This can be seen from the preservation features of fossil material. Furthermore, the gryphaea valves from the Mikhailovsky quarry are well preserved. There are almost no drillings of organisms that could be produced not only in vivo, but also posthumously. The shells are dark grey, and small crusts of pyrite (FeS_2) have been observed on the outer surface.

The valves of the gryphaeas have a light-gray color in the location near the village of Sukhochevo. They contain many perforations, the largest of which belong to the bivalve mollusks, *L. antiquissima*. On the surface of the valves and inside the shell substance, there are traces of ferruginization—sometimes very significant. In this case, the valves become almost completely reddish in color. Some valves are covered with a friable carbonate crust, the origin of which is discussed below (Figure 2f). Possibly, this is one of the surface friable shell layers that was destroyed posthumously.

The valves of the gryphaeas are also light gray in the locality near the town of Roshal. They also have traces of drilling, and in particular, barnacles. Signs of ferruginization are also observed, and some valves have a yellowish-red color or spots on the surface. However, the degree of ferruginization is less than in the locality near the village of Sukhochevo.

3.3. Features of Crystallographic Texture

Only calcite is found in all shells. Other minerals/impurities, including iron minerals—for example, pyrite—are contained in small amounts that do not exceed the resolution of the method. We characterized the crystallographic texture through two measured pole figures for each sample. A pole figure is a two-dimensional function on a unit sphere. For crystallographic texture comparison, we used the maximum values on a pole figure with (0006) and (10–14) Miller indices. Moreover, we also analyzed a pole figures' isoline pattern. As soon as we determined that *G. dilatata* valves could be either convex or flattened, we compared them separately to exclude the sample shape influence.

The sharpest crystallographic texture is observed in *G. dilatata* from the Zheleznogorsk quarry, 2.04 mrd, for the convex left valve of shell. In other cases, it varies from 1.95–1.97 mrd. The differences in these values are minimal. We have previously established [18] that for the mussels, *Mytilus galloprovincialis* Lamarck, 1819, it varies within 1.03 mrd, depending on the habitat. Therefore, these values seem insignificant. The values of the pole density sharpness on the pole figure (10–14) are smaller; they vary from 1.43–1.51 mrd. At the same time, the highest value was revealed for the shells collected in the quarry near the village of Sukhochevo. Furthermore, in this case, the range of variation is very small (Figure 4).

From the point of view of the isoline distribution, the pole figures can be characterized as follows. The distribution of the pole density for both pole figures has a horseshoe shape with two centers of greatest sharpness. There are isolines with a minimal pole density almost at the edges of the pole figure, (0006). On pole figure (10–14), similar isolines are located on the sides. A similar pattern is observed in valves from all three localities (Figure 4). Separately, it is necessary to consider the left valve of *G. dilatata* with a friable carbonate layer on the outer surface from the quarry near the village of Sukhochevo. The maximal sharpness of the pole density is 1.96 mrd (pole figure (0006)) and 1.55 mrd (pole figure (10–14)). In the first case, this falls within the considered range for other valves; in the second case, it slightly exceeds the maximum value by 0.05 mrd (Figure 5).

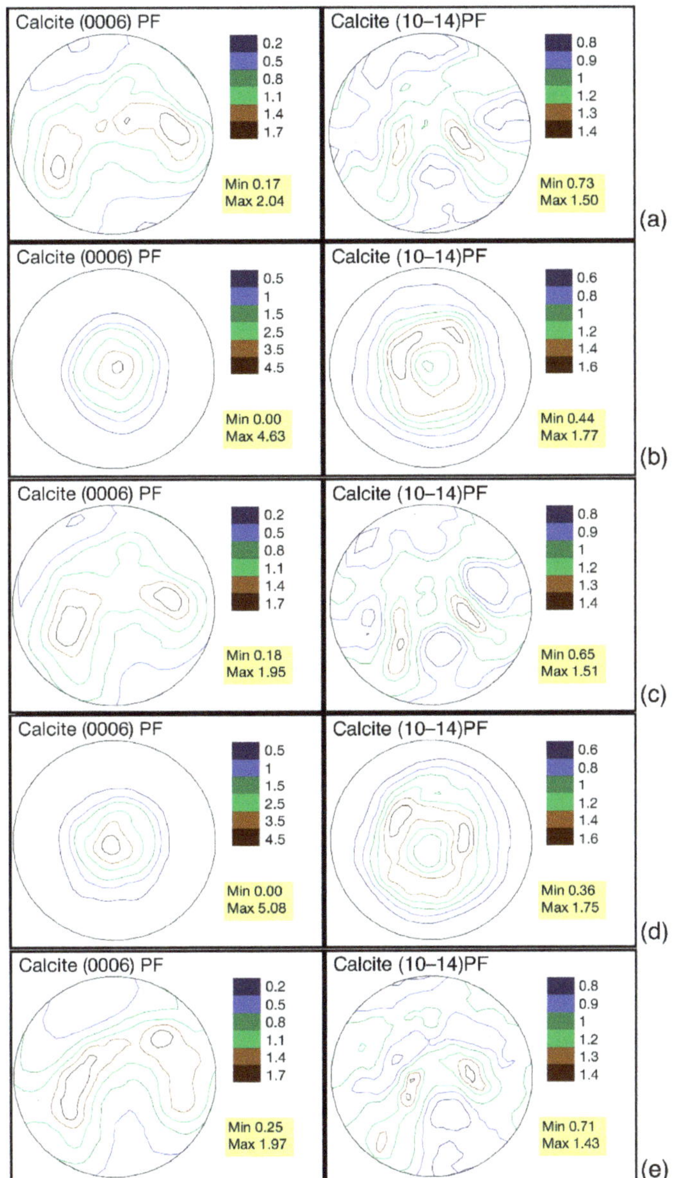

Figure 4. Pole figures of *Gryphaea dilatata* Sowerby, 1816, from different locations: (**a**) Calcite pole figures of the left valve of *Gryphaea dilatata* Sowerby, 1816, Kursk region, near town of Zheleznogorsk, Mikhailovsky quarry, Middle Jurassic, Middle–Upper Callovian; (**b**) Calcite pole figures of the right valve of *Gryphaea dilatata* Sowerby, 1816, Same; (**c**) Calcite pole figures of the left valve of *Gryphaea dilatata* Sowerby, 1816, Oryol region, Kromy district, sand quarry near the village of Sukhochevo, Middle Jurassic, Middle Callovian; (**d**) Calcite pole figures of the right valve of *Gryphaea dilatata* Sowerby, 1816, Same; (**e**) Calcite pole figures of the left valve of *Gryphaea dilatata* Sowerby, 1816, Moscow region, Shatura district, sand quarry on the outskirts of the town of Roshal, Middle Jurassic, Callovian–Upper Jurassic, Lower Oxfordian.

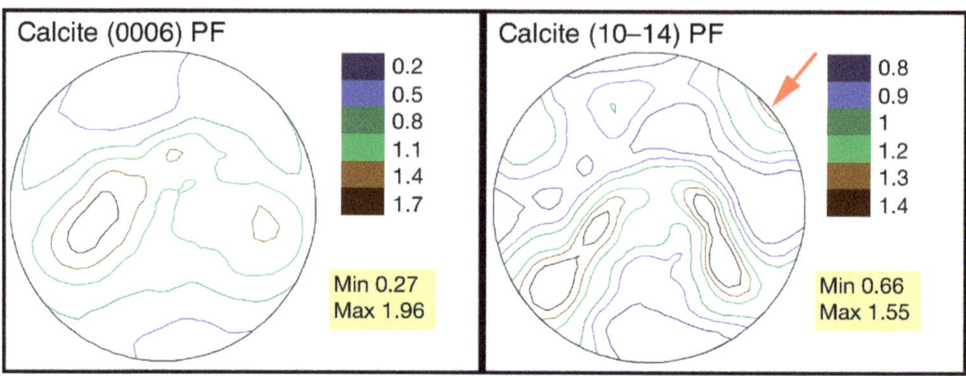

Figure 5. Calcite pole figures for the left valve of *Gryphaea dilatata* Sowerby, 1816, with a friable white surface layer, Oryol region, Kromy district, sand quarry near the village of Sukhochevo.

The red arrow shows the area of anomalous sharpness maxima.

The distribution of the pole density on pole figure (0006) also has a horseshoe shape, but with three centers of greatest sharpness—one of them being more pronounced than the others. Differences are also noted for pole figure (10–14), namely a small maximum with coordinates ($\chi = 90°$, $\varphi = 45°$) is present at the edge of the pole figure, which was not observed for other valves. The microstructure of gryphaeas valves is foliate (Figure 6a,b), and the plates are composed of foliated calcite crystals, as was shown previously [24,25].

When analyzing the outer layer using an electron microscope, it was revealed (Figure 6c) that the friable substance contains elements of the microstructure in the form of separate and grouped plates. There are also clusters of irregularly-shaped globules. The rest of the substance is a continuous mass without any order of elements. Since no fundamental differences are found in the crystallographic texture of the valves, the friable layer on the valve surface may refer to shell material, which was possibly modified to some extent. However, it is absent for most shells; this layer was probably destroyed posthumously. The absence of foulers on the surface of all shells is associated with this fact, although the valves of these mollusks could potentially be a good place for various organisms to settle. During the diagenesis process, the layer where the foulers were located collapsed, and there were not even traces of their presence. However, there are drillings from organisms that could extend to different depths of the shell material, including the inner valve surfaces. Moreover, drillers could perform them after the mollusk's death.

The pole figures of (10–14) for the *G. dilatata* valve with a friable carbonate layer from the Sukhochevo quarry and the valve from the Roshal quarry are very similar in their isoline patterns. However, we confirmed the presence of valve recrystallization by the SEM method in the first case, whereas we can only assume this in the second case.

A different pattern of crystallographic texture was observed for slightly-concave right valves from two localities: the Zheleznogorsk quarry and the quarry near the village of Sukhochevo (Figures 2b,d and 4b,d). The texture sharpness for pole figure (0006) varies from 4.63–5.08 mrd, whereas for pole figure (10–14), it varies between 1.75–1.77 mrd. That is, both values are higher than those of strongly-convex left valves. Given this, the crystals from the right valves are more ordered. This can also be seen by the isoline distributions on the pole figures. The center of the pole density maxima is localized close to the central parts of the pole figures. Most of the isolines are circular. The center of the pole density maxima is bifurcated on pole figure (10–14), and the crystal distribution is very dependent on the valve shape. Given this, it was probably easier for mollusks to arrange calcite crystals in a thick, almost flat, valve than in a thick, strongly-convex one.

Figure 6. Microstructure of valves of the *Gryphaea dilatata* Sowerby, 1816: (**a**,**b**) split of the left valve; (**c**) friable white layer, the white arrow shows the elements of the microstructure, Oryol region, Kromy district, sand quarry near the village of Sukhochevo, Middle Jurassic, Middle Callovian.

A similar pole figure (0006) is known for mussels of the genus, *Mytilus* Linnaeus, 1758. This pole figure looks like deformed circles inscribed in each other, which are located offset from the center. However, its sharpness is much higher, up to 15.72 mrd, for Pleistocene representatives of the species, *M. galloprovincialis* [17] (Figure 7).

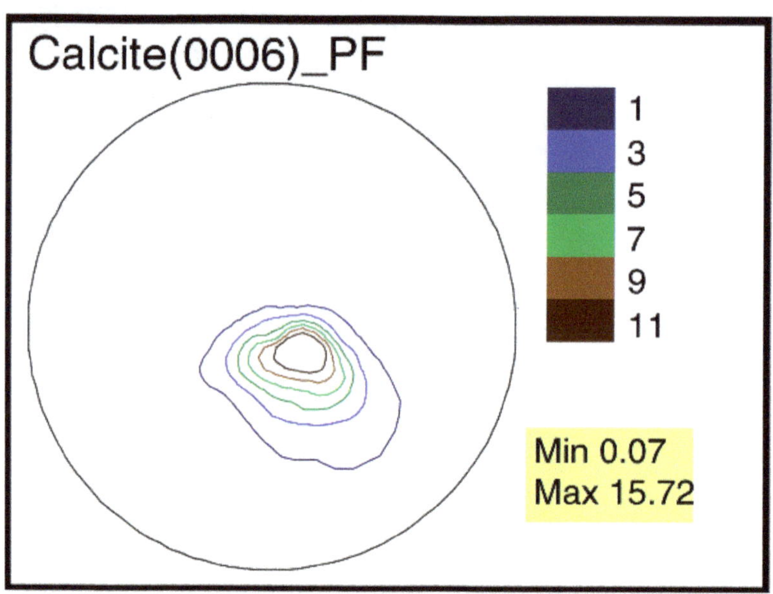

Figure 7. Calcite pole figure (0006) of the valves of *Mytilus galloprovincialis* Lamarck, 1819, Tuzla Spit, Kerch Strait, Pleistocene, Karangat deposits.

It is interesting to compare the crystallographic texture of studied *G. dilatata* valves with the texture of thick valves from closely-related mollusks. To do this, recent mollusks, such as *O. edulis*, from the close family, Ostreidae, and the left valve of the mollusk, *P. mirabilis*— family Gryphaeidae—were taken from the Maastrichtian deposits of the Crimea Peninsula. The left, thick-walled valve of *P. mirabilis* was studied. It is convex in shape (Figure 2g), but not as strongly as the left valves of *G. dilatata*. The shell contains only calcite. Its pole figure (0006) (Figure 8a) is very similar to figure (0006) of the almost-flat right valves of *G. dilatata*, and the calcite crystals are less ordered. The pole density maximum of pole figure (0006) of the *P. mirabilis* left valve is 2.41 mrd, while this value varies for the right *G. dilatata* valves—within 4.63–5.08 mrd. The isoline pattern of pole figure (10–14) is similar to that of the same pole figure of *G. dilatata* right valves. The pole density maximum is 1.39 mrd, which is less than in the left valves of *G. dilatata*. Moreover, the pole density sharpness of both *P. mirabilis* pole figures are much less than one in the right valves of *G. dilatata*, although the pole figures have similarities. Thus, with an increase in the convexity of the thick-walled valves, the sharpness of the crystallographic texture decreases. The pole figures of the left valves differ significantly, although the values of the greatest sharpness either coincide, or are very close, for related genera from the same family.

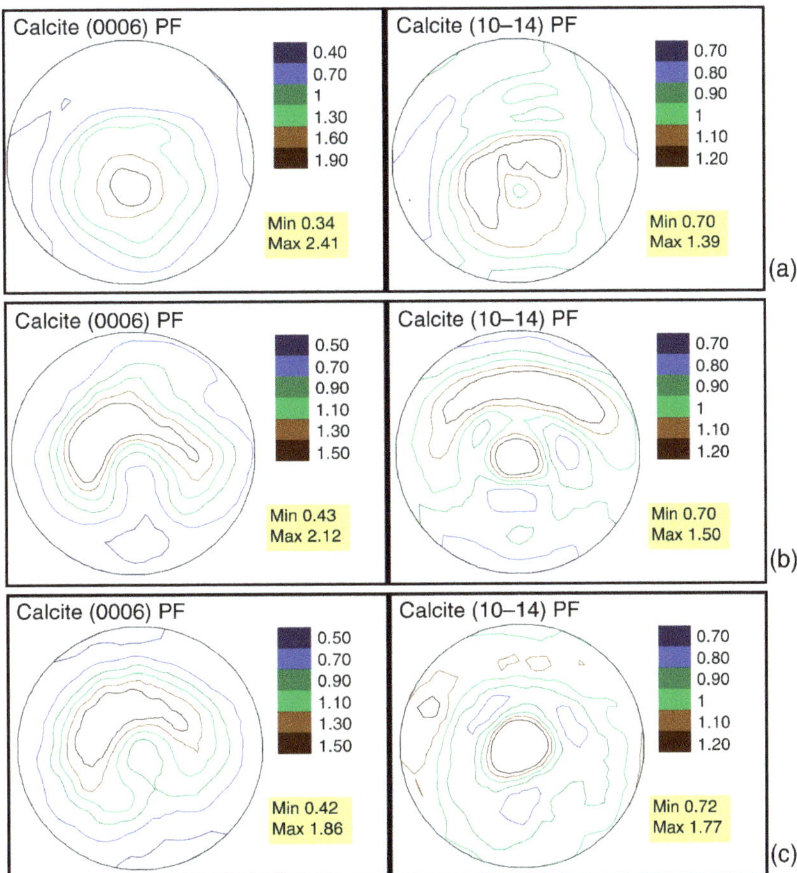

Figure 8. Calcite pole figures of thick-walled valves from fossil bivalves: (**a**) left valve of *Pycnodonte mirabilis* Rousseau, 1842, Crimea Peninsula, Upper Cretaceous, Maastrichtian; (**b**) right valves of *Ostrea edulis* Linnaeus, 1758, the coast of the Arabatsky Gulf of the Azov Sea in the town of Shchelkino (Crimea Peninsula), Pleistocene, Karangat deposits; (**c**) right valves of *Ostrea edulis* Linnaeus, 1758, Taman Peninsula, Chushka Spit, Pleistocene, Karangat deposits.

The shells of recent oysters also consist of calcite and have a variable, flattened, or convex shape with various elements of radial and concentric sculpturing on the shell's surface. The maximum values of pole figure (0006) of the recent oysters, *O. edulis* (Black Sea, Maly Utrish), right valves are 2.49 mrd (Figure 9a), and 2.38 mrd for pole figure (10–14). For the left valve of *O. edulis* (port Lagos, coast of Portugal), the same values are 2.53 and 1.93 mrd, respectively (Figure 9b). The range of sharpness variation, as in the case of gryphaea, is small. For recent oysters, the sharpness values are slightly higher than the ones in the left valves of *G. dilatata,* and significantly lower than the same values for pole figure (0006) of the flat right valves of gryphaeas. For fossil oysters (right valves) of the same species from Pleistocene deposits, the maximum sharpness of pole figure (0006) varies in the range of 1.86–2.12 mrd, while that of pole figure (10–14) varies from 1.5–1.77 mrd (Figure 8b,c). The ranges of variation are also small. The maximal pole figure values are less than those of recent oysters, which has already been noted [18]. The largest sharpness values of the pole figure (0006) for *G. dilatata* left valves (Figure 4) and the Pleistocene *O. edulis* right valves (Figure 8b,c) are almost the same, and for pole figure (10–14) of *G. dilatata* they are slightly lower, although the largest value is smaller than the lower ones of the range for the fossil *O. edulis*. A comparison of the right and left valves of oysters is justified, given that the left and right valves of *O. edulis* differ a little in shape and sculpture. The isoline patterns on the pole figures of both valves are also slightly different.

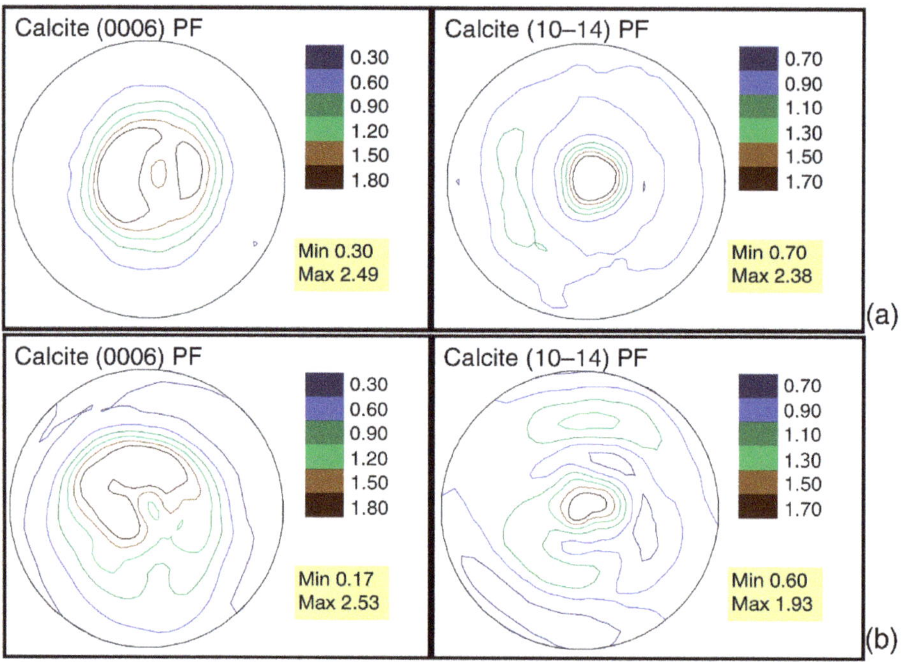

Figure 9. Calcite pole figures of the thick-walled valves of *Ostrea edulis* Linnaeus, 1758: (**a**) right valve, coast near the village of Maly Utrish, Black Sea; (**b**) left valve, Portugal, coast near the port of Lagos.

The calcite pole figures (0006) of the left and right valves of recent oysters remained an arc according to the isoline pattern (Figure 9), but with different curvature degrees. In this way, they are similar to the ones for the pole figures of the *G. dilatata* left valves. However, the latter ones have a much stronger isoline curvature on the pole figure's central part.

The calcite pole figure (0006) of recent oysters from the coast near the village of Maly Utrish is very close to the same pole figure of the *G. dilatata* right valve and the *P. mirabilis* left valve. The isolines of this pole figure have an almost circle shape, but the central part is divided into three unequal sectors, the larger of which is depressed, which gives it the shape of a wide, short arc. The calcite pole figure (10–14) of oysters has several circles or ovals inscribed into each other in the central part (Figure 9).

4. Discussion

The Jurassic *G. dilatata* bivalves turned out to be a very good object for studying the influence of habitat and fossilization conditions on their calcite crystallographic texture. The studied shells were found at a distance of tens (Mikhailovsky and Sukhochevo quarries) and hundreds (Roshal and Sukhochevo, Mikhailovsky and Roshal quarries) of kilometers from each other, that is, the mollusks lived in different water areas under different habitats. *G. dilatata* from the quarries near the village of Sukhochevo and Mikhailovsky could have lived at about the same time, and the mollusks from the quarry near the town of Roshal are slightly younger than the other samples. Since even Callovian *G. dilatata* can come from different substages (Middle and Upper), the habitats at different times are unlikely to be the same. In addition, by the beginning of the Late Jurassic, the sea-bay, which was on the territory of the Russian platform, became more open, connecting with other water areas [26]. This affected the change in habitat for example, resulting in an increase in salinity, or the transition of terrigenous bottom sediments to calc-terrigenous ones. In addition, the presence of a large amount of charred and pyritized wood in the Sukhochevo quarry is evidence of the coastal strip's proximity, which means an increased drift of terrigenous material into the sea basin.

The degree of ferruginization and calcite replacement with iron minerals differs for grypheas from different localities. This diagenetic factor can significantly affect the microstructure of a fossil, as was observed in the case of belemnite rostra from the Kuntsev–Filyovsky Park in Moscow (Upper Jurassic deposits, Volgian stage) (Figure 10).

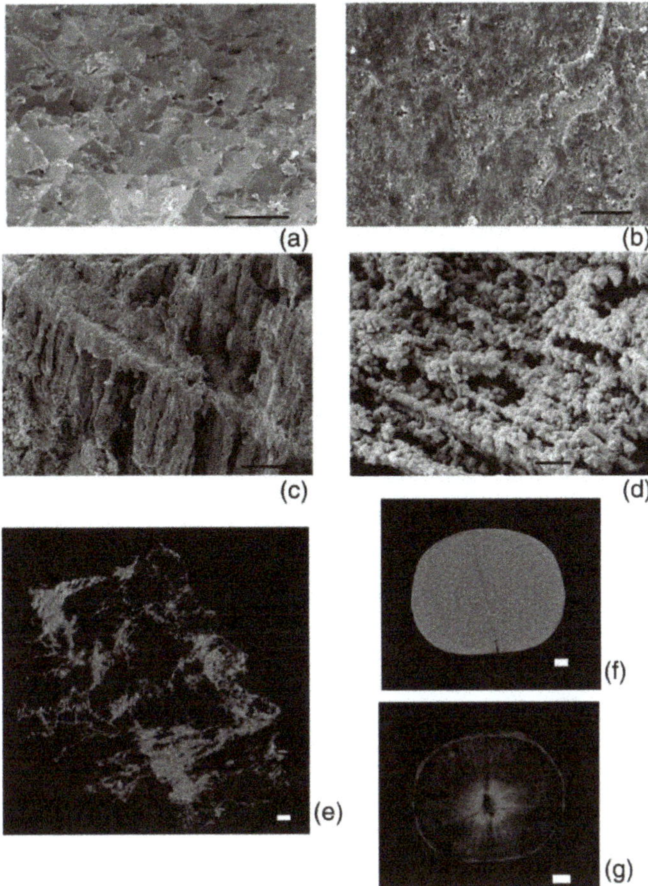

Figure 10. Surface and internal microstructure of belemnite rosters (Upper Jurassic, Upper Volgian; Moscow, Kuntsevo–Filyevsky Park) substituted and unsubstituted with iron minerals: (**a**) transverse split of belemnite rostra not substituted with iron minerals, scale is 100 μm; (**b**) surface of belemnite rostra not substituted with iron minerals, scale is 30 μm; (**c,d**) microstructure of the rostrum substituted with iron minerals, scales are 100 and 10 μm, respectively; (**e**) virtual section of a growth fragment substituted with iron minerals, X-ray microtomography, almost all calcite structures are destroyed inside, scale is 100 μm; (**f**) transverse virtual section of a belemnite roster not substituted with iron minerals, X-ray microtomography, scale is 1 mm; (**g**) transverse virtual section of the belemnite rostrum substituted with iron minerals, X-ray microtomography, scale is 1 mm.

Despite this, the pole figures of all studied *G. dilatata* samples are very similar in terms of the isoline pattern, that is, the habitat and fossilization conditions affected the crystallographic texture very little. Moreover, the pole figures of the samples from different places differ very slightly according to the largest texture sharpness values. This is illustrated by two calcite pole figures, (0006) and (10–14). The most variable sharpness values are observed for pole figure (0006), so it is the most significant for the analysis of the different influences on the crystallographic texture.

The only deviation from the general trend is the left valve of the gryphaea with a friable carbonate surface layer. Despite this, the values of the maximum texture sharpness coincide for this valve and the rest of the left ones. However, there are differences in the pole figures' isoline patterns. An additional peak of maximum sharpness on two pole figures is likely associated with the recrystallization of the valve surface layer, which is also confirmed by microstructure studies. That is, the fact of recrystallization was reflected not in the maximum texture sharpness values, but in the isoline pattern, since the recrystallized layer contains weakly-oriented crystals, and the most strictly-oriented crystals are located in the remaining layers—which all valves have. This is why the values of maximum sharpness are almost the same.

Using the gryphaeas, it is possible to observe how different parts of the skeleton—in this case, the valves—of the same organism have significant differences in crystallographic texture. These differences are associated with different valve shapes. They are reflected both in the difference of the pole figures' isoline patterns, and in the sharpness values. The maximum texture sharpness of the *G. dilatata* right valves is 5.08 mrd. Their pole figures (0006) are very close to axial ones. A similar picture is found for recent and fossil *M. galloprovincialis*. The pole figures (0006) of mussels are almost axial, but the maximum texture sharpness is 15.72 mrd. This situation can also be represented by the example of two peaks: weak and sharp (Figure 11). If one cuts the peaks from their base to the highest point by several sections and looks from above, one gets an axial pole figure, but the cut lines of the different peaks are located at different distances. The crystalline orientations are closer to each other for sharper peaks.

The right valves of gryphaeas, as well as the thick-walled left valves of *P. mirabilis*, have axial pole figures (0006). At the same time, pole figure (10–14) of *P. mirabilis* has an arcuately curved isoline for the maximum of the pole density intensity. A similar situation is observed for oyster's pole figure (0006). In recent oysters, the pole figure (0006) isoline pattern varies from a figure with an arcuate maximum to an almost axial one. Meanwhile, pole figure (10–14) for most of the studied valves has an axial character with a maximum in the center. Only *G. dilatata* has a curved arcuate center maximum. All studied shells were related, i.e., *G. dilatata* and *P. mirabilis* are members of the same family, while *O. edulis* belongs to the close family, Ostreidae. Other similarities are that their shells are thick-walled, and *G. dilatata* and *P. mirabilis* have convex left valves, while other valves (the right gryphaeas valves and oysters) are flattened. At the same time, for the fossil oysters and the recent ones from the coast of Lagos, pole figure (0006) for the left and right valves has a slightly curved maximum, which is similar to the same pole figure of the left *G. dilatata* valves. Furthermore, only for recent Black Sea *O. edulis* does pole figure (0006) of the right valves represent a transitional variant between the axial pole figure and the one with an arcuately-curved isoline for the maximum of the pole density intensity. In this regard, it can be concluded that in the thick-walled shells of the Gryphaeidae and Ostreidae families' bivalve mollusks, the pattern of isolines varies from almost axial to a figure with an arcuately-curved isoline for the maximum of the pole density intensity.

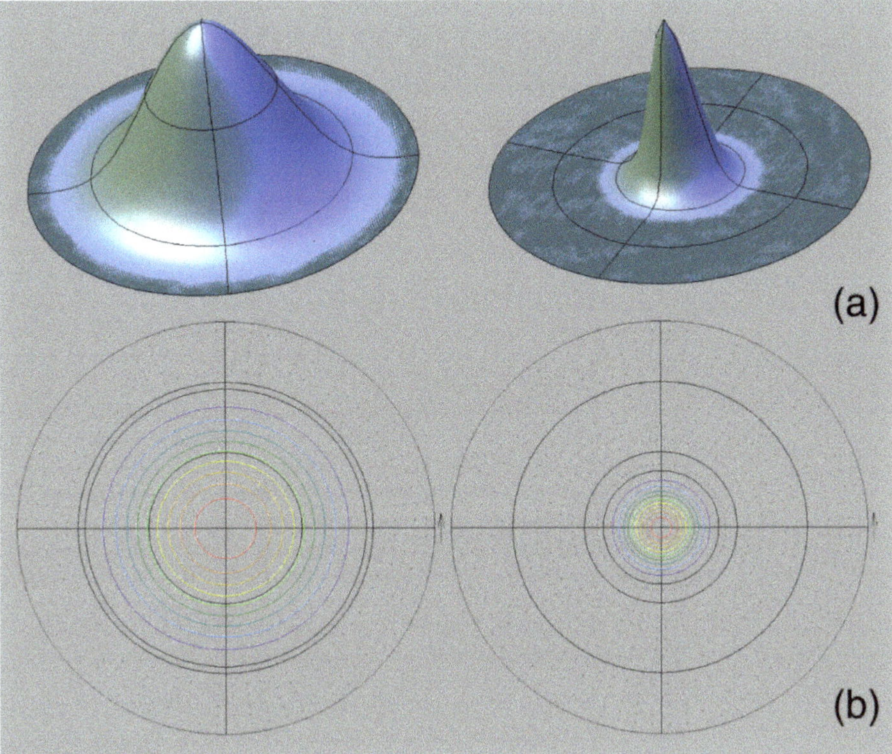

Figure 11. Models of crystal distribution peaks: (**a**) Illustration of the crystallographic texture sharpness; (**b**) its presentation on the pole figures. The circles with different color are isolines with the same pole density values.

5. Conclusions

There are few works concerning the study of the global crystallographic texture of fossil objects. They cannot be fully compared with the results of local crystallographic texture studies fulfilled using X-ray or EBSD diffraction due to the fundamental difference in the studied volumes of shell material. Moreover, in the case of X-rays or EBSD, the specimen is a small piece of the valve, which is not easily related to the coordinate system of the whole valve. Studies of the global crystallographic texture of fossilized shells of the same species from different geological layers and locations was never carried out before. Thus, we can assess the degree of influence of diagenetic and paleoecological factors on the crystallographic texture of calcite in Jurassic shells, since we have not previously found a single factor that would affect the change in such a texture for recent and Pleistocene mollusks. As well, the global crystallographic texture of objects subjected to recrystallization has never been evaluated.

As a result of this study, some important aspects of the calcite crystallographic texture of fossil and recent bivalve mollusk shells have been revealed. Since there are only a few results for whole shell or valve crystallographic texture studies that have been obtained by neutron diffraction, the interpretation of both the numerical values of the pole figures maxima and the isoline pattern is very important.

1. For the first time, the complete pole figures of *G. dilatata* shells were measured using time-of-flight neutron diffraction.

2. The studied *G. dilatata* lived at considerable distances from each other. The mollusks likely did not coincide in their time of habitation. This means that their habitats were different.

3. *G. dilatata* from the three localities almost do not differ in their crystallographic texture maximum sharpness, i.e., the habitat conditions and fossilization did not affect the texture, and it can be considered as a very stable feature.

4. The largest differences of the crystallographic texture sharpness have been noted for the calcite pole figure (0006). Therefore, it is possible to reveal the greatest variability by comparing these figures' sharpness for different objects.

5. The features of *G. dilatata* valve recrystallization, which affect the crystallographic texture, were revealed for the first time using time-of-flight neutron diffraction. They were reflected in the pole figure isolines, but do not appear in the values of maximum sharpness, since the entire valve was not recrystallized and the pole density was low in the recrystallized part. The maximum texture sharpness was determined by the non-recrystallized part of the valve.

6. It has been established that for one mollusk, the crystallographic texture of the left and right valves of various shapes differs: to a greater extent by the pole figures' isoline patterns, to a lesser extent by the values of maximum sharpness.

7. The calcite pole figures and the values of the maximum sharpness of the *G. dilatata* and *O. edulis* left valves are similar. The pole figures isoline patterns and values of the pole density maximum of *G. dilatata* right valves are similar, and along the same parameters as the *P. mirabilis* left valve. All of these species are characterized by either an axial pole figure (or close to it) or a figure with an arcuately-curved sharpness maximum. Transitions between them are found for the recent *O. edulis*. It is possible that these pole figures with have isoline patterns that are characteristic of thick-walled shells.

8. The thick-walled valves of *G. dilatata*, *O. edulis*, and *P. mirabilis* have small values of maximum texture sharpness. They are very close for these mollusks. Perhaps, it is a characteristic feature of all thick-walled valves and reflects an adaptation to shell construction.

Author Contributions: Conceptualization, methodology, all authors; software, D.N. and T.L.; formal analysis, all authors; writing—original draft preparation, A.P.; writing—review and editing, D.N. and T.L.; investigation, visualization, all authors; project administration, T.L. All authors have read and agreed to the published version of the manuscript.

Funding: The APC was funded by Joint Institute for Nuclear Research.

Institutional Review Board Statement: Not applicable.

Informed Consent Statement: Not applicable.

Data Availability Statement: The data supporting reported results can be obtained on request from the article authors.

Acknowledgments: The authors express their heartfelt gratitude to S.V. Bagirov, K.K. Tarasenko, A.V. Guzhov, R.A. Rakitov (PIN RAS), A.A. Kozlova (N.I. Pirogov Russian National Research Medical University), Y.V. Kirzhaev (PJSC «Oryolstroy»), D.S. Yudin (Limited Liability Company «ODSK Nerud») and O.A. Kozlova (GBOU School No. 1502 "Energy").

Conflicts of Interest: The authors declare no conflict of interest.

References

1. Nevesskaya, L.A.; Popov, S.V.; Goncharova, I.A.; Guzhov, A.V.; Yanin, B.T.; Polubotko, I.V.; Byakov, A.S.; Gavrilova, V.A. *Bivalves of Russia and Neighboring Countries in the Phanerozoic*; Scientific World: Moscow, Russia, 2013; 524p.

2. Kennedy, W.J.; Taylor, J.D.; Hall, A. Environmental and Biological Controls on Bivalve Shell Mineralogy. *Biol. Rev.* **1969**, *44*, 499–530. [CrossRef] [PubMed]

3. Carter, J.G.; Clark, G.R. Classification and phylogenetic significance of molluscan shell microstructure. *Ser. Geol. Notes Short Cour.* **1985**, *13*, 50–71. [CrossRef]

4. Zakhera, M.; Kassab, A.; Chinzei, K. Hyotissocameleo, a new Cretaceous oyster subgenus and its shell microstructure, from Wadi Tarfa, Eastern Desert of Egypt. *Paleontol. Res.* **2001**, *5*, 77–86.

5. Nikolayev, D.I.; Lychagina, T.A.; Pakhnevich, A.V. Experimental neutron pole figures of minerals composing bivalve mollusk shells. *Nat. Appl. Sci.* **2019**, *1*, 344. [CrossRef]

6. Chateigner, D.; Hedegaard, C.; Wenk, H.-R. Mollusc shell microstructures and crystallographic textures. *J. Struct. Geol.* **2000**, *22*, 1723–1735. [CrossRef]

7. Nekhoroshkov, P.; Zinicovscaia, I.; Nikolayev, D.; Lychagina, T.; Pakhnevich, A.; Yushin, N.; Bezuidenhout, J. Effect of the elemental content of shells of the bivalve mollusks (*Mytilus galloprovincialis*) from Saldanha Bay (South Africa) on their crystallographic texture. *Biology* **2021**, *10*, 1093. [CrossRef]

8. Bunge, H.J. *Texture Analysis in Material Science, Mathematical Methods*; Butterworths Publ.: London, UK, 1982; 595p.

9. Ouhenia, S.; Chateigner, D.; Belkhir, M.A.; Guilmeau, E. Microstructure and crystallographic texture of Charonia lampas lampas shell. *J. Struct. Biol.* **2008**, *163*, 175–184. [CrossRef]

10. Chateigner, D.; Kaptein, R.; Dupont-Nivet, M. X-ray quantitative texture analysis on Helix aspersa aspera (Pulmonata) shells selected or not for increased weight. *Amer. Malac. Bull.* **2009**, *27*, 157–165. [CrossRef]

11. Frýda, J.; Klicnarová, K.; Frýdová, B.; Mergl, M. Variability in the crystallographic texture of bivalve nacre. *Bull. Geosci.* **2010**, *85*, 645–662. [CrossRef]

12. Hahn, S.; Rodolfo-Metalpa, R.; Griesshaber, E.; Schmahl, W.W.; Buhl, D.; Hall-Spencer, J.M.; Baggini, C.; Fehr, K.T.; Immenhauser, A. Marine bivalve shell geochemistry and ultrastructure from recent low pH environments: Environmental effect versus experimental bias. *Biogeosciences* **2012**, *9*, 1897–1914. [CrossRef]

13. Fitzer, S.C.; Phoenix, V.R.; Cusack, M.; Kamenos, N.A. Ocean acidification impacts mussel control on biomineralization. *Sci. Rep.* **2014**, *4*, 6218. [CrossRef] [PubMed]

14. Almagro, I.; Drzymała, P.; Berent, K.; Saínz-Díaz, C.I.; Willinger, M.G.; Bonarsky, J.; Checa, A.G. New crystallo-graphic relationships in biogenic aragonite: The crossed-lamellar microstructures of mollusks. *Cryst. Growth Des.* **2016**, *16*, 2083–2093. [CrossRef]

15. Kučeráková, M.; Rohlicek, J.; Vratislav, S.; Jarosova, M.; Kalvoda, L.; Lychagina, T.; Nikolayev, D.; Douda, K. Texture and element concentration of the freshwater shells from the Unionidae family collected in the Czech Republic by X-ray, neutron and electron diffraction. *Crystals* **2021**, *11*, 1483. [CrossRef]

16. Pakhnevich, A.; Nikolayev, D.; Lychagina, T.; Balasoiu, M.; Ibram, O. Global Crystallographic Texture of Freshwater Bivalve Mollusks of the Unionidae Family from Eastern Europe Studied by Neutron Diffraction. *Life* **2022**, *12*, 730. [CrossRef] [PubMed]

17. Brokmeier, H.-G. Neutron Diffraction Texture Analysis of Multi-Phase Systems. *Textures Microstruct.* **1989**, *10*, 325–346. [CrossRef]

18. Pakhnevich, A.V.; Nikolayev, D.I.; Lychagina, T.A. Comparison of the Crystallographic Texture of the Recent, Fossil and Subfossil Shells of Bivalves. *Paleontol. J.* **2021**, *55*, 589–599. [CrossRef]

19. Nikolayev, D.I.; Lychagina, T.A.; Nikishin, A.V.; Yudin, V.V. Study of error distribution in measured pole figures. *Solid State Phenom.* **2005**, *105*, 77–82. [CrossRef]

20. Nikolayev, D.I.; Lychagina, T.A.; Nikishin, A.V.; Yudin, V.V. Investigation of measured pole figures errors. *Mater. Sci. Forum* **2005**, *495–497*, 307–312. [CrossRef]

21. Nikolayev, D.I.; Ullemeyer, K. A note on preprocessing of diffraction pole density data. *J. Appl. Cryst.* **1994**, *27*, 517–520. [CrossRef]

22. Gerasimov, P.A. *Guide Fossils of the Mesozoic of the European Central Regions of the USSR. Part 1. Lamellar-Gill, Gastropods, Boat-Footed Mollusks and Brachiopods of the Jurassic Deposits*; Gosgeoltekhizdat: Moscow, Russia, 1955; 379p. (In Russian)

23. Gerasimov, P.A.; Mitta, V.V.; Kochanova, M.D.; Tesakova, E.M. *The Callovian Fossils of Central Russia*; VNIGNI–MosGorSUN: Moscow, Russia, 1996; 127p. (In Russian)

24. Checa, A.G.; Esteban-Delgado, F.J.; Rodríguez-Navarro, A.B. Crystallographic structure of the foliated calcite of bivalves. *J. Structur. Biol.* **2007**, *157*, 393–402. [CrossRef]

25. Checa, A.G.; Harper, E.M.; González-Segura, A. Structure and crystallography of foliated and chalk shell microstructures of the oyster Magallana: The same materials grown under different conditions. *Sci. Rep.* **2018**, *8*, 7507. [PubMed]

26. Nevesskaya, L.A. *Stages of Benthos Development in the Phanerozoic Seas, Mesozoic, Cenozoic*; Transactions of PIN RAS; Rozanov, A.Y., Ed.; "Nauka": Moscow, Russia, 1999; Volume 274, 438p. (In Russian)

Review

Integrative Phylogenetics: Tools for Palaeontologists to Explore the Tree of Life

Raquel López-Antoñanzas [1,2,*], Jonathan Mitchell [3], Tiago R. Simões [4], Fabien L. Condamine [1], Robin Aguilée [5], Pablo Peláez-Campomanes [2], Sabrina Renaud [6], Jonathan Rolland [5] and Philip C. J. Donoghue [7]

[1] Institut des Sciences de l'Évolution (ISE-M, UMR 5554, CNRS/UM/IRD/EPHE), Université de Montpellier, 34090 Montpellier, France

[2] Departamento de Paleobiología, Museo Nacional de Ciencias Naturales-CSIC, 28006 Madrid, Spain

[3] Department of Biology, West Virginia University Institute of Technology, 410 Neville Street, Beckley, WV 25801, USA

[4] Museum of Comparative Zoology & Department of Organismic and Evolutionary Biology, Harvard University, Cambridge, MA 02138, USA

[5] Laboratoire Évolution & Diversité Biologique, Université Paul Sabatier Toulouse III, UMR 5174, CNRS/IRD, 31077 Toulouse, France

[6] Laboratoire de Biométrie et Biologie Evolutive, UMR 5558, CNRS, Université Claude Bernard Lyon 1, 69622 Villeurbanne, France

[7] School of Earth Sciences, University of Bristol, Bristol BS8 1RJ, UK

* Correspondence: raquel.lopez-antonanzas@umontpellier.fr

Citation: López-Antoñanzas, R.; Mitchell, J.; Simões, T.R.; Condamine, F.L.; Aguilée, R.; Peláez-Campomanes, P.; Renaud, S.; Rolland, J.; Donoghue, P.C.J. Integrative Phylogenetics: Tools for Palaeontologists to Explore the Tree of Life. *Biology* **2022**, *11*, 1185. https://doi.org/10.3390/biology11081185

Academic Editors: Mary H. Schweitzer and Ferhat Kaya

Received: 30 June 2022
Accepted: 3 August 2022
Published: 7 August 2022

Publisher's Note: MDPI stays neutral with regard to jurisdictional claims in published maps and institutional affiliations.

Copyright: © 2022 by the authors. Licensee MDPI, Basel, Switzerland. This article is an open access article distributed under the terms and conditions of the Creative Commons Attribution (CC BY) license (https://creativecommons.org/licenses/by/4.0/).

Simple Summary: All life is derived from a single common ancestor, whose descendants coevolved with the planet, shaping the structure of biodiversity and the physical processes that operate on Earth. This complex history cannot be inferred solely by studying the genomes of living organisms, nor through analysis of the fossil remains of their extinct relatives. Only a unified approach integrating living and extinct species and drawing from both genomic and anatomical evidence can achieve this aim. In this review, we highlight recent advances, challenges, and opportunities in this endeavour. These include the development of models for analysis of anatomical data; methods for combined analysis of fossil and living species, as well as anatomical and genomic data; and the combined estimation of evolutionary relationships, geographic range, and evolutionary rates. However, the application of such methods is limited by a shortage of expertise in taxonomy and comparative anatomy, which are skills required for the compilation of anatomical datasets. Whereas there is a common concern for the incompleteness of the fossil record, knowledge with respect to the comparative anatomy of living species is equally incomplete. We anticipate that the increased demand for an integrative phylogenetic approach to reconstruct the tree of life and evolutionary patterns and processes will encourage researchers to overcome these challenges with the aim of elucidating the complexities behind organismal evolution across broad taxonomic and time scales.

Abstract: The modern era of analytical and quantitative palaeobiology has only just begun, integrating methods such as morphological and molecular phylogenetics and divergence time estimation, as well as phenotypic and molecular rates of evolution. Calibrating the tree of life to geological time is at the nexus of many disparate disciplines, from palaeontology to molecular systematics and from geochronology to comparative genomics. Creating an evolutionary time scale of the major events that shaped biodiversity is key to all of these fields and draws from each of them. Different methodological approaches and data employed in various disciplines have traditionally made collaborative research efforts difficult among these disciplines. However, the development of new methods is bridging the historical gap between fields, providing a holistic perspective on organismal evolutionary history, integrating all of the available evidence from living and fossil species. Because phylogenies with only extant taxa do not contain enough information to either calibrate the tree of life or fully infer macroevolutionary dynamics, phylogenies should preferably include both extant and extinct taxa, which can only be achieved through the inclusion of phenotypic data. This integrative phylogenetic approach provides ample and novel opportunities for evolutionary biologists to benefit from palaeontological data to help establish an evolutionary time scale and to test core macroevolutionary hypotheses about the drivers of biological diversification across various dimensions of organisms.

Keywords: taxonomy; morphometrics; phylogeny; evolution; morphological clock; molecular clock; biodiversity; palaeobiogeography; macroevolution

1. Introduction

Establishing an evolutionary time scale is a fundamental yet elusive goal of the Earth and life sciences. Without knowledge of the timing of evolutionary events, it is not possible to test hypotheses of ecological and evolutionary processes over geologic time. The fossil record once constituted the gold standard with respect to attempts to establish evolutionary time scales; however, for more than 50 years [1], that role has been filled by molecular clock approaches for groups with extant representatives. The benefits of analysing and integrating multiple lines of evidence to test hypotheses in science were previously tackled by Kluge [2] in what he called "TOTAL EVIDENCE analysis". This idea was expanded by Nixon and Carpenter [3] in their "simultaneous analysis". Since then, multiple Bayesian methods have been developed to accommodate genomic and/or morphological data.

A molecular clock methodology has also been developed to account for variation in the rate of molecular evolution among lineages and to accommodate the inaccuracies and imprecision inherent in the use of fossil evidence with respect to calibration [4–6], and it is now generally considered to be the most efficient methodology for calibrating evolutionary trees to geologic time. Therefore, evolutionary trees are often built on genomic datasets, putting morphology to one side [7]. However, fossil data provide the key means of clock calibration and are fundamental to the molecular clock methodology [5,8].

Traditionally, molecular clocks use fossil taxa to calibrate the divergences between living lineages (node dating). Nevertheless, the latest methods (tip dating) allow fossil species to be included alongside their living relatives, with the absence of molecular sequence data for fossil taxa remedied by supplementing the sequence alignments for living taxa with phenotype character matrices for both living and fossil taxa in total evidence dating [9,10]. In this way, the temporal constraints on lineage divergence provided by fossil species can be implemented in a more direct manner. Building total-evidence time-calibrated phylogenies is critical to increase the accuracy of the inferences regarding macroevolutionary processes. Tip dating is being increasingly applied with combined datasets, and it has begun to be used in fossils and/or living morphological datasets alone [11] in what has been called the morphological clock. Morphological data are a crucial component of phylogenetic inference, as they are usually the only information available to integrate both living and extinct members of an evolutionary tree [12]. The importance of morphological phylogenetics for dating the tree of life is widely recognized and has been bolstered by recent methodological developments. Statistical techniques, mostly using Bayesian inference, now allow researchers to test and implement variations in clock models, data partitioning, taxon sampling strategies, sampling of ancestors, and tree models (e.g., the fossilized birth–death (FBD) tree model) using morphological data [13–19]. In this way, palaeontologists are now able to achieve more accurate modelling of the diversification process across geological time, a crucial aspect of phylogenies with taxonomic sampling extending into deep time. Over the last years, the concurrent discovery of new fossil sites in previously rarely explored areas, the improvement of dating techniques, and the development of effective and integrative methods in phylogenetics have revitalized the study of speciation and extinction rates, as well as their variation over time and among clades [20–22]. Phylogenetic approaches in macroevolution enable diversification rates to be tied to changes in paleoenvironmental (extrinsic) and/or biotic (intrinsic) factors. These state-of-the-art approaches can be used to establish a time scale for evolution, linking phenotypic evolution with diversification rates and extrinsic phenomena, including causal agents of evolutionary change, such as global climate oscillations [23–25]. Given that phylogenies with extant taxa do not contain enough information for macroevolutionary dynamics to be fully and reliably reconstructed [26,27], phylogenies must include both extant and extinct taxa, which can only be achieved through integration of phenotypic data. In fact, a comprehensive understanding of evolution requires fossil data. Unfortunately, morphological phylogenetic data are lacking for most groups. Moreover, the morphological characteristics of living taxa are usually overlooked, and the data needed to determine the phylogenetic positions of fossil taxa with respect to their present-day relatives are often unavailable for many clades.

Given that establishing evolutionary time scales is a key goal of palaeontology, it is surprising that these phylogenetic methods are not more widely adopted by palaeontologists. Hence, the goal of this paper is to highlight some of the latest methodological advances bridging extinct and extant organismal biology that will help palaeontologists to address key aspects of patterns and processes in evolution.

2. Advances in Integrated Phylogenetics

2.1. Taxonomy

Taxonomy has been marginalized and traditionally treated as a purely descriptive discipline for both extant and extinct organisms [28]. However, the discovery of new fossils from key but underexplored areas of the world and/or key time intervals in the history of life are crucial to evolutionary biology. The study of new fossil taxa can shed light onto phylogenetic relationships in order to infer the time in which anatomical novelties appeared in a given group, as well as the biotic/abiotic factors driving their origin. Taxonomic studies in palaeontology are crucial for tackling all biochronological palaeobiogeographical and macroevolutionary questions. Discovery and description of new species creates generate raw data for further analysis by providing information on character states (and therefore phylogenetic inference), biogeographical locations, and temporal calibrations that are foundational to dating and reconstructing the evolutionary history of life. For instance, the study of the first Neogene micromammals from Zahlé (Bekaa Valley) discovered in Lebanon, one of the only two terrestrial Late Miocene sites in the Arabian Peninsula, has provided relevant data concerning new species situated at pivotal phylogenetic positions. This has allowed for inference of the expected dental morphology of the ancestors of some important lineages of rodents [29], as well as the evolutionary history of such important genera as *Progonomys*, the earliest known murine (Old World mice and rats), which is the first modern representative of the group to spread out of southern Asia [30]. Moreover, these data were relevant with respect to inferring the age of the sites (several million years older than previously thought), as well as the timing and nature of the migration events that took place between Eurasia and Africa via the Arabian plate.

2.2. Morphological Datasets

Modelling the evolution of morphological structures is a complex but crucial task for improving the practice of morphological phylogenetics and for testing evolutionary scenarios. The ability of morphological data to place extinct taxa phylogenetically is widely acknowledged, as sampling fossils for molecular data is typically impossible [31–33]. Fossil data are fundamental to molecular clocks, providing the key means of time calibration, although their commonplace use is far from satisfactory [5,8]. It is essential that the phylogenies of fossil species used in molecular clock calibration be compatible with the phylogenies of the living species that underpin the divergence time analysis. To this end, it is essential that taxonomists gather phenotypic information at the level of individual species, as their molecular counterparts do, instead of as usual, gathering such phenotypic information at a higher level (e.g., genus) [7]. Therefore, an important issue faced by taxonomists is the scarcity of morphological datasets of whole clades at the species level. Surprisingly, sourcing morphological datasets for living species may be more challenging than for their fossil counterparts [34]. Unfortunately, there is a continuously decreasing number of taxonomists able to collect and analyse phenotypic data [7], and even if the need for such expertise is pressing, taxonomical studies are still considered unfashionable instead of being encouraged.

Phylogenetic morphological datasets are frequently composed of discrete characters only or continuous characters discretized into arbitrary categories. However, discrete and continuous characters are jointly evolving, and the latter may contain information concerning gradual variations; ignoring this mutual information may lead to biased parameter estimates [35,36]. There is a long-standing debate in the scientific community concerning the use of quantitative characters for phylogenetic reconstructions, with disagreements concerning their suitability for inferring phylogenies [37,38]. However, continuous traits reduce the subjective bias of discrete characters and represent the full range of interspecific variation; therefore, they can be useful in phylogenetic reconstructions. Geometric and morphometric methods were applied as early as the 1990s to characterize fossil rodent taxa, to assess their relationship with relatives, and quantify evolutionary patterns [39]. Since then, the field has been renewed by the rise of 3D methods [40], enabled by the increasing availability of μCT scanners. A recent avenue of research with involves the joint use of geometric morphometrics and phylogenetic methods to map the evolution of complex structures and test models of evolution [41]. A particular challenge is constituted by the multivariate nature of morphometric data, although phylogenetic models are being developed to accommodate such issues [36,42,43]. It is now possible to use multivariate data directly in divergence time estimation [44].

Another challenge involves integrating developmental constraints on the evolution of morphological character, such as serial homologies and correlated characters, into phylogenetic models [45].

Although further work is needed to solve many of the concerns surrounding continuous data, new approaches are being developed [46–48] to analyse and understand the nature of these characters so that they can be used in support of 'total evidence' analyses.

Recent developments in morphometrics, phylogenetics, and comparative methods have revitalized the use of morphological data by palaeontologists to elucidate the dynamics of evolution over time [13,16,21,49–53]. Therefore, a new era of high-impact and interdisciplinary morphological taxonomy is beginning.

2.3. Calibrating the Tree of Life

Time provides palaeontologists with a unique perspective on phylogenetic history. Maximum parsimony was, until recently, the only way for palaeontologists to analyse morphological datasets. Despite initial attempts to integrate stratigraphic data with parsimony analyses in the 1990s [54,55], the problem faced by palaeontologists had was the considerable time and effort they had to dedicate to manually calibrate the resulting trees. Moreover, subject to the number of taxa included in a dataset, palaeontologists had to infer the distribution of morphological characters without including temporal data, with a consequent loss of information. The recent development of methodological approaches facilitating a posteriori time calibration of phylogenetic trees, such as PaleoTree [56,57] or STRAP (Stratigraphic Tree Analysis for Palaeontology) [58] allows for time calibration of phylogenetic trees resulting from parsimony analyses, as well as assessment of their agreement with the stratigraphic record (stratigraphic congruence) [58–61]. Owing to the development of such methods, the incorporation of stratigraphic data into parsimony analysis has been bolstered, presenting the opportunity to use additional techniques of phylogenetic reconstruction with morphological data [52,58,61,62].

Since the introduction of Bayesian tip-dating phylogenetic methods, which were first applied with simplified clock and tree models [9,10], the inclusion of stratigraphic data into phylogenetic analyses has boomed [52,63,64]. The development of tip dating with more complex mechanistic tree models, such as the fossilized birth–death (FBD) tree [65,66] and its subsequent variations—such as the skyline FBD [13–19], which enables speciation, extinction, and fossilization parameters to be changed piecewise across the tree—allowed fossil species to be analysed in conjunction with and within the same analytical framework as extant taxa using Bayesian phylogenetics. This has been particularly useful for palaeontologists, who have revitalizing the use of morphological data to elucidate the dynamics of evolution across the tree of life [16,17,49,51,52,63,67–69] (Figure 1). These analyses can be carried out with morphological datasets alone, in what is called the morphological clock [16,17,51,52,63,70], which adds another method of reconstructing evolution to the palaeontologist's toolbox. A morphological clock can be applied with data from extinct clades only [17,52,60,63,67,71] or with data from fossils and extant taxa [16,49,51,70,72,73]. Therefore, these recent developments applying Bayesian methods using fossil taxa as tips make it possible to compare phylogenies of extinct taxa obtained by means of evolutionary models with those resulting from maximum parsimony, which remains the most widely applied method for analysis of morphological data. The availability and inclusion of fossils in analysis enables Bayesian tip dating, which may improve the accuracy and precision of divergence time estimates [74,75]. Nevertheless, tip calibration has been shown to lead to 'deep root attraction' [76–78]. However, this artefact can be mitigated by using informative priors for FBD parameters [16,77] or a combination of tip and node calibration (whereby in the absence of tree-internal clade age constraints, the age estimates are unbounded by anything other than the root, leading to ages that become more precise with proximity to the root) [79].

Morphological data can be combined with a molecular matrix either based on a few loci (generated with Sanger techniques) [49] or based on hundreds of loci (generated with next-generation sequencing techniques) [80]. A morphological clock is then integrated, along with several molecular clocks, taking into account rate and clock heterogeneity across the dataset. To date, most total-evidence dating analyses have been carried out with molecular matrices composed of a few loci. However, the bird order Sphenisciformes (penguins), for instance, has been studied using both, a few [49] and hundreds of loci [80] and recovered similar results. Future studies will rely on vast ranges of genomic data combined with morphological datasets to estimate phylogenetic relationships and divergence times. A challenge is to infer to what extent the inclusion of increasingly large molecular datasets in combined analyses could affect to the phylogenetic contribution of morphological data. According to Neumann [81], they will continue to have a strong influence, even when outnumbered by molecular data by thousands of times.

Challenges remain for palaeontologists because for many important groups (e.g., rodents), the number of characters available for computational morphological phylogenies is very limited, and commonly, the relationships between taxa are inferred by hand rather than by computational algorithms. The main issue is that building new phylogenies from new character matrices of morphological data is very time-consuming. Systematics must be revitalized and encouraged more than ever. There is a need for palaeontologists and neontologists capable of encapsulating systematic data to infer testable systematic hypotheses [82].

2.4. Exploring Macroevolution

The combination of evidence from species-level phylogenies (with extinct and extant taxa) with robust estimates of divergence time is thus vital to infer biogeographical and macroevolutionary patterns within and among clades. [7,83]. Moreover, the integration of both morphological and molecular data for Bayesian relaxed clock analysis (total-evidence dating) provides a joint estimate of tree topology, divergence times, and evolutionary rates in a multivariate statistical framework [14,70,84]. Total-evidence dating has also been improved by the development of the FBD process to estimate more accurate priors on times [66,84]. Molecular data improve relationship information among living taxa and help to (re-) optimize the morphological characters, improving their ability to accurately place fossils [7].

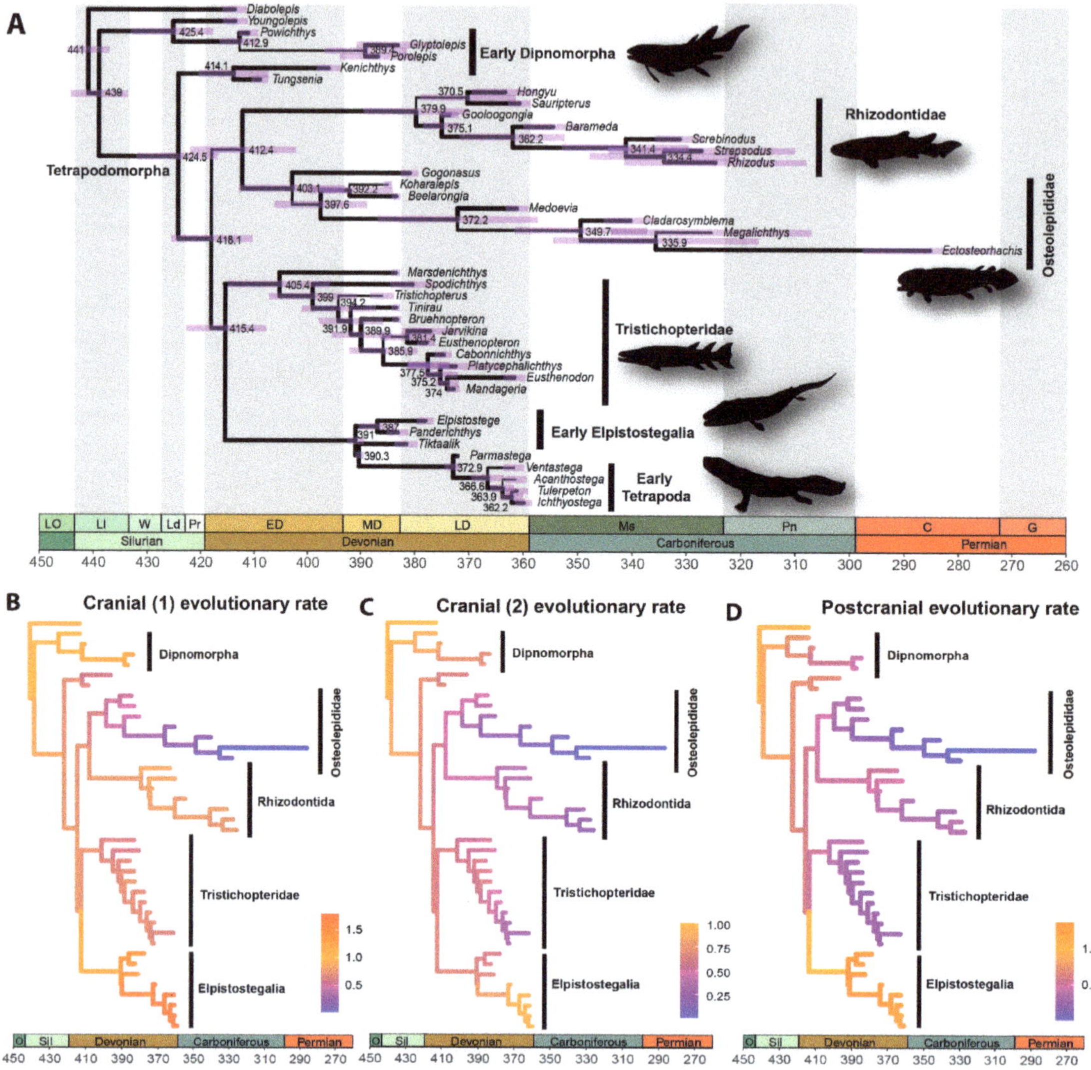

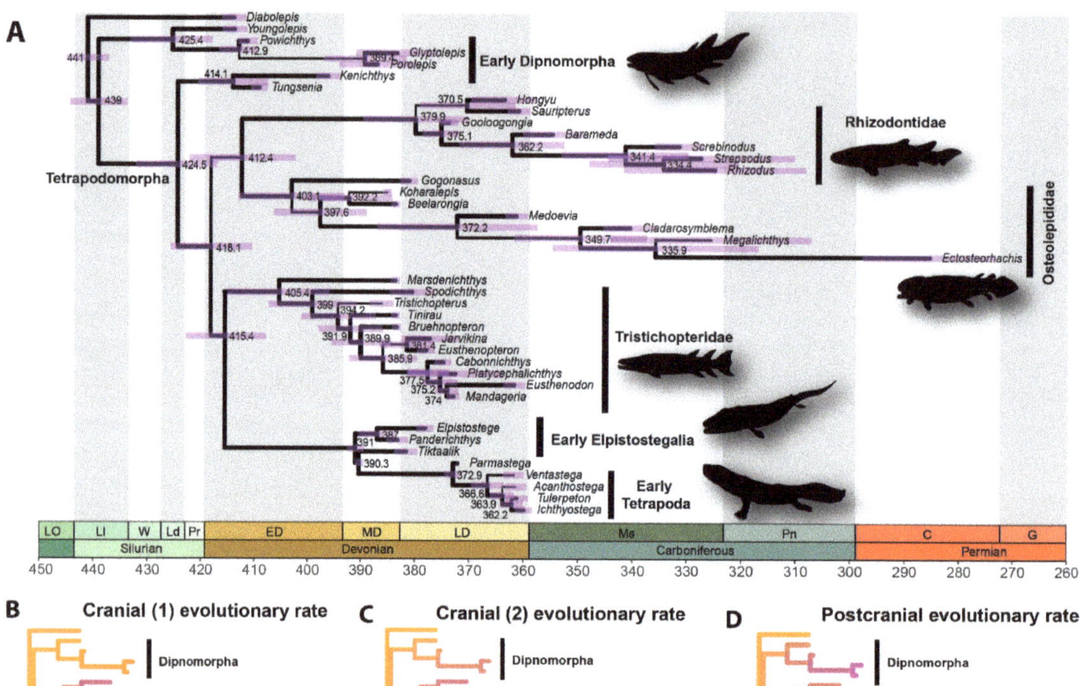

Figure 1. Bayesian evolutionary tree with estimated divergence times and evolutionary rates for the major groups of early tetrapodomorphs (adapted from [67]). (**A**) Divergence times for the fish–tetrapod transition. Node values represent median ages; purple error bars represent the 95% highest posterior density (HPD) intervals; branch thickness is proportional to posterior probabilities. (**B–D**) Relative rates of morphological evolution across subdivisions (partitions) of the phenotype in early tetrapodomorphs: two partitions including cranial characters and one partition including postcranial characters. All silhouettes created by TRS.

Ancestral state estimations represent a central tool for the exploration of trait evolution. They are useful to test hypotheses, such as the biogeographic history and movements of clades through time (e.g., [85,86]), as well as the order and timing of character state changes (e.g., [87,88]). Species distributions are defined by presence or absence in pre-defined geographic units. The most likely biogeographical scenarios at all internal nodes of a given time-calibrated evolutionary tree can be estimated using maximum likelihood or Bayesian approaches, notably with the dispersal–extinction–cladogenesis (DEC) model [85,89–92].

The DEC model and its derivatives [93] allow for investigation of time-calibrated phylogenies with extant and extinct taxa while considering tectonic evolution via the incorporation of time bins in which the connectivity between any two areas can change through time [94,95]. Geological connectivity can be coded as a matrix of connection/disconnection relying on the latest palaeogeographical scenarios (e.g., [96]) for a given region. The DEC model can also incorporate trait-dependent models in which species traits can influence dispersal rates [97]. A DEC model was applied to a dataset of European fossil muroids to reconstruct numerous transitions, revealing the most often utilised migration corridors for these ancient rodents (Figure 2). This analysis exemplifies how the combination of phylogenetic models and fossil data can produce novel insights into the structure of ancient communities and their biogeographic habits.

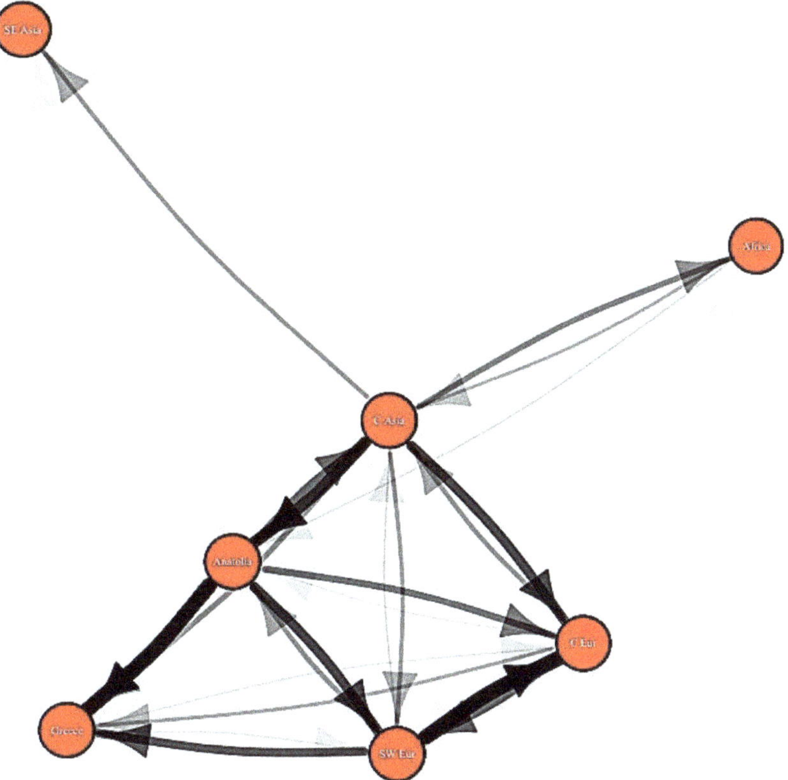

Figure 2. Map of reconstructed immigration and emigration rates for Old World Miocene muroids (work in prep.) based on a DEC analysis run using BioGeoBears in R. Arrows represent reconstructed movement of an individual lineage from one region to another. Arrows are shaded to represent the frequency of a specific transition.

Phylogenetic approaches in macroevolution now allow diversification rates be tied to changes in paleoenvironmental (extrinsic) and/or biotic (intrinsic) factors [98–100]. Relaxed clock Bayesian inference methods under the FBD tree model and its skyline variant, SFBD, allow speciation, extinction, and fossilisation parameters to vary across time bins [13,19]. Such model developments can provide more precise estimates of macroevolutionary parameters, including net diversification, turnover, and fossil sampling rates [17]. Some authors have called into question estimation of speciation and extinction rates from extant trees [26,101,102]. However, even if their limitations have been, in part, overcome by recent methodologies [103], phylogenies that combine palaeontological and neontological data have been proven to provide accurate insights into macroevolutionary dynamics [83,104,105]. This is especially evident in groups with high past diversity but which are currently extinct or only represented by a few taxa [105].

By inferring the diversification rates using several methods with different model assumptions and focusing on clades and lineages with rate shifts that are consistently estimated regardless of method, it is possible to reliably focus on the possible causes of those shifts [22]. Phylogenetically informed hierarchical Bayesian regression [106,107] is a general tool that has only recently begun to be explored. By allowing multiple factors to influence diversification rates and pooling the estimates across time and space, many parameters can be regularised to identify factors that have the largest effect on rate variation. This allows for testing of more nuanced hypotheses; instead of investigating whether abiotic or biotic factors were more important, the relative importance of many different factors can be simultaneously estimated, and their interactions can be investigated [108,109]. Using hierarchical regressions, it is possible to estimate clade-specific values that represent the unmeasured variables and assess how multiple distinct lineages differ in their evolutionary responses to climatic shifts during a given period of time and control for differing geographic locations. Using temporally, geographically, and phylogenetically well-resolved datasets to pool parameter estimates by region and clade allows for exploration of how climatic (Court Jester) and ecological (Red Queen) factors influenced diversification at a level of resolution that has not been achieved to date.

Smits [109] fitted Bayesian hierarchical models to the durations of brachiopod lineages over time and estimated the relative importance of factors such as geographic range, environmental preference, and body size on extinction intensity. This allowed the author to go beyond simply asking whether Court Jester or Red Queen effects predominated but instead to delve into how the relative importance of trait-based or environmentally based factors changed over time in these lineages. Smits was able to use the difference between overall fitness (total duration) and the strength of selection (trait-specific regression coefficients of duration) to show how background and mass extinctions vary in their selective regimes.

Shifts of diversification across a phylogeny can provide lines of evidence for the respective role of biotic and climatic variables in macroevolution [22,110]. A major outstanding question in macroevolution is the extent to which diversification rates are influenced by organismal traits and environmental changes. To differentiate between the Court Jester (extrinsic controls, such as environment and geography) and Red Queen (intrinsic controls, such as traits) hypotheses in shaping the radiation of a given group, palaeontologists have to provide a wealth of data from large geographical areas over long temporal intervals. Smits [108] used a hierarchical Bayesian model to determine which parameters best explain the durations of North American fossil mammal species over the Cenozoic. This flexible approach facilitated estimation of how different factors, ranging from geographic region to locomotor, and dietary categories influenced extinction risk in Cenozoic mammals while accounting for unobserved clade- and species-specific factors.

Repeated convergent trait evolution in clades allows for an examination of the role disparity plays while minimising the effects of phylogenetic pseudoreplication [111,112]. Probabilistic models, such as fossil BAMM [104] or PyRate [113], have been developed to estimate the rates of diversification and preservation using time-calibrated trees and fossil occurrences. Probabilistic models incorporate the distribution of lineage durations (along with the number of fossil occurrences to estimate a preservation rate) to estimate the optimal combination of speciation and extinction events explaining the shape and distribution of branch lengths in a phylogeny.

Uncertainty exists with respect the measured traits, the shape of the phylogeny, and the estimates of the rates themselves. Fuentes-G. [114] recently extended phylogenetic regressions to accommodate these different levels and degrees of uncertainty. In their study [114], they explored how allometric relationships vary in posture across a phylogeny of mammals, although their flexible approach is applicable to any dataset.

Such models are increasingly used in palaeobiological and macroevolutionary studies [108,109,114–118]. Once rates have been estimated using these approaches, hierarchical Bayesian regressions [108] can be used to identify associations between diversification rates and abiotic/biotic variables (such as climate, traits, and local richness) to evaluate the relative importance of each variable in driving diversification. Cole [115] examined the effect of the importance of environmental and habitat-based traits relative to traits focused on feeding mechanisms and selectivity on the extinction propensity of crinoids. This study demonstrated that the same biotic trait (body size) could have opposite effects on extinction depending on the abiotic environment (mixed or siliciclastic) in these crinoids. The ability to test nuanced hypotheses such as that outlined above, whereby environmental conditions influence not only the diversification rates themselves but also the importance and directionality of different biotic traits (and potentially vice-versa) is a recent and exciting development.

3. Conclusions

The statistical techniques mentioned above have only begun to be applied to questions in palaeontology over the past decade but have found extensive applications in phylogenetic comparative analysis, quantitative genetics, and ecology. Complementary methodologies that combine morphological and molecular approaches can provide novel answers to broad evolutionary and deep-time questions with methods to infer the dynamics of speciation and extinction, as well as the variation in species diversification among lineages, using time-calibrated phylogenetic trees. These recent developments provide palaeontologists with a golden opportunity to considerably expand their research toolkit and bridge emerging techniques from evolutionary biology and paleobiology. In paleontological research the many challenges with respect to our understanding of how life has evolved and survived on Earth have to be approached in a collaborative and integrative fashion. Many of the most important problems can now be solved with interdisciplinary teams of scientists using the best available technology. However, it is crucial that all scientist inside and outside the discipline restore the place of palaeontology at the high table of evolutionary biology where it belongs.

Author Contributions: Investigation, R.L.-A., J.M. and T.R.S.; writing—original draft preparation, R.L.-A.; writing—review and editing, R.L.-A., J.M., T.R.S., F.L.C., R.A., P.P.-C., S.R., J.R. and P.C.J.D.; visualization, J.M. and T.R.S.; supervision, R.L.-A.; project administration, R.L.-A. All authors have read and agreed to the published version of the manuscript.

Funding: ANR-AAPG 2022. PGC2018-094122-B-100 (MICU/AEI/FEDER, EU). National Science and Engineering Research Council of Canada (NSERC) postdoctoral fellowship to T.R.S.

Institutional Review Board Statement: Not applicable.

Informed Consent Statement: Not applicable.

Data Availability Statement: Not applicable.

Conflicts of Interest: The authors declare no conflict of interest.

References

1. Kumar, S. Molecular clocks: Four decades of evolution. *Nat. Rev. Genet.* **2005**, *6*, 654–662. [CrossRef]
2. Kluge, A.G. A concern for evidence and a phylogenetic hypothesis of relationships among Epicrates (Boidae, Serpentes). *Syst. Zool.* **1989**, *38*, 7–25. [CrossRef]
3. Nixon, K.C.; Carpenter, J.M. On simultaneous analysis. *Cladistics* **1996**, *12*, 221–241. [CrossRef]
4. Reisz, R.R.; Müller, J. Molecular timescales and the fossil record: A paleontological perspective. *Trends Genet.* **2004**, *20*, 237–241. [CrossRef]
5. Benton, M.J.; Donoghue, P.C.J. Paleontological Evidence to Date the Tree of Life. *Mol. Biol. Evol.* **2007**, *24*, 26–53. [CrossRef]

6.   Donoghue, P.C.J.; Benton, M. Rocks and clocks: Calibrating the Tree of Life using fossils and molecules. *Trends Ecol. Evol.* **2007**, *22*, 424–431. [CrossRef]

7.   Lee, M.S.; Palci, A. Morphological Phylogenetics in the Genomic Age. *Curr. Biol.* **2015**, *25*, R922–R929. [CrossRef]

8.   Parham, J.F.; Donoghue, P.C.J.; Bell, C.J.; Calway, T.D.; Head, J.J.; Holroyd, P.A.; Inoue, J.G.; Irmis, R.B.; Joyce, W.G.; Ksepka, D.T.; et al. Best Practices for Justifying Fossil Calibrations. *Syst. Biol.* **2011**, *61*, 346–359. [CrossRef]

9.   Pyron, R.A. Divergence time estimation using fossils as terminal taxa and the origins of Lissamphibia. *Syst. Biol.* **2011**, *60*, 466–481. [CrossRef]

10.   Ronquist, F.; Teslenko, M.; van der Mark, P.; Ayres, D.L.; Darling, A.; Höhna, S.; Larget, B.; Liu, L.; Suchard, M.A.; Huelsenbeck, J.P. MrBayes 3.2, efficient Bayesian phylogenetic inference and model choice across a large model space. *Syst. Biol.* **2012**, *61*, 539–542. [CrossRef]

11.   Turner, A.H.; Pritchard, A.C.; Matzke, N.J. Empirical and Bayesian approaches to fossil-only divergence times: A study across three reptile clades. *PLoS ONE* **2017**, *12*, e0169885. [CrossRef]

12.   Hunt, G.; Slater, G. Integrating Paleontological and Phylogenetic Approaches to Macroevolution. *Annu. Rev. Ecol. Evol. Syst.* **2016**, *47*, 189–213. [CrossRef]

13.   Gavryushkina, A.; Welch, D.; Stadler, T.; Drummond, A.J. Bayesian inference of sampled ancestor trees for epidemiology and fossil calibration. *PLOS Comput. Biol.* **2014**, *10*, e1003919. [CrossRef]

14.   Warnock, R.; Wright, A. *Understanding the Tripartite Approach to Bayesian Divergence Time Estimation (Elements of Paleontology)*; Cambridge University Press: Cambridge, UK, 2021; pp. 1–46.

15.   Wright, A.; Wagner, P.; Wright, D. *Testing Character Evolution Models in Phylogenetic Paleobiology: A Case Study with Cambrian Echinoderms (Elements of Paleontology)*; Cambridge University Press: Cambridge, UK, 2021; pp. 1–52.

16.   Simões, T.R.; Vernygora, O.; Caldwell, M.W.; Pierce, S.E. Megaevolutionary dynamics and the timing of evolutionary innovation in reptiles. *Nat. Commun.* **2020**, *11*, 3322. [CrossRef]

17.   Simões, T.R.; Caldwell, M.W.; Pierce, S.E. Sphenodontian phylogeny and the impact of model choice in Bayesian morphological clock estimates of divergence times and evolutionary rates. *BMC Biol.* **2020**, *18*, 191. [CrossRef]

18.   Zhang, C. Selecting and averaging relaxed clock models in Bayesian tip dating of Mesozoic birds. *Paleobiology* **2021**, *48*, 340–352. [CrossRef]

19.   Zhang, C.; Stadler, T.; Klopfstein, S.; Heath, T.A.; Ronquist, F. Total-evidence dating under the fossilized birth-death process. *Syst. Biol.* **2016**, *65*, 228–249. [CrossRef]

20.   Benton, M.J. Forth, J.; Langer, M.C. Models for the rise of the dinosaurs. *Curr. Biol.* **2014**, *24*, R87–R95. [CrossRef]

21.   Benton, M.J. Exploring macroevolution using modern and fossil data. *Proc. R. Soc. B* **2015**, *282*, 20150569. [CrossRef]

22.   Condamine, F.; Rolland, J.; Höhna, S.; Sperling, F.; Sanmartin, I. Testing the Role of the Red Queen and Court Jester as Drivers of the Macroevolution of Apollo Butterflies. *Syst. Biol.* **2018**, *67*, 940–964. [CrossRef]

23.   Cantalapiedra, J.L.; FitzJohn, R.G.; Kuhn, T.S.; Hernández-Fernández, M.; DeMiguel, D.; Azanza, B.; Morales, J.; Mooers, A.Ø. Dietary innovations spurred the diversification of ruminants during the Caenozoic. *Proc. R. Soc. B* **2014**, *281*, 20132746. [CrossRef]

24.   Clavel, J.; Morlon, H. Accelerated body size evolution during cold climatic periods in the Cenozoic. *Proc. Natl. Acad. Sci. USA* **2017**, *114*, 4183–4188. [CrossRef]

25.   Lawing, A.M.; Polly, P.D.; Hews, D.K.; Martins, E.P. Including Fossils in Phylogenetic Climate Reconstructions, A Deep Time Perspective on the Climatic Niche Evolution and Diversification of Spiny Lizards (*Sceloporus*). *Am. Nat.* **2016**, *188*, 133–148. [CrossRef]

26.   Louca, S.; Pennell, M.W. Extant timetrees are consistent with a myriad of diversification histories. *Nature* **2020**, *580*, 502–505. [CrossRef]

27.   Louca, S.; Pennell, M.W. Why extinction estimates from extant phylogenies are so often zero. *Curr. Biol.* **2021**, *31*, 3168–3173. [CrossRef]

28.   Agnarsson, I.; Kuntner, M. Taxonomy in a changing world, seeking solutions for a science in crisis. *Syst. Biol.* **2007**, *56*, 531–559. [CrossRef]

29.   López-Antoñanzas, R.; Knoll, F.; Maksoud, S.; Azar, D. Miocene rodent from Lebanon provides the "missing link" between Asian and African gundis (Rodentia, Ctenodactylidae). *Sci. Rep.* **2015**, *5*, 12871. [CrossRef]

30.   López-Antoñanzas, R.; Renaud, S.; Peláez-Campomanes, P.; Azar, D.; Kachacha, G.; Knoll, F. First levantine fossil murines shed new light on the first dispersal of mice. *Sci. Rep.* **2019**, *9*, 11874. [CrossRef]

31.   Wiens, J.J. Speciation and ecology revisited, phylogenetic niche conservatism and the origin of species. *Evolution* **2004**, *58*, 193–197. [CrossRef]

32.   Asher, R.J.; Geisler, J.H.; Sánchez-Villagra, M.R. Morphology, paleontology, and placental mammal phylogeny. *Syst. Biol.* **2008**, *57*, 311–317. [CrossRef]

33.   Springer, M.; Meredith, R.; Eizirik, E.; Teeling, E.; William, M. Morphology and Placental Mammal Phylogeny. *Syst. Biol.* **2008**, *57*, 499–503. [CrossRef] [PubMed]

34.   Guillerme, T.; Cooper, N. Assessment of available anatomical characters for linking living mammals to fossil taxa in phylogenetic analyses. *Biol Lett.* **2016**, *12*, 20151003. [CrossRef] [PubMed]

35.   Revell, L.J. Phytools, an R package for phylogenetic comparative biology (and other things). *Methods Ecol. Evol.* **2012**, *3*, 217–223. [CrossRef]

36. May, M.R.; Moore, B.R. A Bayesian Approach for Inferring the Impact of a Discrete Character on Rates of Continuous-Character Evolution in the Presence of Background-Rate Variation. *Syst. Biol.* **2020**, *69*, 530–544. [CrossRef] [PubMed]
37. Klingenberg, C.P.; Gidaszewski, N.A. Testing and Quantifying Phylogenetic Signals and Homoplasy in Morphometric Data. *Syst. Biol.* **2010**, *59*, 245–261.
38. Adams, D.C.; Cardini, A.; Monteiro, L.R.; O'Higgins, P.; Rohlf, F.J. Morphometrics and phylogenetics: Principal components of shape from cranial modules are neither appropriate nor effective cladistic characters. *J. Hum. Evol.* **2011**, *60*, 240–243. [CrossRef]
39. Renaud, S.; Michaux, J.; Jaeger, J.; Auffray, J. Fourier analysis applied to *Stephanomys* (Rodentia, Muridae) molars: Nonprogressive evolutionary pattern in a gradual lineage. *Paleobiology* **1996**, *22*, 255–265. [CrossRef]
40. Skinner, M.M.; Gunz, P. The presence of accessory cusps in chimpanzee lower molars is consistent with a patterning cascade model of development. *J. Anat.* **2010**, *217*, 245–253. [CrossRef]
41. Alhajeri, B.H.; Steppan, S.J. A phylogenetic test of adaptation to deserts and aridity in skull and dental morphology across rodents. *J. Mammal.* **2018**, *99*, 1197–1216. [CrossRef]
42. Smith, U.E.; Hendricks, J.R. Geometric Morphometric Character Suites as Phylogenetic Data: Extracting Phylogenetic Signal from Gastropod Shells. *Syst. Biol.* **2013**, *62*, 366–385. [CrossRef]
43. Clavel, J.; Escarguel, G.; Merceron, G. mv$_{MORPH}$, an R package for fitting multivariate evolutionary models to morphometric data. *Methods Ecol. Evol.* **2015**, *6*, 1311–1319. [CrossRef]
44. Alvarez-Carretero, S.; Goswami, A.; Yang, Z.; Dos Reis, M. Bayesian estimation of species divergence times using correlated quantitative characters. *Syst. Biol.* **2019**, *68*, 967–986. [CrossRef] [PubMed]
45. Billet, G.; Bardin, J. Serial homology and correlated characters in morphological phylogenetics: Modeling the evolution of dental crests in Placentals. *Syst. Biol.* **2019**, *68*, 267–280. [CrossRef]
46. Brocklehurst, N.; Romano, M.; Fröbisch, J. Principal component analysis as an alternative treatment for morphometric characters: Phylogeny of caseids as a case study. *Palaeontology* **2016**, *59*, 877–886. [CrossRef]
47. Parins-Fukuchi, C. Bayesian placement of fossils on phylogenies using quantitative morphometric data. *Evolution* **2018**, *72*, 1801–1814. [CrossRef] [PubMed]
48. Ascarrunz, E.; Claude, J.; Joyce, W.G. The phylogenetic relationships of geoemydid turtles from the Eocene Messel Pit Quarry: A first assessment using methods for continuous and discrete characters. *PeerJ* **2021**, *9*, e11805. [CrossRef] [PubMed]
49. Gavryushkina, A.; Heath, T.A.; Ksepka, D.T.; Stadler, T.; Welch, D.; Drummond, A.J. Bayesian total-evidence dating reveals the recent crown radiation of penguins. *Syst. Biol.* **2017**, *66*, 57–73. [CrossRef]
50. Parry, L.A.; Edgecombe, G.D.; Eibye-Jacobsen, D.; Vinther, J. The impact of fossil data on annelid phylogeny inferred from discrete morphological characters. *Proc. R. Soc. B* **2016**, *283*, 20161378. [CrossRef]
51. Simões, T.R.; Caldwell, M.W.; Tałanda, M.; Bernardi, M.; Palci, A.; Vernygora, O.; Bernardini, F.; Mancini, L.; Nydam, R.L. The origin of squamates revealed by a Middle Triassic lizard from the Italian Alps. *Nature* **2018**, *557*, 706–709. [CrossRef]
52. López-Antoñanzas, R.; Peláez-Campomanes, P. Bayesian morphological clock versus Parsimony: An insight into the relationships and dispersal events of postvacuum Cricetidae (Rodentia, Mammalia). *Syst. Biol.* **2022**, *71*, 512–525. [CrossRef]
53. Miyashita, T.; Gess, R.W.; Tietjen, K. Non-ammocoete larvae of Palaeozoic stem lampreys. *Nature* **2021**, *591*, 408–412. [CrossRef] [PubMed]
54. Fisher, D.C. Stratocladistics, morphological and temporal patterns and their relation to phylogenetic process. In *Interpreting the Hierarchy of Nature—From Systematic Patterns to Evolutionary Theories*; Grande, L., Rieppel, O., Eds.; Academic Press: New York, NY, USA, 1994; pp. 133–171.
55. Wagner, P.J. Stratigraphic tests of cladistic hypotheses. *Paleobiology* **1995**, *21*, 153–178. [CrossRef]
56. Bapst, D.W. Paleotree, an R package for paleontological and phylogenetic analyses of evolution. *Methods Ecol. Evol.* **2012**, *3*, 803–807. [CrossRef]
57. Bapst, D.W. Assessing the effect of time-scaling methods on phylogeny-based analyses in the fossil record. *Paleobiology* **2014**, *40*, 331–351. [CrossRef]
58. Bell, M.A.; Lloyd, G.T. Strap: An R package for plotting phylogenies against stratigraphy and assessing their stratigraphic congruence. *Palaeontology* **2014**, *58*, 379–389. [CrossRef]
59. O'Connor, A.; Wills, M.A. Measuring stratigraphic congruence across Trees, higher taxa, and time. *Syst Biol.* **2016**, *65*, 792–811. [CrossRef]
60. King, B.; Beck, R. Bayesian Tip-dated Phylogenetics, Topological Effects, Stratigraphic Fit and the Early Evolution of Mammals. *bioRxiv* **2019**. [CrossRef]
61. King, B. Bayesian tip-dated phylogenetics in paleontology: Topological effects and stratigraphic fit. *Syst. Biol.* **2021**, *70*, 283–294. [CrossRef]
62. Sansom, R.S.; Choate, P.G.; Keating, J.N.; Randle, E. Parsimony, not Bayesian analysis, recovers more stratigraphically congruent phylogenetic trees. *Biol. Lett.* **2018**, *14*, 20180263. [CrossRef]
63. King, B.; Qiao, T.; Lee, M.S.; Zhu, M.; Long, J.A. Bayesian morphological clock methods resurrect placoderm monophyly and reveal rapid early evolution in Jawed vertebrates. *Syst. Biol.* **2017**, *66*, 499–516. [CrossRef]
64. King, B.; Rücklin, M. Tip dating with fossil sites and stratigraphic sequences. *PeerJ* **2020**, *8*, e9368. [CrossRef] [PubMed]
65. Stadler, T. Sampling-through-time in birth–death trees. *J. Theor. Biol.* **2010**, *267*, 396–404. [CrossRef] [PubMed]

66. Heath, T.A.; Huelsenbeck, J.P.; Stadler, T. Fossilized birth–death process. *Proc. Natl. Acad. Sci. USA* **2014**, *111*, E2957–E2966. [CrossRef] [PubMed]

67. Simões, T.R.; Pierce, S.E. Sustained high rates of morphological evolution during the rise of tetrapods. *Nat. Ecol. Evol.* **2021**, *5*, 1403–1414. [CrossRef]

68. Paterson, J.R.; Edgecombe, G.D.; Lee, M.S.Y. Trilobite evolutionary rates constrain the duration of the Cambrian explosion. *Proc. Natl. Acad. Sci. USA* **2019**, *116*, 4394–4399. [CrossRef]

69. Jouault, C.; Maréchal, A.; Condamine, F.L.; Wang, B.; Nel, A.; Legendre, F.; Perrichot, V. Including fossils in phylogeny: A glimpse into the evolution of the superfamily Evanioidea (Hymenoptera, Apocrita) under tip-dating and the fossilized birth–death process. *Zool. J. Linn. Soc.* **2022**, *194*, 1396–1423. [CrossRef]

70. Lee, M.S.Y.; Cau, A.; Naish, D.; Dyke, G.J. Morphological clocks in paleontology, and a mid-Cretaceous origin of crown Aves. *Syst. Biol.* **2014**, *63*, 442–449. [CrossRef]

71. Giles, S.; Xu, G.-H.; Near, T.J.; Friedman, M. Early members of 'living fossil' lineage imply later origin of modern ray-finned fishes. *Nature* **2017**, *549*, 265. [CrossRef]

72. Halliday, T.J.D.; dos Reis, M.; Tamuri, A.U.; Ferguson-Gow, H.; Yang, Z.; Goswami, A. Rapid morphological evolution in placental mammals post-dates the origin of the crown group. *Proc. R. Soc. B* **2019**, *286*, 20182418. [CrossRef]

73. Pyron, R.A. Novel Approaches for Phylogenetic Inference from Morphological Data and Total-Evidence Dating in Squamate Reptiles (Lizards, Snakes, and Amphisbaenians). *Syst. Biol.* **2017**, *66*, 38–56. [CrossRef]

74. Luo, A.; Duchêne, D.A.; Chao-Dong Zhu, C.Z.; Ho, S.Y.W.A. Simulation-Based Evaluation of Tip-Dating Under the Fossilized Birth–Death Process. *Syst. Biol.* **2020**, *69*, 325–344. [CrossRef] [PubMed]

75. Mongiardino Koch, N.; Thompson, J.R.; Hiley, A.S.; McCowin, M.F.; Armstrong, A.F.; Coppard, S.E.; Aguilera, F.; Bronstein, O.; Kroh, A.; Mooi, R.; et al. Phylogenomic analyses of echinoid diversification prompt a re-evaluation of their fossil record. *eLife* **2022**, *11*, e72460. [CrossRef] [PubMed]

76. O'Reilly, J.E.; Dos Reis, M.; Donoghue, P.C.J. Dating tips for divergence-time estimation. *Trends Genet.* **2015**, *31*, 637–650. [CrossRef] [PubMed]

77. Ronquist, F.; Lartillot, N.; Phillips, M.J. Closing the gap between rocks and clocks using total-evidence dating. *Phil. Trans. R. Soc. B* **2016**, *371*, 20150136. [CrossRef]

78. Donoghue, P.C.; Yang, Z. The evolution of methods for establishing evolutionary timescales. *Philos. Trans. R. Soc. Lond. B* **2016**, *371*, 20160020. [CrossRef]

79. O'Reilly, J.E.; Donoghue, P.C.J. Tips and nodes are complementary not competing approaches to the calibration of molecular clocks. *Biol. Lett.* **2016**, *12*, 20150975. [CrossRef]

80. Cole, T.L.; Zhou, C.; Fang, M.; Pan, H.; Ksepka, D.; Fiddaman, S.; Emerling, C.; Thomas, D.; Bi, X.; Fang, Q.; et al. Genomic insights into the secondary aquatic transition of penguins. *Nat. Commun.* **2022**, *13*, 3912. [CrossRef]

81. Neumann, J.S.; Desalle, R.; Narechania, A.; Schierwater, B.; Tessler, M. Morphological Characters Can Strongly Influence Early Animal Relationships Inferred from Phylogenomic Data Sets. *Syst Biol.* **2021**, *70*, 360–375. [CrossRef]

82. Lipps, J.H. The future of paleontology—The next 10 years. *Palaeontol. Electr.* **2017**, *10*, 1A.

83. Rabosky, D.L. Extinction rates should not be estimated from molecular phylogenies. *Evolution* **2010**, *64*, 1816–1824. [CrossRef]

84. Zhang, C.; Wang, M. Bayesian tip dating reveals heterogeneous morphological clocks in Mesozoic birds. *R. Soc. Open Sci.* **2019**, *6*, 182062. [CrossRef] [PubMed]

85. Ree, R.H.; Smith, S.A. Maximum Likelihood Inference of Geographic Range Evolution by Dispersal, Local Extinction, and Cladogenesis. *Syst Biol.* **2008**, *57*, 4–14. [CrossRef] [PubMed]

86. Landis, M.J.; Eaton, D.A.R.; Clement, W.L.; Park, B.; Spriggs, E.L.; Sweeney, P.W.; Edwards, E.J.; Donoghue, M.J. Joint phylogenetic estimation of geographic movements and biome shifts during the global diversification of Viburnum. *Syst. Biol.* **2021**, *70*, 67–85. [CrossRef] [PubMed]

87. Schluter, D.; Price, T.; Mooers, A.Ø.; Ludwig, D. Likelihood of Ancestor States in Adaptive Radiation. *Evolution* **1997**, *51*, 1699–1711. [CrossRef]

88. Ackerly, D.D.; Schwilk, D.W.; Webb, C.O. Niche evolution and adaptive radiation, testing the order of trait divergence. *Ecology* **2006**, *87*, S50–S61. [CrossRef]

89. Beeravolu, C.R.; Condamine, F.L. An extended Maximum Likelihood inference of geographic range evolution by dispersal, local extinction and cladogenesis. *bioRxiv* **2016**. [CrossRef]

90. Matzke, N.J. Model selection in historical biogeography reveals that founder-event speciation is a crucial process in island clades. *Syst. Biol.* **2014**, *63*, 951–970. [CrossRef]

91. Matzke, N.J. Statistical Comparison of DEC and DEC+J Is Identical to Comparison of Two Classe Submodels, and Is Therefore Valid. *OSF Prepr.* **2021**. [CrossRef]

92. Ree, R.H. Detecting the historical signature of key innovations using stochastic models of character evolution and cladogenesis. *Evolution* **2005**, *59*, 257–265. [CrossRef]

93. Goldberg, E.E.; Lancaster, L.T.; Ree, R.H. Phylogenetic Inference of Reciprocal Effects between Geographic Range Evolution and Diversification. *Syst. Biol.* **2011**, *60*, 451–465. [CrossRef]

94. Wood, H.M.; Matzke, N.J.; Gillespie, R.G.; Griswold, C.E. Treating Fossils as Terminal Taxa in Divergence Time Estimation Reveals Ancient Vicariance Patterns in the Palpimanoid Spiders. *Syst Biol.* **2013**, *62*, 264–284. [CrossRef] [PubMed]

95. Tejero-Cicuéndez, H.; Simó-Riudalbas, M.; Menéndez, I.; Carranza, S. Ecological specialization, rather than the island effect, explains morphological diversification in an ancient radiation of geckos. *Proc. R. Soc. B* **2021**, *288*, 20211821. [CrossRef] [PubMed]
96. He, Z.; Zhang, Z.; Guo, Z.; Scotese, C.R.; Deng, C. Middle Miocene (~14 Ma) and late Miocene (~6 Ma) paleogeographic boundary conditions. *Paleoceanogr. Paleoclimatol.* **2021**, *36*, e2021PA004298. [CrossRef]
97. Klaus, K.V.; Matzke, N.J. Statistical Comparison of Trait-Dependent Biogeographical Models Indicates That Podocarpaceae Dispersal Is Influenced by Both Seed Cone Traits and Geographical Distance. *Syst. Biol.* **2020**, *69*, 61–75. [CrossRef]
98. Condamine, F.L.; Rolland, J.; Morlon, H. Macroevolutionary perspectives to environmental change. *Ecol. Lett.* **2013**, *16*, 72–85. [CrossRef]
99. Condamine, F.L.; Silvestro, D.; Koppelhus, E.B.; Antonelli, A. The rise of angiosperms pushed conifers to decline during global cooling. *Proc. Natl. Acad. Sci. USA* **2020**, *117*, 28867–28875. [CrossRef] [PubMed]
100. Condamine, F.L.; Romieu, J.; Guinot, G. Climate cooling and clade competition likely drove the decline of lamniform sharks. *Proc. Natl. Acad. Sci. USA* **2019**, *116*, 20584–20590. [CrossRef]
101. Quental, T.; Marshall, C.R. Diversity dynamics: Molecular phylogenies need the fossil record. *Trends Ecol. Evol.* **2010**, *25*, 434–441. [CrossRef]
102. O'Meara, B.C.; Beaulieu, J.M. Potential survival of some, but not all, diversification 661 methods. 2021. Available online: https://ecoevorxiv.org/w5nvd/ (accessed on 6 November 2021).
103. Morlon, H.; Parsons, T.L.; Plotkin, J.B. Reconciling molecular phylogenies with the fossil record. *Proc. Natl. Acad. Sci. USA* **2011**, *108*, 16327–16332. [CrossRef]
104. Mitchell, J.; Etienne, R.S.; Rabosky, D.L. Inferring diversification rate variation from phylogenies with fossils. *Syst. Biol.* **2019**, *68*, 1–18. [CrossRef]
105. Silvestro, D.; Salamin, N.; Schnitzler, J. PyRate, a new program to estimate speciation and extinction rates from incomplete fossil data. *Methods Ecol. Evol.* **2014**, *5*, 1126–1131. [CrossRef]
106. Currie, T.; Meade, A. Keeping yourself updated, Bayesian approaches in phylogenetic comparative methods with a focus on Markov Chain Models of discrete character evolution. In *Modern Phylogenetic Comparative Methods*; Garamszegi, L.Z., Ed.; Springer: Sevilla, Spain, 2014; pp. 263–287.
107. Bürkner, P.-C. Advanced Bayesian Multilevel Modeling with the R Package brms. *R J.* **2018**, *10*, 395–411. [CrossRef]
108. Smits, P.D. Expected time-invariant effects of biological traits on mammal species duration. *Proc. Natl. Acad. Sci. USA* **2015**, *112*, 13015–13020. [CrossRef] [PubMed]
109. Smits, P.D. How macroecology affects macroevolution, The interplay between extinction intensity and trait-dependent extinction in brachiopods. *bioRxiv* **2019**. [CrossRef]
110. Drummond, C.S.; Eastwood, R.J.; Miotto, S.T.S.; Hughes, C.E. Multiple continental radiations and correlates of diversification in *Lupinus* (Leguminosae), testing for key innovation with incomplete taxon sampling. *Syst. Biol.* **2012**, *61*, 443–460. [CrossRef]
111. Maddison, W.P.; FitzJohn, R.G. The Unsolved Challenge to Phylogenetic Correlation Tests for Categorical Characters. *Syst. Biol.* **2015**, *64*, 127–136. [CrossRef]
112. Rabosky, D.L. Goldberg, E.E. FiSSE, A simple nonparametric test for the effects of a binary character on lineage diversification rates. *Evolution* **2017**, *71*, 1432–1442. [CrossRef]
113. Silvestro, D.; Salamin, N.; Antonelli, A.; Meyer, X. Improved estimation of macroevolutionary rates from fossil data using a Bayesian framework. *Paleobiology* **2019**, *45*, 546–570. [CrossRef]
114. Fuentes-G., J.A.; Polly, P.D.; Martins, E.P. A Bayesian extension of phylogenetic generalized least squares: Incorporating uncertainty in the comparative study of trait relationships and evolutionary rates. *Evolution* **2020**, *74*, 311–325. [CrossRef]
115. Cole, S.R. Hierarchical controls on extinction selectivity across the diplobathrid crinoid phylogeny. *Paleobiology* **2021**, *47*, 251–270. [CrossRef]
116. Hagen, O.; Andermann, T.; Quental, T.B.; Antonelli, A.; Silvestro, D. Estimating Age-Dependent Extinction: Contrasting Evidence from Fossils and Phylogenies. *Syst. Biol.* **2018**, *67*, 458–474. [CrossRef] [PubMed]
117. Januario, M.; Quental, T.B. Re-evaluation of the "law of constant extinction" for ruminants at different taxonomical scales. *Evolution* **2021**, *75*, 656–671. [CrossRef] [PubMed]
118. Kostikova, A.; Silvestro, D.; Pearman, P.B.; Salamin, N. Bridging Inter- and Intraspecific Trait Evolution with a Hierarchical Bayesian Approach. *Syst. Biol.* **2016**, *65*, 417–431. [CrossRef] [PubMed]

Review

Chemistry and Analysis of Organic Compounds in Dinosaurs

Mariam Tahoun [1], Marianne Engeser [2], Vigneshwaran Namasivayam [1], Paul Martin Sander [3] and Christa E. Müller [1,*]

[1] PharmaCenter Bonn, Pharmaceutical Institute, Pharmaceutical & Medicinal Chemistry, University of Bonn, D-53121 Bonn, Germany; mtahoun@uni-bonn.de (M.T.); vnamasiv@uni-bonn.de (V.N.)
[2] Kekulé Institute for Organic Chemistry and Biochemistry, University of Bonn, D-53121 Bonn, Germany; marianne.engeser@uni-bonn.de
[3] Institute of Geosciences, Section Paleontology, University of Bonn, D-53113 Bonn, Germany; martin.sander@uni-bonn.de
* Correspondence: christa.mueller@uni-bonn.de

Simple Summary: Fossils of dinosaurs other than birds are at least 66 million years old. Nevertheless, many organic compounds have survived fossilization and can still be found in the fossils. This article describes the discovery of organic molecules in dinosaur fossils. It provides a review of the analytical methods used for their detection and characterization, and presents the wide range of chemical organic compounds, including small molecules and polymers, that have been found in dinosaurs to date. The difficulties in unambiguously confirming the presence of some of the organic molecules in these fossils are also discussed.

Abstract: This review provides an overview of organic compounds detected in non-avian dinosaur fossils to date. This was enabled by the development of sensitive analytical techniques. Non-destructive methods and procedures restricted to the sample surface, e.g., light and electron microscopy, infrared (IR) and Raman spectroscopy, as well as more invasive approaches including liquid chromatography coupled to tandem mass spectrometry (LC-MS/MS), time-of-flight secondary ion mass spectrometry, and immunological methods were employed. Organic compounds detected in samples of dinosaur fossils include pigments (heme, biliverdin, protoporphyrin IX, melanin), and proteins, such as collagens and keratins. The origin and nature of the observed protein signals is, however, in some cases, controversially discussed. Molecular taphonomy approaches can support the development of suitable analytical methods to confirm reported findings and to identify further organic compounds in dinosaur and other fossils in the future. The chemical properties of the various organic compounds detected in dinosaurs, and the techniques utilized for the identification and analysis of each of the compounds will be discussed.

Keywords: fossil; dinosaur; molecular paleontology; paleoproteomics; porphyrin; collagen; melanin; keratin

Citation: Tahoun, M.; Engeser, M.; Namasivayam, V.; Sander, P.M.; Müller, C.E. Chemistry and Analysis of Organic Compounds in Dinosaurs. *Biology* **2022**, *11*, 670. https://doi.org/10.3390/biology11050670

Academic Editors: Mary H. Schweitzer and Ferhat Kaya

Received: 20 March 2022
Accepted: 22 April 2022
Published: 27 April 2022

Publisher's Note: MDPI stays neutral with regard to jurisdictional claims in published maps and institutional affiliations.

Copyright: © 2022 by the authors. Licensee MDPI, Basel, Switzerland. This article is an open access article distributed under the terms and conditions of the Creative Commons Attribution (CC BY) license (https://creativecommons.org/licenses/by/4.0/).

1. Introduction

After an organism's death, microbial decomposition of organic constituents occurs very fast, mostly leaving behind mineralized skeletal remains. If this degradation process is arrested early enough, due to factors related to the burial environment and dependent on the characteristics of the molecular or tissue components [1–3], preservation of "soft tissue" can occur. Such fossils are exceptional and very valuable because they may contain information related to evolution, biology, or the environment that can be revealed by analyzing their composition [2,3]. Preserved soft tissue has been reported from a variety of fossil fish, amphibians, reptiles, dinosaurs, and mammals. This includes cells, organelles, skin, scales, feathers, hair, colored structures, digestive organs, eggshells, and muscles [2]. This mode of preservation is unique because the original organic material is minimally altered. Furthermore, this should be differentiated from fossilization involving alteration of original material, e.g., replacement of organic matter by minerals such as phosphates (phosphatization) or conversion to thin films of carbon (carbonization) [4,5].

Researchers try to understand the factors that hinder decay processes and contribute to the preservation of organic compounds present in soft tissues. These include (but are not limited to) intrinsic properties of the organic molecules, their environment (including metals such as Fe and Mn present) [6–8], and the type of preserved soft tissue. The presence of moisture, microorganisms or enzymes speeds up the decay process [1]. The most labile molecular bonds are first targeted during decomposition. For example, proteins and DNA are susceptible to degradation by hydrolysis of their peptide and phosphoric acid ester bonds, respectively. However, association of organic compounds with minerals (e.g., in bone or teeth) or with macromolecules may isolate and protect them from the external environment.

Oxidative conditions usually lead to faster decay than reductive conditions. Hydrophobic organic compounds are more likely to be preserved than hydrophilic compounds because of their limited water-solubility, which protects them from hydrolysis and other reactions. Polymeric structures may be preserved due to crosslinking and intramolecular interactions. Environmental factors greatly affect fossilization, e.g., by applying pressure on tissue, limiting the mobility of molecules and exposure to water, microbes, and enzymes. Moreover, extremes of temperature, pH, and salinity play a role in molecular preservation by inhibiting microbial activity and affecting the rate of the chemical decomposition process. Taphonomic studies at a molecular scale ("molecular taphonomy") can be used to establish analytical methods for understanding chemical processes that lead to the degradation of organic compounds upon fossilization ([1,2,9–15] and references therein).

Since the first discoveries of microstructures (collagen-like fibrils, vessels, and cells) in a 200-million-year-old dinosaur bone in 1966 [16] there has been an increased interest in studying the large number of available dinosaur fossils for signs of molecular preservation of organic compounds. Such finds provide information about the dinosaurs' biology, including their evolution, eating habits, and environment. Most reports on organic matter in fossilized dinosaurs have been focused on their bones. In recent years, studies on eggshells, cartilage, feathers, and integumentary structures have emerged, albeit mostly discussing in situ analyses, and relying on morphological and microscopic observations due the uniqueness of the studied fossils (reviewed in [2,10] and references therein).

To date, organic compounds have been recovered from a wide array of dinosaur taxa, including the early-branching coelurosaur *Sinosauropteryx* [17], the tyrannosaur *Tyrannosaurus rex* [18–20], the ovirapotorosaurs *Heyuannia huangi* [21,22] and *Citipati osmolskae* [23], the alvarezsaurid *Shuvuuia deserti* [24], the dromaeosaur *Sinornithosaurus* [17], the early-branching avialan *Anchiornis huxleyi* [25], the early-branching sauropodomorph *Lufengosaurus sp.* [3], an unidentified titanosaurid dinosaur [26], the ankylosaur *Borealopelta markmitchelli* [27], the ceratopsian *Psittacosaurus* [28], the hadrosaur *Brachylophosaurus canadensis* [29,30], an indeterminate hadrosaur material [26], and *Hypacrosaurus stebingeri* [31]. Here, we review the chemistry of the organic molecules recovered to date from fossilized non-avian dinosaurs and discuss the analytical methods used for their detection.

2. Analytical Techniques to Investigate Preserved Organic Compounds

The principles of the analytical techniques used in paleontological research, along with their advantages and drawbacks, have recently been reviewed in detail [32,33]. The application of mass spectrometry in proteomic analysis of fossils was specifically discussed by Schweitzer et al. (2019) [34]. The following paragraphs present selected analytical techniques that have been utilized to detect organic compounds in fossilized dinosaurs.

2.1. Microscopy

Initial studies carried out on fossils in search of organic matter included a thorough screening of the fossils' surface or of petrographic thin sections to identify regions in which soft tissues and associated organic compounds could be preserved [35]. Imaging techniques such as optical microscopy (OM), scanning electron microscopy (SEM), and transmission electron microscopy (TEM) have been used for this purpose. Optical microscopy is useful for the visualization of petrographic thin sections to identify preserved cellular structures. Mineralization, diagenetic alteration, and/or microbial contamination of tissues can be detected by means of this technique [35].

Electron microscopy is more powerful due to its much higher resolution. It is therefore used to examine subcellular structures in greater detail. In SEM, electrons are directed onto the surface of the sample, generating and transmitting secondary electrons to a detector. Therefore, SEM is limited to studying the sample surfaces by generating a pseudo-3D gray-scale topographical image without collecting chemical signals [35,36]. However, a technique known as energy-dispersive X-ray spectrometry (EDS is often combined with SEM, which uses high energy X-rays characteristic for a specific element released alongside the secondary electrons [32,35,37]. Integration of the elemental information from EDS into the topographical map from SEM allows the localization of elements to be identified in the sample [35]. Other variations of SEM exist, such as field emission SEM (FESEM) [35] and variable pressure SEM (VPSEM) [38]. VPSEM allows for analysis of uncoated samples within a wider range of beam energies than traditional SEM [35]. VPSEM can also be used without prior sample preparation (e.g., dehydration or drying) in soft samples [39]. Thus, FESEM and VPSEM reduce the risk of sample contamination. Both techniques have been used for the study of soft tissues in dinosaur bones [35,38].

In transmission electron microscopy (TEM), electrons are directed to partially demineralized or very thin-cut sections of a sample in a way that only the electrons that cross through the sample are detected. This feature makes TEM a high-resolution technique that can be used for identifying subcellular structures such as organelles or characteristic structural patterns, e.g., the 67 nm bands of collagen fibers [35].

2.2. Spectroscopy and Spectrometry

2.2.1. UV/Vis Spectroscopy

Ultraviolet/visible light (UV/Vis) spectroscopy is an analytical technique to measure the absorption, transmittance, or reflectance of light by molecules upon irradiation with ultraviolet (190–380 nm) or visible (380–750 nm) light [40,41]. The functional group(s) of the molecule responsible for light absorption is known as the chromophore, e.g., due to conjugated C=C double bonds and/or aromatic rings. The chromophore contains valence electrons having low excitation energy, which become excited and transit to higher energy levels when the molecule is irradiated [41]. The wavelengths at which light is absorbed can be used to identify the structure of a compound. The amount of light absorbed is directly proportional to the concentration of the compound and thus allows for its quantification [42]. UV/Vis spectroscopy is frequently used in molecular paleontology, particularly when analyzing colored fossils, to detect characteristic absorption bands of pigments; it has, for example, been used for detecting heme [18].

2.2.2. Infrared and Raman Spectroscopy

Further studies on fossils use chemical imaging techniques such as Fourier transform infrared spectroscopy (FTIR) and Raman spectroscopy to search for chemical signals, e.g., of functional groups (e.g., amide or carbonyl group) in the samples [9,21,32]. FTIR excites the vibrations of chemical bonds using infrared irradiation. Each type of chemical bond will absorb infrared (IR) waves in a distinct wave number range in the near-IR (12,500–4000 cm^{-1}), mid-IR (4000–400 cm^{-1}), or far-IR (400–10 cm^{-1}) regions. Most of the important chemical signals that are indicative of functional groups will be present in the mid-IR range [43]. FTIR can be combined with light microscopy to identify the location of the detected functional groups in the sample. However, FTIR entails many disadvantages. The wave number ranges can overlap if the sample contains many organic signals leading to frequent misinterpretations of chemical signals, especially if diagenetic changes occurred to the original structure. In addition, any contamination on the surface of the sample will be recorded in the spectra and may not be distinguishable from the sample signals. FTIR has been used to detect characteristic absorption bands of peptide bonds, amide I (C=O bond, ca. 1655 cm^{-1}) and amide II (N-H bond, ca. 1545 cm^{-1}), associated with collagen in cartilage, in addition to peptide bonds specific for melanin (1580 cm^{-1}) [32,34].

Other variants of IR spectroscopy have been used to study fossils. For example, synchrotron-radiation Fourier transformed infrared spectroscopy (SR-FTIR) uses a much brighter light source (synchrotron radiation) ranging from far-IR to near-IR [44] to produce chemical maps. SR-FTIR has a higher resolution and a better signal-to-noise ratio than classical FTIR [45]. In addition, attenuated-total reflection IR (ATR-IR) has been used for the analysis of liquid samples [46].

Raman spectroscopy applies monochromatic laser light (ultraviolet, infrared, or visible) to irradiate the layer directly below the surface of the sample. Some of the light is then scattered with a defined frequency generating a signal that can be detected and plotted as a graph of intensity versus wave number. The observed scattering depends on the type of functional group and its vibration [47]. It can be combined with other microscopic techniques such as confocal microscopy to form a chemical map of the functional groups present in the sample. Raman spectroscopy is currently one of the most preferred methods to search for preserved organic matter and other chemical constituents in fossils because it does not require exhaustive sample preparation. However, in contrast to FTIR, signals present on the outermost surface cannot be detected. The produced signals are weak, often requiring prolonged periods of intense irradiation [48], which can lead to a degradation of thermolabile compounds due to the heat produced by the laser [32,49]. Raman spectroscopy has been used for the detection of heme in dinosaur bones [18] and for the detection of the heme degradation product biliverdin and of its precursor protoporphyrin IX in dinosaur eggshells [21].

2.2.3. Mass Spectrometry

Mass spectrometric techniques are among the most sensitive, reliable methods to detect organic compounds. Soft ionization techniques allow measuring the mass-to-charge ratio of intact molecular ions. In addition, different chemical classes of compounds have characteristic fragmentation patterns observed in mass spectrometry [50]. However, detecting only fragments or only molecular ions is often not sufficient for identification of specific organic molecules [32,35], whereas a combination of both can be highly informative.

Chemical information, especially on molecular fragments, can be obtained by time-of-flight secondary ion mass spectrometry (TOF-SIMS) and pyrolysis coupled to gas chromatography-mass spectrometry (Py-GC-MS). Only fragments can be detected by the latter method because of the harsh ionization conditions used, often leading to a complete destruction of the sample.

TOF-SIMS is a surface imaging technique with ultra-high spatial resolution which directs high energy ionizing beams (e.g., gallium ions) over the sample surface. Molecules are released, ionized and often fragmented [51]. The ions are transmitted to the time-of-flight mass spectrometer and detected according to the time it takes for them to reach the detector. The heavier their masses are, the more time it will take. It can be used for analyzing fragile or small amounts of fossil samples because measurements take place at the surface without the need for extractions. Determination of the location of the signal in the sample is the main advantage of the method, and it is therefore useful for organic compound screening [52]. However, as TOF-SIMS only analyzes the surface, any changes on the surface or contamination will influence the results [32,35]. This method has been used to detect heme [53], melanin [54], protein fragments of β-keratin [24], and collagen [26] in fossils.

To overcome the extensive fragmentation, especially of higher molecular weight ions, a variant of TOF-SIMS known as cluster secondary ion mass spectrometry was developed. Its principle relies on bombardment of the sample using a polyatomic cluster of ions, such as gold (Au$_3$) or a C$_{60}$-based ion cluster, buckminsterfullerene. This allows the detection of intact molecular ions in the range of 1000–3000 D, which was not possible with traditional TOF-SIMS [55]. In addition, spatial resolution beyond the micrometer range can be achieved [56].

Py-GC-MS is a technique that uses intense heat (ca. 400–600 °C) to fragment molecular bonds. The generated fragments are gaseous; they are separated by gas chromatography and detected by mass spectrometry. Unlike TOF-SIMS, the sample is destroyed, and the location of the chemical signal in the original sample cannot be determined. This method does not require sample preparation, and therefore, the risk of detecting artifacts is lowered [57]. It has been used to detect molecular fragments characteristic of proteins, lignin and chitin in fossils [58–60]. Due to the destruction of the sample, Py-GC-MS is not preferred if alternative approaches are possible; therefore, it is only used for analyses of insoluble fossil material which cannot be analyzed otherwise [35].

All of the aforementioned analytical techniques are often not suitable for unambiguously determining the identity of organic compounds. However, they can help narrowing down sample regions that contain organic compounds, which may then be subjected to more invasive mass spectrometry techniques.

On rare or unique fossils, only non-destructive or highly sensitive methods can be applied. Modern mass spectrometry techniques now provide options for analyzing such precious samples since they require only small quantities of material.

In order to identify intact organic compounds, mild ionization methods, such as electrospray ionization, need to be applied [61]. In most cases, samples are extracted, separated by reverse-phase liquid chromatography, which is coupled to tandem mass spectrometry (LC-MS/MS). The prerequisite for this type of analysis is a solution of the analytes; thus, compounds that are insoluble in the typically used solvents (methanol, acetonitrile, water, and their mixtures) cannot be analyzed [62].

The type of mass analyzer used is decisive for mass accuracy and sensitivity of mass spectrometric measurements. Quadrupole, time-of-flight, linear ion-trap, Fourier transform ion cyclotron resonance (FT-ICR) and Orbitrap analyzers are commonly used for organic compounds in fossils. Instruments with high mass accuracy are needed to determine elemental compositions of organic compounds. With ion trap instruments or when two types of mass analyzers are combined in series, more advanced mass spectrometric analyses are possible, known as tandem mass spectrometry. Typical combinations are quadrupole/quadrupole, quadrupole/time-of-flight (q/TOF), and quadrupole or linear ion-trap coupled to Orbitrap. Tandem mass spectrometers allow for a unique type of analysis known as collision-induced dissociation, in which intact ions of a defined mass-to-charge ratio are selected and then deliberately fragmented to analyze the fragments [62]. This method is used to achieve ultra-high sensitivity, and it provides structural information on the molecules of interest. It is the method of choice in proteomics to identify peptide sequences and to obtain information about diagenetic changes to the chemical structure, and to identify post-translational modifications [34,35]. For fossils, LC-MS/MS is one of the most selective, accurate and sensitive methods to identify organic compounds. However, this is often not applicable due to limited sample availability and/or difficulties in extracting the target compounds due to a lack of solubility [33,35].

2.3. Immunological Techniques

Immunological techniques are based on antigen–antibody reactions. Antibodies used in the process are specific to a certain epitope in the target tissue. These sensitive techniques are used to screen for the presence of macromolecules such as proteins or DNA. Using antibodies, sequence determination is not possible, but regions in the sample may be located, in which proteinaceous or genetic material has been preserved, and which can be selected subsequently for mass spectrometric analysis [34]. Immunological techniques include enzyme-linked immunosorbent assays (ELISAs), Western blotting (immunoblotting), and immunohistochemistry/immunostaining procedures.

A prerequisite for the detection of proteins by Western blot and ELISA is a liquid extract containing the protein of interest. ELISA is the more sensitive technique [63]. There are different forms of ELISA: direct ELISA, indirect ELISA, sandwich ELISA and competitive ELISA, which are typically performed in well plates. The first step is to immobilize the antigen of interest by direct adsorption to the surface or through binding to a capture antibody fixed to the plate. Direct and indirect ELISA are used for antigens immobilized directly to the well plate, whereas sandwich ELISA is used for antigens bound to a capture antibody [64]. Direct ELISA uses an enzyme-linked antibody that binds directly to the antigen of interest. Upon washing to remove unbound antibodies, and subsequent addition of the suitable substrate, a color change will occur only in the wells that contain the antigen–antibody complex [65]. Indirect ELISA is used to detect the presence of antibodies rather than antigens. Addition of a sample expected to contain a primary antibody specific to the antigen of interest results in the formation of a complex with the immobilized antigen. A secondary antibody linked to an enzyme and specific to the primary antibody is added. After washing, any unbound antibodies are removed. The substrate is added and the enzymatic reaction occurs to produce a colored product that confirms the presence of the antibody [64,65].

Sandwich ELISA is used to detect the presence of antigens and is the most commonly used form of ELISA. The well surface is first coated with a capture antibody specific to the antigen of interest, onto which the antigen from a sample will be immobilized. A primary antibody specific to the antigen will then be added. If the antigen is present, the primary antibody will bind to it. The next steps are the same as those for indirect ELISA, by which the color change will confirm the presence of the antigen [64,65]. It is worth noting that sandwich ELISA will only be possible for antigens which have two separate epitopes for binding a capture antibody and a primary antibody. Using two antibodies for detection of the same antigen makes sandwich ELISA highly specific [66]. In ELISA, proteins are detected in their natural conformation.

In competitive ELISA, antigens in a sample compete with a reference antigen coated on the surface of a well in binding to a labeled primary antibody of known concentration. The sample is incubated first with the primary antibody. Then this solution is added to the wells. The more antigens are present in the sample, the more primary antibodies will bind to them [67]. Any unbound antibody will then bind to the reference antigen. Following a washing step, an enzyme-linked secondary antibody is added. The substrate for the enzyme is then added, and the intensity of the resulting color is inversely related to the concentration of the antigens present in the sample. If few primary antibodies are bound to the reference antigen, a faint color will be observed, and this indicates a high concentration of antigens in the sample [68].

Western blot is used to identify a protein from a complex mixture [64]. Before performing a Western blot experiment, the mixture of proteins in a sample are separated by polyacrylamide gel electrophoresis according to size [69,70]. There are two main types of gel electrophoresis, depending on the type of additives used: sodium dodecyl sulfate-polyacrylamide gel electrophoresis (SDS-PAGE) and blue native-polyacrylamide gel electrophoresis (BN-PAGE). SDS-PAGE uses the detergent sodium dodecyl sulfate which denatures the proteins, whereas BN-PAGE uses the mild Coomassie blue dye and does not denature the protein of interest [69,71,72]. The bands containing the separated proteins are transferred to an immobilizing nitrocellulose or polyvinylidene difluoride membrane. This is followed by adding a blocking buffer containing non-fat dried milk or 5% bovine serum albumin, in order to prevent binding of antibodies to the membrane [70]. A primary antibody specific to the protein of interest is incubated with the membrane, followed by washing to remove unbound antibodies. Then, a secondary antibody is added that binds specifically to the primary antibody, and which is radiolabeled or linked to an enzyme. Afterwards, either a substrate is added to initiate the enzymatic reaction, or a photographic film for a radio-labeled substrate is used for detection of the target antigen–antibody complex and to locate the protein [64].

To account for diagenetic changes to the original structure, polyclonal antibodies are often used during analysis of fossil extracts; however, problems with poor specificity of antibodies may arise. Both Western blot and ELISA are prone to contamination and/or interference from extraction buffer components [34].

Immunohistochemistry is based on the same principles as ELISA and Western blot, the only difference is that the antibodies are applied in situ on intact tissue instead of utilizing extracts [73]. In situ analyses are preferred to destructive techniques because they minimize the loss of precious sample material and/or degradation of organic material during preparation (e.g., after exposure to chemicals or air) [34]. Suitable microscopic tissue slides containing the epitopes of interest are fixed, usually by formalin, into a polymer or paraffin wax [74]. If the fixation process is known to mask the antigens of interest, an extra step is usually performed by physical (e.g., heat or ultrasound) or chemical (e.g., enzymatic digestion) methods to break any cross-links formed, making the antigens re-accessible to antibodies [75]. The next step is incubation with a blocking buffer such as bovine serum albumin to prevent non-specific binding. This is followed by adding primary antibodies specific to the antigen of interest, then washing to remove unbound antibodies. Fluorescence-labeled or enzyme-linked (e.g., peroxidase or alkaline phosphatase) secondary antibodies are then added [73]. Visualization of positive reactivity takes place by light or fluorescence microscopy, or after addition of substrate and monitoring of the color change due to the enzymatic reaction. This immunological assay allows for the localization of target antigens in tissues, which is not possible with ELISA and Western blot techniques [74].

3. Organic Compounds Found in Dinosaurs

The following sections will describe the evidence and chemistry of organic compounds found to date in non-avian dinosaurs. An overview of the localities and age of the dinosaurs is depicted in Figure 1.

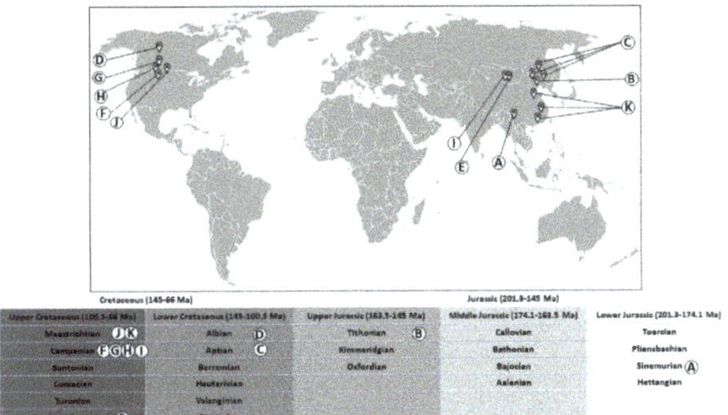

Figure 1. World map showing localities and age of dinosaurs in which organic compounds have been detected to date: (**A**) Dawa, Lufeng County, Yunnan Province, China [3]. (**B**) Yaolugao locality in Jianching County, western Liaoning Province, China [25]. (**C**) Dawangzhangzhi, Lingyuan City, Liaoning Province, China) and Sihetun, Beipiao City, Liaoning Province, China), and Yixian Formation, China [17,28]. (**D**) Suncor Millenium Mine, Fort McMurray, Alberta, Canada [27]. (**E**) Ukhaa Tolgod in southwestern Mongolia [24]. (**F**) Judith River Formation, eastern Montana, USA [29,30]. (**G**) Dinosaur Park Formation, Alberta, Canada [26]. (**H**) Two Medicine Formation, northern Montana, USA [31]. (**I**) Djadokhta Formation, Mongolia [23]. (**J**) Hell Creek Formation, eastern Montana, USA [18–20] (**K**) Chinese provinces (Henan, Jiangxi, and Guangdong) [21,22]. Concept adapted from reference [76]. The world map "BlankMap-World-IOC" by Chanheigeorge (https://commons.wikimedia.org/wiki/File:BlankMap-World-IOC.PNG, accessed on 19 March 2022) from 2008 has been used as a template onto which location markers, lines and letters were added. It is licensed under CC-BY-SA 3.0 (https://creativecommons.org/licenses/by-sa/3.0/legalcode, accessed on 19 March 2022) via Wikimedia Commons.

3.1. Pigments

Pigments are molecules that absorb light of wavelengths in the visible range (ca. 380–750 nm) and, accordingly, are responsible for the colors seen in many organisms and some minerals. Examples of naturally occurring pigments or biochromes are porphyrins, melanins and carotenoids [77]. Recent research has focused on investigating the preservation of pigments that are responsible for colors seen in fossils. Based on molecular analyses, scientists have been able to reconstruct the original color of some dinosaurs, also referred to as paleocolor reconstructions [78,79]. The pigments believed to have been preserved in dinosaur fossils include porphyrins (heme and protoporphyrin IX), their open-chain tetrapyrrole derivatives (biliverdin) and the biopolymer melanin (eumelanin and pheomelanin).

3.1.1. Porphyrins

Porphyrins are a family of organic compounds containing four pyrrole rings connected by methine bridges. Examples are heme (**1**), the iron-complexing main prosthetic group of hemoglobin, and protoporphyrin IX (**2**), the metal-free precursor of heme (see Table 1 for structures). Metabolic degradation products include linear tetrapyrrole derivatives, e.g., biliverdin (**3**) (see Table 1). Porphyrins and their derivatives are relatively stable, even for hundreds of millions of years; they have been recovered from sediments and crude oil extracts, the oldest record being from 1.1-billion-year-old sediments [80]. Porphyrins have also been detected in fossil tissues from dinosaurs [eggshells [22] and trabecular bone [18]] and the abdomen of a female mosquito [53]. The chemistry of porphyrins in fossils has been recently reviewed [81]; the porphyrins detected in fossils derived from dinosaurs are compiled in Table 1.

Table 1. Porphyrins detected in dinosaurs.

Organic Compound	Heme	Protoporphyrin IX	Biliverdin
Structure and exact mass	**Heme** (**1**) Exact Mass: 616.18	**Protoporphyrin IX** (**2**) Exact Mass: 562.26	**Biliverdin** (**3**) Exact Mass: 582.25
Analytical technique	HPLC-UV UV/Vis spectroscopy Raman spectroscopy	LC-ESI-q/TOF-MS Raman spectroscopy	LC-ESI-q/TOF-MS Raman spectroscopy
Dinosaur species and age	*Tyrannosaurus rex* (67 Ma)	*Heyuannia huangi* (66 Ma)	*Heyuannia huangi* (66 Ma)
Location of fossil	Hell Creek formation, eastern Montana, USA	Chinese provinces (Henan, Jiangxi, and Guangdong)	Chinese provinces (Henan, Jiangxi, and Guangdong)
Type of tissue	Extracts of trabecular bone tissues	Extract of eggshells	Extract of eggshells
Reference	[18]	[21,22]	[21,22]

Heme was identified in trabecular bone extracts of *Tyrannosaurus rex* in 1997 [18]. The distinct chemical feature which made it possible to confirm the identity of heme was its chromophore in the ultraviolet/visible light range [18]. The porphyrin ring has a very characteristic band in the ultraviolet range of around 410 nm, known as the *Soret* band, which could be detected using ultraviolet/visible light (UV/Vis) spectroscopy. This band was observed in bone extracts but not in controls, indicating that the signals were derived solely from the bone and not from contaminating factors in the surrounding sandstone sediment or extraction buffers. In addition, four of the six characteristic Raman peaks for hemoglobin (marker bands I, II, IV, and V) were detected with high intensity in the extracts. The six marker bands are found in the following spectral regions: band I (1340–1390 cm^{-1}), band II (1470–1505 cm^{-1}), band III (1535–1575 cm^{-1}), band IV (1550–1590 cm^{-1}), band V (1605–1645 cm^{-1}), and band VI (1560–1600 cm^{-1}) [82]. Resonance Raman spectroscopy analyses on extracts also showed that iron was present in the oxidized ferric state, which indicates a diagenetic alteration of heme (Fe^{2+} complex) to the oxidized hemin form. In addition, proton NMR spectra on the fossil extract were similar to those from degraded hemoproteins containing ferric iron [18].

A further case of heme in the fossil record, although not in dinosaurs, was reported 16 years later, when traces of heme were found in the abdomen of a female fossil mosquito (46 Ma), analyzed in situ by TOF-SIMS [53].

Only recently, the metal-free porphyrin, protoporphyrin IX (**2**), and the linear tetrapyrrole derivative biliverdin (**3**) were detected in extracts of eggshells from the oviraptorid dinosaur *Heyuannia huangi* by liquid-chromatography electrospray ionization-quadrupole-time-of-flight mass spectrometry (LC-q/TOF-MS) [22]. The exact masses were detected with high resolution in the mass spectra as protonated molecular ions, [M + H]$^+$, from three fossil eggshell samples. For confirmation, extant emu eggshell extracts and commercial standards of the two compounds were also analyzed. These peaks were not detected either in the sediment samples or in control samples, indicating that the peaks truly belonged to the analyzed fossil. Protoporphyrin IX (**2**) is more hydrophobic than biliverdin (**3**) and therefore more likely to be preserved due to its resistance to hydrolytic attack. In addition, the ring system of protoporphyrin is more stable than the open chain structure of biliverdin. Based on these results, a reconstruction of eggshell color as blue-green was performed [22]. A year later, protoporphyrin IX and biliverdin were reported using Raman spectroscopy in various fossilized eggshells, including *Heyuannia huangi* [21]. This study has been criticized by experts in Raman spectroscopy because the authors had based their observations only on a single analytical technique; Alleon et al. even argued that the observed signals were due to instrumental artefacts caused by background luminescence, and not due to Raman scattering [83,84].

There appears to be still much potential for future discoveries of porphyrins and their metabolites and degradation products in dinosaurs and other fossils.

3.1.2. Melanins

Melanins are a group of dark-colored biopolymeric structures. Different types of melanin are known: eumelanin (**4**), pheomelanin (**5**), allomelanin, pyomelanin and neuromelanin (see Figure 2). Eumelanin, pheomelanin, and allomelanin are most relevant when studying fossils. Eumelanin (**4**) and pheomelanin (**5**) are nitrogen-containing melanins found in animals. Allomelanin is a nitrogen-free melanin (see Figure 3) which is found in plants, fungi and bacteria; it is relevant when studying fossils because detection of its chemical signals can imply external microbial contamination [25].

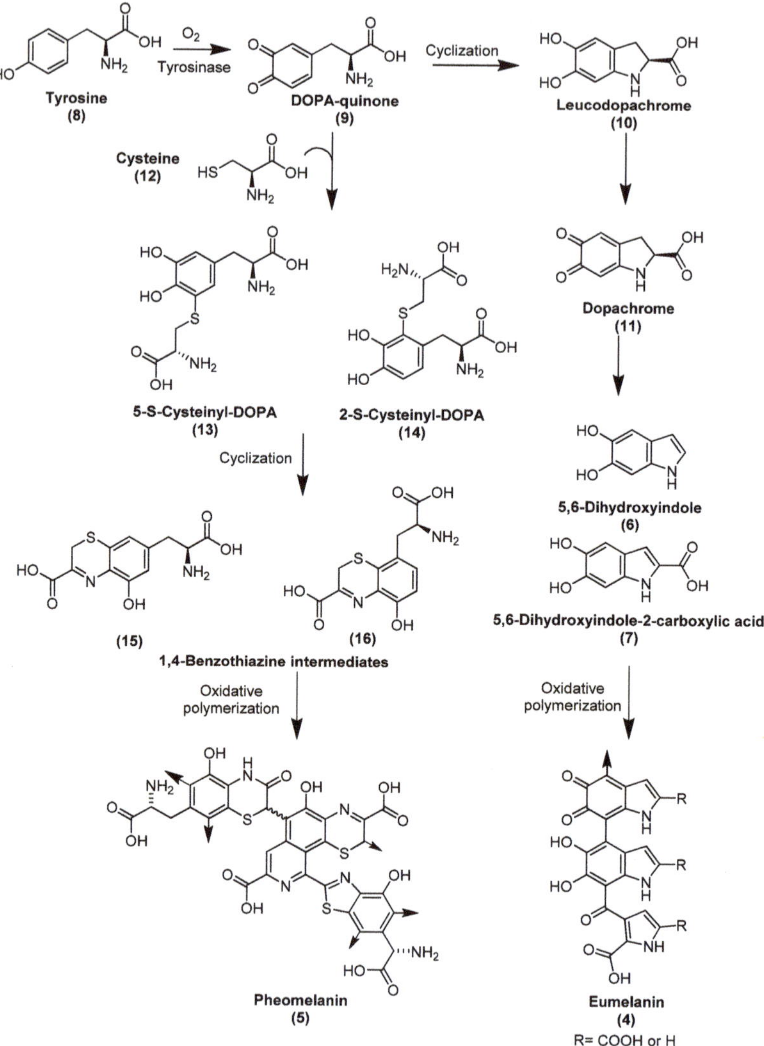

Figure 2. Structures and biosynthesis of eumelanin and pheomelanin. Arrows on structures (4) and (5) show points of polymer expansion [85,86].

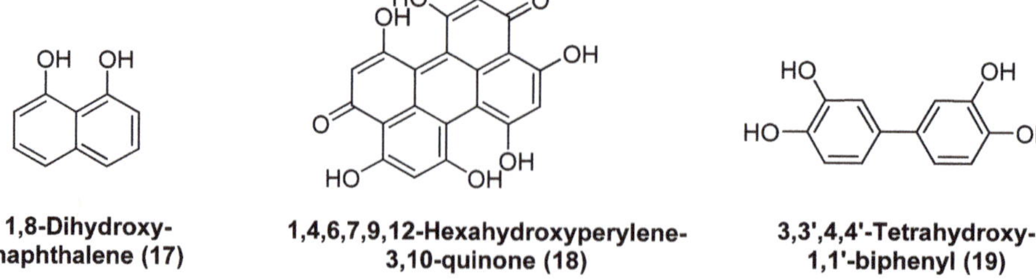

Figure 3. Structures of precursors involved in different types of allomelanin.

The biosynthesis of eumelanin and pheomelanin takes place in melanocytes in the dermis. Figure 2 shows the biosynthesis of eumelanin and pheomelanin, including their intermediates. The synthesized melanins are transported into the keratinocytes, found in the epidermis, in special lysosome-like vesicles known as eumelanosomes and pheomelanosomes [85,87]. Both types of melanosomes are then incorporated into the outer layer of the skin, determining the color of skin, hair, and eyes. Melanins are responsible for absorbing UV light and for scavenging free radicals that can be formed upon exposure to UV light, in order to protect the inner layers of the skin from harmful radiation and radical reactions [86,88].

Eumelanin is brown to black in color and contains repeating units of 5,6-dihydroxyindole (**6**) and 5,6-dihydroxyindole-2-carboxylic acid (**7**). In its biosynthesis, it is derived from the amino acid tyrosine (**8**), which, upon action of tyrosinase, or by oxidation, is converted to DOPA-quinone (**9**), which is then cyclized and decarboxylated to form 5,6-dihydroxyindole (**6**) through the intermediate compounds leucodopachrome (**10**) and dopachrome (**11**) [25,85,86] (for structures see Figure 2). Some indole units may randomly undergo partial oxidative cleavage via formation of an ortho-benzoquinone leading to pyrrole-di-carboxylic acid derivatives, which are incorporated into the polymeric structure of eumelanin [89,90]. Pheomelanin is a reddish-yellow sulfur-containing melanin which contains units of 1,4-benzothiazine and 1,3-benzothiazole [91]. Similar to eumelanin, pheomelanin is derived from tyrosine (**8**), and additionally from cysteine (**12**), that is fused with DOPA-quinone (**9**) to form cysteinyl-DOPA derivatives **13** and **14**, which undergo several oxidation steps to form 1,4-benzothiazine intermediates **15** and **16** [85] (see Figure 2).

Allomelanin has not been studied as much as eumelanin and pheomelanin. However, it is established that several subtypes of allomelanin can be distinguished according to the precursors from which they are derived. The precursors comprise 1,8-dihydroxynapthalene (**17**), 1,4,6,7,9,12-hexahydroxyperylene-3,10-quinone (**18**) and biphenolic dimers such as 3,3′,4,4′-tetrahydroxy-1,1′-biphenyl (**19**), biosynthesized from acetyl-CoA, malonyl-CoA, and catechol, respectively (see Figure 3). Accordingly, three types of allomelanin are distinguished:1,8-dihydroxynapthalene melanin, 1,4,6,7,9,12-hexahydroxyperylene-3,10-quinonemelanin, and catechol-melanin [92].

There is emerging morphological and chemical evidence for eumelanin and pheomelanin detected in a variety of fossils with or without association with melanosomes. Examples are fossilized marine reptiles such as a Paleogene turtle (55 Ma), Cretaceous mosasaur (86 Ma), and Jurassic ichthyosaur (ca. 196–190 Ma) [93]. The compounds were also found in several species of fish (359–366 Ma), amphibians (Ypresian/Lutetian, Eocene, Aquitanian, Miocene, Chattian, Oligocene), birds (56–34 Ma), and mammals (56-34 Ma) [94]. Furthermore, they were detected in dinosaurs (150–112 Ma) [25,27]. A summary of findings on melanins and/or melanosomes in the dinosaur fossil record is compiled in Table 2, along with the analytical methods used.

Imaging studies using SEM in combination with EDS have been used to detect melanin based on the presence and shape of melanosomes in preserved integumentary structures of the theropods *Sinosauropteryx* and *Sinornithosaurus* [17], as well as *Psittacosaurus* [28]. More recently, analytical techniques such as TOF-SIMS and Py-GC-MS have been utilized to confirm the chemical fingerprint of melanin in the early avialan *Anchiornis huxleyi* [25] and the ankylosaur *Borealopelta markmitchelli* [27]. Due to the resemblance between melanosomes of dinosaurs and keratinophilic bacteria on the microscopic level [17,95], a chemical analysis is necessary in order to confirm the presence of melanin [25].

TOF-SIMS analyses of a feather fossil derived from *Anchiornis huxleyi* (150 Ma) showed negative ion spectra characteristic for melanins in the areas where microscopic melanosome-like structures were observed [25]. Compared to spectra of synthetic and natural variants of eumelanin and pheomelanin, many high-intensity mass signals were in common, indicating the presence of eumelanin of animal origin. Absorption bands suggesting the presence of eumelanin as well were detected using infrared spectroscopy. Bacterial contamination was excluded in the examined areas due to the absence of peaks corresponding to peptidoglycans and hopanoids [25]. Peptidoglycans, polymers consisting of sugars and peptides, are cell wall components of Gram-positive and Gram-negative bacteria [96], while hopanoids are cyclic lipophilic triterpenoids that are located in the bacterial cell membrane and have been detected in the fossil record of bacteria [97,98]. The TOF-SIMS spectra of bacteria-derived melanin, namely allomelanin, which does not contain nitrogen (see Figure 3), does not show any of the nitrogen-derived peaks that were found in the fossil (mass-to-charge ratios of 50, 66, 74, 98, 122, and 146). Analysis of the surrounding sediment using the same method showed negative ion spectra corresponding to silicate-rich minerals, but no nitrogen-containing peaks were observed. Signals for sulfur-containing compounds that could originate from pheomelanin were not intense enough to confirm its presence in the fossil [25].

Table 2. Melanin detected in dinosaurs.

	Eumelanosomes and Pheomelanosomes	Eumelanin-like Pigmentation (Black and Yellow)	Eumelanin	Mixture of Pheomelanin and Eumelanin
Analytical technique	SEM imaging combined with EDS	Imaging with digital camera	TOF-SIMS EDS IR micro-spectroscopy	TOF-SIMS Py-GC-MS EDS
Dinosaur, location and age of fossil	*Sinosauropteryx* (125 Ma, Dawangzhangzhi, Lingyuan City, Liaoning Province, China) *Sinornithosaurus* (125 Ma, Sihetun, Beipiao City, Liaoning Province, China)	*Psittacosaurus* (125 Ma) Yixian formation in China	*Anchiornis huxleyi* (150 Ma) Yaolugao locality in Jianching County, western Liaoning, China	*Borealopelta markmitchelli* (112 Ma) Suncor Millenium Mine, Fort McMurray, Alberta, Canada
Type of tissue	Integumentary filaments from the tail	Preserved epidermal scales scattered from head to tail	Filamentous epidermal appendages ("feathers")	Integumentary structures (epidermis and keratinized scales)
Reference	[17]	[28]	[25]	[27]

The preserved integumentary structures of the ankylosaur *Borealopelta markmitchelli* (112 Ma) were analyzed by TOF-SIMS and pyrolysis-GC-MS to investigate the presence of melanin [27]. TOF-SIMS analysis showed negative ions similar to those of melanin in previously reported fossils [93], resembling natural and synthetic melanin. In addition, ions containing sulfur (1,3-benzothiazole) indicative of pheomelanin [93] were detected, suggesting that a mixture of eumelanin and pheomelanin was present [27]. Pyrolysis-GC-MS analysis showed signals corresponding to eumelanin (N- and O-heterocyclic and aromatic compounds), as reported previously in fossils [99,100]. Signals derived from pheomelanin (1,3-benzothiazole) were also present, which were not detected in the surrounding sediment [27].

3.2. Proteins

Although met with controversy, especially when considering chemical instability, there are more and more reports on proteins and their fragments detected in fossils. In the early years, this was backed mainly by morphological examination and the application of vibrational spectroscopy and immunological techniques. In recent years, the field of paleoproteomics has flourished, applying high-resolution mass spectrometry to determine peptide sequences and to map them on the extant versions of the proteins of interest [33,34,101]. Further paleoproteomic research, especially sequencing of proteins by mass spectrometry, would be required to confirm the endogeneity of the detected protein fragments [34]. It has to be kept in mind that cross-contamination remains an important issue when analyzing peptide sequences [102]. Not only can cross-contamination arise from laboratory reagents and controls, it can also occur due to previously analyzed samples. Thus, it is necessary to rule out cross-contamination by suitable measures, such as careful and self-critical approaches, and appropriate controls [102].

Vibrational methods such as infrared spectroscopy have been used to detect proteins, showing characteristic absorption bands of the amide bonds; however, these signals are non-specific and it is not possible to identify the type of protein or its sequence [34]. Early trials to detect proteins utilized amino acid analysis after degradation of the proteins. This method is also insufficient for determining the original peptide sequence [33]. TOF-SIMS employs a harsh ionization method which causes extensive fragmentation of proteins. Therefore, while it cannot be used for sequencing, it is useful for obtaining a chemical map, revealing the regions where amino acid fragments are found in a fossil, which may then be further analyzed [32,34,76].

Immunological techniques including immunohistochemistry, Western blotting, and ELISA rely on positive antigen–antibody reactions, detecting specific epitopes of a protein or nucleic acid. Specificity depends on the employed antibodies, but protein or nucleic acid sequences cannot be determined. These methods can be useful to locate the regions that may contain preserved proteins (or nucleic acids) suitable for subsequent mass spectrometric analysis. In addition, liquid chromatography coupled to electrospray ionization high-resolution mass spectrometry is used for the identification of proteins. The techniques used in paleoproteomics and their limitations were recently reviewed [33,34]. Most of the proteins detected in dinosaur fossils belonged to the most abundant ones including collagen type I (found in bones), collagen type II (found in cartilage), and beta-keratin (found in scales, turtle shell, claws of reptiles, and in avian feathers) [103]. The following section will discuss the evidence for proteins detected to date in dinosaurs and their chemistry.

3.2.1. Collagens

Collagens constitute a family of glycoproteins that are the main components of the extracellular matrix of different tissues. In animals, 29 different types of collagen have been found, but only 3 types (collagen I, II and III) constitute around 80–90% of the total collagen. Collagens are structural proteins in the extracellular matrix which confer mechanical strength especially to connective tissues. They directly interact with other components of the extracellular matrix, such as proteoglycans, fibronectin and laminin. Proteoglycans are glycoproteins that form a gel-like network in the extracellular matrix. Collagens and other fibrous proteins (fibronectin and laminin) are located within this network. Fibronectin and laminin are non-collagenous glycoproteins that form fibrous networks and affect the shape of the extracellular matrix. They possess binding sites important for cell adhesion [104]. In addition, collagens interact with secreted soluble factors such as the von Willebrand factor and interleukin-2, and with cell surface receptors such as integrins [105]. These interactions aim to regulate tissue development and mechanical responses to cell signaling such as cell adhesion, migration and chemotaxis [106,107]. The primary polypeptide structure of collagen is known as the α-chain. All types of collagens share the repeating amino acid sequence [Gly-X-Y], where X and Y are usually proline and hydroxyproline, respectively (see Figure 4). In 12% of collagen sequences, both proline and hydroxyproline are present in their respective positions, while in 44% of the sequences, only one of them is present [108]. The secondary structure of collagen is formed from three α-chains arranged in parallel. They are twisted together to form the tertiary structure, a rope-like triple helix with a molecular weight of ca. 300 kDa, a length of 280 nm and a diameter of 1.4 nm. The abundance of the cyclic amino acids, proline and hydroxyproline, sterically hinders rotation around the peptide bonds in the α-chains which contributes to the stability and rigidity of the triple helix. In addition, two types of hydrogen bonds stabilize the triple helix. The first type of intermolecular hydrogen bonds are formed between the NH of glycine and the carbonyl group of proline residues in neighboring α-chains. The second type are intramolecular hydrogen bonds, formed between the carbonyl or hydroxyl groups of hydroxyproline and the carbonyl group of glycine or hydroxyproline residues in the same α-chain, mediated by a water molecule [108]. Moreover, the X and Y positions in further collagen sequences are occupied by other amino acids, but never contain tryptophan, tyrosine, or cysteine, as these would destabilize the triple helix [109]. Post-translational modifications such as hydroxylation of proline and lysine residues or glycosylation (with galactose or a disaccharide of glucose and galactose) also contribute to stability and are typical of collagen. The more hydroxyproline residues there are, the more thermally stable the triple helix is. Post-translational hydroxylation of proline residues is often used in identification of collagen from fossils, especially since this cannot be performed by bacteria [108]. Hydroxylysine is the point of attachment of the sugars via an O-glycosidic linkage, and this stabilizes the collagen fibrils mechanically by formation of covalent crosslinks [108].

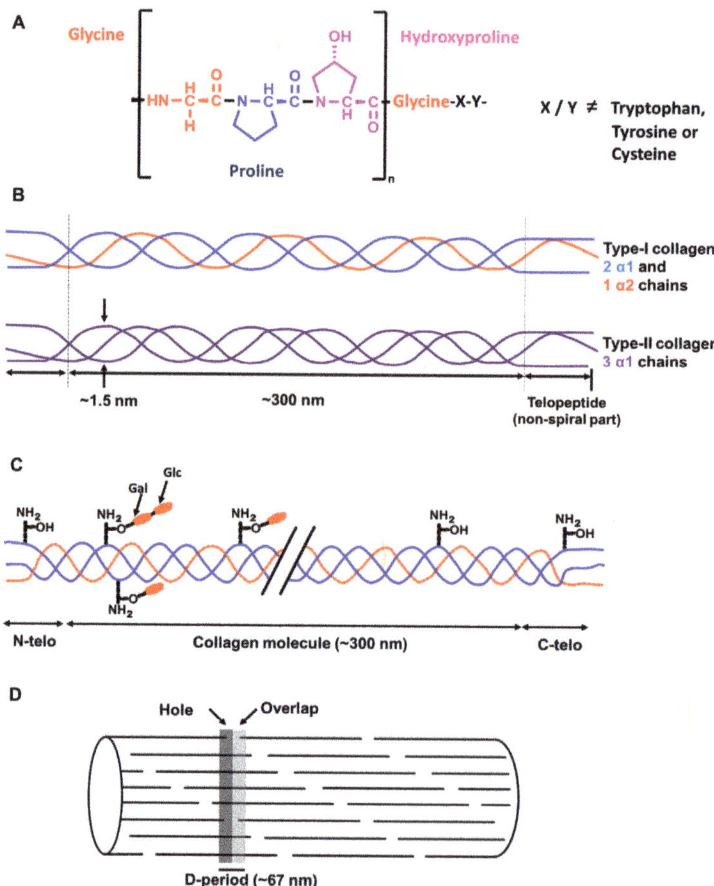

Figure 4. (**A**) The most common repeating sequence present in collagen types I and II. Positions X and Y can be occupied by any amino acid except tryptophan, tyrosine or cysteine. The most common amino acids in positions X and Y are proline and hydroxyproline, respectively. (**B**) Schematic representation of the triple helical structure of collagens type I and II. In collagen type I, there are two α1 chains and one α2 chain, whereas in collagen type II, there are three α1 chains. (**C**) Diagram of a collagen molecule showing the post-translational modifications that occur, which are hydroxylation of lysine residues and glycosylation of hydroxylysine by galactose and glucose. (**D**) The stacked arrangement of collagen fibers, visible under a transmission electron microscope, shows a characteristic staggered pattern known as the D-band or D-period of approximately 67 nm in periodicity. This banding is a unique feature used for identification of collagen fibers under the microscope. Adapted from [110,111].

Collagen Type I

Collagen type I is the major component of bone organic phase, but is also present in skin, tendons, ligaments, lung, blood vessels, cornea, brain and spinal cord [108,112] (see Figure 4 for structures). Collagen I is composed of two $\alpha1(I)$ chains and one $\alpha2(I)$ chain and assembles into elongated fibrils of 500 μm in length and 500 nm in diameter. The fibrils have a characteristic tight arrangement. Every 64–67 nm, there also is a pattern repeating itself, known as *D*-banding. This pattern is visible in the electron microscope and can be utilized for identification of collagen type I [108,113–116]. Studies reporting evidence for collagen type I found in the dinosaur fossil record are compiled in Table 3.

Table 3. Collagen type I and II in the dinosaur fossil record.

Study	Collagen Type I	Analytical Technique(s)	Dinosaur Name, Location and Age	Type of Tissue	Reference
1	Amino acid fragments and peptide sequences (5 from $\alpha1$ chain, 1 from $\alpha2$ chain)	Immuno-histochemistry, ELISA, TOF-SIMS and LC-MS/MS	*Tyrannosaurus rex* (68 Ma) Hell Creek Formation, eastern Montana, USA	Trabecular bone	[19,20]
2	Infrared absorption bands	SR-FTIR and confocal Raman microscopy	*Lufengosaurus* (ca. 195 Ma) Dawa, Lufeng County, Yunnan Province, China	Rib bone (thin sections)	[3]
3	Amino acid fragments (alanine, arginine, glycine, and proline)	TOF-SIMS	Various Dinosauria (75 Ma) Dinosaur Park Formation, Alberta, Canada	Claw, ungual phalanx, astragalus, tibia, rib	[26]
4	Peptide sequences (6 for $\alpha1$ chain, 2 for $\alpha2$ chain)	Immuno-histochemistry, Western blot, ATR-IR, TOF-SIMS, and LC-MS/MS	*Brachylophosaurus canadensis* (80 Ma) Judith River Formation, eastern Montana, USA	Femur from hind limb (4 different samples)	[29]
5	Peptide sequences (6 for $\alpha1$ chain, 2 for $\alpha2$ chain)	Nano-LC-MS/MS and FT-ICR-MS	*Brachylophosaurus canadensis* (80 Ma) Judith River Formation, eastern Montana, USA	Femur from hind limb (4 different samples)	[30]
6	Collagen type II	Immunohisto-chemistry	*Hypacrosaurus stebingeri* (75 Ma) Two Medicine Formation, northern Montana, USA.	Calcified cartilage from supraoccipital	[31]

Attempts to detect collagen type I in dinosaurs were performed on samples of *Tyrannosaurus rex* [19] (see Table 3). Trabecular bone extracts showed positive reactivity in an ELISA employing avian collagen I antibodies. The signal was weaker in the dinosaur as compared to extant emu cortical and trabecular bone, but the signal detected in the fossil was larger than those in buffer controls and in the sediment. The same pattern was observed by in situ immunohistochemistry studies. Antibody binding decreased significantly when the fossil tissue was digested with collagenase I before exposure to the antibodies. TOF-SIMS analysis revealed amino acid residues in the fossil including glycine (highest relative signal intensity), alanine, proline, lysine, leucine and isoleucine [19]. A subsequent study [20] applied a softer mass spectrometric technique to avoid undesired fragmentation, liquid chromatography tandem mass spectrometry (LC-MS/MS). In this study, the dinosaur fossil was compared to similarly treated ostrich and mastodon samples. The mass spectra obtained from *T. rex* bone extracts detected seven collagen peptide sequences, five from the $\alpha1(I)$ chain, one from the $\alpha2(I)$ chain, and one belonging to the $\alpha1(II)$ chain of type II collagen, that were aligned with database sequences from extant vertebrates. Post-translational modifications, especially hydroxylation of proline, lysine and glycine, were detected in the dinosaur fossil as well as in the mastodon and ostrich samples, while no collagen sequences were detected in control samples of the surrounding sediment and the extraction buffers. The sediment contained peptides of bacterial origin, but no collagen [20].

In another study, investigation of the hadrosaurid dinosaur *Brachylophosaurus canadensis* provided evidence for collagen type I [29]. This was confirmed by studies in different laboratories and at different times using different methodology including sample preparation technique, mass spectrometry instrument, and data analysis software [29,30]. Microscopic observation (by field-emission SEM) of fibrous structures in demineralized femur bones was followed up by immunoblot assays. A positive reactivity to antibodies raised against avian collagen type I was observed in whole fossil bone extracts and in intact demineralized fossil bones [29]. In situ immunohistochemistry studies performed on demineralized fossil bones confirmed the results. The extraction buffers and the surrounding sediments showed no reactivity. Antibody binding decreased significantly when the samples were digested with collagenase before exposure to the antibodies, or when exposed to antibodies that had been pre-incubated with excess collagen. Gel electrophoresis studies on samples of the surrounding sediment did not show any visible protein bands. Infrared spectroscopy showed absorption bands of amide bonds (Amide I and Amide II). Analysis using TOF-SIMS indicated fragments of lysine, proline, alanine, glycine, and leucine residues in intact blood vessels and in matrix of demineralized bone. Further experiments using reversed-phase microcapillary liquid chromatography tandem mass spectrometry (linear ion-trap alone or hybridized with Orbitrap mass spectrometry) recovered eight collagen type I sequences, containing a total of 149 amino acids. Six of these sequences were attributed to the α1 chain and two to the α2 chain [29].

High-resolution measurements performed eight years later by LC-tandem mass spectrometry coupled to Fourier-transform ion cyclotron resonance mass spectrometry (FT-ICR-MS) again showed eight collagen type I sequences in the range of 250 kDa, two of which had previously been detected for the α1 chain, in addition to three new α1 chain sequences, and three new sequences for the α2 chain [30]. In both studies, no collagen sequences could be detected in spectra of extraction buffer or samples of the surrounding sediment. In addition, post-translational modification of hydroxylated proline was observed, which is important for the triple helix structure of collagen I and cannot be produced by microbes [29,30].

Amino acid fragments in association with direct observations of fibrous structures showing the 67 nm banding of typical collagen were detected using TOF-SIMS by Bertazzo et al. (2015) in a variety of dinosaur bone samples from the Late Cretaceous Dinosaur Park Formation of Alberta, Canada [26]. The banding indicated that the quaternary structure of collagen may have been preserved. In addition, TOF-SIMS analyses were performed to search for amino acids using thick sections of the fossil bone sample, as well as modern rabbit bone, non-calcified fossil samples, surrounding sediment, and the sample holder made of copper as controls. The fossil dinosaur bone samples and the rabbit bone contained similar amino acid peaks which were neither present in the non-calcified fossil samples nor in the surrounding sediment or in the sample holder. Fragments belonging to glycine, arginine, alanine, and proline were detected only in the permineralized fossil samples [26].

Synchrotron-radiation Fourier transformed infrared spectroscopy (SR-FTIR) and confocal Raman spectroscopy were used to identify characteristic vibrations of chemical bonds at specific absorption bands for each functional group, producing high resolution images and spectra [34]. Infrared absorption bands characteristic for collagen type I were detected in thin sections of the rib bone of a 195-million-year-old *Lufengosaurus*, and early-branching sauropodomorph, and the geologically oldest dinosaur sample analyzed to date. The detection was especially in the regions where vascular canals could microscopically be observed. The infrared absorption bands of the fossil samples were very similar to the reference samples of extant collagen I extracted from calf skin [3].

The published evidence for collagen type I and other proteins and their sequences in dinosaurs should still be treated with caution. For example, TOF-SIMS is not suitable for sequencing but can only help to locate samples for subsequent tandem-mass spectrometry experiments. A combination of different analytical techniques is usually needed, combined with the proper controls. Tandem mass spectrometry is the main technique to prove the presence of peptides and to sequence polypeptides/proteins.

Collagen Type II

Collagen type II is a structural protein mostly present in cartilage, tendons, and in the intervertebral disc [108]. In contrast to collagen type I, it is a homotrimer, composed of 3 α1(II) chains [113]. Similar to collagen type I, it also forms a triple helix of around 1000 amino acids in length and has the repeating amino acid pattern of Gly-X-Y [117], forming an aggregated fibrous structure.

The first report on collagen II associated with preserved calcified cartilage in dinosaurs [31] was from *Hypacrosaurus stebingeri*, a 75-million-year-old hadrosaur nestling discovered in the Two Medicine Formation of northern Montana, USA. Techniques used to chemically characterize the observed chondrocyte-like microstructures were histochemical and immunological techniques, as shown in Table 3. Thin sections of demineralized fossil cartilage exposed to antibodies raised against avian collagen type II showed positive reactivity after visualization by green fluorescence. The observed pattern was interrupted and less intense compared to the homogenous distribution of the binding pattern in extant cartilage from emu (*Dromaius novaehollandiae*), suggesting that either the epitopes are few or that the epitopes recognized by avian collagen I antibodies are not similar to those present in the dinosaur. Collagen II is not produced by bacteria; thus, contamination is less likely to have occurred [31].

Specificity of the antibodies was checked by prior digestion of the thin sections by collagenase II and exposure to the antibodies, after which the binding decreased significantly in both fossil and recent material under the same conditions. This supports the interpretation that collagen II is likely present in the fossil. Antibodies against avian collagen I did not show any binding in both fossil and recent cartilage, which is not expected to be found there [31].

3.2.2. Keratins

Keratins are structural proteins which are the major constituents of hair, nails, feathers, horns, and hooves [118]. They are characterized by a high cysteine content (7–13%). Keratins have several biological functions, including (i) mechanical effects and (ii) altering cellular metabolism. By disassembly and reassembly, keratins provide flexibility to the cytoskeletal structure, making cells and tissues withstand mechanical stress and maintain their shape. Keratins affect the response to cellular signaling by binding to various signaling proteins such as protein kinases and phosphatases. Thus, keratins are involved in the regulation of cell growth, cell differentiation, mitosis, and protein synthesis, which may lead to a change in cellular metabolism [119–121].

Keratins have a molecular weight of 40–70 kDa [119,122]. The amino acids in the primary sequence of keratins are often cysteine, glycine, proline, and serine, and to a lesser extent lysine, histidine, and methionine. Tryptophan is rarely present [118,123]. The secondary structure of keratins is either an α-helix or a β-sheet, depending on the type of amino acids present. Accordingly, two types of keratins can be distinguished: α-keratin and β-keratin [119] (see Figure 5).

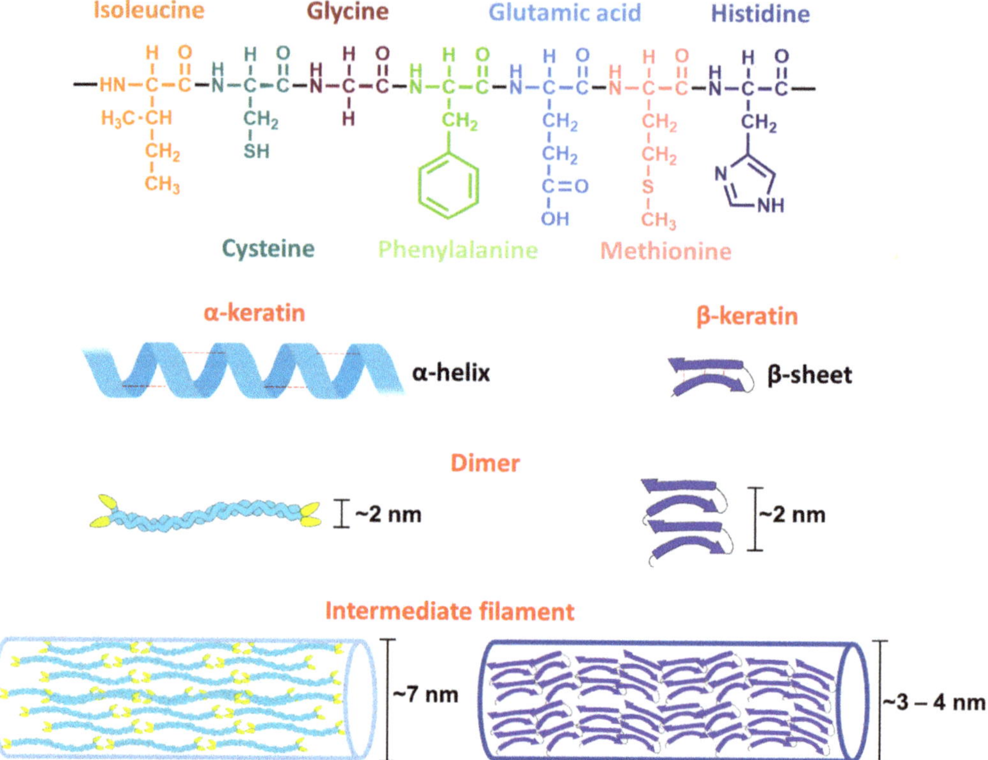

Figure 5. Diagram of the four different levels of keratin structure. The primary sequence of keratin is shown, including the most common amino acids present (the amino acids are L-configurated, but the stereochemistry is not shown). The secondary structure of keratins can be either an α-helix or a β-sheet, classifying them into α-keratins and β-keratins, respectively. The tertiary structures of both keratin types are heterodimers. The quaternary structure is composed of intermediate filaments, that are 7 nm in diameter for α-keratin and 3–4 nm in diameter in β-keratin. Adapted from [124].

The tertiary structure of keratins is composed of a dimer that forms the building block of keratin filaments. It is stabilized by inter- and intra-molecular interactions such as disulfide bridges, hydrogen bonds, hydrophobic interactions, and ionic bonds [118]. Their quaternary structure consists of self-assembling intermediate filaments having a characteristic electron-lucent region of 7–8 nm in diameter observed under the electron microscope. The formation of keratin filaments is affected by pH and osmolarity [119].

Post-translational modifications occur to the secondary structure of keratins, which in turn affect their overall structure, physicochemical properties and functions. Phosphorylation or formation of intra- and interchain covalent bonds (e.g., disulfide bonds) can directly modify the structure. Changes in pH, the types of ions present, and osmolarity can alter the physicochemical properties indirectly, for example, by changing the isoelectric point. Keratins can modify their filaments due to mechanical stress such as tension, compression, and shearing [119].

Keratins are insoluble in water, alkali, weak acids, and organic solvents. They are stable in the presence of proteases such as pepsin and trypsin. The crosslinking via disulfide bonds stabilizes the overall tertiary structure and lowers the water solubility [118,123].

α-Keratin is expressed in all vertebrates [23]. Its structure is better described than that of β-keratin [119]. α-Keratins are classified into two types according to their isoelectric point (pI) range: type I (pI = 4.9–5.4) and type II (pI = 6.5–8.5). α-Keratins with more acidic amino acids are of type I, while those containing more basic amino acids belong to type II [119] (see Figure 5).

β-Keratins are exclusively expressed in reptiles and birds (e.g., claw sheaths and feathers), and differ from the α-keratins in their lower solubility and the high rigidity of their microfibril filaments [24]. β-keratin has a core of 30 amino acids and forms antiparallel β-sheets, joined by regions of β-turns and stabilized by hydrogen bonds. The quaternary structure of β-keratin is characterized by microfibril filaments of 3 nm in diameter [125]. The presence of hydrophobic amino acids in the core, such as valine and proline [125], increases their preservation potential because they will not be readily hydrolyzed [23]. β-Keratins are not expressed in humans or microorganisms; thus, if β-keratins are detected in fossils, exogenous contamination can likely be ruled out [23]. β-Keratin has been detected in fossil dinosaurs mainly by immunohistochemistry techniques as shown in Table 4.

Table 4. Evidence of beta-keratin in the dinosaur fossil record.

	β-Keratin and Its Amino Acid Fragments	β-Keratin Epitopes
Analytical technique(s)	TOF-SIMS Immunohistochemistry	Immunohistochemistry
Dinosaur species Location and age of fossil	*Shuvuuia deserti* (100 Ma) Ukhaa Tolgod in southwestern Mongolia	*Citipati osmolskae* (75 Ma) Djadokhta Formation of Mongolia
Type of tissue	Feather-like epidermal appendages	Original keratinous-like claw sheath
Reference	[24]	[23]

There are some amino acids which are common in the sequence of both types of keratins, such as glycine, serine, valine, leucine, glutamate, cysteine, and alanine [119,126]. Amino acids which are more abundant in α-keratin are methionine, histidine, phenylalanine, and isoleucine [127]. Amino acids that are more abundant in α-keratin are proline and aspartate [119], whereas histidine, methionine, tryptophan, and tyrosine are rarely present [126].

The first characterization of β-keratin in fossil dinosaurs was from feather-like structures of the 100-million-year-old *Shuvuuia deserti* collected at Ukhaa Tolgod in southwestern Mongolia (see Table 4) [24]. Immunohistochemical studies using antibodies raised against avian α- and β-keratins showed a strong reactivity in both fossil and extant (duck feather) tissue samples for β-keratin, and less reactivity for α-keratin. No reactivity was seen in control samples, including incubation with antibodies not specific to β-keratin. Reduced binding was observed when the antibodies against β-keratin were incubated with excess β-keratin before exposure to the tissues, thus confirming the specificity of this approach. Furthermore, TOF-SIMS analysis was performed on isolated fiber structures to search for amino acids to support the immunological findings. Several amino acid fragments, containing glycine, serine, leucine, cysteine, proline, valine and alanine, were detected in the mass spectra. The targeted sampling location supports that these amino acids could belong to the fossil, but sequencing by higher resolution methods would be needed for confirmation [24].

Antibodies raised against β-keratin have shown positive binding to demineralized thin sections of claw sheaths from the 75-million-year-old oviraptorid dinosaur, *Citipati osmolskae*, from the Djadokhta Formation of Mongolia, which showed keratinous-like microstructures [23]. Reference samples of extant emu and ostrich claw sheath were additionally studied. An in situ immunohistochemical approach combined with immunofluorescence and electron microscopy was employed that reaffirmed the previous claims that β-keratin can be preserved over millions of years. However, the available sample material from dinosaur fossils limits sequencing approaches. Yet, a targeted high-resolution mass spectrometric approach has been suggested for further studies based on sampling the regions which exhibited positive reactivity to β-keratin antibodies [23].

4. Conclusions

This review provides a collection of organic compounds identified in dinosaur bone and soft tissues to date, giving insights into their chemistry and the analytical techniques used for their identification. Reports on organic compounds are increasing as more targeted sensitive analytical approaches that use less and less sample material are being developed. Organic compounds detected from dinosaurs so far comprise pigments, such as porphyrins and melanins, and proteins, including collagen type I, collagen type II and β-keratin. The analytical techniques used have been a combination of imaging using microscopy, absorption, reflectance and vibrational spectroscopy. Chemical imaging on the sample surface using time-of-flight secondary ion mass spectrometry, and more invasive techniques, namely liquid chromatography coupled with tandem mass spectrometry were also employed. Yet, even as analytical techniques become more advanced and highly sensitive, it still remains challenging to prove the endogeneity of the detected structures, especially when searching for proteins or DNA. Further development of sample preparation techniques that minimizes contamination is required.

Author Contributions: All authors contributed to writing and editing, and approved the final version. All authors have read and agreed to the published version of the manuscript.

Funding: The authors were funded by the Deutsche Forschungsgemeinschaft (DFG) within the Research Unit FOR 2685 "The Limits of the Fossil Record: Analytical and Experimental Approaches to Fossilization." This is manuscript #45 of FOR 2685.

Institutional Review Board Statement: Not applicable.

Informed Consent Statement: Not applicable.

Data Availability Statement: Not applicable.

Acknowledgments: The authors are grateful for support by the Deutsche Forschungsgemeinschaft (DFG, FOR2685).

Conflicts of Interest: There are no conflict to declare.

Abbreviations

ATR-IR	Attenuated-total reflection infrared spectroscopy
BN-PAGE	Blue native-polyacrylamide gel electrophoresis
EDS	Energy-dispersive X-ray spectrometry
ELISA	Enzyme linked immunosorbent assay
ESI-MS	Electrospray ionization mass spectrometry
FESEM	Field emission scanning electron microscopy
FT-ICR-MS	Fourier transform ion cyclotron resonance mass spectrometry
FTIR	Fourier transform infrared spectroscopy
IR	Infrared spectroscopy
LC-MS/MS	Liquid chromatography tandem mass spectrometry
NMR	Nuclear magnetic resonance
OM	Optical microscopy
Py-GC-MS	Pyrolysis gas-chromatography mass spectrometry
q/TOF-MS	Quadrupole time-of-flight mass spectrometry
SDS-PAGE	Sodium dodecyl sulfate-polyacrylamide gel electrophoresis
SEM	Scanning electron microscopy
SR-FTIR	Synchrotron-radiation Fourier transform infrared spectroscopy
TEM	Transmission electron microscopy
TOF-SIMS	Time-of-flight secondary-ion mass spectrometry
UV/VIS	Ultraviolet-visible light
VPSEM	Variable-pressure scanning electron microscopy

References

1. Eglinton, G.; Logan, G.A. Molecular preservation. *Philos. Trans. R. Soc.* **1991**, *333*, 315–328. [CrossRef]
2. Schweitzer, M.H. Soft tissue preservation in terrestrial Mesozoic vertebrates. *Annu. Rev. Earth Planet. Sci.* **2011**, *39*, 187–216. [CrossRef]
3. Lee, Y.C.; Chiang, C.C.; Huang, P.Y.; Chung, C.Y.; Huang, T.D.; Wang, C.C.; Chen, C.I.; Chang, R.S.; Liao, C.H.; Reisz, R.R. Evidence of preserved collagen in an Early Jurassic sauropodomorph dinosaur revealed by synchrotron FTIR microspectroscopy. *Nat. Commun.* **2017**, *8*, 2–9. [CrossRef]
4. Ji, Q.; Luo, Z.-X.; Yuan, C.-X.; Wible, J.R.; Zhang, J.-P.; Georgi, J.A. The earliest known eutherian mammal. *Nature* **2002**, *416*, 816–822. [CrossRef]
5. Gioncada, A.; Collareta, A.; Gariboldi, K.; Lambert, O.; Di Celma, C.; Bonaccorsi, E.; Urbina, M.; Bianucci, G. Inside baleen: Exceptional microstructure preservation in a late Miocene whale skeleton from Peru. *Geology* **2016**, *44*, 839–842. [CrossRef]
6. Cadena, E.-A. In situ SEM/EDS compositional characterization of osteocytes and blood vessels in fossil and extant turtles on untreated bone surfaces: Different preservational pathways microns away. *PeerJ* **2020**, *8*, e9833. [CrossRef]
7. Surmik, D.; Dulski, M.; Kremer, B.; Szade, J.; Pawlicki, R. Iron-mediated deep-time preservation of osteocytes in a Middle Triassic reptile bone. *Hist. Biol.* **2019**, *33*, 186–193. [CrossRef]
8. Schweitzer, M.H.; Zheng, W.; Cleland, T.P.; Goodwin, M.B.; Boatman, E.; Theil, E.; Marcus, M.A.; Fakra, S.C. A role for iron and oxygen chemistry in preserving soft tissues, cells and molecules from deep time. *Proc. R. Soc. B Biol. Sci.* **2014**, *281*, 20132741. [CrossRef]
9. Wiemann, J.; Fabbri, M.; Yang, T.R.; Stein, K.; Sander, P.M.; Norell, M.A.; Briggs, D.E.G. Fossilization transforms vertebrate hard tissue proteins into N-heterocyclic polymers. *Nat. Commun.* **2018**, *9*, 4741. [CrossRef]
10. Schweitzer, M.H.; Wittmeyer, J.L.; Horner, J.R. Soft tissue and cellular preservation in vertebrate skeletal elements from the Cretaceous to the present. *Proc. R. Soc. B Biol. Sci.* **2007**, *274*, 183–197. [CrossRef]
11. Briggs, D.E.G. Molecular taphonomy of animal and plant cuticles: Selective preservation and diagenesis. *Philos. Trans. R. Soc. Lond. Ser. B Biol. Sci.* **1999**, *354*, 7–17. [CrossRef]
12. Collins, M.J.; Nielsen-Marsh, C.M.; Hiller, J.; Smith, C.I.; Roberts, J.P.; Prigodich, R.V.; Wess, T.J.; Csapò, J.; Millard, A.R.; Turner-Walker, G. The survival of organic matter in bone: A review. *Archaeometry* **2002**, *44*, 383–394. [CrossRef]
13. Briggs, D.E.G.; Mcmahon, S. The role of experiments in investigating the taphonomy of exceptional preservation. *Palaeontology* **2016**, *59*, 1–11. [CrossRef]
14. Keenan, S.W. From bone to fossil: A review of the diagenesis of bioapatite. *Am. Mineral.* **2016**, *101*, 1943–1951. [CrossRef]
15. Butterfield, N.J. Exceptional fossil preservation and the Cambrian Explosion. *Integr. Comp. Biol.* **2003**, *43*, 166–177. [CrossRef]
16. Pawlicki, R.; Korbel, A.; Kubiak, H. Cells, collagen fibrils and vessels in dinosaur bone. *Nature* **1966**, *211*, 655–657. [CrossRef]
17. Zhang, F.; Kearns, S.L.; Orr, P.J.; Benton, M.J.; Zhou, Z.; Johnson, D.; Xu, X.; Wang, X. Fossilized melanosomes and the colour of Cretaceous dinosaurs and birds. *Nature* **2010**, *463*, 1075–1078. [CrossRef]
18. Schweitzer, M.H.; Marshall, M.; Carron, K.; Bohle, D.S.; Busse, S.C.; Arnold, E.V.; Barnard, D.; Horner, J.R.; Starkey, J.R. Heme compounds in dinosaur trabecular bone. *Proc. Natl. Acad. Sci. USA* **1997**, *94*, 6291–6296. [CrossRef]
19. Schweitzer, M.H.; Suo, Z.; Avci, R.; Asara, J.M.; Allen, M.A.; Arce, F.T.; Horner, J.R. Analyses of soft tissue from *Tyrannosaurus rex* suggest the presence of protein. *Science* **2007**, *316*, 277–280. [CrossRef]
20. Asara, J.M.; Schweitzer, M.H.; Freimark, L.M.; Phillips, M.; Cantley, L.C. Protein sequences from mastodon and *Tyrannosaurus rex* revealed by mass spectrometry. *Science* **2007**, *316*, 280–285. [CrossRef]
21. Wiemann, J.; Yang, T.R.; Norell, M.A. Dinosaur egg colour had a single evolutionary origin. *Nature* **2018**, *563*, 555–558. [CrossRef]
22. Wiemann, J.; Yang, T.R.; Sander, P.N.; Schneider, M.; Engeser, M.; Kath-Schorr, S.; Müller, C.E.; Sander, P.M. Dinosaur origin of egg color: Oviraptors laid blue-green eggs. *PeerJ* **2017**, *5*, e3706. [CrossRef]
23. Moyer, A.E.; Zheng, W.; Schweitzer, M.H. Microscopic and immunohistochemical analyses of the claw of the nesting dinosaur, *Citipati osmolskae*. *Proc. R. Soc. B Biol. Sci.* **2016**, *283*, 20161997. [CrossRef]
24. Schweitzer, M.H.; Watt, J.A.; Avci, R.; Knapp, L.; Chiappe, L.; Norell, M.; Marshall, M. Beta-keratin specific immunological reactivity in feather-like structures of the cretaceous alvarezsaurid, *Shuvuuia deserti*. *J. Exp. Zool.* **1999**, *285*, 146–157. [CrossRef]
25. Lindgren, J.; Sjövall, P.; Carney, R.M.; Cincotta, A.; Uvdal, P.; Hutcheson, S.W.; Gustafsson, O.; Lefèvre, U.; Escuillié, F.; Heimdal, J.; et al. Molecular composition and ultrastructure of Jurassic paravian feathers. *Sci. Rep.* **2015**, *5*, 13520. [CrossRef]
26. Bertazzo, S.; Maidment, S.C.R.; Kallepitis, C.; Fearn, S.; Stevens, M.M.; Xie, H.N. Fibres and cellular structures preserved in 75-million-year-old dinosaur specimens. *Nat. Commun.* **2015**, *6*, 7352. [CrossRef]
27. Brown, C.M.; Henderson, D.M.; Vinther, J.; Fletcher, I.; Sistiaga, A.; Herrera, J.; Summons, R.E. An exceptionally preserved three-dimensional armored dinosaur reveals insights into coloration and Cretaceous predator-prey dynamics. *Curr. Biol.* **2017**, *27*, 2514–2521.e3. [CrossRef]
28. Lingham-Soliar, T.; Plodowski, G. The integument of *Psittacosaurus* from Liaoning Province, China: Taphonomy, epidermal patterns and color of a ceratopsian dinosaur. *Naturwissenschaften* **2010**, *97*, 479–486. [CrossRef]
29. Schweitzer, M.H.; Zheng, W.; Organ, C.L.; Avci, R.; Suo, Z.; Freimark, L.M.; Lebleu, V.S.; Duncan, M.B.; Heiden, M.G.V.; Neveu, J.M.; et al. Biomolecular characterization and protein sequences of the Campanian hadrosaur *B. canadensis*. *Science* **2009**, *324*, 626–631. [CrossRef]

30. Schroeter, E.R.; Dehart, C.J.; Cleland, T.P.; Zheng, W.; Thomas, P.M.; Kelleher, N.L.; Bern, M.; Schweitzer, M.H. Expansion for the *Brachylophosaurus canadensis* collagen I sequence and additional evidence of the preservation of Cretaceous protein. *J. Proteome Res.* **2017**, *16*, 920–932. [CrossRef]

31. Bailleul, A.M.; Zheng, W.; Horner, J.R.; Hall, B.K.; Holliday, C.M.; Schweitzer, M.H. Evidence of proteins, chromosomes and chemical markers of DNA in exceptionally preserved dinosaur cartilage. *Natl. Sci. Rev.* **2020**, *7*, 815–822. [CrossRef]

32. Pan, Y.; Hu, L.; Zhao, T. Applications of chemical imaging techniques in paleontology. *Natl. Sci. Rev.* **2019**, *6*, 1040–1053. [CrossRef]

33. Cleland, T.P.; Schroeter, E.R. A comparison of common mass spectrometry approaches for paleoproteomics. *J. Proteome Res.* **2018**, *17*, 936–945. [CrossRef]

34. Schweitzer, M.H.; Schroeter, E.R.; Cleland, T.P.; Zheng, W. Paleoproteomics of Mesozoic dinosaurs and other Mesozoic fossils. *Proteomics* **2019**, *19*, 1800251. [CrossRef]

35. Schweitzer, M.H.; Avci, R.; Collier, T.; Goodwin, M.B. Microscopic, chemical and molecular methods for examining fossil preservation. *Comptes Rendus Palevol* **2008**, *7*, 159–184. [CrossRef]

36. Zhou, W.; Apkarian, R.; Wang, Z.L.; Joy, D. Fundamentals of scanning electron microscopy (SEM). In *Scanning Microscopy for Nanotechnology*; Springer: New York, NY, USA, 2006; pp. 1–40.

37. Adams, F.; Barbante, C. Electron-based imaging techniques. In *Comprehensive Analytical Chemistry*; Elsevier: Amsterdam, The Netherlands, 2015; Volume 69, pp. 269–313.

38. Saitta, E.T.; Liang, R.; Lau, M.C.; Brown, C.M.; Longrich, N.R.; Kaye, T.G.; Novak, B.J.; Salzberg, S.L.; Norell, M.A.; Abbott, G.D.; et al. Cretaceous dinosaur bone contains recent organic material and provides an environment conducive to microbial communities. *Elife* **2019**, *8*, e46205. [CrossRef]

39. Relucenti, M.; Familiari, G.; Donfrancesco, O.; Taurino, M.; Li, X.; Chen, R.; Artini, M.; Papa, R.; Selan, L. Microscopy methods for biofilm imaging: Focus on SEM and VP-SEM pros and cons. *Biology* **2021**, *10*, 51. [CrossRef]

40. Klijn, M.E.; Hubbuch, J. Application of ultraviolet, visible, and infrared light imaging in protein-based biopharmaceutical formulation characterization and development studies. *Eur. J. Pharm. Biopharm.* **2021**, *165*, 319–336. [CrossRef]

41. Picollo, M.; Aceto, M.; Vitorino, T. UV-Vis spectroscopy. *Phys. Sci. Rev.* **2019**, *4*, 20180008. [CrossRef]

42. Akash, M.S.H.; Rehman, K. Ultraviolet-visible (UV-VIS) spectroscopy. In *Essentials of Pharmaceutical Analysis*; Springer: Singapore, 2020; pp. 29–56, ISBN 978-981-15-1547-7.

43. Olcott Marshall, A.; Marshall, C.P. Vibrational spectroscopy of fossils. *Palaeontology* **2015**, *58*, 201–211. [CrossRef]

44. Loutherback, K.; Birarda, G.; Chen, L.; N Holman, H.-Y. Microfluidic approaches to synchrotron radiation-based Fourier transform infrared (SR-FTIR) spectral microscopy of living biosystems. *Protein Pept. Lett.* **2016**, *23*, 273–282. [CrossRef] [PubMed]

45. Wang, M.; Lu, X.; Yin, X.; Tong, Y.; Peng, W.; Wu, L.; Li, H.; Yang, Y.; Gu, J.; Xiao, T.; et al. Synchrotron radiation-based Fourier-transform infrared spectromicroscopy for characterization of the protein/peptide distribution in single microspheres. *Acta Pharm. Sin. B* **2015**, *5*, 270–276. [CrossRef] [PubMed]

46. Baeten, V.; Dardenne, P. Spectroscopy: Developments in instrumentation and analysis. *Grasas Aceites* **2002**, *53*, 45–63. [CrossRef]

47. Lohumi, S.; Kim, M.S.; Qin, J.; Cho, B.-K. Raman imaging from microscopy to macroscopy: Quality and safety control of biological materials. *TrAC Trends Anal. Chem.* **2017**, *93*, 183–198. [CrossRef]

48. Raman, C.V.; Krishnan, K.S. A new type of secondary radiation. *Nature* **1928**, *121*, 501–502. [CrossRef]

49. Marigheto, N.A.; Kemsley, E.K.; Potter, J.; Belton, P.S.; Wilson, R.H. Effects of sample heating in FT-Raman spectra of biological materials. *Spectrochim. Acta Part A Mol. Biomol. Spectrosc.* **1996**, *52*, 1571–1579. [CrossRef]

50. Milman, B.L. General principles of identification by mass spectrometry. *TrAC Trends Anal. Chem.* **2015**, *69*, 24–33. [CrossRef]

51. Altelaar, A.F.M.; Luxembourg, S.L.; McDonnell, L.A.; Piersma, S.R.; Heeren, R.M.A. Imaging mass spectrometry at cellular length scales. *Nat. Protoc.* **2007**, *2*, 1185–1196. [CrossRef]

52. Thiel, V.; Sjövall, P. Using time-of-flight secondary ion mass spectrometry to study Biomarkers. *Annu. Rev. Earth Planet. Sci.* **2011**, *39*, 125–156. [CrossRef]

53. Greenwalt, D.E.; Goreva, Y.S.; Siljeström, S.M.; Rose, T.; Harbach, R.E. Hemoglobin-derived porphyrins preserved in a Middle Eocene blood-engorged mosquito. *Proc. Natl. Acad. Sci. USA* **2013**, *110*, 18496–18500. [CrossRef]

54. Lindgren, J.; Uvdal, P.; Sjövall, P.; Nilsson, D.E.; Engdahl, A.; Schultz, B.P.; Thiel, V. Molecular preservation of the pigment melanin in fossil melanosomes. *Nat. Commun.* **2012**, *3*, 824. [CrossRef] [PubMed]

55. Winograd, N. The Magic of Cluster SIMS. *Anal. Chem.* **2005**, *77*, 142 A–149 A. [CrossRef]

56. Kozole, J.; Winograd, N. Cluster secondary ion mass spectrometry. In *Surface Analysis and Techniques in Biology*; Smentkowski, V.S., Ed.; Springer: Cham, Switzerland, 2014; pp. 71–98.

57. Meier, D.; Faix, O. Pyrolysis-gas chromatography-mass spectrometry. In *Methods in Lignin Chemistry*; Lin, S.Y., Dence, C.W., Eds.; Springer: Berlin/Heidelberg, Germany, 1992; pp. 177–199, ISBN 13:978-3-642-74065-7.

58. Stankiewicz, B.A.; Briggs, D.E.G.; Evershed, R.P.; Flannery, M.B.; Wuttke, M. Preservation of chitin in 25-million-year-old fossils. *Science* **1997**, *276*, 1541–1543. [CrossRef]

59. Stankiewicz, B.A.; Poinar, H.N.; Briggs, D.E.G.; Evershed, R.P.; Poinar, J. Chemical preservation of plants and insects in natural resins. *Proc. R. Soc. B Biol. Sci.* **1998**, *265*, 641–647. [CrossRef]

60. Stankiewicz, B.A.; Mastalerz, M.; Kruge, M.A.; Van Bergen, P.F.; Sadowska, A. A comparative study of modern and fossil cone scales and seeds of conifers: A geochemical approach. *New Phytol.* **1997**, *135*, 375–393. [CrossRef]

61. Fenn, J.B.; Mann, M.; Meng, C.K.; Wong, S.F.; Whitehouse, C.M. Electrospray ionization for mass spectrometry of large biomolecules. *Science* **1989**, *246*, 64–71. [CrossRef]

62. El-Aneed, A.; Cohen, A.; Banoub, J. Mass spectrometry, review of the basics: Electrospray, MALDI, and commonly used mass analyzers. *Appl. Spectrosc. Rev.* **2009**, *44*, 210–230. [CrossRef]

63. Crowther, J.R. Overview of ELISA in relation to other disciplines. In *The ELISA Guidebook. Methods in Molecular Biology*; Walker, J.M., Ed.; Humana Press: New York, NY, USA, 2009; pp. 1–8, ISBN 978-1-60327-254-4.

64. Yu, H.-W.; Halonen, M.J.; Pepper, I.L. Immunological methods. In *Environmental Microbiology*; Pepper, I.L., Gerba, C.P., Gentry, T.J., Eds.; Elsevier: Amsterdam, The Netherlands, 2015; pp. 245–269, ISBN 978-0-12-394626-3.

65. Aydin, S. A short history, principles, and types of ELISA, and our laboratory experience with peptide/protein analyses using ELISA. *Peptides* **2015**, *72*, 4–15. [CrossRef]

66. Stanker, L.H.; Hnasko, R.M. A double-sandwich ELISA for identification of monoclonal antibodies suitable for sandwich immunoassays. In *Methods in Molecular Biology*; Hnasko, R., Ed.; Humana Press: New York, NY, USA, 2015; pp. 69–78, ISBN 978-1-4939-2741-8.

67. Gan, S.D.; Patel, K.R. Enzyme immunoassay and enzyme-linked immunosorbent assay. *J. Investig. Dermatol.* **2013**, *133*, e12. [CrossRef]

68. Makarananda, K.; Weir, L.R.; Neal, G.E. Competitive ELISA. In *Immunochemical Protocols*; Pound, J.D., Ed.; Humana Press: Totowa, NJ, USA, 1998; pp. 155–160, ISBN 978-1-59259-257-9.

69. Gallagher, S.R. SDS-polyacrylamide gel electrophoresis (SDS-PAGE). *Curr. Protoc. Essent. Lab. Tech.* **2008**, *6*, 7-3. [CrossRef]

70. Yang, P.-C.; Mahmood, T. Western blot: Technique, theory, and trouble shooting. *N. Am. J. Med. Sci.* **2012**, *4*, 429. [CrossRef]

71. Crichton, P.G.; Harding, M.; Ruprecht, J.J.; Lee, Y.; Kunji, E.R.S. Lipid, detergent, and Coomassie Blue G-250 affect the migration of small membrane proteins in Blue Native gels. *J. Biol. Chem.* **2013**, *288*, 22163–22173. [CrossRef]

72. Na Ayutthaya, P.P.; Lundberg, D.; Weigel, D.; Li, L. Blue native polyacrylamide gel electrophoresis (BN-PAGE) for the analysis of protein oligomers in plants. *Curr. Protoc. Plant Biol.* **2020**, *5*, e20107. [CrossRef]

73. Gillett, C.E. Immunohistochemistry. In *Breast Cancer Research Protocols*; Brooks, S.A., Harris, A.L., Eds.; Humana Press: Totowa, NJ, USA, 2006; pp. 191–200.

74. Magaki, S.; Hojat, S.A.; Wei, B.; So, A.; Yong, W.H. An introduction to the performance of immunohistochemistry. In *Biobanking: Methods and Protocols*; Yong, W.H., Ed.; Springer: New York, NY, USA, 2019; pp. 289–298, ISBN 978-1-4939-8935-5.

75. D'Amico, F.; Skarmoutsou, E.; Stivala, F. State of the art in antigen retrieval for immunohistochemistry. *J. Immunol. Methods* **2009**, *341*, 1–18. [CrossRef]

76. Alfonso-Rojas, A.; Cadena, E.-A. Exceptionally preserved 'skin' in an Early Cretaceous fish from Colombia. *PeerJ* **2020**, *8*, e9479. [CrossRef]

77. Roy, A.; Pittman, M.; Saitta, E.T.; Kaye, T.G.; Xu, X. Recent advances in amniote palaeocolour reconstruction and a framework for future research. *Biol. Rev.* **2020**, *95*, 22–50. [CrossRef]

78. Vinther, J. A guide to the field of palaeo colour. *BioEssays* **2015**, *37*, 643–656. [CrossRef]

79. Vinther, J. Reconstructing vertebrate paleocolor. *Annu. Rev. Earth Planet. Sci.* **2020**, *48*, 345–375. [CrossRef]

80. Gueneli, N.; Mckenna, A.M.; Ohkouchi, N.; Boreham, C.J.; Beghin, J.; Javaux, E.J.; Brocks, J.J. 1.1-billion-year-old porphyrins establish a marine ecosystem dominated by bacterial primary producers. *Proc. Natl. Acad. Sci. USA* **2018**, *115*, E6978–E6986. [CrossRef]

81. Tahoun, M.; Gee, C.T.; McCoy, V.E.; Sander, P.M.; Müller, C.E. Chemistry of porphyrins in fossil plants and animals. *RSC Adv.* **2021**, *11*, 7552–7563. [CrossRef]

82. Asher, S.A. Resonance Raman spectroscopy of hemoglobin. In *Methods in Enzymology*; Antonini, E., Chiancone, E., Rossi-Bernardi, L., Eds.; Academic Press: New York, NY, USA, 1981; pp. 371–413.

83. Alleon, J.; Montagnac, G.; Reynard, B.; Brulé, T.; Thoury, M.; Gueriau, P. Pushing Raman spectroscopy over the edge: Purported signatures of organic molecules in fossil animals are instrumental artefacts. *BioEssays* **2021**, *43*, 2000295. [CrossRef]

84. Wiemann, J.; Briggs, D.E.G. Raman spectroscopy is a powerful tool in molecular paleobiology: An analytical response to Alleon et al. (https://doi.org/10.1002/bies.202000295). *BioEssays* **2022**, *44*, 2100070. [CrossRef]

85. Nasti, T.H.; Timares, L. MC1R, Eumelanin and Pheomelanin: Their role in determining the susceptibility to skin cancer. *Photochem. Photobiol.* **2015**, *91*, 188–200. [CrossRef]

86. Suzukawa, A.A.; Vieira, A.; Winnischofer, S.M.B.; Scalfo, A.C.; Di Mascio, P.; Ferreira, A.M.D.C.; Ravanat, J.-L.; Martins, D.D.L.; Rocha, M.E.M.; Martinez, G.R. Novel properties of melanins include promotion of DNA strand breaks, impairment of repair, and reduced ability to damage DNA after quenching of singlet oxygen. *Free Radic. Biol. Med.* **2012**, *52*, 1945–1953. [CrossRef]

87. Lindgren, J.; Moyer, A.; Schweitzer, M.H.; Sjövall, P.; Uvdal, P.; Nilsson, D.E.; Heimdal, J.; Engdahl, A.; Gren, J.A.; Schultz, B.P.; et al. Interpreting melanin-based coloration through deep time: A critical review. *Proc. R. Soc. B Biol. Sci.* **2015**, *282*, 20150614. [CrossRef]

88. Ortonne, J.-P. Photoprotective properties of skin melanin. *Br. J. Dermatol.* **2002**, *146*, 7–10. [CrossRef]

89. D'Ischia, M.; Napolitano, A.; Ball, V.; Chen, C.-T.; Buehler, M.J. Polydopamine and Eumelanin: From structure–property relationships to a unified tailoring strategy. *Acc. Chem. Res.* **2014**, *47*, 3541–3550. [CrossRef]

90. Pezzella, A.; Napolitano, A.; D'Ischia, M.; Prota, G.; Seraglia, R.; Traldi, P. Identification of partially degraded oligomers of 5,6-dihydroxyindole-2-carboxylic acid in sepia melanin by matrix-assisted laser desorption/ionization mass spectrometry. *Rapid Commun. Mass Spectrom.* **1997**, *11*, 368–372. [CrossRef]

91. Falk, H.; Wolkenstein, K. Natural product molecular fossils. In *Progress in the Chemistry of Organic Natural Products*; Kinghorn, A.D., Falk, H., Gibbons, S., Kobayashi, J., Eds.; Springer: Cham, Switzerland, 2017; pp. 1–126, ISBN 978-3-319-45618-8.

92. Cao, W.; Zhou, X.; McCallum, N.C.; Hu, Z.; Ni, Q.Z.; Kapoor, U.; Heil, C.M.; Cay, K.S.; Zand, T.; Mantanona, A.J.; et al. Unraveling the structure and function of melanin through synthesis. *J. Am. Chem. Soc.* **2021**, *143*, 2622–2637. [CrossRef]

93. Lindgren, J.; Sjövall, P.; Carney, R.M.; Uvdal, P.; Gren, J.A.; Dyke, G.; Schultz, B.P.; Shawkey, M.D.; Barnes, K.R.; Polcyn, M.J. Skin pigmentation provides evidence of convergent melanism in extinct marine reptiles. *Nature* **2014**, *506*, 484–488. [CrossRef]

94. Colleary, C.; Dolocan, A.; Gardner, J.; Singh, S.; Wuttke, M.; Rabenstein, R.; Habersetzer, J.; Schaal, S.; Feseha, M.; Clemens, M.; et al. Chemical, experimental, and morphological evidence for diagenetically altered melanin in exceptionally preserved fossils. *Proc. Natl. Acad. Sci. USA* **2015**, *112*, 12592–12597. [CrossRef]

95. Moyer, A.E.; Zheng, W.; Johnson, E.A.; Lamanna, M.C.; Li, D.; Lacovara, K.J.; Schweitzer, M.H. Melanosomes or microbes: Testing an alternative hypothesis for the origin of microbodies in fossil feathers. *Sci. Rep.* **2015**, *4*, 4233. [CrossRef]

96. Schleifer, K.H.; Kandler, O. Peptidoglycan types of bacterial cell walls and their taxonomic implications. *Bacteriol. Rev.* **1972**, *36*, 407–477. [CrossRef] [PubMed]

97. Härtner, T.; Straub, K.L.; Kannenberg, E. Occurrence of hopanoid lipids in anaerobic *Geobacter* species. *FEMS Microbiol. Lett.* **2005**, *243*, 59–64. [CrossRef]

98. Belin, B.J.; Busset, N.; Giraud, E.; Molinaro, A.; Silipo, A.; Newman, D.K. Hopanoid lipids: From membranes to plant–bacteria interactions. *Nat. Rev. Microbiol.* **2018**, *16*, 304–315. [CrossRef] [PubMed]

99. Glass, K.; Ito, S.; Wilby, P.R.; Sota, T.; Nakamura, A.; Bowers, C.R.; Vinther, J.; Dutta, S.; Summons, R.; Briggs, D.E.G.; et al. Direct chemical evidence for eumelanin pigment from the Jurassic period. *Proc. Natl. Acad. Sci. USA* **2012**, *109*, 10218–10223. [CrossRef] [PubMed]

100. Glass, K.; Ito, S.; Wilby, P.R.; Sota, T.; Nakamura, A.; Russell Bowers, C.; Miller, K.E.; Dutta, S.; Summons, R.E.; Briggs, D.E.G.; et al. Impact of diagenesis and maturation on the survival of eumelanin in the fossil record. *Org. Geochem.* **2013**, *64*, 29–37. [CrossRef]

101. Smejkal, G.B.; Schweitzer, M.H. Will current technologies enable dinosaur proteomics? *Expert Rev. Proteom.* **2007**, *4*, 695–699. [CrossRef]

102. Buckley, M.; Warwood, S.; van Dongen, B.; Kitchener, A.C.; Manning, P.L. A fossil protein chimera; difficulties in discriminating dinosaur peptide sequences from modern cross-contamination. *Proc. R. Soc. B Biol. Sci.* **2017**, *284*, 20170544. [CrossRef]

103. Sawyer, R.H.; Glenn, T.; French, J.O.; Mays, B.; Shames, R.B.; Barnes, G.L.; Rhodes, W.; Ishikawa, Y. The expression of beta (β) keratins in the epidermal appendages of reptiles and birds. *Am. Zool.* **2000**, *40*, 530–539. [CrossRef]

104. Alberts, B.; Johnson, A.; Lewis, J.; Raff, M.; Roberts, K.; Walter, P. Cell junctions, cell adhesion, and the extracellular matrix. In *Molecular Biology of the Cell*, 4th ed.; Garland Science: New York, NY, USA, 2002; pp. 1065–1126.

105. Koide, T. Triple helical collagen-like peptides: Engineering and applications in matrix biology. *Connect. Tissue Res.* **2005**, *46*, 131–141. [CrossRef] [PubMed]

106. Frantz, C.; Stewart, K.M.; Weaver, V.M. The extracellular matrix at a glance. *J. Cell Sci.* **2010**, *123*, 4195–4200. [CrossRef] [PubMed]

107. Rozario, T.; DeSimone, D.W. The extracellular matrix in development and morphogenesis: A dynamic view. *Dev. Biol.* **2010**, *341*, 126–140. [CrossRef] [PubMed]

108. Sorushanova, A.; Delgado, L.M.; Wu, Z.; Shologu, N.; Kshirsagar, A.; Raghunath, R.; Mullen, A.M.; Bayon, Y.; Pandit, A.; Raghunath, M.; et al. The collagen suprafamily: From biosynthesis to advanced biomaterial development. *Adv. Mater.* **2019**, *31*, 1801651. [CrossRef]

109. Shoulders, M.D.; Raines, R.T. Collagen structure and stability. *Annu. Rev. Biochem.* **2009**, *78*, 929–958. [CrossRef]

110. Curtis, R.W.; Chmielewski, J. A comparison of the collagen triple helix and coiled-coil peptide building blocks on metal ion-mediated supramolecular assembly. *Pept. Sci.* **2021**, *113*, e24190. [CrossRef]

111. Yamauchi, M.; Sricholpech, M. Lysine post-translational modifications of collagen. *Essays Biochem.* **2012**, *52*, 113–133. [CrossRef]

112. Hulmes, D.J.S. Collagen diversity, synthesis and assembly. In *Collagen*; Fratzl, P., Ed.; Springer: Boston, MA, USA, 2008; pp. 15–47.

113. Line, S.; Rhodes, C.; Yamada, Y. Molecular Biology of Cartilage Matrix. In *Cellular and Molecular Biology of Bone*; Noda, M., Ed.; Academic Press: San Diego, CA, USA, 1993; pp. 539–555.

114. Fedarko, N.S. Osteoblast/osteoclast development and function in osteogenesis imperfecta. In *Osteogenesis Imperfecta*; Shapiro, J.R., Ed.; Academic Press: San Diego, CA, USA, 2014; pp. 45–56, ISBN 9780123971654.

115. Henriksen, K.; Karsdal, M.A. Type I collagen. In *Biochemistry of Collagens, Laminins and Elastin*; Karsdal, M.A., Ed.; Elsevier: Amsterdam, The Netherlands, 2019; pp. 1–12, ISBN 9780128170687.

116. Byers, P.H.; Bonadio, J.F. The molecular basis of clinical heterogeneity in osteogenesis imperfecta: Mutations in type I collagen genes have different effects on collagen processing. In *Genetic and Metabolic Disease in Pediatrics*; Lloyd, J.K., Scriver, C.R., Eds.; Butterworth & Co. (Publishers) Ltd.: Bodmin, UK, 1985; pp. 56–90.

117. Bächinger, H.P.; Mizuno, K.; Vranka, J.A.; Boudko, S.P. Collagen formation and structure. In *Comprehensive Natural Products II: Chemistry and Biology*; Liu, H.-W., Mander, L., Eds.; Elsevier: Amsterdam, The Netherlands, 2010; Volume 5, pp. 469–530, ISBN 9780080453828.

118. Shavandi, A.; Silva, T.H.; Bekhit, A.A.; Bekhit, A.E.-D.A. Keratin: Dissolution, extraction and biomedical application. *Biomater. Sci.* **2017**, *5*, 1699–1735. [CrossRef]

119. Bragulla, H.H.; Homberger, D.G. Structure and functions of keratin proteins in simple, stratified, keratinized and cornified epithelia. *J. Anat.* **2009**, *214*, 516–559. [CrossRef]
120. Coulombe, P.A.; Omary, M.B. 'Hard' and 'soft' principles defining the structure, function and regulation of keratin intermediate filaments. *Curr. Opin. Cell Biol.* **2002**, *14*, 110–122. [CrossRef]
121. Gu, L.-H.; Coulombe, P.A. Keratin function in skin epithelia: A broadening palette with surprising shades. *Curr. Opin. Cell Biol.* **2007**, *19*, 13–23. [CrossRef]
122. Sun, T.-T.; Eichner, R.; Nelson, W.G.; Scheffer Tseng, C.G.; Weiss, R.A.; Jarvinen, M.; Woodcock-Mitchell, J. Keratin classes: Molecular markers for different types of epithelial differentiation. *J. Investig. Dermatol.* **1983**, *81*, S109–S115. [CrossRef] [PubMed]
123. Korniłłowicz-Kowalska, T.; Bohacz, J. Biodegradation of keratin waste: Theory and practical aspects. *Waste Manag.* **2011**, *31*, 1689–1701. [CrossRef] [PubMed]
124. Wang, B.; Yang, W.; McKittrick, J.; Meyers, M.A. Keratin: Structure, mechanical properties, occurrence in biological organisms, and efforts at bioinspiration. *Prog. Mater. Sci.* **2016**, *76*, 229–318. [CrossRef]
125. Fraser, R.D.B.; Parry, D.A.D. Molecular packing in the feather keratin filament. *J. Struct. Biol.* **2008**, *162*, 1–13. [CrossRef]
126. Toni, M.; Dalla Valle, L.; Alibardi, L. Hard (beta-)keratins in the epidermis of reptiles: Composition, sequence, and molecular organization. *J. Proteome Res.* **2007**, *6*, 3377–3392. [CrossRef]
127. Perța-Crișan, S.; Ursachi, C.Ș.; Gavrilaș, S.; Oancea, F.; Munteanu, F.-D. Closing the loop with keratin-rich fibrous materials. *Polymers* **2021**, *13*, 1896. [CrossRef]

biology

MDPI

Article

Taphonomic and Diagenetic Pathways to Protein Preservation, Part I: The Case of *Tyrannosaurus rex* Specimen MOR 1125

Paul V. Ullmann [1,*], Kyle Macauley [1], Richard D. Ash [2], Ben Shoup [3] and John B. Scannella [4,5]

[1] Department of Geology, Rowan University, Glassboro, NJ 08028, USA; macauleyk4@students.rowan.edu
[2] Department of Geology, University of Maryland, College Park, MD 20742, USA; rdash@umd.edu
[3] Absaroka Energy & Environmental Solutions, Buffalo, WY 82834, USA; ben.shoup@absarokasolutions.com
[4] Museum of the Rockies, Montana State University, Bozeman, MT 59717, USA; john.scannella@montana.edu
[5] Department of Earth Sciences, Montana State University, Bozeman, MT 59717, USA
* Correspondence: ullmann@rowan.edu

Simple Summary: Contrary to traditional views, fossil bones have been shown to occasionally retain original cells, blood vessels, and structural tissues that are still comprised, in part, by their original proteins. To help clarify how such remarkable preservation occurs, we explored the fossilization history of a famous *Tyrannosaurus rex* specimen previously shown to yield original cells, vessels, and collagen protein sequences. By analyzing the trace element composition of the femur of this tyrannosaur, we show that after death its carcass decayed underwater in a brackish, oxic, estuarine channel and then became buried by sands that quickly cemented around the bones, largely protecting them from further chemical alteration. Other bones yielding original proteins have also been found to have fossilized within rapidly-cementing sediments in oxidizing environments, which strongly suggests that such settings are conducive to molecular preservation.

Citation: Ullmann, P.V.; Macauley, K.; Ash, R.D.; Shoup, B.; Scannella, J.B. Taphonomic and Diagenetic Pathways to Protein Preservation, Part I: The Case of *Tyrannosaurus rex* Specimen MOR 1125. *Biology* **2021**, *10*, 1193. https://doi.org/10.3390/biology10111193

Academic Editor: Douglas S. Glazier

Received: 21 October 2021
Accepted: 14 November 2021
Published: 17 November 2021

Publisher's Note: MDPI stays neutral with regard to jurisdictional claims in published maps and institutional affiliations.

Copyright: © 2021 by the authors. Licensee MDPI, Basel, Switzerland. This article is an open access article distributed under the terms and conditions of the Creative Commons Attribution (CC BY) license (https://creativecommons.org/licenses/by/4.0/).

Abstract: Many recent reports have demonstrated remarkable preservation of proteins in fossil bones dating back to the Permian. However, preservation mechanisms that foster the long-term stability of biomolecules and the taphonomic circumstances facilitating them remain largely unexplored. To address this, we examined the taphonomic and geochemical history of *Tyrannosaurus rex* specimen Museum of the Rockies (MOR) 1125, whose right femur and tibiae were previously shown to retain still-soft tissues and endogenous proteins. By combining taphonomic insights with trace element compositional data, we reconstruct the postmortem history of this famous specimen. Our data show that following prolonged, subaqueous decay in an estuarine channel, MOR 1125 was buried in a coarse sandstone wherein its bones fossilized while interacting with oxic and potentially brackish early-diagenetic groundwaters. Once its bones became stable fossils, they experienced minimal further chemical alteration. Comparisons with other recent studies reveal that oxidizing early-diagenetic microenvironments and diagenetic circumstances which restrict exposure to percolating pore fluids elevate biomolecular preservation potential by promoting molecular condensation reactions and hindering chemical alteration, respectively. Avoiding protracted interactions with late-diagenetic pore fluids is also likely crucial. Similar studies must be conducted on fossil bones preserved under diverse paleoenvironmental and diagenetic contexts to fully elucidate molecular preservation pathways.

Keywords: REE; *Tyrannosaurus rex*; molecular paleontology; geochemical taphonomy; diagenesis; bone; protein; collagen; Hell Creek Formation

1. Introduction

1.1. Biomolecular Preservation in Fossils

Preservation of endogenous biomolecules like proteins and DNA in ancient fossils was once thought implausible by many paleontologists. The plethora of microbial and inorganic agents of decay inherent in fossilization were long expected to inevitably either breakdown labile biomolecules into minute and useless components, or alter them so thoroughly via recombination and crosslinking that they become intractably unrecognizable [1]. Yet, as discussed in detail by several recent reviews [2–7], innovative adaptations of molecular biology techniques toward analyzing fossil samples and significant advances in analytical resolution have proven that recognizable biomolecules persist in many, diverse fossils. Ancient DNA and even entire nuclear and mitochondrial genomes are now being recovered from Pleistocene fossils [8–12], some of which are older than 1 My [13], and proteins such as collagen I, hemoglobin, and β-keratin have been identified within vertebrate fossils dating back to the Jurassic [14–21]. Collagen I peptides have even been recovered from two Cretaceous non-avian dinosaur bones [18,22,23], and cladistic analyses found those peptides to place non-avian dinosaurs in their expected phylogenetic context within Archosauria [18,24], thus corroborating their authenticity. The growing number of these reports by independent research groups using diverse analytical techniques clearly demonstrates that 'surviving' fossilization is not an insurmountable challenge for biomolecules. Yet, this concept remains controversial, in part because there are still many gaps in our understanding of the geochemical processes that result in fossilization in general, let alone the processes that result in exceptional preservation of soft tissues and their component molecules.

"Traditional" explanations attributing the preservation of proteinaceous tissues such as skin, blood vessels, or feathers to simple carbonization (e.g., [25]) or mineral replacement (e.g., via phosphatization [26]) are clearly insufficient, as they cannot explain cases of retention of endogenous molecular signatures in such fossil tissues (e.g., [15,27]). Yet, it is only within the last decade that molecular paleontologists have begun developing alternative hypotheses about preservation mechanisms capable of fostering long-term molecular stability. Schweitzer et al. [28,29] were the first to propose such a mechanism, suggesting that iron-catalyzed free radical reactions could mediate natural tissue fixation by inducing intramolecular crosslinking. Ensuing actualistic experiments by Boatman et al. [30] found support for this hypothesis in that Fenton- and glycation-treated modern chicken collagen samples exhibited the same types of crosslinks present in the walls of fossil blood vessels recovered from a Cretaceous *Tyrannosaurus rex* femur (MOR 555/USNM 555000). To date, only two other studies have directly addressed preservation mechanisms. Schweitzer et al. [31] suggested that concurrent recrystallization of bone hydroxyapatite may encase the molecular condensation products described above as they form, shielding them in a manner similar to intracrystalline proteins (cf. [32]), whereas Edwards et al. [33] alternatively suggested that ternary complexation of biomolecules with dissolved metal cations and mineral crystal surfaces may stabilize them over geologic timescales. As the brevity of this summary shows, attempts to investigate mechanisms of molecular preservation remain rare.

To better elucidate these (and other, as-yet-unrecognized) potential preservation mechanisms, it is imperative to identify geochemical regimes conducive to such "exceptional" preservation and characterize the full suite of physicochemical taphonomic parameters at play. Although certain taphonomic circumstances can reliably be linked to "exceptional" preservation (e.g., rapid burial) [34,35], many geologic and geochemical variables need further study to elucidate their influence(s) on decay at the molecular level. For example, how much do factors such as groundwater chemistry and diffusion history control the preservation potential of biomolecules? Are the depositional setting (e.g., floodplain, seafloor) and geochemical microenvironment of burial primary factors controlling whether or not original soft tissues and their component molecules persist? If so, which sedimentary facies and diagenetic histories are most conducive to biomolecular preservation in fossils, and why? These important questions remain largely unexplored.

1.2. Trace Element Taphonomy of Fossil Bone

Among the many established and emerging analytical methods that could be used to shed light on potential answers to the questions listed above, trace element analyses are arguably the most useful, as they offer unparalleled windows into the geochemical and diagenetic history of fossils. Rare earth elements (REE; lanthanum-lutetium), uranium, scandium, and other trace elements released from sediments into groundwaters are ubiquitously adsorbed by bone hydroxyapatite during fossilization (primarily via surface adsorption and cationic exchange into the crystal lattice, leaving histologic structure unaltered; [36,37]). These elements are negligibly present in bone mineral during life, meaning that their presence in a fossil bone reflects the geochemical and hydrodynamic history of the diagenetic environment [37,38]. The proportions of trace elements adsorbed depends largely on groundwater chemistry [37], and their spatial distributions within fossil bone tissues form records of the pore-fluid interactions a bone experienced through diagenesis [36,39–44]. In short, analyses of REE and other trace elements offer unique insights not only into the magnitude of chemical alteration a specimen has undergone, but also: (1) the chemistry of ancient surface and groundwaters in the burial environment (e.g., [43,45,46]); (2) spatial and temporal trends in redox conditions both within and around biologic remains as they were fossilizing (e.g., [46–48]), and; (3) the number and timing of interactions a specimen had with pore fluids through diagenesis [39].

Given these diverse and critically-relevant utilities, two of us (P.V.U. and R.D.A.) elected to employ trace element analyses in parallel with molecular assays in an initial case study examining the diagenetic history of *Edmontosaurus* bones retaining endogenous collagen I from a mass-death bonebed in the Cretaceous Hell Creek Formation [20,46,49,50]. Although those studies clarified one set of paleoenvironmental, diagenetic, and geochemical circumstances conducive to biomolecular preservation, all of the specimens examined in that project derived from a single locality and taxon. The current study builds upon our prior work by examining the diagenetic history of another non-avian dinosaur preserved under drasticallydifferent paleoenvironmental and diagenetic circumstances within the Hell Creek Formation: *Tyrannosaurus rex* specimen MOR 1125. This specimen became one of the most widely-known fossils in the world when Schweitzer et al. [51,52] and Asara et al. [22] reported, respectively, the preservation of original bone cells, blood vessels, and pliable proteinaceous matrix in its right femur and both tibiae, as well as endogenous collagen I peptides in its right femur. Those studies of MOR 1125 ignited unprecedented interest in the now-growing field of molecular paleontology, so it is due time for the taphonomic and diagenetic history of this specimen to be resolved in comprehensive detail.

2. Taphonomic and Geologic Context

MOR 1125 was collected from exposures of the Maastrichtian Hell Creek Formation northwest of Jordan, MT, and just south of the Fort Peck Reservoir on lands managed by the United States Fish and Wildlife Service (Charles M. Russell Wildlife Refuge) (Figure 1). Its more than 220 skeletal elements were disarticulated but closely associated (Figure 2) within sandstone underneath ~50 ft of overburden (Figure 3A and Figure S1). All preserved cranial elements were found within a 5 m^2 area in the northern corner of the quarry, and nearly all teeth were found dislodged from the jaw elements in two clusters adjacent to the skull bones. Based on data from the 25 m^2 excavated in 2002 (Figure 2), skeletal element abundances range from 0 to 16 specimens/m^2 with an average of 5 specimens/m^2. Field data collected by Museum of the Rockies crews reveal that the bones were stratigraphically separated from one another by as much as 55 cm, though all of the cranial elements were found within a narrower ~15 cm interval. Although this stratigraphic interval is thick enough to accommodate bones stacked on top of one another, almost all specimens were found spatially isolated. Strike measurements acquired for 14 long bones (e.g., limb bones, ribs) during the 2002 field season show a bimodal pattern comprising a northeast-southwest trend and a nearly north-south trend (Figure 3B), implying probable hydraulic orientation of bones within the quarry. Bones range in size from phalanges and cervical ribs up to girdle elements and large limb bones (i.e., bones pertaining to both Voorhies Groups I and II were recovered; cf. [53]), and all major portions of the body are represented. Most of the bones are complete, though some exhibit transverse and longitudinal fractures stemming from minor post-fossilization compaction. None of the bones exhibit any noteworthy signs of weathering or abrasion, and the femur examined by Schweitzer et al. [51,52] and Asara et al. [22] exhibits infilling of the medullary cavity by crushed trabeculae and sedimentary matrix.

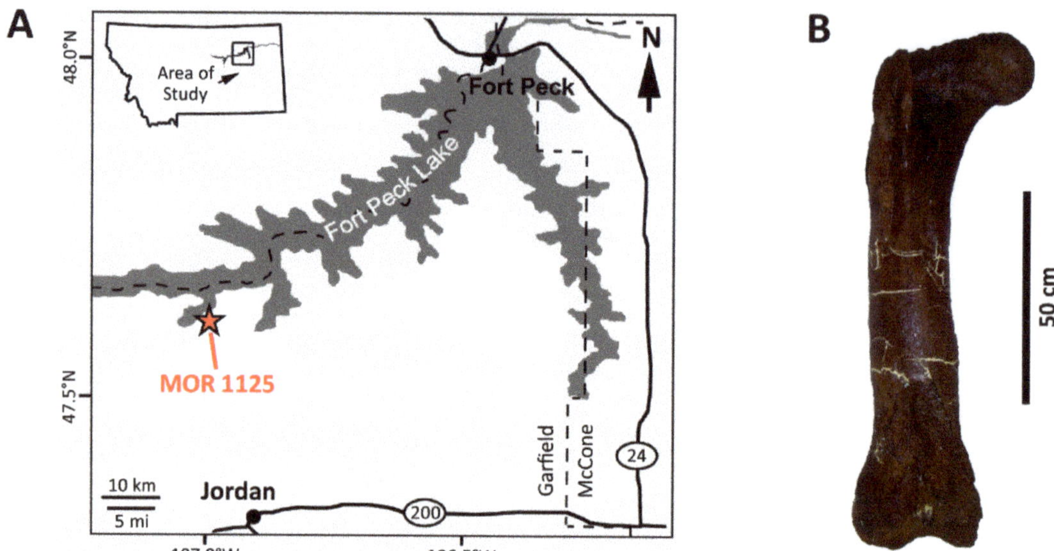

Figure 1. (**A**) Map showing the location of the MOR 1125 quarry in Garfield County, Montana. (**B**) Right femur of MOR 1125 examined in this study. Map modified from [54] and bone photograph modified from [55], each under CC BY-4.0 licenses.

Schweitzer et al. [51] (p. 1953) briefly characterized the lithology from which MOR 1125 was recovered as a "soft, well-sorted sandstone that was interpreted as estuarine in origin", but the stratigraphy and sedimentology of the quarry have never been reported in detail, despite their relevance to interpreting the depositional environment and taphonomic history of this specimen. Below, we briefly summarize the stratigraphy of the quarry and the entire butte encasing this specimen.

The specimen was recovered from the Basal Sand of the Hell Creek Formation (*sensu* Hartman et al. [56] and Fowler [54]), 1.5 m above its basal contact with underlying Colgate tidal flat facies. This places it within the lower unit (L3 *sensu* Horner et al. [57]) of the Hell Creek Formation (Figure S1). Dark, marine deposits of the Bearpaw Shale are exposed near the base of the butte from which the specimen was recovered, and they exhibit a gradational contact with a shallow-marine sequence identified as the Fox Hills Sandstone. The Fox Hills Formation is approximately 11.2 m thick and coarsens subtly up-section from silty, hummocky cross-stratified fine sands at the base to low-angle planar cross-stratified, fine-medium sands near the top (Figure S1). Sublitharenitic, trough cross-stratified, fine to medium-grained sands of the Colgate Sandstone erosively scour into the Fox Hills Formation (Figure S1), as is common across the region (following the stratigraphic definitions of Fowler [54]).

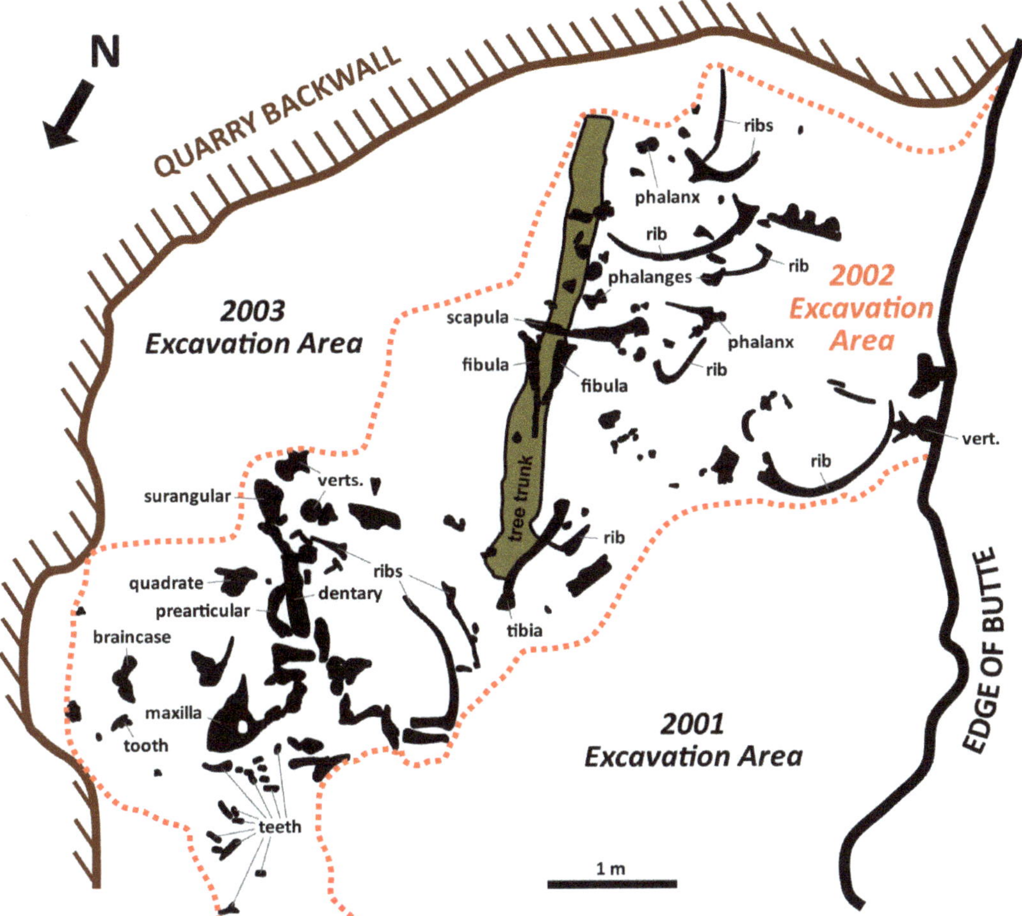

Figure 2. Quarry map from the 2002 field season at the MOR 1125 quarry. Only bones discovered during that field season are shown. Select identifiable bones are identified as labeled. Although the large tree trunk was found above all the bones in the central region of the quarry, it is shown behind them to allow the bones to be better seen. Scale bar as indicated.

Specifically, MOR 1125 was found entombed near the base of a normally-graded, swaley to trough cross-bedded channel lag deposit within a 11.8 m thick section of fine to coarse-grained, trough cross-bedded sandstones (Figure 3A and Figure S1). Pebbles and rounded, silty, rip-up clasts occur immediately beneath the bones, and all three of these larger clast types are supported by a homogenous matrix of medium and coarse sand, all of which are indicative of deposition within a relatively high-energy channel (see Discussion).

Strata exposed in the upper portion of the quarry wall largely consist of cross-bedded fluvial channel sandstones, shaly floodplain/overbank mudstones, and massive crevasse-splay sandstones typical of the middle portion of the Hell Creek Formation [54,55,58–61]. For further details on the sedimentology of the entombing sandstone and strata overlying it, please see Figure S1 and the Supplementary Materials.

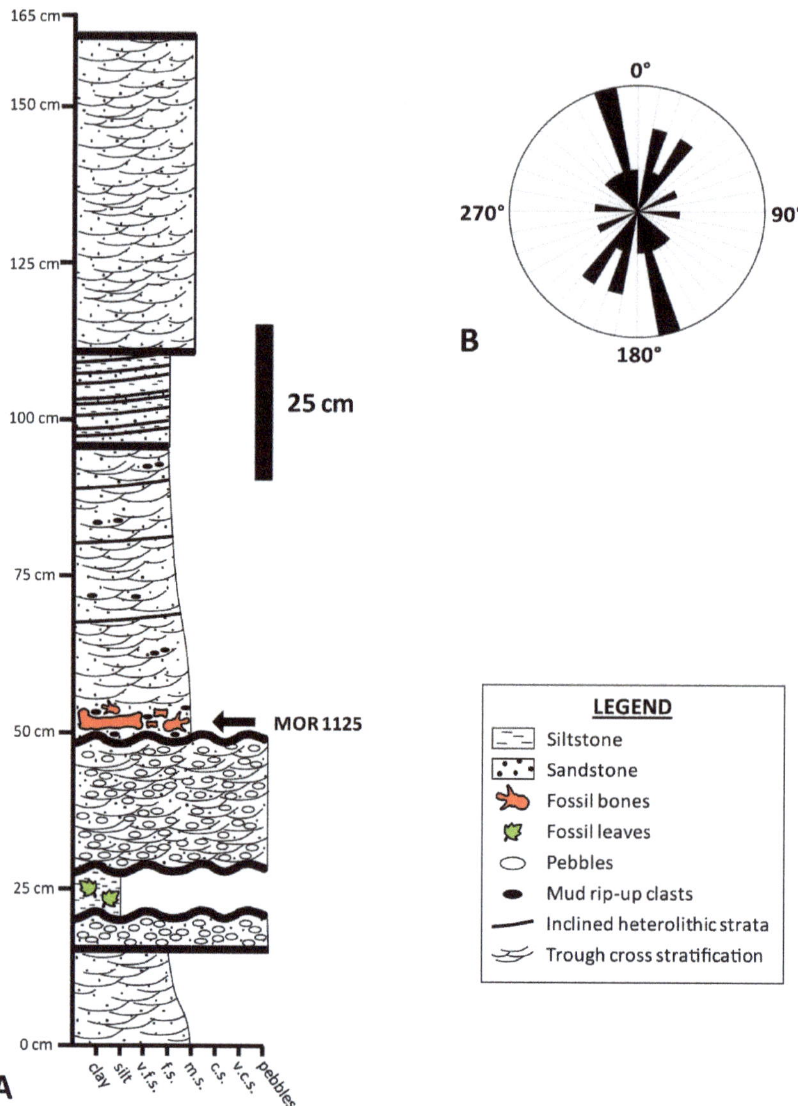

Figure 3. (**A**) Stratigraphic section taken within the MOR 1125 quarry. (**B**) Rose diagram of MOR 1125 bone orientations based on data from 14 long bones collected during the 2002 field season. Presented as an arithmetic plot with 10° bins. Abbreviations: c.s.—coarse sand; f.s.—fine sand; m.s.—medium sand; v.c.s.—very coarse sand; v.f.s.—very fine sand. Scale bar for (**A**) as indicated.

3. Materials and Methods

3.1. Materials

A portion of cortex excised from the midshaft of the right femur of *Tyrannosaurus rex* MOR 1125 was used in this study. Although it is not the exact piece of cortex from which Schweitzer et al. [52] and Asara et al. [22] recovered endogenous protein and peptide sequences, respectively, this fragment is derived from the same region of the midshaft of the same bone, thereby minimizing any potential chemical heterogeneities between the two cortex samples. The fragment comprises nearly the entire cortical thickness of the bone, including the intact external cortex margin but not the internal wall of the medullary cavity (hence, none of the medullary tissue identified by Schweitzer et al. [62] is present in this fragment). Macroscopically, the femur is well-preserved, exhibiting no signs of pre-burial weathering or postmortem abrasion.

3.2. Methods

3.2.1. Sample Preparation

An autoclaved chisel was initially used to isolate a smaller piece of the midshaft fragment of the femur of MOR 1125 for embedding and sectioning. The resulting subsample, which captured the cortical width of the femur, was then embedded under vacuum in Silmar 41^{TM} resin (US Composites). A Hillquist SF-8 trim saw was used to cut a thick section (~3 mm) from the embedded subsample, which was then rinsed with distilled water and allowed to thoroughly dry. For laser ablation-inductively coupled plasma mass spectrometry (LA-ICPMS) analyses, the thick section was placed directly in the laser ablation chamber; no further polishing was necessary or performed.

3.2.2. LA-ICPMS Analyses

We employed the same mass spectrometry methods as Ullmann et al. [46] in this study, and refer the reader to that publication for details. Briefly, LA-ICPMS was used to examine spatial heterogeneity of REE and other pertinent trace elements in the fossil in order to reconstruct the diagenetic history of the specimen and the geochemical regimes to which it was exposed. Iron concentrations are reported in weight percentage (wt. %), while all other concentrations are reported in parts per million (ppm). To enable comparisons to fossil bones from other sites, REE concentrations were normalized against the North American Shale Composite (NASC) using values from Gromet et al. [63] and Haskin et al. [64] (a subscript N denotes shale-normalized values or ratios). Reproducibility, taken as the percent relative standard deviation for all REE in an NIST 610 glass standard, averaged 2% and was at or below 3% for every element except iron (6.6%). For further analytical specifics of the LA-ICPMS runs performed in this study, please see the Supplementary Materials.

4. Results

4.1. Overall REE Composition

At the specimen level (i.e., considering all transect data combined), the femur of MOR 1125 exhibits a $\sum$REE of 596 ppm; this value thus represents the average REE content of the cortex (Table 1). Manganese (Mn) and strontium (Sr) concentrations are the highest of all recorded elements (2439 and 2386 ppm, respectively), with concentrations more than double all the other trace elements and more than an order of magnitude higher than all REE (Table 1). The average concentration of yttrium (Y) is also higher than all REE (1102 ppm), and the average scandium (Sc) concentration is very high (83 ppm) compared to fresh bone. At the whole-bone level, light (LREE) and heavy rare earth element (HREE) concentrations are elevated compared to the middle rare earths (MREE, Sm–Gd), indicating fraction among REE occurred during uptake (see below). Though uranium (U) concentrations exhibit an average (38 ppm) higher than other dinosaur bones recently analyzed from the Hell Creek Formation (2–18 ppm) [46], the femur of MOR 1125 exhibits a comparatively lower amount of iron (0.73 wt. %; compared to 1.23–1.76 wt. % in [46]).

Table 1. Average whole-bone trace element composition of the right femur of *Tyrannosaurus rex* MOR 1125. Numbers presented are averages of all transect data acquired across the cortex. Iron (Fe) is presented in weight percent (wt. %); all other elements are in parts per million (ppm). Absence of $(Ce/Ce^*)_N$, $(Pr/Pr^*)_N$, $(Ce/Ce^{**})_N$, and $(La/La^*)_N$ anomalies occurs at 1.0. $(Ce/Ce^*)_N$, $(Pr/Pr^*)_N$, $(Ce/Ce^{**})_N$, and $(La/La^*)_N$ anomalies, comparing observed (Ce, Pr, La) versus expected (Ce*, Pr*, Ce**, La*) concentrations of each element, are calculated as in the Materials and Methods section of the text. The Y/Ho value reflects this mass ratio.

Element	Concentration
Sc	83.43
Mn	2439
Fe	0.73
Sr	2386
Y	1102
Ba	888
La	66.48
Ce	107.47
Pr	14.24
Nd	63.72
Sm	17.65
Eu	5.45
Gd	34.38
Tb	6.88
Dy	63.45
Ho	18.82
Er	74.13
Tm	12.51
Yb	93.69
Lu	16.68
Th	0.13
U	37.77
ΣREE	596
$(Ce/Ce^*)_N$	0.82
$(Pr/Pr^*)_N$	0.92
$(Ce/Ce^{**})_N$	1.26
$(La/La^*)_N$	2.83
Y/Ho	58.55

4.2. Intra-Bone REE Depth Profiles

All REE exhibit steeply declining concentrations with cortical depth, and certain elements exhibit hints of weak secondary diffusion from within the medullary cavity (e.g., Figure 4A; also see Supplementary Materials). Among the REE, cerium (Ce) concentrations are the highest at the cortical margin (~2200 ppm) and thulium (Tm) exhibits the lowest concentrations at the outer cortex edge (~50 ppm). LREE exhibit the steepest declines (on average from ~1300 ppm near the cortical margin to <15 ppm by 1 cm into the cortex; Figure 4A), generally encompassing a decrease of two orders of magnitude, which is clearly indicative of greater uptake in the external cortex than deeper within the bone. MREE concentration profiles are generally intermediate in slope between those of LREE and HREE, and MREE concentrations are so low throughout the middle and internal cortex (<2 ppm) that they frequently encroach on or fall below detection limit (Data S1). HREE exhibit the flattest profiles among the rare earths (e.g., ytterbium [Yb] in Figure 4A), and unlike LREE and MREE, they exhibit increasing concentrations with depth through the internal cortex (again reflective of fractionation during uptake; see below). For example, Yb concentrations rise from ~30–80 ppm in the middle cortex to ~50–150 ppm in the internal cortex (Data S1). Moreover, HREE only decline from an average of ~250 ppm at the cortical margin to ~40 ppm by 1 cm into the cortex, constituting less than an order of magnitude decrease.

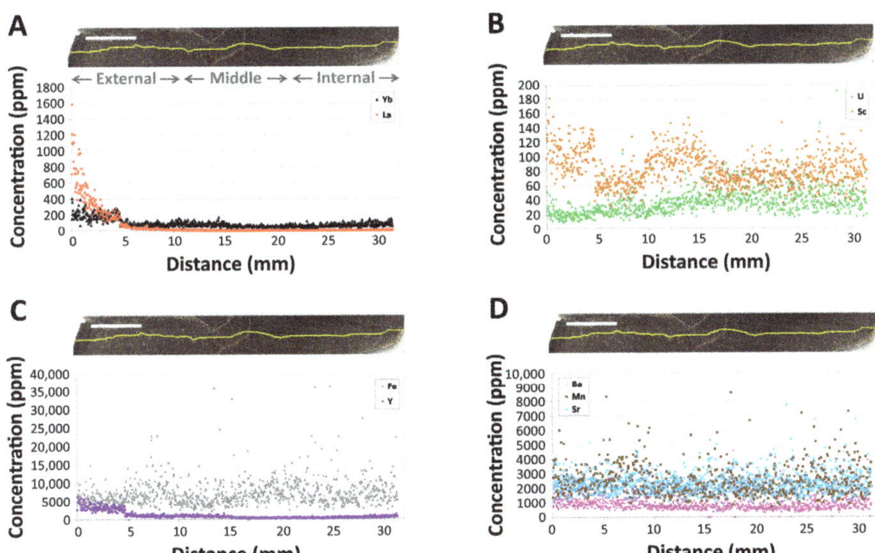

Figure 4. Intra-bone REE concentration gradients of various elements in the femur of MOR 1125. (**A**) Lanthanum (La) and ytterbium (Yb). (**B**) Scandium (Sc) and uranium (U). (**C**) Iron (Fe) and yttrium (Y). (**D**) Barium (Ba), manganese (Mn), and strontium (Sr). Note the different concentration scales for each panel. The laser track is denoted by the yellow line in each bone cross section. Gray text labels in (**A**) span the approximate regions considered as the 'external', 'middle', and 'internal' cortices. Scale bars, in white over bone images, each equal 1 mm.

Although intermittent spikes in REE concentrations are present in osteonal tissue surrounding Haversian canals (indicative of uptake through vascular systems, e.g., brief Yb spike near 14 mm in Figure 4A), there are no obvious signs of a uniform deflection in elemental profiles reflective of significant double medium diffusion effects (cf. [44]). Instead, most trace element profiles exhibit a distinct plateau and then drop in concentrations at ~4.7 mm into the cortex (Figure 4). A fine, open, diagenetic crack passes diagonally across the laser transect at this depth. As shown in Figure 4A, REE concentrations are uniformly higher in the bone on the external side of this crack than on the internal side of it.

Scandium (Sc) is the only element to exhibit a distinct, broad peak in concentrations within the middle cortex (Figure 4B). Uranium (U) also exhibits a depth profile shape distinct from those of all the other elements examined, characterized by an initial, weak decrease in concentration from the cortical margin followed by a slow, steady increase in concentration throughout the middle cortex to a stable plateau of the highest concentrations within the bone (~40–50 ppm) within the internal cortex (Figure 4B). Unlike REE, Sc, and Y, the profile of U does not exhibit any disruption related to the crack at 4.7 mm. Iron (Fe) exhibits a nearly flat profile (Figure 4C) with comparatively less locallyrestricted spikes in concentrations in the external cortex than strontium (Sr), manganese (Mn), and barium (Ba), each of which exhibit high concentrations throughout the cortex (Figure 4D). Yttrium (Y) exhibits the same general profile shape as the HREE in the bone, including a slight decrease in concentrations near 15 mm and slight increase in concentrations toward the internal end of the transect (Figure 4C), indicative of similar uptake behavior for these elements in the femur of MOR 1125.

4.3. NASC-Normalized REE Patterns

Although the external-most cortex of the femur exhibits considerably greater REE enrichment than deeper portions of the cortex (as is common in fossil bones, e.g., [39,65]; Figure 5A), the external 250 µm of the transect still exhibits an NASC-normalized pattern very similar to that of the bone as a whole (compare Figures 5B and 6A), but with lesser relative enrichment in HREE. Both exhibit a modest negative Ce anomaly (visually evident as a downward deflection of the pattern at this element) and relative LREE depletion and HREE enrichment relative to MREE. Relative enrichment in HREE (perhaps resulting from uptake from brackish pore fluids; see Discussion) is also evident in how closely a data point for the whole-bone composition of MOR 1125 plots to the Yb corner of a Nd_N-Gd_N-Yb_N ternary plot (Figure 5C). In the external 250 µm of the cortex, shale-normalized concentrations range from ~30–100 times NASC values.

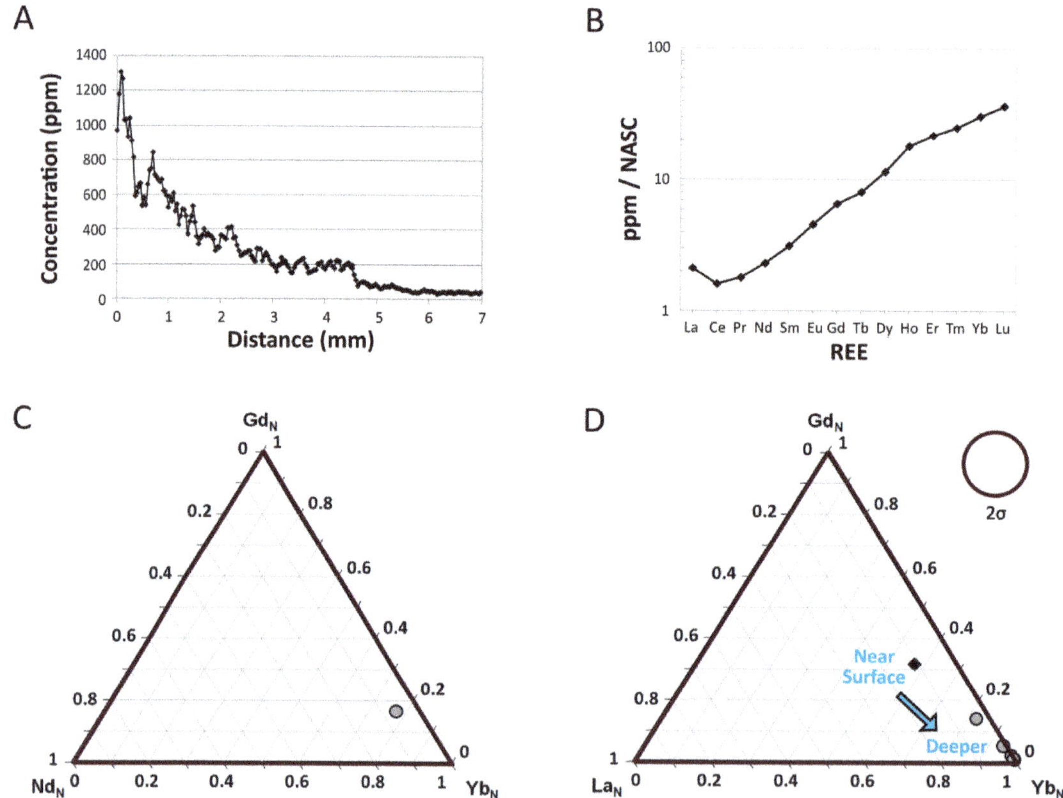

Figure 5. REE composition of the femur of MOR 1125. (**A**) Three-point moving average profile of La concentrations in the outermost 7 mm of the bone. (**B**) Average NASC-normalized REE composition of the fossil specimen as a whole. (**C,D**) Ternary diagrams of NASC-normalized REE. (**C**) Average composition of the bone. (**D**) REE compositions divided into data from each individual laser transect (~5 mm of data each). Compositional data from the transect that included the outer bone edge is denoted by a dark diamond; all other internal transect data are indicated by gray circles. The 2σ circle represents two standard deviations based on ±5% relative standard deviation.

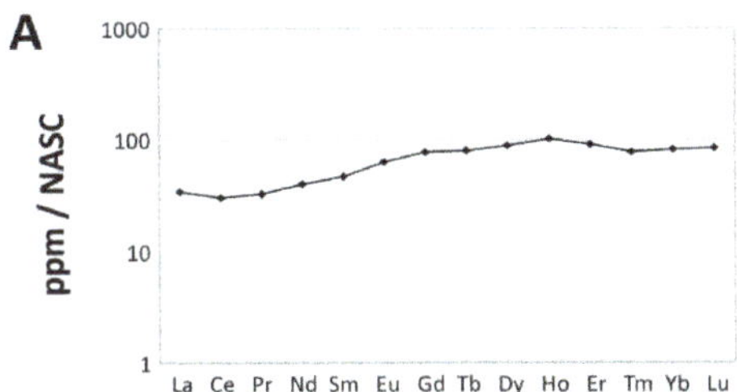
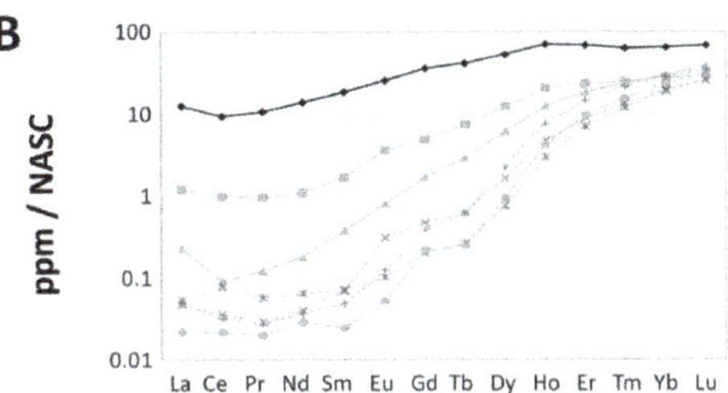

Figure 6. Spider diagrams of intra-bone NASC-normalized REE distribution patterns within the femur of MOR 1125. (**A**) Average composition of the outermost 250 μm of the cortex, demonstrating a similar magnitude of relative LREE depletion in the outermost cortex as seen in the bone as a whole (Figure 5B). (**B**) Variation in compositional patterns by laser transects. The pattern which includes the external margin of the bone is shown in black, those from deepest within the bone by dotted, lightgray lines, and all other analyses in between by solid, darkgray lines.

A ternary plot of La_N-Gd_N-Yb_N (Figure 5D) for each individual laser run compiled into the full transect identifies considerable spatial variation in bone composition (i.e., variation exceeds two standard deviations). This is confirmed by a spider diagram of individual laser runs (Figure 6B) which also shows substantial contrasts in the proportions of REE by laser run, with differences primarily corresponding to cortical depth. As can be seen in these figures, the femur of MOR 1125 generally becomes increasingly enriched in HREE relative to LREE and MREE with increasing depth into the internal cortex, with the bone overall shifting from modestly HREE enriched in the external-most cortex to drastically HREE enriched in the internal cortex and inner half of the middle cortex (roughly the inner half of the transect). Proportionally, this trend shifts from roughly one order of magnitude enrichment of HREE relative to LREE in the external cortex to approximately three orders of magnitude in the internal cortex.

Relative to NASC, middle and internal cortex transects generally exhibit roughly equal depletion in LREE and enrichment in HREE (Figure 6B). The internal-most laser run exhibits uniformly higher REE concentrations than the run immediately external to it. Although the laser run with the lowest concentrations, which crosses the outer portion of the internal cortex, exhibits slightly elevated peaks at neodymium (Nd), Gd, and holmium (Ho), there are no distinct signs of tetrad effects (i.e., 'M'- or 'W'-shaped shale-normalized patterns; [47] and references therein) in other laser runs (Figure 6B) or in the bone as a whole (Figure 5B; also see Supplementary Materials for further discussion on potential tetrad effects in MOR 1125).

4.4. $(La/Yb)_N$ vs. $(La/Sm)_N$ Ratio Patterns

At the whole-bone level, the fibula of MOR 1125 exhibits an $(La/Sm)_N$ value of 0.75 and a $(La/Yb)_N$ of 0.04, reflective of substantial HREE enrichment relative to most environmental water samples, dissolved loads, and sedimentary particulates (Figure 7A). In fact, this extremely low $(La/Yb)_N$ value places the bone within the compositional range of marine pore fluids and outside the ranges of all the other environmental samples examined (see Supplementary Materials for a breakdown of the literature sources used for environmental samples).

Plotting REE ratios for individual laser runs reveals a consistent pattern of decreasing $(La/Yb)_N$ and unchanging $(La/Sm)_N$ with increasing cortical depth (Figure 7B). The laser run including the external margin of the bone exhibits an $(La/Yb)_N$ value more than two orders of magnitude greater than laser runs across the internal cortex. All laser runs across the internal cortex exhibit $(La/Yb)_N$ ratios < 0.003, and those across the middle cortex still remain <0.01. All laser run $(La/Sm)_N$ ratios, regardless of cortical depth, range between 0.6–1.0.

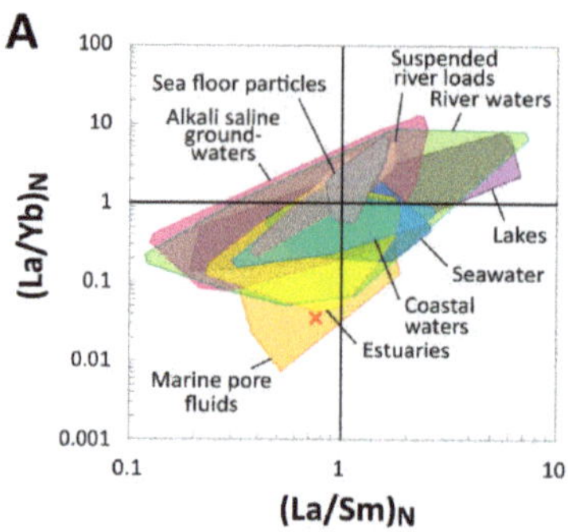

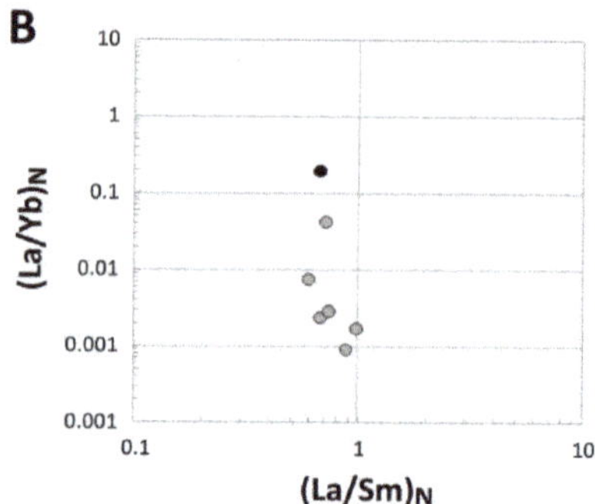

Figure 7. (La/Yb)$_N$ and (La/Sm)$_N$ ratios of the femur of MOR 1125. (**A**) Comparison of the whole-bone average (La/Yb)$_N$ and (La/Sm)$_N$ ratios of the fossil to ratios from various environmental waters and sedimentary particulates. Literature sources for environmental samples are provided in the Supplementary Materials. (**B**) REE compositions of individual laser transects expressed as NASC-normalized (La/Yb)$_N$ and (La/Sm)$_N$ ratios. The transect including the external bone margin is denoted by the black symbol, whereas all other (internal) transects are represented by gray symbols.

4.5. REE Anomalies

(Ce/Ce*)$_N$, (La/La*)$_N$, and La-corrected (Ce/Ce**)$_N$ anomalies, which are based on relative proportions of REE, are essentially absent at the outer cortex edge (Figure S2). However, all three of these anomalies fluctuate between positive and negative values across the length of the transect. Gaps in the data for (Ce/Ce*)$_N$ and (La/La*)$_N$ anomalies become abundant in the internal cortex due to: (1) concentrations of praseodymium (Pr) and Nd falling below lower detection limit in this region of the bone and; (2) occasionally, Nd concentrations are significantly greater than those of Pr (Data S1).

(Ce/Ce*)$_N$ anomalies fluctuate positively and negatively from ~0.4–4.0 across the transect, with values often being slightly negative through much of the external and middle cortices but positive in the internal cortex (Figure S2). There is a distinct ~0.2 magnitude rise in (Ce/Ce*)$_N$ values just interior to the open crack at a cortical depth of 4.7 mm, where the values appear to remain consistently alike those at the cortical margin for roughly the next 3 mm. When averaged for the entire bone (Table 1), MOR 1125 exhibits a weakly negative (Ce/Ce*)$_N$ value of 0.82, which is consistent with the negative inflection of the NASC-normalized pattern at Ce in the whole-bone spider diagram (Figure 5B).

To further aid in differentiating true, redox-related cerium anomalies from apparent anomalies produced by (La/La*)$_N$ anomalies, (Ce/Ce*)$_N$ values were also plotted against (Pr/Pr*)$_N$ values (following [66]). Anomaly values from the inner regions of the cortex occupy a substantially wider range of both (Ce/Ce*)$_N$ and (Pr/Pr*)$_N$ than those from the external-most 1 mm of the bone (Figure 8), signifying a relatively more heterogeneous composition in the middle and internal cortices. All but one external cortex data point plot near the upper margins of fields 2a and 4b, indicative of slightly positive La and Ce anomalies, whereas internal cortex measurements plot in every field (Figure 8). Relatively few data points plot in fields 2b and 4a, generally indicative of negative La and Ce anomalies; most of these pertain to the internal cortex (and all of those that do not pertain to the middle cortex).

Quantitative calculations of (La/La*)$_N$ anomalies and La-corrected (Ce/Ce**)$_N$ anomalies confirm these qualitative inferences. (Ce/Ce**)$_N$ anomalies are generally slightly positive throughout the external cortex but become slightly negative (<1) in the middle and internal cortex (Figure S2). Although there are numerous data gaps (due to the Pr and Nd concentration factors noted above), this pattern is exemplified by the innermost 7 mm of the internal cortex exhibiting a negative (Ce/Ce**)$_N$ average of 0.94. Additionally, fluctuations in (Ce/Ce**)$_N$ values across the middle cortex are considerable, encompassing variation of more than three orders of magnitude. At the whole-bone level, the bone exhibits a slightly positive $(Ce/Ce^{**})_N$ anomaly (1.26; Table 1). When plotted by laser run, $(Ce/Ce^{**})_N$ anomalies display a positive correlation with U concentrations ($r^2 = 0.75$; Figure S3). $(La/La^*)_N$ anomalies are commonly positive in the external cortex, and they decrease steadily to negative values in the inner half of the transect (Figure S2). At the whole-bone level, MOR 1125 exhibits a positive $(La/La^*)_N$ anomaly (average = 2.83; Table 1); however, this value is drastically biased toward readings from the external cortex due to abundant data gaps in the internal cortex (owing to the Pr and Nd concentration factors noted above).

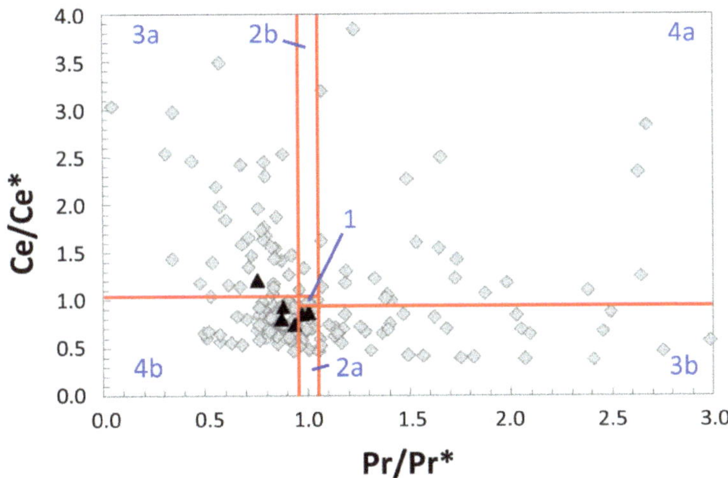

Figure 8. $(Ce/Ce^*)_N$ vs. $(Pr/Pr^*)_N$ plot (after [66]) of five-point averages along the transect across the cortex of MOR 1125 recorded by LA-ICPMS. Separate fields (labeled by blue text) are as follows: 1, neither Ce nor La anomaly; 2a, no Ce and positive La anomaly; 2b, no Ce and negative La anomaly; 3a, positive Ce and negative La anomaly; 3b, negative Ce and positive La anomaly; 4a, negative Ce and negative La anomaly; 4b, positive Ce and positive La anomaly. Measurements from the outer 1 mm of the external cortex are plotted as black triangles, and all measurements from deeper within the bone are plotted as grey diamonds. $(Ce/Ce^*)_N$ and $(Pr/Pr^*)_N$ anomalies, comparing observed (Ce, Pr) versus expected (Ce*, Pr*) concentrations of each element, are calculated as in the Materials and Methods section of the text.

Yttrium/holmium (Y/Ho) ratios are slightly above chondritic (26) [67] near the cortical margin and through most of the external cortex (range ~30–70). Deeper in the bone, Y/Ho anomalies become increasingly positive through the middle cortex, forming a broad peak near 24 mm of ~90–250, then gently decline through the innermost cortex to values of ~60–100 at the end of the transect (Figure S2). When averaged for the entire transect, the Y/Ho anomaly is positive (59; Table 1).

5. Discussion

5.1. Clarifying MOR 1125's Paleoenvironmental and Taphonomic Context

Based on the stratigraphic section recorded across the entire butte in the field (Figure S1), MOR 1125 was recovered from strata comprising the lower unit (L3 of [57] or Reid Coulee unit of [56]) of the Hell Creek Formation, specifically from the Basal Sand [57]. This stratigraphic context, which coincidentally makes MOR 1125 one of the oldest (stratigraphically-lowest) *T. rex* specimen known from the Hell Creek Formation [57], implies that the carcass was likely buried in a low-elevation environment relatively close to the coast (i.e., within 5 km) of the Western Interior Cretaceous Seaway (WIKS; cf. [61,68]). This conclusion is empirically supported by the presence of organic-rich inclined heterolithic strata in the entombing sandstone, as well as a large tree trunk and numerous fossil leaves (likely indicative of subaqeous burial within a low-elevation environment). However, the ratio of sediment to organics in this sandstone is too high for it to have been deposited in a persistent swamp (cf. [69,70]), and it also lacks root traces which would be expected in an abandoned channel or marshy environment [71,72]. Indeed, outside of a significant debris flow event, sandstones would not generally be expected to be deposited in quiescent swamps or marshes [71]. Rather, many aspects of the sedimentology of the entombing sandstone indicate that it was deposited under relatively high energy compared to all other strata observed in the quarry, including its considerable thickness (11.8 m), normally-graded structure, medium to coarse-grained matrix, trough and occasional swaley cross-bedding, and inclusions consisting of pebbles and rounded mud rip-up clasts. Moderate stratigraphic dispersal of the bones of MOR 1125 (by up to 55 cm) is also suggestive of burial occurring in a temporally-persistent, relatively high-energy setting.

Collectively, all of these findings strongly suggest that MOR 1125 was buried within an active channel in a lush, low-elevation environment near the coast. Absence of scales from the freshwater gar *Lepisosteus* in the entombing sandstone, combined with the presence of fragmentary turtle remains and an abundance of organic detritus in this stratum, implies deposition in a brackish estuarine setting, as inferred by Schweitzer et al. [51] (p. 1953); we therefore agree with their interpretation of the depositional setting having likely been an "estuarine" channel. This conclusion is also in agreement with those of Flight [73] and Fowler [54], who also interpreted the Basal Sand of the Hell Creek Formation as representing deposition within estuarine channels. The mean dip direction of cosets within these beds (see Taphonomic and Geologic Context above) indicate that the channel primarily flowed toward the southeast, and an average flow in this general direction is also supported by the predominant south-southeast orientation of long bones within the quarry (inferred to be parallel to the current; Figure 3B).

The cause of death of this tyrannosaur remains unknown. However, we view drought to be an unlikely cause due to the absence of red/oxidized paleosols and caliche (cf. [74,75]). Obrution and miring are also unlikely due to the lack of contorted strata, skeletal articulation, and preferential preservation of bones from the tail, hind limbs, and/or pelvis (cf. [76]).

In contrast, the postmortem history of MOR 1125 is much clearer. Taken together, the spatial distribution of the bones and their well-preserved character indicate that decomposition was allowed to take place long enough to result in complete disarticulation of the skeleton, yet also short enough for the bones to avoid weathering. This combination (as well as marked HREE enrichment, see below) implies that a significant phase of decomposition occurred in an oxic, subaqeous environment (cf. [75]), as the weak temperature swings of constant submersion can thwart significant bone weathering [77]. Based on the close association of the skeletal remains, negligible signs of abrasion, and lack of evidence for significant winnowing (i.e., representation of both Voorhies Groups and all major portions of the skeleton, including small and light-weight bones), burial appears to have taken place very close to the site of skeletonization; the remains can thus be considered modestlyparautochthonous. Given the wealth of lush terrestrial habitats across the floodplains recorded by strata of the Hell Creek Formation [61,78], it is likely the tyrannosaur died at a location slightly upstream from the quarry. Finally, occurrence of all bones in a single bed exhibiting slight normal grading of sediments (Figure 3A) signifies that burial was accomplished by a single depositional event characterized by waning flow competency [79].

We thus conclude the following scenario for the burial history of MOR 1125 based on the available geologic and taphonomic data: this *Tyrannosaurus* perished near a fluvial channel close to the coast of the WIKS. The river/stream carried the carcass downstream, during which time the carcass underwent decomposition primarily underwater. Opening of the fluvial channel into a broader estuary diminished flow competency, which caused the carcass to sink to the floor of a likely brackish estuarine channel where its major soft tissues (e.g., skin and muscle) continued to decay, allowing its skeleton to become disarticulated. A significant, brief rise in flow competency, perhaps fueled by a major rain event/flood, entrained abundant coarse sand, siltstone pebbles, and clay rip-up clasts which buried the remains within the normallygraded deposit now seen at the quarry.

5.2. Reconstructing the Geochemical History of MOR 1125

Having resolved the biostratinomic history of the carcass, we now discuss insight into its diagenetic history after burial. In particular, our trace element data provide informative clues about the geochemical history of MOR 1125 which allowed it to retain endogenous cells, soft tissues, and collagen I.

The femur of MOR 1125 exhibits surface concentrations of LREE (average ~1300 ppm; Data S1) comparable to those of other dinosaur bones previously analyzed from the Hell Creek Formation (~100–1400) [46], but overall modest ΣREE (596 ppm) compared to most other Cretaceous bones tested to date (Table 2), which have been reported to exhibit ΣREE ranging from 1100 ppm to over 25,000 ppm [36,80–83]. Average concentrations of Fe (0.73 wt. %), Sr (2386 ppm), and Ba (888 pm) are also low in this specimen compared to other dinosaur bones recently analyzed from the Hell Creek Formation (1.25–1.76 wt. %, ~2300–3700 ppm, and ~1500–2100 ppm, respectively) [46]. Conversely, MOR 1125 possesses (on average) considerably greater concentrations of Y (1102 ppm), Lu (17 ppm), and U (38 ppm) than other bones from the Hell Creek Formation (7–250 ppm, 0.1–3 ppm, and 2–18 ppm, respectively) [46]. Such comparisons are admittedly coarse as these studies span a wide range of taxa, intra-specimen cortical widths, and depositional environments, but they show that (concerning the vast majority of trace elements considered) the femur of MOR 1125 is only modestly chemically altered for its age. This fact alone may largely account for how this fossil has managed to retain original cells, tissues, and peptide sequences [22,51,52]. The relatively low concentrations of REE and other trace elements in the bone may stem, in part, from a combination of limited availability in the regional surface and groundwaters due to complexation with carbonates and humic acids [83–90] and/or partial earlydiagenetic removal from pore fluids by coprecipitation in secondary phosphates within the surrounding sediments [34,91–95] (see Supplementary Materials for further discussion of these potential sequestration processes).

Table 2. Summary of the REE composition of the right femur of *Tyrannosaurs rex* MOR 1125. Qualitative ΣREE content is based on the value shown in Table 1 (596 ppm) in comparison to values from other Mesozoic bones (as listed in the main text). Abbreviations: DMD—double medium diffusion *sensu* [44]; LREE—light rare earth elements.

Clear DMD Kink for LREE?	Relative Noise in Outer Cortex for La	REE Suggest Flow in Marrow Cavity?	Relative ΣREE Content (Whole Bone)	Relative U Content (Whole Bone)	Relative Porosity of the Cortex
No	Low	No	Low	Low	Low

Although REE concentrations in the femur of MOR 1125 steeply decrease with cortical depth, indicative of a single phase of simple diffusive uptake, considerable HREE enrichment is apparent within the middle and internal cortices (Figure 4A). This pattern of elemental signatures, which greatly contrasts trends of relative LREE and MREE enrichment throughout the cortex of other Hell Creek Formation dinosaur bones we have recently analyzed [46], is consistent with protracted uptake from: (1) relatively HREE-enriched lowland waters (such as those in brackish estuaries or tidallyinfluenced river channels) [36,37,96], and/or; (2) diagenetic pore fluids under oxidizing conditions. At the whole-bone level, the femur exhibits an overall positive $(Ce/Ce^{**})_N$ anomaly (1.26) indicative of overall oxidizing diagenetic conditions. However, a $(Ce/Ce^{**})_N$ anomaly is, on average, absent in the middle cortex (Figure S2), and the internal cortex was actually found to exhibit a slightly negative average of 0.94, indicating that REE uptake in the inner regions of the bone primarily occurred under slightly reducing conditions. We interpret this contrast to reflect the development of a locallyreducing microenvironment within the central medullary cavity and other internal regions of the bone during early diagenesis, during which time the bone was externally exposed to oxidizing conditions (cf. [97]). Three additional chemical attributes identify the external early-diagenetic environment as oxidizing: (1) positive $(Ce/Ce^{**})_N$ anomalies in the external cortex (Figures S2 and S3; (2) high concentrations of U throughout the cortex (Table 1 and Figure 4B) [98]; and (3) high Sc concentrations in the external cortex [99,100].

Though recent oxidation could also partially contribute to positive $(Ce/Ce^{**})_N$ values in the external cortex, we view it likely that any such contribution is minor for three reasons. First, any such influence would likely disproportionately raise anomaly values only near the outermost edge of the transect, not throughout the entire external cortex (cf. [101]). Second, the bone was collected from beneath 50 feet of rock rather than near the modern ground surface. Finally, the surrounding sediment is wellcemented, which would hinder efficient percolation of modern pore fluids to and around the fossil. For these reasons, we infer that spatial variations in the $(Ce/Ce^{**})_N$ anomaly across the width of the specimen's cortex reliably record contrasting early-diagenetic redox conditions within and outside of the bone. Part of this conclusion involves primary, early-diagenetic uptake of REE occurring in an oxidizing environment, which is consistent with the coarse grain sizes of the entombing sediment and the estuarine channel setting inferred from the stratigraphy of the quarry (see above). Thus, our geochemical findings independently support our paleoenvironmental interpretations.

The peculiar "offset" in concentration-depth profiles of many elements (e.g., Sc, Y, REE) at ~4.7 mm, characterized by a brief plateau followed by a dramatic drop in concentrations (e.g., Figure 4A,C), clearly appears to result from diagenetic pore fluid flow along a crack cutting across the laser transect at this cortical depth. The plateau of higher concentrations of each of these elements on the external side of the crack is most likely attributable to relatively protracted uptake from externallyderived pore fluids that entered via this conduit and flowed preferentially through the bone tissue exterior to it. Further, $(Ce/Ce^{*})_N$ and $(Ce/Ce^{**})_N$ anomalies each decrease from oxidizing values external to the crack at 4.7 mm to near-zero values interior to it (Figure S2), indicating that this crack created millimeter-scale spatial contrasts in redox conditions within the external cortex during early diagenesis.

Interestingly, the profiles of Fe, Mn, Sr, Ba, and U do not exhibit any disruption related to the crack at 4.7 mm (Figure 4B–D). For Fe, Mn, and Ba, their generally flat, high concentration-depth profiles imply they are primarily incorporated into common, homogeneouslydistributed, secondary mineral phases (e.g., goethite, Mn oxides, barite) [102]. The flat, high concentration-depth profile of Sr presumably reflects its spatiallyhomogenous substitution for Ca ions in bone apatite, as typically seen in bioapatitic fossils [41,103]. The unique profile shape exhibited by U (Figure 4B) indicates that the majority of U was adsorbed from a pore fluid which diffused through the bone either prior to formation of the crack or during late diagenesis (i.e., well after the bone became a stable fossil). The latter of these two hypotheses would imply that substantial late-diagenetic overprinting occurred to the trace element composition of the bone, which seems unlikely given the low average concentrations of most trace elements with diffusivities in bone similar to that of U ($\sim 1 \times 10^{-14}$ cm^2/s [101]; see Table 1 and Data S1), very weak signs of secondary diffusion from within the medullary cavity (Figure 4), and lack of evidence for leaching or secondary REE incorporation phases (Figures 4 and 5A). Atypical placement of MOR 1125 far from the fields occupied by natural fresh waters and those of estuarine and coastal waters in the $(La/Yb)_N$ vs. $(La/Sm)_N$ plot (Figure 7A) is also inconsistent with incorporation of a significant proportion of the trace element inventory of the bone from a late-diagenetic pore fluid. Thus, most U was likely incorporated into the bone during early diagenesis.

Many chemical attributes of the femur of MOR 1125 suggest that the chemistry of pore fluids percolating through it changed over time through the primary, early-diagenetic phase of diffusive uptake. For example, typical intra-bone fractionation trends (*sensu* [104]) in the $(La/Yb)_N$ vs. $(La/Sm)_N$ plot for this specimen (Figure 7B, and discussed further in the Supplementary Materials relative to prior literature [40,105,106]) clearly demonstrate that pore fluids became relatively depleted in LREE by the time they reached the middle and internal cortices. Similarly, one would expect elements with similar diffusivities, such as U and REE [101], to develop similarlyshaped concentration depth profiles if pore fluid chemistry was held constant by sustained replenishment, but these elements exhibit blatantlydifferent profile shapes in MOR 1125 (compare Figure 4A,B): whereas REE profiles steeply decline from the cortical margin to low concentrations throughout the interior of the bone, U concentrations gradually increase inward from a subtle minimum near 2–3 mm, such that the highest average concentrations found in the bone occur in the outer portion of the internal cortex (ca. 24 mm). This suggests that the availability of REE for uptake by the bone's interior dwindled over time while that of U remained relatively high. Such a situation could arise due to the greater mobility of U complexes than those of REE in oxic fluids [83], and the relatively lower partition coefficients (in apatite) of U than REE could further help maintain high U availability in pore fluids diffusing through the internal regions of the bone [101]. The positive correlation between U concentrations and average $(Ce/Ce^{**})_N$ anomaly for each laser run ($r^2 = 0.75$; Figure S3) further supports this interpretation, as it suggests that Ce and U were incorporated into the bone over similar timescales [98]. Finally, positive Y/Ho anomalies throughout the middle and internal cortices and positive $(La/La^*)_N$ anomalies in the external cortex are also likely products of fractionation during uptake from pore fluids which were changing in composition over time (Table 1, and discussed in the Supplementary Materials). If the majority of uptake with fractionation occurred during early diagenesis, as all evidence appears to support, then the composition of MOR 1125 likely reflects uptake from circum-neutral pH groundwaters (cf. [107]).

To summarize, our cumulative trace element data indicate that the femur of MOR 1125 experienced moderate trace element uptake from a circum-neutral pH, HREE-enriched pore fluid during early diagenesis. Decay produced locallyreducing microenvironments within the medullary cavity and in regions within the cortex while it was externally exposed to oxidizing conditions. Combining these findings with the traditional taphonomic observations discussed above (i.e., minimal weathering and abrasion, disarticulation but close association of the remains) reveals that MOR 1125 underwent skeletonization during fairly prolonged, subaqeous decay in a sandy estuary channel and, following burial, its bones remained exposed to an oxic, potentiallybrackish pore fluid through early diagenesis and fossilization, after which they experienced relatively minimal further chemical alteration (Figure 9).

5.3. Emerging Taphonomic and Diagenetic Themes

To date, this report constitutes only the second study to have geochemically characterized the depositional circumstances of a pre-Cenozoic locality where bones yield original biomolecules, with the other being recent work by some of us (P.V.U. and R.D.A.) on the Standing Rock Hadrosaur Site (SRHS) [21,46]. Despite the somewhat contrasting taphonomic and diagenetic histories of *T. rex* MOR 1125 and *Edmontosaurus* bones from SRHS [46,51], both have been shown to retain endogenous cells, soft tissues [50,51], and collagen I protein [21,22,52]. More importantly, these specimens still share certain geochemical similarities indicative of diagenetic circumstances in common between these two sites that have evidently permitted long-term biomolecular preservation in fossil bones. We therefore now highlight these common themes in a preliminary attempt to constrain diagenetic pathways to molecular preservation.

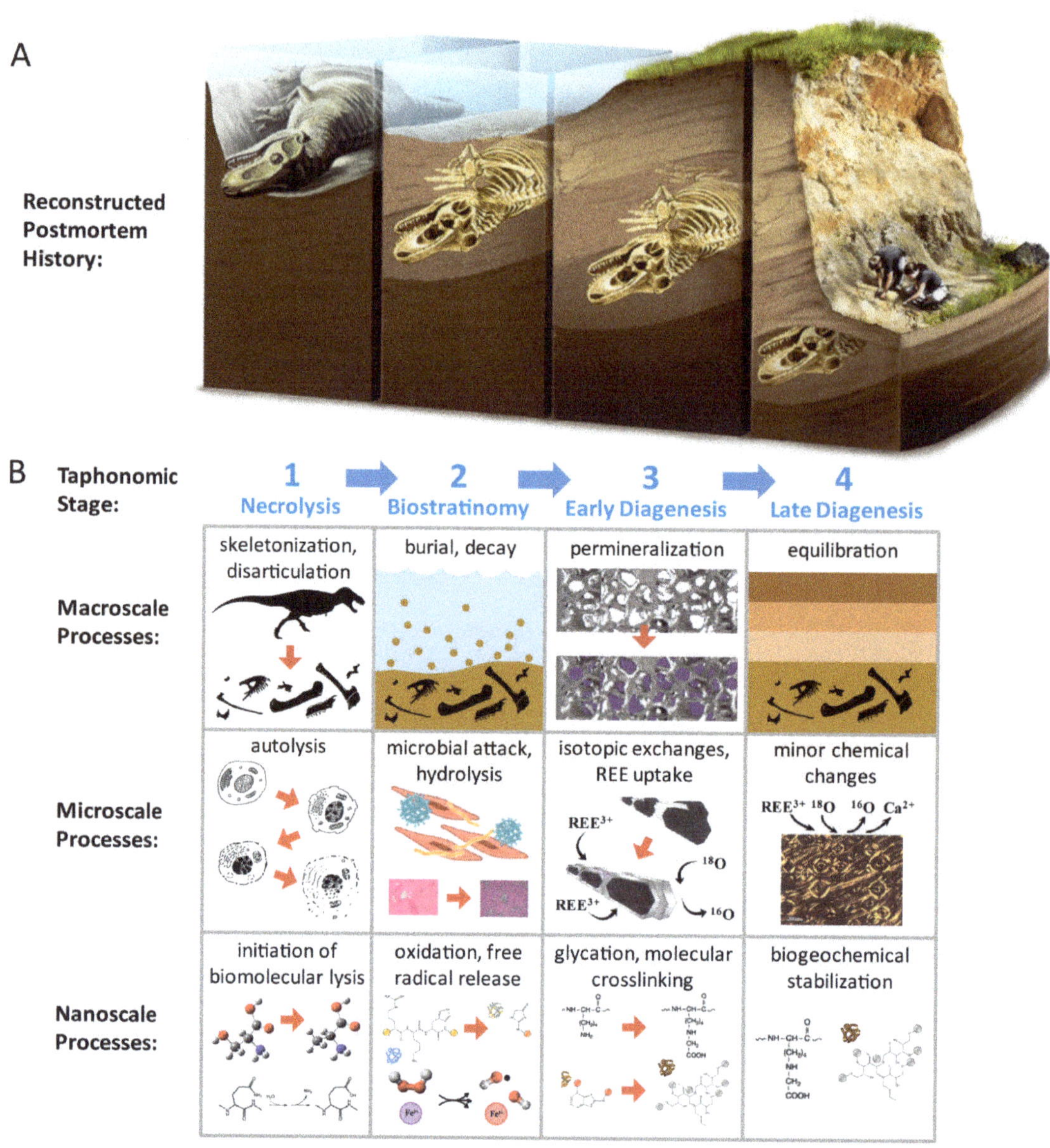

Figure 9. Reconstruction of the taphonomic and diagenetic history of *Tyrannosaurus rex* MOR 1125. (**A**) Generalized postmortem history of the carcass, *sans* disarticulation. Reprinted in modified form with permission from ref. [108]. Copyright 2016 Porto Editora. (**B**) Synopsis of macroscale, microscale, and nanoscale (molecular level) processes inferred to have taken place during each taphonomic stage in the decay of this specimen, as portrayed in (**A**). Nanoscale processes of biomolecular decay and stabilization based on propositions by [29,30,109,110]. Black *Tyrannosaurus* silhouette by Scott Hartman, www.phylopic.org (accessed on 8 August 2021), CC BY-NC-SA-3.0. Autolysis, REE uptake, biomolecular lysis, and glycation images each respectively modified, with permission, from [38,111–114]. Hydrolysis image modified from [115], and oxidation and crosslinking images modified from [109], each under CC BY-4.0 licenses.

Although our cumulative trace element data show that the coarse-grained nature of the entombing sand allowed MOR 1125 to experience overall greater chemical alteration than bones at SRHS [46], fossil specimens from both sites appear minimally altered at the elemental level. For example, they exhibit steeplydeclining REE profiles (Figure 4A, and Figures 1 and 2 of [46]), MREE concentrations which frequently drop below detection limit through the middle cortex (Data S1, and appendix A of [46]), and low $\sum$REE compared to other Cretaceous bones reported in the literature (see above). This similarity bears the obvious implication that minimal chemical alteration permits molecular preservation, as predicted by Trueman et al. [39]. Taphonomically, both assemblages were buried in lowland coastal settings, and the carcasses at each site underwent decay primarily in subaqueous environments ([49] and this study). These conditions could have permitted substantial trace element uptake, but instead it appears that burial soon after disarticulation followed by limited exposure to early-diagenetic pore fluids limited the magnitude of trace element uptake at each site. At SRHS, this was accomplished via rapid burial in a low-permeability mudstone which hindered pore fluid flow in partnership with partial encasement of select bones in early-diagenetic siderite concretions [46,50], and early cementation of the coarse sandstone at the MOR 1125 locality appears to have hampered diagenetic pore fluid flow in a similar fashion.

At the whole-bone level, specimens from both sites also exhibit significant enrichment in Sr compared to most other trace elements (as is common in bioapatitic fossils in general) [41,103], positive Y/Ho and (La/La*)$_N$ anomalies (which are both common in fossil bones due to the slightly faster diffusivities of Y and LREE in bone) [101], high Sc enrichment (likely related to precipitation of secondary minerals under oxidizing conditions) [99,100], and slightly negative (Ce/Ce*)$_N$ anomalies reflective of oxidizing diagenetic conditions (Table 1, and Table 1 of [46]). The last two of these findings agree with those of Wiemann et al. [109], who found that fossils from oxidizing depositional environments are more likely to yield endogenous soft tissues, despite the possibility of oxidative damage to biomolecules [35]. Indeed, in contrast to traditional views like those expressed by Eglinton and Logan [35], recent experiments by Boatman et al. [30] indicate that oxidizing conditions may actually promote biomolecular stabilization through free radical-induced inter- and intramolecular crosslinking.

Both MOR 1125 and bones from SRHS also exhibit clear signs of fractionation of REE during uptake from an early-diagenetic pore fluid, namely steeper concentration-depth profiles for LREE than HREE (Figure 4A, and Figure 2 of [46]) and shifting proportions of LREE and HREE by cortical depth (Figure 6B, and Figure 4 of [46]). Such fractionation patterns are commonly seen in fossil non-avian dinosaur bones possessing a thick cortex [40,43,65,116] due to a 'filtering' effect created by the thick rim of dense bone tissue, which can cause pore fluids to become relatively depleted in LREE by the time they reach the internal cortex [101]. Observation of such patterns in fossil bones is important, because they signify retention of early-diagenetic trace element signatures, which in turn demonstrates that late-diagenetic overprinting (which is nearly universal to some degree) [42,104] has neither completely obfuscated signatures imparted from the initial burial environment nor drastically altered the chemistry of the fossil specimen [36,48,80,117–120]. In other words, the marked fractionation trends seen in the femur of MOR 1125 and bones at SRHS demonstrate that the bulk of the REE inventory in each respective specimen derives from early-diagenetic uptake from a single pore fluid (rather than late-diagenetic overprinting), and that these bones avoided any significant interactions with other pore fluids during late diagenesis. This does not imply that the majority of the term of burial was 'dry', but rather that any exposure(s) to pore fluids after initial fossilization (such as phreatic groundwaters or recent vadose fluids) have not meaningfully influenced the chemistry of these fossils. Avoidance of protracted interactions with pore fluids through later phases of diagenesis thus appears to promote long-term biomolecular stability.

Interestingly, REE ternary diagrams, (Ce/Ce*)$_N$ vs. (Pr/Pr*)$_N$ plots, and anomaly profiles reveal considerable spatial heterogeneity in the degree of chemical alteration of fossil bones at both sites, especially within the internal cortex (Figure 5, Figure 8, Figure S2 and Figures 3d, 6, and A.4 of [46]). Although it is not surprising that external cortices show more homogenous patterns of alteration (because they generally equilibrate with external pore fluids more quickly and to a greater degree) [83,101], recovery of endogenous soft tissues from such randomlyaltered bone tissue could be considered surprising. However, the magnitude of alteration is what likely matters more; as discussed above, the middle and internal cortices of each specimen exhibit lower trace element concentrations than the corresponding external cortex, signifying that even though patterns of alteration are comparatively more erratic internally, these internal regions still constitute the least-altered portion of dense bone tissue in each fossil. Therefore, based on these specimens, we suggest that unless signs of significant secondary diffusion from within the medullary cavity are encountered (e.g., Figure 11 of [120]), future paleomolecular sampling efforts should (where possible) target the middle and internal cortex of bones rather than the external cortex.

Finally, MOR 1125 and SRHS bones also each exhibit relatively high, flat profiles for Fe, Mn, and Ba (Figure 4C,D, and appendix A of [46]). We interpret these profiles to reflect homogenous distributions of minute, secondary mineral phases in the fossilized bone tissues (cf. [121]). Sustained peaks in any of these elements along a laser ablation transect could signify an unfavorable presence of more or larger crystals of secondary minerals (e.g., goethite, Mn oxides, and barite), which could be used alongside REE analyses as a potential screening tool to further direct paleomolecular sampling within the cortex of a fossil bone.

6. Conclusions

By considering the taphonomic and geochemical signatures of the femur of MOR 1125 and bones from SRHS in context with other recent actualistic and analytical studies cited in the Discussion above, we deduce the following about diagenetic pathways to molecular preservation:

1. Our results strengthen the hypothesis that oxidizing depositional environments or, more specifically, oxidizing microenvironmental conditions during early diagenesis, can foster chemical reactions which stabilize bone cells, soft tissues, and their component biomolecules;
2. In general, middle and internal cortical tissues usually experience less chemical alteration onaverage than the external cortex of any given fossil bone, implying that these interior regions should be the primary targets of future paleomolecular sampling;
3. Diagenetic circumstances which restrict exposure to percolating pore fluids (i.e., burial in compact, fine-grained sediments, early/rapid cementation of sediments, encasement in early-diagenetic concretion) elevate molecular preservation potential by minimizing gross chemical alteration;
4. Avoidance of protracted interactions with late-diagenetic pore fluids is almost certainly crucial to sustain long-term stability of biomolecules in a fossil bone.

These findings are intriguing, but it must be remembered that our conclusions remain based on data from only two Cretaceous localities in the same geologic formation, physiographic region, and current climatic regime (the Hell Creek Formation in the U.S. Western Interior). To fully elucidate molecular preservation mechanisms and the taphonomic circumstances which facilitate them, traditional taphonomic and trace element analyses must be conducted alongside paleomolecular assays (e.g., immunoassays or proteomics) at additional localities of varied ages which yield bones preserved under widelydiverse paleoenvironmental and diagenetic contexts.

Supplementary Materials: The following are available online at https://www.mdpi.com/article/10.3390/biology10111193/s1; Supplementary Materials narrative: a DOCX Word file including further details on the sedimentology of the entombing sandstone and overlying Hell Creek Formation strata within the MOR 1125 quarry, additional details on our LA-ICPMS methods, discussion of potential tetrad effects in the right femur of MOR 1125, further discussion of intra-bone fractionation trends in $(La/Yb)_N$ vs, $(La/Sm)_N$ plots, discussion of potential sequestration process that may have limited REE availability to the bones of MOR 1125, sources for environmental data in Figure 7, and Figures S1–S3. Figure S1: Complete stratigraphic section across the entire butte from which MOR 1125 was collected. Figure S2: Intra-bone patterns of $(Ce/Ce^*)_N$, $(Ce/Ce^{**})_N$, and $(La/La^*)_N$ anomalies and Y/Ho ratios within the femur of MOR 1125. Figure S3: Cerium anomaly $(Ce/Ce^{**})_N$ values plotted against uranium (U) concentrations in the femur of MOR 1125. Data S1: Raw transect data acquired and analyzed in this study, provided as an XLSX Excel spreadsheet.

Author Contributions: Conceptualization, P.V.U.; methodology, P.V.U.; formal analysis, P.V.U., K.M., R.D.A. and B.S.; investigation, P.V.U., K.M. and B.S.; resources, J.B.S.; data curation, P.V.U. and J.B.S.; writing—original draft preparation, P.V.U.; writing—review and editing, P.V.U., K.M., B.S. and J.B.S.; visualization, P.V.U.; project administration, P.V.U.; funding acquisition, P.V.U. All authors have read and agreed to the published version of the manuscript.

Funding: This research was funded by a Rowan University Seed Fund to P.V.U.

Institutional Review Board Statement: Not applicable.

Informed Consent Statement: Not applicable.

Data Availability Statement: All data generated by this study are available in this manuscript and the accompanying Supplementary Materials.

Acknowledgments: This study was only made possible by the diligent excavation efforts of Museum of the Rockies field crews for three field seasons from 2001 through 2003, whom we thank for tackling the challenge of recovering a *Tyrannosaurus rex* from beneath 50 ft of rock; one of us who took part in this excavation (P.V.U.) can vouch that rappelling with a jackhammer is no easy feat! Gratitude is also extended to the United States Fish and Wildlife Service (Charles M. Russell Wildlife Refuge). We also thank David Grandstaff, Elena Schroeter, Bob Harmon, Jack Horner, Mary Schweitzer, Laura Wilson, and Nels Peterson for helpful discussions and three reviewers for their helpful comments which greatly improved the manuscript.

Conflicts of Interest: The authors declare no conflict of interest.

References

1. Curry, G.B. Molecular palaeontology: New life for old molecules. *Trends Ecol. Evol.* **1987**, *2*, 161–165. [CrossRef]
2. Schweitzer, M.H. Soft tissue preservation in terrestrial Mesozoic vertebrates. *Annu. Rev. Earth Planet. Sci.* **2011**, *39*, 187–216. [CrossRef]
3. Hofreiter, M.; Collins, M.; Stewart, J.R. Ancient biomolecules in Quaternary palaeoecology. *Quat. Sci. Rev.* **2012**, *33*, 1–13. [CrossRef]
4. Schweitzer, M.H.; Zheng, W.; Cleland, T.P.; Goodwin, M.B.; Boatman, E.; Theil, E.; Marcus, M.A.; Fakra, S.C. A role for iron and oxygen chemistry in preserving soft tissues, cells and molecules from deep time. *Proc. R. Soc. B* **2014**, *281*, 20132741. [CrossRef]
5. Schweitzer, M.H.; Schroeter, E.R.; Cleland, T.P.; Zheng, W. Paleoproteomics of Mesozoic dinosaurs and other Mesozoic fossils. *Proteomics* **2019**, *19*, 1800251. [CrossRef] [PubMed]
6. Thomas, B.; Taylor, S. Proteomes of the past: The pursuit of proteins in paleontology. *Expert Rev. Proteom.* **2019**, *16*, 881–895. [CrossRef] [PubMed]
7. Bailleul, A.M.; Li, Z. DNA staining in fossil cells beyond the Quaternary: Reassessment of the evidence and prospects for an improved understanding of DNA preservation in deep time. *Earth-Sci. Rev.* **2021**, *216*, 103600. [CrossRef]
8. Orlando, L.; Bonjean, D.; Bocherens, H.; Thenot, A.; Argant, A.; Otte, M.; Hanni, C. Ancient DNA and the population genetics of cave bears (*Ursus spelaeus*) through space and time. *Mol. Biol. Evol.* **2002**, *19*, 1920–1933. [CrossRef]
9. Orlando, L.; Ginolhac, A.; Zhang, G.; Froese, D.; Albrechtsen, A.; Stiller, M.; Schubert, M.; Cappellini, E.; Petersen, B.; Moltke, I.; et al. Recalibrating *Equus* evolution using the genome sequence of an early Middle Pleistocene horse. *Nature* **2013**, *499*, 74–78. [CrossRef]
10. Noonan, J.P.; Coop, G.; Kudaravalli, S.; Smith, D.; Krause, J.; Alessi, J.; Chen, F.; Platt, D.; Pääbo, S.; Pritchard, J.K.; et al. Sequencing and analysis of Neanderthal genomic DNA. *Science* **2006**, *314*, 1113–1118. [CrossRef]
11. Oskam, C.L.; Haile, J.; McLay, E.; Rigby, P.; Allentoft, M.E.; Olsen, M.E.; Bengtsson, C.; Miller, G.H.; Schwenninger, J.-L.; Jacomb, C.; et al. Fossil avian eggshell preserves ancient DNA. *Proc. R. Soc. B* **2010**, *277*, 1991–2000. [CrossRef] [PubMed]
12. Barnett, R.; Westbury, M.V.; Sandoval-Velasco, M.; Vieira, F.G.; Jeon, S.; Zazula, G.; Martin, M.D.; Ho, S.Y.W.; Mather, N.; Gopalakrishnan, S.; et al. Genomic adaptations and evolutionary history of the extinct scimitar-toothed cat, *Homotherium latidens*. *Curr. Biol.* **2020**, *30*, 5018–5025. [CrossRef] [PubMed]

13. van der Valk, T.; Pečnerová, P.; Díez-del-Molino, D.; Bergström, A.; Oppenheimer, J.; Hartmann, S.; Xenikoudakis, G.; Thomas, J.A.; Dehasque, M.; Sağlican, E.; et al. Million-year-old DNA sheds light on the genomic history of mammoths. *Nature* **2021**, *591*, 265–269. [CrossRef] [PubMed]

14. Pan, Y.; Zheng, W.; Moyer, A.E.; O'Connor, J.K.; Wang, M.; Zheng, X.; Wang, X.; Schroeter, E.R.; Zhou, Z.; Schweitzer, M.H. Molecular evidence of keratin and melanosomes in feathers of the Early Cretaceous bird *Eoconfuciusornis*. *Proc. Natl. Acad. Sci. USA* **2016**, *113*, 7900–7907. [CrossRef]

15. Pan, Y.; Zheng, W.; Sawyer, R.H.; Pennington, M.W.; Zheng, X.; Wang, X.; Wang, M.; Hu, L.; O'Connor, J.; Zhao, T.; et al. The molecular evolution of feathers with direct evidence from fossils. *Proc. Natl. Acad. Sci. USA* **2019**, *116*, 3018–3023. [CrossRef]

16. Lindgren, J.; Kuriyama, T.; Madsen, H.; Sjövall, P.; Zheng, W.; Uvdal, P.; Engdahl, A.; Moyer, A.E.; Gren, J.A.; Kamezaki, N.; et al. Biochemistry and adaptive colouration of an exceptionally preserved juvenile fossil sea turtle. *Sci. Rep.* **2017**, *7*, 13324. [CrossRef]

17. Lindgren, J.; Sjövall, P.; Thiel, V.; Zheng, W.; Ito, S.; Wakamatsu, K.; Hauff, R.; Kear, B.P.; Engdahl, A.; Alwmark, C.; et al. Soft-tissue evidence for homeothermy and crypsis in a Jurassic ichthyosaur. *Nature* **2018**, *564*, 359–365. [CrossRef]

18. Schroeter, E.R.; DeHart, C.J.; Cleland, T.P.; Zheng, W.; Thomas, P.M.; Kelleher, N.L.; Bern, M.; Schweitzer, M.H. Expansion for the *Brachylophosaurus canadensis* collagen I sequence and additional evidence of the preservation of Cretaceous protein. *J. Proteome Res.* **2017**, *16*, 920–932. [CrossRef]

19. Schweitzer, M.H.; Zheng, W.; Moyer, A.E.; Sjövall, P.; Lindgren, J. Preservation potential of keratin in deep time. *PLoS ONE* **2018**, *13*, e0206569.

20. Bailleul, A.M.; Zheng, W.; Horner, J.R.; Hall, B.K.; Holliday, C.M.; Schweitzer, M.H. Evidence of proteins, chromosomes and chemical markers of DNA in exceptionally preserved dinosaur cartilage. *Natl. Sci. Rev.* **2020**, *7*, 815–822. [CrossRef]

21. Ullmann, P.V.; Voegele, K.K.; Grandstaff, D.E.; Ash, R.D.; Zheng, W.; Schroeter, E.R.; Schweitzer, M.H.; Lacovara, K.J. Molecular tests support the viability of rare earth elements as proxies for fossil biomolecule preservation. *Sci. Rep.* **2020**, *10*, 15566. [CrossRef] [PubMed]

22. Asara, J.M.; Schweitzer, M.H.; Freimark, L.M.; Phillips, M.; Cantley, L.C. Protein sequences from mastodon and *Tyrannosaurus rex* revealed by mass spectrometry. *Science* **2007**, *316*, 280–285. [CrossRef] [PubMed]

23. Schweitzer, M.H.; Zheng, W.; Organ, C.L.; Avci, R.; Suo, Z.; Freimark, L.M.; Lebleu, V.S.; Duncan, M.B.; Vander Heiden, M.G.; Neveu, J.M.; et al. Biomolecular characterization and protein sequences of the Campanian hadrosaur *B. canadensis*. *Science* **2009**, *324*, 626–631. [CrossRef] [PubMed]

24. Organ, C.L.; Schweitzer, M.H.; Zheng, W.; Freimark, L.M.; Cantley, L.C.; Asara, J.M. Molecular phylogenetics of *Mastodon* and *Tyrannosaurus rex*. *Science* **2008**, *320*, 499. [CrossRef] [PubMed]

25. Davis, P.G.; Briggs, D.E.G. Fossilization of feathers. *Geology* **1995**, *23*, 783–786. [CrossRef]

26. Briggs, D.E.G.; Kear, A.J. Fossilization of soft tissue in the laboratory. *Science* **1993**, *259*, 1439–1442. [CrossRef]

27. Cleland, T.P.; Schroeter, E.R.; Zamdborg, L.; Zheng, W.; Lee, J.E.; Tran, J.C.; Bern, M.; Duncan, M.B.; Lebleu, V.S.; Ahlf, D.R.; et al. Mass spectrometry and antibody-based characterization of blood vessels from *Brachylophosaurus canadensis*. *J. Proteome Res.* **2015**, *14*, 5252–5262. [CrossRef]

28. Schweitzer, M.H.; Wittmeyer, J.L.; Horner, J.R. Soft tissue and cellular preservation in vertebrate skeletal elements from the Cretaceous to the present. *Proc. R. Soc. B* **2007**, *274*, 183–197. [CrossRef] [PubMed]

29. Schweitzer, M.H.; Schroeter, E.R.; Goshe, M.B. Protein molecular data from ancient (>1 million years old) fossil material: Pitfalls, possibilities, and grand challenges. *Anal. Chem.* **2014**, *86*, 6731–6740. [CrossRef]

30. Boatman, E.M.; Goodwin, M.B.; Holman, H.-Y.N.; Fakra, S.; Zheng, W.; Gronsky, R.; Schweitzer, M.H. Mechanisms of soft tissue and protein preservation in *Tyrannosaurus rex*. *Sci. Rep.* **2019**, *9*, 15678. [CrossRef]

31. Schweitzer, M.H.; Zheng, W.; Cleland, T.P.; Bern, M. Molecular analyses of dinosaur osteocytes support the presence of endogenous molecules. *Bone* **2013**, *52*, 414–423. [CrossRef]

32. Sykes, G.A.; Collins, M.J.; Walton, D.I. The significance of a geochemically isolated intracrystalline organic fraction within biominerals. *Org. Geochem.* **1995**, *23*, 1059–1065. [CrossRef]

33. Edwards, N.P.; Barden, H.E.; van Dongen, B.E.; Manning, P.L.; Larson, P.L.; Bergmann, U.; Sellers, W.I.; Wogelius, R.A. Infrared mapping resolves soft tissue preservation in 50 million year-old reptile skin. *Proc. R. Soc. B* **2011**, *278*, 3209–3218. [CrossRef]

34. Allison, P.A. Konservat-Lagerstätten: Cause and classification. *Paleobiology* **1988**, *14*, 331–344. [CrossRef]

35. Eglinton, G.; Logan, G.A. Molecular preservation. *Philos. Trans. R. Soc. Lond. B* **1991**, *333*, 315–328.

36. Trueman, C.N. Rare earth element geochemistry and taphonomy of terrestrial vertebrate assemblages. *PALAIOS* **1999**, *14*, 555–568. [CrossRef]

37. Trueman, C.N. Trace element geochemistry of bonebeds. In *Bonebeds: Genesis, Analysis, and Paleobiological Significance*; Rogers, R.R., Eberth, D.A., Fiorillo, A.R., Eds.; University of Chicago Press: Chicago, IL, USA, 2007; pp. 397–435.

38. Trueman, C.N.; Tuross, N. Trace elements in Recent and fossil bone apatite. *Rev. Miner. Geochem.* **2002**, *48*, 489–521. [CrossRef]

39. Trueman, C.N.; Palmer, M.R.; Field, J.; Privat, K.; Ludgate, N.; Chavagnac, V.; Eberth, D.A.; Cifelli, R.; Rogers, R.R. Comparing rates of recrystallisation and the potential for preservation of biomolecules from the distribution of trace elements in fossil bones. *C. R. Palevol* **2008**, *7*, 145–158. [CrossRef]

40. Trueman, C.N.; Kocsis, L.; Palmer, M.R.; Dewdney, C. Fractionation of rare earth elements within bone mineral: A natural cation exchange system. *Palaeogeogr. Palaeocl.* **2011**, *310*, 124–132. [CrossRef]

41. Hinz, E.A.; Kohn, M.J. The effect of tissue structure and soil chemistry on trace element uptake in fossils. *Geochim. Cosmochim. Acta* **2010**, *74*, 3213–3231. [CrossRef]

42. Kocsis, L.; Trueman, C.N.; Palmer, M.R. Protracted diagenetic alteration of REE contents in fossil bioapatites: Direct evidence from Lu-Hf isotope systematics. *Geochim. Cosmochim. Acta* **2010**, *74*, 6077–6092. [CrossRef]

43. Herwartz, D.; Tütken, T.; Jochum, K.P.; Sander, P.M. Rare earth element systematics of fossil bone revealed by LA-ICPMS analysis. *Geochim. Cosmochim. Acta* **2013**, *103*, 161–183. [CrossRef]

44. Kohn, M.J. Models of diffusion-limited uptake of trace elements in fossils and rates of fossilization. *Geochim. Cosmochim. Acta* **2008**, *72*, 3758–3770. [CrossRef]

45. Kocsis, L.; Gheerbrant, E.; Mouflih, M.; Cappetta, H.; Ulianov, A.; Chiaradia, M.; Bardet, N. Gradual changes in upwelled seawater conditions (redox, pH) from the late Cretaceous through early Paleogene at the northwest coast of Africa: Negative Ce anomaly trend recorded in fossil bio-apatite. *Chem. Geol.* **2016**, *421*, 44–54. [CrossRef]

46. Ullmann, P.V.; Grandstaff, D.E.; Ash, R.D.; Lacovara, K.J. Geochemical taphonomy of the Standing Rock Hadrosaur Site: Exploring links between rare earth elements and cellular and soft tissue preservation. *Geochim. Cosmochim. Acta* **2020**, *269*, 223–237. [CrossRef]

47. Grandstaff, D.E.; Terry, D.O. Rare earth element composition of Paleogene vertebrate fossils from Toadstool Geologic Park, Nebraska, USA. *Appl. Geochem.* **2009**, *24*, 733–745. [CrossRef]

48. Bosio, G.; Gioncada, A.; Gariboldi, K.; Bonaccorsi, E.; Collareta, A.; Pasero, M.; Celma, C.D.; Malinverno, E.; Urbina, M.; Bianucci, G. Mineralogical and geochemical characterization of fossil bones from a Miocene marine Konservat-Lagerstätte. *J. S. Am. Earth Sci.* **2021**, *105*, 102924. [CrossRef]

49. Ullmann, P.V.; Shaw, A.; Nellermoe, R.; Lacovara, K.J. Taphonomy of the Standing Rock Hadrosaur Site, Corson County, South Dakota. *PALAIOS* **2017**, *32*, 779–796. [CrossRef]

50. Ullmann, P.V.; Pandya, S.H.; Nellermoe, R. Patterns of soft tissue and cellular preservation in relation to fossil bone microstructure and overburden depth at the Standing Rock Hadrosaur Site, Maastrichtian Hell Creek Formation, South Dakota, USA. *Cretaceous Res.* **2019**, *99*, 1–13. [CrossRef]

51. Schweitzer, M.H.; Wittmeyer, J.L.; Horner, J.R.; Toporski, J.K. Soft-tissue vessels and cellular preservation in *Tyrannosaurus rex*. *Science* **2005**, *307*, 1952–1955. [CrossRef]

52. Schweitzer, M.H.; Suo, Z.; Avci, R.; Asara, J.M.; Allen, M.A.; Arce, F.T.; Horner, J.R. Analyses of soft tissue from *Tyrannosaurus rex* suggest the presence of protein. *Science* **2007**, *316*, 277–280. [CrossRef] [PubMed]

53. Gates, T.A. The Late Jurassic Cleveland-Lloyd Dinosaur Quarry as a drought-induced assemblage. *PALAIOS* **2005**, *20*, 363–375. [CrossRef]

54. Fowler, D. The Hell Creek Formation, Montana: A stratigraphic review and revision based on a sequence stratigraphic approach. *Geosciences* **2020**, *10*, 435. [CrossRef]

55. San Antonio, J.D.; Schweitzer, M.H.; Jensen, S.T.; Kalluri, R.; Buckley, M.; Orgel, J.P.R.O. Dinosaur peptides suggest mechanisms of protein survival. *PLoS ONE* **2011**, *6*, e20381. [CrossRef] [PubMed]

56. Hartman, J.H.; Butler, R.D.; Weiler, M.W.; Schumaker, K.K. Context, naming and formal designation of the Cretaceous Hell Creek Formation lectostratotype, Garfield County, Montana. In *Through the End of the Cretaceous in the Type Locality of the Hell Creek Formation in Montana and Adjacent Areas*; Wilson, G.P., Clemens, W.A., Horner, J.R., Hartman, J.H., Eds.; Geological Society of America Special Paper 503; Geological Society of America: Boulder, CO, USA, 2014; pp. 89–121.

57. Horner, J.R.; Goodwin, M.B.; Myhrvold, N. Dinosaur census reveals abundant *Tyrannosaurus* and rare ontogenetic stages in the Upper Cretaceous Hell Creek Formation (Maastrichtian), Montana, USA. *PLoS ONE* **2011**, *6*, e16574. [CrossRef] [PubMed]

58. Waage, K.M. Cretaceous transitional environments and faunas in central South Dakota. In *A Symposium on Paleoenvironments of the Cretaceous Seaway in the Western Interior, Geological Society of America Rocky Mountain Section 20th Annual Meeting*; Colorado School of Mines: Golden, CO, USA, 1968; pp. 237–266.

59. Frye, C.I. Stratigraphy of the Hell Creek Formation in North Dakota. *N. Dak. Geol. Surv. Bull.* **1969**, *54*, 1–65.

60. Moore, W.L. Stratigraphy and Environments of Deposition of the Cretaceous Hell Creek Formation (Reconnaissance) and the Paleocene Ludlow Formation (Detailed), Southwestern North Dakota. Report of Investigations No. 56. United States. 1976. Available online: https://www.osti.gov/biblio/7346007 (accessed on 3 November 2021).

61. Murphy, E.C.; Hoganson, J.W.; Johnson, K.R. Lithostratigraphy of the Hell Creek Formation in North Dakota. *Geol. Soc. Am. Spec. Pap.* **2002**, *361*, 9–34.

62. Schweitzer, M.H.; Wittmeyer, J.L.; Horner, J.R. Gender-specific reproductive tissue in ratites and *Tyrannosaurus rex*. *Science* **2005**, *308*, 1456–1460. [CrossRef]

63. Gromet, L.P.; Dymek, R.F.; Haskin, L.A.; Korotev, R.L. The "North American shale composite": Its compilation, major and trace element characteristics. *Geochim. Cosmochim. Acta* **1984**, *48*, 2469–2482. [CrossRef]

64. Haskin, L.A.; Haskin, M.A.; Frey, F.A.; Wildeman, T.R. Relative and absolute terrestrial abundances of the rare earths. In *Origin and Distribution of the Elements*; Ahrens, L.H., Ed.; Pergamon: Oxford, UK, 1968; pp. 889–912.

65. Herwartz, D.; Tütken, T.; Munker, C.; Jochum, K.P.; Stoll, B.; Sander, P.M. Timescales and mechanisms of REE and Hf uptake in fossil bones. *Geochim. Cosmochim. Acta* **2011**, *75*, 82–105. [CrossRef]

66. Bau, M.; Dulski, P. Anthropogenic origin of positive gadolinium anomalies in river waters. *Earth Planet. Sc. Lett.* **1996**, *143*, 245–255. [CrossRef]

67. Pack, A.; Russell, S.S.; Shelley, J.M.G.; van Zuilen, M. Geo- and cosmochemistry of the twin elements yttrium and holmium. *Geochim. Cosmochim. Acta* **2007**, *71*, 4592–4608. [CrossRef]
68. Colson, M.C.; Colson, R.O.; Nellermoe, R. Stratigraphy and depositional environments of the upper Fox Hills and lower Hell Creek Formations at the Concordia Hadrosaur Site in northwestern South Dakota. *Rocky Mt. Geol.* **2004**, *39*, 93–111. [CrossRef]
69. Staub, J.R.; Cohen, A.D. The Snuggedy Swamp of South Carolina: A back-barrier estuarine coal-forming environment. *J. Sediment. Petrol.* **1979**, *49*, 133–144.
70. McCabe, P.J. Depositional environments of coal and coal-bearing strata. *Spec. Publ. Int.* **1984**, *7*, 13–42.
71. Reineck, H.-E.; Singh, I.B. *Depositional Sedimentary Environments*; Springer: Berlin, Germany, 1980; pp. 1–439.
72. Kroeger, T.J. Palynology of the Hell Creek Formation (Upper Cretaceous, Maastrichtian) in northwestern South Dakota: Effects of paleoenvironment on the composition of palynomorph assemblages. *Geol. Soc. Am. Spec. Pap.* **2002**, *361*, 457–472.
73. Flight, J.N. Sequence Stratigraphic Analysis of the Fox Hills and Hell Creek Fms (Maastrichtian), Eastern Montana and Its Relationship to Dinosaur Paleontology. Master's Thesis, Montana State University, Bozeman, MT, USA, 2004.
74. Schlesinger, P.M. The formation of caliche in soils of the Mojave Desert, California. *Geochim. Cosmochim. Acta* **1992**, *49*, 57–66. [CrossRef]
75. Rogers, R.R. Taphonomy of three dinosaur bone beds in the Upper Cretaceous Two Medicine Formation of northwestern Montana: Evidence for drought-related mortality. *PALAIOS* **1990**, *5*, 394–413. [CrossRef]
76. Rogers, R.R.; Kidwell, S.M. A conceptual framework for the genesis and analysis of vertebrate skeletal concentrations. In *Bonebeds: Genesis, Analysis, and Paleobiological Significance*; Rogers, R.R., Eberth, D.A., Fiorillo, A.R., Eds.; University of Chicago Press: Chicago, IL, USA, 2007; pp. 1–64.
77. Behrensmeyer, A.K. Taphonomic and ecologic information from bone weathering. *Palaeogeogr. Palaeocl.* **1978**, *4*, 150–162. [CrossRef]
78. Retallack, G.J. Dinosaurs and dirt. In *Dinofest International Proceedings*; Wolberg, D.L., Stump, E., Rosenberg, G.D., Eds.; Academy of Natural Sciences: Philadelphia, PA, USA, 1997; pp. 345–359.
79. Behrensymer, A.K. Vertebrate preservation in fluvial channels. *Palaeogeogr. Palaeocl.* **1988**, *63*, 183–199. [CrossRef]
80. Suarez, C.A.; Suarez, M.B.; Terry, D.O.; Grandstaff, D.E. Rare earth element geochemistry and taphonomy of the Early Cretaceous Crystal Geyser Dinosaur Quarry, east-central Utah. *PALAIOS* **2007**, *22*, 500–512. [CrossRef]
81. Suarez, C.A.; Morschhauser, E.M.; Suarez, M.B.; You, H.; Li, D.; Dodson, P. Rare earth element geochemistry of bone beds from the Lower Cretaceous Zhonggou Formation of Gansu Province, China. *J. Vertebr. Paleontol.* **2019**, *38*, 22–35. [CrossRef]
82. Rogers, R.R.; Fricke, H.C.; Addona, V.; Canavan, R.R.; Dwyer, C.N.; Harwood, C.L.; Koenig, A.E.; Murray, R.; Thole, J.T.; Williams, J. Using laser ablation-inductively coupled plasma-mass spectrometry (LA-ICP-MS) to explore geochemical taphonomy of vertebrate fossils in the Upper Cretaceous Two Medicine and Judith River Formations of Montana. *PALAIOS* **2010**, *25*, 183–195. [CrossRef]
83. Suarez, C.A.; Kohn, M.J. Caught in the act: A case study on microscopic scale physicochemical effects of fossilization on stable isotopic composition of bone. *Geochim. Cosmochim. Acta* **2020**, *268*, 277–295. [CrossRef]
84. Cantrell, K.J.; Byrne, R.H. Rare earth element complexation by carbonate and oxalate ions. *Geochim. Cosmochim. Acta* **1987**, *51*, 597–605. [CrossRef]
85. Luo, Y.-R.; Byrne, R.H. Carbonate complexation of yttrium and the rare earth elements in natural waters. *Geochim. Cosmochim. Acta* **2004**, *68*, 691–699. [CrossRef]
86. Livingstone, D.A. Chemical composition of rivers and lakes. In *Data of Geochemistry*; Fleischer, M., Ed.; USGS: Washington, DC, USA, 1963; pp. G1–G64.
87. Zhang, J.; Létolle, R.; Jusserand, C. Major element chemistry of the Huanghe (Yellow River), China—Weathering processes and chemical fluxes. *J. Hydrol.* **1995**, *168*, 173–203. [CrossRef]
88. Stallard, R.F.; Edmond, J.M. Geochemistry of the Amazon 3. Weathering chemistry and limits to dissolved inputs. *J. Geophys. Res.* **1987**, *92*, 8293–8302. [CrossRef]
89. Berner, R.A. Calcium carbonate concretions formed by the decomposition of organic matter. *Science* **1968**, *159*, 195–197. [CrossRef]
90. Pourret, O.; Davranche, M.; Gruau, G.; Dia, A. Rare earth elements complexation with humic acid. *Chem. Geol.* **2007**, *243*, 128–141. [CrossRef]
91. Föllmi, K.B. The phosphorous cycle, phosphogenesis and marine phosphate-rich deposits. *Earth-Sci. Rev.* **1996**, *40*, 55–124. [CrossRef]
92. Wang, L.; Liu, Q.; Hu, C.; Liang, R.; Qiu, J.; Wang, Y. Phosphorous release during decomposition of the submerged macrophyte *Potamogeton crispus*. *Limnology* **2018**, *19*, 355–366. [CrossRef]
93. Byrne, R.H.; Liu, X.; Schijf, J. The influence of phosphate coprecipitation on rare earth distributions in natural waters. *Geochim. Cosmochim. Acta* **1996**, *60*, 3341–3346. [CrossRef]
94. Liu, X.; Byrne, R.H.; Schijf, J. Comparative coprecipitation of phosphate and arsenate with yttrium and the rare earths: The influence of solution complexation. *J. Solut. Chem.* **1997**, *26*, 1187–1198. [CrossRef]
95. Brunet, R.-C.; Astin, K.B. Variation in phosphorus flux during a hydrological season: The River Ardour. *Water Res.* **1998**, *32*, 547–558. [CrossRef]
96. Elderfield, H.; Upstill-Goddard, R.; Sholkovitz, E.R. The rare earth elements in rivers, estuaries, and coastal seas and their significance to the composition of ocean waters. *Geochim. Cosmochim. Acta* **1990**, *54*, 971–991. [CrossRef]

97. Trueman, C.N.; Benton, M.J.; Palmer, M.R. Geochemical taphonomy of shallow marine vertebrate assemblages. *Palaeogeog. Palaeocl.* **2003**, *197*, 151–169. [CrossRef]
98. Metzger, C.A.; Terry, D.O.; Grandstaff, D.E. Effect of paleosol formation on rare earth element signatures in fossil bone. *Geology* **2004**, *32*, 497–500. [CrossRef]
99. Williams, C.T.; Potts, P.J. Element distribution maps in fossil bones. *Archaeometry* **1988**, *30*, 237–247. [CrossRef]
100. Williams, C.T.; Henderson, P.; Marlow, C.A.; Molleson, T.I. The environment of deposition indicated by the distribution of rare earth elements in fossil bones from Olduvai Gorge, Tanzania. *Appl. Geochem.* **1997**, *12*, 537–547. [CrossRef]
101. Kohn, M.J.; Moses, R.J. Trace element diffusivities in bone rule out simple diffusive uptake during fossilization but explain in vivo uptake and release. *Proc. Natl. Acad. Sci. USA* **2013**, *110*, 419–424. [CrossRef] [PubMed]
102. Wings, O. Authigenic minerals in fossil bones from the Mesozoic of England: Poor correlation with depositional environments. *Palaeogeog. Palaeocl.* **2004**, *204*, 15–32. [CrossRef]
103. Janssens, K.; Vincze, L.; Vekemans, B.; Williams, C.T.; Radtke, M.; Haller, M.; Knochel, A. The non-destructive determination of REE in fossilized bone using synchrotron radiation induced K-line X-ray microfluorescence analysis. *J. Anal. Chem.* **1999**, *363*, 413–420. [CrossRef]
104. Herwartz, D.; Munker, C.; Tütken, T.; Hoffmann, J.E.; Wittke, A.; Barbier, B. Lu-Hf isotope systematics of fossil biogenic apatite and their effects on geochronology. *Geochim. Cosmochim. Acta* **2013**, *101*, 328–343. [CrossRef]
105. Reynard, B.; Lecuyer, C.; Grandjean, P. Crystal-chemical controls on rare-earth element concentrations in fossil biogenic apatites and implications for paleoenvironmental reconstructions. *Chem. Geol.* **1999**, *155*, 233–241. [CrossRef]
106. Trueman, C.N.; Behrensmeyer, A.K.; Potts, R.; Tuross, N. High-resolution records of location and stratigraphic provenance from the rare earth element composition of fossil bones. *Geochim. Cosmochim. Acta* **2006**, *70*, 4343–4355. [CrossRef]
107. Martin, J.E.; Patrick, D.; Kihm, A.J.; Foit, F.F.; Grandstaff, D.E. Lithostratigraphy, tephrochronology, and rare earth element geochemistry of fossils at the classical Pleistocene Fossil Lake area, south central Oregon. *J. Geol.* **2005**, *113*, 139–155. [CrossRef]
108. Costa, I.A.; Barros, J.A.; Malta, L.; Viana, M.A.; Santos, R.P. *Viva a Terra! 7*; Porto Editora: Port, Portugal, 2016; pp. 1–48.
109. Wiemann, J.; Fabbri, M.; Yang, T.-R.; Stein, K.; Sander, P.M.; Norell, M.A.; Briggs, D.E.G. Fossilization transforms vertebrate hard tissue proteins into N-heterocyclic polymers. *Nat. Commun.* **2018**, *9*, 4741. [CrossRef]
110. Schweitzer, M.H. Molecular paleontology: Some current advances and problems. *Ann. Paleontol.* **2004**, *90*, 81–102. [CrossRef]
111. Vermes, I.; Haanen, C. Apoptosis and programmed cell death in health and disease. *Adv. Clin. Chem.* **1994**, *31*, 177–246.
112. Bayoumy, A.M.; Youssif, G.; Elgohary, E.A.; Husien, S.; El Deen, H.S.; Albeltagy, N.M.; Abdelnaby, D.R.; Elhaes, H.; Ibrahim, M.A. Impact of salvation on the geometrical parameters of some amino acids. *Lett. Appl. NanoBioSci.* **2019**, *8*, 567–570.
113. Schroeter, E.R.; Cleland, T.P. Glutamine deamidation: An indicator of antiquity, or preservational quality? *Rapid Commun. Mass Spectrom.* **2016**, *30*, 251–255. [CrossRef] [PubMed]
114. Lima, M.; Baynes, J.W. Glycation. In *Encyclopedia of Biological Chemistry*; Lennarz, W.J., Lane, M.D., Eds.; Elsevier: London, UK, 2013; pp. 405–411.
115. Mello, R.B.; Silva, M.R.R.; Alves, M.T.S.; Evison, M.P.; Guiamarães, M.A.; Francisco, R.A.; Astolphi, R.D.; Iwamura, E.S.M. Tissue microarray analysis applied to bone diagenesis. *Sci. Rep.* **2017**, *7*, 39987. [CrossRef] [PubMed]
116. Ferrante, C.; Cavin, L.; Vennemann, T.; Martini, R. Histology and geochemistry of *Allosaurus* (Dinosauria: Theropoda) from the Cleveland-Lloyd Dinosaur Quarry (Late Jurassic, Utah): Paleobiological implications. *Front. Earth Sci.* **2021**, *9*, 641060. [CrossRef]
117. Kowal-Linka, M.; Jochum, K.P.; Surmik, D. LA-ICP-MS analysis of rare earth elements in marine reptile bones from the Middle Triassic bonebed (Upper Silesia, S Poland): Impact of long-lasting diagenesis, and factors controlling the uptake. *Chem. Geol.* **2014**, *363*, 213–228. [CrossRef]
118. McCormack, J.M.; Bahr, A.; Gerdes, A.; Tütken, T.; Prinz-Grimm, P. Preservation of successive diagenetic stages in Middle Triassic bonebeds: Evidence from in situ trace element and strontium isotope analysis of vertebrate fossils. *Chem. Geol.* **2015**, *410*, 108–123. [CrossRef]
119. Decrée, S.; Herwartz, D.; Mercadier, J.; Miján, I.; de Buffrénil, V.; Leduc, T.; Lambert, O. The post-mortem history of a bone revealed by its trace element signature: The case of a fossil whale rostrum. *Chem. Geol.* **2018**, *477*, 137–150. [CrossRef]
120. McLain, M.; Ullmann, P.V.; Ash, R.D.; Bohnstedt, K.; Nelsen, D.; Clark, R.; Brand, L.R.; Chadwick, A.V. Independent confirmation of fluvial reworking at a Lance Formation (Maastrichtian) bonebed by traditional and chemical taphonomic analyses. *PALAIOS* **2021**, *36*, 193–215. [CrossRef]
121. Goodwin, M.B.; Grant, P.G.; Bench, G.; Holroyd, P.A. Elemental composition and diagenetic alteration of dinosaur bone: Distinguishing micron-scale spatial and compositional heterogeneity using PIXE. *Palaeogeogr. Palaeocl.* **2007**, *253*, 458–476. [CrossRef]

biology

MDPI

Article

Taphonomic and Diagenetic Pathways to Protein Preservation, Part II: The Case of *Brachylophosaurus canadensis* Specimen MOR 2598

Paul V. Ullmann [1,*], Richard D. Ash [2] and John B. Scannella [3,4]

[1] Department of Geology, Rowan University, Glassboro, NJ 08028, USA
[2] Department of Geology, University of Maryland, College Park, MD 20742, USA
[3] Museum of the Rockies, Montana State University, Bozeman, MT 59717, USA
[4] Department of Earth Sciences, Montana State University, Bozeman, MT 59717, USA
[*] Correspondence: ullmann@rowan.edu

Simple Summary: Reports of the recovery of proteins and other molecules from fossils have become so common over the last two decades that some paleontologists now focus almost entirely on studying how biologic molecules can persist in fossils. In this study, we explored the fossilization history of a specimen of the hadrosaurid dinosaur *Brachylophosaurus* which was previously shown to preserve original cells, tissues, and structural proteins. Trace element analyses of the tibia of this specimen revealed that after its bones were buried in a brackish estuarine channel, they fossilized under wet conditions which shifted in redox state multiple times. The successful recovery of proteins from this specimen, despite this complex history of chemical alterations, shows that the processes which bind and stabilize biologic molecules shortly after death provide them remarkable physical and chemical resiliency. By uniting our results with those of similar studies on other dinosaur fossils known to also preserve original proteins, we also conclude that exposure to oxidizing conditions in the initial ~48 h postmortem likely promotes molecular stabilization reactions, and the retention of early-diagenetic trace element signatures may be a useful proxy for molecular recovery potential.

Abstract: Recent recoveries of peptide sequences from two Cretaceous dinosaur bones require paleontologists to rethink traditional notions about how fossilization occurs. As part of this shifting paradigm, several research groups have recently begun attempting to characterize biomolecular decay and stabilization pathways in diverse paleoenvironmental and diagenetic settings. To advance these efforts, we assessed the taphonomic and geochemical history of *Brachylophosaurus canadensis* specimen MOR 2598, the left femur of which was previously found to retain endogenous cells, tissues, and structural proteins. Combined stratigraphic and trace element data show that after brief fluvial transport, this articulated hind limb was buried in a sandy, likely-brackish, estuarine channel. During early diagenesis, percolating groundwaters stagnated within the bones, forming reducing internal microenvironments. Recent exposure and weathering also caused the surficial leaching of trace elements from the specimen. Despite these shifting redox regimes, proteins within the bones were able to survive through diagenesis, attesting to their remarkable resiliency over geologic time. Synthesizing our findings with other recent studies reveals that oxidizing conditions in the initial ~48 h postmortem likely promote molecular stabilization reactions and that the retention of early-diagenetic trace element signatures may be a useful proxy for molecular recovery potential.

Keywords: REE; *Brachylophosaurus*; molecular paleontology; geochemical taphonomy; diagenesis; bone; protein; collagen; Judith River Formation

Citation: Ullmann, P.V.; Ash, R.D.; Scannella, J.B. Taphonomic and Diagenetic Pathways to Protein Preservation, Part II: The Case of *Brachylophosaurus canadensis* Specimen MOR 2598. *Biology* **2022**, *11*, 1177. https://doi.org/10.3390/biology11081177

Academic Editor: Zhifei Zhang

Received: 14 June 2022
Accepted: 3 August 2022
Published: 5 August 2022

Publisher's Note: MDPI stays neutral with regard to jurisdictional claims in published maps and institutional affiliations.

Copyright: © 2022 by the authors. Licensee MDPI, Basel, Switzerland. This article is an open access article distributed under the terms and conditions of the Creative Commons Attribution (CC BY) license (https://creativecommons.org/licenses/by/4.0/).

1. Introduction

1.1. Shifting Views on Molecular Preservation

Fossilization has historically been viewed as a "harsh" process involving forced mineralization coincident with the wholesale loss of organic tissues and their component biomolecules (e.g., [1]). However, an ever-expanding wealth of recent studies employing methods from histochemistry and immunoassays to genomics and proteomics have shattered this 'traditional' paradigm. As reviewed by ourselves and others [2–7], it is now clear that not only is the long-term preservation of endogenous DNA, proteins, and other biomolecules possible in fossils, but select burial circumstances may actually promote molecular preservation in both plants and animals interred in a diverse array of depositional environments. For example, drastic advances in analytical resolution over the last two decades have enabled the recovery of nuclear and mitochondrial genomes and/or proteomes from fossils of a number of Pleistocene mammals, such as cave bears [8], mammoths [9–12], horses [13], saber-toothed cats [14], and ancient hominins [15,16].

Given their greater resilience to decay than DNA [17], structural proteins (e.g., collagen I, actin, and β-keratin) are now known from vertebrate fossils dating all the way back to the Jurassic (e.g., [18–22]). Considerable attention has been paid especially to the protein collagen I due to its sheer abundance in bioapatitic tissues [23] and inferred high preservation potential [18,24–47]. Recent studies have employed numerous independent techniques to identify collagen within fossils, with arguably the most convincing evidence being the identification of original peptide sequences and diagenetiforms (protein remains demonstrably modified by diagenetic alterations [48]) in fossil bones via high-resolution tandem mass spectrometry [18,28,30,42,45]. Remarkably, cladistic analyses incorporating collagen I peptides recovered from two Cretaceous nonavian dinosaur bones confirmed their archosaurian identities and thus endogeneity [45,49]. Such findings unequivocally demonstrate both that biomolecules can 'survive' fossilization and that portions of them can persist over strikingly-long geologic timescales.

However, the idea of molecular preservation in Mesozoic (and possibly even older) fossils remains controversial to some due to our incomplete understanding of soft-tissue fossilization in general and the geochemical reactions which may stabilize biomolecules within cells and tissues over such immense time frames. While it is universally agreed that processes such as rapid burial can facilitate the preservation of soft tissues in fossils [50,51], it largely remains unclear how other taphonomic processes and the physical (e.g., sedimentology and hydrodynamics) and chemical attributes (e.g., aqueous geochemistry) of depositional environments influence decay at the molecular level (but see [52] for an informative initial foray into this subject). It is therefore vital for researchers to not only demonstrate the authenticity of biomolecular remnants in fossils but to also identify the physicochemical factors acting within the diagenetic settings which permitted such cases of "exceptional" preservation. Pioneering actualistic studies by Schweitzer et al. [4] and Boatman et al. [53] demonstrated that iron free radicals in diagenetic pore fluids likely play a role by inducing intra-molecular crosslinking, but are other aspects of groundwater chemistry (i.e., redox state; cf. [54]) and diffusion history (i.e., duration spent saturated; cf. [47]) equally important in determining whether or not biomolecules persist in fossils? Additionally, if they are, which depositional settings and diagenetic histories most favor long-term molecular preservation? In short, we are just beginning to explore these questions.

1.2. Insights from Trace Element Analyses

One of the most effective means of clarifying the geochemical history of a fossil is through studying its trace element composition. After being solubilized from surrounding sediments by percolating groundwaters via oxidation, dissolution, and other processes, trace element ions, including those of the rare earth elements (REE: lanthanum–lutetium), uranium, and scandium, are ubiquitously adsorbed by bone hydroxyapatite during diagenesis [55]. Since these elements are essentially absent in bone tissue in vivo, their presence in fossil bones derives almost entirely from postmortem interactions with surface and groundwaters [55,56]. As a result, the proportions and spatial distributions of trace elements within a fossil bone provide detailed insights into the chemistries of past pore fluids and the geochemical milieus to which a specimen was exposed throughout its term of burial (e.g., [57–64]). Trace element signatures have thus been successfully utilized to: (1) infer the relative degree of chemical alteration a specimen has endured [65,66]; (2) characterize the chemistry of pore fluids in past environments (e.g., [60,63,67]); (3) track spatiotemporal trends in redox conditions within specimens throughout diagenesis (e.g., [62–64]); and (4) clarify the number and relative timing of exposures to pore fluids throughout diagenesis (e.g., [64,68]). REE signatures, in particular, have also been shown to potentially be viable proxies for molecular preservation in fossil bones (i.e., they can help identify the most ideal specimens for paleomolecular investigation [47]).

Several of our recent studies capitalized on these diverse utilities of trace elements to elucidate the paleoenvironmental, geochemical, and diagenetic history of two fossils from the Cretaceous Hell Creek Formation which were each documented to preserve endogenous collagen I [28,47], specifically an *Edmontosaurus* fibula [63] and a femur of *Tyrannosaurus rex* specimen Museum of the Rockies (MOR) 1125 [64] (also see [69] for alternative taxonomic assignment of MOR 1125). These studies provided intriguing insights into taphonomic pathways to protein preservation, but they still merely constitute two case studies in the same geologic formation; the full suite of taphonomic and diagenetic variables at play in molecular preservation remains to be clarified.

In this study, we conducted trace element analyses on the only Mesozoic fossil other than *T. rex* MOR 1125 known to yield endogenous peptide sequences: *Brachylophosaurus canadensis* specimen MOR 2598. Schweitzer et al. [30] and Schroeter et al. [45] each recovered numerous peptides of collagen I from the left femur of this hadrosaur, the authenticity of which were independently corroborated by multiple forms of microscopy, infrared spectroscopy, mass spectrometry, and immunoassays replicated in multiple laboratories by separate researchers each using dedicated equipment and reagents (see each reference for further details). The cumulatively-comprehensive approach undertaken by these studies to demonstrate reproducibility and authenticate the endogeneity of collagen in MOR 2598 set a rigorous standard that has yet to be matched again, despite over a decade of ensuing research on other specimens. Given this great significance of MOR 2598 in providing concrete foundations for the field of molecular paleontology, it is only right to resolve the taphonomic and diagenetic history of this specimen in equally comprehensive detail.

2. Taphonomic and Geologic Context

MOR 2598 consists of an articulated left hind limb of a subadult *Brachylophosaurus canadensis* recovered from an outcrop of the Campanian Judith River Formation north of Malta, Montana, on lands managed by the Montana Department of Natural Resources and Conservation (Figure 1). The specimen was found within a thick sequence (~7 m) of trough cross-stratified channel sandstones exposed along the southern side of Cottonwood Creek. Schweitzer et al. [31] concluded that these strata were deposited in a fluvial channel within the overall lowland fluviodeltaic system of the Judith River ecosystems [70,71]. The tibia, fibula, and pes were collected in the summer of 2006, and the femur was collected in a separate plaster jacket the following year. The articulation of these skeletal elements implies they were still joined by connective tissues (i.e., ligaments) at the time of burial. The incomplete nature of the tibia (see below) and slightly lighter color of this bone than the femur indicate that portions of the tibia were exposed by modern erosion/weathering upon discovery (whereas the femur was not [30]). All of the skeletal elements are brown in color (e.g., Figure 1B), indicating their mineralogy has likely been transformed from hydroxyapatite to fluorapatite, which is typical of bone fossilization [72,73]. As discussed by Schweitzer et al. [30], the femur was collected with 10–12 cm of sediment still encasing it to maintain geochemical equilibrium for as long as possible before examination via demineralization, scanning electron microscopy, multiple immunoassays, and liquid chromatography–tandem mass spectrometry. All of the skeletal elements appear well-preserved morphologically in that they lack any signs of weathering or abrasion, but the tibia (examined herein) is missing its distal end and a section of the posterior and medial portions of the shaft near its proximal end. The tibia is also highly fractured, exhibiting numerous transverse and longitudinal fractures arising from compaction after fossilization. Though the medullary cavity of this bone is partly 'filled' due to compaction and the partial crushing of cancellous trabeculae, there are no signs of permineralization or infilling by the sedimentary matrix (pers. observations, and [74]).

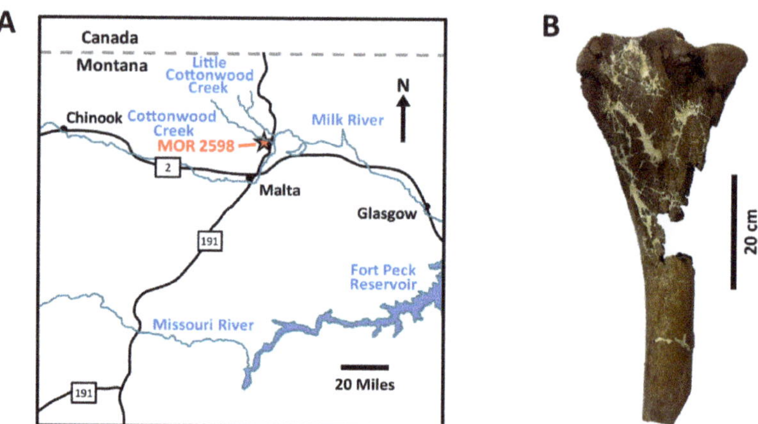

Figure 1. (**A**) Map showing the locality from which MOR 2598 was recovered in Phillips County, Montana. (**B**) Left tibia of MOR 2598 examined in this study, shown in lateral view. Map redrawn and modified from [75].

3. Materials and Methods

3.1. Materials

It was not possible to acquire a sample from the left femur of *Brachylophosaurus canadensis* MOR 2598 for this study as prior histologic and paleomolecular studies of this particular skeletal element of the specimen [30,45,74] and reconstructive efforts undertaken during preparation to maintain permanent stability of the bone left no portion of the cortex easily removable without compromising the integrity of the fossil. Therefore, we instead extracted a fragment of the cortex from the midshaft of the left tibia of MOR 2598. This bone was found in articulation with the left femur in the field and was accordingly buried at the same stratigraphic position within the same stratum as the left femur, so it was almost certainly exposed to the same environmental conditions postmortem and early-diagenetic regime(s) after burial as the left femur. The cortex at the midshaft of tibiae also possesses a similar thickness, density, and histologic microstructure to the midshaft cortex of femora in hadrosaurids (e.g., [76,77]). For these reasons, we are confident that trace element signatures within the left tibia should be very similar to those that would be identified in the left femur examined by prior studies, and that this tibia therefore represents a suitable choice for examining the geochemical history of the hind limb of MOR 2598 as a whole.

The excised cortical fragment encompasses the majority of the cortical thickness of the bone, including the external margin, but fragile cancellous bone within the medullary cavity disintegrated away during preparation. Because of this, we infer that the innermost portion of the internal cortex was not included in our analyses.

3.2. Methods

3.2.1. Sample Preparation

The cortical sample was embedded in Silmar 41^{TM} resin (US Composites, West Palm Beach, FL, USA) under vacuum, then sectioned using a Hillquist SF-8 trim saw

(Hillquist, Arvada, CO, USA). The resulting ~3 mm-thick section was briefly rinsed with distilled water then briefly polished with 600 grit silicon carbide to acquire an evenly smooth surface for ensuing laser ablation–inductively coupled plasma mass spectrometry (LA-ICPMS) analyses.

3.2.2. LA-ICPMS Analyses

We employed the same LA-ICPMS methods as Ullmann et al. [63,64] in this study and refer the reader to the Supplementary Materials and those publications for thorough details. In brief, LA-ICPMS was chosen as a powerful means of examining the spatial distribution of REEs and other minor and trace elements within the tibia of MOR 2598, which can provide unique insights into the diagenetic history of a fossil bone and any geochemical shifts it endured through its fossilization. Concentrations of elements are reported in parts per million (ppm) except for iron, which is reported in weight percent (wt. %). REE concentrations were normalized against the North American Shale Composite (NASC; [78,79]) to facilitate comparisons to other fossil bones from other localities. The use of a subscript $_N$ denotes NASC-normalized values and ratios. The reproducibility of our results was taken as the percent relative standard deviation for all REEs in the NIST 610 glass standard; it averaged 1.5% and was below 3% for all analyzed elements. NASC-normalized REE ratios were used to calculate $(Ce/Ce^*)_N$, $(Ce/Ce^{**})_N$, $(Pr/Pr^*)_N$, and $(La/La^*)_N$ anomalies following Herwartz et al. [60]: $(Ce/Ce^*)_N = Ce_N/(0.5*La_N + 0.5*Pr_N)$, $(Ce/Ce^{**})_N = Ce_N/(2*Pr_N - Nd_N)$, $(Pr/Pr^*)_N = Pr_N/(0.5*Ce_N + 0.5*Nd_N)$, and $(La/La^*)_N = La_N/(3*Pr_N - 2*Nd_N)$.

4. Results

4.1. Overall REE Composition

As a whole (i.e., by summing all transect data), the left tibia of MOR 2598 exhibits a $\sum$REE value of 256 ppm (Table 1). The three most abundant trace elements in the cortex of the bone are iron (Fe), strontium (Sr), and barium (Ba), which exhibit concentrations of 0.94 wt. %, 2499 ppm, and 1448 ppm, respectively (Table 1). All of these elements, as well as manganese (Mn), are present in concentrations approximately one to two orders of magnitude higher than REEs (Table 1). Whereas the average scandium (Sc) enrichment (59 ppm) is around the same magnitude as those of most LREEs (~10–90 ppm), the average yttrium (Y) concentration (190 ppm) is more than double that of the highest REE (89 ppm for cerium, Ce; Table 1). Among REEs, there is substantially greater enrichment in light rare earth elements (LREEs, La–Nd) than middle (MREEs, Sm–Gd) and heavy rare earths (HREEs, Tb–Lu), clearly indicative of fractionation during uptake (see Discussion). The average whole-bone concentration of uranium (U), 51 ppm, is distinctly higher than those of bones from the Hell Creek Formation known to also yield endogenous collagen I (2–38 ppm [63,64]).

Table 1. Average whole-bone trace element composition of the left tibia of *Brachylophosaurus canadensis* MOR 2598. Numbers presented are averages of all transect data acquired across the cortex. Iron (Fe) is presented in weight percent (wt. %); all other elements are in parts per million (ppm). Absence of $(Ce/Ce^*)_N$, $(Pr/Pr^*)_N$, $(Ce/Ce^{**})_N$, and $(La/La^*)_N$ anomalies occurs at 1.0, and these anomalies were calculated as in the Materials and Methods. The Y/Ho value reflects this mass ratio.

Element	Concentration
Sc	59.23
Mn	834
Fe	0.94
Sr	2499
Y	190
Ba	1448
La	40.65
Ce	88.61
Pr	9.86

Table 1. *Cont.*

Element	Concentration
Nd	35.26
Sm	7.41
Eu	2.61
Gd	12.41
Tb	2.09
Dy	17.33
Ho	4.44
Er	15.26
Tm	2.21
Yb	15.42
Lu	2.62
Th	0.18
U	51.13
$\sum$REE	256
$(Ce/Ce^*)_N$	1.04
$(Pr/Pr^*)_N$	0.95
$(Ce/Ce^{**})_N$	1.10
$(La/La^*)_N$	1.12
Y/Ho	42.85

4.2. Intra-Bone Concentration Depth Profiles

Each REE exhibits a steeply-declining concentration profile from the cortical margin. As an example, lanthanum (La) concentrations decrease from ~1000 ppm at the outer edge of the cortex to ~10 ppm at 5 mm, thus constituting an order of magnitude decrease across this distance (Figure 2A). Concentrations of HREEs and the latter half of the MREE series, as well as U, Sc, Y, and lutetium (Lu), all increase toward the internal end of the transect (Figure 2A–C). For example, Yb concentrations increase from ~1–2 ppm at a depth of 25 mm to ~15–20 ppm at the internal end of the transect (Figure 2A, Data S1). Such increases signify secondary diffusion from within the medullary cavity (see the Discussion and Supplementary Materials). Among REEs, Ce exhibits the highest concentration at the cortical margin (~3000 ppm), whereas Lu exhibits the lowest (~10 ppm). LREEs generally exhibit the steepest concentration profiles, reflective of spatially-heterogeneous uptake, whereas HREEs generally exhibit flatter profiles, reflective of comparatively more spatially homogenous uptake. MREE profiles are generally intermediate in steepness, and MREE concentrations commonly fall below the lower detection limit in the middle and internal cortices (Data S1).

Brief spikes in concentrations typically encountered in osteonal tissue around Haversian canals are rare and generally of miniscule magnitude. Although this would seem to imply a lack of major uptake through vascular systems, most REE (e.g., La in Figure 2A) profiles exhibit a subtle deflection near 2.5 mm reflective of uptake via double medium diffusion (*sensu* [80]). Near 1 mm, numerous elements, especially Y, HREE, Sc, and U, alternatively exhibit a roughly 80% increase in concentrations over values at the cortical margin (e.g., Figure 2A–C), perhaps reflective of late diagenetic near-surface leaching (see Discussion).

Fe, Ba, Sr, and Mn each exhibit much flatter profiles at higher concentrations than all other elements (Figure 2C,D), with Fe exhibiting both the highest values and greatest range in variation of concentrations across the transect among these four elements. Unlike all other elements we investigated, Sc and U each exhibit "W-shaped" profiles with increasing cortical depth: after slowly decreasing from the cortical margin, each profile includes a broad, moderate peak in concentrations in the central portion of the middle cortex followed by steadily-increasing concentrations across the internal cortex (Figure 2B). Y exhibits the same profile shape as HREEs in the tibia of MOR 2598 (Figure 2C), indicating similar uptake behavior for these elements in this fossil.

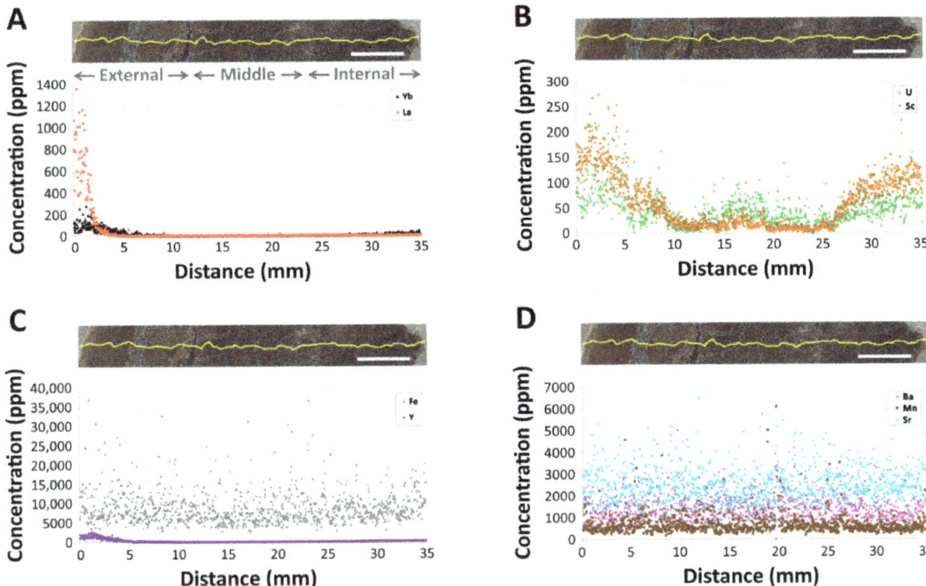

Figure 2. Intra-bone concentration gradients of various elements in the tibia of MOR 2598. (**A**) Lanthanum (La) and ytterbium (Yb). (**B**) Scandium (Sc) and uranium (U). (**C**) Iron (Fe) and yttrium (Y). (**D**) Barium (Ba), manganese (Mn), and strontium (Sr). Note the different concentration scales for each panel. The laser track is denoted by the yellow line in each bone cross section. Gray text labels in (**A**) span the approximate regions considered as the 'external', 'middle', and 'internal' cortices. Scale bars, in white over bone images, each equal 5 mm.

4.3. NASC-Normalized REE Patterns

Spider diagrams of NASC-normalized REE concentrations reveal overall HREE enrichment in the bone as a whole (Figure 3B,C), but significant relative enrichment of LREEs within the external 250 μm of the cortex (Figure 4A). A ternary plot of Nd_N-Gd_N-Yb_N confirms this trend of relative HREE enrichment by revealing that a data point for the specimen as a whole plots more closely to the Yb corner (Figure 3C). Whereas there is no apparent Ce anomaly in the bone as a whole (Figure 3B), the external-most 250 μm of the cortex exhibits a modest positive Ce anomaly (seen as an upward deflection of the pattern at this element; Figure 4A). REE concentrations range from ~25 to 50 times NASC values in the external 250 μm of the cortex.

Substantial spatial heterogeneity in REE composition is evident in both a ternary plot of La_N-Gd_N-Yb_N (Figure 3D) and a spider diagram of individual laser runs compiled into the full transect (Figure 4B). Both of these figures reveal significantly greater LREE content in the external-most laser run compared to all other laser runs (i.e., variation exceeds two standard deviations), signifying variations in composition are largely controlled by cortical depth. In general, the bone becomes increasingly enriched in MREEs and HREEs relative to LREEs with increasing cortical depth, with transects through the middle and internal cortices exhibiting both similar magnitudes of REE enrichment and drastic relative enrichment in HREEs (Figure 4B). Proportionally, transects through these internal regions of the bone exhibit one to two orders of magnitude of enrichment in HREEs over LREEs, compared to just half an order of magnitude of HREE enrichment in the external-most transect.

All laser runs through the middle and internal cortices exhibit isolated peaks at gadolinium (Gd; Figure 4B), likely attributable to isobaric interference effects between LREE oxides and other ions likely present within the fossil (e.g., spectral overlap between Gd^{157} and BaF [81,82]). Most spider diagrams, especially those which separately plot data from the external cortex (e.g., Figure 4), also exhibit subtle peaks at europium (Eu) and holmium (Ho). These peaks impart a weak 'M' shape to the shale-normalized patterns, which most authors (e.g., [83] and references therein) attribute to influences of tetrad effects during uptake (also see Supplementary Materials for further discussion on potential tetrad effects in MOR 2598).

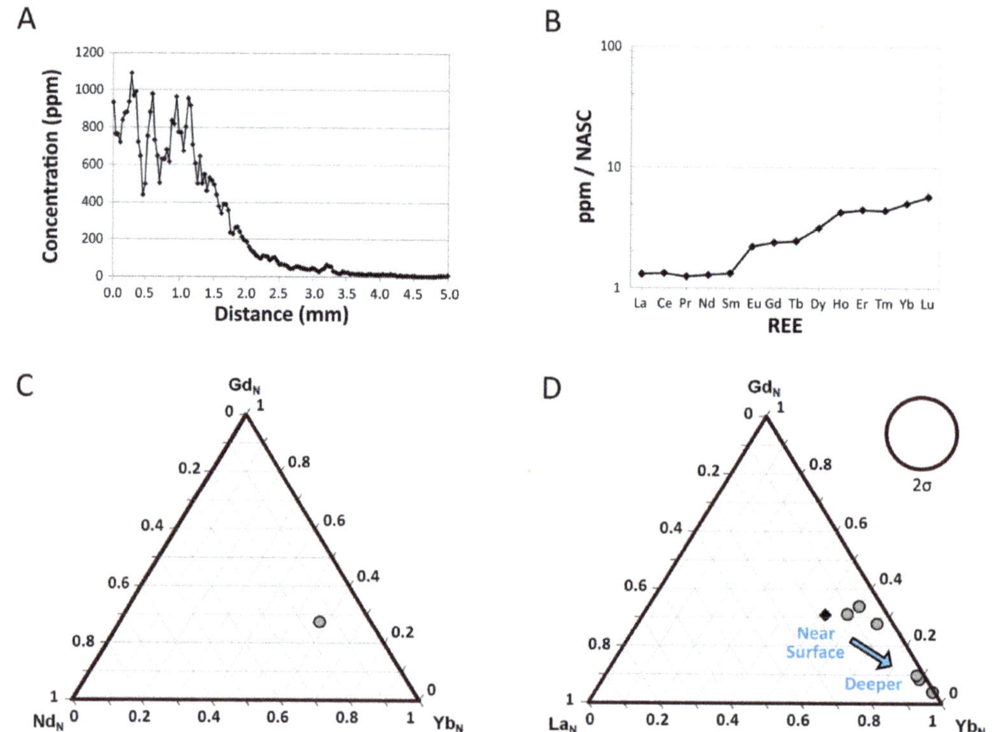

Figure 3. REE composition of the tibia of MOR 2598. (**A**) Three-point moving average profile of La concentrations in the outermost 5 mm of the bone. (**B**) Average NASC-normalized REE composition of the fossil specimen as a whole. (**C,D**) Ternary diagrams of NASC-normalized REE. (**C**) Average composition of the bone. (**D**) REE compositions divided into data from each individual laser transect (~5 mm of data each). Compositional data from the transect that included the outer bone edge is denoted by a dark diamond; all other internal transect data are indicated by gray circles. The 2σ circle represents two standard deviations based on ± 5% relative standard deviation.

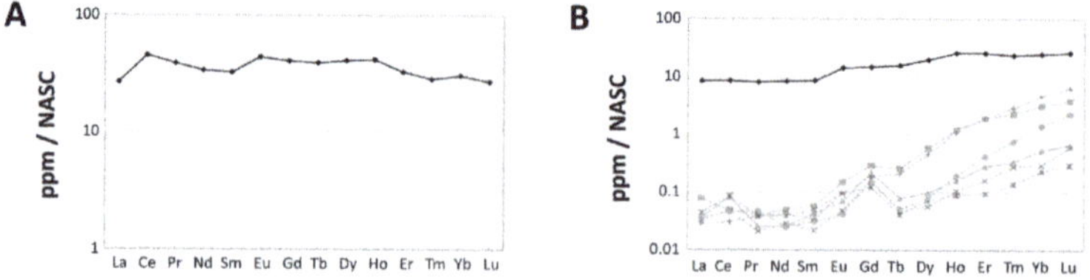

Figure 4. Spider diagrams of intra-bone NASC-normalized REE distribution patterns within the tibia of MOR 2598. (**A**) Average composition of the outermost 250 μm of the cortex, demonstrating substantially greater LREE and MREE content in the outermost cortex compared to in the bone as a whole (**B**). (**B**) Variation in compositional patterns by laser transects. The pattern which includes the external margin of the bone is shown in black, those from deepest within the bone by dotted, light-gray lines, and all other analyses in between by solid, dark-gray lines.

4.4. $(La/Yb)_N$ vs. $(La/Sm)_N$ Ratio Patterns

The tibia of MOR 2598 exhibits a whole-bone average $(La/Sm)_N$ value of 0.99 and a $(La/Yb)_N$ of 0.26. These values signify modest HREE enrichment relative to many environmental water samples, dissolved loads, and sedimentary particulates. Specifically, these values place the bone within the compositional range of river waters, brackish estuary waters, and marine pore fluids (Figure 5A).

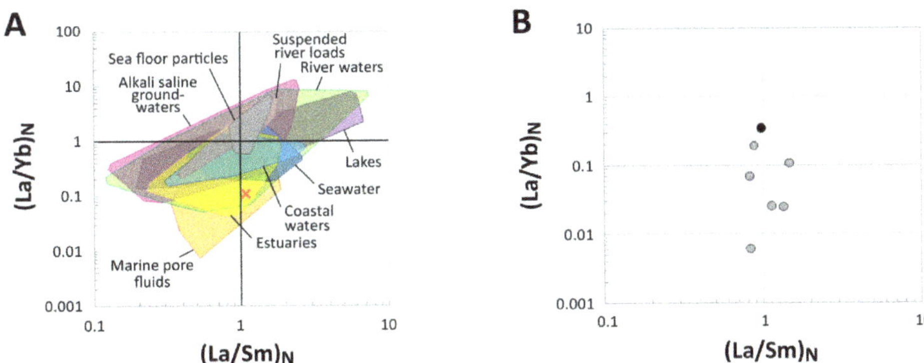

Figure 5. $(La/Yb)_N$ and $(La/Sm)_N$ ratios of the tibia of MOR 2598. (**A**) Comparison of the whole-bone average $(La/Yb)_N$ and $(La/Sm)_N$ ratios of the fossil to ratios from various environmental waters and sedimentary particulates. Literature sources for environmental samples are provided in the Supplementary Materials. (**B**) REE compositions of individual laser transects expressed as NASC-normalized $(La/Yb)_N$ and $(La/Sm)_N$ ratios. The transect including the external bone margin is denoted by the black symbol, whereas all other (internal) transects are represented by gray symbols.

When plotted by individual laser runs (Figure 5B), REE ratios from the middle and internal cortices plot with similar $(La/Sm)_N$ values to those seen in the external cortex but with consistently lower $(La/Yb)_N$ values. This difference encompasses roughly one order of magnitude, on average. The most internal laser run exhibits the lowest $(La/Yb)_N$ ratio (0.006), and all but two laser runs across the middle and internal cortices exhibit $(La/Yb)_N$ ratios < 0.1. $(La/Sm)_N$ ratios range between 0.8 and 1.5 and exhibit no apparent relationship with cortical depth.

4.5. REE Anomalies

Due to the concentrations of many trace elements in the middle cortex being so low that they fall below the lower detection limit (Data S1), every anomaly examined exhibits major gaps in coverage through this region of the bone (Figure S1). Occasional instances of significantly higher neodymium (Nd) than praseodymium (Pr) concentrations also create gaps in the anomaly profiles. Whereas $(Ce/Ce^*)_N$ and La-corrected $(Ce/Ce^{**})_N$ anomalies are absent at the outer cortex edge, $(La/La^*)_N$ anomalies are slightly negative in the external-most ~280 μm (Figure S1). All three of these anomalies exhibit substantial positive and negative fluctuations across the transect.

Although $(Ce/Ce^*)_N$ anomalies fluctuate from ~0.2 to 20 across the transect, they are largely positive throughout most of the internal half of the transect (Figure S1). This trend is not reflected, however, in the whole-bone $(Ce/Ce^*)_N$ average for the tibia, which is essentially absent (1.04; Table 1). $(Ce/Ce^*)_N$ values were also plotted against $(Pr/Pr^*)_N$ values (following [84]) to aid us in differentiating true, redox-related cerium anomalies from apparent anomalies induced by variations in $(La/La^*)_N$ anomalies. The majority of $(Ce/Ce^*)_N$ values from the external 1 mm of the bone plot near the lower margins of fields 3a and 4a (Figure 6), reflective of slightly negative La anomalies in the external cortex (in agreement with Figure S1). In contrast, anomaly values from inner regions of the cortex plot over a broad range encompassing every field of the diagram (Figure 6), indicative of substantial heterogeneity at the sub-millimeter scale in the middle and internal cortices. Within this broad spectrum, there are relatively few data points in fields 1 and 2b (Figure 6); regions of the internal cortex represented by these data points lack a Ce anomaly.

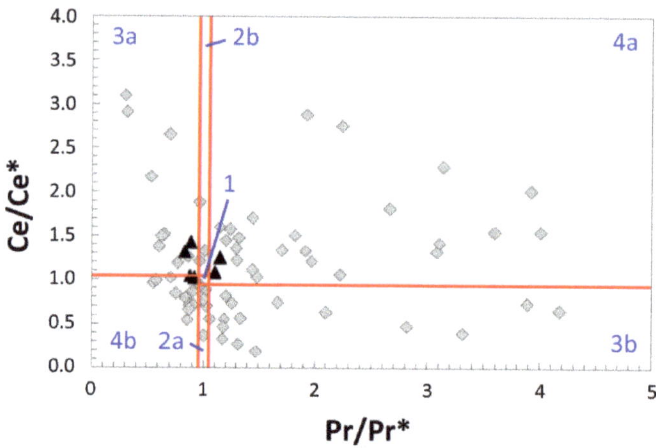

Figure 6. $(Ce/Ce^*)_N$ vs. $(Pr/Pr^*)_N$ plot (after [84]) of five-point averages along the transect across the cortex of MOR 2598 recorded via LA-ICPMS. Separate fields (labeled by blue text) are as follows: 1, neither Ce nor La anomaly; 2a, no Ce and positive La anomaly; 2b, no Ce and negative La anomaly; 3a, positive Ce and negative La anomaly; 3b, negative Ce and positive La anomaly; 4a, negative Ce and negative La anomaly; 4b, positive Ce and positive La anomaly. Measurements from the outer 1 mm of the external cortex are plotted as black triangles, and all measurements from deeper within the bone are plotted as gray diamonds. $(Ce/Ce^*)_N$ and $(Pr/Pr^*)_N$ anomalies calculated as in the Materials and Methods section of the text.

$(La/La^*)_N$ anomalies and La-corrected $(Ce/Ce^{**})_N$ anomalies were also directly calculated (see Methods) to quantitatively assess these qualitative inferences. Unfortunately, as mentioned above, frequent drops in concentrations of LREEs below the detection limit severely limit coverage in these profiles. However, based on Figure S1, it is clear that $(La/La^*)_N$ anomalies are exclusively negative in the internal 29 mm of the cortex. The average $(La/La^*)_N$ value across this region is 0.20. $(Ce/Ce^{**})_N$ anomalies are also almost exclusively negative in the middle and internal cortices, exhibiting a similarly low average (0.65) across this same span. The values of both of these anomalies fluctuate by roughly two orders of magnitude across the transect. As a whole, the tibia of MOR 2598 exhibits slightly positive $(Ce/Ce^{**})_N$ and $(La/La^*)_N$ anomalies (1.10 and 1.12, respectively; Table 1), but these are each clearly biased by overweighting of data from the external cortex (caused by abundant missing data from internal regions of the bone, as discussed above). Plotting $(Ce/Ce^{**})_N$ anomalies against U concentrations for each laser run yielded a poor correlation between these two redox-sensitive signatures ($r^2 = 0.29$; Figure S2).

The yttrium/holmium (Y/Ho) ratios are slightly above chondritic (26; [85]) in the outer ~10 mm of the bone and the innermost ~7 mm of the transect, wherein they range ~20–300. Though data are sporadic through the middle cortex due to very low concentrations, ratios from this region form a broad swale below these peaks in the external and internal cortices (Figure S1). Specifically, the average of the Y/Ho ratios through the central 15 mm of the cortex (10–25 mm along the transect) is 35, and values across this region mostly fall between ~10 and 80. These spatial contrasts partially negate one another when data are averaged for the entire transect, which yields a slightly positive whole-bone average anomaly of 43 (Table 1).

5. Discussion

5.1. MOR 2598's Paleoenvironmental and Taphonomic Context

Though limited taphonomic and stratigraphic data are available, the excellent preservation quality of MOR 2598 suggests it was protected by rapid burial postmortem. The full articulation of the hind limb, negligible signs of (ancient) weathering and abrasion, and excellent histological preservation (Figure 10B of [74]) each support the interpretation that burial took place within a few years postmortem, perhaps even just weeks after death [86]. However, the absence of the remainder of the skeleton implies this hind limb became disarticulated from the remainder of the carcass during brief subaerial decay, as well as the probable short-distance transport of the limb (from an upstream site of death) prior to burial (cf. [87,88]). The recovery of MOR 2598 from a channel sandstone [30] strongly suggests that: (1) decay primarily occurred subaqueously under oxygenated conditions (cf. [89]); (2) fluvial currents likely caused the separation of the limb from the body, and; (3) the carcass likely reached the "bloat" or active decay phase of postmortem decomposition for this to occur (cf. [90,91]). Ultimately, a lull in flow competency induced deposition and burial of the limb within the channel. Based on the great thickness of the succession of channel sandstone horizons from which MOR 2598 was recovered (7 m), it appears burial occurred within a well-established lowland channel rather than a recently-formed avulsion channel. This conclusion is consistent with prior interpretations of the Judith River Formation as generally representing lowland fluvial environments close to the coastline of the Western Interior Cretaceous Seaway (WIKS; [70,71]).

To briefly summarize, the available data reveal that this *Brachylophosaurus canadensis* individual (MOR 2598) died within or near a fluvial channel on the coastal lowlands. Its carcass experienced fairly brief decay within the channel, where currents eventually led to disarticulation of the left hind limb which was carried shortly downstream. Either an obstruction in the channel, a temporary lull in flow competency, or slowing of currents due to gradual channel broadening caused the deposition of the limb on the channel floor where it became quickly buried and fossilized within cross-stratified sands. Our trace element data provide illuminating insights into the ensuing diagenetic history of MOR 2598, which we now characterize in an effort to constrain geochemical pathways to cellular, soft tissue, and biomolecular preservation.

5.2. Reconstructing the Geochemical History of MOR 2598

The tibia of MOR 2598 exhibits low REE concentrations near the cortical margin (e.g., ~800 ppm for La; Data S1) and a low whole-bone ΣREE (256 ppm) compared to many other bones from the Cretaceous period (Table 2), which have been found to possess ΣREE ranging from 1110 to 25,000 ppm [58,63,92–95]. Notably, however, these values each fall within the range of other dinosaur bones we have examined from the Cretaceous Hell Creek Formation [63,64] which have also been found to yield endogenous proteins [28,29,47]. Compared to those specimens from the Hell Creek Formation, MOR 2598 exhibits lower average concentrations of Fe (0.94 wt. %), Mn (834 ppm), and Y (190 ppm), a higher concentration of U (51 ppm), and similar concentrations of Sr (2499 ppm), Ba (1448 ppm), and Lu (3 ppm). Although these comparisons do not take into account differences in taxon, cortical width, histology, or diagenetic regimes, they still reveal that the tibia of MOR 2598 is (for most elements examined) less chemically altered than the majority of fossil bones of similar age. We have previously attributed such cases of minimal alteration to various sequestration processes limiting the availability of trace element ions in early-diagenetic pore fluids (e.g., complexation with humic acids and/or dissolved carbonates [96–99] and coprecipitation with phosphates in entombing sediments [100–103]), and those processes may also account for the modest alteration of MOR 2598 (see Supplementary Materials for further discussion).

Table 2. Summary of the REE composition of the left tibia of *Brachylophosaurus canadensis* MOR 2598. Qualitative $\sum$REE content is based on the value shown in Table 1 (256 ppm) in comparison to values from other Mesozoic bones (as listed in the main text). Abbreviations: DMD, double medium diffusion *sensu* [80]; LREEs, light rare earth elements.

Clear DMD Kink for LREE?	Relative Noise in Outer Cortex for La	REE Suggest Flow in Marrow Cavity?	Relative $\sum$REE Content (Whole Bone)	Relative U Content (Whole Bone)	Relative Porosity of the Cortex
Yes	Moderate	Yes	Low	Moderate	Low

Ba, Fe, and Mn each exhibit flat concentration profiles (Figure 2C,D) probably indicative of incorporation into homogenously distributed, minute, secondary mineral phases, presumably barite, goethite, and Mn oxides [104]. Sr likely exhibits a similarly flat profile shape due to spatially homogenous substitution for Ca in bone hydroxyapatite [105,106]. In contrast to these more abundant elements, all REEs exhibit steep declines in concentrations from the cortical margin, with LREEs exhibiting the steepest declines and HREEs the shallowest (due to crystal–chemical controls based on ionic radius [107]). Meanwhile, MREE concentrations commonly drop below detection limit in the middle and internal cortices (Data S1). These trends are typical of fossil bones which experienced relatively brief uptake largely by simple 'external-to-internal' diffusion and did not equilibrate with external pore fluids during diagenesis (cf. [60,63,64,68,108]). However, clear kinks in concentration profiles for many REEs and substantial, locally-restricted variations in their concentrations in the external cortex (e.g., La in Figure 2A) also indicate at least partial uptake via double medium diffusion (*sensu* [80]) through Haversian canals.

Spider diagrams reveal even proportions of REEs within the external-most cortex (Figure 4A) yet significant relative HREE enrichment throughout the bone as a whole (Figure 3B,C and Figure 4B). Relative HREE enrichment is especially evident in the internal cortex, where, for example, concentrations of Yb rise to more than double those of La (Figure 2A). These signatures are very similar to, but less pronounced than, those observed in a *Tyrannosaurus rex* femur recovered from an estuarine channel sandstone in the Hell Creek Formation [64]. As with that specimen, it is likely that such HREE enrichment reflects protracted trace element uptake from relatively HREE-enriched brackish waters and/or diagenetic pore fluids under oxidizing conditions. This interpretation is supported by the whole-bone composition of this specimen being similar to those of lowland river waters, estuarine waters, and marine pore fluids (Figure 5A) which typically exhibit such relative HREE enrichment [55,92,109]. Interestingly, these findings strongly suggest that the channel in which MOR 2598 was interred was likely tidally influenced, which in turn suggests that it was recovered from an estuarine channel, not a (strictly speaking) fluvial channel—an insight not apparent from the sedimentology/stratigraphy of the quarry.

Regarding redox regimes through diagenesis, at the whole-bone level, the tibia exhibits a slightly positive $(Ce/Ce^{**})_N$ anomaly (1.10) reflective of a weakly oxidizing overall diagenetic history; this is consistent with the inferred burial setting having been an estuarine channel (see above). Generally high U concentrations throughout much of the cortex (Figure 2B and Table 1) and a relatively high average Sc concentration (59 ppm) corroborate this signal, as U and Sc enrichment have each been linked with uptake under oxidizing conditions [110–112]. The plotting of numerous data points from the middle and internal cortices in fields 3b and 4a of the $(Ce/Ce^*)_N$ vs. $(Pr/Pr^*)_N$ plot (Figure 6) also supports the presence of oxidizing conditions within the bone. However, both $(Ce/Ce^*)_N$ and $(Ce/Ce^{**})_N$ anomalies are essentially absent at the cortical margin, whereas their values fluctuate considerably (both positively and negatively) throughout the middle and internal cortices (Figure S1). These contrasts, as well as the broad distribution of data points in Figure 6, demonstrate the presence of considerable spatial heterogeneity in redox conditions throughout the bone through diagenesis, especially in the middle and internal cortices.

Although the trace element anomaly profiles in Figure S1 may seem somewhat stochastic, $(Ce/Ce^*)_N$ and $(Ce/Ce^{**})_N$ values are generally positive and negative, respectively, across the inner half of the transect. If these trends are taken as reliable records of redox regimes during early diagenesis/fossilization, as all indications appear to support (see below), then these anomaly trends would signify that uptake occurred under prevailingly reducing conditions within the interior of the bone. The development of reducing microenvironments within fossil bones is relatively common [104] due to the release of iron and hydrogen sulfide from decaying organics within a dysaerobic, enclosed space [113]. However, high Sc concentrations in the internal cortex (Figure 2B) appear incompatible with this interpretation (as this should be a product of uptake under oxidizing conditions, as discussed above). We attribute this apparent "contrast" to temporal changes in redox conditions in this region of the bone through early diagenesis. Specifically, these conflicting signals could arise via the significant uptake of Sc in the internal cortex under initially oxidizing conditions followed by a shift to reducing conditions, during which latter time the internal cortex secondarily acquired positive $(Ce/Ce^*)_N$ anomalies (and Sc ions remained sequestered by, as in adsorbed to, bone crystallites; also see the Supplementary Materials for further discussion of the peculiar shapes of Sc and U concentration profiles in MOR 2598).

The redox scenario just described would necessitate a supply of significant amounts of Sc to the interior of the bone. This would have to be supplied by a pore fluid percolating through the medullary cavity (after the decay of blood and other internal organics), which would presumably also supply numerous other trace elements to the internal cortex. The concentration profiles of HREEs (e.g., Yb in Figure 2A), U (Figure 2B), Y, Lu, and the latter half of the MREE series (Data S1) each exhibit increases toward the internal end of the transect (in the internal cortex), providing concrete evidence of uptake from a second diffusion front in the interior of the bone. That LREEs exhibit negligible rises in concentrations toward the internal end of the transect (e.g., La in Figure 2A) indicates that the majority of elements supplied by the pore fluid passing through the medullary cavity were mostly those with comparatively-modest to low diffusivities (based on [108]). This bias signifies that the pore fluid must have been a chemically 'evolved', highly-fractionated fluid which, based on the magnitude of select elemental enrichments in the internal cortex (e.g., Figure 2B), either flowed through the medullary cavity for an extended period of time or, more likely due to burial and compaction, became pooled there, allowing protracted uptake. It is also apparent that this pore fluid was likely not simply an HREE-enriched solution passing through during some later phase of late diagenesis because there are no clear signs of similar HREE enrichment in the external-most cortex (e.g., Figure 2A,B). Instead, the majority of elements exhibiting enrichment toward the internal end of the transect (e.g., U, Y, HREE) exhibit a subtle 'plateau' of stable concentrations in the outermost ~1 mm of the cortex followed by a ~80% increase near ~1.5–2 mm (Figure 2A–C). We interpret this pattern to reflect modest leaching of trace elements from the outermost ~1 mm of the external cortex, most likely during late diagenesis and under slightly oxidizing conditions (based on the weakly positive $(Ce/Ce^*)_N$ anomalies in this region; Figure S1). This conclusion may also be supported by: (1) the common presence of negative $(La/La^*)_N$ anomalies in the outermost ~500 µm of the external cortex (potentially reflective of near-surface loss of La; Figure S1), and; (2) a lack of a correlation between U concentrations and $(Ce/Ce^{**})_N$ anomalies for each laser run ($r^2 = 0.29$; Figure S2), implying uptake of U and REEs over differing timescales [112].

The absence of more major signs of leaching or late-diagenetic trace element uptake at the cortical edge (Figure 2), as well as the retention of clear evidence of spatiotemporal changes in pore fluid compositions (described in the last few paragraphs), imply that late-diagenetic overprinting of trace element signatures in MOR 2598 was not substantial, and, therefore, that the tibia at least partially retains early-diagenetic signatures. Indeed, there are numerous signs of relatively brief interaction with pore fluids. For example, a spider diagram (Figure 4B) and ternary plot (Figure 3D) of REE proportions by individual laser runs each reveal clear signs of significant fractionation during uptake from circum-neutral pH surface/groundwaters (cf. [114]) in the form of increasing relative LREE depletion/HREE enrichment with increasing cortical depth (as in, e.g., [59,107,115]). These fractionation effects are also evident from significant spatial variations in $(La/Yb)_N$ ratios (Figure 5B), a positive $(La/La^*)_N$ anomaly and Y/Ho ratio for the bone as a whole (1.12 and 43, respectively; cf. [60]), and differing concentration profile shapes (contrast Figure 2A,B) for U and REE (which should otherwise have similar shapes due to their similar diffusivities [108]). Similarly-high Y/Ho ratios in the external and internal cortices (Figure S1) further imply fractionation also occurred during uptake from the chemically 'evolved' pore fluid pooled in the medullary cavity.

To review, our trace element data thus identify that after brief subaqeous decay and transport down a fluvial stream, the hind limb of MOR 2598 became fossilized after burial in a sandy, oxic, estuarine channel (Figure 7). Its bones experienced relatively brief primary uptake of REEs and other trace elements from circum-neutral pH, HREE-enriched, and potentially brackish channel waters and groundwaters under oxidizing conditions during early diagenesis. Comparatively slower percolation of pore fluids through the medullary cavity of the tibia than around its exterior led to the development of reducing conditions inside the bone. Recent erosion, typical of that in desert/badlands environments from which most fossil bones are recovered, re-exposed MOR 2598 to oxidizing conditions and caused minor leaching of trace elements from the outermost ~1 mm of the cortex but had no significant effects on the overall chemistry of the bone. As a result, the specimen remains modestly altered compared to others of similar age.

5.3. Insights into Molecular Taphonomy from Comparative Geochemistry

MOR 2598 is only the third vertebrate fossil of pre-Cenozoic age to have both yielded endogenous protein and have its geochemical history characterized through trace element analyses. Both other specimens in this short list, namely *Tyrannosaurus rex* MOR 1125 [28,29,64] and *Edmontosaurus annectens* SRHS-DU-231 [47,63], are also large nonavian dinosaurs recovered from Late Cretaceous strata in Montana, USA. Adding MOR 2598 into this comparative framework reveals both geochemical similarities to, and differences from, these two other specimens which: (1) bolster prior hypotheses about protein preservation pathways; and (2) add new insights into the complexity of post-burial diagenetic alterations which biomolecules can withstand.

Taken as a whole, the biostratinomic history of *Brachylophosaurus* MOR 2598 is quite similar to that which we recently explicated for *T. rex* MOR 1125 [64]. After death and a short period of (likely subaqueuous) decay, brief fluvial transport brought each specimen into coastal estuaries along the western coast of the WIKS where they were rapidly buried in sandy estuarine channels, and throughout this history, the bones of each specimen were acquiring trace elements from brackish, HREE-enriched surface and groundwaters (this study and [64]). As at the MOR 1125 quarry [64], early cementation of the sediments entombing MOR 2598 appears to have limited trace element uptake by the fossil bones, allowing them to exhibit minimal alteration at the elemental level. This is evident in the low $\sum$REE of the tibia compared to many other bones of Cretaceous age (as discussed above), as well as its steep declines in REE concentrations from the cortical margin (Figure 2A) and very low concentrations of elements with ionic radii similar to that of Ca^{2+} (i.e., MREE) in the middle cortex (Data S1). Thus, MOR 2598 adds further support to the assertions of Schweitzer [116], Herwartz et al. [59], and Ullmann et al. [64] that: (1) early-diagenetic cementation of sediments can effectively thwart protracted decay and chemical alteration of bones after burial (presumably by minimizing exposure to percolating groundwaters and the exogenous microbes they carry with them), and; (2) this diagenetic pathway also facilitates rapid molecular stabilization (presumably via the iron free-radical-induced molecular crosslinking mechanism elucidated by Boatman et al. [53]). Fossils from the Standing Rock Hadrosaur Site (SRHS; [47,63,117]) demonstrate that rapid burial in fine-grained sediments with low-permeability and/or encasement in early-diagenetic concretion can similarly hinder the decay of endogenous cells, tissues, and their component biomolecules.

As for SRHS-DU-231 [63] and MOR 1125 [64], MOR 2598 exhibits high Sc enrichment and a slightly positive whole-bone $(Ce/Ce^{**})_N$ anomaly (Table 1) reflective of a generally oxidizing diagenetic history. Although this pattern superficially supports the proposition by Wiemann et al. [54] that oxidizing depositional environments may be more favorable settings for molecular preservation than reducing environments (perhaps due to greater release of crosslink-catalyzing iron free radicals; cf. [53]), it is clear that both MOR 2598 and bones from SRHS also experienced reducing conditions during diagenesis. This is evident from consistently-positive $(Ce/Ce^*)_N$ anomalies in the internal cortex of the tibia of MOR 2598 (Figure S1) and external cortices of many SRHS bones (Figure 6 of [63]). Microbial decay of organics (including those within bones) is known to decrease local pH and create dysoxic to anoxic conditions, thereby eliciting the production of reducing conditions [118], especially within confined microenvironments such as those within and around a carcass in compacted or low-permeability sediments (e.g., [50,119]) or within the internal pore spaces of bones (e.g., [104,120,121]). It is also well known that reducing early diagenetic conditions do not preclude exquisite morphologic and biochemical preservation of structural soft tissues, perhaps due to induced dysoxia/anoxia (e.g., [122–125]). Thus, there are reasons that fossil bones preserved under largely-reducing conditions may still yield original molecules.

Though the acidic pH that would have temporarily accompanied reducing conditions within the medullary cavity of this specimen may seem (at face value) preclusive to molecular preservation, weak acidity has been implicated in the rapid nucleation of inert, protective, microcrystalline goethite crystals within 'osteocytes' and 'blood vessels' recovered from fossil bones [4,53]. For this reason, initial redox conditions in the immediate ~48 h after death may ultimately be the most critical, as it has been demonstrated that inter- and intra-molecular crosslinking (i.e., stabilization) reactions can operate in this brief timeframe, promoting equilibration with the early-diagenetic environment which may then persist through fossilization and late diagenesis [53]. However, we must note that actualistic studies examining molecular stability through temporal shifts in redox regimes would be necessary to further evaluate this hypothesis. Regardless, our recognition of varied pH and redox conditions over time within the tibia of MOR 2598 is thus not incompatible with these prior studies, but rather augments them by revealing that biomolecular remains may survive multiple changes in redox regimes through diagenesis (Figure 7).

In particular, as discussed above, the bones of MOR 2598 experienced two comparatively 'extra' diagenetic events which MOR 1125 and bones from SRHS did not [63,64]: (1) protracted trace element uptake from stagnant pore fluids 'pooled' in the central medullary cavity (evidenced by elevated concentrations of HREEs, U, Sc, Y, and Lu in the internal cortex; Figure 2, Data S1), and; (2) modest late-diagenetic leaching of the cortical surface (evidenced by negative $(La/La^*)_N$ anomalies and reduced concentrations of many trace elements in the outermost ~0.5–1 mm; Figure 2 and Figure S1). Despite this complex diagenetic history, cortical bone from the femur of MOR 2598 has yielded numerous microstructures retaining endogenous peptide sequences of multiple proteins (namely collagen, actin, tubulin, histones, myosin, and tropomyosin [18,30,45]). This fact indicates that processes which stabilized diagenetiforms within this specimen in the initial hours to days postmortem imparted remarkable long-term resiliency. Novel experiments by Schwietzer et al. [4] and Boatman et al. [53] have begun to shed light on how this may occur, but testing of more fossils and further actualistic studies are needed to fully resolve the endurance of diagenetiforms under the wide array of physicochemical/thermodynamic regimes of natural diagenetic environments.

Finally, all three specimens examined in this discussion (MOR 2598, SRHS-DU-231, and MOR 1125) also each exhibit abundant signals of fractionation of REEs during uptake, which when combined with their low $\sum$REE content indicate at least partial retention of early-diagenetic trace element signatures. These signs include positive whole-bone Y/Ho anomalies (Table 1, and Table 1 in [63,64]), negative $(La/La^*)_N$ anomalies in the middle and internal cortices (Figures S1 and S2 of [64]), steeper concentration profiles for LREEs than HREEs (Figure 2A, Figure 2 of [63], and Figure 4A of [64]), and increasing relative-HREE enrichment with cortical depth (Figure 4B, Figure 4 of [63], and Figure 6B of [64]). As suggested by Ullmann et al. [64], the retention of such early-diagenetic signatures may constitute a useful proxy for molecular recovery potential because it indicates a specimen has avoided protracted interactions with any late-diagenetic pore fluids (e.g., phreatic groundwaters) which could plausibly cause hydrolysis and other decay processes.

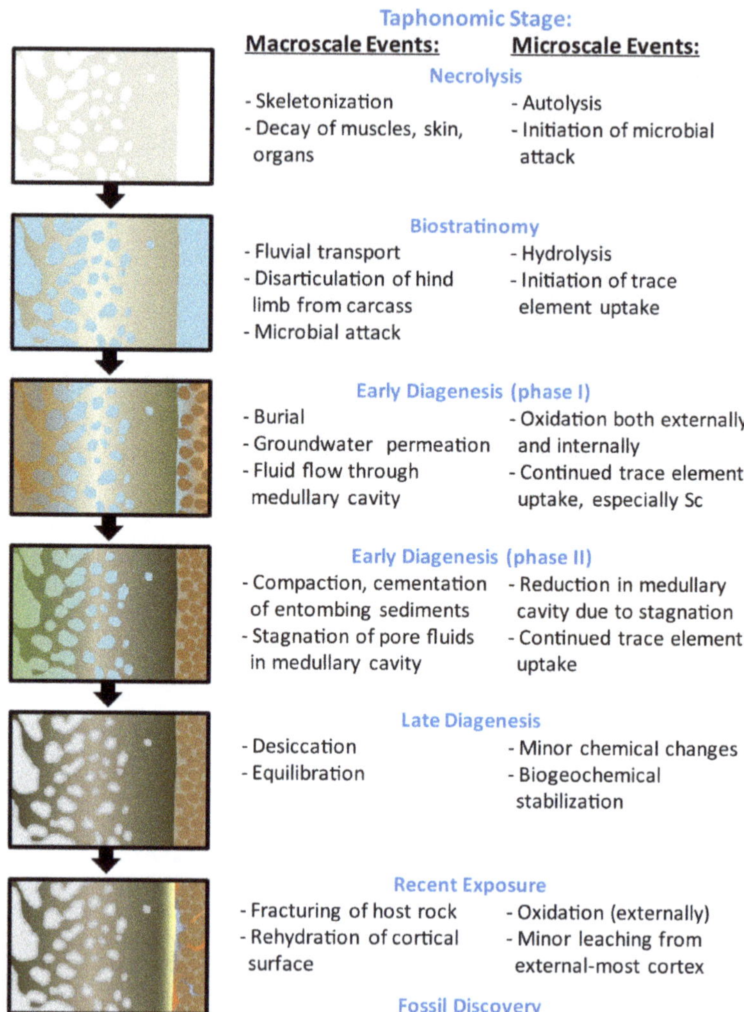

Figure 7. Reconstruction of the diagenetic history of *Brachylophosaurus canadensis* MOR 2598, summarizing macroscale and microscale events effecting its skeletal elements after death. Events are classified into approximate taphonomic stages, with a visual portrayal of the diagenetic processes affecting the bones during each stage shown at left.

6. Conclusions

Synthesizing our results with those of other recent experimental and actualistic studies in molecular taphonomy leads us to conclude the following:

1. By allowing the quick characterization of spatial patterns of diagenetic alteration within a fossil, trace element analyses constitute a useful and effective means of screening fossil tissues prior to paleomolecular analyses;

2. Retention of early-diagenetic trace element signatures may constitute a useful proxy for molecular recovery potential because it indicates a specimen has avoided protracted interactions with any late-diagenetic pore fluids;
3. Although burial in coarse, permeable sediments in oxic environments appears conducive to molecular preservation, it is also possible for fossils preserved in reducing paleoenvironments to yield endogenous molecules. This dichotomy suggests that the presence of oxidizing conditions in the initial ~48 h postmortem may be more key to molecular preservation than the redox state of the final setting of burial;
4. Rapid burial in fine-grained sediments with low permeability, encasement in early-diagenetic concretion, and/or early-diagenetic cementation of entombing sediments can effectively thwart protracted decay and chemical alteration of bones (and their component cells, tissues, and biomolecules) by minimizing exposure to percolating groundwaters and the exogenous microbes they carry with them, and;
5. Biomolecular remains may survive multiple changes in redox regimes through diagenesis, and this indicates that processes which stabilize biomolecules in the initial hours to days postmortem can impart remarkable long-term resiliency.

While these insights are obviously enlightening, it must be reiterated that they primarily derive from fossils from just three Cretaceous localities. In agreement with Schroeter et al. [126], we propose that paleomolecular and trace element analyses on Paleogene and Miocene fossils are direly needed to close the long window of the Cenozoic for which protein preservation has yet to be explored. Based on all data currently available, it seems very likely that future studies will continue to broaden the suite of depositional and diagenetic circumstances known to be conducive to molecular preservation.

Supplementary Materials: The following supporting information can be downloaded at: https://www.mdpi.com/article/10.3390/biology11081177/s1, Supplementary Materials narrative: A DOCX Word file including further details on our LA-ICPMS methods, discussion of potential tetrad effects in the left tibia of MOR 2598, discussion of potential sequestration processes that may have limited REE availability to the bones of MOR 2598, discussion of potential causes for the peculiar concentration profile shapes of U and Sc in the left tibia of MOR 2598, sources for environmental data in Figure 5, and additional data on the trace element composition of the left tibia of MOR 2598 (Figures S1 and S2). Data S1: Raw transect data acquired and analyzed in this study, provided as an XLSX Excel spreadsheet. References [127–189] are cited within the Supplementary Materials narrative.

Author Contributions: Conceptualization, P.V.U.; methodology, P.V.U.; formal analysis, P.V.U. and R.D.A.; investigation, P.V.U.; resources, J.B.S.; data curation, P.V.U. and J.B.S.; writing—original draft preparation, P.V.U.; writing—review and editing, P.V.U. and J.B.S.; visualization, P.V.U.; project administration, P.V.U.; funding acquisition, P.V.U. All authors have read and agreed to the published version of the manuscript.

Funding: This research was funded by Rowan University.

Institutional Review Board Statement: Not applicable.

Informed Consent Statement: Not applicable.

Data Availability Statement: All data generated by this study are available in this manuscript and the accompanying Supplementary Materials.

Acknowledgments: We thank Bob Harmon, Nels Peterson, and the rest of their Museum of the Rockies field crews for their efforts to excavate and collect this specimen over the course of two field seasons in 2006 to 2007. Gratitude is also extended to the Montana Department of Natural Resources and Conservation for access to the land where MOR 2598 was recovered. We also thank Ellen-Thérèse Lamm for specimen sampling and Elena Schroeter, Jack Horner, and Mary Schweitzer for helpful discussions.

Conflicts of Interest: The authors declare no conflict of interest.

References

1. Curry, G.B. Molecular palaeontology: New life for old molecules. *Trends Ecol. Evol.* **1987**, *2*, 161–165. [CrossRef]
2. Schweitzer, M.H. Soft tissue preservation in terrestrial Mesozoic vertebrates. *Annu. Rev. Earth Planet. Sci.* **2011**, *39*, 187–216. [CrossRef]
3. Hofreiter, M.; Collins, M.; Stewart, J.R. Ancient biomolecules in Quaternary palaeoecology. *Quat. Sci. Rev.* **2012**, *33*, 1–13. [CrossRef]
4. Schweitzer, M.H.; Zheng, W.; Cleland, T.P.; Goodwin, M.B.; Boatman, E.; Theil, E.; Marcus, M.A.; Fakra, S.C. A role for iron and oxygen chemistry in preserving soft tissues, cells and molecules from deep time. *Proc. R. Soc. B* **2014**, *281*, 20132741. [CrossRef]
5. Schweitzer, M.H.; Schroeter, E.R.; Cleland, T.P.; Zheng, W. Paleoproteomics of Mesozoic dinosaurs and other Mesozoic fossils. *Proteomics* **2019**, *19*, 1800251. [CrossRef]
6. Thomas, B.; Taylor, S. Proteomes of the past: The pursuit of proteins in paleontology. *Expert Rev. Proteom.* **2019**, *16*, 881–895. [CrossRef]
7. Bailleul, A.M.; Li, Z. DNA staining in fossil cells beyond the Quaternary: Reassessment of the evidence and prospects for an improved understanding of DNA preservation in deep time. *Earth-Sci. Rev.* **2021**, *216*, 103600. [CrossRef]
8. Orlando, L.; Bonjean, D.; Bocherens, H.; Thenot, A.; Argant, A.; Otte, M.; Hänni, C. Ancient DNA and the population genetics of cave bears (*Ursus spelaeus*) through space and time. *Mol. Biol. Evol.* **2002**, *19*, 1920–1933. [CrossRef]
9. Gilbert, M.T.P.; Drautz, D.I.; Lesk, A.M.; Ho, S.Y.W.; Qi, J.; Ratan, A.; Hsu, C.-H.; Sher, A.; Dalén, L.; Götherström, A.; et al. Intraspecific phylogenetic analysis of Siberian woolly mammoths using complete mitochondrial genomes. *Proc. Natl. Acad. Sci. USA* **2008**, *105*, 8327–8332. [CrossRef]
10. Miller, W.; Drautz, D.I.; Ratan, A.; Pusey, B.; Qi, J.; Lesk, A.M.; Tomsho, L.P.; Packard, M.D.; Zhao, F.; Sher, A.; et al. Sequencing the nuclear genome of the extinct woolly mammoth. *Nature* **2008**, *456*, 387–390. [CrossRef]
11. Cappellini, E.; Jensen, L.J.; Szklarczyk, D.; Ginolhac, A.; da Fonseca, R.A.R.; Stafford, T.W., Jr.; Holen, S.R.; Collins, M.J.; Orlando, L.; Willerslev, E.; et al. Proteomic analysis of a Pleistocene mammoth femur reveals more than one hundred ancient bone proteins. *J. Proteome Res.* **2012**, *11*, 917–926. [CrossRef] [PubMed]
12. van der Valk, T.; Pečnerová, P.; Díez-del-Molino, D.; Bergström, A.; Oppenheimer, J.; Hartmann, S.; Xenikoudakis, G.; Thomas, J.A.; Dehasque, M.; Sağlican, E.; et al. Million-year-old DNA sheds light on the genomic history of mammoths. *Nature* **2021**, *591*, 265–269. [CrossRef] [PubMed]
13. Orlando, L.; Ginolhac, A.; Zhang, G.; Froese, D.; Albrechtsen, A.; Stiller, M.; Schubert, M.; Cappellini, E.; Petersen, B.; Moltke, I.; et al. Recalibrating *Equus* evolution using the genome sequence of an early Middle Pleistocene horse. *Nature* **2013**, *499*, 74–78. [CrossRef]
14. Barnett, R.; Westbury, M.V.; Sandoval-Velasco, M.; Vieira, F.G.; Jeon, S.; Zazula, G.; Martin, M.D.; Ho, S.Y.W.; Mather, N.; Gopalakrishnan, S.; et al. Genomic adaptations and evolutionary history of the extinct scimitar-toothed cat, *Homotherium latidens*. *Curr. Biol.* **2020**, *30*, 5018–5025. [CrossRef] [PubMed]
15. Noonan, J.P.; Coop, G.; Kudaravalli, S.; Smith, D.; Krause, J.; Alessi, J.; Chen, F.; Platt, D.; Pääbo, S.; Pritchard, J.K.; et al. Sequencing and analysis of Neanderthal genomic DNA. *Science* **2006**, *314*, 1113–1118. [CrossRef]
16. Welker, F.; Ramos-Madrigal, J.; Gutenbrunner, P.; Mackie, M.; Tiwary, S.; Jersie-Christensen, R.R.; Chiva, C.; Dickinson, M.R.; Kuhlwilm, M.; de Manuel, M.; et al. The dental proteome of *Homo antecessor*. *Nature* **2020**, *580*, 235–238. [CrossRef]
17. Briggs, D.E.G.; Evershed, R.P.; Lockheart, M.J. The biomolecular paleontology of continental fossils. *Paleobiology* **2000**, *26*, 169–193. [CrossRef]
18. Cleland, T.P.; Schroeter, E.R.; Zamdborg, L.; Zheng, W.; Lee, J.E.; Tran, J.C.; Bern, M.; Duncan, M.B.; Lebleu, V.S.; Ahlf, D.R.; et al. Mass spectrometry and antibody-based characterization of blood vessels from *Brachylophosaurus canadensis*. *J. Proteome Res.* **2015**, *14*, 5252–5262. [CrossRef]
19. Pan, Y.; Zheng, W.; Moyer, A.E.; O'Connor, J.K.; Wang, M.; Zheng, X.; Wang, X.; Schroeter, E.R.; Zhou, Z.; Schweitzer, M.H. Molecular evidence of keratin and melanosomes in feathers of the Early Cretaceous bird *Eoconfuciusornis*. *Proc. Natl. Acad. Sci. USA* **2016**, *113*, E7900–E7907. [CrossRef]
20. Pan, Y.; Zheng, W.; Sawyer, R.H.; Pennington, M.W.; Zheng, X.; Wang, X.; Wang, M.; Hu, L.; O'Connor, J.; Zhao, T.; et al. The molecular evolution of feathers with direct evidence from fossils. *Proc. Natl. Acad. Sci. USA* **2019**, *116*, 3018–3023. [CrossRef]
21. Lindgren, J.; Kuriyama, T.; Madsen, H.; Sjövall, P.; Zheng, W.; Uvdal, P.; Engdahl, A.; Moyer, A.E.; Gren, J.A.; Kamezaki, N.; et al. Biochemistry and adaptive colouration of an exceptionally preserved juvenile fossil sea turtle. *Sci. Rep.* **2017**, *7*, 13324. [CrossRef] [PubMed]
22. Lindgren, J.; Sjövall, P.; Thiel, V.; Zheng, W.; Ito, S.; Wakamatsu, K.; Hauff, R.; Kear, B.P.; Engdahl, A.; Alwmark, C.; et al. Soft-tissue evidence for homeothermy and crypsis in a Jurassic ichthyosaur. *Nature* **2018**, *564*, 359–365. [CrossRef] [PubMed]
23. Ambler, R.P.; Daniel, M. Proteins and molecular paleontology. *Philos. Trans. R. Soc. B* **1991**, *333*, 381–389.
24. Baird, R.F.; Rowley, M.J. Preservation of avian collagen in Australian Quaternary cave deposits. *Palaeontology* **1990**, *33*, 447–451.
25. Wojtowicz, A.; Yamauchi, M.; Montella, A.; Bandiera, P.; Sotowski, R.; Ostrowski, K. Persistence of bone collagen cross-links in skeletons of the Nuraghi population living in Sardinia 1500-1200 B.C. *Calcif. Tissue Int.* **1999**, *64*, 370–373. [CrossRef] [PubMed]
26. Tuross, N. Alterations in fossil collagen. *Archaeometry* **2002**, *44*, 427–434. [CrossRef]
27. Avci, R.; Schweitzer, M.H.; Boyd, R.D.; Wittmeyer, J.L.; Arce, F.T.; Calvo, J.O. Preservation of bone collagen from the Late Cretaceous period studied by immunological techniques and atomic force microscopy. *Langmuir* **2005**, *21*, 3584–3590. [CrossRef]

28. Asara, J.M.; Schweitzer, M.H.; Freimark, L.M.; Phillips, M.; Cantley, L.C. Protein sequences from *Mastodon* and *Tyrannosaurus rex* revealed by mass spectrometry. *Science* **2007**, *316*, 280–285. [CrossRef]

29. Schweitzer, M.H.; Suo, Z.; Avci, R.; Asara, J.M.; Allen, M.A.; Arce, F.T.; Horner, J.R. Analyses of soft tissue from *Tyrannosaurus rex* suggest the presence of protein. *Science* **2007**, *316*, 277–280. [CrossRef]

30. Schweitzer, M.H.; Zheng, W.; Organ, C.L.; Avci, R.; Suo, Z.; Freimark, L.M.; Lebleu, V.S.; Duncan, M.B.; Vander Heiden, M.G.; Neveu, J.M.; et al. Biomolecular characterization and protein sequences of the Campanian hadrosaur *B. canadensis*. *Science* **2009**, *324*, 626–631. [CrossRef]

31. Lingham-Soliar, T.; Wesley-Smith, J. First investigation of the collagen D-band ultrastructure in fossilized vertebrate integument. *Proc. R. Soc. B* **2008**, *275*, 2207–2212. [CrossRef] [PubMed]

32. Buckley, M.; Collins, M.; Thomas-Oates, J.; Wilson, J.C. Species identification by analysis of bone collagen using matrix-assisted laser desorption/ionisation time-of-flight mass spectrometry. *Rapid Commun. Mass Spectrom.* **2009**, *23*, 3843–3854. [CrossRef] [PubMed]

33. Buckley, M.; Kansa, S.W.; Howard, S.; Campbell, S.; Thomas-Oates, J.; Collins, M. Distinguishing between archaeological sheep and goat bones using a single collagen peptide. *J. Archaeol. Sci.* **2010**, *37*, 13–20. [CrossRef]

34. Buckley, M.; Larkin, N.; Collins, M. Mammoth and *Mastodon* collagen sequences; survival and utility. *Geochim. Cosmochim. Acta* **2011**, *75*, 2007–2016. [CrossRef]

35. Buckley, M.; Harvey, V.L.; Orihuela, J.; Mychajliw, A.M.; Keating, J.N.; Milan, J.N.A.; Lawless, C.; Chamberlain, A.T.; Egerton, V.M.; Manning, P.L. Collagen sequence analysis reveals evolutionary history of extinct West Indies *Nesophontes* (island-shrews). *Mol. Biol. Evol.* **2020**, *37*, 2931–2943. [CrossRef] [PubMed]

36. Dobberstein, R.C.; Collins, M.J.; Craig, O.E.; Taylor, G.; Penkman, K.E.H.; Ritz-Timme, S. Archaeological collagen: Why worry about collagen diagenesis? *Archaeol. Anthrop. Sci.* **2009**, *1*, 31–42. [CrossRef]

37. Lindgren, J.; Uvdal, P.; Engdahl, A.; Lee, A.H.; Alwmark, C.; Bergquist, K.-E.; Nilsson, E.; Ekström, P.; Rasmussen, M.; Douglas, D.A.; et al. Microspectroscopic Evidence of Cretaceous Bone Proteins. *PLoS ONE* **2011**, *6*, e19445. [CrossRef]

38. Antonio, J.D.S.; Schweitzer, M.H.; Jensen, S.T.; Kalluri, R.; Buckley, M.; Orgel, J.P.R.O. Dinosaur peptides suggest mechanisms of protein survival. *PLoS ONE* **2011**, *6*, e20381. [CrossRef]

39. Zylberberg, L.; Laurin, M. Analysis of fossil bone organic matrix by transmission electron microscopy. *C. R. Palevol* **2011**, *10*, 357–366. [CrossRef]

40. Wang, S.-Y.; Cappellini, E.; Zhang, H.-Y. Why collagens best survived in fossils? Clues from amino acid thermal stability. *Biochem. Biophys. Res. Commun.* **2012**, *422*, 5–7. [CrossRef]

41. Buckley, M. Ancient collagen reveals evolutionary history of the endemic South American 'ungulates'. *Proc. R. Soc. B* **2015**, *282*, 20142671. [CrossRef] [PubMed]

42. Cleland, T.P.; Schroeter, E.R.; Feranec, R.S.; Vashishth, D. Peptide sequences from the first *Castoroides ohioensis* skull and the utility of old museum collections for palaeoproteomics. *Proc. R. Soc. B* **2016**, *283*, 20160593. [CrossRef] [PubMed]

43. Hill, R.C.; Wither, M.J.; Nemkov, T.; Barrett, A.; D'Alessandro, A.; Dzieciatkowska, M.; Hansen, K.C. Preserved proteins from extinct *Bison latifrons* identified by tandem mass spectrometry; hydroxlysine glycosides are a common feature of ancient collagen. *Mol. Cell. Proteom.* **2015**, *14*, 1946–1958. [CrossRef]

44. Lee, Y.-C.; Chiang, C.-C.; Huang, P.-Y.; Chung, C.-Y.; Huang, T.D.; Wang, C.-C.; Chen, C.-I.; Chang, R.-S.; Liao, C.-H.; Reisz, R.R. Evidence of preserved collagen in an Early Jurassic sauropodomorph dinosaur revealed by synchrotron FTIR microspectroscopy. *Nat. Commun.* **2017**, *8*, 14220. [CrossRef]

45. Schroeter, E.R.; DeHart, C.J.; Cleland, T.P.; Zheng, W.; Thomas, P.M.; Kelleher, N.L.; Bern, M.; Schweitzer, M.H. Expansion for the *Brachylophosaurus canadensis* collagen I sequence and additional evidence of the preservation of Cretaceous protein. *J. Proteome Res.* **2017**, *16*, 920–932. [CrossRef] [PubMed]

46. Anné, J.; Edwards, N.P.; Brigidi, F.; Gueriau, P.; Harvey, V.L.; Geraki, K.; Slimak, L.; Buckley, M.; Wogelius, R.A. Advances in bone preservation: Identifying possible collagen preservation using sulfur speciation mapping. *Palaeogeogr. Palaeoclimatol. Palaeoecol.* **2019**, *520*, 181–187. [CrossRef]

47. Ullmann, P.V.; Voegele, K.K.; Grandstaff, D.E.; Ash, R.D.; Zheng, W.; Schroeter, E.R.; Schweitzer, M.H.; Lacovara, K.J. Molecular tests support the viability of rare earth elements as proxies for fossil biomolecule preservation. *Sci. Rep.* **2020**, *10*, 15566. [CrossRef]

48. Cleland, T.P.; Schroeter, E.R.; Colleary, C. Diagenetiforms: A new term to explain protein changes as a result of diagenesis in paleoproteomics. *J. Proteom.* **2021**, *230*, 103992. [CrossRef]

49. Organ, C.L.; Schweitzer, M.H.; Zheng, W.; Freimark, L.M.; Cantley, L.C.; Asara, J.M. Molecular phylogenetics of *Mastodon* and *Tyrannosaurus rex*. *Science* **2008**, *320*, 499. [CrossRef]

50. Allison, P.A. The role of anoxia in the decay and mineralization of proteinaceous macro-fossils. *Paleobiology* **1988**, *14*, 139–154. [CrossRef]

51. Eglinton, G.; Logan, G.A. Molecular preservation. *Philos. Trans. R. Soc. Lond. B* **1991**, *333*, 315–328.

52. Colleary, C.; Lamadrid, H.M.; O'Reilly, S.S.; Dolocan, A.; Nesbitt, S.J. Molecular preservation in mammoth bone and variation based on burial environment. *Sci. Rep.* **2021**, *11*, 2662. [CrossRef] [PubMed]

53. Boatman, E.M.; Goodwin, M.B.; Holman, H.-Y.N.; Fakra, S.; Zheng, W.; Gronsky, R.; Schweitzer, M.H. Mechanisms of soft tissue and protein preservation in *Tyrannosaurus rex*. *Sci. Rep.* **2019**, *9*, 15678. [CrossRef] [PubMed]

54. Wiemann, J.; Fabbri, M.; Yang, T.-R.; Stein, K.; Sander, P.M.; Norell, M.A.; Briggs, D.E.G. Fossilization transforms vertebrate hard tissue proteins into N-heterocyclic polymers. *Nat. Commun.* **2018**, *9*, 4741. [CrossRef] [PubMed]

55. Trueman, C.N. Trace element geochemistry of bonebeds. In *Bonebeds: Genesis, Analysis, and Paleobiological Significance*; Rogers, R.R., Eberth, D.A., Fiorillo, A.R., Eds.; University of Chicago Press: Chicago, IL, USA, 2007; pp. 397–435.

56. Trueman, C.N.; Behrensmeyer, A.K.; Tuross, N.; Weiner, S. Mineralogical and compositional changes in bones exposed on soil surfaces in Amboseli National Park, Kenya: Diagenetic mechanisms and the role of sediment pore fluids. *J. Archaeol. Sci.* **2004**, *31*, 721–739. [CrossRef]

57. Suarez, C.A.; Macpherson, G.L.; González, L.A.; Grandstaff, D.E. Heterogeneous rare earth element (REE) patterns and concentrations in a fossil bone: Implications for the use of REE in vertebrate taphonomy and fossilization history. *Geochim. Cosmochim. Acta* **2010**, *74*, 2970–2988. [CrossRef]

58. Suarez, C.A.; Morschhauser, E.M.; Suarez, M.B.; You, H.; Li, D.; Dodson, P. Rare earth element geochemistry of bone beds from the Lower Cretaceous Zhonggou Formation of Gansu Province, China. *J. Vertebr. Paleontol.* **2019**, *38*, 22–35. [CrossRef]

59. Herwartz, D.; Tütken, T.; Münker, C.; Jochum, K.P.; Stoll, B.; Sander, P.M. Timescales and mechanisms of REE and Hf uptake in fossil bones. *Geochim. Cosmochim. Acta* **2011**, *75*, 82–105. [CrossRef]

60. Herwartz, D.; Münker, C.; Tütken, T.; Hoffmann, J.E.; Wittke, A.; Barbier, B. Lu–Hf isotope systematics of fossil biogenic apatite and their effects on geochronology. *Geochim. Cosmochim. Acta* **2013**, *101*, 328–343. [CrossRef]

61. Kowal-Linka, M.; Jochum, K.P.; Surmik, D. LA-ICP-MS analysis of rare earth elements in marine reptile bones from the Middle Triassic bonebed (Upper Silesia, S Poland): Impact of long-lasting diagenesis, and factors controlling the uptake. *Chem. Geol.* **2014**, *363*, 213–228. [CrossRef]

62. Decrée, S.; Herwartz, D.; Mercadier, J.; Miján, I.; de Buffrénil, V.; Leduc, T.; Lambert, O. The post-mortem history of a bone revealed by its trace element signature: The case of a fossil whale rostrum. *Chem. Geol.* **2018**, *477*, 137–150. [CrossRef]

63. Ullmann, P.V.; Grandstaff, D.E.; Ash, R.D.; Lacovara, K.J. Geochemical taphonomy of the Standing Rock Hadrosaur Site: Exploring links between rare earth elements and cellular and soft tissue preservation. *Geochim. Cosmochim. Acta* **2020**, *269*, 223–237. [CrossRef]

64. Ullmann, P.V.; Macauley, K.; Ash, R.D.; Shoup, B.; Scannella, J.B. Taphonomic and diagenetic pathways to protein preservation, part I: The case of *Tyrannosaurus rex* specimen MOR 1125. *Biology* **2021**, *10*, 1193. [CrossRef]

65. McCormack, J.M.; Bahr, A.; Gerdes, A.; Tütken, T.; Prinz-Grimm, P. Preservation of successive diagenetic stages in Middle Triassic bonebeds: Evidence from in situ trace element and strontium isotope analysis of vertebrate fossils. *Chem. Geol.* **2015**, *410*, 108–123. [CrossRef]

66. Grimstead, D.N.; Clark, A.E.; Rezac, A. Uranium and Vanadium Concentrations as a Trace Element Method for Identifying Diagenetically Altered Bone in the Inorganic Phase. *J. Archaeol. Method. Theory* **2018**, *25*, 689–704. [CrossRef]

67. Kocsis, L.; Gheerbrant, E.; Mouflih, M.; Cappetta, H.; Ulianov, A.; Chiaradia, M.; Bardet, N. Gradual changes in upwelled seawater conditions (redox, pH) from the late Cretaceous through early Paleogene at the northwest coast of Africa: Negative Ce anomaly trend recorded in fossil bio-apatite. *Chem. Geol.* **2016**, *421*, 44–54. [CrossRef]

68. Trueman, C.N.; Palmer, M.R.; Field, J.; Privat, K.; Ludgate, N.; Chavagnac, V.; Eberth, D.A.; Cifelli, R.; Rogers, R.R. Comparing rates of recrystallisation and the potential for preservation of biomolecules from the distribution of trace elements in fossil bones. *C. R. Palevol* **2008**, *7*, 145–158. [CrossRef]

69. Paul, G.S.; Persons, W.S.; Van Raalte, J. The tyrant lizard king, queen and emperor: Multiple lines of morphological and stratigraphic evidence support subtle evolution and probable speciation within the North American genus *Tyrannosaurus*. *Evol. Biol.* **2022**, *49*, 156–179. [CrossRef]

70. Rogers, R.R.; Kidwell, S.M. Associations of vertebrate skeletal concentrations and discontinuity surfaces in terrestrial and shallow marine records: A test in the Cretaceous of Montana. *J. Geol.* **2000**, *108*, 131–154. [CrossRef]

71. Rogers, R.R.; Kidwell, S.M.; Deino, A.L.; Mitchell, J.P.; Nelson, K.; Thole, J.T. Age, correlation, and lithostratigraphic revision of the Upper Cretaceous (Campanian) Judith River Formation in its type area (north-central Montana), with a comparison of low- and high-accommodation alluvial records. *J. Geol.* **2016**, *124*, 99–135. [CrossRef]

72. Elorza, J.; Astibia, H.; Murelaga, X.; Pereda-Superbiola, X. Francolite as a diagenetic mineral in dinosaur and other Upper Cretaceous reptile bones (Laño, Iberian Peninsula): Microstructural, petrological and geochemical features. *Cretaceous Res.* **1999**, *20*, 169–187. [CrossRef]

73. Keenan, S.W. From bone to fossil: A review of the diagenesis of bioapatite. *Am. Miner.* **2016**, *101*, 1943–1951. [CrossRef]

74. Schweitzer, M.H.; Moyer, A.E.; Zheng, W. Testing the hypothesis of biofilm as a source for soft tissue and cell-like structures preserved in dinosaur bone. *PLoS ONE* **2016**, *11*, e0150238. [CrossRef]

75. LaRock, J.W. Sedimentology and Taphonomy of a Dinosaur Bonebed from the Upper Cretaceous (Campanian) Judith River Formation of North Central Montana. Master's Thesis, Montana State University, Bozeman, MT, USA, 2000.

76. Horner, J.R.; de Ricglès, A.; Padian, K. Variation in dinosaur skeletochronlogy indicators: Implications for age assessment and physiology. *Paleobiology* **1999**, *25*, 295–304. [CrossRef]

77. Horner, J.R.; De Ricqlès, A.; Padian, K. Long bone histology of the hadrosaurid dinosaur *Maiasaura peeblesorum*: Growth dynamics and physiology based on an ontogenetic series of skeletal elements. *J. Vertebr. Paleontol.* **2000**, *20*, 115–129. [CrossRef]

78. Haskin, L.A.; Haskin, M.A.; Frey, F.A.; Wildeman, T.R. Relative and absolute terrestrial abundances of the rare earths. In *Origin and Distribution of the Elements*; Ahrens, L.H., Ed.; Pergamon: Oxford, UK, 1968; pp. 889–912.

79. Gromet, L.P.; Dymek, R.F.; Haskin, L.A.; Korotev, R.L. The "North American shale composite": Its compilation, major and trace element characteristics. *Geochim. Cosmochim. Acta* **1984**, *48*, 2469–2482. [CrossRef]

80. Kohn, M. Models of diffusion-limited uptake of trace elements in fossils and rates of fossilization. *Geochim. Cosmochim. Acta* **2008**, *72*, 3758–3770. [CrossRef]

81. Kemp, R.A.; Trueman, C.N. Rare earth elements in Solnhofen biogenic apatite: Geochemical clues to the paleoenvironment. *Sediment. Geol.* **2003**, *155*, 109–127. [CrossRef]

82. Smirnova, E.V.; Mysovskaya, I.N.; Lozhkin, V.I.; Sandimirova, G.P.; Pakhomova, N.N.; Smagunova, A.A. Spectral interferences from polyatomic barium ions in inductively coupled plasma mass spectrometry. *J. Appl. Spectrosc.* **2006**, *73*, 911–917. [CrossRef]

83. Grandstaff, D.; Terry, D. Rare earth element composition of Paleogene vertebrate fossils from Toadstool Geologic Park, Nebraska, USA. *Appl. Geochem.* **2009**, *24*, 733–745. [CrossRef]

84. Bau, M.; Dulski, P. Anthropogenic origin of positive gadolinium anomalies in river waters. *Earth Planet. Sci. Lett.* **1996**, *143*, 245–255. [CrossRef]

85. Pack, A.; Russell, S.S.; Shelley, J.M.G.; van Zuilen, M. Geo- and cosmochemistry of the twin elements yttrium and holmium. *Geochim. Cosmochim. Acta* **2007**, *71*, 4592–4608. [CrossRef]

86. Behrensmeyer, A.K. Taphonomic and ecologic information from bone weathering. *Paleobiology* **1978**, *4*, 150–162. [CrossRef]

87. Hill, A.; Behrensmeyer, A.K. Disarticulation patterns of some modern East African mammals. *Paleobiology* **1984**, *10*, 366–376. [CrossRef]

88. Brand, L.R.; Hussey, M.; Taylor, J. Decay and disarticulation of small vertebrates in controlled experiments. *J. Taphon.* **2003**, *1*, 69–95.

89. Rogers, R.R. Taphonomy of three dinosaur bone beds in the Upper Cretaceous Two Medicine Formation of northwestern Montana: Evidence for erought-related mortality. *Palaios* **1990**, *5*, 394–413. [CrossRef]

90. Payne, J.A. A summer carrion study of the baby pig *Sus scrofa* Linnaeus. *Ecology* **1965**, *46*, 592–602. [CrossRef]

91. Syme, C.E.; Salisbury, S.W. Patterns of aquatic decay and disarticulation in juvenile Indo-Pacific crocodiles (*Crocodylus porosus*), and implications for the taphonomic interpretation of fossil crocodyliform material. *Palaeogeogr. Palaeoclimatol. Palaeoecol.* **2014**, *412*, 108–123. [CrossRef]

92. Trueman, C.N. Rare Earth Element Geochemistry and Taphonomy of Terrestrial Vertebrate Assemblages. *Palaios* **1999**, *14*, 555–568. [CrossRef]

93. Suarez, C.A.; Suarez, M.B.; Terry, D.O.; Grandstaff, D.E. Rare earth element geochemistry and taphonomy of the Early Cretaceous Crystal Geyser Dinosaur Quarry, east-central Utah. *Palaios* **2007**, *22*, 500–512. [CrossRef]

94. Rogers, R.R.; Fricke, H.C.; Addona, V.; Canavan, R.R.; Dwyer, C.N.; Harwood, C.L.; Koenig, A.E.; Murray, R.; Thole, J.T.; Williams, J. Using laser ablation-inductively coupled plasma-mass spectrometry (LA-ICP-MS) to explore geochemical taphonomy of vertebrate fossils in the Upper Cretaceous Two Medicine and Judith River Formations of Montana. *Palaios* **2010**, *25*, 183–195. [CrossRef]

95. Suarez, C.A.; Kohn, M.J. Caught in the act: A case study on microscopic scale physicochemical effects of fossilization on stable isotopic composition of bone. *Geochim. Cosmochim. Acta* **2020**, *268*, 277–295. [CrossRef]

96. Cantrell, K.J.; Byrne, R.H. Rare earth element complexation by carbonate and oxalate ions. *Geochim. Cosmochim. Acta* **1987**, *51*, 597–605. [CrossRef]

97. Zhang, J.; Huang, W.W.; Létolle, R.; Jusserand, C. Major element chemistry of the Huanghe (Yellow River), China—Weathering processes and chemical fluxes. *J. Hydrol.* **1995**, *168*, 173–203. [CrossRef]

98. Luo, Y.-R.; Byrne, R.H. Carbonate complexation of yttrium and the rare earth elements in natural waters. *Geochim. Cosmochim. Acta* **2004**, *68*, 691–699. [CrossRef]

99. Pourret, O.; Davranche, M.; Gruau, G.; Dia, A. Rare earth elements complexation with humic acid. *Chem. Geol.* **2007**, *243*, 128–141. [CrossRef]

100. Byrne, R.H.; Liu, X.; Schijf, J. The influence of phosphate coprecipitation on rare earth distributions in natural waters. *Geochim. Cosmochim. Acta* **1996**, *60*, 3341–3346. [CrossRef]

101. Föllmi, K.B. The phosphorus cycle, phosphogenesis and marine phosphate-rich deposits. *Earth-Sci. Rev.* **1996**, *40*, 55–124. [CrossRef]

102. Liu, X.; Byrne, R.H.; Schijf, J. Comparative coprecipitation of phosphate and arsenate with yttrium and the rare earths: The influence of solution complexation. *J. Solut. Chem.* **1997**, *26*, 1187–1198. [CrossRef]

103. Brunet, R.-C.; Astin, K.B. Variation in phosphorus flux during a hydrological season: The River Ardour. *Water Res.* **1998**, *32*, 547–558. [CrossRef]

104. Wings, O. Authigenic minerals in fossil bones from the Mesozoic of England: Poor correlation with depositional environ-ments. *Palaeogeogr. Palaeoclimatol.* **2004**, *204*, 15–32. [CrossRef]

105. Janssens, K.; Vincze, L.; Vekemans, B.; Williams, C.T.; Radtke, M.; Haller, M.; Knochel, A. The non-destructive determination of REE in fossilized bone using synchrotron radiation induced K-line X-ray microfluorescence analysis. *Anal. Bioanal. Chem.* **1999**, *363*, 413–420. [CrossRef]

106. Hinz, E.A.; Kohn, M.J. The effect of tissue structure and soil chemistry on trace element uptake in fossils. *Geochim. Cosmochim. Acta* **2010**, *74*, 3213–3231. [CrossRef]

107. Trueman, C.N.; Kocsis, L.; Palmer, M.R.; Dewdney, C. Fractionation of rare earth elements within bone mineral: A natural cation exchange system. *Palaeogeogr. Palaeoclim. Palaeoecol.* **2011**, *310*, 124–132. [CrossRef]

108. Kohn, M.J.; Moses, R.J. Trace element diffusivities in bone rule out simple diffusive uptake during fossilization but explain in vivo uptake and release. *Proc. Natl. Acad. Sci. USA* **2013**, *110*, 419–424. [CrossRef]

109. Elderfield, H.; Upstill-Goddard, R.; Sholkovitz, E.R. The rare earth elements in rivers, estuaries, and coastal seas and their significance to the composition of ocean waters. *Geochim. Cosmochim. Acta* **1990**, *54*, 971–991. [CrossRef]

110. Williams, C.T.; Potts, P.J. Research notes and application reports element distribution maps in fossil bones. *Archaeometry* **1988**, *30*, 237–247. [CrossRef]

111. Williams, C.T.; Henderson, P.; Marlow, C.A.; Molleson, T.I. The environment of deposition indicated by the distribution of rare earth elements in fossil bones from Olduvai Gorge, Tanzania. *Appl. Geochem.* **1997**, *12*, 537–547. [CrossRef]

112. Metzger, C.A.; Terry, D.O.; Grandstaff, D.E. Effect of paleosol formation on rare earth element signatures in fossil bone. *Geology* **2004**, *32*, 497–500. [CrossRef]

113. Brett, C.E.; Baird, G.C. Comparative Taphonomy: A Key to Paleoenvironmental interpretation based on fossil preservation. *Palaios* **1986**, *1*, 207–227. [CrossRef]

114. Martin, J.E.; Patrick, D.; Kihm, A.J.; Foit, F.F., Jr.; Grandstaff, D.E. Lithostratigraphy, tephrochronology, and rare earth element geochemistry of fossils at the classical Pleistocene Fossil Lake area, south central Oregon. *J. Geol.* **2005**, *113*, 139–155. [CrossRef]

115. Ferrante, C.; Cavin, L.; Vennemann, T.; Martini, R. Histology and geochemistry of *Allosaurus* (Dinosauria: Theropoda) from the Cleveland-Lloyd Dinosaur Quarry (Late Jurassic, Utah): Paleobiological implications. *Front. Earth Sci.* **2021**, *9*, 641060. [CrossRef]

116. Schweitzer, M. Molecular paleontology: Some current advances and problems. *Ann. Paléontol.* **2004**, *90*, 81–102. [CrossRef]

117. Ullmann, P.V.; Pandya, S.H.; Nellermoe, R. Patterns of soft tissue and cellular preservation in relation to fossil bone micro-structure and overburden depth at the Standing Rock Hadrosaur Site, Maastrichtian Hell Creek Formation, South Dakota, USA. *Cretaceous Res.* **2019**, *99*, 1–13. [CrossRef]

118. Allison, P.A. Decay processes. In *Paleobiology: A Synthesis*; Briggs, D.E.G., Crowther, P.R., Eds.; Blackwell Scientific Publications: Oxford, UK, 1990; pp. 213–216.

119. Martill, D.M. Macromolecular resolution of fossilized muscle tissue from an elopomorph fish. *Nature* **1990**, *346*, 171–172. [CrossRef]

120. Hubert, J.F.; Panish, P.T.; Chure, D.J.; Prostak, K.S. Chemistry, microstructure, petrology, and diagenetic model of Jurassic dinosaur bones, Dinosaur National Monument, Utah. *J. Sediment. Res.* **1996**, *66*, 531–547.

121. Pfretzschner, H.-U. Pyrite in fossil bone. *Neues Jahrb. Für Geol. Und Paläontologie* **2001**, *220*, 1–23. [CrossRef]

122. Franzen, J.L. Exceptional preservation of Eocene vertebrates in the lake deposit of Grube Messel (West Germany). *Philos. Trans. R. Soc. B* **1985**, *311*, 181–186.

123. McNamara, M.E.; Orr, P.J.; Kearns, S.L.; Alcalá, L.; Anadón, P.; Peñalver-Mollá, E. High-fidelity organic preservation of bone marrow in ca. 10 Ma amphibians. *Geology* **2006**, *34*, 641–644. [CrossRef]

124. McNamara, M.E.; Orr, P.J.; Kearns, S.L.; Alcalá, L.; Anadon, P.; Molla, E.P. Soft-tissue preservation in Miocence frogs from Libros, Spain: Insights into the genesis of decay microenvironments. *Palaios* **2009**, *24*, 104–117. [CrossRef]

125. Manning, P.; Morris, P.M.; McMahon, A.; Jones, E.; Gize, A.; Macquaker, J.H.S.; Wolff, G.; Thompson, A.; Marshall, J.; Taylor, K.G.; et al. Mineralized soft-tissue structure and chemistry in a mummified hadrosaur from the Hell Creek Formation, North Dakota (USA). *Proc. R. Soc. B* **2009**, *276*, 3429–3437. [CrossRef]

126. Schroeter, E.R.; Cleland, T.P.; Schweitzer, M.H. Deep time paleoproteomics: Looking forward. *J. Proteome Res.* **2022**, *21*, 9–19. [CrossRef] [PubMed]

127. Bau, M. Controls on the fractionation of isovalent trace elements in magmatic and aqueous systems: Evidence from Y/Ho, Zr/Hf, and lanthanide tetrad effect. *Contrib. Mineral. Petr.* **1996**, *123*, 323–333. [CrossRef]

128. Herwartz, D.; Tütken, T.; Jochum, K.P.; Sander, P.M. Rare earth element systematics of fossil bone revealed by LA-ICPMS analysis. *Geochim. Cosmochim. Acta* **2013**, *103*, 161–183. [CrossRef]

129. Livingstone, D.A. Chemical composition of rivers and lakes. In *Data of Geochemistry*; Fleischer, M., Ed.; USGS: Washington, DC, USA, 1963; pp. G1–G64.

130. Stallard, R.F.; Edmond, J.M. Geochemistry of the Amazon 3. Weathering chemistry and limits to dissolved inputs. *J. Geophys. Res.* **1987**, *92*, 8293–8302. [CrossRef]

131. Berner, R.A. Calcium carbonate concretions formed by the decomposition of organic matter. *Science* **1968**, *159*, 195–197. [CrossRef]

132. Allison, P.A. Konservat-Lagerstätten: Cause and classification. *Paleobiology* **1988**, *14*, 331–344. [CrossRef]

133. Wang, L.; Liu, Q.; Hu, C.; Liang, R.; Qiu, J.; Wang, Y. Phosphorous release during decomposition of the submerged macrophyte *Potamogeton crispus*. *Limnology* **2018**, *19*, 355–366. [CrossRef]

134. Hoyle, J.; Elderfield, H.; Gledhill, A.; Greaves, M. The behaviour of the rare earth elements during mixing of river and sea waters. *Geochim. Cosmochim. Acta* **1984**, *48*, 143–149. [CrossRef]

135. Goldstein, S.J.; Jacobsen, S.B. Rare earth elements in river waters. *Earth Planet. Sci. Lett.* **1988**, *89*, 35–47. [CrossRef]

136. Biddau, R.; Cidu, R.; Frau, F. Rare earth elements in waters from the albitite-bearing granodiorites of Central Sardinia, Italy. *Chem. Geol.* **2002**, *182*, 1–14. [CrossRef]

137. Åström, M.; Corin, N. Distribution of rare earth elements in anionic, cationic and particulate fractions in boreal humus-rich streams affected by acid sulphate soils. *Water Res.* **2003**, *37*, 273–280. [CrossRef]

138. Bwire Ojiambo, S.; Lyons, W.B.; Welch, K.A.; Poreda, R.J.; Johannesson, K.H. Strontium isotopes and rare earth elements as tracers of groundwater-lake water interactions, Lake Naivasha, Kenya. *Appl. Geochem.* **2003**, *18*, 1789–1805. [CrossRef]

139. Tang, J.; Johannesson, K.H. Speciation of rare earth elements in natural terrestrial waters: Assessing the role of dissolved organic matter from the modeling approach. *Geochim. Cosmochim. Acta* **2003**, *67*, 2321–2339. [CrossRef]

140. Centeno, L.M.; Faure, G.; Lee, G.; Talnagi, J. Fractionation of chemical elements including the REEs and ^{226}Ra in stream contaminated with coal-mine effluent. *Appl. Geochem.* **2004**, *19*, 1085–1095. [CrossRef]

141. Johannesson, K.H.; Tang, J.; Daniels, J.M.; Bounds, W.J.; Burdige, D.J. Rare earth element concentrations and speciation in organic-rich blackwaters of the Great Dismal Swamp, Virginia, USA. *Chem. Geol.* **2004**, *209*, 271–294. [CrossRef]

142. Gammons, C.H.; Wood, S.A.; Pedrozo, F.; Varekamp, J.C.; Nelson, B.J.; Shope, C.L.; Baffico, G. Hydrogeochemistry and rare earth element behavior in a volcanically acidified watershed in Patagonia, Argentina. *Chem. Geol.* **2005**, *222*, 249–267. [CrossRef]

143. Gammons, C.H.; Wood, S.A.; Nimick, D.A. Diel behavior of rare earth elements in a mountain stream with acidic to neutral pH. *Geochim. Cosmochim. Acta* **2005**, *69*, 3747–3758. [CrossRef]

144. Barroux, G.; Sonke, J.E.; Boaventura, G.; Viers, J.; Godderis, Y.; Bonnet, M.-P.; Sondag, F.; Gardoll, S.; Lagane, C.; Seyler, P. Seasonal dissolved rare earth element dynamics of the Amazon River main stem, its tributaries, and the Curuaí floodplain. *Geochem. Geophy. Geosy.* **2006**, *7*, Q12005. [CrossRef]

145. Bau, M.; Knappe, A.; Dulski, P. Anthropogenic gadolinium as a micropollutant in river waters in Pennsylvania and in Lake Erie, northeastern United States. *Chem. Erde-Geochem.* **2006**, *66*, 143–152. [CrossRef]

146. Kulaksiz, S.; Bau, M. Contrasting behaviour of anthropogenic gadolinium and natural rare earth elements in estuaries and the gadolinium input into the North Sea. *Earth Planet. Sci. Lett.* **2007**, *260*, 361–371. [CrossRef]

147. Kulaksiz, S.; Bau, M. Anthropogenic gadolinium as a microcontaminant in tap water used as drinking water in urban areas and megacities. *Appl. Geochem.* **2011**, *26*, 1877–1885. [CrossRef]

148. Pokrovsky, O.S.; Viers, J.; Shirokova, L.S.; Shevchenko, V.P.; Filipov, A.S.; Dupré, B. Dissolved, suspended, and colloidal fluxes of organic carbon, major and trace elements in the Severnaya Dvina River and its tributary. *Chem. Geol.* **2010**, *273*, 136–149. [CrossRef]

149. Censi, P.; Sposito, F.; Inguaggiato, C.; Zuddas, P.; Inguaggiato, S.; Venturi, M. Zr, Hf and REE distribution in river water under different ionic strength conditions. *Sci. Total Environ.* **2018**, *645*, 837–853. [CrossRef] [PubMed]

150. Kalender, L.; Aytimur, G. REE geochemistry of Euphrates River, Turkey. *J. Chem.* **2016**, *1*, 1–13. [CrossRef]

151. Smith, C.; Liu, X.-M. Spatial and temporal distribution of rare earth elements in the Neuse River, North Carolina. *Chem. Geol.* **2018**, *488*, 34–43. [CrossRef]

152. Merschel, G.; Bau, M.; Schmidt, K.; Münker, C.; Dantas, E.L. Hafnium and neodymium isotopes and REY distribution in the truly dissolved, nanoparticulate/colloidal and suspended loads of rivers in the Amazon Basin, Brazil. *Geochim. Cosmochim. Acta* **2017**, *213*, 383–399. [CrossRef]

153. Smedley, P.L. The geochemistry of rare earth elements in groundwater from the Carnmenellis area, southwest England. *Geochim. Cosmochim. Acta* **1991**, *55*, 2767–2779. [CrossRef]

154. Johannesson, K.H.; Lyons, W.B.; Stetzenbach, K.J.; Byrne, R.H. The solubility control of rare earth elements in natural terrestrial waters and the significance of PO_4^{3-} and CO_3^{2-} in limiting dissolved rare earth concentrations: A review of recent information. *Aquat. Geochem.* **1995**, *1*, 157–173. [CrossRef]

155. Johannesson, K.H.; Stetzenbach, K.J.; Hodge, V.F. Rare earth elements as geochemical tracers of regional groundwater mixing. *Geochim. Cosmochim. Acta* **1997**, *61*, 3605–3618. [CrossRef]

156. Johannesson, K.H.; Farnham, I.M.; Guo, C.; Stetzenbach, K.J. Rare earth element fractionation and concentration variations along a groundwater flow path within a shallow, basin-fill aquifer, southern Nevada, USA. *Geochim. Cosmochim. Acta* **1999**, *63*, 2697–2708. [CrossRef]

157. Leybourne, M.I.; Goodfellow, W.D.; Boyle, D.R.; Hall, G.M. Rapid development of negative Ce anomalies in surface waters and contrasting REE patterns in groundwaters associated with Zn-Pb massive sulphide deposits. *Appl. Geochem.* **2000**, *15*, 695–723. [CrossRef]

158. Tang, J.; Johannesson, K.H. Controls on the geochemistry of rare earth elements along a groundwater flow path in the Carrizo Sand aquifer, Texas, USA. *Chem. Geol.* **2006**, *225*, 156–171. [CrossRef]

159. Esmaeili-Vardanjani, M.; Shamsipour-Dehkordi, R.; Eslami, A.; Moosaei, F.; Pazand, K. A study of differentiation pattern and rare earth elements migration in geochemical and hydrogeochemical environments of Airekan and Cheshmeh Shotori areas (Central Iran). *Environ. Earth Sci.* **2013**, *68*, 719–732. [CrossRef]

160. Chevis, D.A.; Johannesson, K.H.; Burdige, D.J.; Tang, J.; Moran, S.B.; Kelly, R.P. Submarine groundwater discharge of rare earth elements to a tidally-mixed estuary in Southern Rhode Island. *Chem. Geol.* **2015**, *397*, 128–142. [CrossRef]

161. Liu, H.; Guo, H.; Xing, L.; Zhan, Y.; Li, F.; Shao, J.; Niu, H.; Liang, X.; Li, C. Geochemical behaviors of rare earth elements in groundwater along a flow path in the North China Plain. *J. Asian Earth Sci.* **2016**, *117*, 33–51. [CrossRef]

162. Johannesson, K.H.; Lyons, W.B. Rare-earth element geochemistry of Colour Lake, an acidic freshwater lake on Axel Heiberg Island, Northwest Territories, Canada. *Chem. Geol.* **1995**, *119*, 209–223. [CrossRef]

163. De Carlo, E.H.; Green, W.J. Rare earth elements in the water column of Lake Vanda, McMurdo Dry Valleys, Antarctica. *Geochim. Cosmochim. Acta* **2002**, *66*, 1323–1333. [CrossRef]

164. Sholkovitz, E.R. The geochemistry of rare earth elements in the Amazon River estuary. *Geochim. Cosmochim. Acta* **1993**, *57*, 2181–2190. [CrossRef]

165. Nozaki, Y.; Lerche, D.; Alibo, D.S.; Tsutsumi, M. Dissolved indium and rare earth elements in three Japanese rivers and Tokyo Bay: Evidence for anthropogenic Gd and In. *Geochim. Cosmochim. Acta* **2000**, *64*, 3975–3982. [CrossRef]

166. Rousseau, T.C.C.; Sonke, J.E.; Chmeleff, J.; van Beek, P.; Souhat, M.; Boaventura, G.; Seyler, P.; Jeandel, C. Rapid neodymium release to marine waters from lithogenic sediments in the Amazon estuary. *Nat. Commun.* **2015**, *6*, 7592. [CrossRef] [PubMed]

167. Elderfield, H.; Sholkovitz, E.R. Rare earth elements in the pore waters of reducing nearshore sediments. *Earth Planet. Sci. Lett.* **1987**, *82*, 280–288. [CrossRef]

168. Bayon, G.; Birot, D.; Ruffine, L.; Caprais, J.-C.; Ponzevera, E.; Bollinger, C.; Donval, J.-P.; Charlou, J.-L.; Voisset, M.; Grimaud, S. Evidence for intense REE scavenging at cold seeps from the Niger Delta margin. *Earth Planet. Sci. Lett.* **2011**, *312*, 443–452. [CrossRef]

169. Elderfield, H.; Greaves, M.J. The rare earth elements in seawater. *Nature* **1982**, *296*, 214–219. [CrossRef]

170. de Baar, H.J.W.; Bacon, M.P.; Brewer, P.G. Rare-earth distributions with a positive Ce anomaly in the Western North Atlantic Ocean. *Nature* **1983**, *301*, 324–327. [CrossRef]

171. German, C.R.; Holliday, B.P.; Elderfield, H. Redox cycling of rare earth elements in the suboxic zone of the Black Sea. *Geochim. Cosmochim. Acta* **1991**, *55*, 3553–3558. [CrossRef]

172. Piepgras, D.J.; Jacobsen, S.B. The behavior of rare earth elements in seawater: Precise determination of variations in the North Pacific water column. *Geochim. Cosmochim. Acta* **1992**, *56*, 1851–1862. [CrossRef]

173. Sholkovitz, E.R.; Landing, W.M.; Lewis, B.L. Ocean particle chemistry: The fractionation of rare earth elements between suspended particles and seawater. *Geochim. Cosmochim. Acta* **1994**, *58*, 1567–1579. [CrossRef]

174. German, C.R.; Masuzawa, T.; Greaves, M.J.; Elderfield, H.; Edmond, J.M. Dissolved rare earth elements in the Southern Ocean: Cerium oxidation and the influence of hydrography. *Geochim. Cosmochim. Acta* **1995**, *59*, 1551–1558. [CrossRef]

175. Zhang, J.; Nozaki, Y. Rare earth elements and yttrium in seawater: ICP-MS determinations in the East Caroline, Coral Sea, and South Fiji basins of the western South Pacific Ocean. *Geochim. Cosmochim. Acta* **1996**, *60*, 4631–4644. [CrossRef]

176. Hongo, Y.; Obata, H.; Alibo, D.S.; Nozaki, Y. Spatial variations of rare earth elements in North Pacific surface water. *J. Oceanogr.* **2006**, *62*, 441–455. [CrossRef]

177. Wang, Z.-L.; Yamada, M. Geochemistry of dissolved rare earth elements in the Equatorial Pacific Ocean. *Environ. Geol.* **2007**, *52*, 779–787. [CrossRef]

178. van de Flierdt, T.; Pahnke, K.; Amakawa, H.; Andersson, P.; Basak, C.; Coles, B.; Colin, C.; Crocket, K.; Frank, M.; Frank, N.; et al. GEOTRACES intercalibration of neodymium isotopes and rare earth element concentrations in seawater and suspended particles. Part 1: Reproducibility of results for the international intercomparison. *Limnol. Oceanogr. Methods* **2012**, *10*, 234–251. [CrossRef]

179. Grenier, M.; Jeandel, C.; Lacan, F.; Vance, D.; Venchiarutti, C.; Cros, A.; Cravatte, S. From the subtropics to the central equatorial Pacific Ocean: Neodymium isotopic composition and rare earth element concentration variations. *J. Geophys. Res.-Oceans* **2013**, *118*, 592–618. [CrossRef]

180. Jeandel, C.; Delattre, H.; Grenier, M.; Pradoux, C.; Lacan, F. Rare earth element concentrations and Nd isotopes in the Southeast Pacific Ocean. *Geochem. Geophy. Geosy.* **2013**, *14*, 328–341. [CrossRef]

181. Garcia-Solsona, E.; Jeandel, C.; Labatut, M.; Lacan, F.; Vance, D.; Chavagnac, V.; Pradoux, C. Rare earth elements and Nd isotopes tracing water mass mixing and particle-seawater interactions in the SE Atlantic. *Geochim. Cosmochim. Acta* **2014**, *125*, 351–372. [CrossRef]

182. Abbott, A.N.; Haley, B.A.; McManus, J.; Reimers, C.E. The sedimentary flux of dissolved rare earth elements to the ocean. *Geochim. Cosmochim. Acta* **2015**, *154*, 186–200. [CrossRef]

183. Hathorne, E.C.; Stichel, T.; Brück, B.; Frank, M. Rare earth element distribution in the Atlantic sector of the Southern Ocean: The balance between particle scavenging and vertical supply. *Mar. Chem.* **2015**, *177*, 157–171. [CrossRef]

184. Zheng, X.-Y.; Plancherel, Y.; Saito, M.A.; Scott, P.M.; Henderson, G.M. Rare earth elements (REEs) in the tropical South Atlantic and quantitative deconvolution of their non-conservative behavior. *Geochim. Cosmochim. Acta* **2016**, *177*, 217–237. [CrossRef]

185. Johannesson, K.H.; Palmore, C.D.; Fackrell, J.; Prouty, N.G.; Swarzenski, P.W.; Chevis, D.A.; Telfeyan, K.; White, C.D.; Burdige, D.J. Rare earth element behavior during groundwater-seawater mixing along the Kona Coast of Hawaii. *Geochim. Cosmochim. Acta* **2017**, *198*, 229–258. [CrossRef]

186. Osborne, A.H.; Hathorne, E.C.; Schijf, J.; Plancherel, Y.; Böning, P.; Frank, M. The potential of sedimentary foraminiferal rare earth element patterns to trace water masses in the past. *Geochem. Geophy. Geosy.* **2017**, *18*, 1550–1568. [CrossRef]

187. de Baar, H.J.W.; Bruland, K.W.; Schijf, J.; van Heuven, S.M.A.C.; Behrens, M.K. Low cerium among the dissolved rare earth elements in the central North Pacific Ocean. *Geochim. Cosmochim. Acta* **2018**, *236*, 5–40. [CrossRef]

188. Haley, B.A.; Klinkhammer, G.P.; McManus, J. Rare earth elements in pore waters of marine sediments. *Geochim. Cosmochim. Acta* **2004**, *68*, 1265–1279. [CrossRef]

189. Kim, J.-H.; Torres, M.E.; Haley, B.A.; Kastner, M.; Pohlman, J.W.; Riedel, M.; Lee, Y.J. The effect of diagenesis and fluid migration on rare earth element distribution in fluids of the northern Cascadia accretionary margin. *Chem. Geol.* **2011**, *291*, 152–165. [CrossRef]

biology

MDPI

Article

Soft-Tissue, Rare Earth Element, and Molecular Analyses of *Dreadnoughtus schrani*, an Exceptionally Complete Titanosaur from Argentina

Elena R. Schroeter [1,*], Paul V. Ullmann [2], Kyle Macauley [2], Richard D. Ash [3], Wenxia Zheng [1], Mary H. Schweitzer [1] and Kenneth J. Lacovara [2]

[1] Department of Biological Sciences, North Carolina State University, Raleigh, NC 27695, USA; wzheng2@ncsu.edu (W.Z.); mhschwei@ncsu.edu (M.H.S.)
[2] Department of Geology, Rowan University, Glassboro, NJ 08028, USA; ullmann@rowan.edu (P.V.U.); macauleyk4@students.rowan.edu (K.M.); lacovara@rowan.edu (K.J.L.)
[3] Department of Geology, University of Maryland, College Park, MD 20742, USA; rdash@umd.edu
* Correspondence: easchroe@ncsu.edu

Citation: Schroeter, E.R.; Ullmann, P.V.; Macauley, K.; Ash, R.D.; Zheng, W.; Schweitzer, M.H.; Lacovara, K.J. Soft-Tissue, Rare Earth Element, and Molecular Analyses of *Dreadnoughtus schrani*, an Exceptionally Complete Titanosaur from Argentina. *Biology* **2022**, *11*, 1158. https://doi.org/10.3390/biology11081158

Academic Editor: Zhifei Zhang

Received: 30 June 2022
Accepted: 31 July 2022
Published: 2 August 2022

Publisher's Note: MDPI stays neutral with regard to jurisdictional claims in published maps and institutional affiliations.

Copyright: © 2022 by the authors. Licensee MDPI, Basel, Switzerland. This article is an open access article distributed under the terms and conditions of the Creative Commons Attribution (CC BY) license (https://creativecommons.org/licenses/by/4.0/).

Simple Summary: Although many analytical techniques have shown that organic material can be preserved in fossils for millions of years, the geochemical factors that allow this preservation are not well understood. This is partly because paleomolecular studies often do not include geochemical analyses of the fossil or burial environment from which it came. We conducted in-depth geological, geochemical, and molecular analyses of a specimen of *Dreadnoughtus schrani*, an immense dinosaur from Argentina. We reviewed physical aspects of the sediments in which *Dreadnoughtus* was deposited, then characterize the following features: the structural integrity of the bone microstructure; the amount and type of external mineral that infiltrated the bone; the concentration of elements that are rare in the Earth's crust (REEs) throughout the bone; the preservation of soft-tissue structures (e.g., bone cells and blood vessels); the preservation of bone protein using antibodies that specifically recognize collagen I. Our data show that original bone microstructures and protein are preserved in *Dreadnoughtus*, and that after burial, the specimen was exposed to weakly-oxidizing conditions and groundwaters rich in "light" REEs but experienced little further chemical alteration after this early stage of fossilization. Our findings support the idea that fossils showing lower concentrations of REEs are well suited for molecular analyses.

Abstract: Evidence that organic material preserves in deep time (>1 Ma) has been reported using a wide variety of analytical techniques. However, the comprehensive geochemical data that could aid in building robust hypotheses for how soft-tissues persist over millions of years are lacking from most paleomolecular reports. Here, we analyze the molecular preservation and taphonomic history of the *Dreadnougtus schrani* holotype (MPM-PV 1156) at both macroscopic and microscopic levels. We review the stratigraphy, depositional setting, and physical taphonomy of the *D. schrani* skeletal assemblage, and extensively characterize the preservation and taphonomic history of the humerus at a micro-scale via: (1) histological analysis (structural integrity) and X-ray diffraction (exogenous mineral content); (2) laser ablation-inductively coupled plasma mass spectrometry (analyses of rare earth element content throughout cortex); (3) demineralization and optical microscopy (soft-tissue microstructures); (4) in situ and in-solution immunological assays (presence of endogenous protein). Our data show the *D. schrani* holotype preserves soft-tissue microstructures and remnants of endogenous bone protein. Further, it was exposed to LREE-enriched groundwaters and weakly-oxidizing conditions after burial, but experienced negligible further chemical alteration after early-diagenetic fossilization. These findings support previous hypotheses that fossils that display low trace element uptake are favorable targets for paleomolecular analyses.

Keywords: fossil proteins; molecular paleontology; diagenesis; taphonomy; rare earth elements; soft-tissue preservation; geochemistry

1. Introduction

In the last few decades, the concept of "exceptional preservation" has expanded beyond specimens that retain exquisite morphological details through processes such as phosphatization (e.g., [1]) to include fossils that preserve some of their original organic content. Evidence that soft-tissue structures, original proteins, and other organic material have been preserved in deep time (>1 Ma) has been reported using a wide variety of analytical techniques, including (but not limited to) amino acid analyses (e.g., [2], Raman spectroscopy (e.g., [3,4]), Fourier transform infrared spectroscopy (FT-IR) (e.g., [5–7]), immunology (e.g., [8,9]), time-of-flight secondary ion mass spectrometry (TOF-SIMS) (e.g., [10–12]), and tandem mass spectrometry (LC-MS/MS) (e.g., [13,14]). In spite of the diverse, growing literature that original, endogenous organic material can preserve for millions of years, these reports are often regarded with skepticism (e.g., [15]), in no small part, because the geochemical mechanisms that allow for such preservation are not completely understood. Although hypotheses exist as to geochemical factors that may positively influence preservation (e.g., involvement of iron [16,17], microbial activity ([18]), and/or condensation reactions [19–22]), such studies often examine specific cases in isolation, making it difficult to infer larger scale relationships between the geochemical environment and preservation. As a result, the comprehensive depositional and geochemical data that could aid in building robust, multi-faceted hypotheses about molecular preservation in deep time are lacking from most paleomolecular reports [23,24].

The holotype of *Dreadnoughtus schrani* (MPM-PV 1156) is a specimen that provides the opportunity for in-depth analyses of its preservation and taphonomic history at both the macroscopic and microscopic levels. MPM-PV 1156 represents a massive (59.3 metric tons; 65.4 short tons) titanosaur recovered from the Santa Cruz Province of Argentina [25]. This *D. schrani* holotype has retained 45.3% of its skeleton, including many associated and articulated elements [25]—an extraordinary portion for a sauropod of its size [26]. It has exceptional completeness, coupled with evidence of syndepositional deformation of sedimentary beds, led Lacovara et al. [25] to hypothesize that the individual was entombed during a rapid burial event, such as a crevasse splay. Rapid burial has also been implicated in other types of "exceptional preservation", including soft-tissues, original organics, and other labile features commonly destroyed by diagenesis (e.g., skin, feathers, cells, proteins) [19,27–35]. Given the remarkable preservation of MPM-PV 1156 at the macroscopic level, we hypothesized that the burial event that protected the massive *D. schrani* holotype from extensive biostratinomic processes may have been sufficient to delay early diagenesis at other levels as well, preserving both microstructure and original molecular components, such as proteins, in the fossil tissue.

Here, we review the gross geological context of MPM-PV 1156, including the stratigraphy and depositional sediments of its locality and the physical taphonomy of the *D. schrani* skeletal assemblage as a whole. Then, we perform extensive analyses of the left humerus of the *D. schrani* holotype (MPM-PV 1156-46) to examine its preservation and taphonomic history at a micro-scale. These include: (1) histological analysis and X-ray diffraction (XRD) to assess the structural integrity of the bone microstructure and the extent of its exogenous mineral content; (2) laser ablation inductively coupled plasma mass spectrometry (LA-ICPMS) to assess its geochemical history based on the spatial heterogeneity of rare earth elements (REEs) and other pertinent trace elements throughout the bone cortex; (3) chemical demineralization and optical microscopy of bone tissue to assess the morphological preservation of soft-tissue microstructures; (4) in situ and in-solution immunological assays to assess the presence of endogenous protein.

2. Geologic and Taphonomic Context

Specimen MPM-PV 1156 is the holotype of *Dreadnoughtus schrani* [25] (Figure 1A). It was collected between 2005 and 2008 from outcrops of the Late Cretaceous Cerro Fortaleza Formation, along the east bank of the Río La Leona in Santa Cruz Province, Argentina (Figure 1B). Lacovara et al. [25] originally reported its age to be Campanian–Maastrichtian, based on the prior recovery of Campanian ammonites from the underlying La Anita Formation and Maastrichtian palynomorphs from the overlying La Irene Formation (see [36] and references therein). Recent radiometric dating of detrital zircons from the Cerro Fortaleza Formation in the region of the *Dreadnoughtus* quarry constrain the age of these deposits specifically to the Campanian [37]. The 116 skeletal elements of MPM-PV 1156 were found as partially-articulated and closely associated remains alongside those of a second, smaller individual, *D. schrani* paratype MPM-PV 3546 (Figure 2).

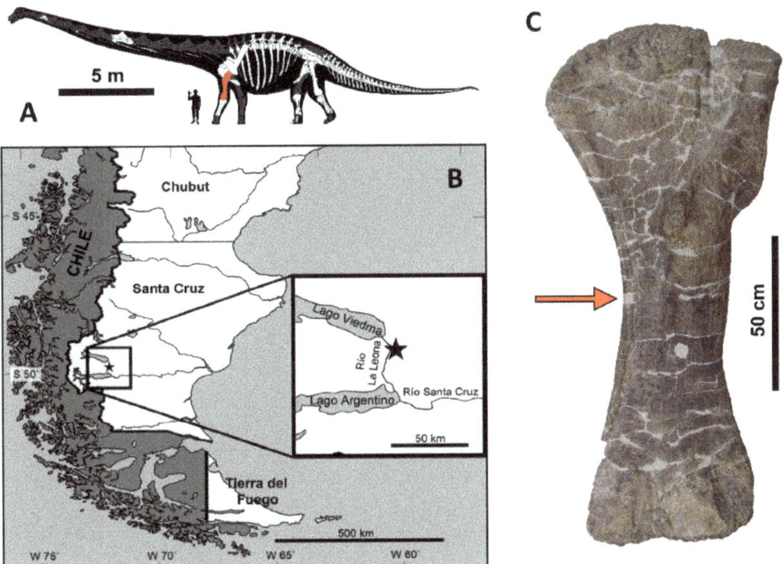

Figure 1. (**A**) Reconstruction of *Dreadnoughtus schrani* holotype (MPM-PV 1156) with preserved elements shown in white. The left humerus (MPM-PV 1156-49; highlighted in red) was examined in this study. (**B**) Map showing the location of the *Dreadnoughtus* quarry in Santa Cruz Province, Patagonia, Argentina. (**C**) Left humerus (MPM-PV 1156-49) of *D. schrani*, shown in anterior view. Red arrow indicates the tissue sampling location for these analyses. Scale bars are as labeled. Image modified from Figures 1–5; of Ullmann and Lacovara [38], with permission.

Strata of the Cerro Fortaleza Formation comprise fluvial channel and overbank facies deposited in the Austral Basin, now east of the Southern Patagonian Andes [39,40]. Bones of both *Dreadnoughtus* individuals were recovered from a mixed lithosome comprised of tan, finely trough cross-bedded, fine to medium-grained sandstone and gray, homogenous mudstone containing abundant plant remains. Lacovara et al. [25] interpreted these deposits to represent a crevasse splay horizon deposited onto a fluvial floodplain, based on an abundance of large-scale (primary), convoluted bedforms interpreted to have formed either (1) via liquefaction during rapid deposition, or (2) by sediment redistribution around the large, heavy skeletal remains of the *Dreadnoughtus* holotype at the time of burial, as it subsided into a thick, "soft" substrate. Abundant silicified wood [41] and palynomorphs [42] in strata of the Cerro Fortaleza Formation demonstrate that fluvial floodplains across the region hosted diverse forests.

Evidence for the authocthonous nature of MPM-PV 1156 includes: (1) most of the tail and the left femur, tibia, and fibula remained in articulation within the south-central portion of the quarry; (2) signs of abrasion to the bones are absent [43,44], and; (3) the enormous size of the individual likely hindered any long-distance transport (Figure 2). Although the bones exhibit negligible weathering [43,44], from the disarticulated nature of large portions of the body (e.g., the ribcage, left forelimb, and dorsal series (Figure 2), we infer the carcass underwent a protracted phase of decay prior to burial. Recovery of several shed cf. *Orkoraptor burkei* teeth within the quarry, along with bite marks on a caudal vertebra of the paratype (MPM-PV 3546) and a dorsal vertebra (belonging to either the holotype or the paratype), further support this conclusion. Preferential preservation of appendicular elements from the left side indicates the animal was likely lying on its left side at the time of burial [25].

Figure 2. Map of the *Dreadnoughtus schrani* quarry. Numbers denote individual bone specimen numbers. Bones pertaining to paratype individual MPM-PV 3546 are shaded in gray, lignite/wood specimens are shaded in brown, and left humerus (MPM-PV 1156-49) examined in this study is shaded in red. Some overlapping elements and a few specimens found eroding from the outcrop beyond the area depicted in this map are not shown. Select bones are identified as labeled. Grid is shown in 2 m increments; scale bar = 1 m. (Map by J. DiGnazio).

Based on the quarry map (Figure 2), skeletal element abundances range from 0 to 18 specimens per 4 m^2 area, with an average of 6 specimens per 4 m^2. During excavation, multiple bones of both *Dreadnoughtus* individuals were found at strongly plunging angles within the quarry, including the left femur MPM-PV 3546-21 of the paratype, which was found standing vertically (i.e., perpendicular to bedding) [25]. Such plunge angles, in combination with numerous occurrences of bones stacked three deep (e.g., left scapula MPM-PV 1156-48, left radius MPM-PV 1156-51, and dorsal rib MPM-PV 1156-77 near the eastern edge of the quarry; Figure 2), demonstrate that burial occurred in sediment with little bearing strength, in agreement with our prior interpretation of the host sediments deriving from a large-scale crevasse splay event [25]. The general orientations of elongate skeletal elements (e.g., limb bones, ribs, the articulated caudal series) within the quarry form a distinct bimodal pattern, composed of roughly north–south and east–west trends (Figure 2). This might result from the hydraulic alignment of numerous skeletal elements parallel and perpendicular to a paleocurrent flowing along one of these bearings. Individual elements of MPM-PV 1156 range in size from an isolated tooth and distal caudal vertebrae up to girdle elements and stylopodial limb bones (i.e., bones pertaining to both Voorhies Groups I and II were recovered; cf. [45]). All major portions of the body are represented, further supports interpretation of the remains as being autochthonous. Though most of the bones are complete, many exhibit transverse and longitudinal fractures and/or plastic deformation arising from post-fossilization compaction, especially ribs and the pelvic elements; on the left humerus (MPM-PV 1156-49), a distinct fracture through the proximal end is observed (Figure 1C).

3. Methods

Brief descriptions of the experimental methods are described below. Comprehensive details of these analyses are provided in the Supplementary Materials, as specified in each section.

3.1. Sample Collection

Fossil and sediment samples for molecular assays were collected using aseptic techniques as follows: using nitrile gloves, diaphyseal cortical bone fragments were removed from the humerus of MPM-PV 1156 (Figure 2) immediately upon the opening of its protective field jacket. A sediment sample was taken from within the plaster jacket containing the humerus, from an area between the bone surface and the inner wall of the jacket. Both samples were wrapped in autoclaved foil and stored in an autoclaved glass container under desiccation at room temperature until analyses. During processing, specimens were handled within a fume hood. Bench surfaces were thoroughly cleaned using 95% ethanol, followed by 10% bleach, before sterile foil was laid down over the work area. Nitrile gloves, shoe covers, a hair net, a facemask, and a lab coat (all permanently kept in the ancient-isolated environment) were worn throughout all procedures. For REE analyses, nitrile gloves were used to extract a second fragment of the midshaft of MPM-PV 1156-49 for embedding and sectioning.

Extant controls for molecular assays, which included extant alligator (*Alligator mississippiensis*) and chicken (*Gallus gallus*) limb bones, were processed separately, and in isolation. All limb elements were separately defleshed, then degreased in 10% Zout (Dial Co., Scottsdale, AZ, USA) or 10% Shout (Johnson Co., New York, NY, USA). Following degreasing, diaphyseal portions of all limb elements were sectioned into pieces, rinsed in Epure water, then wrapped in foil and stored at −80 °C until analyses. The alligator humerus was used for immunofluorescence. All other techniques used the alligator femur.

3.2. Assessments of Preservational Quality and Geochemical History

3.2.1. Histological Analysis

Embedding of *D. schrani* bone tissue samples followed methods described in Green et al. [46], Boyd et al. [47], and Cleland et al. [48]. A fragment of the humerus from MPM-PV 1156 was embedded in Silmar 41^{TM} resin (US Composites, West Palm Beach, FL, USA) and a 2 mm thick, transverse section was cut with a Isomet 1000 precision wafer saw (Buehler, Lake Bluff, IL, USA). The section was then mounted on a glass slide using clear Loctite epoxy (Henkel Adhesives, Rocky Hill, CT, USA) and ground/polished to transparency using a Ecomet 4000 grinder (Buehler, Lake Bluff, IL, USA). Ground sections were imaged with a transmitted light microscope (10x; Axioplan, Zeiss, White Plains, NY, USA) fitted with circularly polarized light filters and a motorized XYZ stage (Marzhauser, Wetzlar, Germany) with output to StereoInvestigator software (MBF Bioscience, Williston, VT, USA) for automated image montaging.

3.2.2. X-ray Diffraction (XRD)

D. schrani cortical fragments (~1–2 mg) were powdered in a tungsten carbide Mixer-Mill (SPEX SamplePrep 8000, Cole-Parmer, Vernon Hills, IL, USA) to ~10 μm. Analyses were performed on a X'Pert diffractometer (#DY1738, Philips, Amsterdam, The Netherlands) using Cu Kα radiation (λ = 1.54178 Å), and operating at 45 kV and 40 mA. Diffraction patterns were measured from 5–75° 2θ with a step size of 0.017° 2θ and a time of 1.3 s per step (~0.77 degrees/min). HighScore Plus software version 3.0e (Philips) was used to interpret diffraction traces.

3.2.3. Laser Ablation Inductively Coupled Plasma Mass Spectrometry (LA-ICPMS)

We employed the same mass spectrometry methods as Ullmann et al. [49]. REE concentrations were normalized against the North American Shale Composite (NASC) to enable comparisons to fossil bones from other sites, using values from Gromet et al. [50] and Haskin et al. [51]; a subscript N denotes shale-normalized values or ratios. NASC-normalized REE ratios were used to calculate $(Ce/Ce^*)_N$, $(Ce/Ce^{**})_N$, $(Pr/Pr^*)_N$, and $(La/La^*)_N$ anomalies (sensu [52]) as follows: $(Ce/Ce^*)_N = Ce_N/(0.5^*La_N + 0.5^*Pr_N)$, $(Ce/Ce^{**})_N = Ce_N/(2^*Pr_N - Nd_N)$, $(Pr/Pr^*)_N = Pr_N/(0.5^*Ce_N + 0.5^*Nd_N)$, and $(La/La^*)_N = La_N/(3^*Pr_N - 2^*Nd_N)$. See Supplemental Methods Section S1.1 for more details.

3.3. Assessments of Soft Tissue Preservation

To prevent exogenous contamination, all sample preparation and analyses performed on fossil materials and all negative controls were conducted using instruments, buffers, and chemicals reserved for and dedicated to fossil analyses, in a laboratory where no extant vertebrate remains were ever housed or analyzed. Preparation and analyses of extant material (i.e., positive controls) occurred in a separate laboratory, and no interchange of solutions, instruments, or other materials occurred.

3.3.1. Demineralization and Evaluation of Morphological Structures

Fragments of *D. schrani* humerus were incubated in 0.5 M disodium ethylenediaminetetraacetic acid (EDTA) (pH 8.0) for two weeks. Pliable demineralization products were transferred to a glass slide, treated directly with acetone (to rule out all possibility of glue contamination), and imaged in transmitted light and cross-polarized light using a Zeiss Axiocam MRC5 camera mounted to a Zeiss Axioskop 2 Plus biological microscope and a Zeiss Axioskop petrographic polarizing microscope, respectively. Low magnification images were acquired on a Zeiss Stemi 2000-C dissecting microscope. See Supplemental Methods Section S1.2 for more details.

3.3.2. Immunofluorescence (IF)

Demineralized tissue from *D. schrani*, chicken, and alligator were embedded in resin, thinly sectioned (220 nm), and subjected to immunofluorescence assays using polyclonal rabbit anti-chicken collagen I antibodies. For detailed descriptions of IF procedures and the numerous antibody specificity tests performed in conjunction with them, see Supplemental Methods Section S1.4.

3.3.3. Enzyme-Linked Immunosorbent Assay (ELISA)

Bone and sediment fragments were demineralized with hydrochloric acid, followed by protein solubilization with guanidine hydrochloride (GuHCl). Buffer samples (empty tubes that received reagent but no tissue; "blanks") were analyzed simultaneously with fossil and sediment samples to serve as an additional negative control. GuHCl extracts of all tissues and negative controls were subjected to ELISA with polyclonal chicken specific anti-collagen I. For a detailed description of the chemical extraction and ELISA procedures, see Supplemental Methods Section S1.5.

4. Results

4.1. Preservational Quality

4.1.1. Histological Analysis

The microstructure of the humeral sample from MPM-PV 1156-49 shows primary fibrolamellar bone at the periosteal surface transitioning to densely remodeled secondary bone deeper at the medullary cavity (Figure 3). Structures consistent with fungal tunneling or MDF were not observed in any of the tissue examined, indicating that microbial alteration of this fossil sample was minimal [53]. Under cross-polarized light (Figure 3), deviation from the expected pattern of birefringence in fibrolamellar bone was not apparent, consistent with a lack of alteration of the mineral phase of bone. The lack of indicators of extensive microbial alteration, coupled with the well-preserved microstructure of the observed primary and secondary bone tissue, give this fossil sample an integrity rank of 5 (out of 5) on the Histology Index [54].

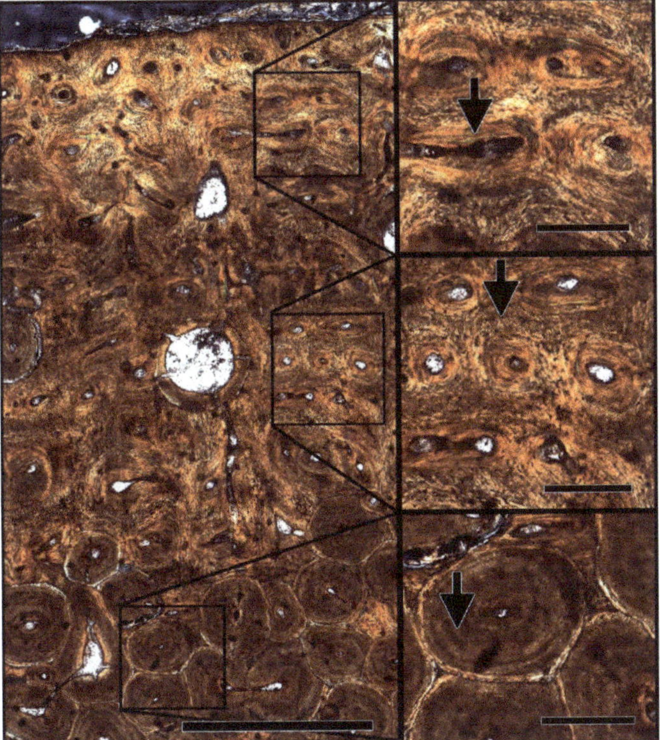

Figure 3. Transverse ground section through a sample of MPM-PV 1156-49, collected from the medial side of the midshaft (complete section is not represented) and imaged under cross-polarized light. The tissue microstructure contains well-preserved primary fibrolamellar bone (FLB) at the periosteal surface, with a birefringence pattern consistent with extant FLB: primary osteons surrounded by concentric layers of lamellar bone (top inset, arrow) and interstitial spaces with unorganized woven bone (middle inset, arrow). Secondary osteons with distinct lamellae (bottom inset, arrow) are also observed. Microscopic destructive foci (MDF) [53] are not apparent, indicating that this sample has not been extensively altered by microbes. Large black spots scattered throughout image represent bubbles in the epoxy adhering the section to the slide. Microstructural preservation of the FLB and Haversian tissues in this fossil sample, and the lack of abundant indicators of microbial attack, suggests that the sample taken from *D. schrani* is an appropriate target for molecular study. Scale bar = 1 mm, inset scale bars = 200 μm.

4.1.2. X-ray Diffraction (XRD)

To determine whether the humeral sample extracted from MPM-PV 1156-49 has been structurally compromised by extensive replacement with exogenous minerals, we performed XRD analysis on a portion of this tissue. Diffraction peaks (Figure S1) recovered in the analysis corresponded with four minerals, including dolomite and three variants of isomorphous [55] apatite (Table 1). Semi-quantitative analysis of the data used gave the approximate percentage of the mineral composition of the fossil tissue as 5% calcite/dolomite and 95% apatite (Table 1), indicating that the replacement and/or inclusion of exogenous minerals was minimal in this sample.

Table 1. Results of X-ray diffraction (XRD) analysis of the humeral sample from MPM-PV 1156-49. Semi-quantitative analysis of the data (HighScore Plus, Phillips) determined that the mineral composition of the fossil tissue is 95% isomorphous variations of apatite, indicating minimal permineralization with exogenous mineral.

Ref. Code	Score	Mineral	Displacement [°2Th.]	Scale Factor	Chemical Formula	Semi-Quantitative Analysis
01-071-1663	43	Dolomite	0.000	0.334	$Mg_{0.1}Ca_{0.9}CO_3$	5%
01-084-1999	29	Chloroapatite	0.000	0.437	$Ca_5(PO_4)_3F_{0.09}C_{10.88}$	17%
01-071-0880	62	Fluoroapatite	0.000	0.934	$Ca_5(PO_4)_3F$	41%
01-073-1731	67	Hydroxyapatite	0.000	0.866	$Ca_5(PO_4)_3(OH)$	36%

4.1.3. REE Analyses

LA-ICPMS was used to examine spatial heterogeneity of REE and other pertinent trace elements in the thick bone section. At the whole-bone level, MPM-PV 1156-49 exhibits a $\sum$REE of 1964 ppm (Table 2). Manganese (Mn; 3233 ppm) and strontium (Sr; 1747 ppm) concentrations are the highest of all recorded elements, with concentrations more than double all the other trace elements examined (Table 2). Yttrium (Y; 505 ppm) is enriched to a similar level as light REE (LREE) elements, whereas the average scandium (Sc) concentration is low (5 ppm). On average, LREE concentrations are an order of magnitude higher than those of the heavy rare earth elements (HREE); the middle rare earths (MREE; Sm–Gd) generally exhibit intermediate concentrations. The average concentrations of uranium (U; 19 ppm) and iron (Fe; 0.19 wt%) are each quite low, owing to low concentrations throughout much of the middle and internal cortices (see below).

Intra-Bone REE Depth Profiles: Concentrations of all REE decline steeply with cortical depth, with profiles for, essentially, every REE forming smoothly decreasing exponential curves (e.g., Figure 4A; see also Supplementary Materials). Among the REE, cerium (Ce) concentrations are the highest at the cortical margin (~4800 ppm) and thulium (Tm) concentrations are the lowest (~10 ppm). LREE exhibit by far the steepest declines in terms of magnitude (on average from ~2300 ppm near the cortical margin to ~100 ppm by 1 cm into the cortex; Figure 4A), generally constituting a decrease of approximately 1.5 orders of magnitude. All MREE concentration profiles exhibit moderate slopes in between those of LREE and HREE. LREE and MREE concentrations are so low throughout much of the middle and internal cortices (<2 ppm) that they frequently encroach on or fall below the lower detection limit (Data S1). In contrast, HREE concentrations generally stay above the detection limit throughout the internal cortex. As exemplified by ytterbium (Yb) in Figure 4A, HREE exhibit the shallowest profiles among the rare earths. The comparatively "shallow" decline in HREE profiles is evident in their magnitude of decrease from the cortical margin: whereas LREE concentrations decrease by two to three orders of magnitude by half way across the width of the cortex, HREE concentrations only decrease by approximately 50% across this same distance (Data S1).

Table 2. Average whole-bone trace element composition of the left humerus of *Dreadnoughtus schrani* (MPM-PV 1156-49). Numbers presented are averages of all transect data acquired across the cortex. Iron (Fe) is presented in weight percent (wt%), all other elements are in parts per million (ppm). Absence of $(Ce/Ce^*)_N$, $(Pr/Pr^*)_N$, $(Ce/Ce^{**})_N$, and $(La/La^*)_N$ anomalies occurs at 1.0. The Y/Ho value reflects this mass ratio.

Element	Concentration
Sc	4.96
Mn	3233
Fe	0.19
Sr	1747
Y	505
Ba	447
Th	0.23
U	18.64
Light Rare Earth Elements (LREEs)	
La	398.95
Ce	762.75
Pr	90.67
Nd	371.52
Middle Rare Earth Elements (MREEs)	
Sm	78.35
Eu	16.91
Gd	84.46
Heavy Rare Earth Elements (HREEs)	
Tb	10.67
Dy	63.76
Ho	13.32
Er	36.67
Tm	4.69
Yb	27.38
Lu	4.23
ΣREE	1964
$(Ce/Ce^*)_N$	0.94
$(Pr/Pr^*)_N$	0.92
$(Ce/Ce^{**})_N$	1.22
$(La/La^*)_N$	1.75
Y/Ho	37.90

Even though nearly the entire cortex is composed of densely vascularized Haversian tissue formed by multiple generations of overlapping secondary osteons (Figure 3) [25], there are no obvious spikes in REE concentrations in osteonal tissue surrounding Haversian canals (Figure 4A) nor clear signs of kinks in REE profiles reflective of uptake via double medium diffusion (cf. [56]). Of the elements examined, only iron (Fe) exhibits brief spikes in concentrations in osteonal tissue surrounding a few Haversian canals, primarily within the middle and internal cortices (e.g., at ~15.3 mm; Figure 4C).

Uranium (U) is the only element to exhibit a broad peak in concentrations within the internal cortex (Figure 4B), though concentrations in this region vary substantially. Uranium is also the only element for which concentrations drop to a minimum in the outer portion of the middle cortex. Scandium (Sc) and yttrium (Y) exhibit the same general profile shapes as the MREE in the bone, characterized by a slow, steady decrease in concentrations from the cortical margin (Figure 4C). Strontium (Sr), manganese (Mn), and barium (Ba) each exhibit high and relatively constant concentrations throughout the cortex, with very slightly greater enrichment in the external-most ~3 mm (Figure 4D).

NASC-Normalized REE Patterns: Because concentrations of LREE in the external cortex are considerably higher than through the rest of the cortex, the external 250 μm of the transect exhibits a more LREE-enriched NASC-normalized pattern than the bone as a whole (compare Figures 5B and 6A). Additionally, whereas the external 250 μm plot exhibit a slightly positive Ce anomaly (visually evident as an upward deflection of the pattern at this element), there is none in the spider diagram for the entire bone. However, both plots exhibit a general trend of relative LREE and MREE enrichment relative to HREE. This trend is also evident in how a data point representing the whole-bone composition of MPM-PV 1156-49 plots near the Nd-Gd edge of a Nd_N-Gd_N-Yb_N ternary plot (Figure 5C). Shale-normalized concentrations range from ~20–100 times NASC values in the external 250 μm of the cortex.

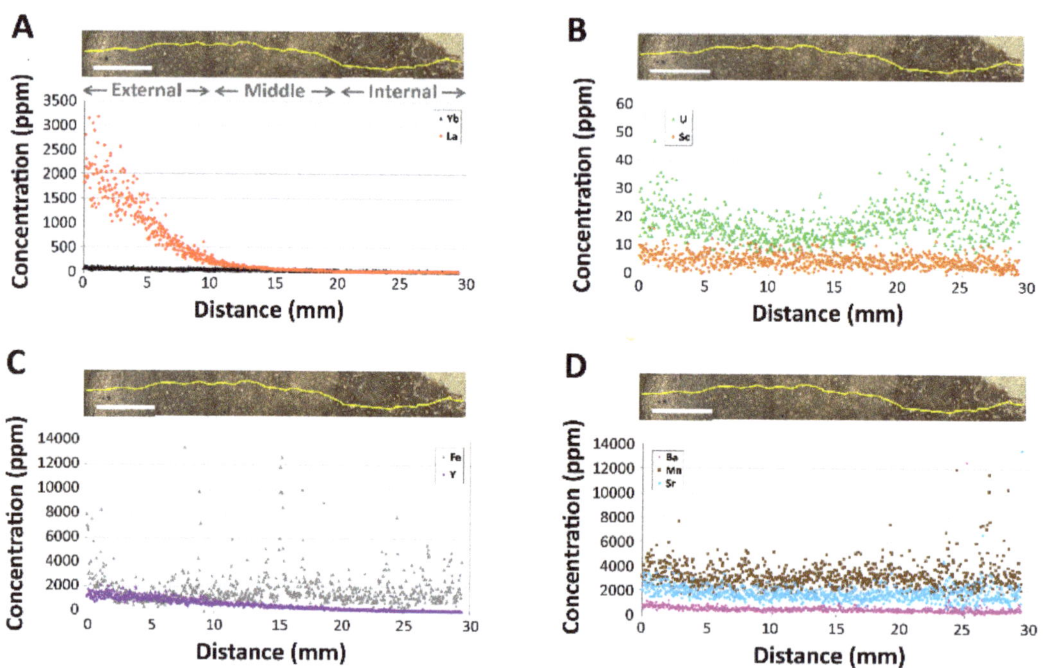

Figure 4. Intra-bone REE concentration gradients of various elements in the left humerus of *Dreadnoughtus* (MPM-PV 1156-49). (**A**) Lanthanum (La) and ytterbium (Yb). (**B**) Scandium (Sc) and uranium (U). (**C**) Iron (Fe) and yttrium (Y). (**D**) Barium (Ba), manganese (Mn), and strontium (Sr). Note that each panel has different concentration scales. Yellow line at the top of each panel depicts the track of the laser across the thick bone section during analyses. Gray text labels in (**A**) indicate the approximate regions of the cortex categorized as 'external', 'middle', and 'internal'. Scale bars (in white over bone images) each equal 1 mm.

Plotting each individual laser run compiled into the full transect into a La_N-Gd_N-Yb_N ternary plot (Figure 5D) shows that there is a tremendous spatial variation in bone composition (i.e., variation greatly exceeds two standard deviations). A spider diagram of individual laser runs (Figure 6B) further confirms this pattern, revealing substantial contrasts in the REE proportions by cortical depth. As these figures show, MPM-PV 1156-49 becomes substantially enriched in HREE relative to LREE and MREE with increasing depth into the cortex. Overall, the bone shifts from being modestly LREE and MREE enriched in the external-most cortex to drastically HREE enriched in the middle and internal cortices (approximately the inner two-thirds of the transect). In the internal-most laser run, HREE are proportionally enriched relative to LREE by over two orders of magnitude (Figure 6B).

Laser runs across the middle and internal cortices generally exhibit roughly equal depletion in LREE and enrichment in HREE relative to the NASC (Figure 6B). The three most internal laser runs exhibit a slight peak at Nd, and all runs across the middle and internal cortices exhibit a slightly negative Ce anomaly. The internal-most laser run exhibits slightly higher LREE and MREE concentrations than the run immediately external to it. Although there are no clear signs of tetrad effects (i.e., 'M'- or 'W'-shaped shale-normalized patterns; [57] and references therein) in individual laser runs (Figure 6B), subtle peaks at Nd, gadolinium (Gd), and holmium (Ho) in the whole-bone and external-most 250 μm spider diagrams (Figures 5B and 6A) may reflect slight tetrad effects (see Supplementary Section S3.1 for further discussion of potential tetrad effects in MPM-PV 1156-49).

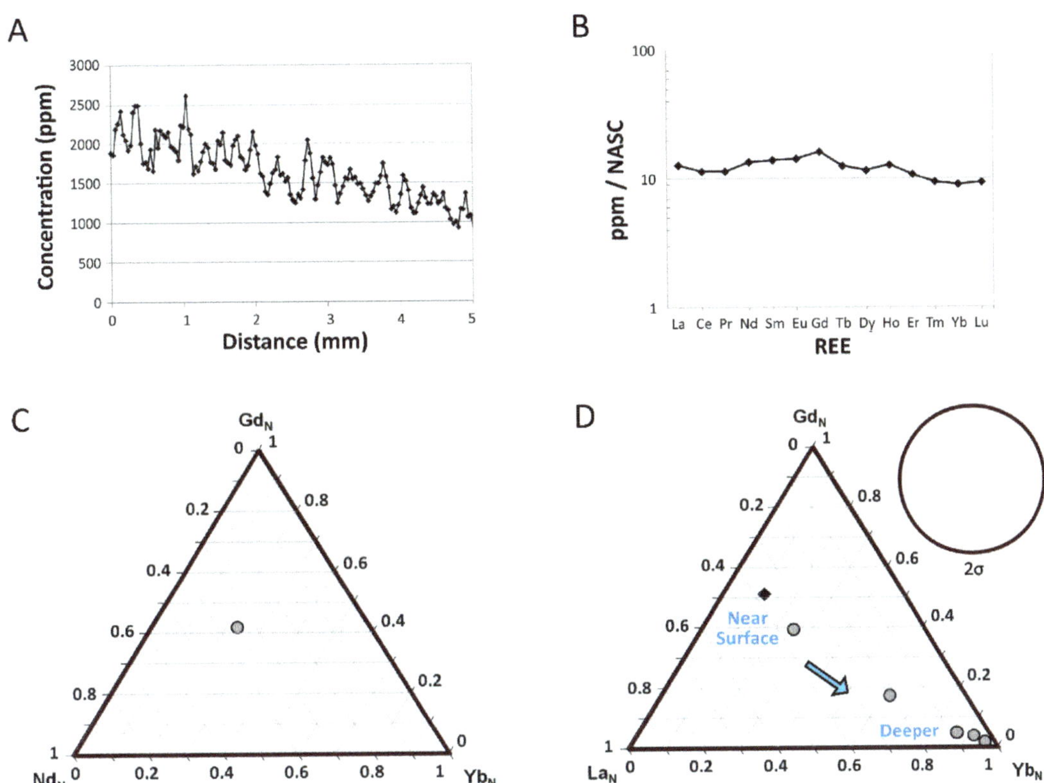

Figure 5. REE composition of the left humerus of *Dreadnoughtus* (MPM-PV 1156-49). (**A**) Three-point moving average profile of La concentrations in the outermost 5 mm of the bone. (**B**) Average NASC-normalized REE composition of the fossil specimen as a whole. (**C**,**D**) Ternary diagrams of NASC-normalized REE. (**C**) Average composition of the bone. (**D**) REE compositions divided into data from each individual laser transect (~5 mm of data each). Compositional data from the transect that included the outer bone edge is denoted by a dark diamond; all other internal transect data are indicated by gray circles. The 2σ circle depicts the area on the plot that represents two standard deviations based on ±5% relative standard deviation.

(La/Yb)$_N$ vs. (La/Sm)$_N$ Ratio Patterns: At the specimen level, MPM-PV 1156-49 exhibits a high (La/Sm)$_N$ value of 3.29 and a (La/Yb)$_N$ of 0.88, indicative of substantial LREE enrichment compared to most environmental water samples, dissolved loads, and sedimentary particulates (Figure 7A). This combination of values places the bone just outside the compositional ranges of freshwater in modern rivers and lakes (see Supplementary Materials, Section S2.2 for the literature sources used for environmental samples).

When REE ratios are plotted by individual laser runs, the bone exhibits a consistent pattern of decreasing (La/Yb)$_N$ and increasing (La/Sm)$_N$ as cortical depth progresses

(Figure 7B). The laser run including the external margin of the bone exhibits an $(La/Yb)_N$ value nearly two orders of magnitude greater than laser runs across the internal cortex. All laser runs across the internal cortex exhibit $(La/Yb)_N$ ratios < 0.05, and those across the middle cortex remain <0.4. Laser run $(La/Sm)_N$ ratios range roughly from 0.7–10.0.

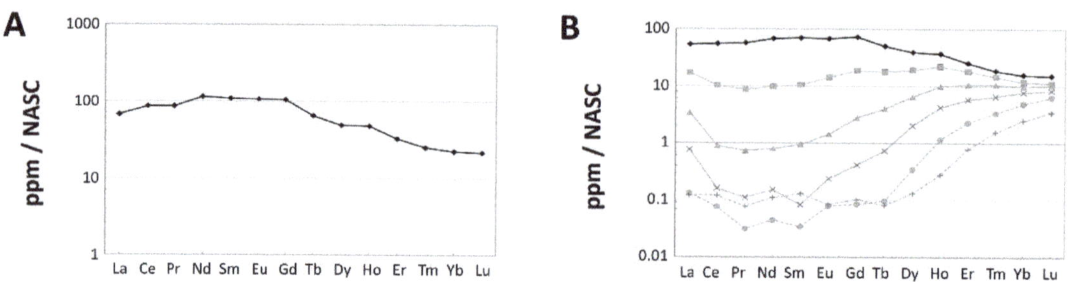

Figure 6. Spider diagrams of intra-bone NASC-normalized REE distribution patterns within the left humerus of *Dreadnoughtus* (MPM-PV 1156-49). (**A**) Average composition of the outermost 250 μm of the cortex, demonstrating a similar degree of relative LREE enrichment in the outermost cortex as seen in the bone as a whole (Figure 5B). (**B**) Variation in compositional patterns by laser transects. The pattern which includes the external margin of the bone is shown in black, those from deepest within the bone by dotted, light-gray lines, and all other analyses in between by solid, dark-gray lines.

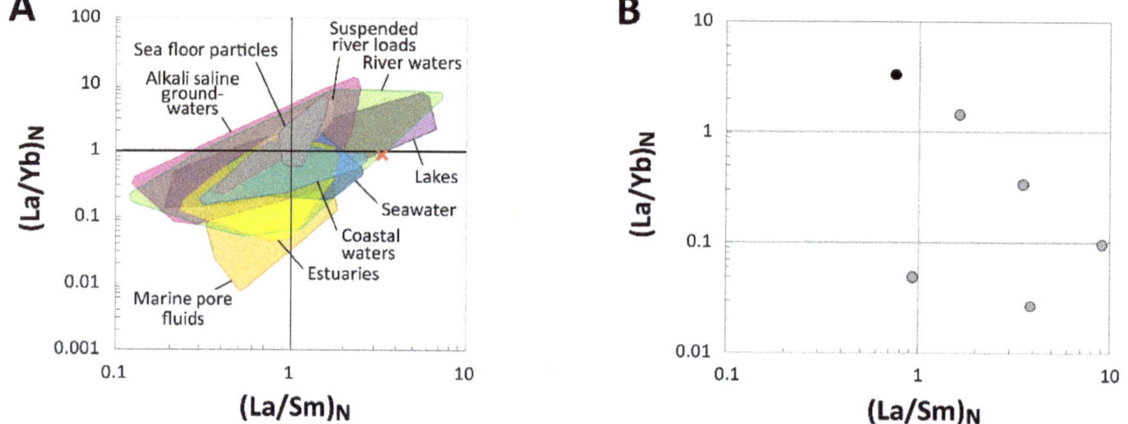

Figure 7. $(La/Yb)_N$ and $(La/Sm)_N$ ratios of the humerus of *Dreadnoughtus* (MPM-PV 1156-49). (**A**) Comparison of the whole-bone average $(La/Yb)_N$ and $(La/Sm)_N$ ratios of the fossil to ratios from various environmental waters and sedimentary particulates. Literature sources for environmental samples are provided in the Supplementary Materials (Section S2.2). (**B**) REE compositions of individual laser transects expressed as NASC-normalized $(La/Yb)_N$ and $(La/Sm)_N$ ratios. The transect including the external bone margin is denoted by the black symbol whereas all other (internal) transects are represented by gray symbols.

REE Anomalies: $(Ce/Ce^*)_N$ and La-corrected $(Ce/Ce^{**})_N$ anomalies are essentially absent in the external-most 5 mm of the cortex (Figure 8A). Values of each of these anomalies, as well as those of $(La/La^*)_N$, fluctuate both positively and negatively across the transect. Abundant data gaps occur for $(Ce/Ce^{**})_N$ and $(La/La^*)_N$ anomalies in the middle and internal cortices due to concentrations of praseodymium (Pr) and Nd commonly falling below the lower detection limit and, occasionally, Nd concentrations significantly exceeding those of Pr (Data S1).

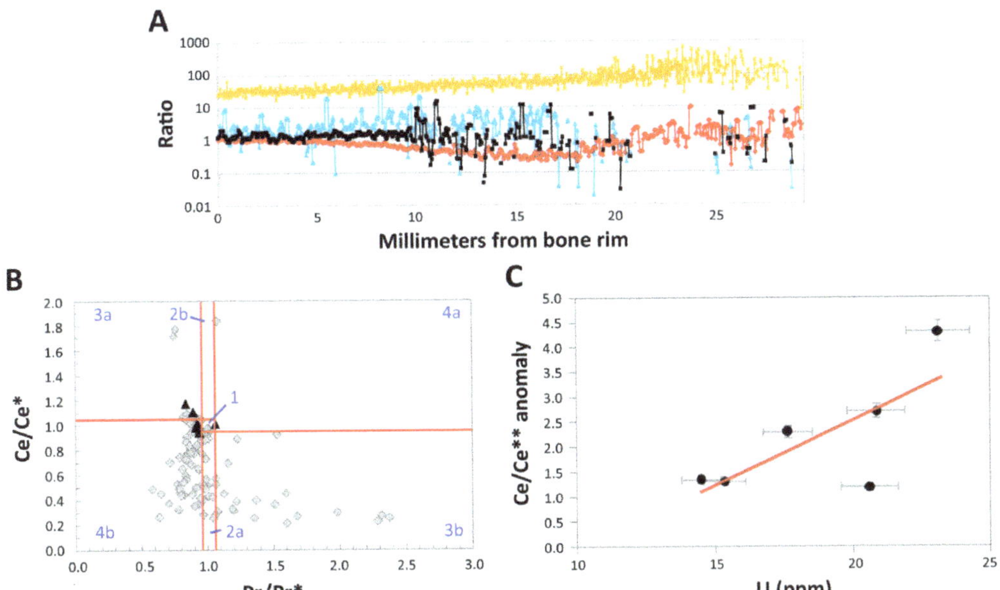

Figure 8. REE anomalies within the left humerus of *Dreadnoughtus* (MPM-PV 1156-49). (**A**) Intra-bone patterns of $(Ce/Ce^*)_N$ (red curve), $(Ce/Ce^{**})_N$ (black curve), and $(La/La^*)_N$ (blue curve) anomalies and Y/Ho ratios (orange curve). (**B**) $(Ce/Ce^*)_N$ vs. $(Pr/Pr^*)_N$ plot (after Bau and Dulski [58]) of five-point averages along the transect across the cortex of MPM-PV 1156-49 recorded by LA-ICPMS. Separate fields (labeled by blue text) are as follows: 1, neither Ce nor La anomaly; 2a, no Ce and positive La anomaly; 2b, no Ce and negative La anomaly; 3a, positive Ce and negative La anomaly; 3b, negative Ce and positive La anomaly; 4a, negative Ce and negative La anomaly; 4b, positive Ce and positive La anomaly. Measurements from the outer 1 mm of the external cortex are plotted as black triangles, and all measurements from deeper within the bone are plotted as gray diamonds. (**C**) Cerium anomaly $(Ce/Ce^{**})_N$ values plotted against uranium (U) concentrations. Error bars, in gray, are based on analytical reproducibility of $\pm 5\%$. There is a weak positive correlation ($r^2 = 0.54$), shown by the red trendline. Anomaly values were calculated as outlined in the Methods. Absence of $(Ce/Ce^*)_N$, $(Ce/Ce^{**})_N$, and $(La/La^*)_N$ anomalies occurs at 1.0.

Of all of the anomalies considered, $(Ce/Ce^*)_N$ exhibits the most variable patterns with cortical depth. Specifically, $(Ce/Ce^*)_N$ values slowly decrease to a negative peak of ~0.3 in the middle cortex (~16 mm), then steadily rebound to positive values throughout most of the internal cortex (Figure 8A). The highest values recorded along the transect (~9.6) occur near 24 mm, forming the apex of a subtle, positive peak in the outer portion of the internal cortex. However, when averaged across the entire transect (Table 2), these fluctuations cancel each other out; as a whole, MPM-PV 1156-49 exhibits essentially no $(Ce/Ce^*)_N$ anomaly (value = 0.94). This finding agrees with the lack of any clear inflection of the NASC-normalized pattern at Ce in the whole-bone spider diagram (Figure 5B).

Following Bau and Dulski [58], $(Ce/Ce^*)_N$ values were also plotted against $(Pr/Pr^*)_N$ values in order to differentiate true, redox-related cerium anomalies from apparent anomalies produced by $(La/La^*)_N$ anomalies. In this plot (Figure 8B), both $(Ce/Ce^*)_N$ and $(Pr/Pr^*)_N$ values exhibit significantly greater variation in the middle and internal cortices than is seen in the external-most 1 mm of the bone. This pattern indicates that the internal regions of the bone are relatively more heterogeneous in composition than the external-most portion of the cortex. All but one data point from the external cortex plot near the right margins of fields 3a and 4b, indicative of slightly positive Ce anomalies and variable

La anomalies (Figure 8B). The vast majority of data points from the middle and internal cortices plot in fields 2a, 3b, and 4b, generally indicative of positive La anomalies.

To quantitatively assess these qualitative trends, we also calculated $(La/La^*)_N$ anomalies and La-corrected $(Ce/Ce^{**})_N$ anomalies (see Methods). $(Ce/Ce^{**})_N$ anomalies are consistently slightly positive throughout most of the external cortex (i.e., the external-most 10 mm exhibits an average anomaly value of 1.40), after which they fluctuate considerably between positive and negative values around an average of zero in the middle cortex (Figure 8A). Unfortunately, minimal data are available from the internal cortex due to concentrations of these LREE often falling below the detection limit. Variations in $(Ce/Ce^{**})_N$ values across the middle cortex encompass a range of more than two orders of magnitude. At the specimen level, the whole-bone exhibits a slightly positive $(Ce/Ce^{**})_N$ anomaly (1.22; Table 2). $(Ce/Ce^{**})_N$ anomaly values are only weakly correlated with U concentrations ($r^2 = 0.54$) when they are plotted by the laser run (Figure 8C). $(La/La^*)_N$ anomalies are commonly positive through the external 17 mm of the bone, after which point data become too sparse to reliably interpret (Figure 8A). This is also evident by the positive $(La/La^*)_N$ anomaly average for the entire specimen (1.75; Table 2); however, again, this signature is almost entirely derived from the external and middle cortices.

Yttrium/holmium (Y/Ho) ratios are essentially chondritic (26; [59]) near the cortical margin, but they steadily become increasingly positive with depth (Figure 8A). Y/Ho anomalies form a broad peak in the internal cortex (whose apex is near 23 mm) at values from ~50–600, then gently decline to values from ~60–150 at the internal end of the transect. For the entire specimen (i.e., all transect data averaged together), the Y/Ho anomaly is slightly positive (38; Table 2).

4.2. Soft Tissue Preservation

4.2.1. Demineralization/Morphological Structures

Demineralized cortical bone yielded soft, flexible structures morphologically consistent with vessels, the fibrous collagenous matrix, and osteocytes (Figure 9A–F), and we hereafter refer to them as such for brevity and clarity. During demineralization in ethylenediaminetetraacetic acid (EDTA), vessels were observed emerging directly from the fossil tissue (Figure 9A) as well as in solution. Isolated samples of these hollow, flexible tubes did not dissolve after multiple treatments with acetone, refuting the hypothesis that they are casts formed by the in-filling of acetone-soluble glues and/or field consolidates. The tapering bifurcation pattern observed in modern vessels and in reported soft tissues from other ancient vertebrates (e.g., [60,61]) was also observed among the recovered vessels from *D. schrani* (Figure 9B). These vessels are inconsistent with fungal hyphae, which are cylindrical and grow from an apical tip extension [62,63] and, thus, do not taper when branching. Hyphae are also generally an order of magnitude smaller than the structures we observed (e.g., Figure 2 in [30]). When imaged under cross-polarized light, vessels displayed minimum birefringence, regardless of stage orientation (Figure 9C). Any birefringence that was observed was limited to small, isolated areas along the vessel wall, leaving the majority of the structure dark. Because most minerals are anisotropic and display birefringence under cross-polarized light [64,65], the lack of birefringence in these structures, as well as their pliability, are features more consistent with an amorphous material of organic origin than a mineralized pseudomorph [19,66].

Matrix recovered from *D. schrani* was soft, pliable, and fibrous in appearance, with encased elongated osteocytes oriented with their long axes parallel to one another (Figure 9D). In general, osteocytes displayed shorter, blunted filopodia-like structures in comparison to those observed on putative osteocytes from other extinct taxa [13,60,61,66–68]. It is not known if this is a result of tissue degradation during the time between the exhumation of the fossil and analyses, or an artifact of preservation caused by the specific depositional environment in which the specimen was entombed. Regardless, we recovered examples of these elongated structures bearing the hallmark filopodia-like extensions of modern osteocytes (Figure 9E: black arrows). Similar to the vessels, isolated osteocytes did not dissolve in acetone and demonstrated very little birefringence when observed under cross-polarized light (Figure 9F), precluding their origin as either glue or mineral-infill of osteocyte lacunae, respectively [19,66].

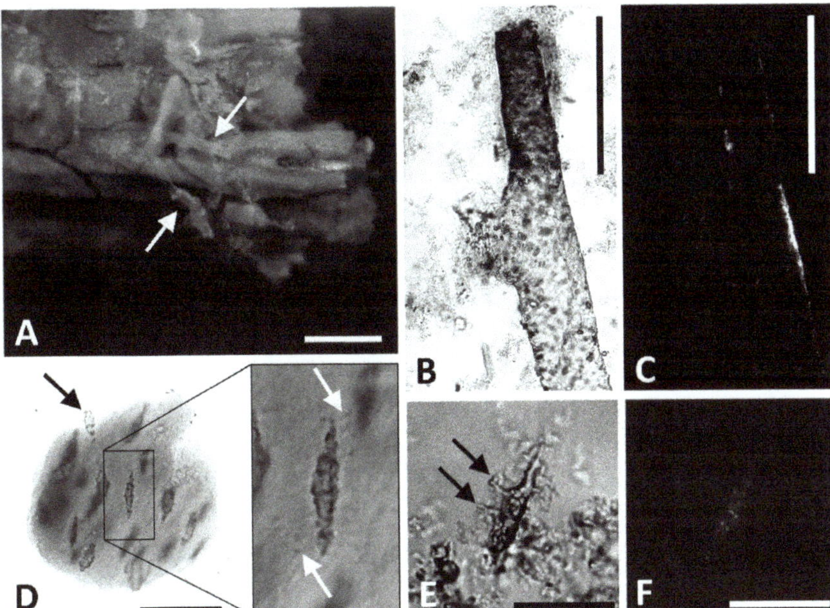

Figure 9. Structures recovered from MPM-PV 1156-49 that are morphologically consistent with soft-tissues. (**A**) White arrows indicate pliable, hollow, transparent tubes emerging from a humeral fragment as the mineral phase is dissolved in EDTA. Scale bar = 500 µm. (**B**) Vessel directly treated three times with acetone, then imaged under transmitted light. Scale bar = 200 µm. (**C**) When imaged under cross-polarized light, the vessel (**B**) showed minimal birefringence, indicating it is not a mineralized structure. Scale bar = 200 µm. (**D**) Fibrous matrix with embedded osteocytes. Osteocytes can be observed emerging from the matrix (black arrow), indicating that these structures are three-dimensional. Inset of the osteocyte shows distinct, lateral projections from the cell-like structure into the surrounding matrix (white arrows), which may represent preserved filopodia or empty canaliculi. Scale bar = 50 µm. (**E**) Isolated osteocyte imaged in transmitted light. Note the narrow, branching structures consistent with filopodia (black arrows). Scale bar = 25 µm. (**F**) Osteocyte imaged under cross-polarized light. Minimal birefringence indicates this structure is not a mineral in-fill of a lacuna. Scale bar = 25 µm.

4.2.2. Immunofluorescence (IF)

When exposed to commercial polyclonal antibodies raised against chicken collagen I, both chicken (*Gallus gallus*) sections (Figure 10A,D,G,J) and alligator (*Alligator mississippiensis*) sections (Figure 10B,E,H,K) showed a clear signal of immunoreactivity. Binding intensity was greatest in chicken sections (Figure 10D), consistent with our use of antibodies raised against chicken collagen I. Binding was diminished in alligator sections (Figure 10E), which is expected given the phylogenetic distance between alligators and the chicken collagen used as an immunogen [69,70]. Despite the lower signal intensity in alligator, positive binding for this tissue indicates that many of the same collagen epitopes have been retained in both species despite their divergence 237 million years in the past [71]. Thus, we can conclude these antibodies are appropriate for analyses of sauropod bone, as Sauropoda is bracketed phylogenetically by these two extant archosaurs [72] and should, therefore, share the archosaur collagen I epitopes possessed by both *Gallus* and *Alligator*.

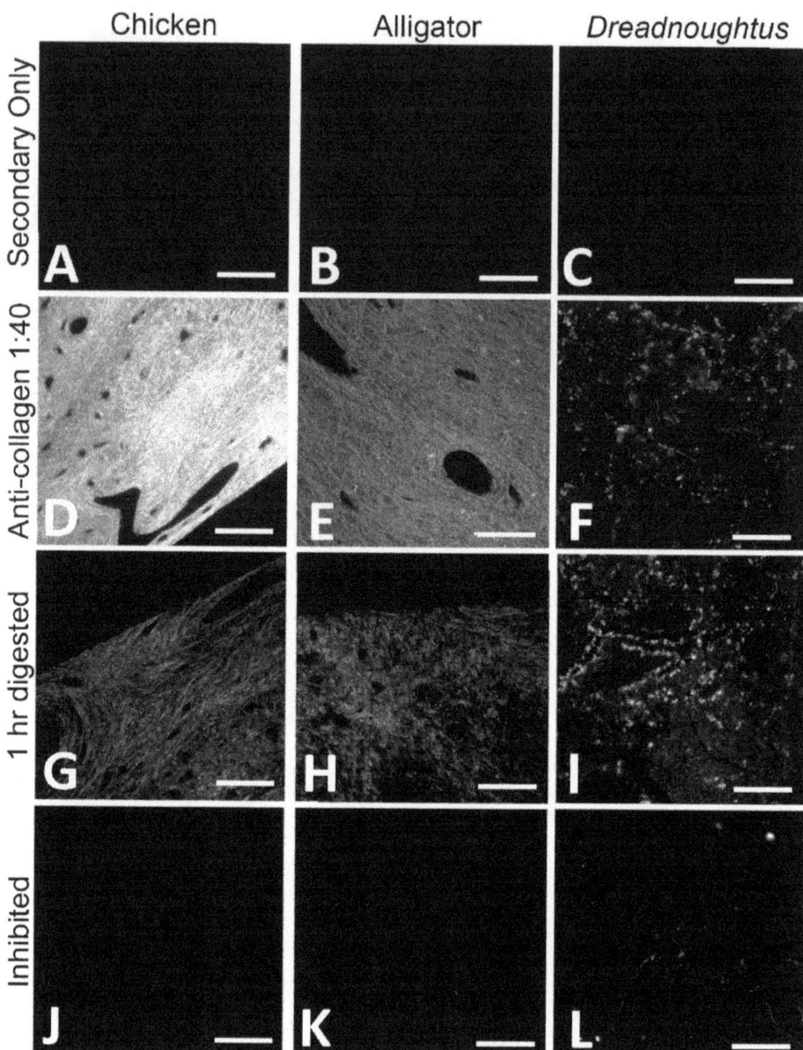

Figure 10. In situ localization of collagen I in demineralized chicken (**A,D,G,J**), alligator (**B,E,H,K**), and *Dreadnoughtus schrani* (**C,F,I,L**) cortical bone fragments using immunofluorescence, imaged at 150 ms. (**A–C**). Sections that have been exposed to secondary antibodies without incubation in collagen I antibodies to control for non-specific binding of the secondary antibody and background fluorescence (i.e., "secondary only" control). Lack of signal indicates that the observed fluorescence in other sections cannot be attributed to non-specific binding of the secondary. (**D–F**) Sections incubated with anti-collagen I antibodies (diluted 1:40) showed a positive signal for collagen I in all tested specimens. (**G–I**) Tissues that have been digested with collagenase prior to exposure to collagen I antibodies show a reduction in signal in the (**G**) chicken and (**H**) alligator sections, and an increase in signal in the (**I**) *D. schrani* section. This indicates that the targeted digestion of collagen I in these tissues directly affect the binding pattern of the antibody, supporting its specificity for collagen I epitopes. (**J–L**) Inhibition of the collagen I antibody with chicken collagen shows reduced binding for all taxa, again supporting the specificity of this antibody. Scale bars = 50 μm.

Indeed, the collagen I signal was also observed in *D. schrani* tissue sections (Figure 10C,F,I,L), consistent in pattern as with the extant controls (Figure 10F), though the intensity of the signal was greatly reduced relative to both chicken and alligator. Antibody binding was detected within visible sections of tissue, but not in void regions of the sections that contained only resin.

To confirm that the signal observed in *Dreadnoughtus* tissue sections did not result from non-specific binding of the primary antibody, we employed multiple controls. Some sections of each tissue were enzymatically digested with collagenase A (Roche) before incubation with anti-collagen I antibodies [73,74]. This protease specifically cleaves the X-Gly bond in the sequence Pro-X-Gly-Pro, which occurs at a high frequency in collagen but is rare in other proteins [74]. Because collagenase targets a bond abundant in collagen, but rare in non-collagenous proteins to which antibodies may bind non-specifically, collagenase treatment of the tissues in situ provides a specificity control for collagen I antibodies. If the antibodies are binding to collagen, then collagenase will substantially alter the immunoreactive response observed in the tissue; either diminished binding as collagen I is destroyed [75], or increased binding as digestion exposes more epitopes for binding [76,77]. If antibodies are binding non-specifically to proteins other than collagen, then collagenase should have no effect. In both chicken and alligator tissues, the fluorescent signal was greatly reduced after digestion in collagenase for 1 h (Figure 10G,H), indicating that the primary antibody is binding to collagen I specifically, as the targeted destruction of collagen has direct impact on the binding pattern [75].

In contrast to the extant controls, digestion for 1 h in collagenase increased the intensity of signal in *D. schrani* tissue (Figure 10I). It has been suggested that condensation reactions, such as Amadori rearrangements and Maillard reactions, may help protect proteins across geologic time through the generation of intermolecular cross-links, forming insoluble aggregates that are resistant to degradation, but which also block reactive epitopes [19–22]. If such processes have played a role in the preservation of these molecules, brief digestion in collagenase may promote antigen retrieval in fossilized tissues by breaking apart proteinaceous aggregates, exposing more epitopes for binding, and increasing signal intensity [78]. An analogous effect has been observed in extant tissues, as many immunofluorescence protocols optimize the signal by initial incubation of sections in proteinase K, which breaks cross-links generated by formalin during tissue fixation to increase epitope exposure and improve the signal [76]. To test this hypothesis, we extended the duration of collagenase digestion (Figure 11). While sections digested for only 1 h showed a slight increase in antibody binding (Figure 11B), sections incubated in collagenase overnight (Figure 11C) or longer (Figure 11D) demonstrated a near total loss of signal. Additionally, tissue sections themselves became smaller after longer digestions, a further indication they are comprised of material that is directly affected by collagenase.

In addition to the tissue digestion experiments, we also inhibited the collagen I antibody with an excess of collagen I. In a specific antibody, collagen I blocks all binding sites on the antibody, making them unavailable to bind to epitopes in the tissue [79]. After inhibition, antibody binding to chicken and alligator tissue was completely absent (Figure 10J,K), supporting the specificity of our antibody. *Dreadnoughtus schrani* sections incubated with inhibited antibodies showed a sharp reduction in the fluorescent signal, though inhibition was not as complete as in the extant material (Figure 10L).

4.2.3. Enzyme-Linked Immunosorbent Assay (ELISA)

ELISA is a method that can identify the presence of protein by quantifying the intensity of color-change generated by antigen–antibody interactions linked to chromogenic substrates [80]. Unlike the in situ immunofluorescence assays, ELISA works by tagging epitopes in chemical extracts of tissue. Because it is performed on chemically extracted antigens that have been solubilized (as opposed to antigen embedded within whole tissue), ELISA is generally an order of magnitude more sensitive than in situ assays [81] and, thus, provides a valuable complement for results obtained through immunofluorescence.

Additionally, unlike assays that require extracted proteins to be denatured for analysis (e.g., electrophoresis), ELISA is capable of identifying native epitopes [82]. When incubated with anti-chicken collagen I antibodies, GuHCl extraction products from *D. schrani* tissue showed an absorbance value more than 300% above the value of its secondary-only negative control (Figure 12, columns 1 and 2). This level of absorbance surpasses the standard criteria for a 'positive' detection of an antigen, which is twice the background signal (e.g., [83–85]). Conversely, extraction products from sediment and buffer controls displayed slightly negative absorbance values, indicating no immunoreactivity was detected in these samples, and that these controls show no evidence of collagen I (Figure 12, columns 3–6). This suggests that the immunoreactivity detected in the *D. schrani* extractions is not the result of contamination derived from the laboratory or burial environments, as these negative controls were subject to the same locations, conditions, and reagents, but contained no reactive material.

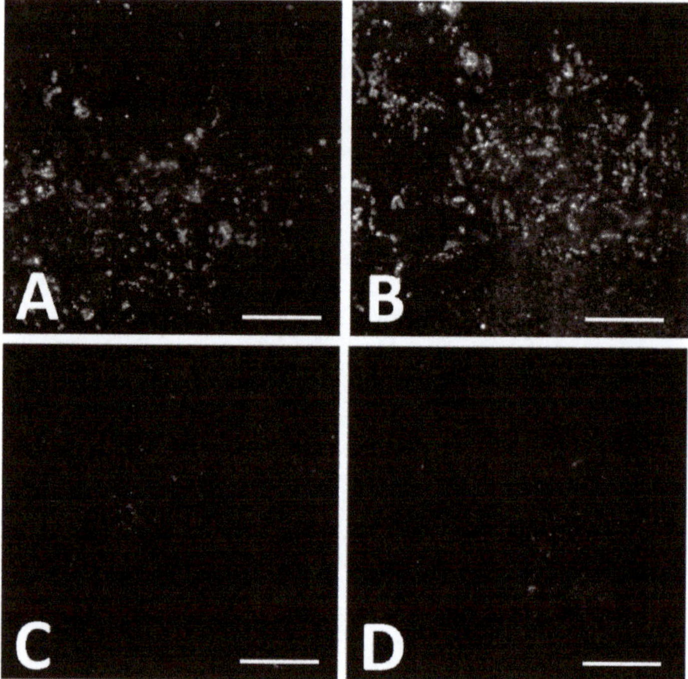

Figure 11. Reduction in immunofluorescent signal in *Dreadnoughtus schrani* tissue with prolonged collagenase digestion prior to incubation with anti-chicken collagen I antibodies. All sections were imaged at an exposure of 150 ms. Sections were digested for (**A**) 0 h (undigested), (**B**) 1 h, (**C**) overnight, and (**D**) 24 h. Compared to the undigested tissue (**A**), immunoreactivity slightly increased with tissue that was digested in collagenase for 1 h prior to incubation with primary antibodies (**B**). This effect is also seen in some extant tissues, as partial digestion in proteinase K increases exposure of epitopes [76]. Extending digestion times to overnight or longer (**C**,**D**) resulted in substantial loss of signal, supporting the specificity of the antibody. Scale bars = 50 μm.

The detected signal for *D. schrani* was greatly reduced compared to values obtained for modern controls, which tended to reach saturation (absorbance = 3.0+) within the first 90 min of reading, even with sample concentrations as low as 500 ng/well (Figure S5). After an equivalent amount of time, extraction products from 150 mg of *D. schrani* fossil showed absorbance values approximately 50 times lower than the values obtained for 500 ng of extracted chicken protein.

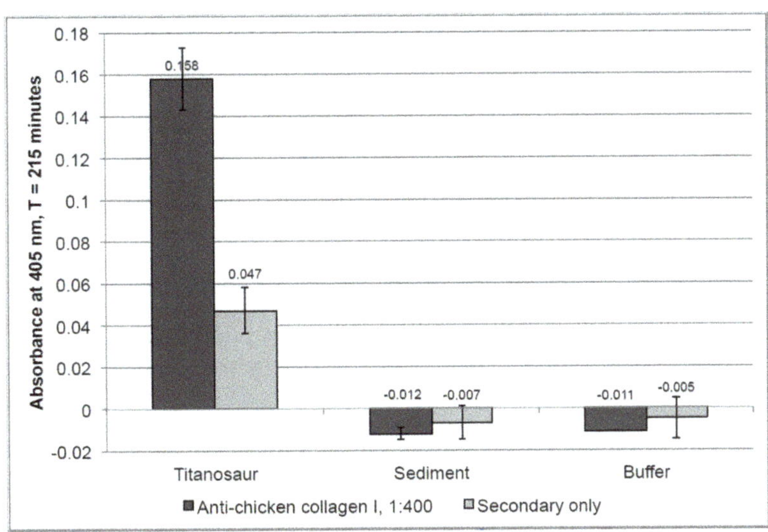

Figure 12. Enzyme—linked immunosorbent assay (ELISA) results of chemical extractions from *Dreadnoughtus schrani* bone, sediment, and laboratory reagents (this trial used the extraction products from 150 mg of bone/sediment per well). Dark gray columns (left) represent absorbance (405 nm, time = 215 min) obtained for extractions incubated with anti-collagen I antibodies (1:400). Light gray columns (right) represent background control (i.e., sample exposed only to secondary antibodies). The absorbance value obtained for *D. schrani* tissue is more than three times its corresponding background control. Negative absorbance values were obtained for extraction products from sediment and buffer samples, corresponding to a lack of immunoreactivity in these samples and indicating that the results obtained for *D. schrani* are not the result of contamination from the entombing environment or lab reagents. Error bars represent one standard deviation above and below mean absorbance values for each sample.

5. Discussion

The holotype of *Dreadnoughtus schrani* (MPM-PV 1156) represents one of the most morphologically and skeletally complete titanosaur species known [25]. To determine whether the taphonomic conditions that resulted in its extraordinary preservation at the gross morphological level also allowed molecular preservation, we performed extensive analyses of its tissues, including the characterization of its histology, geochemical history, and molecular content. First, the bone microstructure and mineral content was evaluated. Thin sections of the humerus displayed well-preserved primary and secondary bone tissues and lack indicators of microbial attack (fungal tunneling or MDF) or microstructural alteration, giving this specimen a 5/5 on the Histologic Index [54]. XRD analysis showed little exogenous mineralization, with ~95% of the bone mineral comprised of apatite (Table S10). Such a high level of structural integrity is generally correlated with preserved protein content in archaeological bone [21,86–88], indicating that MPM-PV 1156 is an appropriate specimen for more in-depth geochemical and molecular characterization.

5.1. Reconstructing the Geochemical History of MPM-PV 1156

Our trace element data elucidate the geochemical history of the *Dreadnoughtus* holotype, as well as clarify the early and late diagenetic conditions to which its left humerus (MPM-PV 1156-49) was exposed, and which allowed it to retain endogenous cells, soft tissues, and collagen I. MPM-PV 1156-49 appears to preserve original, early diagenetic signatures that have not been meaningfully obfuscated by late diagenetic overprinting. This is supported by: (1) a lack of oversteepened elemental profiles that would indicate significant, late uptake (Figure 4); (2) the absence of signs of trace element leaching (i.e., most elemental concentrations increase rather than decrease toward the cortical margin; Figure 4), and; (3) an REE composition most similar in appearance to circum-neutral pH (cf., [89]) rivers and lake freshwaters rather than alkali groundwaters (Figure 7A), which is inconsistent with the incorporation of a major portion of the trace element inventory of the bone from late diagenetic fluids. Overall, the bone exhibits high surface concentrations of LREE (average ~2300 ppm; Data S1), yet its $\sum$REE value of 1964 ppm is comparable to that of most other Mesozoic bones reported in the prior literature (Table 3), which exhibit $\sum$REE ranging from ~300 ppm to over 25,000 ppm [90–100], as are its concentrations of Y (505 ppm), Lu (4 ppm), and U (19 ppm). In contrast, Fe (0.19 wt%), Sr (1747 ppm), and Ba (447 ppm) each exhibit low average concentrations compared to other protein-bearing dinosaur bones we have recently analyzed (0.73–1.76 wt%, ~2300–3700 ppm, and ~900–2100 ppm, respectively) [49,100], perhaps reflecting low abundance of these elements within early diagenetic pore fluids at this site. Although these comparisons are rough due to the specimens deriving from diverse taxa and ranging widely in cortical width and depositional circumstances, they demonstrate that the humerus of MPM-PV 1156 generally exhibits average chemical alteration for its age.

Table 3. Summary of the REE composition of the left humerus of *Dreadnoughtus schrani* MPM-PV 1156-49. Qualitative $\sum$REE content is based on the value shown in Table 2 (1964 ppm) in comparison to values from other Mesozoic bones (as listed in the main text). Abbreviations: DMD, double medium diffusion *sensu* Kohn ([56]); LREE, light rare earth elements.

Clear DMD Kink for LREE?	Relative Noise in Outer Cortex for La	REE Suggest Flow in Marrow Cavity?	Relative $\sum$REE Content (Whole Bone)	Relative U Content (Whole Bone)	Relative Porosity of the Cortex
No	Low	No	Moderate	Low	Low

Concentrations of REE and Y steeply decline from the cortical margin of the bone (Figure 4A,C), indicating that primary trace element uptake occurred via a single phase of simple diffusion (*sensu* [101]) of one pore fluid. Such concentration depth profiles may arise either via brief uptake from a pore fluid which is not being replenished (e.g., [49]) or as a result of fractionation during protracted uptake from a continually replenished solution (e.g., [102]). We rule out the first of these alternatives due to the significant magnitude of REE enrichment throughout the cortex (e.g., Table 2, Data S1) and heterolithic nature of the entombing sediments, which would have allowed for sustained pore fluid to flow through the remains throughout the early diagenetic, primary phase of trace element uptake. In addition, the cortex of humerus MPM-PV 1156-49 is both extremely thick (~3 cm) and dense—attributes which are known to cause a significant 'filtering' effect on trace element diffusion [52,99,100,102,103]. For these reasons, we instead interpret the stark differences in REE enrichment and composition with cortical depth to have arisen via fractionation during protracted, additive uptake (see below). In comparison, the flatter profiles of Fe, Ba, and Mn (Figure 4C,D) suggest that as pore fluids percolated through the bone, these elements were gradually incorporated into homogenously distributed secondary mineral phases, most likely barite, goethite, and Mn oxides [104]. The similarly flat concentration depth profile of Sr is attributed to spatially homogenous substitution for Ca ions in the bone apatite, which commonly occurs during fossilization of bioapatitic tissues [105,106].

Elevated concentrations of REE and other trace elements in the external cortex of MPM-PV 1156-49 could reflect either brief uptake from trace element enriched surface and groundwaters (cf., [107]) or protracted uptake from pore fluids possessing low concentrations of these elements (cf., [108]). We find the former of these possibilities unlikely for three reasons. First, most natural waters in fluviodeltaic environments (such as the environment interpreted for MPM-PV 1156 [25]) possess low concentrations of REE, Y, U, and other trace elements [90,109–111]. This is because they form a complex with carbonates and humic acids (e.g., [112,113]) and/or may be partially removed in early diagenesis by coprecipitation in secondary phosphates within sediments (e.g., [114]). Second, REE concentration depth profiles within the humerus form typical "simple diffusion" gradients (e.g., La in Figure 4A), which reflect sustained diffusion; they do not show the oversteepened curves that would reflect either preferential uptake from a trace element enriched pore fluid within the external cortex or major late diagenetic uptake (cf., [101]). Finally, many trace element concentrations with relatively slow diffusivities [108], including those of HREE (e.g., Yb), remain >2 ppm throughout most of the internal cortex (see Data S1). This is also consistent with protracted uptake, not brief uptake, because these internal regions are farthest from the external pore fluid source. Thus, we conclude that the bones of MPM-PV 1156 interacted with pore fluids for a longer period of time than other Cretaceous specimens which have been found to retain original protein [49,100].

Numerous trace element signatures within the humerus of MPM-PV 1156 demonstrate that the composition of pore fluids percolating through the specimen changed over time through early diagenesis. This is particularly apparent from the spider diagram of REE proportions by transect (Figure 6B), the ternary diagram of REE ratios (Figure 5D), and $(La/Yb)_N$ vs. $(La/Sm)_N$ plot (Figure 7B), each of which show obvious signs of substantial intra-bone fractionation (*sensu* [52]) occurring during uptake. Specifically, each of these figures show that pore fluids became significantly depleted in LREE by the time they reached the middle and internal cortices. Similarly, despite the similar diffusivities of REE and U in bone [108], they exhibit highly contrasting concentration depth profile shapes (compare Figure 4A,B): REE profiles decline steeply from the cortical margin to low concentrations throughout the interior of the bone, while those of U steadily increase from the internal portion of the middle cortex to a maximum in the internal cortex. Further, the weak correlation between U concentrations and $(Ce/Ce^*)_N$ values for each laser run (Figure 8C) suggests these were incorporated into the bone over similar timescales [115]. Taken together, it is apparent that the availability of REE diminished as pore fluids percolated deeper into the bone whereas that of U remained high. As discussed by Suarez and Kohn [97] and Kohn and Moses [108], such patterns commonly arise during uptake from oxic fluids, caused by the relatively lower partition coefficients in apatite for U than REE, and greater mobility of U complexes than REE complexes under oxic conditions. Increasingly positive Y/Ho anomalies with cortical depth and positive $(La/La^*)_N$ anomalies in the external and middle cortices (Figure 8A) are also likely products of fractionation [100].

The lack of significant $(Ce/Ce^*)_N$ or $(Ce/Ce^{**})_N$ anomalies at the whole-bone level (Table 2) and near the cortical margin (Figure 8A) indicates that the early diagenetic environment was neither strongly oxidizing nor reducing. However, slightly negative $(Ce/Ce^*)_N$ values throughout the middle cortex and the internal portion of the external cortex (Figure 8A) indicate that weak oxidizing conditions prevailed in these regions through the timeframe of uptake. $(Ce/Ce^{**})_N$ anomalies were found to be generally slightly positive in the external cortex, indicative of slightly oxidizing conditions (Figure 8A), which is supported by the presence of high Sc and moderate U concentrations in this region (Figure 4B). Conversely, data points for this region of the bone in the $(Ce/Ce^*)_N$ vs. $(Pr/Pr^*)_N$ plot (Figure 8B) fall within fields indicative of slightly reducing conditions. Figure 8A demonstrates that this disagreement arises from the presence of slightly positive $(La/La^*)_N$ anomalies in the external cortex, which bias calculations of traditional $(Ce/Ce^*)_N$ anomalies [52].

In summary, our cumulative trace element data indicate that the humerus of MPM-PV 1156 experienced protracted trace element uptake from a circum-neutral pH pore fluid during early diagenesis. The similar REE composition of the bone to freshwaters in lakes and rivers indicates that this pore fluid was predominantly fed from surficial sources rather than groundwater sources. By combining these geochemical insights with sedimentologic and taphonomic observations by Lacovara et al. ([25]; i.e., partial articulation, recovery of several cf. *Orkoraptor burkei* teeth within the quarry, burial within a mixed lithosome), we conclude that the carcass of MPM-PV 1156 experienced decay and scavenging for a moderate length of time in close proximity to a fluvial channel on a dry floodplain, after which it became buried by a major crevasse splay event. Its remains were exposed to LREE-enriched groundwaters under weakly-oxidizing conditions for a considerable time following burial, but after early diagenetic fossilization they experienced negligible further chemical alteration.

5.2. Evaluating the Preservation of Soft Tissues

The soft, pliable textures of the observed matrix, osteocytes, and vessels are not consistent with mineral in-filling of vessel canals or osteocyte lacunae, and the resistance of these structures to acetone washes precludes glue or consolidate in-filling. All of the osteocytes isolated from the demineralization solution were elongated and flattened in morphology, consistent with osteocytes found in mature bone (i.e., lamellar bone, osteons) [66]. This is consistent with the source fossil, as the humerus of *D. schrani* is composed primarily of remodeled secondary osteons, a trait commonly seen in titanosaurs even before an individual reaches skeletal maturity (e.g., [116–118]). Additionally, the vessels recovered are not septate at any point along their length (Figure 9B), distinguishing them from the majority of fungal hyphae [119]. It has been suggested that similar structures previously reported as vessels (e.g., [60]) may represent biofilm endocasts as opposed to original tissues [120]. However, this hypothesis has not been, and is not, supported. Although it has been suggested that biofilms may play a role in endogenous molecular preservation [18], there is no direct evidence that microbes are capable of producing biofilms that can perfectly replicate these structures in fine detail, nor that biofilms are able to retain a three-dimensional shape in solution [67]. Nevertheless, we do not consider morphological similarity, on its own, conclusive evidence that the structures described in this paper are original and endogenous. To that end, we conducted additional experimentation to investigate the endogeneity of the ostensible matrix recovered from *D. schrani*.

Both in situ and in-solution immunological tests supported the preservation of collagen I in *D. schrani* tissues. Immunofluorescent assays displayed in situ antibody binding in fossil sections that was consistent with assays of modern bone (Figure 10). Furthermore, the immunological response of fossil tissues in assays to test for antibody specificity were also consistent with what was observed for modern bone; binding was inhibited when collagen I antibodies were "blocked" by pre-absorption chicken collagen I and was directly affected by the targeted enzymatic digestion of tissue with collagenase. Interestingly, the specific response of the fossil tissue to collagenase is congruous with the presence of diagenetically cross-linked collagen and correlates with an initial increase in signal caused by the exposure of additional epitopes, followed by the subsequent decline in binding as digestion progresses to more extensively degrade the present collagen I. ELISA data showed antibody binding to chemical extraction products that were consistent with a substantial, but not complete, protein loss in fossilized bone compared to recent tissues, and absence of binding in the sediment or laboratory reagents.

Although it has been suggested that immunoreactivity in fossils could be the result of contamination with collagen-like proteins that are produced by bacteria [120] or fungi [121], or non-specific binding to soil microorganisms [122], it has not been explained how these microbial proteins could at once be so ubiquitous as to contaminate specimens from drastically different burial environments and localities, but not ubiquitous enough to be present in the very sediment entombing those specimens, as shown in our ELISA data. Further, the antibody specificity tests conducted on both extant and ancient archosaurian tissue in situ (Figures 9 and 10) support the specificity of the antibody used for ELISA and IF testing to collagen I, indicating that our immunological data are neither the result of cross-reaction with other molecules that may be present in fossilized or non-fossilized bone (digestion control), nor spurious binding with extraneous paratopes in our polyclonal antibody (inhibition control). Thus, the hypothesis that the collagen I signal we have retrieved from *D. schrani* is of exogenous origin is not supported.

Our molecular analyses have demonstrated the presence of soft-tissue and collagen I preservation in the holotype of *Dreadnoughtus schrani* through three independent techniques: (1) morphological identification; (2) in situ localization of antibody–antigen complexes; (3) immunoreactivity to chemical extracts. These assays universally support the identification of collagenous matrix (microscopy), or collagen I specifically (IF, ELISA), and have failed to show evidence of similar molecular content in the entombing sediment or the laboratory environment. Any alternative hypotheses must identify a contaminating agent that contains epitopes recognized by specific antibodies raised against (and inhibited by) archosaur collagen I, which is degraded by collagenase, that is completely absent in the surrounding sediment, and that does not leave histological evidence of microbial/fungal tunneling or destructive foci, and that can contaminate bone tissue both in situ and in-solution without simultaneously contaminating negative control samples that are conducted in tandem. We conclude that the most parsimonious explanation for these results is that original, endogenous collagen I has been preserved in the holotype of *Dreadnoughtus schrani*.

5.3. Implications for Future Paleomolecular Studies

That the holotype of *Dreadnoughtus* exhibits average alteration for its age could be viewed as a surprising finding, as most prior reports of endogenous biomolecule recovery from pre-Cenozoic fossil bones derive from specimens exhibiting comparatively less chemical alteration (e.g., *Tyrannosaurus rex* MOR 1125 [100] and *Edmontosaurus* bones from the Standing Rock Hadrosaur Site [49]). However, the cortex sample of MPM-PV 1156-49 found herein to yield original cells, blood vessels, fibrous matrix, and endogenous collagen I consisted primarily of middle and internal cortical tissues, each of which are far less altered than the external cortex of the specimen. Thus, the hypothesis advanced by Trueman et al. [101] and Ullmann et al. [123] that regions of bones exhibiting low uptake of REE are likely the best targets for paleomolecular analyses is supported, even though MPM-PV 1156-49 exhibits 'average' alteration at the whole-bone level. The pattern of far greater alteration to the external cortex of MPM-PV 1156-49 than in its more internal regions also conforms to the recommendation by Ullmann et al. [100] for future paleomolecular studies to concentrate sampling efforts on middle and internal cortices rather than the external cortex of fossil specimens.

Similar to other Cretaceous bones we have analyzed, which are documented to retain endogenous collagen I (see [49,100]), MPM-PV 1156-49 appears to have been preserved under (slightly, in this case) oxidizing conditions (Table 2, Figure 8A). This finding agrees with recent claims by Wiemann et al. [3] and Boatman et al. [7], in that oxidizing depositional environments promote soft tissue and biomolecular preservation (by inducing free radical-mediated molecular condensation reactions). That prolonged early diagenetic exposure to moist conditions did not lead to the complete loss of original organics in MPM-PV 1156-49 is an encouraging finding from the perspective of a molecular paleontologist. In particular, it implies: (1) that other 'typical' fossil bones exhibiting 'average' levels of alteration might also still retain original biomolecules, and, perhaps more importantly; (2) that 'normal' diagenetic pathways to fossilization, such as concurrent recrystallization and permineralization (e.g., [124]), may permit molecular preservation (at least in fluviodeltaic environments). If 'normal' bone fossilization processes do not reduce molecular preservation potential to zero, then the pool of fossil specimens that may yield biomolecular material is drastically larger than previously thought (indeed, if this is the case, molecular preservation might not actually be 'exceptional'). Although recrystallization and permineralization have each been hypothesized to possibly promote molecular preservation in fossil bones (via mineral encapsulation [16,101,125–127] and hindrance of microbial infiltration [18,19,128], respectively), it remains premature to claim that 'average' fossil bones constitute favorable paleomolecular samples because this outlook remains based on a sample size of one: *Dreadnoughtus* humerus MPM-PV 1156-49. All other protein-bearing, pre-Cenozoic fossil bones whose trace element inventories have been characterized to date exhibit less REE enrichment [49,100], and the REE content of all other specimens documented to yield original molecules (e.g., those analyzed by Tuross [125] and Schweitzer et al. [13]) remain unknown. Therefore, numerous other 'typically-altered' fossil bones (e.g., possessing $\sum$REE > 1500 ppm) must be tested via immunoassays or paleoproteomics to evaluate the true molecular potential of 'average' specimens.

6. Conclusions

Our assembled molecular and diagenetic data show that, in addition to its exceptional skeletal completeness, the fossil tissue of the *D. schrani* holotype preserved soft-tissue microstructures and remnants of endogenous bone protein. This preservation occurred in a geochemical setting in which its bones were exposed to LREE-enriched groundwaters and weak oxidizing conditions for an extended period after burial. However, following early diagenetic fossilization, the bones experienced negligible further chemical alteration. These findings support the hypotheses advanced by Trueman et al. [101], Ullmann et al. [123], and Gatti et al. [129] that bones exhibiting low trace element uptake are favorable targets for paleomolecular analyses. Moving forward, we encourage the paleomolecular community to include more extensive geochemical analyses as part of their molecular testing routine, as such data will ultimately hold the key to unraveling the complex relationship between diagenesis and 'exceptional' preservation.

Supplementary Materials: The following supporting information can be downloaded at: https://www.mdpi.com/article/10.3390/biology11081158/s1, Figure S1: XRD diffractogram, Figure S2: Intra-bone REE concentration gradients of various elements in *Dreadnoughtus*, Figure S3: ELISA results of chicken bone compared to *Dreadnoughtus*; Table S1: Fossil demineralization protocol, Table S2: Extant bone defleshing and degreasing protocols, Table S3: Extant bone fixation and demineralization protocols, Table S4: Tissue embedding protocol, Table S5: Immunofluorescence protocol, Table S6: Digestion protocol, Table S7: Protein extraction protocol, Table S8: ELISA protocol [49,52,56,77,89,103,106,109–111,124,130–187]; Data File S1: DreadREE_Data S1.

Author Contributions: Conceptualization, E.R.S. and P.V.U.; Data curation, E.R.S. and P.V.U.; Formal analysis, E.R.S., P.V.U., K.M., R.D.A. and W.Z.; Funding acquisition, E.R.S., P.V.U., M.H.S. and K.J.L.; Investigation, E.R.S., P.V.U., K.M., W.Z. and K.J.L.; Methodology, E.R.S., P.V.U., W.Z. and M.H.S.; Project administration, P.V.U., M.H.S. and K.J.L.; Resources, M.H.S. and K.J.L.; Supervision, M.H.S. and K.J.L.; Visualization, E.R.S. and P.V.U.; Writing—original draft, E.R.S., P.V.U. and K.M.; Writing—review and editing, E.R.S., P.V.U., W.Z., M.H.S. and K.J.L. All authors have read and agreed to the published version of the manuscript.

Funding: Funding for this research was provided by the NSF Graduate Research Fellowship (DGE 1002809), the Jurassic Foundation, and the Biological Sciences Department of North Carolina State University to ERS, and Rowan University Seed Funding to PVU. The APC was funded (waived) by MDPI.

Institutional Review Board Statement: Ethical review and approval were not required; no animals were killed or harmed for the purpose of this study. Bone tissues were exclusively obtained from animals that were priorly deceased via circumstances unrelated to this research: chicken bone was obtained from a local grocery store, alligator bone was kindly donated by L. Ibiricu (Drexel University) and C. Berger/SportMan Gold, and of course, *Dreadnoughtus* bone was harvested from an individual that perished millions of years prior to this study.

Informed Consent Statement: Not applicable.

Data Availability Statement: The data presented in this manuscript, as well as extended details on methodology, are available in the supplementary file and the supplementary data file (Data S1), which have been made available for download.

Acknowledgments: We thank: T. Cleland for assistance with benchwork; M. Lamanna for assistance with figures; J. DiGnazio for creating the quarry map; C. Berger, SportMan Gold, and L. Ibiricu for fresh alligator limb elements; E. Vaughn-Cleland, A. Moyer, E. Johnson, K. Voegle, Z. Boles, and A. Carter for logistical assistance. We thank three reviewers whose comments greatly improved this manuscript.

Conflicts of Interest: The authors declare no conflict of interest.

References

1. Gioncada, A.; Collareta, A.; Gariboldi, K.; Lambert, O.; Di Celma, C.; Bonaccorsi, E.; Urbina, M.; Bianucci, G. Inside baleen: Exceptional microstructure preservation in a late Miocene whale skeleton from Peru. *Geology* **2016**, *44*, 839–842. [CrossRef]
2. McCoy, V.E.; Gabbott, S.E.; Penkman, K.; Collins, M.J.; Presslee, S.; Holt, J.; Grossman, H.; Wang, B.; Solórzano, M.M.; Delclòs, X.; et al. Ancient amino acids from fossil feathers in amber. *Sci. Rep.* **2019**, *9*, 6420. [CrossRef] [PubMed]
3. Wiemann, J.; Fabbri, M.; Yang, T.R.; Stein, K.; Sander, P.M.; Norell, M.A.; Briggs, D.E.G. Fossilization transforms vertebrate hard tissue proteins into N-heterocyclic polymers. *Nat. Commun.* **2018**, *9*, 4741. [CrossRef]
4. Wiemann, J.; Crawford, J.M.; Briggs, D.E.G. Phylogenetic and physiological signals in metazoan fossil biomolecules. *Sci. Adv.* **2020**, *6*, eaba6883. [CrossRef] [PubMed]
5. Lee, Y.C.; Chiang, C.C.; Huang, P.Y.; Chung, C.Y.; Huang, T.D.; Wang, C.C.; Chen, C.I.; Chang, R.S.; Liao, C.H.; Reisz, R.R. Evidence of preserved collagen in an Early Jurassic sauropodomorph dinosaur revealed by synchrotron FTIR microspectroscopy. *Nat. Commun.* **2017**, *8*, 2–9. [CrossRef] [PubMed]
6. Reisz, R.R.; Huang, T.D.; Roberts, E.M.; Peng, S.; Sullivan, C.; Stein, K.; Leblanc, A.R.H.; Shieh, D.; Chang, R.; Chiang, C.; et al. Embryology of Early Jurassic dinosaur from China with evidence of preserved organic remains. *Nature* **2013**, *496*, 210–214. [CrossRef]
7. Boatman, E.M.; Goodwin, M.B.; Holman, H.Y.N.; Fakra, S.; Zheng, W.; Gronsky, R.; Schweitzer, M.H. Mechanisms of soft tissue and protein preservation in Tyrannosaurus rex. *Sci. Rep.* **2019**, *9*, 15678. [CrossRef] [PubMed]
8. Pan, Y.; Zheng, W.; Moyer, A.E.; O'Connor, J.K.; Wang, M.; Zheng, X.; Wang, X.; Schroeter, E.R.; Zhou, Z.; Schweitzer, M.H. Molecular evidence of keratin and melanosomes in feathers of the Early Cretaceous bird Eoconfuciusornis. *Proc. Natl. Acad. Sci. USA* **2016**, *113*, E7900–E7907. [CrossRef]
9. Pan, Y.; Zheng, W.; Sawyer, R.H.; Pennington, M.W.; Zheng, X.; Wang, X.; Wang, M.; Hu, L.; O'Connor, J.; Zhao, T.; et al. The molecular evolution of feathers with direct evidence from fossils. *Proc. Natl. Acad. Sci. USA* **2019**, *116*, 3018–3023. [CrossRef] [PubMed]
10. Jiang, B.; Zhao, T.; Regnault, S.; Edwards, N.P.; Kohn, S.C.; Li, Z.; Wogelius, R.A.; Benton, M.J.; Hutchinson, J.R. Cellular preservation of musculoskeletal specializations in the Cretaceous bird Confuciusornis. *Nat. Commun.* **2017**, *8*, 14779. [CrossRef] [PubMed]
11. Lindgren, J.; Sjövall, P.; Thiel, V.; Zheng, W.; Ito, S.; Wakamatsu, K.; Hauff, R.; Kear, B.P.; Engdahl, A.; Alwmark, C.; et al. Soft-tissue evidence for homeothermy and crypsis in a Jurassic ichthyosaur. *Nature* **2018**, *564*, 359–365. [CrossRef]
12. Surmik, D.; Boczarowski, A.; Balin, K.; Dulski, M.; Szade, J.; Kremer, B.; Pawlicki, R. Spectroscopic studies on organic matter from triassic reptile bones, Upper Silesia, Poland. *PLoS ONE* **2016**, *11*, e0151143. [CrossRef] [PubMed]
13. Schweitzer, M.H.; Zheng, W.; Organ, C.L.; Avci, R.; Suo, Z.; Freimark, L.M.; Lebleu, V.S.; Duncan, M.B.; Heiden, M.G.V.; Neveu, J.M.; et al. Biomolecular characterization and protein sequences of the campanian hadrosaur b. canadensis. *Science* **2009**, *324*, 626–631. [CrossRef] [PubMed]
14. Schroeter, E.R.; DeHart, C.J.; Cleland, T.P.; Zheng, W.; Thomas, P.M.; Kelleher, N.L.; Bern, M.; Schweitzer, M.H. Expansion for the Brachylophosaurus canadensis Collagen I Sequence and Additional Evidence of the Preservation of Cretaceous Protein. *J. Proteome Res.* **2017**, *16*, 920–932. [CrossRef] [PubMed]
15. Buckley, M.; Warwood, S.; van Dongen, B.; Kitchener, A.C.; Manning, P.L. A fossil protein chimera; difficulties in discriminating dinosaur peptide sequences from modern cross-contamination. *Proc. R. Soc. B Biol. Sci.* **2017**, *284*, 20170544. [CrossRef]
16. Schweitzer, M.H.; Zheng, W.; Cleland, T.P.; Goodwin, M.B.; Boatman, E.; Theil, E.; Marcus, M.A.; Fakra, S.C. A role for iron and oxygen chemistry in preserving soft tissues, cells and molecules from deep time. *Proc. R. Soc. Ser. B* **2014**, *281*, 20132741. [CrossRef] [PubMed]
17. Greenwalt, D.E.; Goreva, Y.S.; Siljeström, S.M.; Rose, T.; Harbach, R.E. Hemoglobin-derived porphyrins preserved in a Middle Eocene blood-engorged mosquito. *Proc. Natl. Acad. Sci. USA* **2013**, *110*, 18496–18500. [CrossRef] [PubMed]
18. Peterson, J.E.; Lenczewski, M.E.; Scherer, R.P. Influence of microbial biofilms on the preservation of primary soft tissue in fossil and extant archosaurs. *PLoS ONE* **2010**, *5*, e13334. [CrossRef] [PubMed]
19. Schweitzer, M.H. Soft tissue preservation in terrestrial Mesozoic vertebrates. *Annu. Rev. Earth Planet. Sci.* **2011**, *39*, 187–216. [CrossRef]
20. van Klinken, G.J.; Hedges, R.E.M. Experiments on Collagen-Humic Interactions: Speed of Humic Uptake, and Effects of Diverse Chemical Treatments. *J. Archaeol. Sci.* **1995**, *22*, 263–270. [CrossRef]
21. Hedges, R.E.M. Bone diagenesis: An overview of processes. *Archaeometry* **2002**, *44*, 319–328. [CrossRef]
22. Nielsen-Marsh, C.M.; Richards, M.P.; Hauschka, P.V.; Thomas-Oates, J.E.; Trinkaus, E.; Pettitt, P.B.; Karavanić, I.; Poinar, H.; Collins, M.J. Osteocalcin protein sequences of Neanderthals and modern primates. *Proc. Natl. Acad. Sci. USA* **2005**, *102*, 4409–4413. [CrossRef] [PubMed]
23. Cleland, T.P.; Schroeter, E.R. A Comparison of Common Mass Spectrometry Approaches for Paleoproteomics. *J. Proteome Res.* **2018**, *17*, 936–945. [CrossRef] [PubMed]
24. Schroeter, E.R.; Cleland, T.P.; Schweitzer, M.H. Deep Time Paleoproteomics: Looking Forward. *J. Proteome Res.* **2022**, *21*, 9–19. [CrossRef] [PubMed]

25. Lacovara, K.J.; Lamanna, M.C.; Ibiricu, L.M.; Poole, J.C.; Schroeter, E.R.; Ullmann, P.V.; Voegele, K.K.; Boles, Z.M.; Carter, A.M.; Fowler, E.K.; et al. A gigantic, exceptionally complete titanosaurian sauropod dinosaur from southern Patagonia, Argentina. *Sci. Rep.* **2014**, *4*, 6196. [CrossRef]

26. Curry Rogers, K. Titanosauria: A phylogenetic overview. In *The Sauropods: Evolution and Paleobiology*; Curry Rogers, K., Wilson, J.A., Eds.; University of California Press: Berkeley, CA, USA, 2005; pp. 50–103. ISBN 978-0-520-93233-3.

27. Lingham-Soliar, T. A unique cross section through the skin of the dinosaur Psittacosaurus from China showing a complex fibre architecture. *Proc. R. Soc. B Biol. Sci.* **2008**, *275*, 775–780. [CrossRef] [PubMed]

28. Schweitzer, M.H.; Horner, J.R.H. Intravascular microstructures in trabecular bone tissues of Tyrannosaurus rex. *Ann. Paleontol.* **1999**, *85*, 179–192. [CrossRef]

29. Eglinton, G.; Logan, G.A. Molecular preservation. *Philos. Trans.-R. Soc. Lond. B* **1991**, *333*, 315–328. [CrossRef]

30. Chin, K.; Eberth, D.A.; Schweitzer, M.H.; Rando, T.A.; Sloboda, W.J.; Horner, J.R. Remarkable Preservation of Undigested Muscle Tissue within a Late Cretaceous Tyrannosaurid Coprolite from Alberta, Canada. *Palaios* **2003**, *18*, 286–294. [CrossRef]

31. Zhou, Z.; Barrett, P.M.; Hilton, J. An exceptionally preserved lower cretaceous ecosystem. *Nature* **2003**, *421*, 807–814. [CrossRef] [PubMed]

32. Wegweiser, M.; Hartman, S.; Lovelace, D. Duckbill dinosaur chin skin scales: Ups, downs and arounds of surficial morphology of Upper Cretaceous Lance Formation dinosaur skin. *New Mex. Mus. Nat. Hist. Sci. Bull.* **2006**, *35*, 119–126.

33. Young, G.A.; Rudkin, D.M.; Dobrzanski, E.P.; Robson, S.P.; Nowlan, G.S. Exceptionally preserved late Ordovician biotas from Manitoba, Canada. *Geology* **2007**, *35*, 883–886. [CrossRef]

34. Manning, P.L.; Morris, P.M.; McMahon, A.; Jones, E.; Gize, A.; Macquaker, J.H.S.; Wolff, G.; Thompson, A.; Marshall, J.; Taylor, K.G.; et al. Mineralized soft-tissue structure and chemistry in a mummified hadrosaur from the Hell Creek Formation, North Dakota (USA). *Proc. R. Soc. B Biol. Sci.* **2009**, *276*, 3429–3437. [CrossRef] [PubMed]

35. Davis, M. Census of dinosaur skin reveals lithology may not be the most important factor in increased preservation of hadrosaurid skin. *Acta Palaeontol. Pol.* **2014**, *59*, 601–605. [CrossRef]

36. Schroeter, E.R.; Egerton, V.M.; Ibiricu, L.M.; Lacovara, K.J. Lamniform shark teeth from the late cretaceous of southernmost South America (Santa Cruz Province, Argentina). *PLoS ONE* **2014**, *9*, e104800. [CrossRef] [PubMed]

37. Sickmann, Z.T.; Schwartz, T.M.; Graham, S.A. Refining stratigraphy and tectonic history using detrital zircon maximum depositional age: An example from the Cerro Fortaleza Formation, Austral Basin, southern Patagonia. *Basin Res.* **2018**, *30*, 708–729. [CrossRef]

38. Ullmann, P.V.; Lacovara, K.J. Appendicular osteology of *Dreadnoughtus schrani*, a giant titanosaurian (Sauropoda, Titanosauria) from the Upper Cretaceous of Patagonia, Argentina. *J. Vertebr. Paleontol.* **2016**, *36*, e1225303. [CrossRef]

39. Macellari, C.E.; Barrio, C.A.; Manassero, M.J. Upper Cretaceous to Paleocene depositional sequences and sandstone petrography of southwestern Patagonia (Argentina and Chile). *J. S. Am. Earth Sci.* **1989**, *2*, 223–239. [CrossRef]

40. Marenssi, S.A.; Casadio, S.; Santillana, S.N. Estratigrafía y sedimentología de las unidades del Cretácico superior-Paleógeno aflorantes en la margen sureste del lago Viedma, provincia de Santa Cruz, Argentina. *Rev. Asoc. Geol. Argent.* **2003**, *58*, 403–416.

41. Egerton, V.M.; Williams, C.J.; Lacovara, K.J. A new Late Cretaceous (late Campanian to early Maastrichtian) wood flora from southern Patagonia. *Palaeogeogr. Palaeoclimatol. Palaeoecol.* **2016**, *441*, 305–316. [CrossRef]

42. Rombola, C.F.; Greppi, C.D.; Pujana, R.R.; García Massini, J.L.; Bellosi, E.S.; Marenssi, S.A. Brachyoxylon fossil woods with traumatic resin canals from the Upper Cretaceous Cerro Fortaleza Formation, southern Patagonia (Santa Cruz Province, Argentina). *Cretac. Res.* **2022**, *130*, 105065. [CrossRef]

43. Voegele, K.K.; Ullmann, P.V.; Lamanna, M.C.; Lacovara, K.J. Appendicular myological reconstruction of the forelimb of the giant titanosaurian sauropod dinosaur *Dreadnoughtus schrani*. *J. Anat.* **2020**, *237*, 133–154. [CrossRef] [PubMed]

44. Voegele, K.K.; Ullmann, P.V.; Lamanna, M.C.; Lacovara, K.J. Myological reconstruction of the pelvic girdle and hind limb of the giant titanosaurian sauropod dinosaur *Dreadnoughtus schrani*. *J. Anat.* **2021**, *238*, 576–597. [CrossRef] [PubMed]

45. Gates, T.A. The Late Jurassic Cleveland-Lloyd dinosaur quarry as a drought-induced assemblage. *Palaios* **2005**, *20*, 363–375. [CrossRef]

46. Green, J.L.; Schweitzer, M.H.; Lamm, E.T. Limb bone histology and growth in Placerias hesternus (Therapsida: Anomodontia) from the Upper Triassic of North America. *Palaeontology* **2010**, *53*, 347–364. [CrossRef]

47. Boyd, C.A.; Cleland, T.P.; Novas, F. Osteogenesis, homology, and function of the intercostal plates in ornithischian dinosaurs (Tetrapoda, Sauropsida). *Zoomorphology* **2011**, *130*, 305–313. [CrossRef]

48. Cleland, T.P.; Stoskopf, M.K.; Schweitzer, M.H. Histological, chemical, and morphological reexamination of the "heart" of a small Late Cretaceous Thescelosaurus. *Naturwissenschaften* **2011**, *98*, 203–211. [CrossRef] [PubMed]

49. Ullmann, P.V.; Grandstaff, D.E.; Ash, R.D.; Lacovara, K.J. Geochemical taphonomy of the Standing Rock Hadrosaur Site: Exploring links between rare earth elements and cellular and soft tissue preservation. *Geochim. Cosmochim. Acta* **2020**, *269*, 223–237. [CrossRef]

50. Gromet, L.P.; Haskin, L.A.; Korotev, R.L.; Dymek, R.F. The "North American shale composite": Its compilation, major and trace element characteristics. *Geochim. Cosmochim. Acta* **1984**, *48*, 2469–2482. [CrossRef]

51. Haskin, L.A.; Haskin, M.A.; Frey, F.A.; Wildeman, T.R. Relative and absolute terrestrial abundances of the rare earths. In *Origin and Distribution of the Elements*; Ahrens, L.H., Ed.; Pergamon: Oxford, UK, 1968; pp. 889–912.

52. Herwartz, D.; Tütken, T.; Jochum, K.P.; Sander, P.M. Rare earth element systematics of fossil bone revealed by LA-ICPMS analysis. *Geochim. Cosmochim. Acta* **2013**, *103*, 161–183. [CrossRef]

53. Jans, M.M.E.; Nielsen-Marsh, C.M.; Smith, C.I.; Collins, M.J.; Kars, H. Characterisation of microbial attack on archaeological bone. *J. Archaeol. Sci.* **2004**, *31*, 87–95. [CrossRef]

54. Hedges, R.E.M.; Millard, A.R.; Pike, A.W.G. Measurements and Relationships of Diagenetic Alteration of Bone from Three Archaeological Sites. *J. Archaeol. Sci.* **1995**, *22*, 201–209. [CrossRef]

55. Nielsen-marsh, C.; Gernaey, A.; Turner-Walker, G.; Hedges, R.; Pike, A.; Collins, M. The Chemical Degradation of Bone. In *Human Osteology: In Archaeology and Forensic Science*; Cambridge University Press: Cambridge, UK, 2000; pp. 439–454.

56. Kohn, M.J. Models of diffusion-limited uptake of trace elements in fossils and rates of fossilization. *Geochim. Cosmochim. Acta* **2008**, *72*, 3758–3770. [CrossRef]

57. Grandstaff, D.E.; Terry, D.O. Rare earth element composition of Paleogene vertebrate fossils from Toadstool Geologic Park, Nebraska, USA. *Appl. Geochem.* **2009**, *24*, 733–745. [CrossRef]

58. Bau, M.; Dulski, P. Distribution of yttrium and rare-earth elements in the Penge and Kuruman iron-formations, Transvaal Supergroup, South Africa. *Precambrian Res.* **1996**, *79*, 37–55. [CrossRef]

59. Pack, A.; Russell, S.S.; Shelley, J.M.G.; van Zuilen, M. Geo- and cosmochemistry of the twin elements yttrium and holmium. *Geochim. Cosmochim. Acta* **2007**, *71*, 4592–4608. [CrossRef]

60. Schweitzer, M.H.; Wittmeyer, J.L.; Horner, J.R. Soft tissue and cellular preservation in vertebrate skeletal elements from the Cretaceous to the present. *Proc. R. Soc. Lond. B* **2007**, *274*, 183–197. [CrossRef]

61. Schweitzer, M.H.; Wittmeyer, J.L.; Horner, J.R.; Toporski, J.K. Soft-tissue vessels and cellular preservation in Tyrannosaurus rex. *Science* **2005**, *307*, 1952–1955. [CrossRef]

62. Bartnicki-Garcia, S.; Bracker, C.E.; Gierz, G.; López-Franco, R.; Haisheng, L. Mapping the growth of fungal hyphae: Orthogonal cell wall expansion during tip growth and the role of turgor. *Biophys. J.* **2000**, *79*, 2382–2390. [CrossRef]

63. Harris, S.D. Branching of fungal hyphae: Regulation, mechanisms and comparison with other branching systems. *Mycologia* **2008**, *100*, 823–832. [CrossRef]

64. Nesse, W. Optical mineralogy. In *Introduction to Mineralogy*; Oxford University Press: New York, NY, USA, 2000; pp. 114–159.

65. Perkins, D.; Henke, K.R. *Minerals in Thin Section*; Prentice Hall: Upper Saddle River, NJ, USA, 2003.

66. Cadena, E.A.; Schweitzer, M.H. Variation in osteocytes morphology vs. bone type in turtle shell and their exceptional preservation from the Jurassic to the present. *Bone* **2012**, *51*, 614–620. [CrossRef]

67. Schweitzer, M.H.; Zheng, W.; Cleland, T.P.; Bern, M. Molecular analyses of dinosaur osteocytes support the presence of endogenous molecules. *Bone* **2013**, *52*, 414–423. [CrossRef] [PubMed]

68. Armitage, M.H.; Anderson, K.L. Soft sheets of fibrillar bone from a fossil of the supraorbital horn of the dinosaur Triceratops horridus. *Acta Histochem.* **2013**, *115*, 603–608. [CrossRef] [PubMed]

69. Lowenstein, J.M.; Scheuenstuhl, G. Immunological methods in molecular palaeontology. *Philos. Trans.-R. Soc. Lond. B* **1991**, *333*, 375–380. [CrossRef]

70. Frank, S.A. Classification by antigenicity and phylogeny. In *Immunology and Evolution of Infectious Disease*; Princeton University Press: Princeton, NJ, USA, 2002; pp. 175–187.

71. Hedges, S.B.; Dudley, J.; Kumar, S. TimeTree: A public knowledge-base of divergence times among organisms. *Bioinformatics* **2006**, *22*, 2971–2972. [CrossRef]

72. Brochu, C.A. Progress and future directions in archosaur phylogenetics. *J. Paleontol.* **2001**, *75*, 1185–1201. [CrossRef]

73. Ramos-Vara, J.A.; Beissenherz, M.E. Optimization of immunohistochemical methods using two different antigen retrieval methods on formalin-fixed, paraffin-embedded tissues: Experience with 63 markers. *J. Vet. Diagn. Investig.* **2000**, *12*, 307–311. [CrossRef]

74. Harper, E. Collagenases. *Annu. Rev. Biochem.* **1980**, *49*, 1063–1078. [CrossRef]

75. Harper, E.; Kang, A.H. Studies on the specificity of bacterial collagenases. *Biochem. Biophys. Res. Commun.* **1970**, *41*, 482–487. [CrossRef]

76. Avci, R.; Schweitzer, M.H.; Boyd, R.D.; Wittmeyer, J.L.; Arce, F.T.; Calvo, J.O. Preservation of bone collagen from the Late Cretaceous period studied by immunological techniques and atomic force microscopy. *Am. Chem. Soc.* **2005**, *21*, 3584–3591. [CrossRef]

77. Werner, M.; Chott, A.; Fabiano, A.; Battifora, H. Effect of formalin tissue fixation and processing on immunohistochemistry. *Am. J. Surg. Pathol.* **2000**, *24*, 1016–1019. [CrossRef] [PubMed]

78. Schweitzer, M.H.; Avci, R.; Collier, T.C.; Goodwin, M.B. Microscopic, chemical and molecular methods for examining fossil preservation. *Comptes Rendus Palevol* **2008**, *7*, 159–184. [CrossRef]

79. Saper, C.B. A guide to the perplexed on the specificity of antibodies. *J. Histochem. Cytochem.* **2009**, *57*, 1–5. [CrossRef]

80. Paulie, S.; Perlmann, H.; Perlmann, P. Enzyme-linked Immunosorbent Assay. *Encycl. Life Sci.* **2006**, *21*, 165–209. [CrossRef]

81. Avci, R.; Schweitzer, M.; Boyd, R.D.; Wittmeyer, J.; Steeled, A.; Toporski, J.; Beech, I.; Arce, F.T.; Spangler, B.; Cole, K.M.; et al. Comparison of antibody-antigen interactions on collagen measured by conventional immunological techniques and atomic force microscopy. *Langmuir* **2004**, *20*, 11053–11063. [CrossRef] [PubMed]

82. Lechtzier, V.; Hutoran, M.; Levy, T.; Kotler, M.; Brenner, T.; Steinitz, M. Sodium dodecyl sulphate-treated proteins as ligands in ELISA. *J. Immunol. Methods* **2002**, *270*, 19–26. [CrossRef]

83. Appiah, A.S.; Amoatey, H.M.; Klu, G.Y.; Afful, N.T.; Azu, E.; Owusu, G.K. Spread of African cassava mosaic virus from cassava (Manihot esculenta Crantz) to physic nut (*Jatropha curcas* L.) in Ghana. *J. Phytol.* **2012**, *4*, 31–37.

84. Ostlund, E.N.; Crom, R.L.; Pedersen, D.D.; Johnson, D.J.; Williams, W.O.; Schmitt, B.J. Equine West Nile encephalitis, United States. *Emerg. Infect. Dis.* **2001**, *7*, 665–669. [CrossRef] [PubMed]

85. Tabatabai, L.B.; Deyoe, B.L. Specific enzyme-linked immunosorbent assay for detection of bovine antibody to Brucella abortus. *J. Clin. Microbiol.* **1984**, *20*, 209–213. [CrossRef]

86. Smith, C.I.; Craig, O.E.; Prigodich, R.V.; Nielsen-Marsh, C.M.; Jans, M.M.E.; Vermeer, C.; Collins, M.J. Diagenesis and survival of osteocalcin in archaeological bone. *J. Archaeol. Sci.* **2005**, *32*, 105–113. [CrossRef]

87. Nielsen-Marsh, C.M.; Hedges, R.E.M. Patterns of diagenesis in bone I: The effects of site environments. *J. Archaeol. Sci.* **2000**, *27*, 1139–1150. [CrossRef]

88. Nielsen-Marsh, C.M.; Hedges, R.E.M. Patterns of diagenesis in bone II: Effects of acetic acid treatment and the removal of diagenetic $CO_3{}^{2-}$. *J. Archaeol. Sci.* **2000**, *27*, 1151–1159. [CrossRef]

89. Martin, J.E.; Patrick, D.; Kihm, A.J.; Foit, F.F.; Grandstaff, D.E. Lithostratigraphy, tephrochronology, and rare earth element geochemistry of fossils at the classical pleistocene Fossil Lake area, south central Oregon. *J. Geol.* **2005**, *113*, 139–155. [CrossRef]

90. Trueman, C.N. Rare earth element geochemistry and taphonomy of terrestrial vertebrate assemblages. *Palaios* **1999**, *14*, 555–568. [CrossRef]

91. Suarez, C.A.; Suarez, M.B.; Terry, D.O.; Grandstaff, D.E. Rare earth element geochemistry and taphonomy of the early Cretaceous Crystal Geyser Dinosaur Quarry, east-central Utah. *Palaios* **2007**, *22*, 500–512. [CrossRef]

92. Titus, A.L.; Knoll, K.; Sertich, J.J.W.; Yamamura, D.; Suarez, C.A.; Glasspool, I.J.; Ginouves, J.E.; Lukacic, A.K.; Roberts, E.M. Geology and taphonomy of a unique tyrannosaurid bonebed from the upper Campanian Kaiparowits Formation of southern Utah: Implications for tyrannosaurid gregariousness. *PeerJ* **2021**, *9*, e11013. [CrossRef]

93. Rogers, R.R.; Fricke, H.C.; Addona, V.; Canavan, R.R.; Dwyer, C.N.; Harwood, C.L.; Koenig, A.E.; Murray, R.; Thole, J.T.; Williams, J. Using laser ablation-inductively coupled plasma-mass spectrometry (LA-ICP-MS) to explore geochemical taphonomy of vertebrate fossils in the upper cretaceous two medicine and Judith River formations of Montana. *Palaios* **2010**, *25*, 183–195. [CrossRef]

94. Kowal-Linka, M.; Jochum, K.P.; Surmik, D. LA-ICP-MS analysis of rare earth elements in marine reptile bones from the Middle Triassic bonebed (Upper Silesia, S Poland): Impact of long-lasting diagenesis, and factors controlling the uptake. *Chem. Geol.* **2014**, *363*, 213–228. [CrossRef]

95. McCormack, J.M.; Bahr, A.; Gerdes, A.; Tütken, T.; Prinz-Grimm, P. Preservation of successive diagenetic stages in Middle Triassic bonebeds: Evidence from in situ trace element and strontium isotope analysis of vertebrate fossils. *Chem. Geol.* **2015**, *410*, 108–123. [CrossRef]

96. Suarez, C.A.; Morschhauser, E.M.; Suarez, M.B.; You, H.; Li, D.; Dodson, P. Rare earth element geochemistry of bone beds from the Lower Cretaceous Zhonggou Formation of Gansu Province, China. *J. Vertebr. Paleontol.* **2018**, *38*, 22–35. [CrossRef]

97. Suarez, C.A.; Kohn, M.J. Caught in the act: A case study on microscopic scale physicochemical effects of fossilization on stable isotopic composition of bone. *Geochim. Cosmochim. Acta* **2020**, *268*, 277–295. [CrossRef]

98. Botfalvai, G.; Csiki-Sava, Z.; Kocsis, L.; Albert, G.; Magyar, J.; Bodor, E.R.; Tabara, D.; Ulyanov, A.; Makádi, L. 'X' marks the spot! Sedimentological, geochemical and palaeontological investigations of Upper Cretaceous (Maastrichtian) vertebrate fossil localities from the Valioara valley (Densus-Ciula Formation, Haţeg Basin, Romania). *Cretac. Res.* **2021**, *123*, 104781. [CrossRef]

99. Ferrante, C.; Cavin, L.; Vennemann, T.; Martini, R. Histology and Geochemistry of Allosaurus (Dinosauria: Theropoda) from the Cleveland-Lloyd Dinosaur Quarry (Late Jurassic, Utah): Paleobiological Implications. *Front. Earth Sci.* **2021**, *9*, 225. [CrossRef]

100. Ullmann, P.V.; Macauley, K.; Ash, R.D.; Shoup, B.; Scannella, J.B. Taphonomic and diagenetic pathways to protein preservation, part i: The case of tyrannosaurus rex specimen mor 1125. *Biology* **2021**, *10*, 1193. [CrossRef]

101. Trueman, C.N.; Palmer, M.R.; Field, J.; Privat, K.; Ludgate, N.; Chavagnac, V.; Eberth, D.A.; Cifelli, R.; Rogers, R.R. Comparing rates of recrystallisation and the potential for preservation of biomolecules from the distribution of trace elements in fossil bones. *Comptes Rendus-Palevol* **2008**, *7*, 145–158. [CrossRef]

102. Herwartz, D.; Tütken, T.; Münker, C.; Jochum, K.P.; Stoll, B.; Sander, P.M. Timescales and mechanisms of REE and Hf uptake in fossil bones. *Geochim. Cosmochim. Acta* **2011**, *75*, 82–105. [CrossRef]

103. Trueman, C.N.; Kocsis, L.; Palmer, M.R.; Dewdney, C. Fractionation of rare earth elements within bone mineral: A natural cation exchange system. *Palaeogeogr. Palaeoclimatol. Palaeoecol.* **2011**, *310*, 124–132. [CrossRef]

104. Wings, O. Authigenic minerals in fossil bones from the Mesozoic of England: Poor correlation with depositional environments. *Palaeogeogr. Palaeoclimatol. Palaeoecol.* **2004**, *204*, 15–32. [CrossRef]

105. Janssens, K.; Vincze, L.; Vekemans, B.; Williams, C.T.; Radtke, M.; Haller, M.; Knöchel, A. The non-destructive determination of REE in fossilized bone using synchrotron radiation induced K-line X-ray microfluorescence analysis. *Fresenius. J. Anal. Chem.* **1999**, *363*, 413–420. [CrossRef]

106. Hinz, E.A.; Kohn, M.J. The effect of tissue structure and soil chemistry on trace element uptake in fossils. *Geochim. Cosmochim. Acta* **2010**, *74*, 3213–3231. [CrossRef]

107. Suarez, C.A.; Macpherson, G.L.; González, L.A.; Grandstaff, D.E. Heterogeneous rare earth element (REE) patterns and concentrations in a fossil bone: Implications for the use of REE in vertebrate taphonomy and fossilization history. *Geochim. Cosmochim. Acta* **2010**, *74*, 2970–2988. [CrossRef]

108. Kohn, M.J.; Moses, R.J. Trace element diffusivities in bone rule out simple diffusive uptake during fossilization but explain in vivo uptake and release. *Proc. Natl. Acad. Sci. USA* **2013**, *110*, 419–424. [CrossRef]

109. Elderfield, H.; Upstill-Goddard, R.; Sholkovitz, E.R. The rare earth elements in rivers, estuaries, and coastal seas and their significance to the composition of ocean waters. *Geochim. Cosmochim. Acta* **1990**, *54*, 971–991. [CrossRef]

110. Johannesson, K.H.; Tang, J.; Daniels, J.M.; Bounds, W.J.; Burdige, D.J. Rare earth element concentrations and speciation in organic-rich blackwaters of the Great Dismal Swamp, Virginia, USA. *Chem. Geol.* **2004**, *209*, 271–294. [CrossRef]

111. Smith, C.; Liu, X.M. Spatial and temporal distribution of rare earth elements in the Neuse River, North Carolina. *Chem. Geol.* **2018**, *488*, 34–43. [CrossRef]

112. Luo, Y.R.; Byrne, R.H. Carbonate complexation of yttrium and the rare earth elements in natural waters. *Geochim. Cosmochim. Acta* **2004**, *68*, 691–699. [CrossRef]

113. Pourret, O.; Davranche, M.; Gruau, G.; Dia, A. Rare earth elements complexation with humic acid. *Chem. Geol.* **2007**, *243*, 128–141. [CrossRef]

114. Byrne, R.H.; Liu, X.; Schijf, J. The influence of phosphate coprecipitation on rare earth distributions in natural waters. *Geochim. Cosmochim. Acta* **1996**, *60*, 3341–3346. [CrossRef]

115. Metzger, C.A.; Terry, D.O.; Grandstaff, D.E. Effect of paleosol formation on rare earth element signatures in fossil bone. *Geology* **2004**, *32*, 497–500. [CrossRef]

116. Company, J. Bone histology of the titanosaur Lirainosaurus astibiae (Dinosauria: Sauropoda) from the Latest Cretaceous of Spain. *Naturwissenschaften* **2011**, *98*, 67–78. [CrossRef] [PubMed]

117. Klein, N.; Christian, A.; Sander, P.M. Histology shows that elongated neck ribs in sauropod dinosaurs are ossified tendons. *Biol. Lett.* **2012**, *8*, 1032–1035. [CrossRef] [PubMed]

118. Stein, K.; Csiki, Z.; Curry Rogers, K.; Weishampel, D.B.; Redelstorff, R.; Carballido, J.L.; Sander, P.M. Small body size and extreme cortical bone remodeling indicate phyletic dwarfism in Magyarosaurus dacus (Sauropoda: Titanosauria). *Proc. Natl. Acad. Sci. USA* **2010**, *107*, 9258–9263. [CrossRef] [PubMed]

119. Griffin, D.H. Introduction to the Fungi. In *Fungal Physiology*; Griffin, D.H., Ed.; Wiley-Liss, Inc.: New York, NY, USA, 1994; pp. 1–23.

120. Kaye, T.G.; Gaugler, G.; Sawlowicz, Z. Dinosaurian soft tissues interpreted as bacterial biofilms. *PLoS ONE* **2008**, *3*, e2808. [CrossRef]

121. Buckley, M.; Wadsworth, C. Proteome degradation in ancient bone: Diagenesis and phylogenetic potential. *Palaeogeogr. Palaeoclimatol. Palaeoecol.* **2014**, *416*, 69–79. [CrossRef]

122. Brandt, E.; Weichmann, I.; Grupe, G. Howe reliable are immunological tools for the detection of ancient proteins in fossil bones? *Int. J. Osteoarchaeol.* **2002**, *12*, 307–316. [CrossRef]

123. Ullmann, P.V.; Voegele, K.K.; Grandstaff, D.E.; Ash, R.D.; Zheng, W.; Schroeter, E.R.; Schweitzer, M.H.; Lacovara, K.J. Molecular tests support the viability of rare earth elements as proxies for fossil biomolecule preservation. *Sci. Rep.* **2020**, *10*, 15566. [CrossRef] [PubMed]

124. Hubert, J.F.; Panish, P.T.; Chure, D.J.; Prostak, K.S. Chemistry, microstructure, petrology, and diagenetic model of jurassic dinosaur bones, dinosaur national monument, Utah. *J. Sediment. Res.* **1996**, *66*, 531–547.

125. Tuross, N. Albumin preservation in the Taima-taima mastodon skeleton. *Appl. Geochem.* **1989**, *4*, 255–259. [CrossRef]

126. Grupe, G. Preservation of collagen in bone from dry, sandy soil. *J. Archaeol. Sci.* **1995**, *22*, 193–199. [CrossRef]

127. Koenig, A.E.; Rogers, R.R.; Trueman, C.N. Visualizing fossilization using laser ablation-inductively coupled plasma-mass spectrometry maps of trace elements in Late Cretaceous bones. *Geology* **2009**, *37*, 511–514. [CrossRef]

128. Schweitzer, M.H. Molecular paleontology: Some current advances and problems. *Ann. Paleontol.* **2004**, *90*, 81–102. [CrossRef]

129. Gatti, L.; Lugli, F.; Sciutto, G.; Zangheri, M.; Prati, S.; Mirasoli, M. Combining elemental and immunochemical analyses to characterize diagenetic alteration patterns in ancient skeletal remains. *Sci. Rep.* **2022**, *12*, 5112. [CrossRef] [PubMed]

130. Abbott, A.N.; Haley, B.A.; McManus, J.; Reimers, C.E. The sedimentary flux of dissolved rare earth elements to the ocean. *Geochim. Cosmochim. Acta* **2015**, *154*, 186–200. [CrossRef]

131. Åström, M.; Corin, N. Distribution of rare earth elements in anionic, cationic and particulate fractions in boreal humus-rich streams affected by acid sulphate soils. *Water Res.* **2003**, *37*, 273–280. [CrossRef]

132. Barroux, G.; Sonke, J.E.; Boaventura, G.; Viers, J.; Godderis, Y.; Bonnet, M.-P.; Sondag, F.; Gardoll, S.; Lagane, C.; Seyler, P. Seasonal dissolved rare earth element dynamics of the Amazon River main stem, its tributaries, and the Curuaí floodplain. *Geochem. Geophys. Geosyst.* **2006**, *7*, Q12005. [CrossRef]

133. Bau, M.; Dulski, P. Anthropogenic origin of positive gadolinium anomalies in river waters. *Earth Planet. Sci. Lett.* **1996**, *143*, 245–255. [CrossRef]

134. Bau, M.; Knappe, A.; Dulski, P. Anthropogenic gadolinium as a micropollutant in river waters in Pennsylvania and in Lake Erie, northeastern United States. *Geochemistry* **2006**, *66*, 143–152. [CrossRef]

135. Bayon, G.; Birot, D.; Ruffine, L.; Caprais, J.-C.; Ponzevera, E.; Bollinger, C.; Donval, J.-P.; Charlou, J.-L.; Voisset, M.; Grimaud, S. Evidence for intense REE scavenging at cold seeps from the Niger Delta margin. *Earth Planet. Sci. Lett.* **2011**, *312*, 443–452. [CrossRef]

136. Biddau, R.; Cidu, R.; Frau, F. Rare earth elements in waters from the albitite-bearing granodiorites of Central Sardinia, Italy. *Chem. Geol.* **2002**, *182*, 1–14. [CrossRef]

137. Bwire Ojiambo, S.; Lyons, W.B.; Welch, K.A.; Poreda, R.J.; Johannesson, K.H. Strontium isotopes and rare earth elements as tracers of groundwater-lake water interactions, Lake Naivasha, Kenya. *Appl. Geochem.* **2003**, *18*, 1789–1805. [CrossRef]

138. Censi, P.; Sposito, F.; Inguaggiato, C.; Zuddas, P.; Inguaggiato, S.; Venturi, M. Zr, Hf and REE distribution in river water under different ionic strength conditions. *Sci. Total Environ.* **2018**, *645*, 837–853. [CrossRef] [PubMed]

139. Centeno, L.M.; Faure, G.; Lee, G.; Talnagi, J. Fractionation of chemical elements including the REEs and 226Ra in stream contaminated with coal-mine effluent. *Appl. Geochem.* **2004**, *19*, 1085–1095. [CrossRef]

140. Chevis, D.A.; Johannesson, K.H.; Burdige, D.J.; Tang, J.; Moran, S.B.; Kelly, R.P. Submarine groundwater discharge of rare earth elements to a tidally-mixed estuary in Southern Rhode Island. *Chem. Geol.* **2015**, *397*, 128–142. [CrossRef]

141. de Baar, H.J.W.; Bruland, K.W.; Schijf, J.; van Heuven, S.M.A.C.; Behrens, M.K. Low cerium among the dissolved rare earth elements in the central North Pacific Ocean. *Geochim. Cosmochim. Acta* **2018**, *236*, 5–40. [CrossRef]

142. de Baar, H.J.W.; Bacon, M.P.; Brewer, P.G. Rare-earth distributions with a positive Ce anomaly in the Western North Atlantic Ocean. *Nature* **1983**, *301*, 324–327. [CrossRef]

143. De Carlo, E.H.; Green, W.J. Rare earth elements in the water column of Lake Vanda, McMurdo Dry Valleys, Antarctica. *Geochim. Cosmochim. Acta* **2002**, *66*, 1323–1333. [CrossRef]

144. Elderfield, H.; Greaves, M.J. The rare earth elements in seawater. *Nature* **1982**, *296*, 214–219. [CrossRef]

145. Elderfield, H.; Sholkovitz, E.R. Rare earth elements in the pore waters of reducing nearshore sediments. *Earth Planet. Sci. Lett.* **1987**, *82*, 280–288. [CrossRef]

146. Esmaeili-Vardanjani, M.; Shamsipour-Dehkordi, R.; Eslami, A.; Moosaei, F.; Pazand, K. A study of differentiation pattern and rare earth elements migration in geochemical and hydrogeochemical environments of Airekan and Cheshmeh Shotori areas (Central Iran). *Environ. Earth Sci.* **2013**, *68*, 719–732. [CrossRef]

147. Gammons, C.H.; Wood, S.A.; Pedrozo, F.; Varekamp, J.C.; Nelson, B.J.; Shope, C.L.; Baffico, G. Hydrogeochemistry and rare earth element behavior in a volcanically acidified watershed in Patagonia, Argentina. *Chem. Geol.* **2005**, *222*, 249–267. [CrossRef]

148. Gammons, C.H.; Wood, S.A.; Nimick, D.A. Diel behavior of rare earth elements in a mountain stream with acidic to neutral pH. *Geochim. Cosmochim. Acta* **2005**, *69*, 3747–3758. [CrossRef]

149. Garcia-Solsona, E.; Jeandel, C.; Labatut, M.; Lacan, F.; Vance, D.; Chavagnac, V.; Pradoux, C. Rare earth elements and Nd isotopes tracing water mass mixing and particle-seawater interactions in the SE Atlantic. *Geochim. Cosmochim. Acta* **2014**, *125*, 351–372. [CrossRef]

150. German, C.R.; Masuzawa, T.; Greaves, M.J.; Elderfield, H.; Edmond, J.M. Dissolved rare earth elements in the Southern Ocean: Cerium oxidation and the influence of hydrography. *Geochim. Cosmochim. Acta* **1995**, *59*, 1551–1558. [CrossRef]

151. German, C.R.; Holliday, B.P.; Elderfield, H. Redox cycling of rare earth elements in the suboxic zone of the Black Sea. *Geochim. Cosmochim. Acta* **1991**, *55*, 3553–3558. [CrossRef]

152. Goldstein, S.J.; Jacobsen, S.B. Rare earth elements in river waters. *Earth Planet. Sci. Lett.* **1988**, *89*, 35–47. [CrossRef]

153. Grenier, M.; Jeandel, C.; Lacan, F.; Vance, D.; Venchiarutti, C.; Cros, A.; Cravatte, S. From the subtropics to the central equatorial Pacific Ocean: Neodymium isotopic composition and rare earth element concentration variations. *J. Geophys. Res.-Oceans* **2013**, *118*, 592–618. [CrossRef]

154. Haley, B.A.; Klinkhammer, G.P.; McManus, J. Rare earth elements in pore waters of marine sediments. *Geochim. Cosmochim. Acta* **2004**, *68*, 1265–1279. [CrossRef]

155. Hathorne, E.C.; Stichel, T.; Brück, B.; Frank, M. Rare earth element distribution in the Atlantic sector of the Southern Ocean: The balance between particle scavenging and vertical supply. *Mar. Chem.* **2015**, *177*, 157–171. [CrossRef]

156. Hongo, Y.; Obata, H.; Alibo, D.S.; Nozaki, Y. Spatial variations of rare earth elements in North Pacific surface water. *J. Oceanogr.* **2006**, *62*, 441–455. [CrossRef]

157. Hoyle, J.; Elderfield, H.; Gledhill, A.; Greaves, M. The behaviour of the rare earth elements during mixing of river and sea waters. *Geochim. Cosmochim. Acta* **1984**, *48*, 143–149. [CrossRef]

158. Jeandel, C.; Delattre, H.; Grenier, M.; Pradoux, C.; Lacan, F. Rare earth element concentrations and Nd isotopes in the Southeast Pacific Ocean. *Geochem. Geophys. Geosyst.* **2013**, *14*, 328–341. [CrossRef]

159. Johannesson, K.H.; Palmore, C.D.; Fackrell, J.; Prouty, N.G.; Swarzenski, P.W.; Chevis, D.A.; Telfeyan, K.; White, C.D.; Burdige, D.J. Rare earth element behavior during groundwater-seawater mixing along the Kona Coast of Hawaii. *Geochim. Cosmochim. Acta* **2017**, *198*, 229–258. [CrossRef]

160. Johannesson, K.H.; Farnham, I.M.; Guo, C.; Stetzenbach, K.J. Rare earth element fractionation and concentration variations along a groundwater flow path within a shallow, basin-fill aquifer, southern Nevada, USA. *Geochim. Cosmochim. Acta* **1999**, *63*, 2697–2708. [CrossRef]

161. Johannesson, K.H.; Stetzenbach, K.J.; Hodge, V.F. Rare earth elements as geochemical tracers of regional groundwater mixing. *Geochim. Cosmochim. Acta* **1997**, *61*, 3605–3618. [CrossRef]

162. Johannesson, K.H.; Lyons, W.B.; Stetzenbach, K.J.; Byrne, R.H. The solubility control of rare earth elements in natural terrestrial waters and the significance of PO43- and CO32- in limiting dissolved rare earth concentrations: A review of recent information. *Aquat. Geochem.* **1995**, *1*, 157–173. [CrossRef]

163. Johannesson, K.H.; Lyons, W.B. Rare-earth element geochemistry of Colour Lake, an acidic freshwater lake on Axel Heiberg Island, Northwest Territories, Canada. *Chem. Geol.* **1995**, *119*, 209–223. [CrossRef]

164. Kalender, L.; Aytimur, G. REE geochemistry of Euphrates River, Turkey. *J. Chem.* **2016**, *2016*, 1012021. [CrossRef]

165. Kim, J.-H.; Torres, M.E.; Haley, B.A.; Kastner, M.; Pohlman, J.W.; Riedel, M.; Lee, Y.J. The effect of diagenesis and fluid migration on rare earth element distribution in fluids of the northern Cascadia accretionary margin. *Chem. Geol.* **2011**, *291*, 152–165. [CrossRef]
166. Kulaksiz, S.; Bau, M. Anthropogenic gadolinium as a microcontaminant in tap water used as drinking water in urban areas and megacities. *Appl. Geochem.* **2011**, *26*, 1877–1885. [CrossRef]
167. Kulaksiz, S.; Bau, M. Contrasting behaviour of anthropogenic gadolinium and natural rare earth elements in estuaries and the gadolinium input into the North Sea. *Earth Planet. Sci. Lett.* **2007**, *260*, 361–371. [CrossRef]
168. Leybourne, M.I.; Goodfellow, W.D.; Boyle, D.R.; Hall, G.M. Rapid development of negative Ce anomalies in surface waters and contrasting REE patterns in groundwaters associated with Zn-Pb massive sulphide deposits. *Appl. Geochem.* **2000**, *15*, 695–723. [CrossRef]
169. Liu, H.; Guo, H.; Xing, L.; Zhan, Y.; Li, F.; Shao, J.; Niu, H.; Liang, X.; Li, C. Geochemical behaviors of rare earth elements in groundwater along a flow path in the North China Plain. *J. Asian Earth Sci.* **2016**, *117*, 33–51. [CrossRef]
170. Merschel, G.; Bau, M.; Schmidt, K.; Münker, C.; Dantas, E.L. Hafnium and neodymium isotopes and REY distribution in the truly dissolved, nanoparticulate/colloidal and suspended loads of rivers in the Amazon Basin, Brazil. *Geochim. Cosmochim. Acta* **2017**, *213*, 383–399. [CrossRef]
171. Nozaki, Y.; Lerche, D.; Alibo, D.S.; Tsutsumi, M. Dissolved indium and rare earth elements in three Japanese rivers and Tokyo Bay: Evidence for anthropogenic Gd and In. *Geochim. Cosmochim. Acta* **2000**, *64*, 3975–3982. [CrossRef]
172. Osborne, A.H.; Hathorne, E.C.; Schijf, J.; Plancherel, Y.; Böning, P.; Frank, M. The potential of sedimentary foraminiferal rare earth element patterns to trace water masses in the past. *Geochem. Geophys. Geosyst.* **2017**, *18*, 1550–1568. [CrossRef]
173. Piepgras, D.J.; Jacobsen, S.B. The behavior of rare earth elements in seawater: Precise determination of variations in the North Pacific water column. *Geochim. Cosmochim. Acta* **1992**, *56*, 1851–1862. [CrossRef]
174. Pokrovsky, O.S.; Viers, J.; Shirokova, L.S.; Shevchenko, V.P.; Filipov, A.S.; Dupré, B. Dissolved, suspended, and colloidal fluxes of organic carbon, major and trace elements in the Severnaya Dvina River and its tributary. *Chem. Geol.* **2010**, *273*, 136–149. [CrossRef]
175. Reynard, B.; Lecuyer, C.; Grandjean, P. Crystal-chemical controls on rare-earth element concentrations in fossil biogenic apatites and implications for paleoenvironmental reconstructions. *Chem. Geol.* **1999**, *155*, 233–241. [CrossRef]
176. Rousseau, T.C.C.; Sonke, J.E.; Chmeleff, J.; van Beek, P.; Souhat, M.; Boaventura, G.; Seyler, P.; Jeandel, C. Rapid neodymium release to marine waters from lithogenic sediments in the Amazon estuary. *Nat. Commun.* **2015**, *6*, 7592. [CrossRef] [PubMed]
177. Sholkovitz, E.R. The geochemistry of rare earth elements in the Amazon River estuary. *Geochim. Cosmochim. Acta* **1993**, *57*, 2181–2190. [CrossRef]
178. Sholkovitz, E.R.; Landing, W.M.; Lewis, B.L. Ocean particle chemistry: The fractionation of rare earth elements between suspended particles and seawater. *Geochim. Cosmochim. Acta* **1994**, *58*, 1567–1579. [CrossRef]
179. Smedley, P.L. The geochemistry of rare earth elements in groundwater from the Carnmenellis area, southwest England. *Geochim. Cosmochim. Acta* **1991**, *55*, 2767–2779. [CrossRef]
180. Tang, J.; Johannesson, K.H. Controls on the geochemistry of rare earth elements along a groundwater flow path in the Carrizo Sand aquifer, Texas, USA. *Chem. Geol.* **2006**, *225*, 156–171. [CrossRef]
181. Tang, J.; Johannesson, K.H. Speciation of rare earth elements in natural terrestrial waters: Assessing the role of dissolved organic matter from the modeling approach. *Geochim. Cosmochim. Acta* **2003**, *67*, 2321–2339. [CrossRef]
182. Trueman, C.N.; Behrensmeyer, A.K.; Potts, R.; Tuross, N. High-resolution records of location and stratigraphic provenance from the rare earth element composition of fossil bones. *Geochim. Cosmochim. Acta* **2006**, *70*, 4343–4355. [CrossRef]
183. van de Flierdt, T.; Pahnke, K.; Amakawa, H.; Andersson, P.; Basak, C.; Coles, B.; Colin, C.; Crocket, K.; Frank, M.; Frank, N.; et al. GEOTRACES intercalibration of neodymium isotopes and rare earth element concentrations in seawater and suspended particles. Part 1: Reproducibility of results for the international intercomparison. *Limnol. Oceanogr. Methods* **2012**, *10*, 234–251. [CrossRef]
184. Wang, Z.-L.; Yamada, M. Geochemistry of dissolved rare earth elements in the Equatorial Pacific Ocean. *Environ. Geol.* **2007**, *52*, 779–787. [CrossRef]
185. Zhang, J.; Nozaki, Y. Rare earth elements and yttrium in seawater: ICP-MS determinations in the East Caroline, Coral Sea, and South Fiji basins of the western South Pacific Ocean. *Geochim. Cosmochim. Acta* **1996**, *60*, 4631–4644. [CrossRef]
186. Zheng, W.; Schweitzer, M.H. Chemical Analyses of Fossil Bone. In *Forensic Microscopy for Skeletal Tissues: Methods and Protocols*; Bell, L.S., Ed.; Humana Press: Totowa, NJ, USA, 2012; pp. 153–172, ISBN 978-1-61779-976-1.
187. Zheng, X.-Y.; Plancherel, Y.; Saito, M.A.; Scott, P.M.; Henderson, G.M. Rare earth elements (REEs) in the tropical South Atlantic and quantitative deconvolution of their non-conservative behavior. *Geochim. Cosmochim. Acta* **2016**, *177*, 217–237. [CrossRef]

biology

MDPI

Article

Independent Evidence for the Preservation of Endogenous Bone Biochemistry in a Specimen of *Tyrannosaurus rex*

Jennifer Anné [1,*], Aurore Canoville [2], Nicholas P. Edwards [3], Mary H. Schweitzer [4,5,6] and Lindsay E. Zanno [4,5]

[1] The Children's Museum of Indianapolis, Indianapolis, IN 46208, USA
[2] Stiftung Schloss Friedenstein Gotha, 99867 Gotha, Germany
[3] Stanford Synchrotron Radiation Light Source, SLAC National Accelerator Laboratory, Menlo Park, CA 94025, USA
[4] Department of Biological Sciences, Campus Box 7617, North Carolina State University, Raleigh, NC 27695, USA
[5] Paleontology, North Carolina Museum of Natural Sciences, 11 W. Jones St., Raleigh, NC 27601, USA
[6] Department of Geology, Lund University, Sölvegatan 12, 223 62 Lund, Sweden
*   Correspondence: janne@childrensmuseum.org

Simple Summary: Our understanding of what can preserve in the fossil record, and for how long, is constantly evolving with the use of new scientific techniques and exceptional fossil discoveries. In this study, we examine the state of preservation of a *Tyrannosaurus rex* that died about 66 million years ago. This specimen has previously been studied using a number of advanced methods, all of which have indicated preservation of original soft tissues and bone biomolecules. Here, we use synchrotron—a type of particle accelerator—analyses to generate data identifying and quantifying elements that constitute this fossil bone. We show that trace elements incorporated by the living animal during bone deposition and remodeling, such as zinc, are preserved in the fossil bone in a pattern similar to what is seen in modern bird bones. This pattern is not observed in a microscopically well preserved, but molecularly more degraded dinosaur, a herbivorous *Tenontosaurus*. These data further support the preservation of original biological material in this *T. rex*, suggesting new possibilities for deciphering extinct species life histories. This study also highlights that preservation of original biochemistry in fossils is specimen-specific and cannot be determined by pristine appearance alone.

Citation: Anné, J.; Canoville, A.; Edwards, N.P.; Schweitzer, M.H.; Zanno, L.E. Independent Evidence for the Preservation of Endogenous Bone Biochemistry in a Specimen of *Tyrannosaurus rex*. *Biology* **2023**, *12*, 264. https://doi.org/10.3390/biology12020264

Academic Editor: Zhifei Zhang

Received: 29 December 2022
Revised: 3 February 2023
Accepted: 4 February 2023
Published: 7 February 2023

Copyright: © 2023 by the authors. Licensee MDPI, Basel, Switzerland. This article is an open access article distributed under the terms and conditions of the Creative Commons Attribution (CC BY) license (https://creativecommons.org/licenses/by/4.0/).

Abstract: Biomolecules preserved in deep time have potential to shed light on major evolutionary questions, driving the search for new and more rigorous methods to detect them. Despite the increasing body of evidence from a wide variety of new, high resolution/high sensitivity analytical techniques, this research is commonly met with skepticism, as the long standing dogma persists that such preservation in very deep time (>1 Ma) is unlikely. The Late Cretaceous dinosaur *Tyrannosaurus rex* (MOR 1125) has been shown, through multiple biochemical studies, to preserve original bone chemistry. Here, we provide additional, independent support that deep time bimolecular preservation is possible. We use synchrotron X-ray fluorescence imaging (XRF) and X-ray absorption spectroscopy (XAS) to investigate a section from the femur of this dinosaur, and demonstrate preservation of elements (S, Ca, and Zn) associated with bone remodeling and redeposition. We then compare these data to the bone of an extant dinosaur (bird), as well as a second non-avian dinosaur, *Tenontosaurus tilletti* (OMNH 34784) that did not preserve any sign of original biochemistry. Our data indicate that MOR 1125 bone cortices have similar bone elemental distributions to that of an extant bird, which supports preservation of original endogenous chemistry in this specimen.

Keywords: synchrotron; bone remodeling; elemental analysis; molecular paleontology; diagenetic alteration

1. Introduction

The recent application of chemical and molecular techniques in paleontological research has resulted in a re-examination of the preservation potential of original biological chemistry in deep time. Such studies document the preservation of organic molecules and structures, ranging, for example, from elemental and microstructural evidence of color to ancient endogenous proteins, e.g., [1–5]. These discoveries push the boundaries of what was thought (or assumed) to preserve in the fossil record and have caused paleontologists to reconsider the canonical narrative that all original material is replaced during the fossilization process. Nonetheless, extraordinary claims require extraordinary evidence, and published studies documenting the preservation of organic molecules in fossils (particularly Paleozoic and Mesozoic fossils [1–25]) have, not surprisingly, been met with skepticism, e.g., [26–28]. One mechanism for increasing confidence in previous claims of endogenous chemistry in ancient fossils is the application of multiple and independent techniques to retest previous results in a non-destructive manner. Here, we review prior molecular studies, conducted over almost 20 years by multiple investigators, on *Tyrannosaurus rex* skeletal elements (two tibiae and one femur; Museum of the Rockies; MOR 1125) recovered from the Upper Cretaceous Hell Creek Formation of Montana, and use synchrotron X-ray fluorescence (XRF) to further test claims of original biochemistry preservation and limited diagenetic alteration in this specimen. We then contrast data collected from MOR 1125 with those collected from the tibia of the ornithopod *Tenontosaurus tilletti* OMNH 34784 from the Lower Cretaceous Cloverly Formation, and an extant avian radius (*Cacatua moluccensis*; NCSM 17977) showing similar cortical bone tissues.

1.1. Previous Research on MOR 1125

The *T. rex* femur of MOR 1125, was first studied in 2005 for signs of possible medullary bone [25], a sex-specific, estrogen-sensitive, ephemeral bone tissue produced by extant birds in lay [29]. This tissue identification was based on similarities in microstructure and location in both mineralized and demineralized tissues, when compared to those of extant laying ratites (ostrich and emu; [25]). Since then, 20 different techniques have been applied to this specimen, ranging from mass spectrometry to immunohistochemistry using antibody–antigen recognition (Table 1). Results from these studies showed exceptional micromorphological preservation extending to the cellular level (eg. osteocytes, possible endothelial cells; ([17,18,22–25], Schweitzer et al. in prep) as well as several biochemical signatures of bone-specific proteins (e.g., collagen; [17–19,23]). Additionally, these studies confirmed the observed molecular signatures differed from various diagenetically-induced morphologies/chemistries (e.g., biofilms; [14]) and were comparable to similar morphologies/chemistries of extant archosaurs. A summary of all molecular/chemical techniques applied to specimen MOR 1125 and the results obtained is provided in Table 1.

Accumulated data from multiple studies point to an endogenous source for these molecular signals. Here, we add to this body of evidence by applying new, highly sensitive and high-resolution methods to fully characterize the organic remains within this specimen. In this study, we apply Synchrotron XRF and XAS to further test the hypothesis that original biochemistry is preserved in this specimen.

1.2. Synchrotron XRF and XAS

Synchrotron radiation has many advantages over commercial X-ray analyses based on X-ray tubes, such as being monochromatic, high flux, and tunable [30]. These properties result in high sensitivity to dilute elemental concentrations (1 ppm) and the ability to significantly reduce data acquisition time. Additionally, samples can be analyzed under ambient atmosphere and temperatures, with no special preparation requirements. In some synchrotron XRF imaging stations, decimeter scale samples can also be accommodated, which removes the need for subsampling (e.g., [4]).

Table 1. List of various techniques applied and results obtained from molecular/chemical studies of the *T. rex* MOR 1125.

Analytical Technique	MOR 1125 Tissue Type	Results and Citation
Transmitted light microscopy (LM)	Demineralized endosteal bone (MB) and cortical bone (CB)	Flexible, collagen-like fibrous matrix, transparent, hollow vessel-like structures, osteocyte-like microstructures with filopodia [24,25]
Scanning electron microscopy (SEM)	Demineralized Medullary Bone (MB) and Cortical Bone (CB)	Fibers arranged as bundles, consistent with collagen [24]
SEM	Isolated microstructures consistent with osteocytes (hereafter 'osteocytes' to conserve space) and microstructures consistent with blood vessels (hereafter vessels) liberated from demineralized CB	Isolated, free-floating 3-D osteocyte morphs, with long, extensive filipodia and intracellular contents, consistent in morphology and location to osteocytes from extant vertebrates; interconnected, hollow, flexible and transparent microstructures consistent with extant vessels [18,23,25]
SEM	Biofilm grown in extant deproteinated bone	Patchy distribution after demineralization of bone substrate, unable to maintain shape, differed morphologically from structures observed in MOR 1125; no visible osteocyte structures [14]
Atomic force microscopy (AFM)	Demineralized CB and MB	Matrix fibers demonstrating periodicity of ~70 nm consistent with collagen [22]
Transmitted electron microscopy (TEM), localized electron diffraction	Ultrathin sections of CB and MB	Mineral phase identified as biogenic hydroxylapatite (HA) [22]
Transmitted electron microscopy (TEM), localized electron diffraction	Isolated vessels and osteocyte liberated from CB	Iron intimately associated with vessel walls and osteocyte surfaces. Vessel walls under TEM retain differentiated structures and layers, consistent with extant vessels ([17,18,22,23], Schweitzer et al., in prep)
Electron energy loss spectroscopy (EELS)	Vessels and osteocytes liberated from demineralized CB	Iron localized to vascular walls and osteocytes, but not seen in fibrous matrix [17]
Micro-X-ray fluorescence mapping of Fe	Vessels liberated from demineralized CB	Confirmation of iron localizing to vessel walls [17]
Micro-X-ray Fe XANES	Vessels liberated from demineralized CB	Diagenetic goethite Fe within vessels and vessel walls, also seen in treated extant vessels [17]
Immunohistochemistry (IHC), antibodies against chicken collagen I (Col I) and osteocalcin (OC)	Demineralized CB and MB	Localized binding of antibodies to tissue, supporting preservation of Col I and OC epitopes, using multiple controls [23]
IHC to actin antibodies	Isolated vessels and osteocytes from demineralized CB	Localized antibody binding in patterns seen in extant homologues. Different binding patterns in each tissue/cell type support endogeneity, along with negative controls [17,18]

Table 1. *Cont.*

Analytical Technique	MOR 1125 Tissue Type	Results and Citation
IHC using antibodies to chicken phosphoendopeptidase (PHEX)	Osteocytes liberated from demineralized CB	PHEX antibodies localized to osteocyte-like structures, but not to matrix. Binding was specific and controls were negative. Antibody specificity was confirmed to be avian specific; no binding was seen to extant crocodile osteocytes, only bird [18]
ICH using antibodies to DNA backbone	Osteocytes liberated from demineralized CB	Antibodies localized in a single spot internal to some dinosaur osteocytes. No binding was seen to filopodia or other materials [18]
IHC using antibodies to keratan sulfate	Demineralized CB and MB	Binding in a globular pattern in extant and dinosaur MB, no binding in adjacent CB from the same specimens [13]
Histochemical localization using DNA intercalating stains propidium iodide (PI) and 4′, 6′-diamidino-2-phenylindole dihydrochloride (DAPI)	Osteocytes liberated from demineralized CB	Binding followed pattern seen with anti-DNA backbone antibodies, specific to internal region of cell, and only a single spot. No staining was observed on filopodia or other structures [18]
Histochemical localization using Alcian blue.	Demineralized CB and MB	Differential staining, with MB much more intensely stained than CB, consistent with MB in extant avian bone [13]
Histochemical localization with High-Iron Diamine (HID)	Demineralized CB and MB	Differential staining, with MB staining much more intensely relative to CB, consistent with extant avians [13]
Time-of-flight secondary ion mass spectrometry (ToF-SIMS)	Demineralized CB and MB	Identification of amino acids glycine (gly) and alanine (ala) support presence of Col I [23]
Mass spectrometry (MS)	Demineralized and extracted vessels, matrix and osteocytes	Multiple collagen and actin peptide sequences [18,20,21]
Peptide mapping	Peptides recovered from demineralized and extracted bone matrix	Fossil-derived peptides mapped to monomers 2,3, and 4 on extant collagen models [19]
Laser ablation- inductively coupled plasma mass spectrometry (LAICP-MS)	Sectioned CB	Average concentration of exogenous elements from diagenesis (e.g., Fe) lower than in other Hell Creek specimens, with only moderate alteration [7]

Monochromatic X-rays allow the selection of a very narrow bandwidth of X-ray energy/wavelengths which facilitates X-ray absorption spectroscopy (XAS). XAS is one of the most common and most powerful techniques at synchrotron sources as it provides information about the atomic structure of the absorbing atom, allowing the determination of elemental speciation [30]. In the case of studying fossil tissues, this information is critical in determining whether a detected element is derived from organic or inorganic processes. Furthermore, XRF imaging and XAS can be combined to produce maps of elemental species, which adds a further layer of our capability to tease out physiological processes from diagenetic ones in extinct organisms.

1.3. Previous Synchrotron Work on Fossils

Synchrotron XRF and XAS has been used to examine and interpret the biochemistry of fossils for over 10 years, in specimens recovered from a wide range of geologic ages (~400 mya to recent), tissue types (non-biomineralized (e.g., skin) and mineralized (e.g., bones, teeth) and taxa (invertebrates, vertebrates and plants) [4,8,10–12,15,16,31–34]. These studies revealed biological structures that cannot be observed in visible light, as well as the fractionation of elements within discrete biological structures that can be compared with similar tissues in living organisms. Resulting data have led to the identification of specific elemental biomarkers for a number of biosynthetic pathways, including those involved in bone remodeling, repair and deposition [10,11,16,31–34].

In previous synchrotron analyses involving bone, zinc (Zn) and, in most cases, strontium (Sr), were found to correlate within areas of active ossification, including fracture calli, growth plates and around secondary osteons [10,11,16,32–40]. Sr was also shown to be correlated to diagenetic processes, being isolated to the Haversian canal, and was differentiated from organically-derived Sr through differences in distribution patterns and chemical coordination [11]. Calcium (Ca) distributions were found to be similar to Zn and concentrated in areas of ossification as well as the cutting cones of secondary osteons [10,11,16,32]. Although distributions and concentrations varied slightly depending on species, these patterns were seen in both extant and extinct vertebrates, with the oldest preservation of these patterns seen in a 150 mya dinosaur phalanx [16].

1.4. Bone Elemental Biomarkers

In extant vertebrates, elemental biomarkers of bone physiology are most often concentrated on areas of active ossification [32–40]. Zn is one of these, because it plays a structural role when incorporated into the hydroxyapatite (HAP) lattice, forming Zn–Fluorine (F) complexes [35,39]. Zn is also critical for osteoblastogenesis, stimulating bone formation and inhibiting bone resorption [35,39]. The expression of Zn has been found to be highest in osteocytes, which express alkaline phosphatase (ALP) and osteocalcin (OCN) in mineralized tissue [36]. Elevated concentrations of Zn localize to zones between mineralized and unmineralized tissue within osteons, suggesting a role in bone mineralization and cellular regulation [34,40]. Finally, Zn is associated with matrix metalloproteinases (MMPs), which are important for cartilage degradation [41]. In fracture healing, Zn is expressed between the first and second week in rats, where it increases the expression of OCN needed for hard callus formation [35].

Other key elements in bone physiology include sulfur (S), phosphorus (P), Ca and Sr. Sulfur comprises both the organic and inorganic constituents of bone, aiding in both collagen (e.g., cysteine and methionine; [42]) and HAP structure (sulfate; [43]). Sulfur is also an important constituent of glycoproteins found in various skeletal tissues types (e.g., keratan sulfate in medullary bone and other fast growing skeletal tissues; [13,44]). Sr increases osteoblast activity by increasing osteoblastic marker expression in ALPs, bone sialoprotein (BSP), and OCN, all of which affect osteoblast proliferation and differentiation [37,38]. It also interferes with osteoclast dissolution of bone mineral, by disrupting the actin cytoskeleton organized in the sealing zone, a thick band of actin needed for osteoclast apical-basal polarization and resorption [37].

Ca is most associated with bone formation and metabolism, with a majority of vertebrate Ca found in the HAP structure of bones and teeth [44]. In the skeleton, Ca bound in apatite ($[Ca_3(PO_4)_2]_3Ca(OH)_2$) lattice serves as structural support (rigidity, strength and elasticity) and as a reservoir for Ca needed for other metabolic processes throughout the body. Like Ca, P is predominantly found in bone and teeth [45]. It is primarily associated with Ca, both in HAP and amorphous calcium phosphate. Inorganic P is one of the most important components for HAP formation during the mineralization of the extracellular matrix [45].

1.5. Elements Associated with Fossilization and Diagenetic Alteration

Elements observed in fossilized skeletons are either endogenous, or diagenetic, the latter being influenced by the depositional environment. For fossils from the Hell Creek Formation, the most common and abundant diagenetic element seen in fossil specimens is iron (Fe). In some specimens, the amount of Fe is so high that it masks any possible biological signatures [Anné, unpublished work]. In other specimens, Fe is constrained to the mineral infill of the fossil cavities (e.g., medullary cavity and vascular spaces). Other elements usually associated with diagenetic alteration or mineral infill are silicon (Si) and Ca, derived from the various silicate and carbonate minerals associated with fossilized bone ([10,16,33]; Anné, unpublished work).

Here, we test the hypothesis that the biochemical signatures detectable in MOR 1125 are endogenous, and were present in the once-living animal. We predict that if original biochemistry is preserved in MOR 1125, then the distributional patterns of elements crucial to bone formation and remodeling will mirror those found in extant bone, with elevated and localized concentrations of Ca, Zn and possibly S and Sr as described above.

We also compared results from MOR 1125 to extant bone, as a positive control, and to those we derived from the Cretaceous dinosaur *Tenontosaurus tilleti* (OMNH 34784). The latter introduces taxonomic and depositional variables, allowing us to compare the preservation potential between two fossil bones that differ in geological age and depositional environments.

2. Materials and Methods

2.1. Biologic Samples

Our extant bone sample (NCSM 17977) derives from the radius of the bird *Cacatua moluccensis*. This specimen bears mostly remodeled cortical tissue encased by a fracture callus and reactive bone tissue from a possible infection (Figure 1A,B). The "normal" cortical bone of the radius is mostly composed of secondary tissue with large secondary osteons. The fracture callus is represented by trabecular tissue on both the periosteal and endosteal surfaces. The endosteal pathological tissue completely fills the medullary cavity, while the periosteal tissue more than doubles the diameter of the element (Figure 1A,B).

Fossils are represented by a femoral fragment of *Tyrannosaurus rex* (MOR 1125) recovered from the lower Hell Creek Formation (~67–68 my [46]) and a tibial fragment of *Tenontosaurus tilletti* OMNH 34784 from the Cloverly Formation (~124–98 my [47]). Both specimens have been previously studied histologically [24,48,49] and both were described as containing an unusual endosteal tissue consistent with reproductive ("avian") medullary bone [29]. The bone fragment of MOR 1125 considered in this study consists of portions of the deep cortex adjacent to this endosteal spongious tissue. The deep cortex consists mostly of a dense Haversian tissue with several generations of secondary osteons. The associated spongiosa is formed of secondary trabeculae (Figure 1C,D). Between some trabeculae, there is a fine and crushed bone tissue that is similar in location and microstructure to reproductive medullary bone (Figure 1D) and previous molecular studies supported such identification [13,25].

The OMNH 34784 tibial fragment we analyzed is composed of fibrolamellar bone with secondary remodeling marked by several secondary osteons in the deep cortex, some of which are overlapping (Figure 1E,F). Radially-oriented endosteal trabecular bone is present within the medullary cavity. This tissue has been interpreted as potential medullary bone [48].

2.2. Synchrotron Analyses

XRF and XAS were performed at the Stanford Synchrotron Radiation Lightsource (SSRL) at beam lines 2-3, 7-2 and 14-3 (specifics for each beam line listed below). At these beam lines, XRF imaging is performed by mounting the sample on high precision encoded stages and rastering the sample relative to the incident X-ray beam in a continuous scan mode at a 45° angle to the incident beam. The XRF signal is detected with a Hitachi Vortex silicon drift diode detector positioned at 90° to the incident beam. The detector is coupled to a Quantum Detectors Xspress3 multi-channel analyzer system. Data collection is achieved by the stage moving in a continuous horizontal motion, with an image pixel defined by the cumulative XRF counts binned as a function of distance traveled (step size) and velocity over this distance (dwell time). At the end of each horizontal motion, data acquisition stops, the data is read from the detector system, and the stages move to the beginning of the next line. The fluorescence energies for 16 user selectable elements are collected, as well as the full XRF spectrum per pixel. The 16 elements can be displayed live during data collection and are shown here. The full XRF spectrum data is used in cases where deeper interrogation of the XRF data is required. Further details on signal processing of the imaging system are provided in [50,51]. XAS spectra were collected in fluorescence mode. XRF image data were processed using the MicroAnalysis Toolkit software and XAS spectra were processed using the SIXpack [52,53].

2.2.1. Beam Line 7-2

Beam line 7-2 is a hard X-ray wiggler beam line optimized for XRF imaging and XAS of high Z elements (Ca and heavier). It has an energy range of ~5–16 keV with a Si(111) monochromator. The X-ray spot size on the sample is achieved with XOS polycapillary focusing optics with a ~35 μm or ~65 μm spot size available. In this study, an incident beam energy of 11 keV and a spot size of 35 μm was used. XRF signal was detected with a 4 element Hitachi Vortex silicon drift diode detector. 7-2 has a regular scan range of ~400 × 300 mm which allows for multiple samples (such as thin sections) to be mounted simultaneously or can accommodate larger scale samples.

2.2.2. Beam Line 2-3

Beam line 2-3 is a hard X-ray bending magnet beam line optimized for XRF imaging and XAS of high Z elements (Ca and higher) at a higher resolution than 7-2 (but smaller scan range, 25 × 25 mm). It has an energy range of ~5–19 keV with a Si(111) monochromator. The X-ray spot size on the sample is achieved with Sigray axially symmetric focusing optics with 5 μm or 1 μm available. In this study, an incident energy of 13.5 keV and a spot size of 5 μm was used. XRF signal was detected with a 1 element Hitachi Vortex silicon drift diode detector.

2.2.3. Beam Line 14-3

Beam line 14-3 is a tender X-ray bending magnet beamline optimized for XRF imaging and XAS of low Z elements (S, P, Si, Cl). It has an energy range of 2.1–5 keV with two Si(111) monochromators (phi 0 or phi 90). The X-ray spot size on the sample is achieved with Sigray axially symmetric focusing optics with 5 μm or 1 μm available. In this study, a spot size of 5 μm was used. The sample is placed in a helium purged atmosphere to minimize X-ray attenuation from air. In this study, XRF signal was detected with a 4 element Hitachi Vortex silicon drift diode detector in earlier experiments, but in later experiments the system was upgraded to a 7 element detector.

Specimens analyzed at beamline 14-3 were first scanned at 2500 eV to detect a range of elements. For these maps, only total S can be visualized. To identify variations in S species, several scans were acquired at energies that correlate peaks in the XAS spectra known to be characteristic of specific S species, organic and inorganic, associated with the matrix of interest, in this case, bone. For this study, those energies correlate to ~2473, ~2476, ~2478, ~2480 and ~2482 eV. Principle Component Analysis (PCA) of the images is then performed which produces images that assist in highlighting the distribution of S species. These PCA images are then used to select target locations for S XANES, which are then used to identify the specific S species. An example of the step by step process from total S to S XANES is highlighted for MOR 1125 in Figure 2.

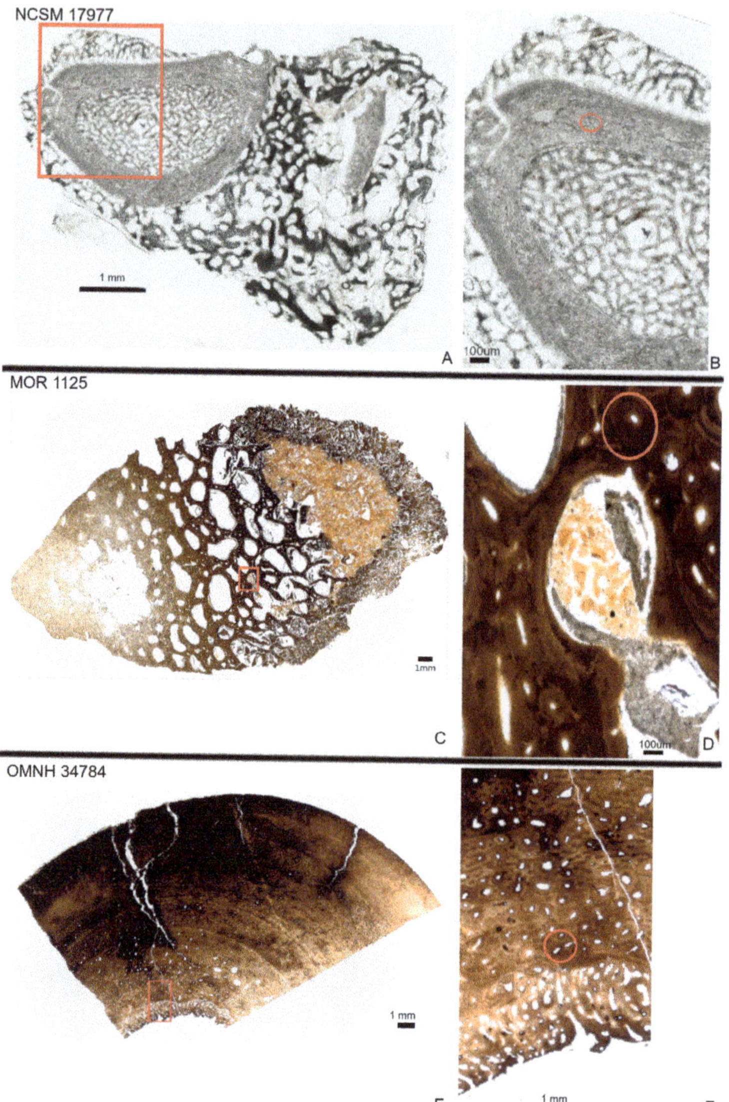

Figure 1. Histology of NCSM 17977 (radius; (**A,B**)), MOR 1125 (femur fragment; (**C,D**)) and OMNH 34784 (tibia fragment; **E,F**). Red boxes indicate where areas of higher magnification were taken. Areas of higher magnification represent those areas scanned for synchrotron XRF. NCSM 17977 is represented by a completed transverse section (**A**). The normal radius wall is composed of cortical bone with large secondary osteons (highlighted with red circle; (**B**)). Pathological tissues on both the periosteal and endosteal surfaces consist of trabeculae. The femur fragment of MOR 1125 consists of a dense Haversian tissues identified by multigenerational secondary osteons (highlighted with red circle) and trabeculae, with crushed bone between some trabeculae (**D**). OMNH 34784 consists of both a "normal" cortex and an unusual endosteal tissue deposited along the endosteal margin and filling part of the medullary cavity (**E**). The cortex is comprised of fibrolamellar bone that is more remodeled towards the endosteal margin as seen by multigenerational secondary osteons (highlighted with red circle; (**F**)).

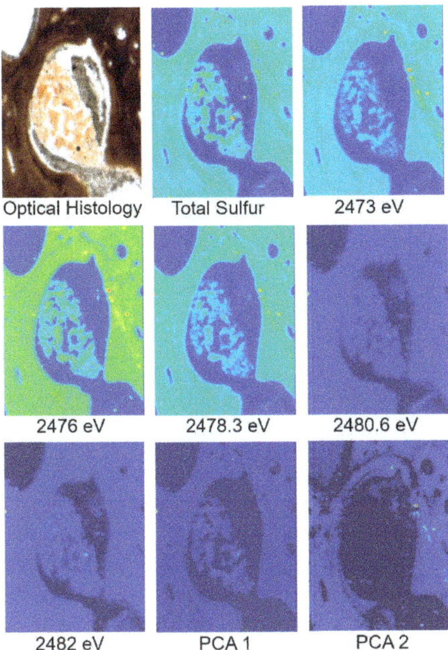

Figure 2. Optical histology, total S, S species, and PCA of S species maps from MOR 1125 (*T. rex*). Optical histology identifies tissues of interest for scanning, in this case Haversian tissue (remodeling) and possible medullary bone (reproductive). Total S represents the distribution of all S species within the area of interest. Various S maps are taken at known energies of important S species within bone, both organic and inorganic. PCA analyses show the greatest difference between the maps, usually indicating differences in concentration and species. XANES are taken based on highlighted areas of species differences in PCA 2 to identify exact S species.

3. Results

3.1. XRF Imaging

Heterogeneous distributions of the elements S, Ca and Zn were observed within discrete histological features of the extant sample, NCSM 17997 (Figures 1 and 3). Ca and Zn highlight areas of remodeling, with elevated concentrations (rings of brighter areas) associated with secondary osteons (Figures 1 and 3). Zn is also elevated within the pathological tissues. Fe is concentrated in spaces between the trabeculae of the pathologic tissues. Sr is elevated in the thicker cortical tissue compared to the finer trabeculae present in pathologic tissues (Figure 4). PCA 2 of S species highlighted potential different species between normal cortical bone and the pathological tissues as well as at the periosteal surface (see XANES results).

Elemental distributions for MOR 1125 were similar to those seen in the extant sample, with Ca and Zn highlighting areas of remodeling (Figures 1 and 3). Fe was highly localized to vascular spaces and along the edges of some bone trabeculae (where some sediment was still present); it was not detected in the adjacent fossil bone tissue, consistent with chemical sequestration and minimal diagenetic alteration of MOR 1125 bone tissue (Figures S1, 1 and 4). Differences in the distribution of S species between different bone tissues were highlighted by PCA of S species maps (Figure 2; see also XANES results). No other distinctions were seen in other elements (Figure S2).

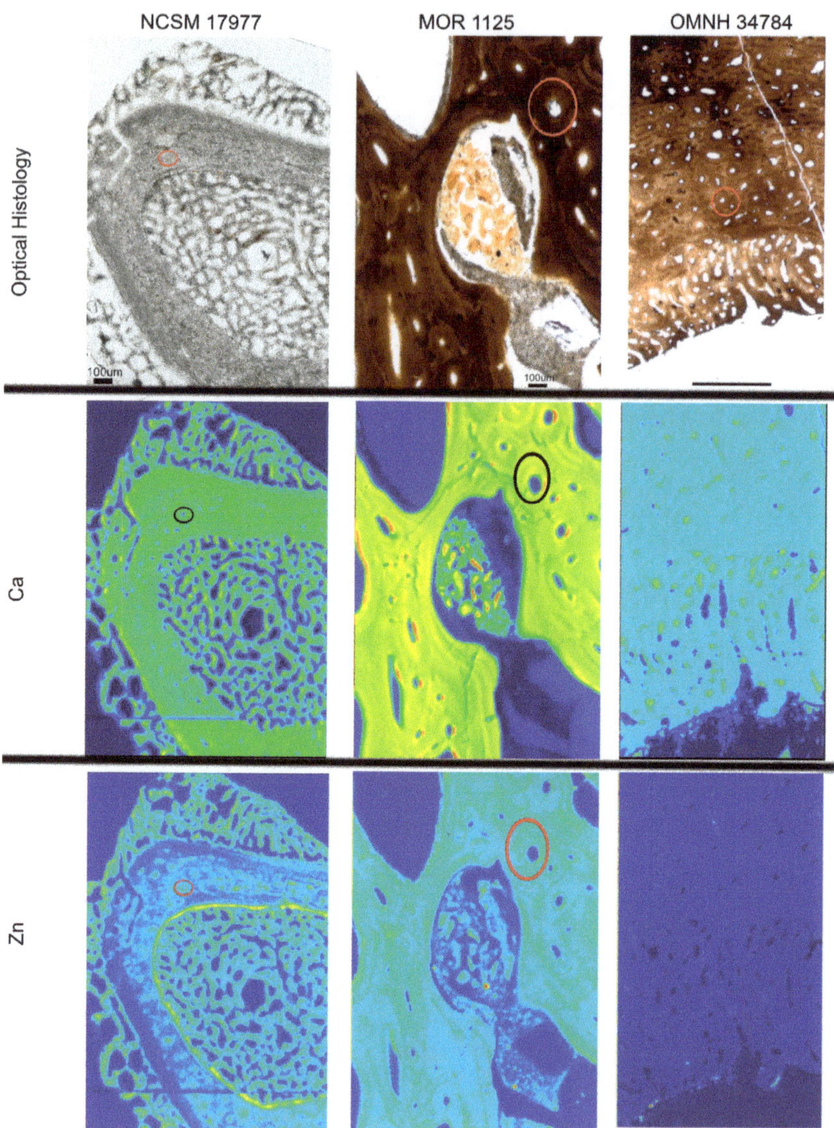

Figure 3. Optical histology and XRF maps taken from beamline 2-3 of Ca and Zn taken for all specimens. Warmer colors correspond to higher concentrations. Secondary osteons are highlighted in Ca and Zn (example osteon highlighted by circles) in NCSM 17977 and MOR 1125. OMNH 34784 shows uniform distribution of Zn, with concentrated Ca in pores between tissues.

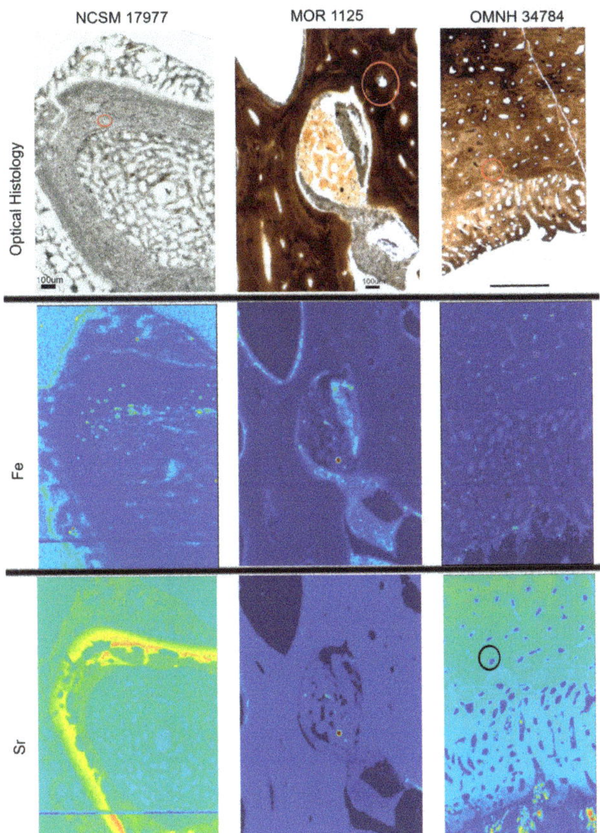

Figure 4. Optical histology and XRF maps taken from beamline 2-3 of Fe and Sr for all specimens. Warmer colors correspond to higher concentrations. Fe is concentrated in spaces between pathologic tissues in NCSM 17977 and within areas of infill between tissues in MOR 1125 and OMNH 34784. Sr is elevated in thicker cortical bone versus the thin trabeculae of the pathologic tissues in NCSM 17977. Sr is relatively uniform in MOR 1125. In OMNH 34784, secondary osteons are highlighted in Sr (example osteon circled).

In contrast to MOR 1125 (*T. rex*), OMNH 34784 (*T. tilletti*) displays uniform distributions of Ca and Zn in both cortical and endosteal bone, with higher concentration of Ca in some, but not all vascular canals, consistent with diagenetic deposition of calcite or another Ca-rich mineral in these areas (Figures 1 and 3). Sr showed elevated concentrations in the cortical bone matrix relative to the endosteal bone (Figure 4), and is also present in lower concentrations in the secondary bone tissue forming some of the deep cortical secondary osteons, when compared to adjacent primary cortical tissue. Some bright spots of Sr are also visible within the medullary cavity. Fe is concentrated within the vascular canals and to some extent within the medullary cavity (Figure 4), similar to that seen in MOR 1125. PCA 2 of S species showed only differential distribution between the bone and diagenetic infill (see XANES results).

3.2. Zn XANES

All Zn XANES spectra from NCSM 17977 (extant sample) exhibit peaks associated with HAP at ~9665 and ~9675 eV, with the most distinct peak at ~9675 eV (Figure 5A; [54–56]). These peaks were also observed in MOR 1125 with the exception of spectrum 4 (Figure 5B).

Spectrum 4 was taken in a cavity containing sediments and smaller bone fragments and differs in several ways from all other spectra including unique peaks at 9668 eV, which is associated with some Zn silicates [54]. This is the only spectrum taken in an area that visually does not appear to be bone tissue, but rather mineral infill (Figure 5A). There is also a unique peak in spectra taken from cortical bone (spectra 1 and 2) that is not seen in the potential medullary bone tissue (3 and 5) at 9676 eV.

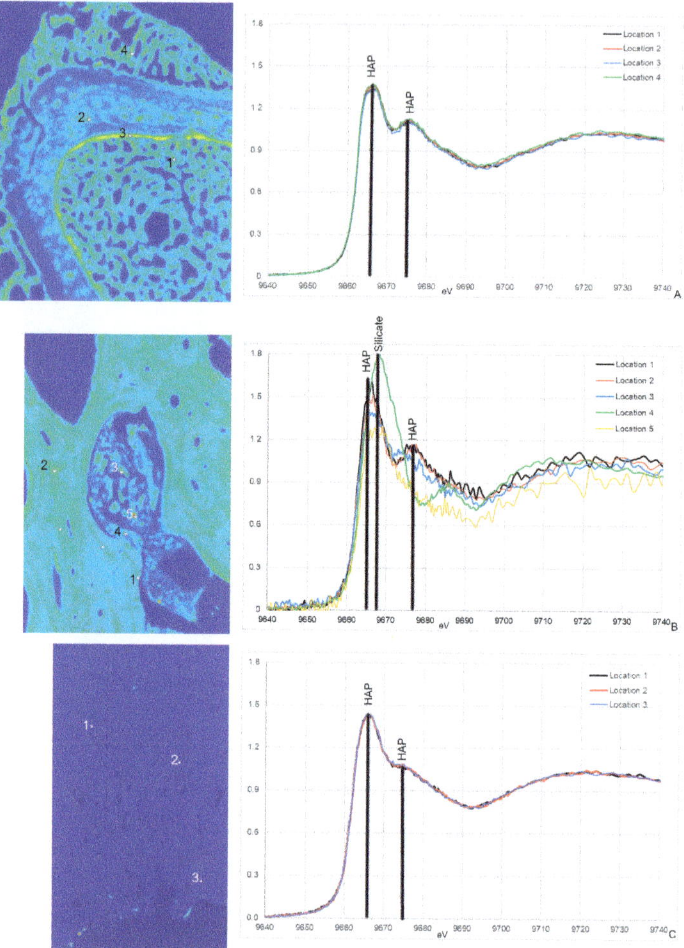

Figure 5. Zn XANES from NCSM 17977 (*C. moluccensis* radius; (**A**)), MOR 1125 (*T. rex* femur; (**B**)) and OMNH 34784 (*T. tilletti* tibia; (**C**)) and with peaks labeled using peak designations based on [54–56]. The locations for the XANES spectra are labeled on the corresponding Zn elemental maps. All spectra show peaks associated with hydroxyapatite (HAP) at ~9665 and ~9675 eV, respectively. For MOR 1125, an additional peak at 9678 eV in spectrum 4 is associated with silicates.

Zn XANES spectra from OMNH 34784 are similar and only exhibit peaks for HAP at ~9665 and ~9675 eV (5C). Both HAP peaks are broader compared to NCSM 17977 and MOR 1125, with the second peak at ~9675 eV reduced to almost a shoulder.

3.3. S XANES

NCSM 17977 exhibits both organic and inorganic S species, with peaks associated with organic disulfides and organic and inorganic sulfides (Figure 6A; [57,58]). Many of these peaks are also seen in MOR 1125, including the disulfide peak at ~2473.5 eV and the sulfate peak at ~2482 eV (Figure 6B) thus consistent with an endogenous source in this dinosaur. NCSM 17977 shows additional organic S species peaks for sulfoxides, as well as some for Native S, which is possibly powder used for thin section grinding. Peak intensity is much lower in endosteal tissue, with broader peaks. OMNH 34784 S XANES does show the dominate sulfate peak at ~2482 eV (Figure 6C). However, all remaining peaks correlate to inorganic sulfates and sulfides, including peaks associated with Fe, Zn, Mg and silicates (Figure 6C).

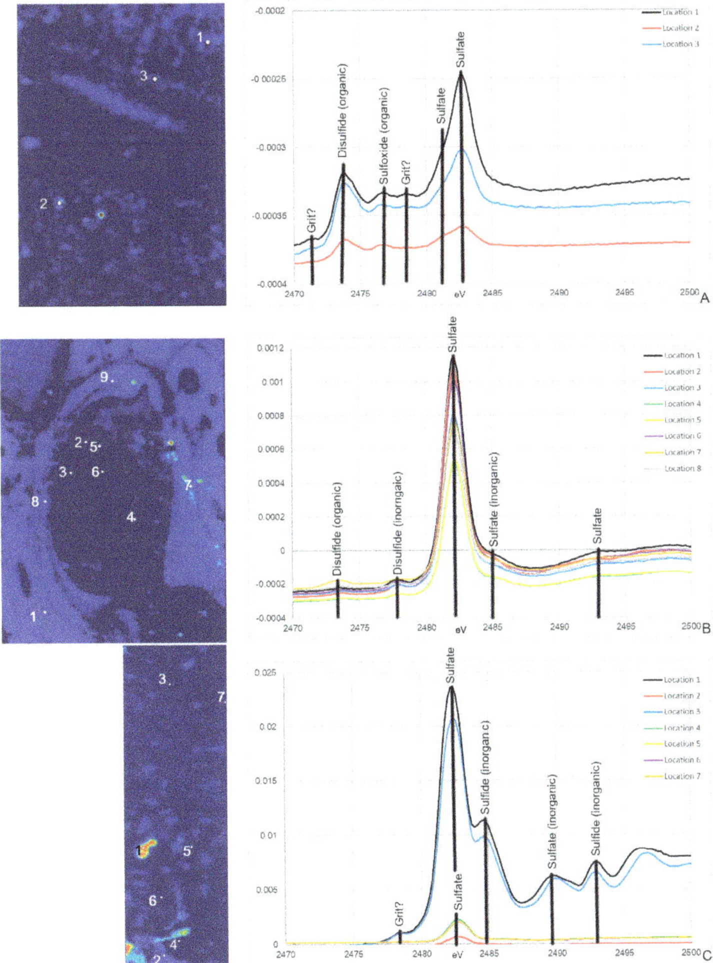

Figure 6. S XANES from NCSM 17977 (*C. moluccensis* radius; (**A**)), MOR 1125 (*T. rex* femur; (**B**)), OMN 34784 (*T. tilletti* tibia; (**C**)) with peaks labeled using peak designations based on [57,58]. The locations for the XANES spectra are labeled on the corresponding S PCA2 maps. Both NCSM 17977 and MOR 1125 exhibit organic disulfide and sulfate peaks. MOR 1125 and OMNH 34784 also have inorganic S peaks for disulfides, sulfates and for OMNH 34784, sulfides.

4. Discussion

4.1. Evidence for Endogenous Biochemistry of MOR 1125

Synchrotron XRF imaging and XAS spectroscopy supports previous multidisciplinary studies showing evidence for retention of original biochemistry in the femur of MOR 1125 (*T. rex*). The distribution of elements Ca and Zn, as well as the suite of detected S species, and the correlation of these to specific histological features associated with bone remodeling, repair, and active ossification (Figures 1–3) are similar to those seen in the extant bird, NCSM 17977, and support an endogenous source and cannot be readily explained by diagenetic overprint [10,11,16,31–34,59–61]. In contrast, despite exceptional histological preservation, OMNH 34784 results varied widely from the *T. rex* and extant bird specimens, showing more uniform distributions of Ca and Zn associated with diagenetic alteration, and a lack of organic S species such as the disulfide peak at ~2473.5 eV (Figures 1, 3 and 6).

The results of this study demonstrate the ability of synchrotron XRF imaging and XAS analyses to identify, spatially resolve, and characterize elements affiliated with discrete biochemical processes in various bone tissue types. This ability may be used in future studies as a way to distinguish between tissue types such as medullary versus pathologic tissues.

4.2. Endogenous Elements in MOR 1125- Ca, Zn and S

MOR 1125 showed Ca, Zn and S elemental distributions and species that correlated to biological processes in (Figures 2, 3, 5 and 6). Elevated Ca within the secondary osteon radius may be due to osteoblastic regulation of mineralization during bone deposition, or by an increase of free Ca released from the bone during the resorption stage of remodeling [10,11,32]. Elevated Zn around secondary osteons has been demonstrated by others on analysis of human osteons, where Zn is known to pool in forming osteons, especially within osteoid at the mineralization front [34–36]. Both extant and fossil specimens exhibit the typical HAP peaks of bone tissue in Zn XANES at ~9665 and ~9675 eV (Figure 5, [54–56]).

Speciation maps of S revealed that different species within MOR 1125 are distributed differently in areas of remodeling and within endosteal (medullary-like) tissue (Figures 2 and 6). Difference in S species within areas of remodeling may be due to S associated with bone mineral (e.g., inorganic and organic sulfate) versus bone collagen (e.g., organic disulfide). Sulfur XANES confirmed that both NCSM 17977 and MOR 1125 exhibit peaks that correspond to multiple organic species of S, notably disulfides and sulfates (Figure 6; ~2473 eV and ~2482 eV). Sulfate is found in the HAP structure of bone in both organic and inorganic phases [55,62]. Organic sulfate is also associated with keratan sulfate (KS), which is found in various skeletal tissues in low concentrations, but is also associated with the deposition of medullary bone in avians at much higher levels (e.g., [43]). MOR 1125 also exhibited peaks for organic disulfides at ~2473 eV [31,58]. Disulfides such as methionine have been found to preserve in association with osteons in archaeological and fossil bone [31].

4.3. Diagenesis in MOR 1125—Fe and S

These data, together with those from previous studies, strongly indicate that original chemistry is preserved in the bone of *T. rex* MOR 1125. However, this specimen also shows evidence of alteration. In both fossil specimens, elevated Fe is confined to diagenetic mineral infill areas between the bone tissue (Figures S1 and 4) that correspond to regions where Fe can be introduced during fossilization (i.e., spaces in the medullary cavity, vascular canals). In addition to peaks associated with bone bioapatite, both MOR 1125 and OMNH 34784 exhibit inorganic S species peaks that correlate to geological input, including inorganic sulfides (e.g., pyrite), sulfates (e.g., gypsum) and sulfites (e.g., $MgSO_3$; Figure 6). The mixture of organic and inorganic S species in MOR 1125 allows us to tease apart primary biochemistry from diagenetic influence in these specimens and supports our interpretation of only limited diagenesis in MOR 1125.

4.4. Differences in Preservation between Fossil Specimens

Despite a shared histological integrity, OMNH 34784 differs from MOR 1125 in the degree of chemical alteration, with little to no sign of endogenous material preserved. This fails to support the hypothesis that histological integrity can be directly correlated to organic preservation. None of the elemental distributions for OMNH 34784 overlap with MOR 1125 or extant material with the exception of the diagenetic Fe (Figures 2 and 4). When comparing XANES, Zn XANES for OMNH 34784 did show the classic HAP peaks at ~9665 and ~9675 eV, though these peaks are broader in OMNH 34784 than in the other two specimens (Figure 5). Broadening may be due to a mixture of different Zn species or concentrations of Zn within the HAP structure, both which may be caused by diagenetic alteration [62]. For S XANES, OMNH 34784 lacks any species of organic S (Figure 6).

5. Conclusions

The XRF and XAS results of this study support previous findings on MOR 1125. Histological analyses, both traditional and CT, confirm excellent morphological preservation of bone tissue at the microstructural and nanostructural levels [24,25]. We then correlated these micromorphological patterns to obtained elemental maps, to show the specific biological features associated with bone remodeling, which has also been seen in extant organisms in both this study and others [10,11,16,31–34]. Synchrotron results strongly support previous studies claiming endogenous preservation at the molecular level, including immunohistolochemical studies confirming the presence of an original biomolecule signal in MOR 1125 and mapped these signals to skeletal regions (i.e., medullary bone). This presents promising implications for the use of synchrotron XRF to differentiate between pathological bone, and reproductive (medullary) bone in MOR 1125, specifically. Although some differences in element distribution are seen between MOR 1125 and the extant specimen, OMNH 34784 shows evidence of far greater chemical alteration, to the exclusion of endogenous material. This highlights that preservation of endogenous bone biochemistry in fossils is specimen-specific and likely tied to variation in depositional history and fossilization processes, although this relationship is far from being fully understood. We conclude that the lack of original molecular/chemical preservation in one specimen cannot be used to eliminate the possibility that it may be preserved in other specimens, regardless of age, type or morphology.

Supplementary Materials: The following supporting information can be downloaded at: https://www.mdpi.com/article/10.3390/biology12020264/s1, Figure S1 shows gross elemental maps for specimens MOR 1125 taken at beamline 7-2. Figure S2 shows additional XRF maps of MOR 1125 that did not show any correlations with specific histological features.

Author Contributions: J.A.—helped design experiments, processed data, conducted data interpretation, wrote the manuscript and made figures. A.C.—helped design experiment, supplied specimens, made thin sections of biological material, aided in data interpretation and edited manuscript. N.P.E.—conducted all synchrotron experiments, aided in data interpretation and edited manuscript. M.H.S.—supplied specimens, aided in data interpretation and edited manuscript. L.E.Z.—helped design experiment, supplied specimens, aided in data interpretation and edited manuscript. All authors have read and agreed to the published version of the manuscript.

Funding: National Science Foundation award #1552328 to LEZ and MHS.

Institutional Review Board Statement: Not applicable.

Informed Consent Statement: Not applicable.

Data Availability Statement: Data available on request due to restrictions.

Acknowledgments: We would like to thank the Stanford Synchrotron Radiation Lightsource (proposal 5643). Use of the Stanford Synchrotron Radiation Lightsource, SLAC National Accelerator Laboratory, is supported by the U.S. Department of Energy, Office of Science, Office of Basic Energy Sciences under Contract No. DE-AC02-76SF00515. The SSRL Structural Molecular Biology Program is supported by the DOE Office of Biological and Environmental Research, and by the National

Institutes of Health, National Institute of General Medical Sciences (P30GM133894). The contents of this publication are solely the responsibility of the authors and do not necessarily represent the official views of NIGMS or NIH. We would like to thank the National Science Foundation for funding (#1552328). We would also like to thank John Scannella from the Museum of the Rockies (MOR), the Sam Noble Oklahoma Museum of Natural History (OMNH), and John Gerwin and Brian O'Shea from the North Carolina Museum of Natural Sciences (NCSM) for loans of the specimens.

Conflicts of Interest: The authors declare no conflict of interest.

References

1. McNamara, M.; Rossi, V.; Slater, T.; Rogers, C.; Ducrest, A.-L.; Dubey, S.; Roulin, A. Decoding the Evolution of Melanin in Vertebrates. *Trends Ecol. Evol.* **2021**, *36*, 430–443. [CrossRef] [PubMed]
2. Wiemann, J.; Yang, T.R.; Sander, P.N.; Schneider, M.; Engeser, M.; Kath-Schorr, S.; Müller, C.E.; Sander, P.M. Dinosaur origin of egg color: Oviraptors laid blue-green eggs. *PeerJ* **2017**, *5*, e3706. [CrossRef] [PubMed]
3. Lindgren, J.; Sjövall, P.; Carney, R.M.; Uvdal, P.; Gren, J.A.; Dyke, G.; Schultz, B.P.; Shawkey, M.D.; Barnes, K.R.; Polcyn, M.J. Skin pigmentation provides evidence of convergent melanism in extinct marine reptiles. *Nature* **2014**, *506*, 484–488. [CrossRef]
4. Wogelius, R.A.; Manning, P.L.; Barden, H.E.; Edwards, N.P.; Webb, S.M.; Sellers, W.I.; Taylor, K.G.; Larson, P.L.; Dodson, P.; You, H.; et al. Trace metals as biomarkers for eumelanin pigment in the fossil record. *Science* **2011**, *333*, 1622–1626. [CrossRef] [PubMed]
5. Schweitzer, M.H.; Zheng, W.; Organ, C.L.; Avci, R.; Suo, Z.; Freimark, L.M.; Lebleu, V.S.; Duncan, M.B.; Vander Heiden, M.G.; Neveu, J.M.; et al. Biomolecular characterization and protein sequences of the Campanian hadrosaur *B. canadensis. Science* **2009**, *324*, 626–631. [CrossRef]
6. Buckley, M.; Lawless, C.; Rybczynski, N. Collagen sequence analysis of fossil camels, *Camelops* and *cf Paracamelus*, from the Arctic and sub-Arctic of Plio-Pleistocene North America. *J. Proteom.* **2019**, *194*, 218–225. [CrossRef]
7. Ullmann, P.V.; Macauley, K.; Ash, R.D.; Shoup, B.; Scannella, J.B. Taphonomic and diagenetic pathways to protein preservation; Part I: The case of *Tyrannosaurus rex* specimen MOR 1125. *Biology* **2021**, *10*, 1193. [CrossRef]
8. Miyashita, T.; Coates, M.I.; Farrar, R.; Larson, P.; Manning, P.L.; Wogelius, R.A.; Edwards, N.P.; Anné, J.; Bergmann, U.; Palmer, A.R.; et al. Hagfish from the Cretaceous Tethys Sea and a reconciliation of the morphological–molecular conflict in early vertebrate phylogeny. *Proc. Natl. Acad. Sci. USA* **2019**, *116*, 2146–2151. [CrossRef]
9. Pan, Y.; Zheng, W.; Sawyer, R.H.; Pennington, M.W.; Zheng, X.; Wang, X.; Wang, M.; Hu, L.; O'Connor, J.; Zhao, T.; et al. The molecular evolution of feathers with direct evidence from fossils. *Proc. Natl. Acad. Sci. USA* **2019**, *116*, 3018–3023. [CrossRef]
10. Anné, J.; Wogelius, R.A.; Edwards, N.P.; van Veelen, A.; Buckley, M.; Sellers, W.I.; Bergmann, U.; Sokaras, D.; Alonso-Mori, R.; Harvey, V.L.; et al. Morphological and chemical evidence for cyclic bone growth in a fossil hyaena. *J. Anal. At. Spectrom.* **2018**, *33*, 2062–2069. [CrossRef]
11. Anné, J.; Wogelius, R.A.; Edwards, N.P.; van Veelen, A.; Ignatyev, K.; Manning, P.L. Chemistry of bone remodelling preserved in extant and fossil Sirenia. *Metallomics* **2016**, *8*, 508–513. [CrossRef] [PubMed]
12. Edwards, N.P.; van Veelen, A.; Anné, J.; Manning, P.; Bergmann, U.; Sellers, W.; Egerton, V.; Sokaras, D.; Alonso-Mori, R.; Wakamatsu, K.; et al. Elemental characterisation of melanin in feathers via synchrotron X-ray imaging and absorption spectroscopy. *Sci. Rep.* **2016**, *6*, 34002. [CrossRef] [PubMed]
13. Schweitzer, M.H.; Zheng, W.; Zanno, L.; Werning, S.; Sugiyama, T. Chemistry supports the identification of gender-specific reproductive tissue in Tyrannosaurus rex. *Sci. Rep.* **2016**, *6*, 23099. [CrossRef] [PubMed]
14. Schweitzer, M.H.; Moyer, A.E.; Zheng, W. Testing the hypothesis of biofilm as a source for soft tissue and cell-like structures preserved in dinosaur bone. *PLoS ONE* **2016**, *11*, e0150238. [CrossRef] [PubMed]
15. Egerton, V.M.; Wogelius, R.A.; Norell, M.A.; Edwards, N.P.; Sellers, W.I.; Bergmann, U.; Sokaras, D.; Alonso-Mori, R.; Ignatyev, K.; van Veelen, A.; et al. The mapping and differentiation of biological and environmental elemental signatures in the fossil remains of a 50 million year old bird. *J. Anal. At. Spectrom.* **2015**, *30*, 627–634. [CrossRef]
16. Anné, J.; Edwards, N.P.; Wogelius, R.A.; Tumarkin-Deratzian, A.R.; Sellers, W.; van Veelen, A.; Bergmann, U.; Sokaras, D.; Alonso-Mori, R.; Ignatyev, K.; et al. Synchrotron imaging reveals bone healing and remodelling strategies in extinct and extant vertebrates. *J. R. Soc. Interface* **2014**, *11*, 20140277. [CrossRef]
17. Schweitzer, M.H.; Zheng, W.; Cleland, T.P.; Goodwin, M.B.; Boatman, E.; Theil, E.; Marcus, M.A.; Fakra, S.C. A role for iron and oxygen chemistry in preserving soft tissues; cells and molecules from deep time. *Proc. Biol. Sci.* **2014**, *281*, 20132741. [CrossRef]
18. Schweitzer, M.H.; Zheng, W.; Cleland, T.P.; Bern, M. Molecular analyses of dinosaur osteocytes support the presence of endogenous molecules. *Bone* **2013**, *52*, 414–423. [CrossRef]
19. San Antonio, J.D.; Schweitzer, M.H.; Jensen, S.T.; Kalluri, R.; Buckley, M.; Orgel, J.P. Dinosaur peptides suggest mechanisms of protein survival. *PLoS ONE* **2011**, *6*, e20381. [CrossRef]
20. Organ, C.L.; Schweitzer, M.H.; Zheng, W.; Freimark, L.M.; Cantley, L.C.; Asara, J.M. Molecular phylogenetics of mastodon and *Tyrannosaurus rex. Science* **2008**, *320*, 499. [CrossRef]
21. Asara, J.M.; Schweitzer, M.H.; Freimark, L.M.; Phillips, M.; Cantley, L.C. Protein sequences from mastodon and *Tyrannosaurus rex* revealed by mass spectrometry. *Science* **2007**, *316*, 280–2855. [CrossRef]

22. Schweitzer, M.H.; Suo, Z.; Avci, R.; Asara, J.M.; Allen, M.A.; Arce, F.T.; Horner, J.R. Analyses of soft tissue from *Tyrannosaurus rex* suggest the presence of protein. *Science* **2007**, *316*, 277–280. [CrossRef]
23. Schweitzer, M.H.; Wittmeyer, J.L.; Horner, J.R. Soft tissue and cellular preservation in vertebrate skeletal elements from the Cretaceous to the present. *Proc. Biol. Sci.* **2007**, *274*, 183–197. [CrossRef] [PubMed]
24. Schweitzer, M.H.; Wittmeyer, J.L.; Horner, J.R.; Toporski, J.K. Soft-tissue vessels and cellular preservation in *Tyrannosaurus rex*. *Science* **2005**, *307*, 1952–1955. [CrossRef] [PubMed]
25. Schweitzer, M.H.; Wittmeyer, J.L.; Horner, J.R. Gender-specific reproductive tissue in ratites and *Tyrannosaurus rex*. *Science* **2005**, *308*, 1456–1460. [CrossRef] [PubMed]
26. Saitta, E.T.; Fletcher, I.; Martin, P.; Pittman, M.; Kaye, T.G.; True, L.D.; Norell, M.A.; Abbott, G.D.; Summons, R.E.; Penkman, K.; et al. Preservation of feather fibers from the Late Cretaceous dinosaur *Shuvuuia deserti* raises concern about immunohistochemical analyses on fossils. *Org. Geochem.* **2018**, *25*, 142–151. [CrossRef]
27. Buckley, M.; Warwood, S.; van Dongen, B.; Kitchener, A.C.; Manning, P.L. A fossil protein chimera; difficulties in discriminating dinosaur peptide sequences from modern cross-contamination. *Proc. Biol. Sci.* **2017**, *284*, 20170544. [CrossRef]
28. Kaye, T.G.; Gaugler, G.; Sawlowicz, Z. Dinosaurian soft tissues interpreted as bacterial biofilms. *PLoS ONE* **2008**, *3*, e2808. [CrossRef]
29. Dacke, C.G.; Arkle, S.; Cook, D.J.; Wormstone, I.M.; Jones, S.; Zaidi, M.; Bascal, Z.A. Medullary bone and avian calcium regulation. *J. Exp. Biol.* **1993**, *184*, 63–88. [CrossRef]
30. Wiedemann, H. Synchrotron radiation. In *Particle Accelerator Physics*; Springer: Berlin/Heidelberg, Germany, 2003; pp. 647–686.
31. Anné, J.; Edwards, N.P.; Brigidi, F.; Gueriau, P.; Harvey, V.L.; Geraki, K.; Slimak, L.; Buckley, M.; Wogelius, R.A. Advances in bone preservation: Identifying possible collagen preservation using sulfur speciation mapping. *Palaeogeogr Palaeoclim. Palaeoecol* **2019**, *520*, 181–187. [CrossRef]
32. Anné, J.; Edwards, N.P.; van Veelen, A.; Egerton, V.M.; Manning, P.L.; Mosselmans, J.F.W.; Parry, S.; Sellers, W.I.; Buckley, M.; Wogelius, R.A. Visualisation of developmental ossification using trace element mapping. *JAAS* **2017**, *32*, 967–974. [CrossRef]
33. Pemmer, B.; Roschger, A.; Wastl, A.; Hofstaetter, J.G.; Wobrauschek, P.; Simon, R.; Thaler, H.W.; Roschger, P.; Klaushofer, K.; Streli, C. Spatial distribution of the trace elements zinc, strontium and lead in human bone tissue. *Bone* **2013**, *57*, 184–193. [CrossRef] [PubMed]
34. Gomez, S.; Rizzo, R.; Pozzi-Mucelli, M.; Bonucci, E.; Vittur, F. Zinc mapping in bone tissues by histochemistry and synchrotron radiation-induced x-ray emission: Correlation with the distribution of alkaline phosphates. *Bone* **1999**, *25*, 33–38. [CrossRef] [PubMed]
35. Yamaguchi, M. Role of nutritional zinc in the prevention of osteoporosis. *Mol. Cell. Biochem.* **2010**, *338*, 241–254. [CrossRef] [PubMed]
36. Hadley, K.B.; Newman, S.M.; Hunt, J.R. Dietary zinc reduces osteoclast resorption activities and increases markers of osteoblast differentiation, matrix maturation, and mineralization in the long bones of growing rats. *J. Nutr. Biochem.* **2010**, *21*, 297–303. [CrossRef]
37. Bonnelye, E.; Chabadel, A.; Saltel, F.; Jurdic, P. Dual effect of strontium ranelate: Stimulation of osteoblast differentiation and inhibition of osteoclast formation and resorption in vitro. *Bone* **2008**, *42*, 129–138. [CrossRef] [PubMed]
38. Barbara, A.; Delannoy, P.; Denis, B.G.; Marie, P.G. Normal matrix mineralization induced by strontium renelate in MC3T3-E1 osteogenic cells. *Metabolis* **2004**, *53*, 532–537. [CrossRef] [PubMed]
39. Lowe, N.M.; Fraser, W.D.; Jackson, M.J. Is there a potential therapeutic value of copper and zinc for osteoporosis? *Proc. Nutr. Soc.* **2002**, *61*, 181–185. [CrossRef]
40. Haumont, S. Distribution of zinc in bone tissue. *J. Histochem. Cyctochem* **1960**, *9*, 141–145. [CrossRef] [PubMed]
41. Bode, M. Characterization of Type I and Type II Collagens in Human Tissues. Ph.D. Thesis, Faculty of Medicine, University of Oulu, Oulu, Finland, 2000.
42. Tran, L.K.; Stepien, K.R.; Bollmeyer, M.M.; Yoder, C.H. Substitution of sulfate in apatite. *Am. Min.* **2017**, *102*, 1971–1976. [CrossRef]
43. Canoville, A.; Zanno, L.E.; Zheng, W.; Schweitzer, M.H. Keratan sulfate as a marker for medullary bone in fossil vertebrates. *J. Anat.* **2021**, *238*, 1296–1311. [CrossRef] [PubMed]
44. Ross, A.C.; Taylor, C.L.; Yaktine, A.L.; Del Valle, H.B. (Eds.) *Dietary Reference Intakes for Calcium and Vitamin D in Institute of Medicine (US) Committee to Review Dietary Reference Intakes for Vitamin D and Calcium*; National Academies Press (US): Washington, DC, USA, 2011.
45. Penido, M.G.; Alon, U.S. Phosphate homeostasis and its role in bone health. *Pediatr. Nephrol.* **2012**, *27*, 2039–2048. [CrossRef]
46. Fowler, D. The Hell Creek Formation, Montana: A stratigraphic review and revision based on a sequence stratigraphic approach. *Geosciences* **2020**, *10*, 435. [CrossRef]
47. D'Emic, M.; Foreman, B.; Jud, N.; Britt, B.; Schmitz, M.; Crowley, J. Chronostratigraphic Revision of the Cloverly Formation (Lower Cretaceous, Western Interior, USA). *Bull Peabody Mus. Nat. Hist.* **2019**, *60*, 3–40. [CrossRef]
48. Lee, A.H.; Werning, S. Sexual maturity in growing dinosaurs does not fit reptilian growth models. *Proc. Natl. Acad. Sci. USA* **2008**, *105*, 582–587. [CrossRef]
49. Horner, J.R.; Padian, K. Age and growth dynamics of *Tyrannosaurus*. *Proc. Biol. Sci.* **2004**, *271*, 1875–1880. [CrossRef]

50. Edwards, N.P.; Bargar, J.R.; van Campen, D.; van Veelen, A.; Sokaras, D.; Bergmann, U.; Webb, S.M. A new μ-high energy resolution fluorescence detection microprobe imaging spectrometer at the Stanford Synchrotron Radiation Lightsource beamline 6-2. *Rev. Sci. Instrum.* **2022**, *93*, 083101. [CrossRef]
51. Edwards, N.P.; Webb, S.M.; Krest, C.M.; van Campen, D.; Manning, P.L.; Wogelius, R.A.; Bergmann, U. A new synchrotron rapid-scanning X-ray fluorescence (SRS-XRF) imaging station at SSRL beamline 6-2. *J. Synchrotron Rad.* **2018**, *25*, 1565–1573. [CrossRef]
52. Webb, S.M. The MicroAnalysis Toolkit: X-ray Fluorescence Image Processing Software. *AIP Conf. Proc.* **2011**, *1365*, 196–199. [CrossRef]
53. Webb, S.M. SIXPACK: A graphical user interface for XAS analysis. In *International Tables for Crystallography*; Chantler, C.T., Boscherini, F., Bunker, B., Eds.; Wiley Online Library: Hoboken, NJ, USA, 2020; Volume 1. [CrossRef]
54. Castorina, E.; Ingall, E.D.; Morton, P.L.; Tavakoli, D.A.; Lai, B. Zinc K-edge XANES spectroscopy of mineral and organic standards. *J. Synchrotron Rad.* **2019**, *26*, 1302–1309. [CrossRef] [PubMed]
55. Hu, W.; Ma, J.; Wang, J.; Zhang, S. Fine structure study on low concentration zinc substituted hydroxyapatite nanoparticles. *Mater Sci. Eng. C* **2012**, *32*, 2404–2410. [CrossRef]
56. Matsunaga, K.; Murata, H.; Mizoguchi, T.; Nakahira, A. Mechanism of incorporation of zinc into hydroxyapatite. *Acta Biomater.* **2010**, *6*, 2289–2293. [CrossRef] [PubMed]
57. Debret, B.; Andreani, M.; Delacour, A.; Rouméjon, S.; Trcera, N.; Williams, H. Assessing sulfur redox state and distribution in abyssal serpentinites using XANES spectroscopy. *EPSL* **2017**, *466*, 1–11. [CrossRef]
58. Koudouna, E.; Veronesi, G.; Patel, I.I.; Cotte, M.; Knupp, C.; Martin, F.L.; Quantock, A.J. Chemical composition and sulfur speciation in bulk tissue by X-ray spectroscopy and X-ray microscopy: Corneal development during embryogenesis. *Biophys. J.* **2012**, *103*, 357–364. [CrossRef]
59. Clarke, B. Normal bone anatomy and physiology. *Clin. J. Am. Soc. Nephrol.* **2008**, *3* (Suppl. S3), S131–S139. [CrossRef]
60. Watts, N.B. Clinical utility of biochemical markers of bone remodeling. *Clin. Chem.* **1999**, *45*, 1359–1368. [CrossRef] [PubMed]
61. Tang, Y.; Chappell, H.F.; Dove, M.T.; Reeder, R.J.; Lee, Y.J. Zinc incorporation into hydroxylapatite. *Biomaterials* **2009**, *30*, 2864–2872. [CrossRef]
62. Raisz, L.G. Physiology and pathophysiology of bone remodeling. *Clin. Chem.* **1999**, *45*, 1353–1358.

Disclaimer/Publisher's Note: The statements, opinions and data contained in all publications are solely those of the individual author(s) and contributor(s) and not of MDPI and/or the editor(s). MDPI and/or the editor(s) disclaim responsibility for any injury to people or property resulting from any ideas, methods, instructions or products referred to in the content.

biology

Article

Fossil Biomarkers and Biosignatures Preserved in Coprolites Reveal Carnivorous Diets in the Carboniferous Mazon Creek Ecosystem

Madison Tripp [1,*], Jasmina Wiemann [2,3], Jochen Brocks [4], Paul Mayer [5], Lorenz Schwark [1,6] and Kliti Grice [1,*]

[1] Western Australian Organic and Isotope Geochemistry Centre, The Institute for Geoscience Research, School of Earth and Planetary Sciences, Curtin University, Kent Street, Bentley, WA 6102, Australia
[2] Department of Earth & Planetary Sciences, Yale University, 210 Whitney Avenue, New Haven, CT 06511, USA
[3] Division of Geological and Planetary Sciences, California Institute of Technology, 1200 E. California Blvd., Pasadena, CA 91125, USA
[4] Research School of Earth Sciences, The Australian National University, Canberra, ACT 2601, Australia
[5] The Field Museum, 1400 S Lake Shore Dr., Chicago, IL 60605, USA
[6] Organic Geochemistry Unit, Institute of Geoscience, Christian-Albrechts-University, 24118 Kiel, Germany
*   Correspondence: madison.tripp@postgrad.curtin.edu.au (M.T.); k.grice@curtin.edu.au (K.G.)

Citation: Tripp, M.; Wiemann, J.; Brocks, J.; Mayer, P.; Schwark, L.; Grice, K. Fossil Biomarkers and Biosignatures Preserved in Coprolites Reveal Carnivorous Diets in the Carboniferous Mazon Creek Ecosystem. *Biology* **2022**, *11*, 1289. https://doi.org/10.3390/biology11091289

Academic Editors: Mary H. Schweitzer and Ferhat Kaya

Received: 2 July 2022
Accepted: 26 August 2022
Published: 30 August 2022

Publisher's Note: MDPI stays neutral with regard to jurisdictional claims in published maps and institutional affiliations.

Copyright: © 2022 by the authors. Licensee MDPI, Basel, Switzerland. This article is an open access article distributed under the terms and conditions of the Creative Commons Attribution (CC BY) license (https://creativecommons.org/licenses/by/4.0/).

Simple Summary: Coprolites (fossilised faeces) can preserve important dietary information through geological time, offering insights into extinct animal diets. When digestion of dietary items leaves no unambiguous morphology to reconstruct the food spectrum of a coprolite producer, preserved biomolecular information can offer unique perspectives into the individual dietary composition and trophic relationships in ancient ecosystems. In this study we combine a uniquely diverse array of chemical techniques to demonstrate that biomarkers and macromolecular biosignatures from Carboniferous coprolites can reveal the dietary spectrum and trophic position of their extinct producers: an overwhelming abundance of cholesteroids, biomarkers of animal cholesterol, and an animal-affinity of the preserved macromolecular phase revealed by the statistical analysis of *in situ* Raman spectra, indicate a likely carnivorous diet for the coprolite producer. The presence of intact primary metabolites, such as sterols, and informative fossilization products of biopolymers, demonstrates the significance of siderite (iron carbonate) concretions in the exceptional preservation of biomolecular information in deep time, facilitated by the rapid encapsulation and remineralisation of organic matter within days to months.

Abstract: The reconstruction of ancient trophic networks is pivotal to our understanding of ecosystem function and change through time. However, inferring dietary relationships in enigmatic ecosystems dominated by organisms without modern analogues, such as the Carboniferous Mazon Creek fauna, has previously been considered challenging: preserved coprolites often do not retain sufficient morphology to identify the dietary composition. Here, we analysed $n = 3$ Mazon Creek coprolites in concretions for dietary signals in preserved biomarkers, stable carbon isotope data, and macromolecular composition. Cholesteroids, metazoan markers of cholesterol, show an increased abundance in the sampled coprolites (86 to 99% of the total steranes) compared to the surrounding sediment, indicating an endogenous nature of preserved organics. Presence of unaltered 5α-cholestan-3β-ol and coprostanol underline the exceptional molecular preservation of the coprolites, and reveal a carnivorous diet for the coprolite producer. Statistical analyses of *in situ* Raman spectra targeting coprolite carbonaceous remains support a metazoan affinity of the digested fossil remains, and suggest a high trophic level for the coprolite producer. These currently oldest, intact dietary stanols, combined with exquisitely preserved macromolecular biosignatures in Carboniferous fossils offer a novel source of trophic information. Molecular and biosignature preservation is facilitated by rapid sedimentary encapsulation of the coprolites within days to months after egestion.

Keywords: steroids; diet; coprolites

1. Introduction

The reconstruction of ancient trophic networks is pivotal to our understanding of ecosystem function and change through time. However, inferring dietary relationships in enigmatic ecosystems dominated by organisms without modern analogues, such as the Carboniferous Mazon Creek fauna, has previously been considered challenging. Coprolites, fossil faecal materials, can offer unique insights into the diets and trophic relationships of extinct life forms in deep time. However, fossil faecal matter can be difficult to interpret, due to the digestion and thus substantial degradation of organismal morphologies. Coprolites can be linked to a producer based on their shape, mineralogy, and geological context of the specimen as well as the presence of any identifiable remains, e.g., [1–7]. Recently, methods of identification have been expanded to include $^{13}C/^{12}C$, $^{15}N/^{14}N$ and DNA analysis [8,9]. However, many coprolite specimens still remain of ambiguous origin and composition and may not contain DNA remains, especially in samples from deep time.

The preservation of biomolecules in deep time is primarily dictated by the early diagenetic chemo-environment, e.g., [10–12]. Delicate details in soft tissues are only known from Konservat Lagerstätten [13–16]; Lagerstätten that preserve soft tissues offer generally more representative insights into extinct biodiversity than fossil sites biased towards the preservation of only hard tissues. The conditions, which result in soft-tissue preservation, are often conducive to the preservation of detailed molecular information, which is otherwise lost during heterotrophic reworking prior to preservation. Carbonate concretions are known to frequently contain carbonaceous fossil soft tissues, e.g., [17], and are thus inferred to form through rapid authigenic mineralisation, which halts significant destruction of the specimen [14,18,19]. During fossilization, lipids are transformed into stable derivatives preserving their original hydrocarbon skeleton. The resulting biomarkers are commonly used to identify original sources of organic matter input into sediments [20].

Biomolecules such as sterols are present in most eukaryotes and contain structural features specific to their function within a group of organisms [21]. Lipids such as sterols tend to be better preserved in sediments compared to other classes of biomolecules (e.g., carbohydrates, proteins) which are often completely degraded under anoxic conditions [22]. Through early diagenesis, biological steroids are defunctionalised as a result of the activity of microbes and clay catalysts, while typically retaining most of their isomeric characteristics [23]. These compounds can include stanols, sterenes, diasterenes and A/B/C-ring monoaromatic steroids. Further diagenetic and catagenetic reactions will result in isomerisation and aromatisation to form more thermodynamically stable configurations [23–25], primarily steranes and triaromatic steroids, which can reside in the sediments for up to hundreds of millions of years. A diagenetic continuum of such steroid hydrocarbons was identified in a study of a Devonian calcite concretion from the Gogo Formation, an exceptional Konservat Lagerstätte in the Canning Basin of Western Australia, by Melendez et al. [26]. Here, microbially mediated eogenetic processes were determined to have resulted in the parallel preservation of stenols and their diagenetic products encompassing steranes and triaromatic steroids. Such instances of preservation offer the opportunity to study not only the characteristic biomolecules from the fossilised specimen, but also the post-depositional processes, which transform these biomolecules and the broader taphonomic history of coprolites.

In contrast, structural biomolecules shared among all organisms, such as proteins, tend to crosslink with oxidation products of lipids and sugars via advanced glycoxidation and lipoxidation reaction schemes to form insoluble complex organic matter composed of heteroatom-rich polymers. The resulting insoluble organic matter retains chemically altered, but not unrecognisable evidence of original biosignatures.

This compositional complexity and potential for chemical transformation through digestive and diagenetic processes requires the combination of complementary chemical analyses. Coprolites have been known to preserve lipid biomarkers, e.g., [27–31], which can be used to reconstruct various molecular inputs including direct dietary information and the processes, which alter these molecules through the digestive tract of the producer [28,32].

In studies of faecal samples, 5β-stanols are common reduced products of dietary sterols including cholesterol, campesterol, sitosterol and stigmasterol [27]. There is a predominance of 5β-cholestan-3β-ol (coprostanol) in human faeces [33], while herbivorous mammal faeces contain mostly 5β-campestanol and 5β-stigmastanol [27,33].

In addition, insoluble organic matter resulting from the diagenetic crosslinking of structural biomolecules has been shown to preserve biologically informative heterogeneities for fossils from the Mazon Creek locality [34]. Extracting dietary information from molecular biomarkers and biosignatures preserved in coprolites opens up new opportunities for tracing trophic networks through time. For example, the presence of high amounts of cholestane in coprolite samples has been attributed to a predominantly carnivorous or omnivorous diet, e.g., [28,29], due to ubiquity of cholesterol in animal tissue. In contrast, the presence of an array of phytosterols and long-chain *n*-alkanes derived from leaf waxes can indicate a herbivorous diet, e.g., [29,33,35]. A fraction of the organic material is chemically transformed into insoluble complex organic matter following ingestion, digestion and egestion, potentially preserving dietary tissue information in form of informative heterogeneities. On the other hand, primary dietary information can also be obtained from indigestible components, which pass through the gastrointestinal system without chemical modification.

The accuracy and detail of the molecular dietary reconstruction is reliant on the degree of biomolecule and biomarker preservation, and diagenetic transformation. The processes, which are responsible for preservation of soft tissue (e.g., rapid burial, rapid mineral growth [14]) are also those which have been demonstrated to preserve biomarkers of low maturity and intact biomolecules, e.g., [11,12,26]. Therefore, exceptionally preserved coprolites that were sealed from environmental influence during earliest diagenesis, such as those from the Mazon Creek Lagerstätte, are likely to preserve biomolecular information alongside soft tissues and thus are ideal candidates for ancient dietary reconstructions.

Palaeoenvironmental Setting

The Mazon Creek Konservat Lagerstätte is one of the most productive fossil Lagerstätten worldwide, with over 350 species of plants and 465 of animals identified [36,37]. The site is renowned for its preservation of delicate soft tissue fossils within iron carbonate (siderite) concretions, of which there are many hundreds of thousands collected. Palynological and palaeobotanical studies of the Lagerstätte, e.g., [36,38,39] have determined its age to be Middle Pennsylvanian (Westphalian D) (306–311 Ma). The siderite concretions are found in the lower 3–8 m of the Francis Creek Shale Member of the Lower Carbondale Formation, overlying the Colchester Coal (No. 2) Member [40–44]. The Mazon Creek site was interpreted as representing a river delta system, e.g., [41–45], wherein some large-scale events such as storms or flash flooding caused the rapid burial of massive amounts of organisms, e.g., [42]. More recently the palaeoenvironment has been re-evaluated to represent a low energy, brackish marine environment where under anaerobic conditions input from peaty forests provided a means of rapid burial of organisms [43]. The abundance of siderite concretions and its occasional co-occurrence with pyrite indicates minimal marine input and low sulfate concentrations, where iron initially reacted with H_2S (where available) to form pyrite. Under anoxic conditions the presence and activity of methanogenic archaea and methanotrophic bacteria initiated the precipitation of iron carbonate, e.g., [42,46–48]. This has also been supported in geochemical studies of $^{34}S/^{32}S$, $^{13}C/^{12}C$ and $^{18}O/^{16}O$ isotopic compositions of the Mazon Creek concretions [43]. Preservation occurs during early microbial decay of deposited organisms, wherein a 'proto-concretion' is formed by the breakdown of degrading organic matter into fatty acids, resulting in mineral precipitation of primarily siderite in an iron-rich environment with limited sulfate [43]. Growth and decay experiments, e.g., [49–52] have demonstrated that this process is initiated within the weeks after deposition. Studies [53,54] have identified that these processes occurred rapidly (within weeks to months) in carbonate concretions formed around decaying soft tissues of tusk-shells.

Here, we analyze $n = 3$ Mazon Creek coprolites preserved in carbonate concretions for dietary signals in preserved biomarkers, stable carbon isotope data, insoluble fossil organic matter composition, and preserved morphology, to provide comprehensive and complementary insights into the trophic structure of the enigmatic Mazon Creek ecosystem and improve our understanding of the role of concretion formation in the preservation of biomolecules in deep time.

2. Materials and Methods

2.1. Sample Preparation and Extraction

Three coprolite fossils in siderite concretions were prepared for analysis (Figure 1). The samples used in this study were obtained from the Chicago Field Museum. Samples had previously been collected from the Pit 11 coal strip mine in Kankakee County, Illinois, from the Francis Creek Shale Member of the Carbondale Formation.

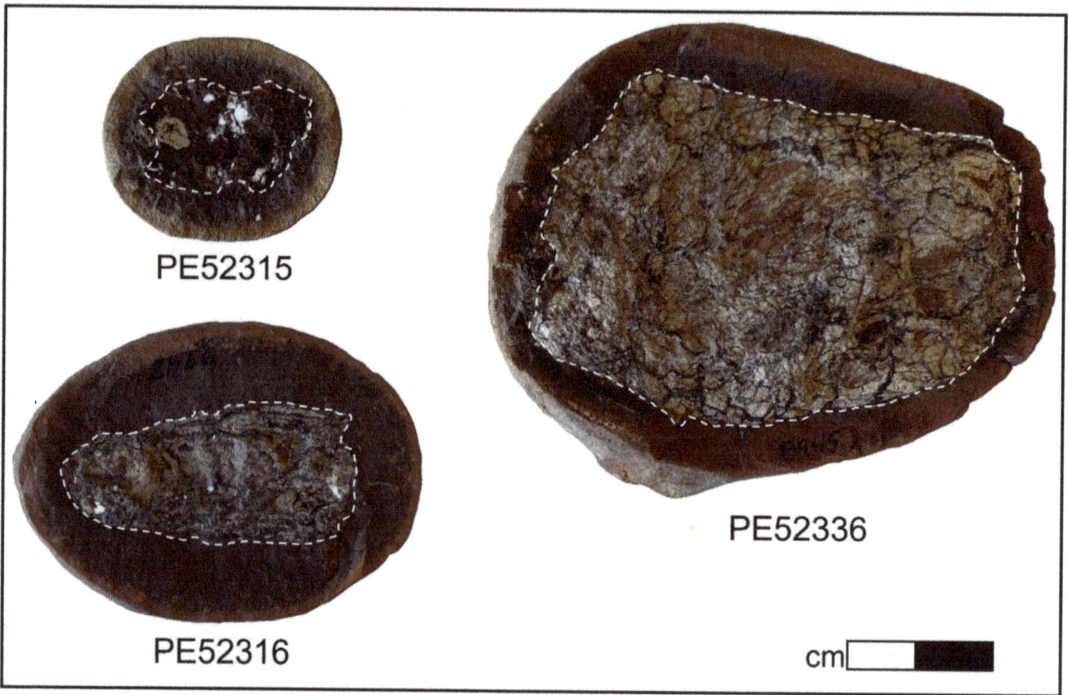

Figure 1. Images of the three samples used in this study. Regions are defined as the 'Fossil' and the 'Matrix' corresponding to the coprolite fossil region and the concretionary region, respectively, as demonstrated by white dashed lines. Subsamples of each region were used for geochemical analysis. 'Matrix' is used throughout to refer to the concretionary region, which is assumed to consist of primarily concretionary material plus potential external organic matter inputs.

One half of each concretion was cut using a handheld Dremel rotary tool with a diamond blade (previously cleaned by sonicating in a mixture of dichloromethane (DCM) and methanol (MeOH) (9:1 v/v) for 15 min intervals, separating the central coprolite fossil (referred to herein as the 'fossil') from the surrounding rock ('matrix') for each sample. The different sections of each concretion were washed by repeated sonication in 15 min intervals to trace removal of external contamination in a mixture of DCM:MeOH (9:1 v/v), before being ground using pre-annealed (450 °C for 3 h) ceramic mortars and pestles. In between sample treatments, pre-annealed sand was ground to clean mortars.

The ground sample material was Soxhlet extracted in individual pre-extracted cellulose thimbles (Soxhlet extracted three times for 24 h using a mixture of DCM:MeOH (9:1 v/v)) for 72 h. A procedural blank of a pre-extracted thimble was run alongside each extraction. The samples were filtered through activated copper powder to remove elemental sulfur. Small scale column chromatography (5 cm × 0.5cm i.d.) using silica gel activated at 160 °C for 24 h was used to separate the total lipid extracts into aliphatic (4 mL n-hexane), aromatic (4 m n-hexane:DCM (9:1 v/v)), porphyrin (4 mL n-hexane:DCM (7:3 v/v)) and polar (4 mL DCM:MeOH (1:1)) fractions for analysis.

2.2. Gas Chromatography-Mass Spectrometry (GC-MS)

Full scan gas chromatography-mass spectrometry analysis (GC-MS) was performed on the aliphatic fractions using an Agilent 7890B GC with a DB-1MS UI capillary column (J and W Scientific, 60 m, 0.25 mm i.d., 0.25 μm film thickness) coupled to an Agilent 5977B MSD. Aromatic fractions were analysed on an Agilent 6890N GC with a DB-5MS UI capillary column (J and W Scientific, 60 m, 0.25 mm i.d., 0.25 μm film thickness) coupled to an Agilent 5975B MSD. The GC oven was ramped from 40 °C to 325 °C at a rate of 3 °C/min with initial and final hold times of 1 min and 30 min, respectively.

Saturated and aromatic steroids and hopanoids were quantified using GC-MS analyses on an Agilent 6890 GC coupled to a Micromass Autospec Premier double sector MS (Waters Corporation, Milford, MA, USA). The GC was equipped with a 60 m DB-5 capillary column (0.25 mm i.d., 0.25 μm film thickness; Agilent J&W Scientific, Agilent Technologies, Santa Clara, CA, USA), and helium was used as the carrier gas at a constant flow of 1 mL/min. Samples were injected in splitless mode into a Gerstel PTV injector at 60 °C (held for 0.1 min) and heated at 260 °C min^{-1} to 300 °C. The MS source was operated at 260 °C in EI mode at 70 eV ionization energy and 8000 V acceleration voltage. All samples were injected in n-hexane to avoid deterioration of chromatographic signals by FeCl$_2$ build-up in the MS ion source through use of halogenated solvents [55]. The GC oven was programmed from 60 °C (held for 4 min) to 315 °C at 4 °C min^{-1}, with a total run time of 100 min. Saturated steranes and hopanes were quantified using metastable reaction monitoring (MRM) in M$^+$ → 217 and M$^+$ → 191 transitions, respectively. Mono- and triaromatic steroids were detected using selected ion recording (SIR) under magnet control of base ions m/z 253 and 231, respectively. All ratios and abundance proportions are reported uncorrected for differences in MS-response. Saturated and aromatic steroid hydrocarbons as measured using MRM are shown in Figures 2 and 3.

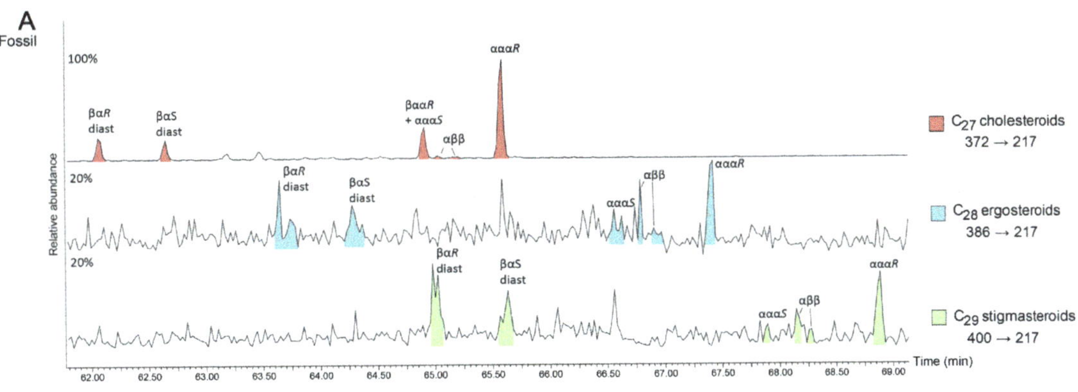

Figure 2. *Cont.*

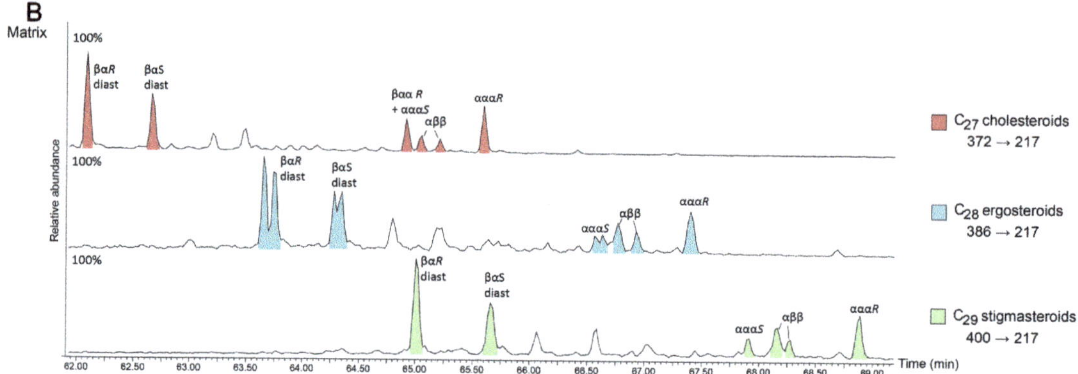

Figure 2. Metastable Reaction Monitoring (MRM) chromatograms of $M^+ \rightarrow 217$ precursor to product transitions of C_{27-29} steranes of PE 52316 Fossil (**A**) and Matrix (**B**). Peaks are coloured according to the number of carbon atoms in the sterane. α and β nomenclature refers to the stereochemistry of hydrogen at C-5, C-14 and C-17 for regular steranes and C-13 and C-17 for diasteranes, while S/R refers to stereochemistry at the C-20 position. Percentages represent the relative abundance of the most abundant peak in each transition.

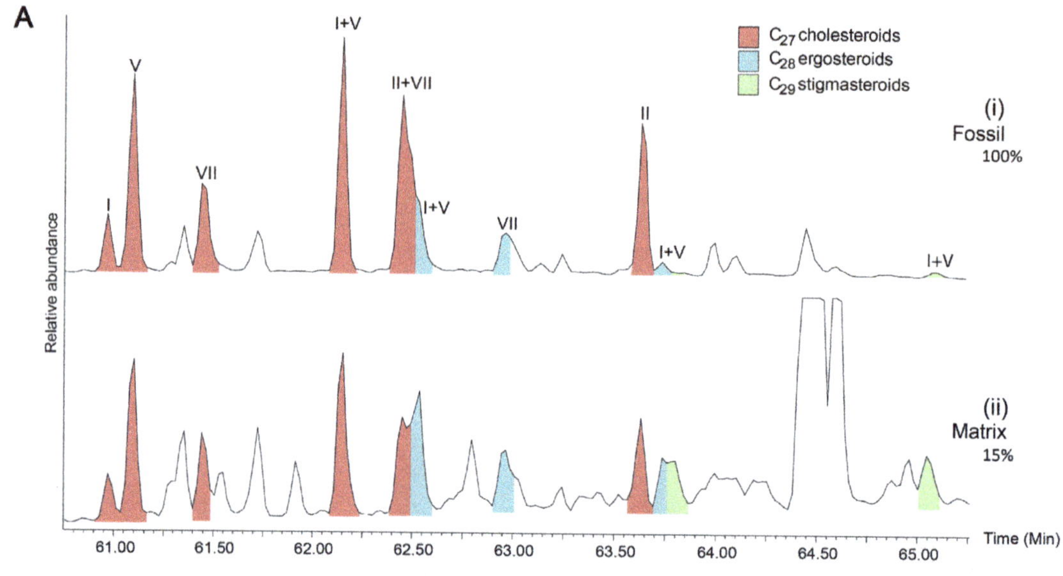

Figure 3. *Cont.*

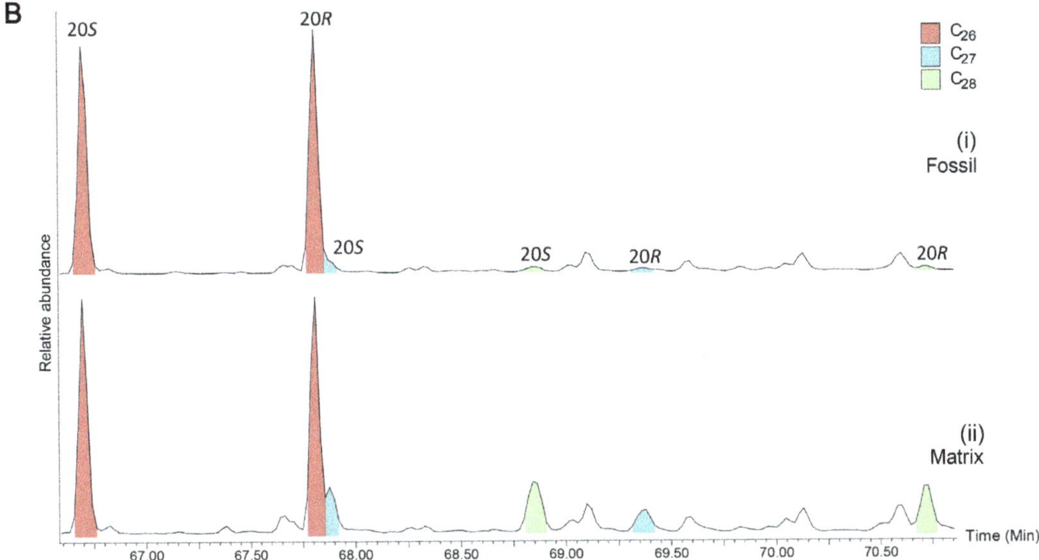

Figure 3. (**A**) Selected Ion Recording (SIR) $m/z = 253$ chromatogram of C_{27-29} monoaromatic steroids of PE 52336 Fossil (**i**) and Matrix (**ii**). I = 5β(H),10β(CH$_3$); II = 5α(H),10β(CH$_3$); V = 5β(CH$_3$),10β(H); VII = 5α(CH$_3$),10α(H). Percentages represent the relative abundance of the most abundant peak in each transition. (**B**) SIM $m/z = 231$ chromatogram of C_{26-28} triaromatic steroids of PE 52336 Fossil (**i**) and Matrix (**ii**). S/R refers to the stereochemistry at C-20.

2.3. Gas Chromatography-Isotope Ratio-Mass Spectrometry (GC-irMS)

Stable isotope ratio mass spectrometric analyses of individual compounds were performed on the aliphatic fractions of each sample to determine δ^{13}C, using a Thermo Scientific Trace GC Ultra coupled to a Thermo Scientific Delta V Advantage mass spectrometer via a GC isolink and a Conflo IV. The reactors consisted of a combustion interface containing a ceramic tube lined with NiO and filled with NiO and CuO, held at 1000 °C. The programs used for the GC column, carrier gas, injector conditions and oven temperatures were identical to those used for GC-MS analysis as described above. The gas chromatography-isotope ratio-mass spectrometer (GC-irMS) measures the δ^{13}C values by monitoring the CO_2 produced by the sample and measuring the response of ions of m/z 44, m/z 45 and m/z 46, relative to the reference gas of known δ^{13}C content.

2.4. Bulk Stable Carbon Isotopes

Ground residue from lipid biomarker extractions were treated with hydrochloric acid (4M) to remove carbonate mineral via repeated addition of fresh acid solution, stirring and heating at 50 °C until gas production ceased. Samples were subsequently washed with Milli-Q water and freeze-dried to remove excess water. δ^{13}C analyses were performed using a Thermo Flash 2000 HT elemental analyser (EA) connected to a Delta V Advantage isotope ratio monitoring mass spectrometer (irMS) via a Conflo IV. Samples were weighed (approximately 6 mg) in triplicate into tin cups (SerCon) and combusted to CO_2 in the nitrogen-carbon reactor (1020 °C). CO_2 passed through the Conflo IV interface into the irMS, which measured m/z 44, 45 and 46. δ^{13}C values were calculated by Thermo Isodat software and normalised to the international VPDB scale by multi-point normalisation using the standard reference materials NBS 19 (+1.95‰) and L-SVEC (−46.60‰) [56]. The standard reference material IAEA-600 was measured to evaluate the accuracy of the normalization. The normalized δ^{13}C values of IAEA-600 from these measurements were within ±0.1‰ of the reported value of −27.77‰ [56].

2.5. Polar Compound Analysis

Polar fractions were analysed at Leeder Analytical (Victoria, Australia). Fractions were dried and internal standard (^{13}C-cholesterol) added, and were then combined with *N*,*O*-bis(trimethylsilyl)fluoroacetamide and trimethylchlorosilane (99:1) and heated (60 °C) for 20 min. Samples were dissolved in toluene (500 µL) before analysis. Gas chromatography-tandem mass spectrometry (GC-MS/MS) was performed on an Agilent 7890B Gas Chromatograph with a DB-5MS UI column (30 m × 0.25 mm 0.25 um film) coupled to an Agilent 7000D Triple Quadruple Mass Spectrometer. Results were quantified against sterol trimethylsilyl-derivative standards.

2.6. In Situ Raman Microspectroscopy and ChemoSpace Analysis of Spectral Data

The set of analysed coprolites was microscopically screened for evidence of preserved carbonaceous matter characterized by a dark brown-to-black colouration (Figure 1), and two coprolites (FMNH PE 52316, FMNH PE 52336) with suitable preservation were identified. A total of $n = 35$ carbonaceous vertebrate, annelid, non-annelid invertebrate, and plant fossils from the Mazon Creek locality (Supplementary Table S1: specimen details) and the two carbonaceous coprolites (FMNH PE 52316, FMNH PE 52336) were surface-cleaned with 70% EtOH, and subjected to high-resolution *in situ* Raman microspectroscopy in the Department of Earth and Planetary Sciences at Yale University. Raman microspectroscopy was performed using a Horiba LabRam HR800 with 532 nm excitation (holographic notch filter; 20 mW at the sample surface). The spectra were obtained in LabSpec 5 software, and the instant processing included only a standard spike removal. Raman scattering was detected by an electron multiplying charge-coupled device (EM-CCD) following dispersal with an 1800 grooves/mm grating and passing through a 200 µm slit (hole size 300 µm). The spectrometer was calibrated using the first order Si band at 520.7 cm^{-1}. Ten spectra were accumulated in the 500–1800 cm^{-1} region, also known as 'organic fingerprint region', for 10 s exposure time each, at 32× magnification. All spectra were analyzed in an identical fashion in SpectraGryph 1.24 spectroscopic software: A conservative adaptive baseline (30%) was fitted, no baseline offset was imposed, and all spectra were normalized to the common highest peak. Relative intensities ($n = 53$; arbitrary units) at pre-selected informative band positions ([34,57]; listed in the Supplementary Information) were exported using the 'Multicursor' function in SpectraGryph 1.2. The resulting variance-covariance matrix was exported into PAST 3 (file available as Source Data), and two separate sample identification strategies were applied: One data matrix contains the extracted spectroscopic signals and binary characters identifying carbonaceous plant, vertebrate, annelid, non-annelid invertebrate remains and coprolites as separate tissue types. A second matrix uses a more agnostic approach and contains, in addition to the spectroscopic signals, only binary characters identifying samples as plants, annelids, non-annelid invertebrates, and vertebrates; in this data matrix coprolite tissue affinity was coded as 'unknown'. Both data matrices are available as Source Data. Endogeneity of organic matter in the carbonaceous films associated with fossil morphology has previously been demonstrated [57], and is here separately assessed using lipid biomarkers. A Canonical Correspondence Analysis of the first data matrix revealed the diagnostic molecular features distinguishing the fossil coprolites from other fossil soft tissues from the Mazon Creek locality (Figure 4A). A second Canonical Correspondence Analysis allowed the coprolite samples to locate in the compositional space (ChemoSpace) based on the affinity of the digested, fossil tissues (Figure 4B). Canonical Correspondence Analysis is a discriminant multivariate analysis that distinguishes previously identified groups of samples in a ChemoSpace, and, if samples of unknown affinity are included, reveals their affinity (which is here translated into the dietary spectrum of a coprolite producer). Due to the discriminant, comparative nature of the analysis and spectroscopic data, axis loadings do not have a dedicated unit. The impact of all $n = 53$ extracted relative intensities is represented by colour-coded (see caption) ChemoSpace vector arrows in Figure 5 (Figure 5B corresponding to the ChemoSpace shown in Figure 4A, and Figure 5C corresponding to the ChemoSpace shown in Figure 4B).

Functional groups were identified using Lambert et al. [58], and have been previously published [34,57]; select functional groups that are characteristically enriched in coprolites and reveal the affinity of their digested tissues are labelled in Figure 5.

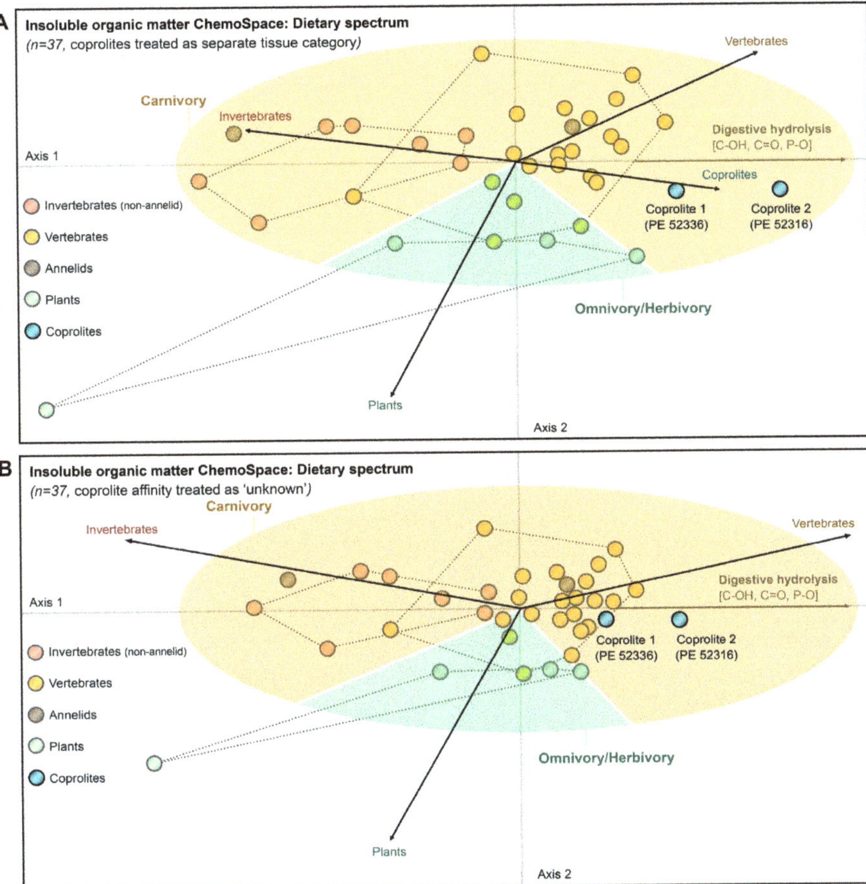

Figure 4. Dietary ChemoSpace analysis of tissue type signals preserved in carbonaceous coprolites from the Mazon Creek locality. A total of n = 37 Mazon Creek carbonaceous fossils and coprolites were spectroscopically fingerprinted (n = 1 biological replicate and n = 10 mean-averaged technical replicates per data point), identified, and analysed by means of a Canonical Correspondence Analysis (CCA; discriminant analysis). (**A**) ChemoSpace resulting from a CCA that treated coprolites as a separate tissue category (black arrows); corresponding trajectories of functional groups in the ChemoSpace are plotted in Figure 5B. (**B**) ChemoSpace resulting from a CCA that treated the tissue affinity of coprolites as unknown; corresponding trajectories of functional groups in the ChemoSpace are plotted in Figure 5C. ChemoSpaces in both, (**A**) (clustered in n = 5 categories, each contained in a dotted, convex hull; except from annelids and coprolites due to small sample size) and (**B**) (clustered in n = 4 categories, each contained in a dotted, convex hull, except from annelids), reveal that fossil coprolites are compositionally distinct from undigested fossil soft tissues due to digestive hydrolysis of macromolecules prior to fossilisation (detailed differences are plotted in Figure 5B), and contain predominantly digested tissues of vertebrate prey items. Thus, carnivory (orange circle fraction, contrary to omnivory/herbivory = green circle fraction) can be inferred for both coprolite producers. Source Data are available for the CCAs in (**A**,**B**).

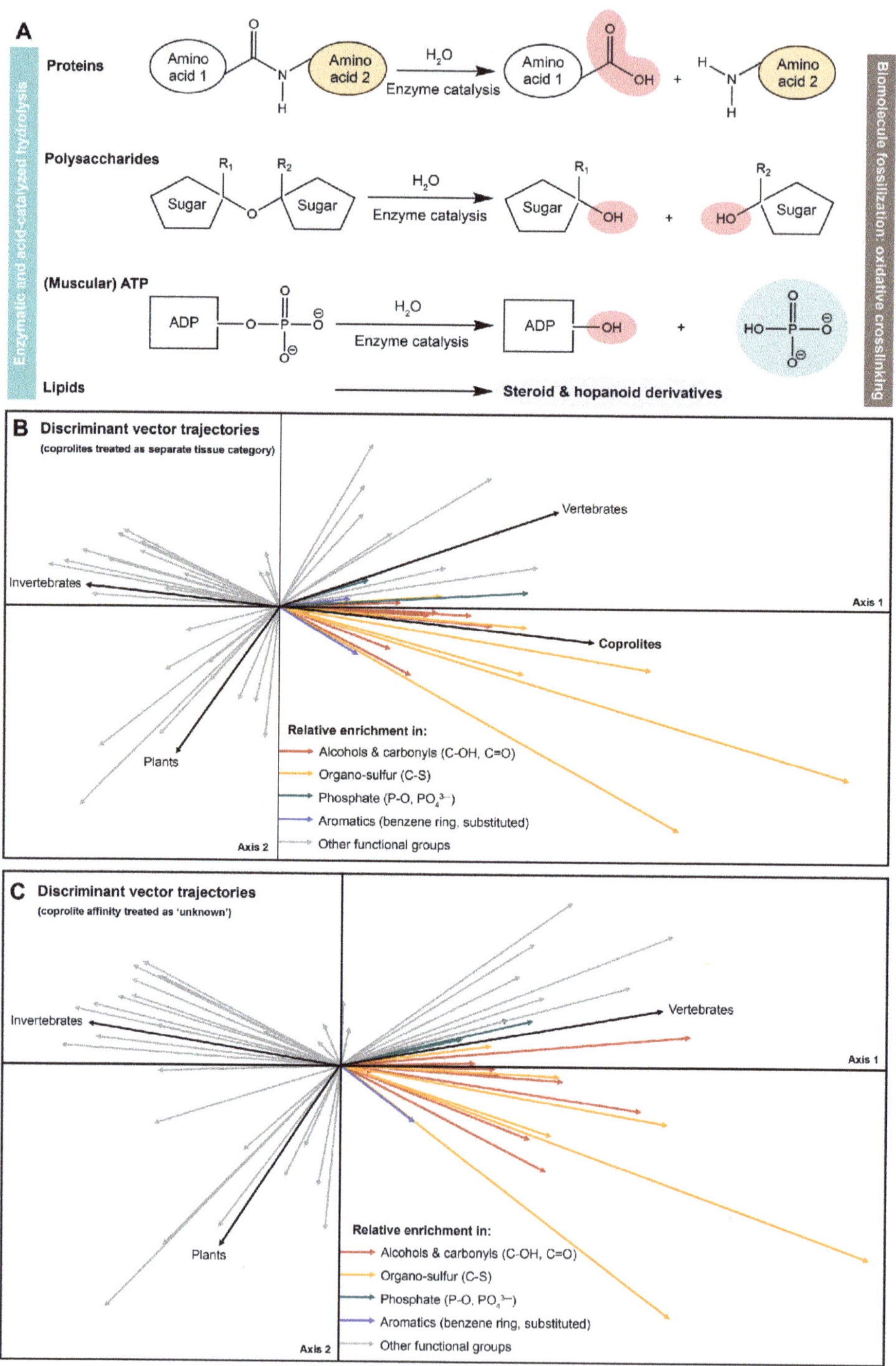

Figure 5. *In vivo* digestive hydrolysis reactions of key macromolecules and functional group trajectories

in the compositional spaces plotted in Figure 4. (**A**) Functional group chemistry during digestive hydrolysis reactions of proteins, polysaccharides, adenosine triphosphate (ATP), and lipids: stomach acid and digestive enzymes catalyse the hydrolytic cleavage of amide, glycosidic, and ester bonds and yield a relative increase in the relative abundance of alcohols and carbonyls (red), phosphate (teal), and aromatic compounds (dark blue); Depending on the dietary source, faecal matter can be enriched in organo-sulfur moieties (orange; S-bearing amino acids: cystein, methionine). Compositional differences associated with digestive hydrolysis survive (as shown in Figure 4) biomolecule fossilization through oxidative crosslinking. ADP = adenosine diphosphate; ATP = adenosine triphosphate. (**B**) Discriminant vector arrows for tissue categories (black, $n = 5$) and functional groups ($n = 53$ extracted relative intensities) in the Canonical Correspondence Analysis (CCA) shown in Figure 4A. Functional groups are colour-coded (corresponding to functional group labels in (**A**)) for vectors that explain the ChemoSpace placement of the two analyzed coprolites. This CCA treated coprolites ($n = 2$) as a separate tissue category to reveal the unique compositional features distinguishing them from other types of carbonaceous soft tissues from the Mazon Creek locality ($n = 35$). (**C**) Discriminant vector arrows for tissue categories (black, $n = 4$) and functional groups ($n = 53$ extracted relative intensities) in the CCA shown in Figure 4B. Functional groups are colour-coded (corresponding to functional group labels in (**A**)) for vectors that explain the ChemoSpace placement of the two analysed coprolites (see Figure 4B). This CCA treated coprolites ($n = 2$) as samples with unknown tissue affinity to identify the primary source of fossil faecal matter ($n = 35$).

2.7. X-ray Diffraction

Powdered samples were analysed using a Bruker-AXS (Karlsruhe, Germany) D8 Advance Powder Diffractometer with a CuKα radiation source (40 kV, 40 mA) and a LynxEye detector. The scan ranged from 5° to 90° 2θ with a step size of 0.015° and a collection time of 0.7 s per step. Crystalline phases were identified by using the Search/Match algorithm, DIFFRAC.EVA 5.2 (Bruker-AXS, Karlsruhe, Germany) to search the International Center for Diffraction Data (ICDD) Powder Diffraction File (PDF4+ 2021 edition).

2.8. Elemental Analysis

Elemental analyses were performed at Source Certain International (Western Australia). Samples (0.25 g) were accurately weighed and digested in nitric acid (16 mL, 65 wt%), perchloric acid (4 mL, 70 wt%) and hydrofluoric acid (10 mL, 50 wt%) at approximately 180 °C for a minimum of 16 h under reflux. The acids were removed by evaporation at approximately 220 °C. Once the residue reached incipient dryness it was dissolved in hydrochloric acid (0.75 mL, 32 wt%), nitric acid (0.25 mL, 65 wt%) and DI-water (20 mL, >16.4 MΩ cm). The solution was suitably diluted for the instrument. Samples were analysed using an Agilent 5110 ICP-AES and an Agilent 7700 ICP-MS.

2.9. Total Organic Carbon

The rock samples were ground to a fine powder and digested in acid (HCl) to remove the carbonate minerals. The remaining residues were analysed using a LECO Carbon-Sulfur Analyser (CS-230). The CO_2 produced was measured with an infra-red detector, and values calculated according to standard calibration.

3. Results and Discussion

Results are discussed in general terms of all three samples, unless a particular sample is specified. Samples subject to the same analysis showed generally consistent results as reflected in biomarker ratios presented in Tables 1–4. Figures present the sample which best demonstrates compositional features.

Table 1. Steroid distributions in extracted organic matter.

		PE52315		PE52316		PE52336	
		Fossil	Matrix	Fossil	Matrix	Fossil	Matrix
[1] Steranes (%)	C_{27}	86.2	69.1	92.8	50.1	99.0	78.0
	C_{28}	7.3	16.3	3.5	25.0	0.6	9.8
	C_{29}	6.5	14.6	3.7	24.9	0.4	12.2
[2] Reg steranes/hopanes		1.9	0.9	4.8	0.5	80.1	1.1
[3] Monoaromatic steroids (%)	C_{27}	54.8	41.8	68.3	38.9	82.2	56.1
	C_{28}	31.7	27.7	25.1	25.9	15.2	22.7
	C_{29}	13.5	30.5	6.6	35.1	2.5	21.2
[4] Triaromatic steroids (%)	C_{27}	43.6	26.7	84.6	41.4	94.8	68.4
	C_{28}	19.4	27.2	6.1	19.9	2.7	11.5
	C_{29}	37.0	46.1	9.3	38.7	2.5	20.1

[1] Distributions of regular steranes were computed using the sum of regular steranes and diasteranes in MRM $M^{+} \rightarrow$ 217 transitions. Steranes: C_{27-29} $\alpha\alpha\alpha$- and $\alpha\beta\beta$-20($S + R$)-steranes and C_{27} $\beta\alpha\alpha$ 20R; $\alpha\alpha\alpha = 5\alpha$(H),14$\alpha$(H),17$\alpha$(H); $\alpha\beta\beta = 5\alpha$(H),14$\beta$(H),17$\beta$(H); $\beta\alpha\alpha = 5\beta$(H),14 α(H),17 α(H); diasteranes: $\beta\alpha$-22($S + R$)-diasteranes; 13β(H),17α(H). [2] Reg steranes/hopanes = [$\sum$(C_{27-29} steranes)]/[$\sum$(C_{27-35} hopanes)], steranes: as above; hopanes: C_{27} Ts, Tm 17α, Tm 17β; C_{29} $\alpha\beta$, Ts, $\beta\alpha$; C_{30} $\alpha\beta$, $\beta\alpha$; C_{31-35} $\alpha\beta$-22($S + R$); $\alpha\beta = 17\alpha$(H),21β(H); $\beta\alpha = 17\beta$(H),21α(H); C_{27} Ts = 18α-22,29,30-trisnor*neo*hopane; C_{27} Tm = 17α-22,29,30-trisnorhopane; C_{29} Ts = 18α-30-*Norneo*hopane. [3] Distributions of monoaromatic steroids (MAS) were computed using the sum of S and R I + V isomers of MAS C_{27-29} homologs in m/z 253 mass chromatograms. [4] Triaromatic steroid distributions (TAS) were computed using the sum of C_{26-28} homologs in the m/z 231 mass chromatograms.

Table 2. Values of sterol compounds reported as total amount in micrograms (µg) total in each fraction, quantified against external sterol standards in 500 mL, as an average of two measurements.

	PE 52315		PE 52316		PE 52336		Procedural Blank
	Fossil	Matrix	Fossil	Matrix	Fossil	Matrix	
5α-Cholestan-3β-ol	-	-	0.43	-	0.16	-	-
Cholesterol	0.07	0.06	0.05	0.17	0.07	0.14	0.07
Coprostan-3-ol	0.03	-	-	-	-	-	-

Table 3. Biomarker parameters in extracts of coprolite samples.

	PE 52315		PE 52316		PE 52336	
	Fossil	Matrix	Fossil	Matrix	Fossil	Matrix
[1] Dia/reg C_{27} steranes	0.64	1.03	0.30	1.50	0.32	0.76
[2] Dia/reg C_{28} steranes	1.61	2.35	0.74	1.87	0.44	2.09
[3] Dia/reg C_{29} steranes	1.25	1.52	0.89	1.47	*	1.46
C_{29} 20S/(20S + 20R) steranes	0.28	0.30	0.18	0.27	0.21	0.25
[4] Ts/(Ts + Tm)	0.52	0.55	0.47	0.52	0.50	0.44
$\beta\alpha$/($\beta\alpha$ + $\alpha\beta$) C_{30} hopane	0.11	0.10	0.08	0.09	0.09	0.11
[5] Pr/Ph	1.00	0.99	0.85	1.03	1.65	1.42

[1] Dia/Reg C_{27} steranes = [$\sum$($\beta\alpha$-22($S + R$)-diacholestane)]/[$\sum$($\alpha\alpha\alpha$- and $\alpha\beta\beta$-20($S + R$)-cholestane)]; [2] Dia/Reg C_{28} steranes = [$\sum$($\beta\alpha$-22($S + R$)-diaergostane)]/[$\sum$($\alpha\alpha\alpha$- and $\alpha\beta\beta$-20($S + R$)-ergostane)]; [3] Dia/Reg C_{29} steranes = [$\sum$($\beta\alpha$-22($S + R$)-diastigmastane)]/[$\sum$($\alpha\alpha\alpha$- and $\alpha\beta\beta$-20($S + R$)-stigmastane)]; [4] Ts = 18α-22,29,30-trisnor*neo*hopane; Tm = 17α-22,29,30-trisnorhopane. Steranes and hopanes identified using the MRM $M^{+} \rightarrow$ 217 and $M^{+} \rightarrow$ 191 transitions, respectively. [5] Pristane and phytane integrated using total ion chromatogram. * Value omitted—coelution with abundant C_{27} steroid hydrocarbons was such that value was not able to be determined.

Table 4. Average δ^{13}C values (bulk residue and compound specific) for PE 52316 and PE 52336, given in per mil (‰).

	PE 52316		PE 52336	
	Fossil	**Matrix**	**Fossil**	**Matrix**
$\delta^{13}C_{org}$	−23.6 (0.07)[3]	−23.8 (0.03)[2]	−23.9 (0.12)[3]	−23.7 (0.03)[3]
$\delta^{13}C_{17}$	−28.4 (0.65 *)[3]	−30.0 (0.24)[3]	−29.4 (0.54 *)[3]	−29.0 (0.62 *)[3]
$\delta^{13}C_{Pristane}$	−28.8 (0.82 *)[2]	−29.2 (0.30)[3]	−28.4 (0.53 *)[3]	−29.3 (0.51 *)[3]
$\delta^{13}C_{18}$	−30.4 (0.41 *)[3]	−31.7 (0.31)[3]	−31.5 (0.13)[3]	−30.7 (0.40)[3]
$\delta^{13}C_{Phytane}$	−33.0 (0.11)[2]	−29.5 (0.23)[3]	−35.1 (0.39)[3]	−33.1 (0.31)[3]
[1]$\delta^{13}C_{20-25}$	−30.5	−31.0	−30.8	−30.7
$\delta^{13}C_{cholestane}$	−32.9 (0.39)[3]	-	−32.6 (0.29)[3]	-

* Number in brackets indicates standard deviation; superscript refers to number of analyses used in average. Values with standard deviations greater than 0.4‰ are marked with an asterisk (*). [1]δ^{13}C values for C_{20-25} *n*-alkanes represent an average of the δ^{13}C values for each *n*-alkane from C_{20} to C_{25}, which were each determined from the average of three isotopic measurements. Standard deviations ranged from 0.02–0.60.

3.1. Inorganic Composition

The coprolite components of samples PE 52316 and PE 52336 are approximately 2 cm and 3.5 cm in length, respectively and appear to be similar in composition (Figure 1). Three primary mineral regions were identified by X-ray diffraction (XRD). The coprolites are preserved in three dimensions as calcium phosphate with cracks filled with sphalerite. The concretionary material is siderite with minor amounts of quartz. Total organic carbon ranged from 0.31 to 0.78 wt% of rock in the fossils and 0.32 to 0.51 wt% in the matrices.

Phosphatic preservation is a characteristic component of carnivore coprolites, e.g., [1,4,59]. It has been suggested that rapid precipitation (within weeks—e.g., [14,19,60]) of dietary calcium phosphate can result in the preservation of fine morphological information [4,61]. Much like the formation of carbonate concretions, the remineralisation of calcium phosphate from faecal material has been demonstrated as autochthonous [7], occurring rapidly after deposition and prior to diagenetic permineralisation.

Elemental analyses revealed enrichment of rare earth elements (REE) in the coprolite fossils compared to the siderite concretion. Phosphatic minerals such as apatite are able to incorporate REE during early fossilisation [62] via substitution for calcium, e.g., [63]. REE are commonly used in palaeontological, palaeoenvironmental and palaeoredox studies, particularly those focused on vertebrate bones, e.g., [64–68] as well as on coprolites fossilised as apatite, e.g., [69,70].

3.2. Lipid Biomarkers of Coprolites

The coprolite fossil regions are characterised by a predominance of cholesterol-derived steroidal hydrocarbons (e.g., Figures 2 and 3, Table 1). C_{27} cholestanes make up 86 to 99% of the total steranes composition of the aliphatic fraction (Table 1). Similarly, aromatic cholesteroids make up the majority of the monoaromatic and triaromatic steroid distributions of each of the fossils (Figure 3, Table 1). Cholesterol and its derivatives have been considered in past studies of faecal material as indicators of an animal diet, e.g., [28,29]. Cholesterol is generally known to be synthesised by animals, while ergosteroids can be found in fungi and some groups of algae [71,72] and sitosterol and stigmasterol are commonly made by higher plants [71]. As such, the relative proportions of related sterane biomarkers can help distinguish the contributions of these inputs in sedimentary organic matter input [73]. However, cholesteroids can also be produced by herbivores via modification of other sterol analogues, particularly phytosterols [74] or synthesised *de novo*, e.g., [74–76] and are also present in almost all eukaryotic cell walls [77]. While the exceptionally high abundance of cholesteroid biomarkers suggests a primarily animal diet for the coprolite producer, minor contributions from plant material as a dietary component or due to occasional grazing cannot be completely ruled out.

The diagenetic fate of biomarkers is typically controlled by factors such as burial depth, redox conditions, mineral and porewater chemistry and geological time [20]. The low diagenetic transformation of the cholesteroids in the coprolite specimens is likely related to favourable preservation conditions during aromatisation. Similarly immature biomolecular signals have been found in previous studies of carbonate concretions where rapid encapsulation, burial and mineralisation supported the preservation of soft tissues and intact biomolecules, e.g., [10,12,26].

Diagenetic processes promote rearrangement of $\alpha\alpha\alpha$ 20R stereoisomers to the more thermodynamically stable $\alpha\alpha\alpha$ 20S and $\alpha\beta\beta$ 20R + S compounds [20,25]. Within the cholestanes distribution the biologically derived cholestane 20R isomer is most dominant (e.g., Figure 2A). Additionally, present is C_{27} $\alpha\alpha\alpha$ 20S cholestane coeluting with coprostane (C_{27} $\beta\alpha\alpha$ 20R). Here, the predominance of the C_{27} $\alpha\alpha\alpha$ 20R over the 20S isomer, (i.e., C_{27} $\alpha\alpha\alpha$ 20S/(20S + 20R) < 0.2), support a low diagenetic conversion of the steroids inside the fossil.

In contrast, the side chain isomerisation of C_{27} to C_{29} steranes as well as the transformation of $\alpha\alpha\alpha$ to $\alpha\beta\beta$-steranes progressed much further in the concretion matrix (Figure 2). This indicates that diagenetic conversion of labile steroids was hampered by early cementation and the lack of clay mineral catalysis, e.g., [11,78,79] in the coprolite, allowing the labile steroids to persevere.

Due to coelution, the proportion of $\beta\alpha\alpha$ 20R compared to $\alpha\alpha\alpha$ 20S was not precisely determined; however, it is presumed that a large area of this peak comprises C_{27} $\beta\alpha\alpha$ 20R based on the low diagenetic conversion of steroids within the fossil and generally low maturity. This was confirmed by the mass spectrum which showed the presence of two coeluting peaks, one with m/z 149 fragment (C_{27} $\alpha\beta\beta$ 20S) and the other with m/z 151 fragment (C_{27} $\beta\alpha\alpha$ 20R). Steroids with $\beta\alpha\alpha$ stereochemistry can be produced by reduction of sterols in the intestine of mammals depending on the primary steroid components of their diet [27] or may also form in sediment through microbial reduction of Δ^5 sterols [80].

Δ^2-sterenes and $\Delta^{3,5}$-steradienes are products formed in early diagenesis via dehydration of $5\alpha(H)$-stanols and Δ^5-sterols, respectively [23,81]. Δ^2-sterenes can also undergo isomerisation and rearrangement to Δ^4- and Δ^5-sterenes and subsequently to diasterenes [23,24,82,83]. A cholestadiene compound has been tentatively identified by comparison of its mass spectrum with that of a 3,5-cholestadiene [81], based on predominant fragments at m/z 213 and 368, in sample PE 52336. This compound is present in a higher relative abundance in the fossil than in the matrix, suggesting that it is likely derived from the original cholesterol content in the coprolite sample. Diasterenes in low abundance were also identified, although in the fossil of PE 52336 only, and are shown in the Supplementary Materials (Figure S1).

A- and B-ring monoaromatic steroids are derived directly from sterol compounds during early diagenesis, e.g., [84–86] while C-ring monoaromatic steroids have previously been attributed to later-stage diagenesis, e.g., [87], which can then be aromatised to form triaromatic hydrocarbons via loss of a methyl group, with increasing thermal maturity [88]. However, it has been suggested, e.g., [26,89] that aromatisation of sterene hydrocarbons to C-ring monoaromatic and triaromatic steroids can occur during early diagenesis via microbially mediated processes. Both the monoaromatic and triaromatic steroids show abundant cholesteroids compared to ergosteroids and stigmasteroids, with cholesteroids comprising between 55 to 82% of the total monoaromatic steroids distribution and 43 to 94% of total triaromatic steranes (Table 1). A predominance of the cholesteroid analogues in all steroid classes demonstrates that diagenetic transformation, while altering the cholesterol predominance somewhat, does not eliminate the primary dietary composition of steroids. Based on conventional models of triaromatic steroid formation, presence of triaromatic steroids in the fossil suggests a catagenetic history, which is inconsistent with the immaturity of the sample. It is therefore plausible that triaromatic steroids were also formed by microbially mediated reaction mechanisms, e.g., [26,89].

5α-cholestan-3β-ol was identified in the fossil of PE 52336 and PE 52316 (Table 2). This stanol may be either from a direct biological input or from early diagenesis by reduction of cholesterol [23,90]. The presence of 5α-cholestan-3β-ol in several of the fossils and its absence in any of the matrices supports that this is derived from the fossil coprolite itself. Coprostan-3-ol (5β-cholestan-3β-ol) was identified, *albeit* in low abundance in the PE 52315 fossil only (Table 2). This is a possible precursor sterol of coprostane that can be formed via reduction of cholesterol in the gut of many higher mammals, e.g., [27] as well as through microbial reduction in sediment [80]. Cholesterol was also identified in the fossil and matrix of all samples; however, this must be considered with caution, as it is also present in comparable concentrations in the procedural blank (Table 2).

3.3. Early Diagenetic Transformation of Dietary Sterols

A series of diagenetic products derived from cholesterol, such as cholestane and triaromatic steroids are preserved in all fossil coprolites. These components support a primarily carnivorous dietary source as they would support a high amount of cholesterol in the original faecal material, which has been rearranged and partially preserved. The cholesteroid compounds identified are summarised in Figure 6. Cholesterol can undergo reduction to 5α- and 5β-stanols [23], such as 5α-cholestan-3β-ol as present in the fossil coprolites. Further reduction would yield cholestane; specifically, the 5α-cholestane, which is the abundant sterane identified in each of the fossils. Cholesterol can also undergo reduction directly to Δ^2-sterenes and $\Delta^{3,5}$-steradienes [23]. While sterenes were not observed, a single cholestadiene was identified in the PE 52336 fossil, which is formed from cholesterol and may be further converted into monoaromatic (Figure 3A) and triaromatic (Figure 3B) steroids.

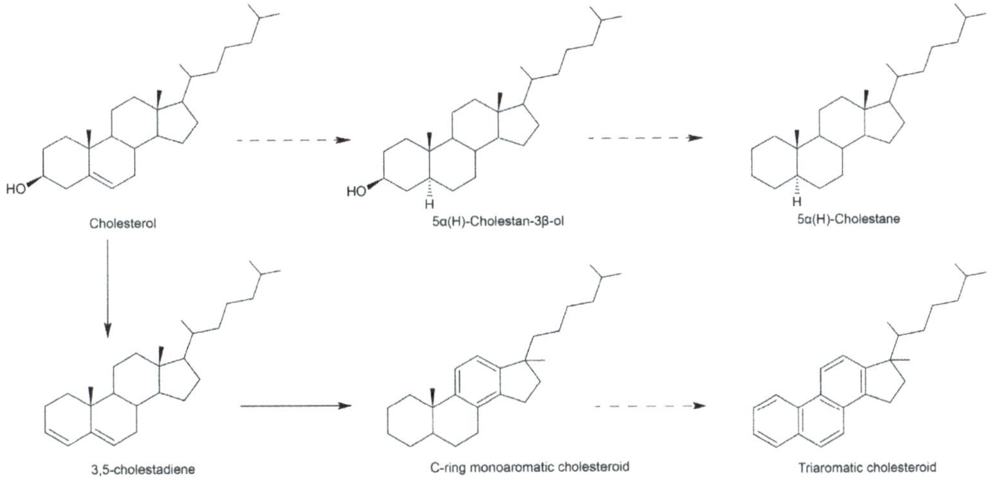

Figure 6. General schematic demonstrating diagenetic rearrangements of cholesteroid compounds, as identified in fossil coprolites in the current study. Dashed lines represent multiple possible intermediates or pathways. In both fossil and matrix of all samples, *n*-alkanes are abundant and consistently range from *n*-C$_{15}$ to *n*-C$_{26}$ with a maximum at *n*-C$_{23}$ but with no odd or even carbon number preference (Figure 7). The predominance of mid-chain (C$_{20}$-C$_{25}$) *n*-alkanes and lack of high-molecular-weight *n*-alkanes supports input from aquatic organic material in a limnic or deltaic environment receiving a minimal input of land plant material, e.g., [91,92]. The freshwater-influenced *n*-alkane distribution is thus in agreement with a geologically inferred large delta setting, e.g., [41–45]. The regular isoprenoids pristane (Pr) and phytane (Ph) are also present with Pr/Ph ratios ranging from 0.85–1.65 (Table 2), ratios supporting fluctuating redox conditions within the environment [73].

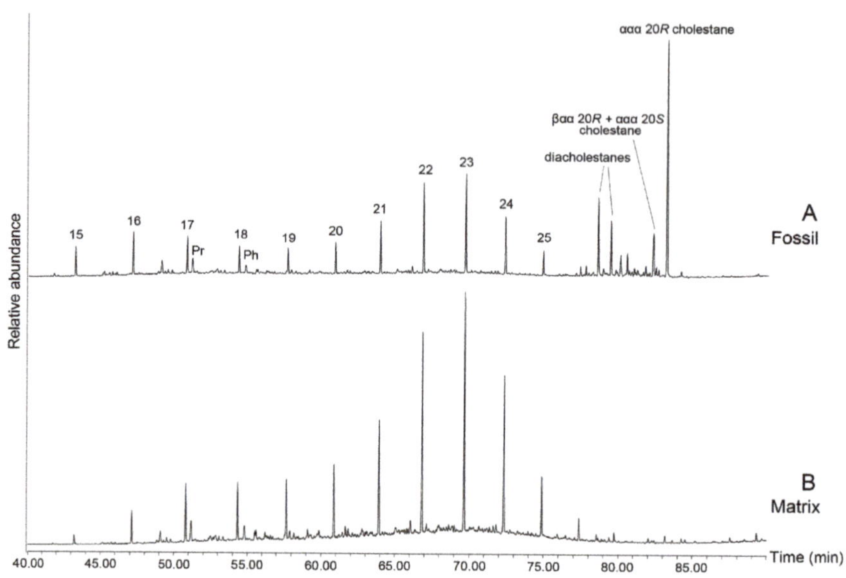

Figure 7. GC-MS total ion chromatogram of PE 52336 showing differences in *n*-alkanes distribution and cholestanes abundances in the Fossil (A) versus the Matrix (B). Numbers represent the carbon number of the hydrocarbon chain. Pr = pristane, Ph = phytane.

3.4. Raman ChemoSpace

The Raman ChemoSpace is here used to complement compositional insights from the analysis of biomarkers with additional information on the molecular makeup and preserved biosignatures in the macromolecular fraction of Mazon Creek coprolites (FMNH PE 52316, FMNH PE 52336). Focus of the analysis are *in situ* Raman spectra collected for carbonaceous fossil remains. The discriminant analysis (CCA) of spectral data collected for all *n* = 37 samples, shown in Figure 4A, identifies key compositional characteristics of fossil vertebrate, non-annelid invertebrate, annelid, plant, and coprolite samples from the Mazon Creek locality. Among these samples, only minimal overlap between the clusters of the diverse array of fossil metazoan tissues is observed; the cluster of carbonaceous plant tissues is separated from the metazoan tissue clusters (Figure 4A). Using the spread of samples in the compositional space (=ChemoSpace), the circular fractions including all metazoan (orange) and plant (green) tissues were plotted in the CCA (Figure 4A,B). It is the nature of the CCA that separates different tissue categories (black vector arrows) radiating from the origin into the ChemoSpace, therefore allowing to constrain the distribution of samples through circular fractions. The trajectories of all tissue vectors and their associated, characteristic molecular features are plotted in Figure 5. Figure 4A locates the two coprolite samples within the circular fraction occupied by metazoan tissues, and reveals as characteristic molecular features a relatively increased abundance of alcohol (C-OH) and carbonyl groups (C=O), aromatics, inorganic phosphate (P-O, PO_4^{3-}), and organo-sulfur moieties (C-S; Figure 5B). The majority of these characteristic molecular features that discriminate fossil coprolites from undigested fossil tissues coincide with the key chemical modifications of macromolecules experienced prior to fossilisation, during *in vivo* enzymatic digestion (Figure 5A): enzyme- and acid-catalysed ($HCl_{(aq)}$, stomach acid) hydrolysis cleaves primarily peptide bonds (amides), glucosidic bonds, esters (including phosphoesters), and breaks down lipids into their aliphatic and aromatic building blocks (please see steroid and hopanoid analyses in this study). Hydrolytic cleavage introduces alcohol groups, which can be converted into carbonyls during subsequent alteration. The molecular fingerprint of hydrolytic digestion preserves in the sampled Mazon Creek coprolites even after oxidative alteration (crosslinking reactions) acting during fossilisation [34,57]. Digestion moves data points in the CCA ChemoSpace along the vectors associated with functional groups introduced during hydrolytic cleavage (colored vectors in Figure 5B,C). The relative increase in organo-sulfur moieties is not related to digestive hydrolysis and thus relates either to the gastro-intestinal concentration processes or post-egestion alteration.

Figure 4B shows a complementary CCA ChemoSpace which, contrary to the analysis in Figure 4A, did not treat coprolites as a separate tissue category, but instead allowed them to group with the other $n = 4$ tissue categories based on the affinity of the contained, digested remains. Both coprolite samples plot within the circular fraction occupied by metazoan samples, and one tangentially overlaps with the vertebrate tissue cluster. Considering that digestion shifts samples in the ChemoSpace CCE (shown in Figures 4A and 5B,C) towards the right, we thus infer a metazoan affinity of the digested faecal matter and identify the coprolite producer as a carnivore rather than omnivore or herbivore (no primary contribution of plants to the diet).

The carnivorous diet, potentially based on vertebrates (as suggested by the proximity to the vertebrate cluster in Figure 4A,B), suggests that the producer of the coprolites represented a high trophic level in the ancient Mazon Creek ecosystem and was potentially an apex predator. Future experimental work on the ChemoSpace effects of different metazoan digestive processes has the potential to allow for the detailed quantification of the contributions of vertebrate, plant, non-annelid invertebrate, and annelid food items to the diet of coprolite producers.

The independent data collected for the soluble organic phase (biomarker analysis, stable isotope fractionation) and the insoluble organic phase (ChemoSpace of carbonaceous fossils) converge in their result: a carnivorous diet is inferred for the producers of the analysed coprolites.

3.5. Lipid Biomarkers of Matrix (Palaeoenvironmental Signal)

Terpane and hopane compounds were more dominant relative to steranes in the concretion matrix compared to the coprolite, as reflected in the regular sterane/hopane ratios (1.9–80 in the fossils compared to 0.54–1.1 in matrices) (Table 3). The Ts/(Ts + Tm) ratio, traditionally considered to be indicative of thermal maturity and clay catalysis, indicates the end of the diagenetic or the onset of the catagenetic stage of thermal transformation for both the coprolite and concretion matrix, which stands in contrast to the presence of a thermally labile steradiene and stanol (Table 3). C_{29} $20S/(20S + 20R)$ steranes ratios (Table 3) are also low overall but consistently higher in the fossil than in the matrix. The Ts/(Ts + Tm) ratios are consistent between the fossil and matrix while the diasterane/regular steranes ratios vary, supporting that different mechanisms are responsible for formation of tris*norneo*hopane than for the isomerization of steranes. Dia/regular sterane ratios for all steroid homologs show a higher proportions of rearranged steroids in the matrix than in the fossil. In general, even in the matrix the C_{27} dia/regular sterane ratio indicates lower maturity; however, the C_{27} sterane ratios are likely strongly influenced by source input.

3.6. Stable Carbon Isotopes

Carbon isotopes of the bulk organic matter were measured on the residue of extracted sediment from samples PE 52316 and PE 52336 after treatment with hydrochloric acid to remove carbonate. Both fossil and matrix of each sample showed $\delta^{13}C$ values around $-23.8‰$ (Table 4).

Compound specific carbon isotopes were measured for the *n*-alkanes of these two samples, as well as for the C_{27} $\alpha\alpha\alpha$ $20S$ sterane in the fossil portions (Table 3). $\delta^{13}C$ values of cholestane ($-32.8‰$) were depleted by approximately 2.0‰ compared to the average isotopic values of *n*-C_{20-25} ($-30.8‰$) and *n*-C_{18} ($-31.1‰$), and by approximately 9.0‰ compared to the bulk organic matter $\delta^{13}C$ values.

Sterane hydrocarbons are typically enriched compared to linear hydrocarbons from the same source by up to 8‰ [93]. Cholestane is here interpreted as derived from dietary cholesterol from the coprolite, while slightly more ^{13}C enriched n-alkanes represent input from external sources, most likely palaeoenvironmental signals. The depletion of δ^{13}C value of cholestane in the two coprolites by approximately 2‰ compared to the average of the abundant mid-chain n-alkanes is therefore consistent with input from two different sources. The isotopic consistency (approximately −30.8‰) of the n-alkanes in the fossil and the matrix is indicative of being derived from a common source, while the depletion of cholestane suggests it has a source different to the n-alkanes. The n-alkanes in the samples are therefore likely to originate from freshwater producers (e.g., algae, aquatic macrophytes, mosses), while cholesteroids intrinsic to the coprolite derive from animal sources.

Phytol is also typically ^{13}C enriched with respect to fatty acids synthesised in phytoplanktonic cells by 2–5‰ [93]. A similar 2–5‰ distinction between phytane and straight-chain hydrocarbons (and pristane) would therefore be expected of phytane derived from phytol, synthesised by chlorophyll *a* [93]. In these samples phytane was depleted in ^{13}C by approximately 2.5–3.5‰ compared to n-C$_{18}$ and 2.5–4‰ compared to mid-chain n-alkanes (Table 3) which is more typical of an origin within a methane cycle, wherein phytane is derived from ether lipids of methanotrophs and not phytol, e.g., [94,95]. Methanogenic archaea and methanotrophs are considered important components of the microbial growth mechanisms of siderite concretions, e.g., [95].

4. Conclusions

This study demonstrates that:

(1)   Cholesteroids including intact 5α-cholestan-3β-ol and coprostanol have been preserved in siderite concretions hosting 306 million-year-old coprolites.
(2)   The molecular data obtained by GC-MS, GC-MRM, GC-irMS and Raman microspectroscopy supports a primarily carnivorous diet and suggest an elevated trophic position for the coprolite producer.
(3)   The preservation of intact dietary sterols and macromolecular biosignatures is attributed to rapid encapsulation of the coprolites within days to months after egestion.
(4)   Siderite (FeCO$_3$) concretions (of Carboniferous age) seem to preserve intact and modified, but not unrecognisable, biomolecules much like calcium carbonate concretions of Jurassic and Devonian ages.

The results demonstrate that molecular information preserved within fossils can provide important ancient dietary insights either alongside or independent of traditional mineralogical or morphological studies of coprolites. Intact dietary sterols present in fossils support the significance of rapid encapsulation and organo-templated growth of carbonate concretions in the preservation of biomolecules in geological time. Carbonate concretions, which are host to soft-tissue fossil preservation are evidently important samples for molecular studies and represent each a unique opportunity to study extinct species and past environments.

Supplementary Materials: The following supporting information can be downloaded at: https://www.mdpi.com/article/10.3390/biology11091289/s1, Figure S1: GC-MS *m/z* 257 chromatogram of aliphatic fraction showing presence of diasterenes. A: 10α, Δ13(17) diacholestene 20*S*; B: 10α, Δ13(17) diacholestene 20*R*; C: 10α, Δ13(17) 24-methyldiacholestene 20*S* (*Tentatively identified based on location of *R*-isomer and mass spectrum); D:10α, Δ13(17) 24-methyldiacholestene 20*R*; E: 10α, Δ13(17) 24-ethyldiacholestene 20*S*; F: 10α, Δ13(17) 24-ethyldiacholestene 20*R*. Compounds were tentatively identified based on comparison of retention time and peak patterns with sample where compounds had been previously identified (e.g., [26]), confirmed using mass spectrum of peaks.; Table S1: A list of all specimens analyzed for the Figures 4 and 5, including catalogue numbers, institutional and museum identifiers, and sampled tissue types. In addition, raw ChemoSpace Canonical Correspondence Analyses (CCA) are plotted (in PAST 3) for the Figures 4 and 5 (separate plots: (1) specimen data points and tissue trajectories for Figure 4, and (2) corresponding discriminant functional group vectors in the CCA space for Figure 5).

Author Contributions: M.T. and K.G. designed the experiments and overall project concept. M.T. performed all laboratory on the fossil concretions except Raman microspectroscopy and GC-MRM. J.W. performed Raman microspectroscopy and J.B. performed GC-MRM. M.T. wrote the manuscript with contributions from all co-authors. K.G. supervised PhD scholar M.T. and P.M. provided Mazon Creek samples and hosted K.G. and M.T. at The Field Museum for this study. L.S. contributed with interpretive work and written contributions. All authors have read and agreed to the published version of the manuscript.

Funding: This research was funded by Australian Research Council (ARC) for an ARC-Laureate Fellowship grant (FL210100103) and ARC infrastructure grants (LE110100119; LE100100041; LE0882836; LE0668345; LE0775551).

Institutional Review Board Statement: Not applicable.

Informed Consent Statement: Not applicable.

Data Availability Statement: Source data for ChemoSpace analyses (Figures 4 and 5) are available in the Supplementary Materials. All other data can be made available upon request from corresponding authors.

Acknowledgments: We thank Peter Hopper, Alex Holman and Janet Hope for their technical support with GC-MS and stable isotope analyses. Tripp thanks Curtin University for a Research Training Postgraduate award, The Australian Institute of Nuclear Science and Engineering for an AINSE Postgraduate Research Award and support from The Institute for Geoscience Research for Star-fish Soxhlet extraction apparatus. Part of this research was undertaken using the XRD instrumentation at the John de Laeter Centre, Curtin University. Tripp thanks Veronica Avery and Matthew Rowles for technical assistance with XRD analysis. We thank Scott Lidgard (Field Museum, Chicago) for providing samples. We thank the three anonymous reviewers for their constructive comments that helped improve this manuscript.

Conflicts of Interest: The authors declare no conflict of interest. The funders had no role in the design of the study; in the collection, analyses, or interpretation of data; in the writing of the manuscript; or in the decision to publish the results.

References

1. Hollocher, K.T.; Alcober, O.A.; Colombi, C.E.; Hollocher, T.C. Carnivore Coprolites from the Upper Triassic Ischigualasto Formation, Argentina: Chemistry, Mineralogy, and Evidence for Rapid Initial Mineralization. *Palaios* **2005**, *20*, 51–63. [CrossRef]
2. Chin, K. The Paleobiological Implications of Herbivorous Dinosaur Coprolites from the Upper Cretaceous Two Medicine Formation of Montana: Why Eat Wood? *Palaios* **2007**, *22*, 554–566. [CrossRef]
3. Gill, F.L.; Crump, M.P.; Schouten, R.; Bull, I.D. Lipid analysis of a ground sloth coprolite. *Quat. Res.* **2009**, *72*, 284–288. [CrossRef]
4. Hollocher, K.T.; Hollocher, T.C.; Rigby, J.K. A Phosphatic Coprolite Lacking Diagenetic Permineralization from the Upper Cretaceous Hell Creek Formation, Northeastern Montana: Importance of Dietary Calcium Phosphate in Preservation. *Palaios* **2010**, *25*, 132–140. [CrossRef]
5. Khosla, A.; Chin, K.; Alimohammadin, H.; Dutta, D. Ostracods, plant tissues, and other inclusions in coprolites from the Late Cretaceous Lameta Formation at Pisdura, India: Taphonomical and palaeoecological implications. *Palaeogeogr. Palaeoclimatol. Palaeoecol.* **2015**, *418*, 90–100. [CrossRef]
6. Northwood, C. Early Triassic coprolites from Australia and their palaeobiological significance. *Palaeontology* **2005**, *48*, 49–68. [CrossRef]
7. Bajdek, P.; Owocki, K.; Niedźwiedzki, G. Putative dicynodont coprolites from the Upper Triassic of Poland. *Palaeogeogr. Palaeoclimatol. Palaeoecol.* **2014**, *411*, 1–17. [CrossRef]
8. Boast, A.P.; Weyrich, L.S.; Wood, J.R.; Metcalf, J.L.; Knight, R.; Cooper, A. Coprolites reveal ecological interactions lost with the extinction of New Zealand birds. *Proc. Natl. Acad. Sci. USA* **2018**, *115*, 1546–1551. [CrossRef]
9. Witt, K.E.; Yarlagadda, K.; Allen, J.M.; Bader, A.C.; Simon, M.L.; Kuehn, S.R.; Swanson, K.S.; Cross, T.-W.L.; Hedman, K.M.; Ambrose, S.H.; et al. Integrative analysis of DNA, macroscopic remains and stable isotopes of dog coprolites to reconstruct community diet. *Sci. Rep.* **2021**, *11*, 3113. [CrossRef]
10. Lengger, S.K.; Melendez, I.M.; Summons, R.E.; Grice, K. Mudstones and embedded concretions show differences in lithology-related, but not source-related biomarker distributions. *Org. Geochem.* **2017**, *113*, 67–74. [CrossRef]
11. Melendez, I.; Grice, K.; Trinajstic, K.; Ladjavardi, M.; Greenwood, P.; Thompson, K. Biomarkers reveal the role of photic zone euxinia in exceptional fossil preservation: An organic geochemical perspective. *Geology* **2013**, *41*, 123–126. [CrossRef]

12. Plet, C.; Grice, K.; Pagès, A.; Verrall, M.; Coolen, M.J.L.; Ruebsam, W.; Rickard, W.D.A.; Schwark, L. Palaeobiology of red and white blood cell-like structures, collagen and cholesterol in an ichthyosaur bone. *Sci. Rep.* **2017**, *7*, 13776. [CrossRef] [PubMed]

13. Allison, P.A. Konservat-Lagerstatten: Cause and Classification. *Paleobiology* **1988**, *14*, 331–344. [CrossRef]

14. Briggs, D.E.G. The role of decay and mineralization in the preservation of soft-bodied fossils. *Annu. Rev. Earth Planet. Sci.* **2003**, *31*, 275–301. [CrossRef]

15. Parry, L.A.; Smithwick, F.; Nordén, K.K.; Saitta, E.T.; Lozano-Fernandez, J.; Tanner, A.R.; Caron, J.B.; Edgecombe, G.D.; Briggs, D.E.; Vinther, J. Soft-Bodied Fossils Are Not Simply Rotten Carcasses—Toward a Holistic Understanding of Exceptional Fossil Preservation. *Bioessays* **2018**, *40*, 1700167. [CrossRef]

16. Seilacher, A. Begriff und bedeutung der Fossil-Lagerstätten: Neues Jarhbuch für Geologie und Paläontologie. *Neues Jahrb. Geol. Paläontologie Mon.* **1970**, 34–39.

17. Grice, K.; Holman, A.I.; Plet, C.; Tripp, M. Fossilised Biomolecules and Biomarkers in Carbonate Concretions from Konservat-Lagerstätten. *Minerals* **2019**, *9*, 158. [CrossRef]

18. Mccoy, V.E. Concretions as agents of soft-tissue preservation: A review. *Paleontol. Soc. Pap.* **2014**, *20*, 147–162. [CrossRef]

19. Sagemann, J.; Bale, S.J.; Briggs, D.E.; Parkes, R.J. Controls on the formation of authigenic minerals in association with decaying organic matter: An experimental approach. *Geochim. Cosmochim. Acta* **1999**, *63*, 1083–1095.

20. Peters, K.E.; Walters, C.C.; Moldowan, J.M. Volume I: Biomarkers and Isotopes in the Environment and Human History. In *The Biomarker Guide*; Cambridge University Press: Cambridge, UK, 2005.

21. Volkman, J.K. A review of sterol markers for marine and terrigenous organic matter. *Org. Geochem.* **1986**, *9*, 83–99. [CrossRef]

22. Huang, W.Y.; Meinschein, W.G. Sterols as source indicators of organic materials in sediments. *Geochim. Cosmochim. Acta* **1976**, *40*, 323–330.

23. Mackenzie, A.S.; Brassel, S.C.; Eglinton, G.; Maxwell, J.R. Chemical fossils: The geological fate of steroids. *Science* **1982**, *217*, 491–504. [CrossRef] [PubMed]

24. Brassell, S.C.; McEvoy, J.; Hoffmann, C.F.; Lamb, N.A.; Peakman, T.M.; Maxwell, J.R. Isomerisation, rearrangement and aromatisation of steroids in distinguishing early stages of diagenesis. *Org. Geochem.* **1984**, *6*, 11–23. [CrossRef]

25. Mackenzie, A.S.; Lamb, N.A.; Maxwell, J.R. Steroid hydrocarbons and the thermal history of sediments. *Nature* **1982**, *295*, 223–226. [CrossRef]

26. Melendez, I.; Grice, K.; Schwark, L. Exceptional preservation of Palaeozoic steroids in a diagenetic continuum. *Sci. Rep.* **2013**, *3*, 2768. [CrossRef]

27. Bull, I.D.; Lockheart, M.J.; Elhmmali, M.M.; Roberts, D.J.; Evershed, R.P. The origin of faeces by means of biomarker detection. *Environ. Int.* **2002**, *27*, 647–654. [CrossRef]

28. Gill, F.L.; Bull, I.D. Lipid analysis of vertebrate coprolites. *N. M. Mus. Nat. Hist. Sci. Bull.* **2012**, *57*, 93–98.

29. Umamaheswaran, R.; Prasad, G.V.R.; Rudra, A.; Dutta, S. Biomarker Signatures in Triassic Coprolites. *Palaios* **2019**, *34*, 458–467. [CrossRef]

30. Weber, D.J.; Lawler, G.C. Lipid Components of the Coprolites. Geology, Paleontology and Paleoecology of a Late Triassic Lake, Western New Mexico. *Brigh. Young Univ. Geol. Stud.* **1978**, *25*, 75–87.

31. Zatoń, M.; Niedźwiedzki, G.; Marynowski, L.; Benzerara, K.; Pott, C.; Cosmidis, J.; Krzykawski, T.; Filipiak, P. Coprolites of Late Triassic carnivorous vertebrates from Poland: An integrative approach. *Palaeogeogr. Palaeoclimatol. Palaeoecol.* **2015**, *430*, 21–46. [CrossRef]

32. Harrault, L.; Milek, K.; Jardé, E.; Jeanneau, L.; Derrien, M.; Anderson, D.G. Faecal biomarkers can distinguish specific mammalian species in modern and past environments. *PLoS ONE* **2019**, *14*, e0211119. [CrossRef] [PubMed]

33. Leeming, R.; Ball, A.; Ashbolt, N.; Nichols, P. Using faecal sterols from humans and animals to distinguish faecal pollution in receiving waters. *Water Res.* **1996**, *30*, 2893–2900. [CrossRef]

34. McCoy, V.E.; Wiemann, J.; Lamsdell, J.C.; Whalen, C.D.; Lidgard, S.; Mayer, P.; Petermann, H.; Briggs, D.E. Chemical signatures of soft tissues distinguish between vertebrates and invertebrates from the Carboniferous Mazon Creek Lagerstatte of Illinois. *Geobiology* **2020**, *18*, 560–565. [CrossRef]

35. Sistiaga, A.; Berna, F.; Laursen, R.; Goldberg, P. Steroidal biomarker analysis of a 14,000 years old putative human coprolite from Paisley Cave, Oregon. *J. Archaeol. Sci.* **2014**, *41*, 813–817. [CrossRef]

36. Clements, T.; Purnell, M.; Gabbott, S. The Mazon Creek Lagerstätte: A diverse late Paleozoic ecosystem entombed within siderite concretions. *J. Geol. Soc.* **2018**, *176*, 1–11. [CrossRef]

37. Wittry, J. *The Mazon Creek Fossil Fauna*; ESCONI: Downers Grove, IL, USA, 2012.

38. Peppers, R.A. *Palynological Correlation of Major Pennsylvanian (Middle and Upper Carboniferous) Chronostratigraphic Boundaries in the Illinois and Other Coal Basins*; The Geological Society of America, Inc.: Boulder, CO, USA, 1996.

39. Pfefferkorn, H.W. High diversity and stratigraphic age of the Mazon Creek flora. In *Mazon Creek Fossils*; Nitecki, M.H., Ed.; Academic Press, Inc.: New York, NY, USA, 1979; pp. 129–142.

40. Baird, G.C. Lithology and fossil distribution, Francis Creek Shale in northeastern Illinois. In *Mazon Creek Fossils*; Nitecki, M.H., Ed.; Academic Press, Inc.: New York, NY, USA, 1979; pp. 41–67.

41. Baird, G.C.; Shabica, C.W.; Anderson, J.L.; Richardson, E.S. Biota of a Pennsylvanian muddy coast: Habitats within the Mazonian delta complex, northeast Illinois. *J. Paleontol.* **1985**, *59*, 253–281.

42. Baird, G.C.; Sroka, S.D.; Shabica, C.W.; Keucher, G.J. Taphonomy of Middle Pennsylvanian Mazon Creek area fossil localities, Northeast Illinois; significance of exceptional fossil preservation in syngenetic concretions. *Palaios* **1986**, *1*, 271–285. [CrossRef]
43. Cotroneo, S.; Schiffbauer, J.D.; McCoy, V.E.; Wortmann, U.G.; Darroch, S.A.F.; Peng, Y.; Laflamme, M. A new model of the formation of Pennsylvanian iron carbonate concretions hosting exceptional soft-bodied fossils in Mazon Creek, Illinois. *Geobiology* **2016**, *14*, 543–555. [CrossRef]
44. Shabica, C.W. Pennsylvanian sedimentation in Northern Illinois: Examination of delta models. In *Mazon Creek Fossils*; Nitecki, M.H., Ed.; Academic Press, Inc.: New York, NY, USA, 1979.
45. Baird, G.C.; Sroka, S.D.; Shabica, C.W.; Beard, T.L. Mazon Creek-type fossil assemblages in the U.S. midcontinent Pennsylvanian: Their recurrent character and palaeoenvironmental significance. *Philos. Trans. R. Soc. Lond. B Biol. Sci.* **1985**, *311*, 87–99.
46. Berner, R.A. A new geochemical classification of sedimentary environments. *J. Sediment. Petrol.* **1981**, *51*, 359–365.
47. Berner, R.A. Sulphate reduction, organic matter decomposition and pyrite formation. *Philos. Trans. R. Soc. Lond. Ser. A Math. Phys. Sci.* **1985**, *315*, 25–38.
48. Claypool, G.E.; Kaplan, I.R. The Origin and Distribution of Methane in Marine Sediments. In *Natural Gases in Marine Sediments*; Maine Science; Kaplan, I.R., Ed.; Springer: Boston, MA, USA, 1974; Volume 3, pp. 99–139.
49. Allison, P.A. The role of anoxia in the decay and mineralization of proteinaceous macro-fossils. *Paleobiology* **1988**, *14*, 139–154. [CrossRef]
50. Berner, R.A. Rate of concretion growth. *Geochim. Cosmochim. Acta* **1968**, *32*, 477–483. [CrossRef]
51. Briggs, D.E.; Kear, A.J. Decay and preservation of polychaetes: Taphonomic thresholds in soft-bodied organisms. *Paleobiology* **1993**, *19*, 107–135. [CrossRef]
52. McCoy, V.E.; Young, R.T.; Briggs, D.E.G. Sediment Permeability and the Preservation of Soft-Tissues in Concretions: An Experimental Study. *Palaios* **2015**, *30*, 608–612. [CrossRef]
53. Yoshida, H.; Ujihara, A.; Minami, M.; Asahara, Y.; Katsuta, N.; Yamamoto, K.; Sirono, S.; Maruyama, I.; Nishimoto, S.; Metcalfe, R. Early post-mortem formation of carbonate concretions around tusk-shells over week-month timescales. *Sci. Rep.* **2015**, *5*, 14123. [CrossRef]
54. Yoshida, H.; Yamamoto, K.; Minami, M.; Katsuta, N.; Sin-ichi, S.; Metcalfe, R. Generalized conditions of spherical carbonate formation around decaying organic matter in early diagenesis. *Sci. Rep.* **2018**, *8*, 6308. [CrossRef]
55. Brocks, J.J.; Hope, J.M. Tailing of chromatographic peaks in GC-MS caused by interaction of halogenated solvents with the ion source. *J. Chromatogr. Sci.* **2014**, *52*, 471–475. [CrossRef]
56. Coplen, T.B.; Brand, W.A.; Gehre, M.; Gröning, M.; Meijer, H.A.J.; Toman, B.; Verkouteren, R.M. New Guidelines for δ^{13}C Measurements. *Anal. Chem.* **2006**, *78*, 2439–2441. [CrossRef]
57. Wiemann, J.; Crawford, J.M.; Briggs, D.E.G. Phylogenetic and physiological signals in metazoan fossil biomolecules. *Sci. Adv.* **2020**, *6*, eaba6883. [CrossRef] [PubMed]
58. Lambert, J.B.; Shurvell, H.F.; Lightner, D.A.; Cooks, R.G.; Stout, G.H. *Introduction to Organic Spectroscopy*; Macmillan: New York, NY, USA, 1987.
59. Chin, K.; Tokaryk, T.T.; Erickson, G.M.; Calk, L.C. A king-sized theropod coprolite. *Nature* **1998**, *393*, 680–682. [CrossRef]
60. Briggs, D.E.; Wilby, P.R. The role of the calcium carbonate-calcium phosphate switch in the mineralization of soft-bodied fossils. *J. Geol. Soc.* **1996**, *153*, 665–668. [CrossRef]
61. Briggs, D.E.; Kear, A.J.; Martill, D.M.; Wilby, P.R. Phosphatization of soft-tissue in experiments and fossils. *J. Geol. Soc.* **1993**, *150*, 1035–1038. [CrossRef]
62. Trueman, C.N. Rare Earth Element Geochemistry and Taphonomy of Terrestrial Vertebrate Assemblages. *Palaios* **1999**, *14*, 555–568. [CrossRef]
63. Fleet, M.E.; Pan, Y. Site preference of rare earth elements in fluorapatite. *Am. Mineral.* **1995**, *80*, 329–335. [CrossRef]
64. Suarez, C.A.; Macpherson, G.L.; González, L.A.; Grandstaff, D.E. Heterogeneous rare earth element (REE) patterns and concentrations in a fossil bone: Implications for the use of REE in vertebrate taphonomy and fossilization history. *Geochim. Cosmochim. Acta* **2010**, *74*, 2970–2988. [CrossRef]
65. Trueman, C.N.; Benton, M.J.; Palmer, M.R. Geochemical taphonomy of shallow marine vertebrate assemblages. *Palaeogeogr. Palaeoclimatol. Palaeoecol.* **2003**, *197*, 151–169. [CrossRef]
66. Trueman, C.N.; Behrensmeyer, A.K.; Potts, R.; Tuross, N. High-resolution records of location and stratigraphic provenance from the rare earth element composition of fossil bones. *Geochim. Cosmochim. Acta* **2006**, *70*, 4343–4355. [CrossRef]
67. Trueman, C.N.; Palmer, M.R.; Field, J.; Privat, K.; Ludgate, N.; Chavagnac, V.; Eberth, D.A.; Cifelli, R.; Rogers, R.R. Comparing rates of recrystallisation and the potential for preservation of biomolecules from the distribution of trace elements in fossil bones. *Comptes Rendus Palevol* **2008**, *7*, 145–158. [CrossRef]
68. Trueman, C.N.; Tuross, N. Trace Elements in Recent and Fossil Bone Apatite. *Rev. Mineral. Geochem.* **2002**, *48*, 489–521. [CrossRef]
69. Niedźwiedzki, G.; Bajdek, P.; Owocki, K.; Kear, B.P. An Early Triassic polar predator ecosystem revealed by vertebrate coprolites from the Bulgo Sandstone (Sydney Basin) of southeastern Australia. *Palaeogeogr. Palaeoclimatol. Palaeoecol.* **2016**, *464*, 5–15. [CrossRef]
70. Owocki, K.; Niedźwiedzki, G.; Sennikov, A.G.; Golubev, V.K.; Janiszewska, K.; Sulej, T. Upper Permian vertebrate coprolites from Vyazniki and Gorokhovets, Vyatkian Regional Stage, Russian Platform. *Palaios* **2012**, *27*, 867–877. [CrossRef]

71. Brocks, J.J.; Grice, K. Biomarkers (Molecular Fossils). In *Encyclopedia of Geobiology*; Reitner, J., Thiel, V., Eds.; Springer: Dordrecht, The Netherlands, 2011; pp. 147–167.

72. Volkman, J.K. Sterols in microorganisms. *Appl. Microbiol. Biotechnol.* **2003**, *60*, 495–506. [CrossRef] [PubMed]

73. Peters, K.E.; Walters, C.C.; Moldowan, J.M. Volume 2: Biomarkers and Isotopes in Petroleum Exploration and Earth History. In *The Biomarker Guide*; Cambridge University Press: Cambridge, UK, 2005.

74. Grice, K.; Klein Breteler, W.C.M.; Schouten, S.; Grossi, V.; de Leeuw, J.W.; Damsté, J.S.S. Effects of zooplankton herbivory on biomarker proxy records. *Paleoceanography* **1998**, *13*, 686–693. [CrossRef]

75. Goad, L.J. Sterol biosynthesis and metabolism in marine invertebrates. *Pure Appl. Chem.* **1981**, *53*, 837–852. [CrossRef]

76. Kanazawa, A. Sterols in marine invertabrates. *Fish. Sci.* **2001**, *67*, 997–1007. [CrossRef]

77. Prost, K.; Birk, J.J.; Lehndorff, E.; Gerlach, R.; Amelung, W. Steroid Biomarkers Revisited—Improved Source Identification of Faecal Remains in Archaeological Soil Material. *PLoS ONE* **2017**, *12*, e0164882. [CrossRef]

78. Nabbefeld, B.; Grice, K.; Schimmelmann, A.; Summons, R.E.; Troitzsch, U.; Twitchett, R.J. A comparison of thermal maturity parameters between freely extracted hydrocarbons (Bitumen I) and a second extract (Bitumen II) from within the kerogen matrix of Permian and Triassic sedimentary rocks. *Org. Geochem.* **2010**, *41*, 78–87. [CrossRef]

79. Van Kaam-Peters, H.M.E.; Köster, J.; Van Der Gaast, S.J.; Dekker, M.; De Leeuw, J.W.; Sinninge Damsté, J.S. The effect of clay minerals on diasterane/sterane ratios. *Geochim. Cosmochim. Acta* **1998**, *62*, 2923–2929. [CrossRef]

80. Peakman, T.M.; De Leeuw, J.W.; Rijpstra, C. Identification and origin of $\Delta^{8(14)}5\alpha$- and $\Delta^{14}5\alpha$-sterenes and related hydrocarbons in an immature bitumen from the Monterey Formation, California. *Geochim. Cosmochim. Acta* **1992**, *56*, 1223–1230. [CrossRef]

81. Gagosian, R.B.; Farrington, J.W. Sterenes in surface sediments from the southwest African shelf and slope. *Geochim. Cosmochim. Acta* **1978**, *42*, 1091–1101. [CrossRef]

82. Brassell, S.C.; Murchison, D.G.; Mason, R.; Durand, B.; Eglinton, G.; Comet, P.A.; Curtis, C.D.; Bada, J.; de Leeuw, J.W. Molecular Changes in Sediment Lipids as Indicators of Systematic Early Diagenesis [and Discussion]. *Philos. Trans. R. Soc. Lond. Ser. A Math. Phys. Sci.* **1985**, *315*, 57–75.

83. Rubinstein, I.; Sieskind, O.; Albrecht, P. Rearranged sterenes in a shale: Occurrence and simulated formation. *J. Chem. Soc. Perkin Trans. 1* **1975**, *19*, 1833–1836. [CrossRef]

84. Hussler, G.; Albrecht, P. C27–C29 Monoaromatic anthrasteroid hydrocarbons in Cretaceous black shales. *Nature* **1983**, *304*, 262–263. [CrossRef]

85. Hussler, G.; Chappe, B.; Wehrung, P.; Albrecht, P. C27–C29 ring A monoaromatic steroids in Cretaceous black shales. *Nature* **1981**, *294*, 556–558. [CrossRef]

86. Schüpfer, P.; Finck, Y.; Houot, F.; Gülaçar, F.O. Acid catalysed backbone rearrangement of cholesta-2,4,6-triene: On the origin of ring A and ring B aromatic steroids in recent sediments. *Org. Geochem.* **2007**, *38*, 671–681. [CrossRef]

87. Riolo, J.; Hussler, G.; Albrecht, P.; Connan, J. Distribution of aromatic steroids in geological samples: Their evaluation as geochemical parameters. *Org. Geochem.* **1986**, *10*, 981–990. [CrossRef]

88. Mackenzie, A.S. Organic Reactions as Indicators of the Burial and Temperature Histories of Sedimentary Sequences. *Clay Miner.* **1984**, *19*, 271–286. [CrossRef]

89. Ling, Y.C.; Moreau, J.; Berwick, L.; Tulipani, S.; Grice, K.; Bush, R. Distribution of iron- and sulfate-reducing bacteria across a coastal acid sulfate soil (CASS) environment: Implications for passive bioremediation by tidal inundation. *Front. Microbiol.* **2015**, *6*, 624. [CrossRef]

90. Gaskell, S.J.; Eglinton, G. Sterols of a contemporary lacustrine sediment. *Geochim. Cosmochim. Acta* **1976**, *40*, 1221–1228. [CrossRef]

91. Cranwell, P.A. Lipid geochemistry of sediments from Upton Broad, a small productive lake. *Org. Geochem.* **1984**, *7*, 25–37. [CrossRef]

92. Ficken, K.J.; Li, B.; Swain, D.L.; Eglinton, G. An *n*-alkane proxy for the sedimentary input of submerged/floating freshwater aquatic macrophytes. *Org. Geochem.* **2000**, *31*, 745–749. [CrossRef]

93. Schouten, S.; Klein Breteler, W.C.M.; Blokker, P.; Schogt, N.; Rijpstra, W.I.C.; Grice, K.; Baas, M.; Damsté, J.S.S. Biosynthetic effects on the stable carbon isotopic compositions of algal lipids: Implications for deciphering the carbon isotopic biomarker record. *Geochim. Cosmochim. Acta* **1998**, *62*, 1397–1406. [CrossRef]

94. Summons, R.E.; Jahnke, L.L.; Roksandic, Z. Carbon isotopic fractionation in lipids from methanotrophic bacteria: Relevance for interpretation of the geochemical record of biomarkers. *Geochim. Cosmochim. Acta* **1994**, *58*, 2853–2863. [CrossRef]

95. Bojanowski, M.J.; Clarkson, E.N.K. Origin of Siderite Concretions in Microenvironments of Methanogenesis Developed in a Sulfate Reduction Zone: An Exception or a Rule? *J. Sediment. Res.* **2012**, *82*, 585–598. [CrossRef]

 *biology*

Article

Soft Tissue and Biomolecular Preservation in Vertebrate Fossils from Glauconitic, Shallow Marine Sediments of the Hornerstown Formation, Edelman Fossil Park, New Jersey

Kristyn K. Voegele [1,*], Zachary M. Boles [1,2], Paul V. Ullmann [1], Elena R. Schroeter [3], Wenxia Zheng [3] and Kenneth J. Lacovara [1,2]

[1] Department of Geology, Rowan University, Glassboro, NJ 08028, USA; bolesz@rowan.edu (Z.M.B.); ullmann@rowan.edu (P.V.U.); lacovara@rowan.edu (K.J.L.)
[2] Jean and Ric Edelman Fossil Park, Rowan University, Mantua Township, NJ 08080, USA
[3] Department of Biological Sciences, North Carolina State University, Raleigh, NC 27695, USA; easchroe@ncsu.edu (E.R.S.); wzheng2@ncsu.edu (W.Z.)
* Correspondence: voegele@rowan.edu

Simple Summary: Original organics and soft tissues are known to persist in the fossil record. To date, these discoveries derive from a limited number of ancient environments, (e.g., rivers, floodplains), and fossils from rarer environments remain largely unexplored. We studied Cretaceous–Paleogene fossils from a peculiar marine environment (glauconitic greensand) from Jean and Ric Edelman Fossil Park in Mantua Township, NJ. Twelve samples were demineralized in acid to remove the mineral component of bone. This treatment frequently yielded products that are visually consistent with bone cells, blood vessels, and bone matrix from modern animals. Fossil specimens that are dark in color exhibit excellent microscopic bone preservation and yielded a greater recovery of original soft tissues, whereas light-colored specimens exhibit poor microscopic preservation and yielded few to no soft tissues. Additionally, a well-preserved femur of a marine crocodilian was found to retain original bone protein by reactions with antibodies. Our results: (1) corroborate previous findings that original soft tissue and proteins can be recovered from fossils preserved in marine environments, and (2) expand the range of ancient environments documented to preserve original organics and soft tissues. This broadens the suite of fossils that may be fruitful to examine in future paleomolecular studies.

Citation: Voegele, K.K.; Boles, Z.M.; Ullmann, P.V.; Schroeter, E.R.; Zheng, W.; Lacovara, K.J. Soft Tissue and Biomolecular Preservation in Vertebrate Fossils from Glauconitic, Shallow Marine Sediments of the Hornerstown Formation, Edelman Fossil Park, New Jersey. *Biology* **2022**, *11*, 1161. https://doi.org/10.3390/biology11081161

Academic Editor: Zhifei Zhang

Received: 18 June 2022
Accepted: 22 July 2022
Published: 2 August 2022

Publisher's Note: MDPI stays neutral with regard to jurisdictional claims in published maps and institutional affiliations.

Copyright: © 2022 by the authors. Licensee MDPI, Basel, Switzerland. This article is an open access article distributed under the terms and conditions of the Creative Commons Attribution (CC BY) license (https://creativecommons.org/licenses/by/4.0/).

Abstract: Endogenous biomolecules and soft tissues are known to persist in the fossil record. To date, these discoveries derive from a limited number of preservational environments, (e.g., fluvial channels and floodplains), and fossils from less common depositional environments have been largely unexplored. We conducted paleomolecular analyses of shallow marine vertebrate fossils from the Cretaceous–Paleogene Hornerstown Formation, an 80–90% glauconitic greensand from Jean and Ric Edelman Fossil Park in Mantua Township, NJ. Twelve samples were demineralized and found to yield products morphologically consistent with vertebrate osteocytes, blood vessels, and bone matrix. Specimens from these deposits that are dark in color exhibit excellent histological preservation and yielded a greater recovery of cells and soft tissues, whereas lighter-colored specimens exhibit poor histology and few to no cells/soft tissues. Additionally, a well-preserved femur of the marine crocodilian *Thoracosaurus* was found to have retained endogenous collagen I by immunofluorescence and enzyme-linked immunosorbent assays. Our results thus not only corroborate previous findings that soft tissue and biomolecular recovery from fossils preserved in marine environments are possible but also expand the range of depositional environments documented to preserve endogenous biomolecules, thus broadening the suite of geologic strata that may be fruitful to examine in future paleomolecular studies.

Keywords: soft tissues; molecular preservation; collagen; Hornerstown Formation; shallow marine; glauconite

1. Introduction

Numerous molecular paleontological investigations have been conducted on geologically ancient fossils, (i.e., >1 Ma years old; [1]), many of which have demonstrated strong evidence for the preservation of endogenous biomolecules and soft tissues in deep time. These studies have yielded the recovery of cellular and tissue structures morphologically similar to their extant counterparts, (e.g., [2–14]), identified the presence of proteins in fossil soft cells and soft tissues via a variety of techniques, (e.g., [4–6,10,15–20]), and even recovered protein sequences from mass spectrometry, (e.g., [4,21–24]). The wealth of such discoveries, and the number of fossil taxa and preservational environments investigated and found to yield soft tissue and biomolecular preservation continues to grow. Initially, only terrestrial environments were investigated, as it was thought that hydrolysis from continuous exposure to water would be inconducive to molecular preservation [25,26]. More recently, a few studies examined aquatic taxa or fossil specimens preserved in marine environments, (e.g., [6,16,27]). However, the number of geologic formations and types of host lithologies, and, therefore, the types of depositional paleoenvironments investigated, still remain limited.

To date, the majority of marine specimens that have been analyzed derive from siliciclastic deposits of various geologic ages (Triassic to Neogene), degree of consolidation (unconsolidated vs. consolidated), and grain size (see Table 1). A few samples have also been examined from shallow marine limestones [16] and chalk [6]. Wiemann et al. [10] also analyzed an indeterminate crocodilian specimen from glauconite containing sands of the Cretaceous Navesink Formation. Only this crocodilian, one of the *Nothosaurus* specimens examined by Surmik et al. [16], and the *Ichthyosaurus* specimen examined by Wiemann et al. [10] were reported to not yield preserved soft tissues. Surmik et al. [16] and Lindgren et al. [6] investigated a suite of specimens from marine environments further for the identification of proteins and found organic compounds, including fragments of amino acids, and endogenous collagen I, respectively. These studies demonstrate that marine fossils from various environments can also preserve soft tissues and biomolecules.

Herein, we examine marine fossils collected from the Jean and Ric Edelman Fossil Park (EFP) at Rowan University in Mantua Township, New Jersey. All specimens were recovered from the Late Cretaceous (Maastrichtian)–early Paleogene (Danian) Hornerstown Formation. All but two of the examined specimens derive from the same assemblage: the Main Fossiliferous Layer (MFL). The MFL is a rich, 10 cm thick, multitaxic bonebed that begins approximately 20 cm above the base of the Formation [28]. As in the prior paleomolecular studies of marine bones discussed above, the Hornerstown Formation is a siliciclastic deposit that was laid down on an organic-rich, shallow-marine shelf [29]. Specifically, the Hornerstown Formation is comprised entirely of heavily bioturbated, unconsolidated, glauconitic greensands [29,30]. It contains comparatively more of this iron- and phosphate-rich, diagenetic mineral, glauconite, than the underlying Navesink Formation, from which a crocodilian specimen was investigated by Wiemann et al. [10]. The wealth of vertebrate fossils in the MFL within the Hornerstown Formation [29,31] allowed us to further investigate the possibility of soft tissue and biomolecular preservation in glauconite-forming shallow marine environments. Additionally, the multitaxic character of the MFL assemblage permitted analysis of specimens from several clades [28], thus allowing depositional and diagenetic effects to be controlled for when making comparisons among taxa. Voegele et al. [28] also demonstrated that the diversity and abundance of fossils within the MFL, as well as their stratigraphic distribution and taphonomic attributes, indicate that the MFL likely represents a mass-death assemblage, meaning that time averaging among specimens is minimal (also see [30]).

We considered the Hornerstown Formation as a lithosome potentially favorable for the preservation of endogenous organics because glauconite is rich in iron, which has been suggested to aid in molecular preservation. Specifically, it has been hypothesized that dissolved iron in diagenetic pore fluids reacts with peroxides sourced from decaying lipids to form iron free radicals, which in turn can induce chemical chain reactions resulting in crosslinking of biomolecules, their decay products, metal cations in solution, and dissolved humics [3,32,33]. Therefore, we examined twelve specimens of five taxa from the MFL and higher within the Formation for soft tissue and biomolecular preservation. This included analyzing one specimen of the marine crocodile *Thoracosaurus* for the preservation of the primary structural protein collagen I using molecular assays.

Table 1. Summary of fossil specimens collected from marine sediments that have previously been examined for cellular and/or soft tissue preservation by demineralization.

Citation	Specimen Number	Taxon	Age	Formation	Rock Type
Barker et al. [34]	PAS11-02	Cetacea indet.	Miocene	Pungo River	Unconsolidated Silt and Clay
Barker et al. [34]	PAS11-03	Cetacea indet.	Miocene	Pungo River	Unconsolidated Silt and Clay
Barker et al. [34]	PAS11-01	Cetacea indet.	Miocene	Calvert	Unconsolidated Silt and Clay
Barker et al. [34]	PAS11-04	Dugongidae indet.	Miocene	Torreya	Unconsolidated Silt and Clay
Cadena and Schweitzer [35]	CCMFC01	Pelomedusoides indet.	Paleogene	Los Cuervos	Sandstone
Lindgren et al. [6]	IRSNB 1624	*Prognathodon*	Cretaceous	Ciply	Phosphatic Chalk
Macauley et al. [36]	RU-CAL-00001	Testudines indet.	Miocene	Calvert	Unconsolidated Silt and Clay
Schweitzer et al. [3]	WCBa-20	*Balaenoptera*	Miocene	Pisco	Mudstone
Surmik et al. [16]	WNoZ/s/7/166	*Nothosaurus*	Triassic	Gogolin	Limestone
Surmik et al. [16]	SUT-MG/F/Tvert/15	*Nothosaurus*	Triassic	Gogolin	Limestone
Surmik et al. [16]	SUT-MG/F/Tvert/2	*Protanystropheus*	Triassic	Gogolin	Limestone
Wiemann et al. [10]	YPM 656	Crocodylia indet.	Cretaceous	Navesink	Unconsolidated Greensand
Wiemann et al. [10]	YMP VP 059362	*Ichthyosaurus*	Jurassic	Posidonia Shale	Mudstone

2. Materials and Methods

2.1. Materials

Fossil specimens examined herein were collected from the Hornerstown Formation between 2011 and 2019; most were collected as part of a systematic grid excavation of the MFL [37]. In the field, all but one sample (*Thoracosaurus* scute RU-EFP-00006-8, drawn from collections) were set aside immediately after discovery for molecular analyses by wrapping each in sterilized aluminum foil and placing them in autoclave-sterilized mason jars with silica gel desiccant beads. Collection was performed while wearing nitrile gloves to limit contamination, and specimens were stored in climate-controlled buildings, thus minimizing temperature and humidity fluxes after collection. Sediment that was adjacent to, but not in contact with, each fossil was collected for use as a negative control and stored as above. To avoid the possibility that glue residues could be mistaken for biological structures, no glues or stabilizing agents were used during either excavation or preparation of specimens. Eight specimens were recovered from the MFL and two were recovered from the upper Hornerstown Formation. Taxonomically, the specimens derive from the turtles *Euclastes*, *Taphrosyphs*, and two other unidentified turtles, the gavialoid crocodilian *Thoracosaurus*, and another unidentified crocodilian. Further details about each specimen are provided in Table 2.

Table 2. An overview of the 12 samples analyzed in this study. Abbreviation: Fm = Formation.

Specimen RU-EFP-#	Taxon	Element	Source Horizon/Fm	Osteocyte Recovery	Vessel Recovery	Fibrous Matrix Recovery	Color	Histology Assessed
00002-2	*Taphrosphys*	Peripheral	MFL	Abundant	Frequent	Frequent	Thin Light Rim	Yes
00006-8	*Thoracosaurus*	Scute	MFL	Frequent	Absent	Rare	Thin Light Rim	Yes
00006-11	*Thoracosaurus*	Femur	MFL	Abundant	Rare	Uncommon	Thin Light Rim	Yes
00018	*Euclastes*	Costal	MFL	Abundant	Absent	Absent	Dark	Yes
00030-1	Crocodylia Indet.	Tibia	MFL	Absent	Absent	Absent	Light	Yes
00030-2	Crocodylia Indet.	Humerus	MFL	Rare	Absent	Absent	Medium Color	No
02175	*Taphrosphys*	Peripheral	MFL	Rare	Absent	Absent	Light	Yes
02222	*Taphrosphys*	Costal	MFL	Abundant	Absent	Absent	Dark	Yes
02245	Pan- Cheloniidae Indet.	Peripheral	MFL	Rare	Absent	Absent	Thin Light Rim	Yes
02295	Testudines Indet.	Femur	MFL	Abundant	Absent	Absent	Dark	No
04161-8	Testudines Indet.	Zeugopodial	Hornerstown	Abundant	Rare	Uncommon	Dark	No
04162	*Taphrosphys*	Costal	Hornerstown	Abundant	Absent	Absent	Dark	No

represents the generic place holder for the numbers above.

Specimen RU-EFP-00006 is a partially articulated skeleton of *Thoracosaurus neocesariensis*. The right femur (Figure 1A) was chosen to analyze for collagen I remnants because this bone is generally well-preserved in terms of gross morphology (Figure S1). Additionally, being a large limb bone, it possesses a greater amount of cortical tissue for analyses. These attributes imply this fossil could be a favorable candidate for molecular retention [25,38,39]. Stylopodial and zeugopodial elements of a juvenile American alligator (*Alligator mississippiensis*) served as a modern positive control for biomolecular assays (see Supp for treatment of extant bones prior to analyses below). Both modern and fossil molecular analyses were conducted with solely cortical tissue.

2.2. Methods

All fossil analyses were performed in a permanently dedicated, fossil-only laboratory at North Carolina State University (NCSU; see Schroeter et al. [24] for additional details on lab sterilization protocols). For the biomolecular analyses, modern control trials were conducted in a separate lab at NCSU. Over the duration of this project, demineralization was carried out by separate project personnel in three separate labs: one sample (*Thoracosaurus* femur RU-EFP-00006-11) was demineralized at NCSU, seven samples were demineralized in a designated, sterilized, fossil-only chemical fume hood at Drexel University, and four were demineralized in a similarly dedicated, sterilized chemical fume hood at Rowan University. Negative controls, which included buffer-only solutions and sediments demineralized and/or co-extracted in the same reagents, were evaluated in tandem with the fossil for all assays. For immunoassays, specificity controls were also conducted (see below). Procedures for all controls and modern/fossil samples were performed in exactly the same manner unless otherwise noted. Three replicates were completed of each assay.

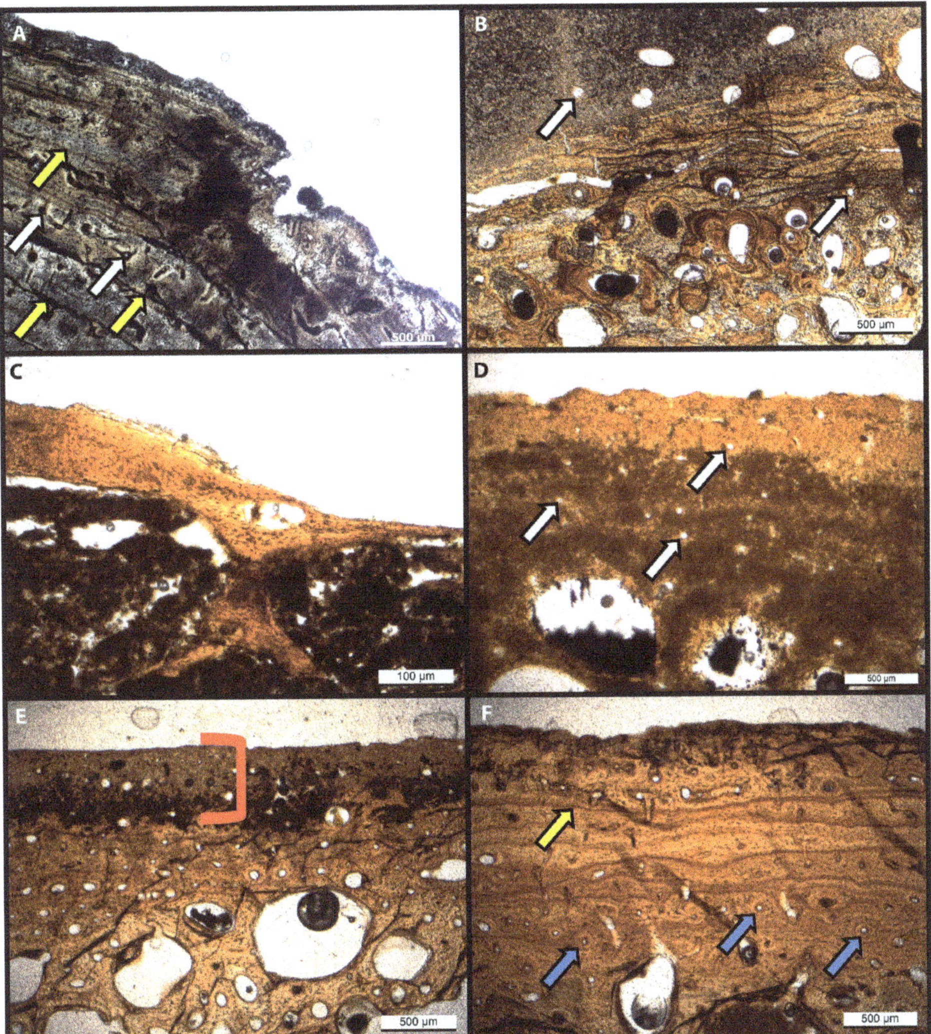

Figure 1. Thin sections of specimens used for demineralization illustrate the varying degrees of bone microstructure preservation. (**A**) *Thoracosaurus* femur (RU-EFP-00006-11) femur showing well-preserved bone microstructure including Sharpey's fibers (Histologic Index [HI] = 5). (**B**) Indeterminate crocodilian tibia (RU-EFP-00030) showing little bone microstructure preserved (HI = 1). (**C**) *Euclastes* costal prong (RU-EFP-00018) showing excellent preservation (HI = 5). (**D**) *Taphrosphys* peripheral (RU-EFP-02175) showed very poorly preserved microstructure (HI in this region = 1; HI for the entire thin section = 2). (**E**) Indet. Pan-Cheloniid peripheral (RU-EFP-02245) showed well preserved internal cortex but a highly altered outer cortex (HI = 4). (**F**) Juvenile *Taphrosphys* costal (RU-EFP-02222) with well-preserved microstructure (HI of this region and the entire thin section = 4–5). White arrows indicate vascular canals, yellow arrows indicate lines of arrested growth, and blue arrows indicate primary osteons. The red bracket corresponds to external rims of light color and poor histological integrity.

Collagen I, a structural protein abundant in bone tissue, was selected as the primary target for biomolecular investigation in this study because of its high preservation potential.

Many factors contribute to this high preservation potential, including: (1) that it is the most abundant protein in bone [40–43]; (2) its primary, secondary, and tertiary structures make it highly stable [25,44,45] and resistant to many proteases [46], and (3) its intimate association with hydroxyapatite crystallites within the tissue structure of bone, (e.g., [25,40,41,47,48]).

2.2.1. Histology

Standard petrographic thin sections were prepared for histological analysis (see [49]). Histological analysis was performed on eight of the twelve specimens that were demineralized (three crocodilian and five turtle samples; see Table 2 for specimen specifics) to acquire a representative dataset of histological variability among specimens from EFP. Bone fragments were embedded in Silmar 41 resin (U.S. Composites), thick sectioned, and mounted to frosted glass slides with Loctite® Heavy Duty 60 Minute epoxy (Westlake, OH, USA). Thin sections were then ground to an appropriate thickness (~100 µm) and polished. The Histological Index (HI) of Hedges and Millard [42] was used to qualify the degree of microstructural preservation.

2.2.2. X-ray Diffraction (XRD)

XRD analyses were conducted using a Phillips X'Pert diffractometer (#DY1738) at the Department of Earth and Environmental Science at the University of Pennsylvania. Approximately 1–2 g of two fossil bone samples (one light tan, one dark purple-brown) and Hornerstown Formation sediment were each mechanically powderized to less than 10 µm in a SPEX tungsten carbide Mixer-Mills (model #8000). Each powdered sample was then loaded into a sample holder and analyzed using Cu Kα radiation (λ = 1.54178A°) and operating at 45 kV and 40 mA. Diffraction patterns were measured from 5–75° 2θ with a step size of 0.017° 2θ and 1.3 s per step (=0.77 degrees per minute). Phillips proprietary software HighScore Plus v. 3.0e was used to interpret the resulting diffraction patterns.

2.2.3. Demineralization

A roughly 0.5 cm^3 fragment of each fossil was initially submerged in 0.5 M disodium ethylenediaminetetraacetic acid (EDTA) to chelate calcium. This solution was changed every other day for approximately four weeks, then weekly for another 1–2 months, as needed. In some cases, demineralization of fossils from this locality was found to occur at a drastically slower rate than we have previously encountered for other similarly aged specimens, including others from marine sediments [11,34,36]. After 1–2 months in EDTA, if demineralization was still mostly ineffective, then as necessary samples were demineralized by acid dissolution in 0.2 M or 0.5 M hydrochloric acid (HCl) for several days to two weeks, with solution changes made every 1–2 days, depending on the resilience of the fossil bone matrix. Demineralization products were transferred to glass slides and visualized using a Zeiss Stemi 2000-C reflected light microscope (Oberkochen, Germany) with a linked AxioCam 506 camera (Zeiss, Oberkochen, Germany). Abundances of osteocytes, vessels, and fibrous matrix fragments from each specimen were assigned to five relative categories based on Ullmann et al. [11]: absent, rare, uncommon, frequent, and abundant. Modern alligator and Hornerstown Formation sediment samples were also demineralized following the same procedures, to serve as positive and negative controls, respectively.

2.2.4. Scanning Electron Microscopy (SEM) and Energy Dispersive X-ray Spectroscopy (EDX)

Osteocytes isolated by demineralization in 0.2 M HCl from RU-EFP-00006-11 were collected using a 1 µm filter (Millipore), which was then placed on an aluminum stub and allowed to dry. The resulting uncoated sample was imaged using a field emission-scanning electron microscope (FE-SEM Zeiss Supra 50VP, Oberkochen, Germany) at Drexel University, operating at a working distance of 6.6 mm and at an accelerating voltage of 1 kV. Elemental spot analyses were collected as a standardless assay using a coupled Oxford model 7430 EDS INCAx microanalyzer (Oxford Instruments, Abingdon, UK).

2.2.5. Protein Extraction

Herein, we used the same protein extraction protocol as Ullmann et al. [20]. In brief, ground bone and sediments, along with a buffer/"blank" control of only the extraction solutions, were each separately extracted with 0.6 M HCl. The resulting supernatants were collected (here called the "HCl fraction"), then the remaining pellet was incubated in 0.05 M ammonium bicarbonate (AMBIC). The resulting supernatants were then also collected (here called the "AMBIC fraction"). HCl fractions were precipitated with trichloroacetic acid (TCA) to form pellets, which were then allowed to dry in a laminar flow hood. AMBIC fractions were dried in a speed vacuum. Samples were stored at either 4 °C, −20 °C, or −80 °C depending on the length of time until analysis. Concentrations are reported based on amounts of pre-extracted bone rather than absolute masses because post-extraction yields from this protocol contain salts from extraction buffers in addition to varying amounts of protein [50]. Fresh *Alligator* cortical bone was extracted in the same manner as the fossil using the same reagents and the same protocol, but in a separate lab at NCSU.

2.2.6. Polyacrylamide Gel Electrophoresis (PAGE) with Silver-Staining

To test for the presence of organics within the extracts from *Thoracosaurus* femur RU-EFP-00006-11, we followed the silver-staining protocols of Zheng and Schweitzer [51], as also detailed in Ullmann et al. [20], but without treatment with the iron chelator pyridoxal isonicotinic hydrazide as it was not found to be necessary (refer to the Supplementary Materials for details).

2.2.7. Enzyme-Linked Immunosorbant Assay (ELISA)

Herein, we used the same ELISA protocol as Ullmann et al. [20]. This included plating of AMBIC extracts at a concentration equivalent to 200 mg of pre-extracted bone per well, blocking of non-specific binding, and incubation with rabbit anti-*Alligator* purified-skin collagen antibodies. Extracts from an extant *Alligator* were analyzed using the same protocols (in a separate lab at NCSU) as a modern control (see the Supplementary Materials for further details).

2.2.8. Immunofluorescence

We followed the fossil tissue embedding and immunofluorescence procedures of Schweitzer et al. [4,15], which are detailed in Zheng and Schweitzer [51]. This included embedding of RU-EFP-00006-11 demineralization products in LR White™ resin (Electron Microscopy Sciences, Hatfield, PA, USA), gathering of 200 nm sections with an ultramicrotome, and incubation with rabbit anti-chicken collagen I antibodies. Inhibition and collagenase digestion specificity controls, as well as a modern *Alligator* control, were also conducted (see the Supplementary Materials for further details).

3. Results

3.1. Histology

Histological investigation of the eight sectioned specimens revealed variable preservation of original bone microstructure among and within samples. The gross morphology of all fossil specimens examined herein is well preserved. However, the external surface of nearly half of the specimens is light in color while that of the other half is dark (Table 2). Variation in color even occurs among bones of the same individuals, such as crocodile RU-EFP-00030 (Table 2). Areas of disrupted microstructure correspond to areas of bone that are light in color under gross morphological investigation, which comprise varying amounts of the total bone thickness/volume (see below). This light alteration is common in fossil bones from this locality and has been previously noted by others ([52]). In long bones, these areas of discoloration occur on the outer cortical layer, (e.g., *Thoracosaurus* femur RU-EFP-00006-11; Figure 2) or extend the entire way across the cortex, (e.g., crocodile tibia RU-EFP-00030-1). Tabular-shaped bones exhibit the same pattern of surficial discoloration and disrupted microstructure; as in long bones, these alteration effects may be restricted to near the cortical margin, (e.g., *Thoracosaurus* scute RU-EFP-00006-8) or pervasive through the entire cross-section of a specimen, (e.g., *Taphrosphys* peripheral RU-EFP-02175). Within these discolored regions, occasional Wedel tunnels are present and the original microstructure is completely or almost completely obliterated, (e.g., Figure 1B,D). In contrast, the darker-colored areas of fossil bones exhibit well-preserved microstructure, with primary and secondary osteons, lines of arrested growth, and osteocyte lacunae clearly visible, (e.g., Figure 1A,C,F).

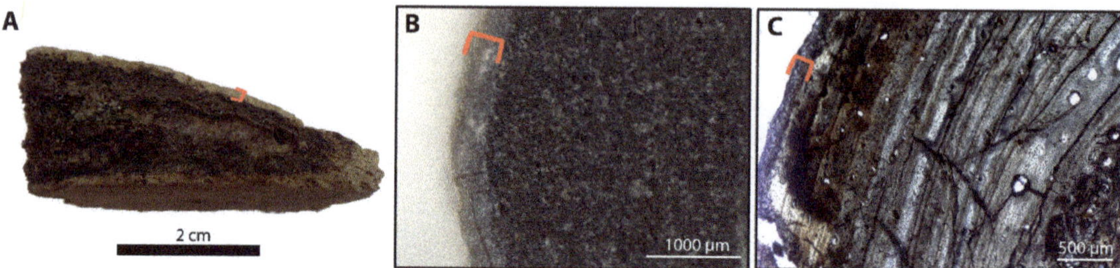

Figure 2. Examples of the two bone colors observed in specimens from EFP: an outer rim of light-colored bone (indicated by red brackets) and a darker interior. (**A**) Hand sample of a bone fragment from EFP. Note the light-coloration of the lower surface of the bone (directly above the scale bar). (**B,C**) *Thoracosaurus neocesariensis* femur (RU-EFP-00006-11) in (**B**) thick section and (**C**) thin section. Images (**B,C**) are not from the same position along the circumference of this bone, and the light-colored rim is not a consistent thickness along the entire circumference. Scale bars as indicated.

The microstructure of RU-EFP-00006-11 is well preserved (HI = 5) except for its outermost 50–250 μm (Figures 1A and 2). In this outer rim, the bone is light in color and the microstructure is completely obliterated (HI = 0). A few Wedl tunnels extend from this poorly preserved rim into the underlying, darker, better-preserved cortex (Figure S2A). The rest of the 5 mm of total cortical thickness is dark in color and well-preserved histologically. The outer cortex is composed of lamellar-zonal bone with at least 17 lines of arrested growth present. The innermost cortex is composed of compact coarse-cancellous bone. Osteocyte lacunae with short branching canaliculi are visible throughout the dark regions of the cortex. Small, longitudinal vascular canals are also present and in some instances are partially infilled with authigenic gypsum and pyrite.

Microstructure preservation among specimens varied, with some being well preserved, (e.g., RU-EFP-00006-11) and others poorly preserved (HI = 1 or 2). Preservation at the microscopic level does not correlate with macroscopic preservation, (i.e., well-preserved gross morphology does not always equate to high histologic integrity) or cortical thickness of a specimen. Specimens RU-EFP-02245, RU-EFP-00006-8, and RU-EFP-00002-2 exhibit a thin rim of this lighter, histologically degraded bone. In these specimens, as in RU-EFP-00006-11, the remainder of the cortex is dark in color and retains high histologic integrity. In contrast, crocodile tibia RU-EFP-00030-1 and *Taphrosphys* peripheral RU-EFP-02175 are entirely light in color with poor microstructure throughout, including large areas of bone tissue that appear completely obliterated by microbial destructions (possibly microscopical focal destructions [MFDs]; Figure S2B; [53]). These alterations are invisible to the naked eye; the gross morphology of RU-EFP-00030-1 is well-preserved with little evidence of bioerosion or damage to the periosteal surface to indicate its microstructure is destroyed. Histological analysis has yet to be conducted on crocodile humerus RU-EFP-00030-2, but thick sections of this bone (made for another study) exhibit an overall moderate level of discoloration between that of the light bone with poor microstructures and dark well-preserved bone seen in other specimens. This is in contrast to the costal prong of RU-EFP-00018, which is dark throughout even though it has a very thin external cortex and an expansive cancellous region.

3.2. X-ray Diffraction

The diffractogram of a sample of RU-EFP-00006-11 identifies it to be composed of fluorapatite (Figure 3A), indicating it has been minorly altered from its original hydroxyapatite composition. Based on these results, combined with the generally well-preserved gross morphology and histology of RU-EFP-00006-11 and its production of relatively abundant demineralization products, we hypothesized that this specimen could be a favorable candidate for biomolecular analyses. The diffractogram of a light-colored, indeterminate crocodilian bone fragment from the upper Hornerstown Formation reveals that altered bone is also composed of fluorapatite. Both bone samples were distinct from Hornerstown Formation sediments, which were identified as glauconite.

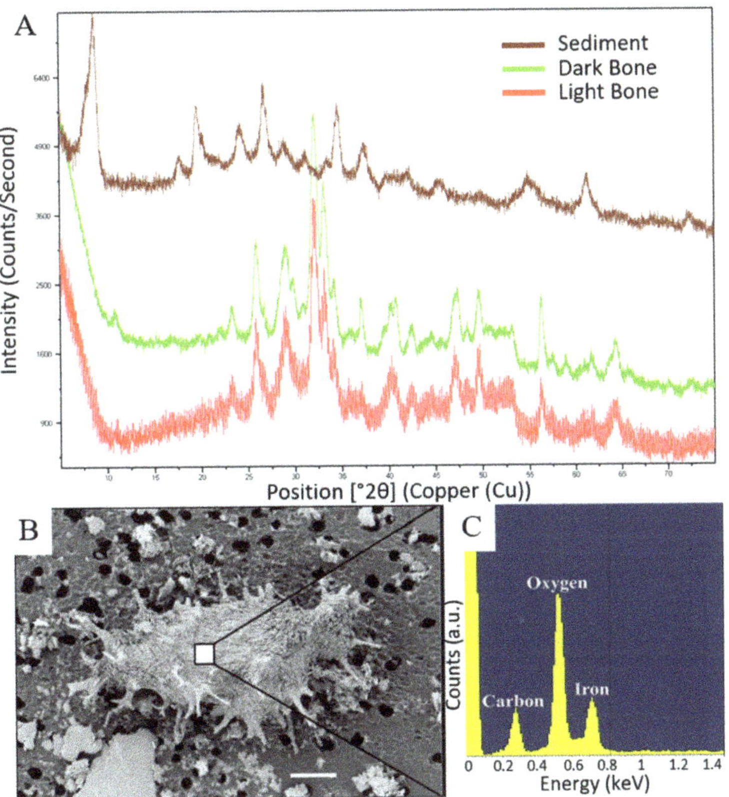

Figure 3. Mineralogical analysis of sediment, dark bone, and light bone samples from EFP and SEM-EDS of a structure morphologically consistent with an osteocyte from RU-EFP-00006-11. (**A**) Diffractogram of XRD results showing distinct differences between the sediment and fossils, which were identified as glauconite and fluorapatite, respectively; (**B**) SEM micrograph of a structure morphologically consistent with an osteocyte; (**C**) EDS spectrum from center of osteocyte (square area) in (**B**). The large peak surrounding 0 keV is an artifact of the very low count rates from the osteocyte. Scale bar 4 μm in (**B**).

3.3. Demineralization

A few bone fragments, (e.g., RU-EFP-00002-2) were demineralized within a month in EDTA, but most remained relatively hard and brittle after six weeks and required further treatment in HCl before they could be imaged. In general, light-colored bone fragments demineralized more quickly than darker bone fragments and would develop a comparatively more granular and crumbly texture early in the process of breaking down (Figure S3B). Both demineralizations with EDTA and HCl yielded structures morphologically consistent with osteocytes, vessels, and fibrous matrix. Although further molecular testing is needed to confirm that these structures represent original cells/tissues, we hereafter refer to them as osteocytes and vessels for brevity. Our antibody assays identified the presence of endogenous collagen I in pieces of fibrous matrix (see below), supporting the endogenous nature of these tissue fragments. Osteocytes were observed fully embedded, partially isolated, or fully isolated from mineralized bone matrix. No sediment samples yielded any structures morphologically consistent with vertebrate cells or tissues; only, small, subrounded to angular glauconite and quartz grains were observed (Figure S3A). Several forms of osteocytes were recovered from fossil bone samples, varying in both cell body shape and complexity of branching of filopodia (Figure 4I–L). Morphologically, these osteocytes correspond to the stellate and flattened oblate morphologies as defined by Cadena and Schweitzer [27]. All morphotypes retain filopodia, some of which possess tertiary, quaternary, or even more ramifications in better preserved specimens, as present in osteocytes isolated from modern bone (Figure 4A). Most recovered "osteocytes" were red-brown in color, (e.g., Figure 4I–L), but a few samples, (e.g., RU-EFP-02175, RU-EFP-04161-8) yielded primarily dark brown to black osteocytes (from both EDTA and HCl demineralization, e.g., Figure S3C).

Following the qualitative categories of Ullmann et al. [11], seven of the twelve demineralized fossils produced Abundant osteocytes, and RU-EFP-00006-8 produced Frequent osteocytes. In the four remaining samples, three (RU-EFP-00030-2, RU-EFP-02175, and RU-EFP-02245) produced only rare osteocytes, and osteocytes were absent in RU-EFP-00030-1. In fact, RU-EFP-00030-1 did not yield any cellular or soft tissue microstructures (osteocytes, vessels, or fibrous matrix). Among all the specimens tested, only three (RU-EFP-00006-11, RU-EFP-04161-8, and RU-EFP-00002-2) yielded vessels, ranging from ~20–50 μm in diameter, and in all three cases their recovery was either frequent or rare. These same three specimens, as well as RU-EFP-00006-8, also yielded frequent to the rare fibrous matrix. Unlike many EFP samples, RU-EFP-00006-11 yielded all three types of microstructures. The fibrous matrix was more plentiful when this bone was demineralized with HCl, but both HCl and EDTA yielded only a few vessels.

3.4. Scanning Electron Microscopy and Energy Dispersive X-ray Spectroscopy

SEM of microfiltered RU-EFP-00006-11 demineralization products revealed three-dimensional osteocytes with rough, fibrous surface texture over each cell body and short, branching filopodia (Figure 3B). Broken bases of filopodia appeared to exhibit brittle fractures and solid cross sections (not hollow). Regarding surface texture, similar rough, crisscrossing grooves on the surface of isolated osteocytes have previously been attributed to either the impression of collagen fibrils or microbial degradation [8,11,54]. EDX spot analyses found these microstructures to be composed of oxygen, iron, and carbon (Figure 3C), similar to the results of many other EDX studies of fossil bone demineralization products, (e.g., [8,11,55]. Under SEM, osteocytes from RU-EFP-00006-11 are indistinguishable from those of extant reptiles, (e.g., [27,56]).

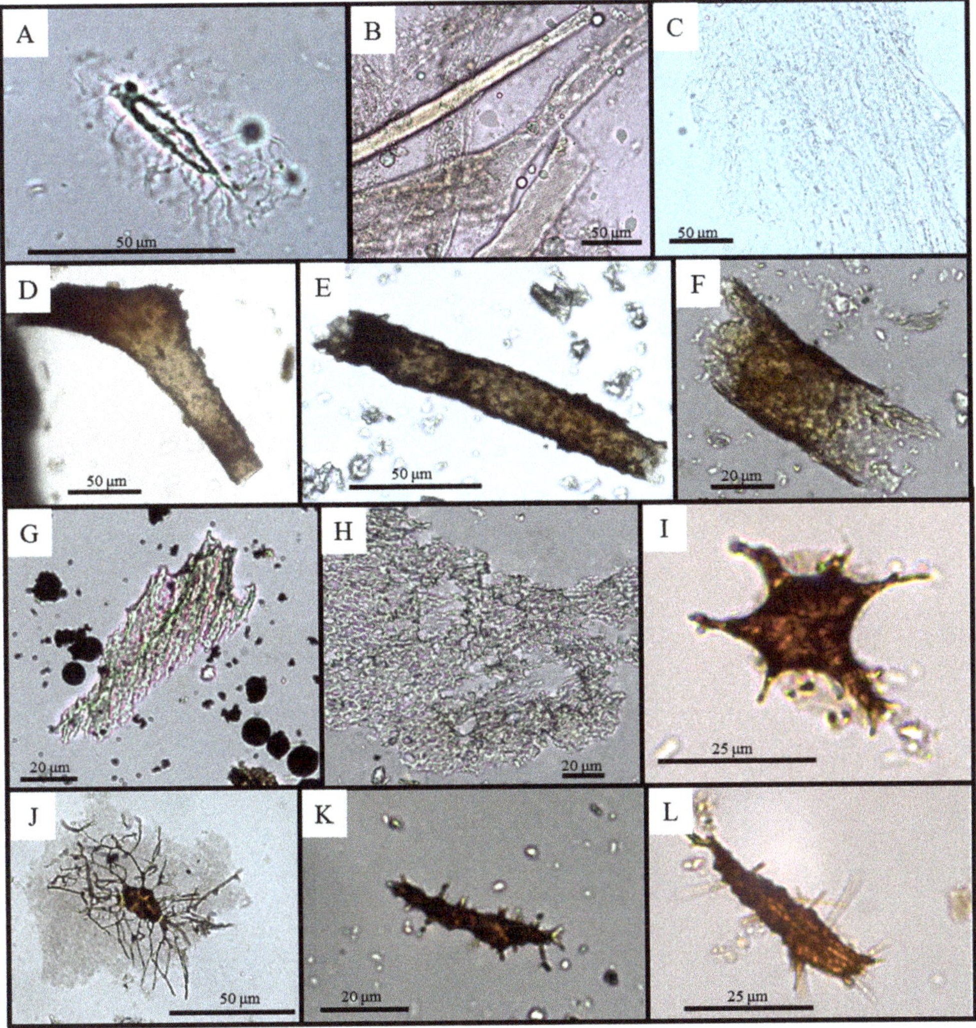

Figure 4. Modern and fossil demineralization products. (**A**) Modern American alligator osteocyte. (**B**) Alligator blood vessels. (**C**) Alligator collagen matrix. (**D,E**) *Thoracosaurus* (RU-EFP-00006-11) blood vessels. (**F**) *Taphrosphys* (RU-EFP-00002-2) blood vessel. (**G**) *Thoracosaurus* (RU-EFP-00006-8) collagen matrix. (**H**) *Taphrosphys* (RU-EFP-00002-2) collagen matrix. (**I**) *Taphrosphys* (RU-EFP-02222) stellate osteocyte. (**J**) Testudines indet. (RU-EFP-04161-8) stellate osteocyte. (**K**) *Taphrosphys* (RU-EFP-04162) flattened oblate osteocyte. (**L**) *Euclastes* (RU-EFP-00018) flattened oblate osteocyte.

3.5. Polyacrylamide Gel Electrophoresis with Silver-Staining

After electrophoresis, the movement of the AMBIC fossil sample through the gel colored the lanes a light yellow-brown (Figure 5A) prior to the application of chemical staining. Incubation in fixing solution (50% methanol) reduced but did not eliminate this "pre-staining" coloration (see Supplementary Materials). Despite this, a drastic increase in color intensity was readily apparent across the entire examined a range of molecular weights after silver-staining (Figure 5B), indicating the presence of organics within the fossil bone AMBIC extracts. "Pre-staining" coloration was decreased by resuspending the portion of the bone pellet that did not dissolve into the solution used for the fossil lane and adding it as a second sample in another lane. Resuspended bone pellet lanes exhibited little to no "pre-staining" and a clear increase in signal intensity after development with silver nitrate (Figure 5A,B), again indicative of the presence of organics within this sample.

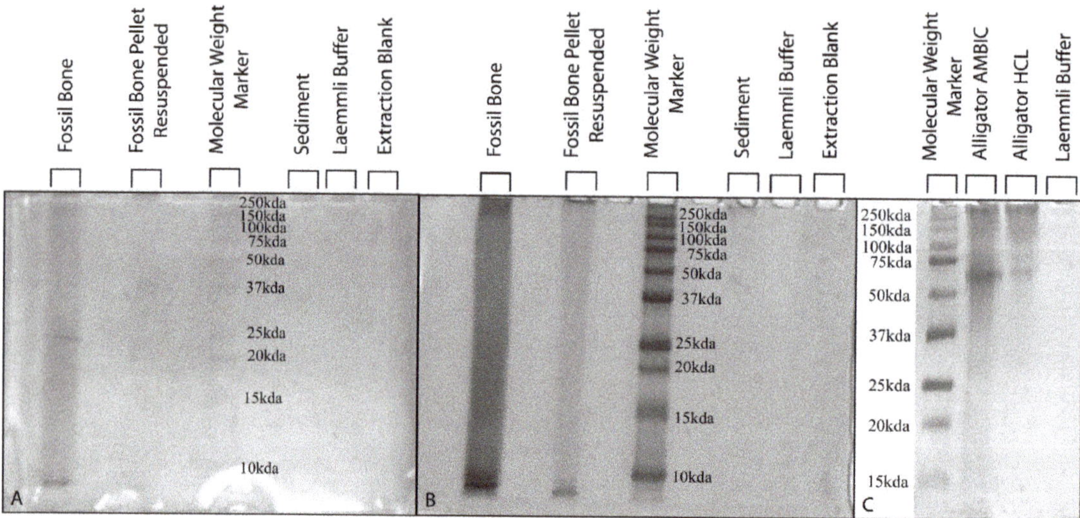

Figure 5. Polyacrylamide gel electrophoresis (PAGE) with silver-staining of fossil and modern samples. (**A**) PAGE of fossil samples prior to silver-staining, exhibiting pre-staining in the fossil bone lane. Fossil samples were loaded at 50mg of pre-extracted bone per lane; (**B**) same gel as (**A**) after development with silver nitrate. Both fossil sample lanes exhibit increased staining, whereas no binding is visible in sediment, Laemmli buffer, and extraction blank lanes; (**C**) silver-staining of modern *Alligator* samples, loaded at 20 µg/lane.

Extraction buffer blanks and Laemmli buffer controls exhibited no staining. Sediment controls exhibited weak "post-development" staining (after incubation with silver nitrate) only at the highest and lowest molecular weights examined (Figure 5). In a subset of replicates, sediment lanes exhibited a faint band near 50 kDa; no banding or increase in staining was discernible in any fossil or other control lane at this molecular weight, indicating it represents a type(s) of organics present only within the sediment controls. In separate gels, modern *Alligator* extracts imparted no "pre-staining" and yielded clear, distinct banding patterns with minimal smearing after development via silver staining (Figure 5C).

3.6. Enzyme-Linked Immunosorbent Assay

In all replicates, a positive signal for collagen I was identified in fossil AMBIC extracts by absorbance readings well over twice the background, (e.g., [57,58]). At 240 min, AMBIC extracts of RU-EFP-00006-11 reached an absorbance of 0.855 (Figure 6). At this same time point, sediment and extraction blank controls exhibited negligible absorbance. In all time point readings, the signal from the fossil bone was at least an order of magnitude higher than in all the negative controls. RU-EFP-00006-11 AMBIC extracts exhibited a significantly reduced signal relative to modern *Alligator* AMBIC extracts, which reached saturation (2.85) at ~150 min (Figure S4). At this same time point in the best replicate, the *Thoracosaurus* AMBIC sample reached an absorbance of 0.53.

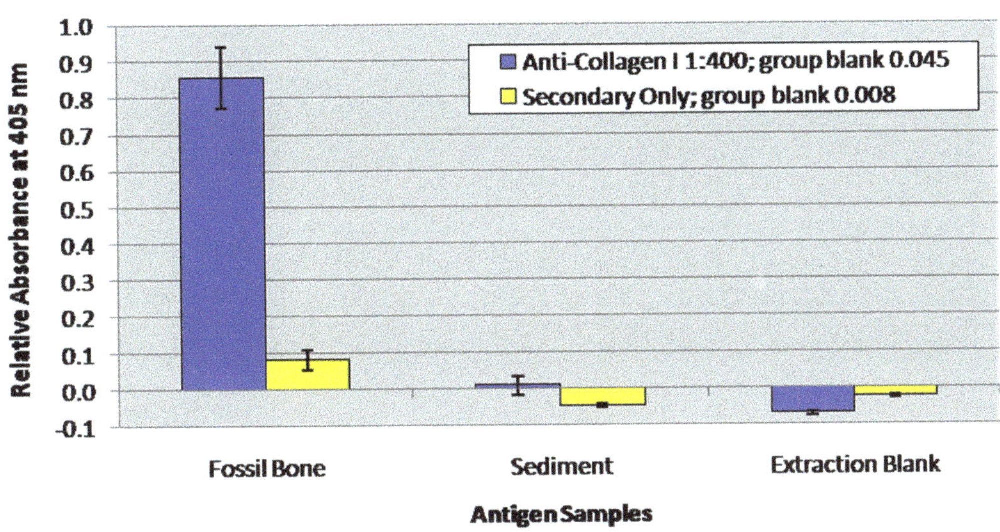

Figure 6. Enzyme-linked immunosorbent assay results of fossil and sediment chemical extracts. Extraction buffers were also run as a negative control to test for contaminants. Fossil and sediment samples were loaded at 200 mg of pre-extracted weight. Blue columns represent absorbance at 405 nm at 240 min, with incubation in anti-chicken collagen I antibodies at a concentration of 1:400. Yellow columns represent the absorbance of the secondary-only controls for each sample. The fossil bone sample is over three times its secondary control and the sediment and extraction blank show low absorbance, indicating that they are not a source for the positive signal in the fossil sample. Error bars represent one standard deviation of the mean absorbance of each sample.

3.7. Immunofluorescence

Fossil demineralization products reacted positively with polyclonal anti-chicken collagen I antibodies. Fluorescence was only observed in primary antibody-incubated tissue sections and was well above the negligible background fluorescence of secondary-only controls (Figure 7A–C). Fluorescence was restricted to tissue pieces, appearing morphologically as a spotty pattern. Specificity controls show decreased to essentially no signal when primary antibodies were inhibited prior to incubation (a control for non-specific paratopes in the polyclonal primary antibodies; Figure 7D) and when tissue sections were digested with collagenase A for 3–6 h prior to exposure to primary antibodies (a control for non-specific binding of the primary antibodies to molecules other than the target protein; Figure 7E). As also found by Schroeter [59] and Ullmann et al. [20], and references therein), digestion with collagenase A for 1 h initially increased signal slightly.

Unexpectedly, modern *Alligator* bone demineralized with HCl exhibited relatively dimmer fluorescence than previous *Alligator* samples demineralized with EDTA and treated with the same primary antibodies (Figure 7F–H; cf. [59], Figure 4D). The expected binding pattern, showing visible bone tissue structures, (e.g., Haversian systems), also appeared patchier than has been observed in previous studies using this antibody (cf. [59], Figure 4D). This patchy binding obliterated most Haversian systems and other recognizable histologic features, leaving instead an irregular fluorescence pattern (Figure 7G,H) somewhat similar to that observed in the fossil *Thoracosaurus* bone (Figure 7B,C). However, despite these apparent artifacts of protocol-induced degradation (see Discussion), the modern *Alligator* tissue sections still exhibited stronger fluorescence than did the fossil samples. Inhibition and digestion controls for the modern samples (see Supplementary Materials), also each dramatically diminished the signal (including after only 1 h of digestion; Figure 7I,J), confirming the specificity of antibody binding in our assays.

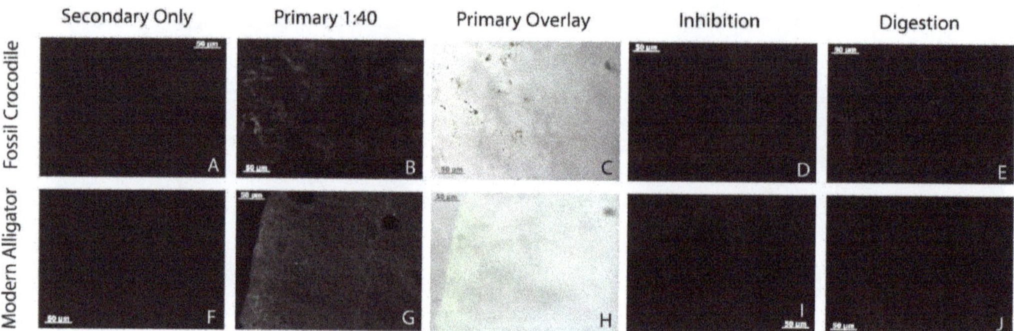

Figure 7. In situ immunofluorescence results for fossil and modern *Alligator* samples. All sections imaged at 200 ms exposure. (**A,F**) Secondary only negative-control tissue sections never exposed to primary antibodies; (**B,G**) tissue sections incubated with anti-chicken collagen I antibodies at 1:40 concentration; (**C,H**) overlays of fluorescence images in (**B,G**) on light microscope images, showing localization of fluorescence to tissue within sections (fluorescence shown as green coloration); (**D,I**) tissue sections treated with the same anti-chicken antibodies but exposed to *Alligator* collagen prior to incubation with sections ("inhibition control"); (**E,J**) tissues sections incubated with collagenase for 3 h prior to incubations with the same primary antibodies. Negative controls and specificity controls exhibit reduced or absent signals, indicating a lack of spurious binding and the presence of endogenous collagen I in each bone sample. Scale bars as indicated.

4. Discussion

Our findings cumulatively support the recovery of endogenous soft tissues and biomolecules from 63–66-million-year-old vertebrate fossils from EFP. As osteocytes were found only in fossil samples and still embedded within bony tissues, it is parsimonious to infer that these structures originated from the fossil material and not the environment. Additionally, multiple assays found only fossil samples to yield organics that reacted positively with antibodies raised against *Alligator* and chicken collagen I. Therefore, the unique depositional environment recorded by sediments of the Hornerstown Formation is the latest in the expanding list of paleoenvironmental settings shown to allow molecular preservation over geologic time.

Wiemann et al. [10] previously demineralized a vertebra from an indeterminate crocodilian (YPM 656) that was collected from the Navesink Formation, which underlies the Hornerstown Formation. In stark contrast to our results, demineralization of YPM 656 failed to yield any cells or soft tissues [10]. There are several possible reasons for this difference. No vertebrae were analyzed in the present study, but even the cancellous turtle shell bones examined herein yielded demineralization products (Table 2), implying that it is unlikely that factors related to skeletal element type alone could account for the lack of recovery of soft tissues from YPM 656 [10]. Although the Navesink Formation also contains iron-rich glauconite, it is in a lower percentage than the Hornerstown Formation [30]. It is thus possible that sediments of the Navesink Formation did not form as conducive of a diagenetic environment for molecular preservation as those of the Hornerstown Formation (cf., [33]). It is also possible that, as observed in the disparity of microstructural preservation between light and dark fossils recovered from the Hornerstown Formation, preservation in the Navesink may be variable as well. Histological examination of YPM 656 may show that its microstructure was poorly preserved despite its intact gross morphology, similar to the light tan fossils of the Hornerstown Formation that also often failed to yield demineralization products herein (Table 2). Additionally, Wiemann et al. [10] acquired their sample from collections, meaning it was not collected fresh for the purposes of paleomolecular analyses. Though historical specimens in collections have yielded results, (e.g., [6]), it has been previously suggested that fresh samples yield better soft tissue and biomolecular recovery ([60] and references therein). However, at this time, it is not possible to concretely resolve which of these processes are responsible for the lack of recovery reported by Wiemann et al. [10].

At EFP, the majority of the Hornerstown Formation currently lies below the natural water table, (e.g., the MFL is positioned ~20 ft beneath it), and studies of regional sequence stratigraphy [61] imply that fossils within the Hornerstown Formation at this locality have likely spent tens of millions of years under saturated conditions. Despite this, we successfully recovered cells and soft tissues from numerous MFL fossils and collagen I from RU-EFP-00006-11. Thus, our findings corroborate other recent studies [3,6,10,16,35] which refute the traditional hypothesis that marine paleoenvironments are inconducive to biomolecular preservation due to hydrolysis caused by constant exposure to water [25,26]. As iron is hypothesized to aid in molecular preservation in many cases [3,32,33], it is possible that the high concentration of glauconite, a mineral rich in iron, aided in the preservation of these endogenous organics, (e.g., via iron free radical-induced molecular crosslinking [33]). It is also possible that the presence of abundant dissolved iron was responsible for some of the analytical challenges encountered herein (see below). Other authors [62] have alternatively suggested that iron may precipitate around cells and other soft tissue microstructures preserved via other pathways during early diagenesis, (e.g., alumino-silicification), forming a mineralized coating later in diagenesis. At this time, there is no evidence that these preservation pathways are mutually exclusive; indeed, at some localities, (e.g., [62]) both may contribute to long-term preservation at the molecular level.

4.1. Overall Preservation

Our histological analyses found light bone to be heavily degraded, having largely lost its original microstructure (HI = 0–2). This histologic alteration appears responsible for light bones yielding few cells and soft tissues upon demineralization (Table 2). For example, the only specimen to not yield any soft tissue products (crocodile tibia RU-EFP-00030-1) is entirely light in cross-section. The other specimen with this condition (*Taphrosphys* peripheral RU-EFP-02175) only yielded rare osteocytes, and crocodile humerus RU-EFP-00030-2 also only yielded rare osteocytes likely because it appears to be preserved in a state of partial degradation. Additionally, when present, the osteocytes recovered from light bones were typically poorly preserved, (i.e., with only short, stubby filopodia) compared to those from dark bones retaining well-preserved histology (HI = 4–5), which more frequently retain long, branching filopodia with multiple ramifications. The only other specimen to yield rare osteocytes (and no vessels or fibrous matrix) was turtle peripheral RU-EFP-02245. Although this specimen only exhibits a thin external rim of light bone, it is primarily composed of highly porous cancellous bone; its accordingly low volume of bone material may have reduced the chances of recovering soft tissues from this specimen, regardless of its level of preservation. All other specimens with a thin layer of light bone or composed entirely of dark bone yielded a consistently greater recovery of cellular and soft-tissue microstructures (Table 2).

Based on our results, there appears to be no association (for these 12 specimens) between soft tissue preservation and either taxon, skeletal element, quality of macroscopic preservation, cortical thickness, or bone tissue type, (i.e., cortical vs. cancellous bone). Of the limb bones sampled, only two yielded abundant cellular and soft tissue microstructures even though all are well-preserved at the gross-morphology level. At this locality, and based on current information, the best predictor of histologic and soft tissue preservation appears to be bone color: bone tissues that are dark are histologically well-preserved bones and generally yield far greater soft tissue recovery upon demineralization. A correlation between morphologic quality and molecular preservation has been suggested previously [25,38,39,63,64], and our results further support this hypothesis. However, it remains unclear what is causing differential degradation among bones preserved within the same probable mass death assemblage [28,30] in the same horizon of the same geologic stratum. It is possible that the light degradation is a late-diagenetic artifact resulting from the modern acidic groundwater of southern New Jersey. Alternatively, these regions of poorly preserved histology within bone could result from microbial activity, as suggested by the presence in some specimens of potential MFDs and Wedl tunnels extending from the light outer layer into the underlying, better-preserved portions of dark bone (cf., [65–67]). This taphonomic question is currently under investigation as part of a separate study.

4.2. Soft Tissue Preservation

Demineralization in EDTA in this study proceeded more slowly and generally yielded fewer microstructures than from other similarly aged fossil bones we have tested [11,59]. Slow demineralization in EDTA was also found by Norris et al. [68] for a Permian *Dimetrodon* bone. As in our study, these authors were able to successfully demineralize their specimen in a solution containing HCl. The cause of such cases of slow demineralization in EDTA remains unknown. Powder XRD analyses identified RU-EFP-00006-11 as being primarily composed of fluorapatite (Figure 3A), indicating that micro-scale permineralization by secondary mineral phases containing non-divalent cations, (e.g., iron oxides with Fe^{3+} ions or quartz with Si^{4+} ions) which cannot be chelated by EDTA is an unlikely explanation. It is possible that significant substitution of trivalent cations for divalent Ca^{2+} ions in bone apatite could hinder the demineralization process (as EDTA only chelates divalent cations; [50]), but further analyses, (e.g., trace element analyses) would be required to evaluate this potential cause. HCl provides harsher, more acidic conditions for demineralization; therefore, demineralization trials employing this solution required significantly less time and yielded a greater recovery of soft tissues. However, as HCl incubation is also a step in our protein extraction protocol (see Supplementary Materials), incubation of fossil bone fragments in this acid may result in solubilization of proteins and other biomolecules which lack divalent cations. Although microstructures isolated by demineralization with HCl did not appear under transmitted light to visually differ from those recovered using EDTA, the effects of HCL versus EDTA demineralization warrants further investigation.

Osteocytes were the most abundant microstructures recovered from our EFP specimens. The common lack of recovery of microstructure morphologically consistent with blood vessels, at least in our crocodilian samples, may be due to the rarity of vessels in the original cortical/cancellous bone tissue. The cortex in modern crocodilians is not as vascular as that of non-avian dinosaurs [69–71] or modern birds; as a result, fossil crocodile bone would not be expected to yield as many structures morphologically consistent with vessels as would dinosaur bone. It should also be noted that although the external cortex of femur RU-EFP-00006-11 has been perforated by occasional Wedl tunnels, the vessel structures we recovered do not appear to represent biofilm coatings of these tunnels because they retained their structure after demineralization, and each possess a clear lumen [72]. Low recovery of the fibrous matrix may relate to the depositional environment, but more testing would be required to elucidate such a causal connection.

4.3. Biomolecular Preservation

Taken together, our molecular assays support the conclusion that the soft tissues and collagen I recovered from this specimen are endogenous. Though a PAGE with silver-stain assay is not a specific test for collagen I or proteins, it can identify the presence and molecular weights of organic compounds in a sample, making it an informative screening assay for paleomolecular studies [4,20,73]. Both fossil AMBIC extract lanes showed a distinct increase in coloration after silver-staining, whereas all negative control lanes (sediment, extraction blank, and Laemmli buffer) exhibited little to no coloration (Figure 5A,B). However, the prestaining of the gel likely points to diagenetic humic substances being present in this sample [74]. Humics can bind silver [75,76], which may have caused the increase in staining. Though this pattern of silver nitrate binding cannot directly support the presence of protein in the sample, this pattern does not falsify the presence of protein in this fossil (as no staining would). Thus, we continued to analyze this sample with assays of greater specificity that are not susceptible to humics. As predicted, the modern *Alligator* extracts exhibited less smearing and better banding than the fossil extracts (Figure 5C; see Supplementary Materials for further discussion of our silver-staining results).

Given the recognition of unique, albeit non-specific, organics in RU-EFP-00006-11 by silver-staining, we next performed ELISA and in situ immunofluorescence as independent assays to identify if the primary structural protein collagen I was present in both whole-bone extracts and demineralization products, respectively. These assays complement one another as ELISA is approximately an order of magnitude more sensitive than immunoflu-orescence [73,77], whereas immunofluorescence allows in situ localization of epitopes in native tissues [15,73]. Fossil extracts in ELISA assays yielded well over double the ab-sorbance of the sediment and extraction buffer controls (Figure 6), supporting the presence of endogenous collagen I in RU-EFP-00006-11. Negative absorbance values for sediment and extraction controls indicate these samples were less reactive than the group blank (PBS only), demonstrating the collagen signal in bone wells is not attributable to these sources of possible contamination. Additionally, the absorbance values of all samples incubated with only secondary antibodies were negligible, (i.e., significantly less than twice the absorbance of the fossil samples incubated with both primary and secondary antibodies), removing non-specific secondary antibody binding (such as to humic substances) as a cause for the high positive signal for the fossil bone extracts. Thus, this assay supports the presence of endogenous collagen I in the femur of this *Thoracosaurus*.

Modern *Alligator* tissue samples analyzed by in situ immunofluorescence exhibited fluorescence when exposed to polyclonal antibodies raised against chicken collagen I (Figure 7G,H). Together with the reduced signals observed in our specificity controls (Figure 7I,J; Supplementary Materials), these results agree with those of Schroeter [59] who concluded that the collagen I epitopes in these two extant archosaurs are conserved enough in structure to each be recognized by the primary antibodies used herein. Because these antibodies can successfully detect collagen I in modern *Alligator*, they were predicted to also bind to fossil crocodilian collagen I. The fluorescence signal exhibited by the fossil tissues was lower in intensity compared to the modern *Alligator* (Figure 7B,C), yet still far brighter than in secondary-only controls (Figure 7A). Fluorescence in fossil tissue sections decreased in all specificity controls (Figure 7D,E; Supplementary Materials), again supporting the specific binding of the primary antibodies to epitopes of collagen I. Both modern and fossil tissues exhibited patchy binding patterns, possibly due to tissue degradation from demineralization with HCl or its ability to solubilize proteins (as in our extraction protocol), as well as decay inherent in fossilization for RU-EFP-00006-11.

5. Conclusions

Our paleomolecular investigations of fossils preserved in the glauconitic, shallow-marine depositional environment of the Hornerstown Formation at EFP add a unique paleoenvironment to the growing list of those which have yielded biomolecular and soft tissue preservation within fossil bones. This depositional environment was rich in iron at the time of burial and remained rich in iron due to the glauconite-dominated composition of the sediments and influx of dissolved iron from recent groundwaters. We hypothesize that this abundant supply of iron may have facilitated soft tissue and biomolecular preser-vation in these specimens via the free radical-induced crosslinking reactions elucidated by Boatman et al. [33]. Soft tissues were recovered by demineralization with HCL and EDTA, though HCL treatments took less time and were required for some samples. Though all specimens we examined are well-preserved in terms of gross morphology, multiple exhibit varying amounts of alteration in the form of light tan-colored bone which was found to be histologically degraded and thus yielded minimal cells and soft tissues upon demineraliza-tion. Association of poor histological preservation with poor soft tissue recovery is logical, and it appears that at EFP this alteration may have occurred, at least in part, by microbial degradation. All biomolecular assays completed on RU-EFP-00006-11 support the presence of endogenous collagen I in this *Thoracosaurus* femur. PAGE with silver-staining exhibited evidence for organics in this fossil, and ELISA and in situ immunofluorescence results each independently support retention of collagen I in this specimen. Collectively, our results and those of previous authors [3,6,10,16,35] support the conclusions of Nielsen-Marsh and Hedges [78] and Hedges [38] that relatively constant immersion in water may not preclude endogenous molecular preservation in fossil bones. As a result, there are many additional paleoenvironments yet to be explored that may preserve endogenous biomolecules and soft tissues.

Supplementary Materials: The following are available online at https://www.mdpi.com/article/10.3390/biology11081161/s1, Supplementary Materials narrative: a DOCX Word file including additional description of methods and continued discussion. Figure S1: Gross morphology of representative fossil specimens examined herein. Figure S2: Examples of microbial bioerosion in histologic thin section. Figure S3: Additional demineralization results. Figure S4: Comparison of enzyme-linked immunosorbent assay results of fossil bone, sediment, and modern *Alligator* chemical extracts. References [79–84] are cited in the Supplementary Materials narrative.

Author Contributions: Conceptualization, K.K.V. and K.J.L.; methodology, K.K.V. and E.R.S.; formal analysis, K.K.V., Z.M.B. and P.V.U.; investigation, K.K.V., Z.M.B., P.V.U. and W.Z.; data curation, K.K.V., Z.M.B. and P.V.U.; writing—original draft preparation, K.K.V.; writing—review and editing, K.K.V., Z.M.B., P.V.U. and E.R.S.; visualization, K.K.V. and Z.M.B.; project administration, K.K.V.; funding acquisition, K.K.V. and P.V.U. All authors have read and agreed to the published version of the manuscript.

Funding: This research was funded by NSF GRFP, DGE Award 1002809, and Rowan University.

Institutional Review Board Statement: Not applicable.

Informed Consent Statement: Not applicable.

Data Availability Statement: All data generated by this study are available in this manuscript and the accompanying Supplementary Materials.

Acknowledgments: Thank you to M. Schweitzer for fruitful discussions during planning and data collection as well as access to facilities and J. Tabacco for assistance in data collection. Additionally, we are grateful for the support and assistance provided by the Inversand Mining Company (Clayton, NJ, USA) for the collection of specimens prior to 2016. We would also like to thank the editor and four reviewers for their suggested improvements to the manuscript.

Conflicts of Interest: The authors declare no conflict of interest.

References

1. Schroeter, E.R.; Cleland, T.P.; Schweitzer, M.H. Deep Time Paleoproteomics: Looking Forward. *J. Proteome Res.* **2021**, *21*, 9–19. [CrossRef]
2. Schweitzer, M.H.; Wittmeyer, J.L.; Horner, J.R.; Toporski, J.K. Soft-tissue vessels and cellular preservation in *Tyrannosaurus rex*. *Science* **2005**, *307*, 1952–1955. [CrossRef] [PubMed]
3. Schweitzer, M.H.; Wittmeyer, J.L.; Horner, J.R. Soft tissue and cellular preservation in vertebrate skeletal elements from the Cretaceous to the present. *Proc. R. Soc. B* **2007**, *274*, 183–197. [CrossRef] [PubMed]
4. Schweitzer, M.H.; Zheng, W.; Organ, C.L.; Avci, R.; Suo, Z.; Freimark, L.M.; Lebleu, V.S.; Duncan, M.B.; Heiden, M.G.V.; Neveu, J.M.; et al. Biomolecular characterization and protein sequences of the Campanian Hadrosaur *B. canadensis*. *Science* **2009**, *324*, 626–631. [CrossRef] [PubMed]
5. Schweitzer, M.H.; Zheng, W.; Cleland, T.P.; Bern, M. Molecular analyses of dinosaur osteocytes support the presence of endogenous molecules. *Bone* **2013**, *52*, 414–423. [CrossRef]
6. Lindgren, J.; Uvdal, P.; Engdahl, A.; Lee, A.H.; Alwmark, C.; Bergquist, K.E.; Nilsson, E.; Ekström, P.; Rasmussen, M.; Douglas, D.A.; et al. Microspectroscopic Evidence of Cretaceous Bone Proteins. *PLoS ONE* **2011**, *6*, e19445. [CrossRef] [PubMed]
7. Armitage, M.H.; Anderson, K.L. Soft sheets of fibrillar bone from a fossil of the supraorbital horn of the dinosaur *Triceratops horridus*. *Acta Histochem.* **2013**, *115*, 603–608. [CrossRef]
8. Cadena, E.A. Microscopical and elemental FESEM and phenom ProX-SEM-EDS analysis of osteocyte- and blood vessel-like microstructures obtained from fossil vertebrates of the eocene messel pit, germany. *PeerJ* **2016**, *4*, e1618. [CrossRef]
9. Cadena, E.A. In situ SEM/EDS compositional characterization of osteocytes and blood vessels in fossil and extant turtles on untreated bone surfaces; different preservational pathways microns away. *PeerJ* **2020**, *8*, e9833. [CrossRef]

10. Wiemann, J.; Fabbri, M.; Yang, T.R.; Stein, K.; Sander, P.M.; Norell, M.A.; Briggs, D.E.G. Fossilization transforms vertebrate hard tissue proteins into N-heterocyclic polymers. *Nat. Commun.* **2018**, *9*, 4741. [CrossRef] [PubMed]

11. Ullmann, P.V.; Pandya, S.H.; Nellermoe, R. Patterns of soft tissue and cellular preservation in relation to fossil bone microstructure and overburden depth at the Standing Rock Hadrosaur Site, Maastrichtian Hell Creek Formation, South Dakota, USA. *Cretac. Res.* **2019**, *99*, 1–13. [CrossRef]

12. Bailleul, A.M.; Zheng, W.; Horner, J.R.; Hall, B.K.; Holliday, C.M.; Schweitzer, M.H. Evidence of proteins, chromosomes and chemical markers of DNA in exceptionally preserved dinosaur cartilage. *Natl. Sci. Rev.* **2020**, *7*, 815–822. [CrossRef] [PubMed]

13. Van der Reest, A.J.; Currie, P.J. Preservation frequency of tissue-like structures in vertebrate remains from the upper Campanian of Alberta: Dinosaur Park Formation. *Cretac. Res.* **2020**, *109*, 104370. [CrossRef]

14. Wiersma, K.; Läbe, S.; Sander, P.M. Organic Phase Preservation in Fossil Dinosaur and Other Tetrapod Bone from Deep Time. In *Fossilization: Understanding the Material Nature of Ancient Plants and Animals*; Gee, C.T., McCoy, V.E., Sander, P.M., Eds.; Johns Hopkins University Press: Baltimore, MD, USA, 2021; pp. 16–54.

15. Schweitzer, M.H.; Suo, Z.; Avci, R.; Asara, J.M.; Allen, M.A.; Arce, F.T.; Horner, J.R. Analyses of soft tissue from *Tyrannosaurus rex* suggest the presence of protein. *Science* **2007**, *316*, 277–280. [CrossRef] [PubMed]

16. Surmik, D.; Boczarowski, A.; Balin, K.; Dulski, M.; Szade, J.; Kremer, B.; Pawlicki, R. Spectroscopic studies on organic matter from Triassic reptile bones, Upper Silesia, Poland. *PLoS ONE* **2016**, *11*, e0151143. [CrossRef]

17. Pan, Y.; Zheng, W.; Moyer, A.E.; O'Connor, J.K.; Wang, M.; Zheng, X.; Wang, X.; Schroeter, E.R.; Zhou, Z.; Schweitzer, M.H. Molecular evidence of keratin and melanosomes in feathers of the Early Cretaceous bird *Eoconfuciusornis*. *Proc. Natl Acad. Sci. USA* **2016**, *113*, 7900–7907. [CrossRef]

18. Pan, Y.; Zheng, W.; Sawyer, R.H.; Pennington, M.W.; Zheng, X.; Wang, X.; Wang, M.; Hu, L.; O'Connor, J.; Zhao, T.; et al. The molecular evolution of feathers with direct evidence from fossils. *Proc. Natl Acad. Sci. USA* **2019**, *116*, 3018–3023. [CrossRef]

19. Lee, Y.C.; Chiang, C.C.; Huang, P.Y.; Chung, C.Y.; Huang, T.D.; Wang, C.C.; Chen, C.I.; Chang, R.S.; Liao, C.H.; Reisz, R.R. Evidence of preserved collagen in an Early Jurassic sauropodomorph dinosaur revealed by synchrotron FTIR microspectroscopy. *Nat. Commun.* **2017**, *8*, 14220. [CrossRef] [PubMed]

20. Ullmann, P.V.; Voegele, K.K.; Grandstaff, D.E.; Ash, R.D.; Zheng, W.; Schroeter, E.R.; Schweitzer, M.H.; Lacovara, K.J. Molecular tests support the viability of rare earth elements as proxies for fossil biomolecule preservation. *Sci. Rep.* **2020**, *10*, 15566. [CrossRef]

21. Asara, J.M.; Schweitzer, M.H.; Freimark, L.M.; Phillips, M.; Cantley, L.C. Protein Sequences from *Mastodon* and *Tyrannosaurus Rex* Revealed by Mass Spectrometry. *Science* **2007**, *316*, 280–285. [CrossRef]

22. Cappellini, E.; Jensen, L.J.; Szklarczyk, D.; Ginolhac, A.; da Fonseca, R.A.; Stafford, T.W., Jr.; Holen, S.R.; Collins, M.J.; Orlando, L.; Willerslev, E.; et al. Proteomic analysis of a pleistocene mammoth femur reveals more than one hundred ancient bone proteins. *J. Proteome Res.* **2012**, *11*, 917–926. [CrossRef] [PubMed]

23. Cleland, T.P.; Schroeter, E.R.; Zamborg, L.; Zheng, W.; Lee, J.E.; Tran, J.C.; Bern, M.; Duncan, M.B.; Lebleu, V.S.; Ahlf, D.R.; et al. Mass spectrometry and antibody-based characterization of blood vessels from *Brachylophosaurus canadensis*. *J. Proteome Res.* **2015**, *14*, 5252–5262. [CrossRef] [PubMed]

24. Schroeter, E.R.; DeHart, C.J.; Cleland, T.P.; Zheng, W.; Thomas, P.M.; Kelleher, N.L.; Bern, M.; Schweitzer, M.H. Expansion for the *Brachylophosaurus canadensis* collagen I sequence and additional evidence of the preservation of Cretaceous protein. *J. Proteome Res.* **2017**, *16*, 920–932. [CrossRef] [PubMed]

25. Eglinton, G.; Logan, G. Molecular Preservation. *Philos. Trans. R. Soc. B* **1991**, *333*, 315–328.

26. Collins, M.J.; Westbroek, P.; Muyzer, G.; de Leeuw, J.W. Experimental evidence for condensation reactions between sugars and proteins in carbonate skeletons. *Geochim. Cosmochim. Acta* **1992**, *56*, 1539–1544. [CrossRef]

27. Cadena, E.A.; Schweitzer, M.H. Variation in osteocytes morphology vs. bone type in turtle shell and their exceptional preservation from the Jurassic to the present. *Bone* **2012**, *51*, 614–620. [CrossRef] [PubMed]

28. Voegele, K.K.; Ullmann, P.V.; Lonsdorf, T.; Christman, Z.; Heierbacher, M.; Kibelstis, B.J.; Putnam, I.; Boles, Z.; Walsh, S.; Lacovara, K.J. Microstratigraphic analysis of fossil distribution in the lower Hornerstown and upper Navesink formations at the Edelman Fossil Park, NJ. *Front. Earth Sci.* **2021**, *9*, 756655. [CrossRef]

29. Gallagher, W.B. The Cretaceous/Tertiary mass extinction event in the northern Atlantic Coastal Plain. *Mosasaur* **1993**, *5*, 75–154.

30. Obasi, C.C.; Terry, D.O., Jr.; Myer, G.H.; Grandstaff, D.E. Glauconite composition and morphology, shocked quartz, and the origin of the Cretaceous(?) Main Fossiliferous Layer (MFL) in southern New Jersey, U.S.A. *J. Sediment. Res.* **2011**, *81*, 479–494. [CrossRef]

31. Gallagher, W.B. Oligotrophic oceans and minimalist organisms: Collapse of the Maastrichtian marine ecosystem and Paleocene recovery in the Cretaceous-Tertiary sequence of New Jersey. *Neth. J. Geosci.* **2003**, *82*, 225–231. [CrossRef]

32. Schweitzer, M.H.; Zheng, W.; Cleland, T.P.; Goodwin, M.B.; Boatman, E.; Theil, E.; Marcus, M.A.; Fakra, S.C. A role for iron and oxygen chemistry in preserving soft tissues, cells and molecules from deep time. *Proc. R. Soc. B* **2014**, *281*, 20132741. [CrossRef] [PubMed]

33. Boatman, E.M.; Goodwin, M.B.; Holman, H.-Y.N.; Fakra, S.; Zheng, W.; Gronsky, R.; Schweitzer, M.H. Mechanisms of soft tissue and protein preservation in *Tyrannosaurus rex*. *Sci. Rep.* **2019**, *9*, 15678. [CrossRef] [PubMed]

34. Barker, K.; Terry, D.O.; Ullmann, P.V. Recovering endogenous cells and soft tissues from fossil bones: Does the depositional environment matter? In Proceedings of the GSA Connect, Portland, Oregon, 10–13 October 2021; 53, p. 225-10.

35. Cadena, E.A.; Schweitzer, M.H. A Pelomedusoid turtle from the Paleocene–Eocene of Colombia exhibiting preservation of blood vessels and osteocytes. *J. Herpetol.* **2014**, *48*, 461–465. [CrossRef]

36. Macauley, K.W.; Ash, R.D.; Ullmann, P.V. Relating rare earth element uptake to soft tissue preservation in Cretaceous and Miocene fossil bones. In Proceedings of the Joint 69th Annual Southeastern/55th Annual Northeastern GSA Section Meeting—2020, Reston, Virginia, 20–22 March 2020; The Geological Society of America: Boulder, CO, USA; 52, p. 59-5.

37. Boles, Z.M. Taphonomy and Paleoecology of a Cretaceous-Paleogene Marine Bone Bed. Ph.D. Thesis, Drexel University, Philadelphia, PA, USA, 2016.

38. Hedges, R.E.M. Bone diagenesis: An overview of processes. *Archaeometry* **2002**, *44*, 319–328. [CrossRef]

39. Schweitzer, M.H. Molecular paleontology: Some current advances and problems. *Ann. Paléontologie* **2004**, *90*, 81–102. [CrossRef]

40. Amber, R.P.; Daniel, M. Proteins and molecular paleontology. *Philos. Trans. R. Soc. B* **1991**, *333*, 381–389.

41. Child, A.M. Microbial taphonomy of archaeological bone. *Stud. Conserv.* **1995**, *40*, 19–30.

42. Hedges, R.E.M.; Millard, A.R. Measurements and relationships of diagenetic alteration of bone from three archaeological sites. *J. Archaeol. Sci.* **1995**, *22*, 201–209. [CrossRef]

43. Sweeney, S.M.; Orgel, J.P.; Fertala, A.; McAuliffe, J.D.; Turner, K.R.; Di Lullo, G.A.; Chen, S.; Antipova, O.; Perumal, S.; Ala-Kokko, L.; et al. Candidate cell and matrix interaction domains on the collagen fibril, the predominant protein of vertebrates. *J. Biol. Chem.* **2008**, *283*, 21187–21197. [CrossRef] [PubMed]

44. San Antonio, J.D.; Schweitzer, M.H.; Jensen, S.T.; Kalluri, R.; Buckley, M.; Orgel, J.P.R.O. Dinosaur peptides suggest mechanisms of protein survival. *PLoS ONE* **2011**, *6*, e20381. [CrossRef] [PubMed]

45. Wang, S.Y.; Cappellini, E.; Zhang, H.Y. Why collagens best survived in fossils? Clues from amino acid thermal stability. *Biochem. Biophys. Res. Commun.* **2012**, *422*, 5–7. [CrossRef] [PubMed]

46. Perumal, S.; Antipova, O.; Orgel, J.P.R.O. Collagen fibril architecture, domain organization, and triple-helical conformation govern its proteolysis. *Proc. Natl. Acad. Sci. USA* **2008**, *105*, 2824–2829. [CrossRef] [PubMed]

47. Collins, M.J.; Nielsen-Marsh, C.M.; Hiller, J.; Smith, C.I.; Roberts, J.P. The survival of organic matter in bone: A review. *Archaeometry* **2002**, *44*, 383–394. [CrossRef]

48. Trueman, C.N.; Martill, D.M. Long Term Survival of Bone: The Role of Bioerosion. *Archaeometry* **2002**, *44*, 371–382. [CrossRef]

49. Chinsamy, A.; Raath, M.A. Preparation of fossil bone for histological examination. *Palaeontol. Afr.* **1992**, *29*, 39–44.

50. Cleland, T.P.; Voegele, K.K.; Schweitzer, M.H. Empirical evaluation of methodology related to the extraction of bone proteins. *PLoS ONE* **2012**, *7*, e31443. [CrossRef]

51. Zheng, W.; Schweitzer, M.H. Chemical Analyses of Fossil Bone. In *Forensic Microscopy for Skeletal Tissues*; Bell, L.S., Ed.; Humana Press: Burnaby, BC, Canada, 2012; pp. 153–172.

52. Pellegrini, R.A.; Callahan, W.R.; Hastings, A.K.; Parris, D.C.; McCauley, J.D. Skeletochronology and Paleohistology of Hyposaurus rogersii (Crocodyliformes, Dyrosauridae) from the Early Paleogene of New Jersey, USA. *Animals* **2021**, *11*, 3067. [CrossRef] [PubMed]

53. Jans, M.M.E. Microbial bioerosion of bone—A review. In *Current Developments in Bioerosion*; Wisshak, M., Tapanila, L., Eds.; Springer: Berlin/Heidelberg, Germany, 2008; pp. 397–413.

54. Pawlicki, R. Histochemical demonstration of DNA in osteocytes from dinosaur bones. *Folia Histochem. Chytobiologica* **1995**, *33*, 183–186.

55. Surmik, D.; Rothschild, B.M.; Pawlicki, R. Unusual intraosseous fossilized soft tissues from the Middle Triassic *Nothosaurus* bone. *Sci. Nat.* **2017**, *104*, 25. [CrossRef] [PubMed]

56. Schweitzer, M.H. Soft tissue preservation in terrestrial Mesozoic vertebrates. *Annu. Rev. Earth Planet. Sci.* **2011**, *39*, 187–216. [CrossRef]

57. Ostlund, E.N.; Crom, R.L.; Pedersen, D.D.; Johnson, D.J.; Williams, W.O.; Schmitt, B.J. Equine west nile encephalitis, United States. *Emerg. Infect. Dis.* **2001**, *7*, 665–669. [CrossRef] [PubMed]

58. Appiah, A.S.; Amoatey, H.M.; Klu, G.Y.P.; Afful, N.T.; Azu, E.; Owusu, G.K. Spread of African cassava mosaic virus from cassava (*Manihot esculenta* Crantz) to physic nut (*Jatropha curcas* L.) in Ghana. *J. Phytol.* **2012**, *4*, 31–37.

59. Schroeter, E.A.R. The Morphology, Histology, and Molecular Preservation of an Exceptionally Complete Titanosaur from Southernmost Patagonia. Ph.D. Thesis, Drexel University, Philadelphia, PA, USA, 2013.

60. Cleland, T.P.; Schroeter, E.R.; Feranec, R.S.; Vashishth, D. Peptide sequences from the first Castoroides ohioensis skull and the utility of old museum collections for palaeoproteomics. *Proc. R. Soc. B Biol. Sci.* **2016**, *283*, 20160593. [CrossRef] [PubMed]

61. Olsson, R.K.; Miller, K.G.; Browning, J.V.; Wright, J.D.; Cramer, B.S. Sequence stratigraphy and sea-level change across the Cretaceous-Tertiary boundary. In *Catastrophic Events and Mass Extinctions: Impacts and Beyond*; Koeberl, C., MacLeod, K.G., Eds.; Geological Society of America: Boulder, CO, USA, 2002; Volume 356, pp. 97–108.

62. Zheng, X.; Bailleul, A.M.; Li, Z.; Wang, X.; Zhou, Z. Nuclear preservation in the cartilage of the Jehol dinosaur *Caudipteryx. Commun. Biol.* **2021**, *4*, 1125. [CrossRef] [PubMed]

63. Hagelberg, E.; Bell, L.S.; Allen, T.; Boyde, A.; Jones, S.J.; Clegg, J.B. Analysis of ancient bone DNA: Techniques and applications. *Philos. Trans. R. Soc. Lond. Ser. B Biol. Sci.* **1991**, *333*, 399–407.

64. Turner-Walker, G.; Jans, M. Reconstructing taphonomic histories using histological analysis. *Palaeogeogr. Palaeoclimatol. Palaeoecol.* **2008**, *266*, 227–235. [CrossRef]

65. Fernandez-Jalvo, Y.; Andrews, P.; Pesquero, D.; Smith, C.; Marin-Monfort, D.; Sanchez, B.; Geigl, E.M.; Alonso, A. Early bone diagenesis in temperate environments Part I: Surface features and histology. *Palaeogeogr. Palaeoclimatol. Palaeoecol.* **2010**, *288*, 62–81. [CrossRef]

66. Danise, S.; Cavalazzi, B.; Dominici, S.; Westall, F.; Monechi, S.; Guioli, S. Evidence of microbial activity from a shallow water whale fall (Voghera, northern Italy). *Palaeogeogr. Palaeoclimatol. Palaeoecol.* **2012**, *317*, 13–26. [CrossRef]

67. Eriksen, A.M.H.; Nielsen, T.K.; Matthiesen, H.; Carøe, C.; Hansen, L.H.; Gregory, D.J.; Turner-Walker, G.; Collins, M.J.; Gilbert, M.T.P. Bio biodeterioration—The effect of marine and terrestrial depositional environments on early diagenesis and bone bacterial community. *PLoS ONE* **2020**, *15*, e0240512. [CrossRef] [PubMed]

68. Norris, S.J.; Paredes, A.M.; Immega, N.; Temple, D.; Bakker, R.T. Release of osteocytes and lamellae through chemical dissolution reveals the intricate, anastomosing circumferential pattern of blood vessels in Dimetrodon neural spines. *Soc. Vertebr. Paleontol. Programs Abstr.* **2017**, 169.

69. Chinsamy, A. Dinosaur bone histology: Implications and inferences. *Paleontol. Soc. Spec. Publ.* **1997**, *7*, 213–227. [CrossRef]

70. Padian, K.; Horner, J.R. Dinosaur Physiology. In *The Dinosauria*, 2nd ed.; Weishampel, D.B., Dodson, P., Osmólska, H., Eds.; University of California Press: Berkeley, CA, USA, 2004; pp. 660–672.

71. Woodward, H.N.; Horner, J.R.; Farlow, J.O. Quantification of intraskeletal histovariability in *Alligator mississippiensis* and implications for vertebrate osteohistology. *PeerJ* **2014**, *2*, e422. [CrossRef] [PubMed]

72. Schweitzer, M.H.; Moyer, A.E.; Zheng, W. Testing the hypothesis of biofilm as a source for soft tissue and cell-like structures preserved in dinosaur bone. *PLoS ONE* **2016**, *11*, e0150238. [CrossRef] [PubMed]

73. Schweitzer, M.H.; Avci, R.; Collier, T.; Goodwin, M.B. Microscopic, chemical and molecular methods for examining fossil preservation. *Comptes Rendus Palevol* **2008**, *7*, 159–184. [CrossRef]

74. Schroeter, E.R.; Blackbum, K.; Goshe, M.B.; Schweitzer, M.H. Proteomic method to extract, concentrate, digest and enrich peptides from fossils with coloured (humic) substances for mass spectrometry analyses. *R. Soc. Open Sci.* **2019**, *6*, 181433. [CrossRef] [PubMed]

75. Dunkelog, R.; Rüttinger, H.-H.; Peisker, K. Comparative study for the separation of aquatic humic substances by electrophoresis. *J. Chromatogr. A.* **1997**, *777*, 355–362. [CrossRef]

76. Evans, R.D.; Villeneuve, J.Y. A method for characterization of humic and fulvic acids by gel electrophoresis laser ablation inductively coupled plasma mass spectrometry. *J. Anal. At. Spectrom.* **1999**, *15*, 157–161. [CrossRef]

77. Avci, R.; Schweitzer, M.H.; Boyd, R.D.; Wittmeyer, J.L.; Arce, F.T.; Calvo, J.O. Preservation of bone collagen from the Late Cretaceous period studied by immunological techniques and atomic force microscopy. *Langmuir* **2005**, *21*, 3584–3590. [CrossRef] [PubMed]

78. Nielsen-Marsh, C.M.; Hedges, R.E.M. Patterns of diagenesis in bone I: The effects of site environments. *J. Archaeol. Sci.* **2000**, *27*, 1139–1150. [CrossRef]

79. Schweitzer, M.H.; Hill, C.L.; Asara, J.M.; Lane, W.S.; Pincus, S.H. Identification of immunoreactive material in mammoth fossils. *J. Mol. Evol.* **2002**, *55*, 696–705. [CrossRef] [PubMed]

80. Tuross, N. Alteration in fossil collagen. *Archaeometry.* **2002**, *44*, 427–434. [CrossRef]

81. Cleland, T.P.; Schroeter, E.R.; Colleary, C. Diagenetiforms: A new term to explain protein changes as a result of diagenesis in paleoproteomics. *Journal of Proteomics.* **2021**, *230*, 103992. [CrossRef]

82. Knauber, W.R.; Krotzky, A.J.; Schink, B. Gradient gel electrophoretic characterization of humic substances and of bound residues of the herbicide bentazon. *Soil Biol. and Biochem.* **1998**, *30*, 969–973. [CrossRef]

83. van Klinken, G.J.; Hedges, R.E.M. Experiements on Collagen-Humic Interactions: Speed of humic uptake, and effects of diverse chemical treatments. *J. Archaeol. Sci.* **1995**, *22*, 263–270. [CrossRef]

84. Bada, J.L.; Xueyun, S.W.; Hamilton, H. Preservation of key biomolecules in the fossil record: Current knowledge and future challenges. *Phil. Trans. R. Soc. Lond. B.* **1999**, *354*, 77–87. [CrossRef]

Article

Environmental Factors Affecting Feather Taphonomy

Mary Higby Schweitzer [1,2,3,4,*], **Wenxia Zheng** [1] **and Nancy Equall** [5]

[1] Department of Biological Sciences, North Carolina State University, Raleigh, NC 27606, USA; wzheng2@ncsu.edu
[2] North Carolina Museum of Natural Sciences, Raleigh, NC 27601, USA
[3] Department of Geology, Lund University, 223 62 Lund, Sweden
[4] Museum of the Rockies, Montana State University, Bozeman, MT 59717, USA
[5] ICAL Facility, Montana State University, Bozeman, NC 59717, USA; equall4@msn.com
[*] Correspondence: schweitzer@ncsu.edu

Simple Summary: This study seeks to test the effect of burial/exposure, sediment type, the addition of feather-degrading microbes, and the addition of minerals on feather preservation, and for the first time, compares these states in ambient vs. elevated CO_2 atmospheres to test the effect of CO_2 on degradation and/or preservation under various depositional settings.

Abstract: The exceptional preservation of feathers in the fossil record has led to a better understanding of both phylogeny and evolution. Here we address factors that may have contributed to the preservation of feathers in ancient organisms using experimental taphonomy. We show that the atmospheres of the Mesozoic, known to be elevated in both CO_2 and with temperatures above present levels, may have contributed to the preservation of these soft tissues by facilitating rapid precipitation of hydroxy- or carbonate hydroxyapatite, thus outpacing natural degradative processes. Data also support that that microbial degradation was enhanced in elevated CO_2, but mineral deposition was also enhanced, contributing to preservation by stabilizing the organic components of feathers.

Keywords: feather; taphonomy; degradation; keratin; microbes; CO_2; apatite; melanin

Citation: Schweitzer, M.H.; Zheng, W.; Equall, N. Environmental Factors Affecting Feather Taphonomy. *Biology* **2022**, *11*, 703. https://doi.org/10.3390/biology11050703

Academic Editor: Andreas Wagner

Received: 27 February 2022
Accepted: 29 April 2022
Published: 3 May 2022

Publisher's Note: MDPI stays neutral with regard to jurisdictional claims in published maps and institutional affiliations.

Copyright: © 2022 by the authors. Licensee MDPI, Basel, Switzerland. This article is an open access article distributed under the terms and conditions of the Creative Commons Attribution (CC BY) license (https://creativecommons.org/licenses/by/4.0/).

1. Introduction

Feathers are arguably the most complex integumentary structures in the entire animal kingdom. The evolutionary origins of feathers are still debated, but growing evidence from both molecular studies in extinct theropods [1–8] and living birds (e.g., [9–18]), as well as numerous fossil discoveries of structures morphologically consistent with feathers (e.g., [4,19–25]) indicate that feathers arose from filamentous structures first identifed in some theropod dinosaurs and birds more than 160 million years ago (e.g., [2,26,27]). However, some data suggest that integumentary structures similar to those from which feathers derived may have been present at the base of Dinosauria [28,29] or perhaps, the base of Archosauria ([30,31] and references therein). Because modern feathers are not biomineralized in life (contra [32,33]) their persistence in the fossil record is counterintuitive, but critical. The impressions of feathers in sediments surrounding skeletal elements led to the identification of *Archaeopteryx* as the first bird [34,35], but there was no organic trace with this specimen to suggest that any original material remained. However, the first specimen attributed to *Archaeopteryx* was a single, isolated feather [36]. This specimen presented differently from feather impressions surrounding the skeletal remains, instead visualized as a carbonized trace clearly distinct from the embedding sediments, suggesting that taphonomic processes resulting in preservation differed between the isolated feather and the skeletal specimen. The environmental factors resulting in these different modes of preservation remain relatively unexplored.

For any organic remains to be preserved in deep time, they must be stabilized before they can degrade [37]. Although many taphonomic modes may result in the preservation of feathers (e.g., carbonized film [36,38], sediment impressions [34,35], three-dimensional filaments [3,39], amber preservation [40,41], bacterial mediation [42,43], or other stabilizing processes (e.g., [42,44–46])), few of these have been subjected to rigorous experimentation, particularly at the molecular level (but see [47]). It is likely that preservation processes differ for every fossil element and/or environment and arise from a complex interplay between the molecular composition of the original structures and the geochemical properties of the surrounding depositional matrix. Factors contributing to feather preservation are testable by approximating naturally occurring conditions in laboratory experiments.

Here, we examine multiple environmental factors that may, to varying degrees, affect the preservation of feathers in the fossil record. We discontinued the experiments after six weeks, because previous experiments (feathers degraded in sandy settings with added microbes) have shown that within this time period, the degradation of feathers was almost complete (unpublished data). We tested feathers buried vs. unburied; soaked in natural pondwater, sterile water, or pondwater after incubation with feather-degrading bacteria; the addition of solubilized hydroxyapatite (HA) to waters; and burial in sand vs. mud. Finally, we repeated the experiment in atmospheres elevated in CO_2. Although the value we employed is higher than proposed for the entire Mesozoic, it is consistent with values proposed for the Ypresian [48]. Thus, this value is in line with what has occurred without human intervention, allowing us to test the direct effect of CO_2 in a shorter time period. We observed that in at least one case, there was no material remaining to be tested at six weeks; degradation was complete.

2. Methods

We subjected the black and white feathers of an extant magpie (*Pica hudsonia*) to these environmental conditions to test their effect on preservation and/or degradation (see Table 1). These feathers allowed us to also test the hypothesis that melanized feather regions would show greater preservation than un-melanized ones [49]. See supplemental documents for additional experimental details.

Table 1. Experimental conditions to test their effect on preservation and/or degradation.

Ambient Atmosphere	Elevated CO_2 Atmosphere
Pondwater, buried (PB)	Pondwater, buried (CPB)
Pondwater, exposed (PE)	Pondwater, exposed (CPE)
Pondwater + *Bacillus licheniformis*, buried (PBB)	Pondwater + *B. licheniformis*, buried (CPBB)
Pondwater + *B. licheniformis*, exposed (PBE)	Pondwater + *B. licheniformis*, exposed (CPBE)
Pondwater + Hydroxylapatite, buried (PMB)	Pondwater + HA, buried (CPMB)
Pondwater + HA, exposed (PME)	Pondwater + HA, exposed (CPME)
E-Pure water, sand + HA, buried (ESMB)	E-Pure water, sand + HA, buried (CESMB)
E-Pure water, sand + HA, exposed (ESME)	E-Pure water, sand + HA, exposed (CESME)
E-Pure water, sand, buried (ESB)	E-Pure water, sand, buried (CESB)
E-Pure water, sand, exposed (ESE)	E-Pure water, sand, exposed (CESE)

We hypothesized that rate, degree, and pattern of the degradation of feathers might differ in atmospheres elevated in CO_2 relative to today's levels, so we divided all experimental conditions into "elevated CO_2" and "ambient" atmospheres. Within these two environments, we tested the effect of burial vs. surface exposure; native microbial populations in surface waters (pondwater) or the addition of feather-degrading microbes (*B. licheniformis*) to those microbes naturally present in pondwater; clean porous sand (No. 1113 Premium Play Sand, Quikrete) vs. natural pond sediments, which were a silty-to-clay mix; and an addition of 6 mM calcium hydroxylapatite (HA; $Ca_{10}(PO_4)_6(OH)_2$), (0.1 M$CaCl_2$ solution mixed with NaH_2PO_4 (0.1 M) with volume ratio 10:6)) [50] to mimic the higher concentration of this mineral within pore waters that occurs under elevated CO_2 and concurrent lower pH [51–53]. This would be more likely to occur in Mesozoic than present-day waters because of the increased acidification and resulting mineral solubilization brought on by elevated CO_2 [52–55]. To test the role of microbes in mediating HA mobilization, we added this mineral to both E-Pure water and pondwater before adding to feathers. Although we used E-Pure water in the last two conditions as controls, some microbes were present, arising from the natural sediments and/or the feathers themselves.

Table 1 illustrates the conditions tested and the abbreviations used in further descriptions. Feathers were cut into sections, marked for periodic recovery, and sampled every two weeks (see supplemental information and Figure S1 for more information). Only data from week six are shown herein, except for buried feathers with added microbes in elevated CO_2 (CPBB). Under this condition, no material remained for testing; we show data from the four-week timepoint for this condition in all figures except SEM images for elevated CO_2 conditions. When we could discern original color, it was noted in the figure legend.

In addition to micromorphological changes, we tested the effect of degradation on antibody binding using a polyclonal antibody raised against extant mature white feathers, which are comprised almost completely of feather corneous β-protein (CβK [39]). In all cases, there was a bacterial component to the experimental conditions, including the E-Pure water conditions, because feathers still rested on unsterilized sediments infiltrated with bacteria in pondwaters or present on the feathers themselves. This is appropriate, as there are no naturally occurring environments that are devoid of bacterial influence. Normal microbial flora would be expected to be present in waters, sediment/sand, and the feathers themselves in all conditions.

3. Results

3.1. Transmitted Light Microscopy (LM)

Figure 1 shows feathers subjected to the above conditions and subsequently imaged using LM under ambient atmosphere (normal air, at room temperature), subjected to the above conditions. Images were taken after six weeks of degradation. For each condition (row), the two left-most panels represent buried feathers, and the right two panels are feathers exposed at the surface.

Under ambient conditions, feathers showed little obvious degradation after two weeks (not shown), but at six weeks, in most cases, degradation was obvious in both buried and exposed feathers with natural pondwater, and greater when *B. licheniformis* was added. In the former (PB, PE; Figure 1A–D), white feather regions that were buried showed more extensive fraying than observed for black feathers, and the buried feathers were more degraded than those exposed at the surface. Both black and white regions of the feathers were significantly degraded in the buried condition (Figure 1A,B), but in the unburied condition, the black regions of the feather were generally less degraded than the white regions (Figure 1C,D). In some regions of the white buried feather, no barbs could be seen. The white regions of the rachis were crumbling and covered in a fine, crystalline material. When *B. licheniformis* was added to the feather setup (Figure 1E–H), degradation was increased over the pondwater-only condition. In general, degradation, as measured by fraying, absence of barbs, and loss of integrity, was greater when feathers were buried, and visual inspection showed that white feather regions suffered greater degradation than did black ones, although white rachises were relatively intact.

When HA was added to pondwaters, preservation was enhanced in both buried (PMB) and unburied (PME) states (Figure 1I–L). In all cases, the barbs and barbules were intact and color was still discernible. Similarly, there was virtually no degradation visible in feathers exposed to E-Pure water with added HA (ESMB/E), even after six weeks (Figure 1M–P). The barbs, barbules, and vane structures were intact in both the white and black regions, both buried (M,N) and unburied (O,P), despite the presence of native microbes in the sediments and water. This was also seen in the feathers incubated with E-Pure water without added minerals (ESB/E, Figure 1Q–T), although some fraying of the white barbs was visible in the exposed regions (Figure 1T).

Figure 1. Feathers degraded in ambient atmospheres after 6 weeks. (**A,B**) Buried and (**C,D**) exposed in natural pondwater; (**E,F**) buried and (**G,H**) exposed feathers with pondwater and added microbes; (**I,J**) buried and (**K,L**) exposed feathers with pondwater and added HA; (**M,N**) buried and (**O,P**) exposed feathers with E-Pure water and HA; (**Q,R**) buried and (**S,T**) exposed feathers with E-Pure water only. Scale is 500 μm for the 1st and 3rd columns, and 200 μm for the 2nd and 4th columns.

Figure 2 shows an experimental set-up identical to Figure 1, but degradation was conducted in an elevated CO_2 atmosphere (5000 ppm), reflecting estimates of some of the highest levels of naturally occurring (i.e., non-anthropogenic) atmospheric CO_2 of the Phanerozoic (e.g., [48]). In each row of images taken after six weeks (except panel E,F, at four weeks), the first two panels were buried and the second two exposed on the surface, as above.

Generally, under light microscopy, degradation appeared more intense in the elevated CO_2 atmosphere than ambient in CP or CPB conditions. Buried feathers were degraded to completion in CPBB and CPBE, possibly indicating an upregulation of urease or carbonic anhydrase enzyme production by these microbes under elevated CO_2 [56]. Because no feathers were recovered at the six-week endpoint, here we include the four-week buried data point (Figure 2E,F). In all other cases, both black and white regions of the feather could still be differentiated under transmitted light; barbs and barbules appeared frayed by the sixth week but were still intact. Buried feathers appeared slightly more degraded than unburied ones, as seen in ambient atmospheres as well, and black regions of feathers in

CPB/CPE (Figure 2A–D) and CPBB/CPBE (Figure 2E,H) were slightly less degraded than white feathers.

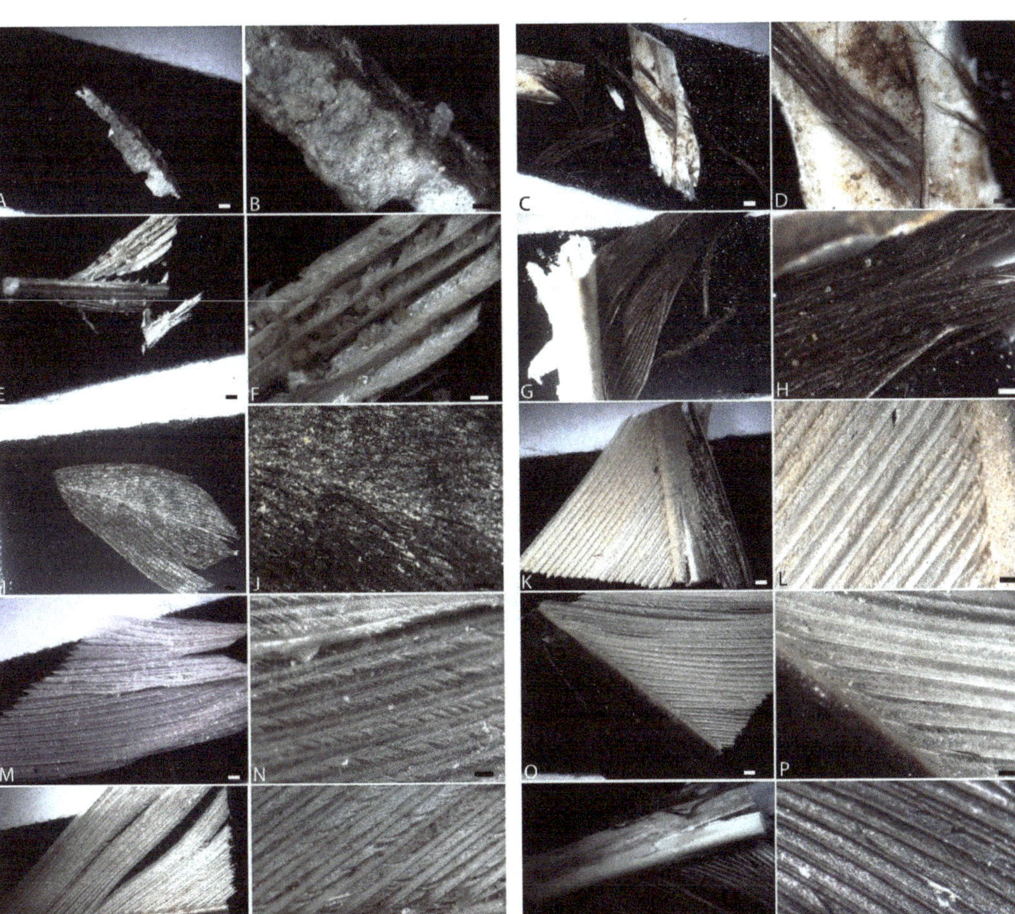

Figure 2. Feathers degraded in elevated CO_2 atmospheres after 6 weeks. (**A,B**) Buried and (**C,D**) exposed with natural pondwater; (**E,F**) buried (4 weeks) and (**G,H**) exposed feathers with pondwater and *B. licheniformis*; (**I,J**) buried and (**K,L**) exposed feathers with pondwater and added HA; (**M,N**) buried and (**O,P**) exposed feathers with E-Pure water and HA; (**Q,R**) buried and (**S,T**) exposed feathers with E-Pure water only. Scale bar for 1st and 3rd columns is 500 µm, and for the 2nd and 4th, 200 µm.

3.2. Scanning Electron Microscopy (SEM)

We used SEM to test the hypothesis that black feathers are more resistant than white feathers under the conditions described above. The SEM images in Figures 3 and 4 were taken after six weeks of degradation. The conditions under which data were collected followed what is shown in Figures 1 and 2. Feathers were differentiated according to color, when possible to discern. Figure 3 represents degradation in ambient atmosphere; Figure 4 is feathers degraded in elevated CO_2 atmospheres.

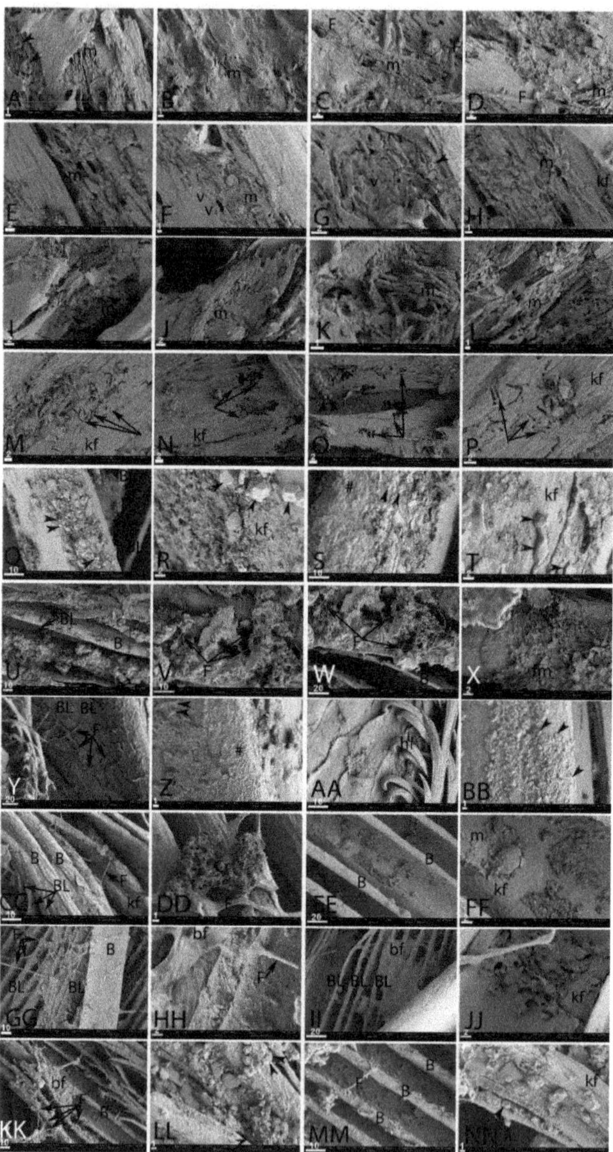

Figure 3. Feathers degraded under ambient conditions. (**A–D**) Buried in pondwater; (**A,B**) are black and (**C,D**) are white. (**E–H**) are feathers exposed at the surface; colors are not discernable. (**I–L**) Feathers buried in pondwater to which *B. licheniformis* has been added. (**M–P**) Surface-exposed feathers in pondwater with *B. licheniformis*; color is not discernible in (**I–P**). Arrows show microbial tunneling. (**Q–T**) represent feathers buried in pondwater with added HA; (**Q,R**) are black and (**S,T**) are white. A film (**#**) covers some regions; also (**U–X**) are exposed feathers with pondwater and HA; colors are not discernible. (**Y–BB**) are feathers buried with E-Pure water and HA; (**Y,Z**) are black and (**AA,BB**) are white. (**CC–FF**) are exposed feathers in E-Pure water with added HA; (**CC,DD**) are black and (**EE,FF**) are white. Finally, (**GG–JJ**) are buried and (**KK–NN**) are exposed in E-Pure water only; (**GG,HH,KK,LL**) are black and (**II,JJ,MM,NN**) are white. Scale as indicated in μm. kf, keratin filaments; bf, biofilm; m, microbial bodies; v, voids; F, fungal hyphae.

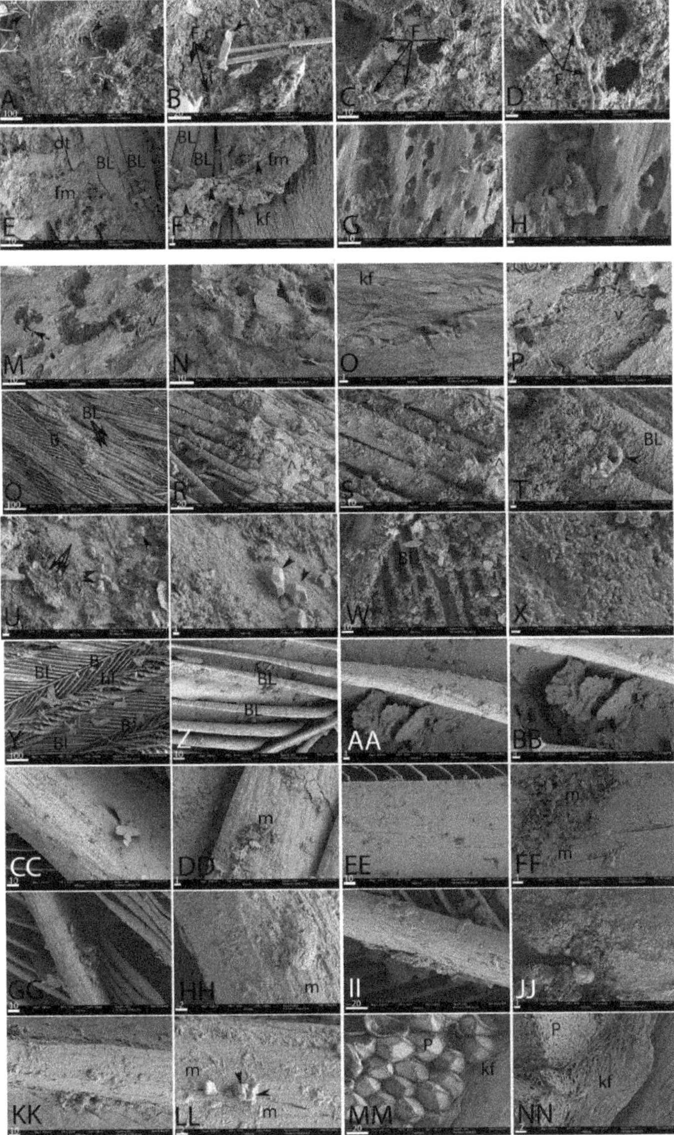

Figure 4. Feathers degraded for 6 weeks under elevated CO$_2$. 4 (**A–D**) are buried in pondwater, color is not discernible. (**E–H**) are surface-exposed; (**E,F**) are black and (**G,H**) are white. Feathers buried in pondwater to which *B. licheniformis* was added did not persist to 6 weeks and no data are shown. (**M–P**) Surface-exposed feathers in pondwater with *B. licheniformis*. Color is only discernible in (**N**), which is black. (**Q–T**) represent feathers buried in pondwater with added HA; all feathers shown are black. (**U–X**) are surface-exposed with pondwater and HA; (**U,V**) are black and (**W,X**) are white. (**Y–BB**) are feathers buried with E-Pure water and HA; all are white. (**CC–FF**) are exposed feathers in E-Pure water with added HA; (**CC,DD**) are black and (**EE,FF**) are white. (**GG–NN**) Incubated in E-Pure water only; (**GG–JJ**) are buried and (**KK–NN**) are surface-exposed. (**GG,HH,KK,LL**) are black, (**II,JJ**) are white, and (**MM,NN**) are indeterminate. Scales as indicated, in µm. kf, keratin filaments; bf, biofilm; m, microbial bodies; v, voids; F, fungal hyphae; ^, amorphous film.

In ambient atmosphere, feathers degraded for six weeks in pondwater showed little difference between buried (PB, Figure 3A–D) and exposed (PE, Figure 3E–H). Highly aligned, confluent microbial mats (m) could be seen on the surface of the fragmented keratinous outer cover in both black (3A–B) and white (3C–D) feathers in both PB and PE samples. Fungal hyphae (F) could also be seen. Occasionally, mineral crystals (Figure 3A, arrowheads) could be seen.

When feathers were exposed on the surface (Figure 3E–H), degradation appeared to be slightly less, but confluent, ordered microbial bodies (m) were seen on the surface of the feathers in all cases. The keratinous outer sheaths were less fragmented, but invasive microbial populations were still visible. Microbial impressions, or "voids" (V), were visible on the surface of the keratinous sheath (Figure 3F,G). Figure 3H shows a bundle of fraying keratin fibers (kf). Mineral crystals and/or diatoms were also visible (Figure 3G, arrowheads). The keratin sheath was densely pitted (Figure 3F, left).

When feathers were soaked in pure cultures of *B. licheniformis* and then added to the normal pondwater flora, degradation was greatly intensified in both the buried (PBB, I–L) and unburied (PBE, M–P) feathers. The keratinous material from the buried samples was fragmented and riddled with holes. Microbial bodies (m) were visible in all panels. In both PBB and PBE, degradation followed a different pattern, clearly visible as microbial tunneling (arrows, Figure 3M–P). This was not observed unless *B. licheniformis* were added, and may be specific to this microbial group. Microbial bodies were not as readily visible in the exposed feathers relative to the buried samples. Bundles of keratin fibers (kf) were loose and ropy. Small regions of apparent crystal growth (arrowheads) can be seen in in direct association with the microbes in Figure 3I.

Buried feathers in natural pondwaters to which HA had been added (Figure 3Q–T) showed better preservation than previous conditions. Barbs (B) and barbules (BL) were still visible (Figure 3Q) and accumulations of geometric crystal growth (arrowheads) were seen on all feather surfaces, both black (Figure 3Q,R) and white (Figure 3S,T). In some cases, an amorphous film could also be seen coating feather surfaces (Figure 3S,Y,Z, (#)). However, deep in the crystal growth and amorphous film, fibers of keratin were still aligned and appeared intact. The unburied feathers with added HA (Figure 3U–X) also showed better preservation than the previous conditions. Feather barbs (B) and barbules (BL) were visible, and showed virtually no degradation; they appeared to be coated in mineral crystals, which may have stabilized the organic structures [57–59]. Fungal hyphae could be seen (Figure 3V,W (F)) but microbial bodies were rare in this condition. Figure 3X shows a fibrous mat of material (fm), the identity of which is uncertain. It was not possible to differentiate black and white feathers.

Buried feathers incubated with E-Pure water with added HA (ESMB) under ambient conditions showed virtually no degradation (Figure 3Y–BB). In some regions of these feathers, fungal hyphae were prevalent (Figure 3Y, (f)), and the same amorphous deposits (#) and mineral crystals (arrowheads) were visible on the feather surfaces. In Figure 3Z, this amorphous and slightly crystalline film (#) appeared to completely coat regions of the buried feather. Barbs, barbules, and hooklets (hl) were visible (Figure 3AA). Degradation did not differ measurably between black (Figure 3Y–Z) and white (Figure 3AA,BB) feathers.

Feathers exposed on the surface and incubated with E-Pure water and added HA (ESME) showed exceptional preservation, with virtually no degradation, although mineral deposition was noted on the surfaces of some regions. There was no obvious difference in preservation between black (Figure 3CC,DD) and white (Figure 3EE,FF) feathers. Unaltered barbs and barbules (BL) could be seen, and filaments of keratin comprising these were visible as aligned fibers (KF). Flattened fungal hyphae (F) were visible, in some cases with outgrowths of bushy, finely crystalline structures (Figure 3DD, (Cr)). These "floret-like" structures were also visible in other regions of unburied feathers and were associated with fungal hyphae. Very few microbial bodies (m) were observed in either buried or unburied feathers in this condition, but a few could be seen in association with this floret-like structure, and they appeared both as bacilliform and coccoid structures (Figure 3FF, (m)).

Feather preservation was also greater when incubated in E-Pure water in both buried (ESB, Figure 3GG–JJ) and unburied (ESE, Figure 3KK–NN) states, consistent with LM data, and virtually no differences were seen between black (Figure 3GG,HH) and white (Figure 3II,JJ) regions. Fungal hyphae (F) were prevalent in buried feathers (and appeared, in some cases, to be associated with a thin amorphous coating tentatively identified as biofilm (bf)) (Figure 3HH,II). This amorphous film appeared to coat buried feather barbs almost completely in some regions.

Some regions of amorphous film (bf) were associated with fungal hyphae in the unburied feathers (Figure 3KK–NN) as well. Feather barbs and barbules were clearly visible, and although crystals had deposited on the surface (arrowheads), the structure of the feather was intact. Even though no excess HA was added to these burial conditions, in some areas, mineral crystals were still visible. The fibrous texture of the keratin comprising the barbs was intact (Figure 3NN).

Figure 4 shows feathers exposed to the same conditions as Figure 3, but conducted in an elevated CO_2 atmosphere. When feathers were buried and incubated with natural pondwaters (CPB, Figure 4A–D), virtually no feather structure remained after six weeks, but fungal hyphae were evident (F) and the feathers revealed highly degraded, pitted surfaces (e.g., Figure 4C,D) colonized by fungi (F). Needle-like crystals (arrowheads) were observed on the surface (Figure 4A,B). The original color of the feathers could not be discerned.

The unburied feathers in this elevated CO_2 pondwater environment (CPE) fared only slightly better (Figure 4E–H). Some of the barbules (Figure 4E–F, (BL)) remained intact, but in parts they were covered with a regional overgrowth of fibrous material (fm). Mineral crystals (arrowheads) and occasionally diatoms (dt) could be seen. However, in other regions, the surface was riddled with holes (Figure 4G,H). Ropey keratin filaments (Figure 4F, (kf)) could be seen. In contrast to the comparable ambient condition, microbial bodies were rarely visible.

In a high CO_2 environment, after six weeks no buried feathers were recovered from the CPBB condition (missing data, Figure 4I–L). However, the feathers exposed at the surface (CPBE) are shown in Figure 4M–P. Microbial tunneling (Figure 4M, arrows) was visible in this condition, as was seen in the corresponding ambient condition. Figure 4M is the only sample where original color could be determined; this feather was black. Figure 4N shows a flaky material, possibly representing degraded keratin or, alternatively, fine clay grains. Although Figure 4O reveals aligned keratin fibers (kf), in most samples the outermost surface of keratin was largely degraded. Contrary to the ambient condition, microbial bodies were difficult to see, but their impressions within the degraded keratin were visible as aligned voids (Figure 4M,P; (v)).

When HA was added to pondwater in the elevated CO_2 environment, preservation was improved in both the buried (CPMB, Figure 4Q–T) and unburied (CPME, Figure 4U–X) conditions. Fine feather structure, including barbs and barbules, remained visible, but a coating of either granular or smooth material (^) could be seen on the buried feathers, that is smoother and less granular than shown in Fig. 3; as a result we used a different symbol. A similar heterogenous material covered the surface of the unburied feathers (Figure 4U–X) to an even greater extent, but beneath this layer, feather barbs appeared intact. Diatoms and other structures were also seen. Geometric mineral crystals (arrowheads) could be seen interspersed with small, round, concave structures approximately 2 μm in diameter (arrows). In Figure 4W, mineral overgrowth on barbules was visible, and thin, plate-like features that could be degraded keratin or clay grains were visible.

Preservation was greatly enhanced in feathers in both buried (CESMB, Figure 4Y–BB) and exposed (CESME, Figure 4CC–FF) conditions in elevated CO_2 when HA was added. Feather structure was almost perfectly preserved, but interspersed with the barbules in the buried feathers were small plate-like structures. These can be seen more clearly in higher-magnification images (Figure. 4BB). In the unburied feathers (Figure 4CC–FF), preservation was virtually perfect, and no alteration of structure was visible in low-magnification images

(Figure 4CC,EE). At higher magnifications, small pockets of microbes appeared on the surface (Figure 4DD,FF, m). Thin, plate-like structures could also be seen.

In the high-CO_2 environment, feathers buried with E-Pure water alone were almost as well preserved as the preceding condition (CESB, Figure 4GG–JJ). Barbs and barbules were preserved with no evidence of fraying or breakdown, although at higher magnifications, some regions appeared to be covered in a crystalline coating (Figure 4JJ), and rarely, microbodies were seen on the surface (Figure 4HH). In the exposed feathers incubated with E-Pure water only, virtually no degradation was seen (Figure 4KK–NN) and barbs and barbules were intact. Small crystals of mineral precipitate were occasionally seen on the surface of the feather (Figure 4LL, arrowhead), and the presence of a few microbial bodies (m) were also noted on the feather surface. Figure 4MM–NN shows a region where keratin filaments (kf) could be seen extending to, and surrounding, a region of geometric pith (P).

3.3. Transmission Electron Microscopy (TEM)

Transmission EM allowed us to study degradation with greater resolution. With ultrathin sections (~90 nm thickness), it is impossible to discern whether black or white regions of the embedded feathers were sectioned. In most cases, where electron-opaque melanosomes were not visualized and no voids were present, we assumed these to be white regions. In all cases, melanosomes, when visible, were most abundant in the feather barbules. They were uniformly opaque to electrons and always embedded in the keratinous matrix, not displayed on the outer surface, as we have previously shown [60]. They did not appear to overlap within the keratin matrix, but were well separated.

Figure 5 shows the various conditions in ambient atmospheres after six weeks of degradation; the first two panels in each row were buried and the second two exposed at the surface. Degradation in pondwater was greater in the buried feathers (PB, Figure 5A,B) than the exposed ones (PE), with a keratinous matrix developing holes. In both buried and exposed feathers, the melanosomes were relatively intact, although some degradation was seen. When *B. licheniformes* were present with pondwater (PBB, PBE; Figure 5E–H), degradation was much more advanced. The keratin matrix was highly degraded in PBB feathers (Figure 5E,F), and the melanosomes showed a great loss of integrity. In the PBE feathers (Figure 5G,H), the keratin matrix showed a loss of smoothness, taking on a "bubbly" texture, but was relatively intact. Melanosomes were not prevalent. This could be because of the region of the feather imaged, as barbules contain more melanosomes than the rest of the feather [61]. When HA was added to the pondwater (PMB, PME; Figure 5I–L), degradation was much less than in previous conditions, and mineral crystals could be seen associated with feather surfaces. Melanosomes and matrix were intact in both cases, although some fraying and loss of integrity was seen in the PME feathers (Figure 5L). Feathers incubated in E-Pure water to which HA was added (ESMB, ESME; Figure 5M–P) showed loss of mineral integrity. No melanosomes were visualized, probably indicating that this was a white feather region. Finally, Figure 5Q–T show feathers incubated in only E-Pure water. Degradation was minimal, but greater than seen when HA was added. Melanosomes were round and almost completely solid; however, in the exposed feathers, some cracking of the keratin matrix was visible.

Figure 6 shows TEM images of feathers degraded in an atmosphere elevated in CO_2. In most conditions, degradation was advanced compared to that seen in ambient conditions. Figure 6A–B, depicting CPB buried feathers, show advanced degradation, with many melanosomes completely degraded, and those remaining showing loss of integrity. Exposed feathers (CPE) fared better, with elongated and round melanosomes present within the keratin matrix. When *B. licheniformis* was added to the pondwater, all feathers were completely degraded at six weeks, so as above, images (Figure E–H) were taken after four weeks. The buried feathers (CPBB) showed greater degradation, with most melanosomes no longer intact. In the exposed feathers (CPBE; Figure 6G,H), voids can also be seen where melanosomes once resided. In high-CO_2 environments, when HA was added to pondwater, degradation was greatly decreased. Melanosomes showed slight degradation in the buried condition (CPMB; Figure 6I,J), with some apparently lysed. Electron-opaque, needle-like mineral deposits (mc) were seen outside of the keratin matrix. Exposed feathers (CPME; Figure 6K–L) showed better preservation, with little degradation of either melanosomes or keratin. There was an unidentified growth on the outer surface of the keratin that was most likely organic, based on its electron translucent character.

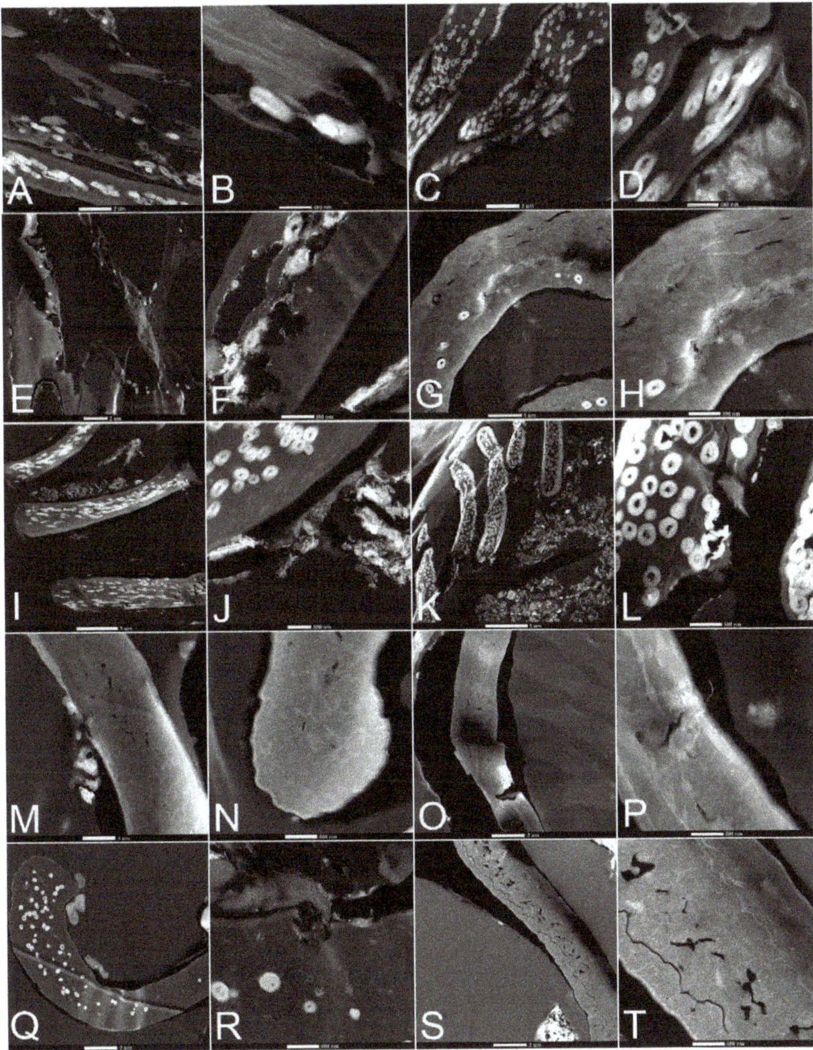

Figure 5. Feathers degraded in ambient conditions. (**A–D**) Exposed to pondwater; (**A,B**) buried and (**C,D**) exposed. (**E–H**) Pondwater with *B. licheniformis* added; (**E,F**) buried and (**G,H**) exposed. (**I,L**) are feathers incubated with pondwater and added HA; (**I,J**) buried and (**K,L**) exposed. Electron-opaque, needle-like mineral crystal (mc) deposits are seen exterior to the keratin of the feather barbules. (**M–P**) show feathers incubated in E-Pure water to which HA has been added. No melanosomes are seen, probably indicating that these are white feather regions, although early degradation is a less likely explanation. (**Q–T**) show feathers incubated in E-Pure water only. (**S**) low magnification, (**T**) higher magnification showing some cracking of the feather matrix. Scale bars as indicated.

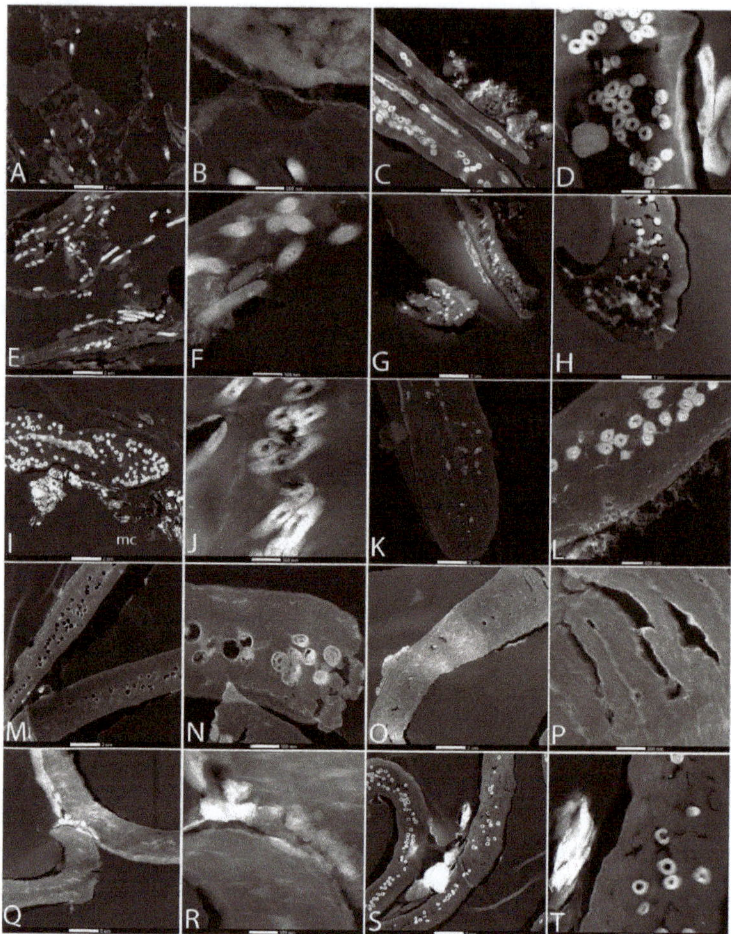

Figure 6. Transmission electron microscopy (TEM) of feathers degraded in elevated CO_2 atmosphere. (**A,B**) are buried and (**C,D**) are exposed in natural pondwaters. (**E–H**) show feathers in natural pondwater with *B. licheniformis* added; (**E,F**) are buried (after 4 weeks) and (**G,H**) are exposed. (**I–L**) Pondwater to which HA has been added; (**I,J**) are buried and (**K,L**) are exposed at the surface. (**M,N**) are buried and (**O,P**) are exposed feathers degraded in E-Pure water to which HA had been added. Finally, (**Q–T**) are feathers in elevated CO_2, degraded in E-Pure water only; (**Q,R**) are buried and (**S,T**) are exposed. Scale bars as indicated.

Buried feathers incubated in elevated CO_2 with E-Pure water to which HA had been added (CESMB; Figure 6M,N) showed relatively advanced degradation of melanosomes, leaving voids, but with little degradation of the keratin matrix in which they were embedded. Exposed feathers (CESME; Figure 6O,P) showed similar degradation. The keratin matrix was relatively unaltered. Melanosomes were rare, but empty voids were present, supporting the idea that these feathers contained pigment organelles.

Feathers incubated in E-Pure water only showed little overall degradation. Buried feathers (CESB; Figure 6Q,R) showed little keratin degradation but no melanosomes or voids, and therefore probably represent white feather regions. Undegraded, hollow melanosomes were visible embedded deep within the keratin matrix in the exposed condition (CESE; Figure 6S,T).

3.4. In Situ Immunohistochemistry (IHC)

We tested the hypothesis that degradation influences antibody binding. Figure 7 shows the response of feathers in different degradation conditions when exposed to a polyclonal antiserum raised against feathers [62]. All images were taken under identical parameters. In the ambient condition, no reduction in intensity of binding was seen, even in the feathers showing the greatest morphological damage (Figure 7A,B,E,F), consistent with preservation of epitopes after structural integrity was lost. Antibody specificity is supported by the lack of binding to the embedding material, or to regions where tissues were completely missing. Negative controls of no primary antibody applied, but all other steps being the same (not shown), indicates that the signal did not arise from non-specific binding of the secondary antibody or fluorescent label. Melanized feather regions (Figure 7A,C,I,K) were more easily visualized in transmitted LM than those regions apparently lacking melanin (Figure 7E,G,M,O,Q,S), but antibodies bound with equal intensity with both types degraded for the duration of this experiment.

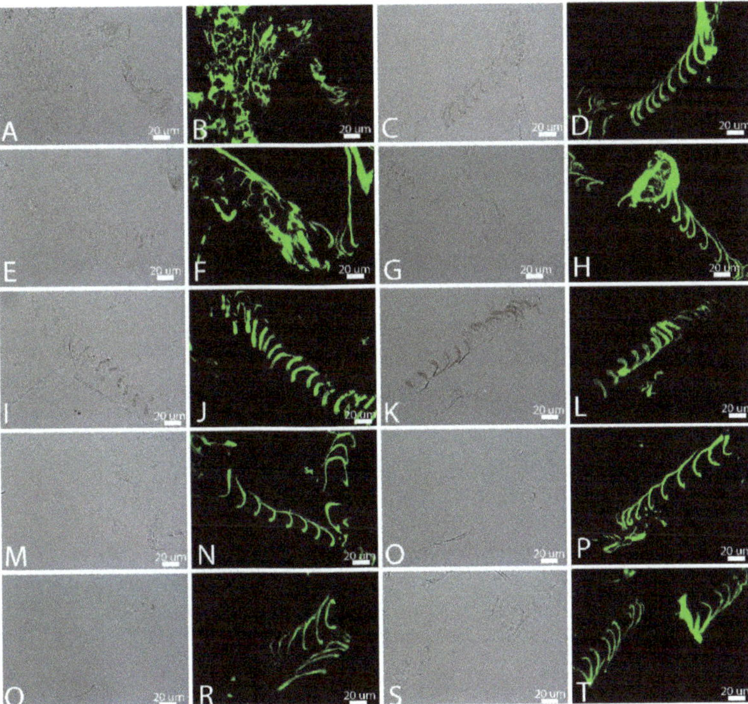

Figure 7. In situ immunochemistry of feathers degraded in ambient atmosphere, then exposed to a polyclonal antiserum raised against whole feather extract and visualized in transmitted light (1st and 3rd image of each row) or using an FITC filter (2nd and 4th image). (**A,B**) are buried and (**C,D**) are exposed in natural pondwaters. (**E–H**) show feathers in natural pondwater with added *B. licheniformis* at 4 weeks; (**E,F**) are buried and (**G,H**) are exposed. (**I–L**) Pondwater to which HA has been added; (**I,J**) are buried and (**K,L**) are exposed at the surface. (**M,N**) are buried and (**O,P**) are exposed feathers degraded in E-Pure water to which HA has been added. Finally, (**Q–T**) are feathers degraded in E-Pure water only; (**Q,R**) are buried and (**S,T**) are surface-exposed. Transmitted light images are modified to increase contrast; see Supplementary Materials.

When feathers were degraded under the same conditions as in Figure 7, but in elevated CO_2 atmospheres, the results were different. Buried feathers exposed to pondwater only (CPB; Figure 8A,B) showed a highly degraded area covering a more intact region. Polyclonal antibodies bound with high avidity, but no melanized regions could be seen with certainty in the transmitted LM (Figure 8A). The exposed feathers (CPE) with pondwater demonstrated better microscopic integrity. Regions of melanized barbules (Figure 8C) could be seen in LM and antibodies bound with the same level of intensity (Figure 8D), as seen in Figure 7. When feathers were buried and incubated with pondwater and *B.licheniformis* in elevated CO_2 (CPBB), no feathers remained for analyses; data are shown for this condition in feathers after four weeks degradation (Figure 8E,F). Integrity was greatly reduced, despite melanin in the visible barbs (Figure 8E). Surface-exposed feathers (CPBE; Figure 8G,H) showed minimal degradation, despite containing less or no melanin. Figure 8I–L shows buried (Figure 8I,J) and exposed (Figure 8K,L) feathers incubated with pondwater with added HA. Both buried and exposed feathers showed melanized regions that remained virtually intact, with no visible degradation, and both bound antibodies with equal and intense avidity. Similarly, Figure 8M–P shows buried (M,N) and exposed (O,P) feathers in E-Pure water to which HA had been added. Virtually no degradation of epitopes was seen, and no substantial differences between lightly melanized and buried (M,N) or unmelanized and exposed (O,P) were seen. Non-melanized buried (Figure 8Q,R) and melanized, exposed (Figure 8S,T) feathers in E-Pure water only were comparable, showing no tissue degradation and no reduction in antibody binding under these conditions.

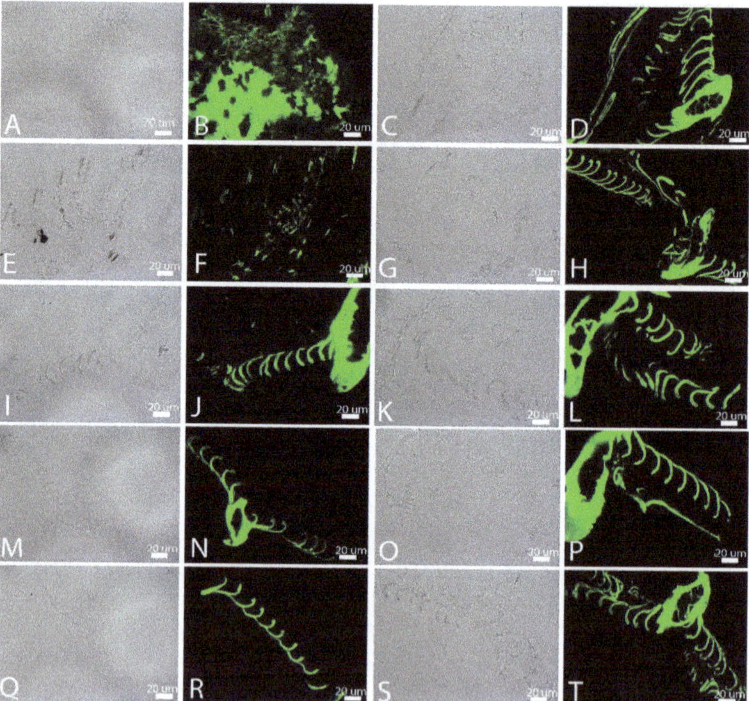

Figure 8. In situ immunochemistry of feathers degraded in elevated CO_2 atmosphere, then exposed to a polyclonal antiserum raised against whole feather extract and visualized in transmitted light (1st and 3rd image of each row) or using an FITC filter (2nd and 4th image). (**A,B**) are buried and (**C,D**) are exposed in natural pondwaters. (**E–H**) show feathers in natural pondwater with added *B. licheniformis*; (**E,F**) are buried (after 4 weeks) and (**G,H**) are exposed. (**I–L**) Pondwater to which $CaPO_4$ has been added; (**I,J**) are buried and (**K,L**) are exposed at the surface. (**M,N**) are buried and (**O,P**) are exposed feathers degraded in E-Pure water to which HA had been added. Finally, (**Q–T**) are feathers in elevated CO_2, degraded in E-Pure water only; (**Q,R**) are buried and (**S,T**) are surface-exposed.

4. Discussion

We showed, using multiple methods, that the degradation of organics is qualitatively and/or quantitatively different in atmospheres elevated to Mesozoic CO_2 levels (or higher) than what is observed in ambient atmospheres. Some microbes are known to precipitate CO_2 as carbonate minerals [63–65], and we sought to determine whether this precipitation could be increased in atmospheres elevated in CO_2. We chose hydroxyapatite because microbes are known to precipitate this mineral [66], and because the solubility of HA rises with increased CO_2 [67]. Because it has been shown that organic molecules can be preserved preferentially in intracrystalline regions of bone [68], we predicted that we may see better molecular preservation in extant samples subjected to elevated CO_2 through rapid stabilization by mineral precipitation. We used both melanized and non-melanized feathers to test the effect of pigment on preservation [47,69]. We added feather-degrading microbes to natural microbial biomass to see whether the action of these organisms was increased in high CO_2 atmospheres, and we tested the effect that increasing the concentration of HA would have on preservation and degradation in varying atmospheres. Finally, we compared preservation when feathers were exposed to waters containing a natural microbial population vs. E-Pure water.

In ambient atmospheres, degradation was greatest when feather-degrading bacteria were added to feathers in the natural microbiota of pondwaters. In general, buried feathers were more highly degraded than surface-exposed feathers, and black regions retained more structural integrity than did white feather regions. The greater degradation observed for buried feathers was not predicted, but we hypothesize that burial gave greater access to microbes by keeping the feathers uniformly damp. In the other conditions, there was little difference, but importantly, adding HA to the waters containing microbes greatly increased preservation, perhaps by slowing the microbially mediated degradation undertaken by microbes naturally present in feathers. In E-Pure water, with or without minerals, virtually no degradation was seen, and black feathers did not differ from white in preservation, supporting a specific role for microbially mediated precipitation in preservation. Feathers in elevated CO_2 showed much greater degradation, whether buried or exposed, in both natural pondwaters and with added microbes; however, in the presence of added HA, degradation was essentially halted, and preservation was relatively greater in elevated CO_2. SEM data for both natural waters and waters with added microbes demonstrated multiple holes and tunneling not seen in ambient atmospheres, suggesting upregulation of microbial enzymes of degradation in elevated CO_2. Increased microbial activity in response to elevated CO_2 was noted in other studies as well [70–73]. The increase in preservation in both atmospheric conditions when HA was added suggests the deposition of mineral-arrested degradation in both buried and unburied states, as suggested elsewhere [74,75], either by adsorbing enzymes of degradation and inactivating them (e.g., [76], by encasing organics, thus making them inaccessible [77]), or both, but this process was relatively enhanced in elevated CO_2. Where color could be detected, there was no measurable difference in preservation/degradation in most conditions. Keratin filaments and intact barbs, barbules, and sheaths showed little to no degradation under the elevated CO_2 + mineral condition. Fungal hyphae, prevalent in the ambient conditions, were not discerned in elevated CO_2 except in the pondwater-only condition.

TEM illustrated high levels of degradation of the keratin sheath in pondwater conditions, both buried and unburied, but melanosomes remained relatively intact, though some degradation of these bodies was visible. However, in buried feathers with added *B. licheniformis*, melanosome and keratin degradation were both advanced over other conditions. Crystal deposits were seen when HA was added, and in these conditions, the feathers appeared unaltered. Patterns were similar in the elevated CO_2 atmospheres for both pondwater and with added microbes, again with datapoints representing only four weeks in the buried condition. In elevated CO_2 with added HA, melanosomes appeared more degraded than did keratin, leaving many voids in the buried feathers.

In ambient atmosphere, antibody binding was intense in all cases, with little difference noted between all conditions shown here. Burial state, added microbes, and presence or absence of melanin did not appear to affect the recognition of keratin epitopes by these polyclonal antibodies, and binding occurred even when structural integrity was not well maintained. Although no material persisted to the six-week timepoint in buried feathers with added *B. licheniformis* in elevated CO_2, the only case where antibody binding was detectably reduced was in the buried feathers after four weeks. Both structural integrity and epitope recognition were greatly diminished.

Processes resulting in the preservation of normally labile and easily degraded organic materials have been of significant interest to the paleontological community, because examples of hair, skin, feathers, and internal organs do not follow standard models of fossilization. It is significant that incidences of exceptional soft tissue preservation are not evenly distributed through time, but occur more frequently in pre-Cenozoic deposits [78,79] during time periods when atmospheric CO_2 was elevated over today's levels [80,81]. In addition, Cenozoic lagerstätte are few, but generally correlate to short periods of elevated CO_2 [82–84]. We propose that this uneven distribution of exceptionally preserved deposits may be due in part to the response of microbes to elevations in atmospheric CO_2 and the subsequent acidification of pore waters, making mineral ions more available for precipitation. This preliminary study demonstrates the need for further investigation into the role of both microbes and CO_2 in preserving organic remains in fossils.

5. Conclusions

Degradation by all measures was greatest in buried feathers in elevated CO_2. However, the addition of HA slowed degradation, whether with natural pondwaters or E-Pure water, and this effect was greater, relative to level of degradation without minerals, in elevated CO_2. The rapid precipitation of minerals on organics outpaced decay in these conditions, illustrating a possible role in exceptional preservation, as suggested by [85–88] and others. Similarly, antibody recognition as a proxy for molecular preservation was still present, even in feathers showing microstructural degradation. The implications for fossilization support the hypothesis that degradation proceeded differently in the elevated CO_2 atmospheres of the Mesozoic. Although degradation was more rapid and more complete in buried feathers, antibody binding was not greatly affecting during the time interval of this experiment, suggesting a lack of direct correlation between histological integrity and antibody binding. In most cases, keratin preservation was equal to or greater than melanosomes under the conditions of this experiment, whatever the atmospheric conditions. We tested the role of apatite because, in atmospheres elevated in CO_2, as in the Mesozoic, warmer waters and the acidification of pore waters would allow an increase in the concentration of this solubilized mineral. Experimental taphonomy designed to address questions of degradation in the past, particularly in deep time intervals, will be misleading unless atmospheric composition and the effects of greenhouse gases on degrading microbes are taken into account.

Supplementary Materials: The following supporting information can be downloaded at https://www.mdpi.com/article/10.3390/biology11050703/s1: Figure S1: Experimental Setup.

Author Contributions: M.H.S. designed the studies, provided the feathers, collected the pondwater and sediment samples, designed the figures, and wrote most of the paper. W.Z. conducted experimental setups; collected LM, TEM, and IHC data; contributed to the paper, and wrote most of the Materials and Methods section. N.E. collected the SEM data. All authors have read and agreed to the published version of the manuscript.

Funding: This research received no external funding.

Institutional Review Board Statement: Not applicable.

Informed Consent Statement: Not applicable.

Data Availability Statement: All data dealing with this study are reported in the paper.

Acknowledgments: This work was performed in part at the Analytical Instrumentation Facility (AIF) at North Carolina State University, which is supported by the State of North Carolina and the National Science Foundation (award number ECCS-2025064). The AIF is a member of the North Carolina Research Triangle Nanotechnology Network (RTNN), a site in the National Nanotechnology Coordinated Infrastructure (NNCI). SEM images were collected at the Image and Chemical Analysis Laboratory (ICAL), Department of Physics, Montana State University. This work was also supported by the NSF (award number 1934844).

Conflicts of Interest: The authors declare no conflict of interest.

References

1. Pan, Y.; Zheng, W.; Moyer, A.E.; O'Connor, J.K.; Wang, M.; Zheng, X.; Wang, X.; Schroeter, E.R.; Zhou, Z.; Schweitzer, M.H. Molecular evidence of keratin and melanosomes in feathers of the Early Cretaceous bird Eoconfuciusornis. *Proc. Natl. Acad. Sci. USA* **2016**, *113*, E7900–E7907. [CrossRef] [PubMed]
2. Pan, Y.; Zheng, W.; Sawyer, R.H.; Pennington, M.W.; Zheng, X.; Wang, X.; Wang, M.; Hu, L.; O'Connor, J.; Zhao, T.; et al. The molecular evolution of feathers with direct evidence from fossils. *Proc. Natl. Acad. Sci. USA* **2019**, *116*, 3018–3023. [CrossRef] [PubMed]
3. Schweitzer, M.H.; Watt, J.A.; Avci, R.; Knapp, L.; Chiappe, L.; Norell, M.; Marshall, M. Beta-keratin specific immunological reactivity in feather-like structures of the cretaceous alvarezsaurid, Shuvuuia deserti. *J. Exp. Zool.* **1999**, *285*, 146–157. [CrossRef]
4. Lindgren, J.; Sjövall, P.; Carney, R.M.; Cincotta, A.; Uvdal, P.; Hutcheson, S.W.; Gustafsson, O.; Lefèvre, U.; Escuillié, F.; Heimdal, J.; et al. Molecular composition and ultrastructure of Jurassic paravian feathers. *Sci. Rep.* **2015**, *5*, 13520. [CrossRef]
5. Alonso, J.; Arillo, A.; Barron, E.; Corral, J.C.; Grimalt, J.; Lopez, J.F.; Lopez, R.; Martinez-Delclos, X.; Ortuno, V.; Penalver, E.; et al. A new fossil resin with biological inclusions in Lower Cretaceous deposits from Alava (northern Spain, Basque-Cantabrian Basin). *J. Paleontol.* **2000**, *74*, 158–178. [CrossRef]
6. Lindgren, J.; Sjövall, P.; Carney, R.M.; Cincotta, A.; Uvdal, P.; Hutcheson, S.W.; Gustafsson, O.; Lefèvre, U.; Escuillié, F.; Heimdal, J.; et al. Molecular composition and ultrastructure of jurassic paravian feathers. *Lundqua Thesis* **2018**, *2018*, 59–71.
7. Gren, J.A.; Sjovall, P.; Eriksson, M.E.; Sylvestersen, R.L.; Marone, F.; Sigfridsson, C.K.G.V. Molecular and microstructural inventory of an isolated fossil bird feather from the Eocene Fur Formation of Denmark. *Palaeontology* **2016**, *60*, 73–90. [CrossRef]
8. Manning, P.L.; Edwards, N.P.; Wogelius, R.A.; Bergmann, U.; Barden, H.P.E.; Larson, P.L.; Schwarz-Wings, D.; Egerton, V.M.; Sokaras, D.; Mori, R.A.; et al. Synchrotron-based chemical imaging reveals plumage patterns in a 150 million year old early bird. *J. Anal. At. Spectrom.* **2013**, *28*, 1024–1030. [CrossRef]
9. Chuong, C.M.; Ping, W.; Zhang, F.C.; Xu, X.; Yu, M.; Widelitz, R.B.; Jiang, T.X.; Hou, L. Adaptation to the Sky: Defining the Feather with Integument Fossils from Mesozoic China and Experimental Evidence from Molecular Laboratories. *J. Exp. Zool. Part B Mol. Dev. Evol.* **2003**, *298*, 42–56. [CrossRef]
10. Alibardi, L. Review: Cornification, morphogenesis and evolution of feathers. *Protoplasma* **2017**, *254*, 1259–1281. [CrossRef]
11. Greenwold, M.J.; Bao, W.; Jarvis, E.D.; Hu, H.; Li, C.; Gilbert, M.T.P.; Zhang, G.; Sawyer, R.H. Dynamic evolution of the alpha (α) and beta (β) keratins has accompanied integument diversification and the adaptation of birds into novel lifestyles. *BMC Evol. Biol.* **2014**, *14*, 249. [CrossRef] [PubMed]
12. Greenwold, M.J.; Sawyer, R.H. Linking the molecular evolution of avian beta (β) keratins to the evolution of feathers. *J. Exp. Zool. Part B Mol. Dev. Evol.* **2011**, *316*, 609–616. [CrossRef] [PubMed]
13. Dhouailly, D. A new scenario for the evolutionary origin of hair, feather, and avian scales. *J. Anat.* **2009**, *214*, 587–606. [CrossRef] [PubMed]
14. Prum, R.O.; Brush, A.H. The Evolutionary Origin and Diversification of Feathers. *Q. Rev. Biol.* **2002**, *77*, 261–295. [CrossRef]
15. Greenwold, M.J.; Sawyer, R.H. Genomic organization and molecular phylogenies of the beta keratin multigene family in the chicken (Gallus gallus) and zebra finch (Taeniopygia guttata): Implications for feather evolution. *BMC Evol. Biol.* **2010**, *10*, 148. [CrossRef]
16. Sawyer, R.H.; Knapp, L.W. Avian Skin Development and the Evolutionary Origin of Feathers. *J. Exp. Zool.* **2003**, *298*, 57–72. [CrossRef]
17. Sawyer, R.H.; Washington, L.D.; Salvatore, B.A.; Glenn, T.; Knapp, L.W. Origin of archosaurian integumentary appendages: The bristles of the wild turkey beard express feather-type b-keratins. *J. Exp. Zool. B Mol. Devel. Evol.* **2003**, *279*, 27–34. [CrossRef]
18. Sawyer, R.H.; Salvatore, B.A.; Potylicki, T.T.F.; French, J.O.; Glenn, T.C.; Knapp, L.W. Origin of Feathers: Feather Beta (β) Keratins Are Expressed in Discrete Epidermal Cell Populations of Embryonic Scutate Scales. *J. Exp. Zool. Part B Mol. Dev. Evol.* **2003**, *295*, 12–24. [CrossRef]
19. Hu, D.; Hou, L.; Zhang, L.; Xu, X.; Xing, X.; Xu, X. A pre-Archaeopteryx troodontid theropod from China with long feathers on the metatarsus. *Nature* **2009**, *461*, 640–643. [CrossRef]
20. Qiang, J.; Currie, P.J.; Norell, M.A.; Shu-an, J. Two feathered dinosaurs from northeastern China. *Nature* **1998**, *393*, 753–761. [CrossRef]
21. Kellner, A.W.A. A review of avian Mesozoic fossil feathers. In *Mesozoic Birds: Above the Heads of Dinosaurs*; Chiappe, L.M., Witmer, L.M., Eds.; UCal Press: Berkeley, CA, USA; Los Angeles, CA, USA; London, UK, 2002; p. 529. ISBN 0-520-20094-2.

22. Kundrát, M.; Rich, T.H.; Lindgren, J.; Sjövall, P.; Vickers-Rich, P.; Chiappe, L.M.; Kear, B.P. A polar dinosaur feather assemblage from Australia. *Gondwana Res.* **2020**, *80*, 1–11. [CrossRef]

23. Martin, L.D.; Czerkas, S.A. The fossil record of feather evolution in the mesozoic. *Am. Zool.* **2000**, *40*, 687–694. [CrossRef]

24. Xu, X.; Zhao, Q.; Norell, M.; Sullivan, C.; Hone, D.; Erickson, G.; Wang, X.; Han, F.; Guo, Y. A new feathered maniraptoran dinosaur fossil that fills a morphological gap in avian origin. *Chin. Sci. Bull.* **2009**, *54*, 430–435. [CrossRef]

25. Wang, X.; Pittman, M.; Zheng, X.; Kaye, T.G.; Falk, A.R.; Hartman, S.A.; Xu, X. Basal paravian functional anatomy illuminated by high-detail body outline. *Nat. Commun.* **2017**, *8*, 14576. [CrossRef] [PubMed]

26. Xu, X.; Zheng, X.; Sullivan, C.; Wang, X.; Xing, L.; Wang, Y.; Zhang, X.; O'Connor, J.K.; Zhang, F.; Pan, Y. A bizarre Jurassic maniraptoran theropod with preserved evidence of membranous wings. *Nature* **2015**, *521*, 70–73. [CrossRef] [PubMed]

27. Zhang, F.; Zhou, Z.; Xing, X.; Wang, X.; Sullivan, C. A bizarre Jurassic maniraptoran from China with elongate ribbon-like feathers. *Nature* **2008**, *455*, 1105–1108. [CrossRef] [PubMed]

28. Lingham-Soliar, T.; Godefroit, P.; Sinitsa, S.M.; Dhouailly, D.; Bolotsky, Y.L.; Sizov, A.V.; McNamara, M.E.; Benton, M.J.; Spagna, P. A Jurassic ornithischian dinosaur from Siberia with both feathers and scales. *Science* **2014**, *345*, 451–455. [CrossRef]

29. Zheng, X.-T.; You, H.-L.; Xu, X.; Dong, Z.-M. An Early Cretaceous heterodontosaurid dinosaur with filamentous integumentary structures. *Nature* **2009**, *458*, 333–336. [CrossRef]

30. Yang, Z.; Jiang, B.; McNamara, M.E.; Kearns, S.L.; Pittman, M.; Kaye, T.G.; Orr, P.J.; Xu, X.; Benton, M.J. Pterosaur integumentary structures with complex feather-like branching. *Nat. Ecol. Evol.* **2019**, *3*, 24–30. [CrossRef]

31. Benton, M.J.; Dhouailly, D.; Jiang, B.; McNamara, M. The Early Origin of Feathers. *Trends Ecol. Evol.* **2019**, *34*, 856–869. [CrossRef]

32. Saitta, E.T.; Rogers, C.S.; Brooker, R.A.; Vinther, J. Experimental taphonomy of keratin: A structural analysis of early taphonomic changes. *Palaios* **2017**, *32*, 647–657. [CrossRef]

33. Vinther, J.; Nicholls, R.; Lautenschlager, S.; Pittman, M.; Kaye, T.G.; Rayfield, E.; Mayr, G.; Cuthill, I.C. 3D Camouflage in an Ornithischian Dinosaur. *Curr. Biol.* **2016**, *26*, 2456–2462. [CrossRef] [PubMed]

34. Von Meyer, H. Archaeopteryx lithographica (Vogel-Feder) und Pterodactylus von Solnhofen. *Geol. Palaeontol.* **1861**, *29*, 678–679.

35. Owen, R. On the Archaeopteryx of von Meyer, with a description of the fossil remains of a long-tailed species, from the lithographic limestone of Solenhofen. *Philos. Trans. R. Soc. London* **1863**, *153*, 33–47.

36. Von Meyer, H. Vogel-Feder und Palpipes priscus von Solenhofen. *Neues Jahrb. Miner. Geogn. Geol. Petrefakten-Kd.* **1861**, *561*, 1861a.

37. Briggs, D.E.G.G. The role of decay and mineralization in the preservation of soft-bodied fossils. *Annu. Rev. Earth Planet. Sci.* **2003**, *31*, 275–301. [CrossRef]

38. Griffiths, P.J. The isolated Archaeopteryx feather. *Archaeopteryx* **1996**, *14*, 1–26.

39. Moyer, A.E.; Zheng, W.; Schweitzer, M.H. Keratin durability has implications for the fossil record: Results from a 10 year feather degradation experiment. *PLoS ONE* **2016**, *11*, e157699. [CrossRef]

40. McKellar, R.C.; Chatterton, B.D.E.; Wolfe, A.P.; Currie, P.J. A Diverse Assemblage of Late Cretaceous Dinosaur and Bird Feathers from Canadian Amber. *Science* **2011**, *333*, 1619–1622. [CrossRef]

41. Amber, T.M.; Xing, L.; Mckellar, R.C.; Xu, X.; Tseng, K.; Ran, H.; Currie, P.J. Report A Feathered Dinosaur Tail with Primitive Plumage Trapped in Mid-Cretaceous Amber. *Curr. Biol.* **2016**, *26*, 3352–3360. [CrossRef]

42. Davis, P.G.; Briggs, D.E.G. Fossilization of feathers. *Geology* **1995**, *23*, 783–786. [CrossRef]

43. Briggs, D.E.G. The role of biofilms in the fossilization of non-mineralized tissues. In *Fossil and Recent Biofilms*; Krumbein, W.E., Ed.; Springer Science and Business Media: Dordrecht, The Netherlands, 2003; pp. 281–290.

44. Parry, L.A.; Smithwick, F.; Nordén, K.K.; Saitta, E.T.; Lozano-Fernandez, J.; Tanner, A.R.; Caron, J.B.; Edgecombe, G.D.; Briggs, D.E.G.; Vinther, J. Soft-Bodied Fossils Are Not Simply Rotten Carcasses—Toward a Holistic Understanding of Exceptional Fossil Preservation: Exceptional Fossil Preservation Is Complex and Involves the Interplay of Numerous Biological and Geological Processes. *BioEssays* **2018**, *40*, 1–11. [CrossRef] [PubMed]

45. Babcock, L.E.; Leslie, S.A.; Elliot, D.H.; Stigall, A.L.; Ford, L.A.; Briggs, D.E.G. The "Preservation Paradox": Microbes as a Key to Exceptional Fossil Preservation in the Kirkpatrick Basalt (Jurassic), Antarctica. *Sediment. Rec.* **2006**, *4*, 4–8. [CrossRef]

46. Wilby, P.R.; Briggs, D.E.G.G.; Bernier, P.; Gaillard, C. Role of microbial mats in the fossilization of soft tissues. *Geology* **1996**, *24*, 787–790. [CrossRef]

47. Slater, T.S.; McNamara, M.E.; Orr, P.J.; Foley, T.B.; Ito, S.; Wakamatsu, K. Taphonomic experiments resolve controls on the preservation of melanosomes and keratinous tissues in feathers. *Palaeontology* **2020**, *63*, 103–115. [CrossRef]

48. Fletcher, B.J.; Brentnall, S.J.; Anderson, C.W.; Berner, R.A.; Beerling, D.J. Atmospheric carbon dioxide linked with Mesozoic and early Cenozoic climate change. *Nat. Geosci.* **2008**, *1*, 43–48. [CrossRef]

49. Justyn, N.M.; Peteya, J.A.; D'Alba, L.; Shawkey, M.D. Preferential attachment and colonization of the keratinolytic bacterium Bacillus licheniformis on black- A nd white-striped feathers. *Auk* **2017**, *134*, 466–473. [CrossRef]

50. Li, J.; Liu, X.; Zhang, J.; Zhang, Y.; Han, Y.; Hu, J.; Li, Y. Synthesis and characterization of wool keratin/hydroxyapatite nanocomposite. *J. Biomed. Mater. Res. Part B Appl. Biomater.* **2012**, *100*, 896–902. [CrossRef]

51. Doney, S.C.; Balch, W.M.; Fabry, V.J.; Feely, R.A. Ocean acidification: A critical emerging problem for the ocean sciences. *Oceanography* **2009**, *22*, 16–25. [CrossRef]

52. Weiss, L.C.; Pötter, L.; Steiger, A.; Kruppert, S.; Frost, U.; Tollrian, R. Rising pCO2 in Freshwater Ecosystems Has the Potential to Negatively Affect Predator-Induced Defenses in Daphnia. *Curr. Biol.* **2018**, *28*, 327–332.e3. [CrossRef]

53. Oh, N.H.; Richter, D.D. Soil acidification induced by elevated atmospheric CO_2. *Glob. Chang. Biol.* **2004**, *10*, 1936–1946. [CrossRef]

54. Zeebe, R.E. History of seawater carbonate chemistry, atmospheric CO_2, and ocean acidification. *Annu. Rev. Earth Planet. Sci.* **2012**, *40*, 141–165. [CrossRef]

55. Hasler, C.T.; Jeffrey, J.D.; Schneider, E.V.C.; Hannan, K.D.; Tix, J.A.; Suski, C.D. Biological consequences of weak acidification caused by elevated carbon dioxide in freshwater ecosystems. *Hydrobiologia* **2018**, *806*, 1–12. [CrossRef]

56. Stock, S.C.; Köster, M.; Dippold, M.A.; Nájera, F.; Matus, F.; Merino, C.; Boy, J.; Spielvogel, S.; Gorbushina, A.; Kuzyakov, Y. Environmental drivers and stoichiometric constraints on enzyme activities in soils from rhizosphere to continental scale. *Geoderma* **2019**, *337*, 973–982. [CrossRef]

57. Trębacz, H.; Wójtowicz, K. Thermal stabilization of collagen molecules in bone tissue. *Int. J. Biol. Macromol.* **2005**, *37*, 257–262. [CrossRef]

58. McMahon, S.; Anderson, R.P.; Saupe, E.E.; Briggs, D.E.G. Experimental evidence that clay inhibits bacterial decomposers: Implications for preservation of organic fossils. *Geology* **2016**, *44*, 867–870. [CrossRef]

59. Collins, M.J.; Nielsen-Marsh, C.M.; Hiller, J.; Smith, C.I.; Roberts, J.P.; Prigodich, R.V.; Wess, T.J.; Csapò, J.; Millard, A.R.; Turner-Walker, G.; et al. The survival of organic matter in bone: A review. *Archaeometry* **2002**, *44*, 383–394. [CrossRef]

60. Moyer, A.E.; Zheng, W.; Johnson, E.A.; Lamanna, M.C.; Li, D.D.Q.; Lacovara, K.J.; Schweitzer, M.H. Melanosomes or microbes: Testing an alternative hypothesis for the origin of microbodies in fossil feathers. *Sci. Rep.* **2014**, *4*, 4233. [CrossRef]

61. Field, D.J.; D'Alba, L.; Vinther, J.; Webb, S.M.; Gearty, W.; Shawkey, M.D. Melanin Concentration Gradients in Modern and Fossil Feathers. *PLoS ONE* **2013**, *8*, e59451. [CrossRef]

62. Schweitzer, M.H.; Moyer, A.E.; Zheng, W. Testing the hypothesis of biofilm as a source for soft tissue and cell-like structures preserved in dinosaur bone. *PLoS ONE* **2016**, *11*, e150238. [CrossRef]

63. Chafetz, H.S.; Guidry, S.A. Bacterial shrubs, crystal shrubs, and ray-crystal shrubs: Bacterial vs. abiotic precipitation. *Sediment. Geol.* **1999**, *126*, 57–74. [CrossRef]

64. Cheng, L.; Cord-Ruwisch, R.; Shahin, M.A. Cementation of sand soil by microbially induced calcite precipitation at various degrees of saturation. *Can. Geotech. J.* **2013**, *50*, 81–90. [CrossRef]

65. Mortensen, B.M.; Haber, M.J.; Dejong, J.T.; Caslake, L.F.; Nelson, D.C. Effects of environmental factors on microbial induced calcium carbonate precipitation. *J. Appl. Microbiol.* **2011**, *111*, 338–349. [CrossRef] [PubMed]

66. Pan, H.; Darvell, B.W. Effect of carbonate on hydroxyapatite Solubility. *Cryst. Growth Des.* **2010**, *10*, 845–850. [CrossRef]

67. Keenan, S.W. From bone to fossil: A review of the diagenesis of bioapatite. *Am. Miner.* **2016**, *101*, 1943–1951. [CrossRef]

68. Sykes, G.A.; Collins, M.J.; Walton, D.I. The significance of a geochemically isolated intracrystalline organic fraction within biominerals. *Org. Geochem.* **1995**, *23*, 1059–1065. [CrossRef]

69. Gunderson, A.R.; Frame, A.M.; Swaddle, J.P.; Forsyth, M.H. Resistance of melanized feathers to bacterial degradation: Is it really so black and white? *J. Avian Biol.* **2008**, *39*, 539–545. [CrossRef]

70. Sadowsky, M.J.; Schortemeyer, M. Soil microbial responses to increased concentrations of atmospheric CO2. *Glob. Chang. Biol.* **1997**, *3*, 217–224. [CrossRef]

71. Blagodatskaya, E.; Blagodatsky, S.; Dorodnikov, M.; Kuzyakov, Y. Elevated atmospheric CO2 increases microbial growth rates in soil: Results of three CO2 enrichment experiments. *Glob. Chang. Biol.* **2010**, *16*, 836–848. [CrossRef]

72. Yu, T.; Chen, Y. Effects of elevated carbon dioxide on environmental microbes and its mechanisms: A review. *Sci. Total Environ.* **2019**, *655*, 865–879. [CrossRef]

73. Jin, J.; Tang, C.; Robertson, A.; Franks, A.E.; Armstrong, R.; Sale, P. Increased microbial activity contributes to phosphorus immobilization in the rhizosphere of wheat under elevated CO2. *Soil Biol. Biochem.* **2014**, *75*, 292–299. [CrossRef]

74. Crosby, C.H.; Bailey, J.V. The role of microbes in the formation of modern and ancient phosphatic mineral deposits. *Front. Microbiol.* **2012**, *3*, 241. [CrossRef] [PubMed]

75. Plet, C.; Grice, K.; Pagès, A.; Ruebsam, W.; Coolen, M.J.L.; Schwark, L. Microbially-mediated fossil-bearing carbonate concretions and their significance for palaeoenvironmental reconstructions: A multi-proxy organic and inorganic geochemical appraisal. *Chem. Geol.* **2016**, *426*, 95–108. [CrossRef]

76. Romanowski, G.; Lorenz, M.G.; Wackernagel, W. Adsorption of Plasmid DNA to Mineral Surfaces and Protection against Dnase-I. *Appl. Environ. Microbiol.* **1991**, *57*, 1057–1061. [CrossRef]

77. Playter, T.; Konhauser, K.; Owttrim, G.; Hodgson, C.; Warchola, T.; Mloszewska, A.M.; Sutherland, B.; Bekker, A.; Zonneveld, J.P.; Pemberton, S.G.; et al. Microbe-clay interactions as a mechanism for the preservation of organic matter and trace metal biosignatures in black shales. *Chem. Geol.* **2017**, *459*, 75–90. [CrossRef]

78. Muscente, A.D.; Schiffbauer, J.D.; Broce, J.; Laflamme, M.; O'Donnell, K.; Boag, T.H.; Meyer, M.; Hawkins, A.D.; Huntley, J.W.; McNamara, M.; et al. Exceptionally preserved fossil assemblages through geologic time and space. *Gondwana Res.* **2017**, *48*, 164–188. [CrossRef]

79. Schiffbauer, J.D.; Laflamme, M. Lagersttten through time: A collection of exceptional preservational pathways from the terminal neoproterozoic through today. *Palaios* **2012**, *27*, 275–278. [CrossRef]

80. Retallack, G.J. Permian and Triassic greenhouse crises. *Gondwana Res.* **2013**, *24*, 90–103. [CrossRef]

81. Breecker, D.O.; Sharp, Z.D.; McFadden, L.D. Atmospheric CO_2 concentrations during ancient greenhouse climates were similar to those predicted for A.D. 2100. *Proc. Natl. Acad. Sci. USA* **2010**, *107*, 576–580. [CrossRef]

82. Gienapp, P.; Teplitsky, C.; Alho, J.S.; Mills, J.A.; Merilä, J. Climate change and evolution: Disentangling environmental and genetic responses. *Mol. Ecol.* **2008**, *17*, 167–178. [CrossRef]

83. Mills, B.J.W.; Krause, A.J.; Scotese, C.R.; Hill, D.J.; Shields, G.A.; Lenton, T.M. Modelling the long-term carbon cycle, atmospheric CO2, and Earth surface temperature from late Neoproterozoic to present day. *Gondwana Res.* **2019**, *67*, 172–186. [CrossRef]
84. Zhang, Y.G.; Pagani, M.; Liu, Z.; Bohaty, S.M.; Deconto, R. A 40-million-year history of atmospheric CO 2 A -million-year history of atmospheric CO. *Phil. Trans. R. Soc. B* **2013**, *271*, 20130096. [CrossRef] [PubMed]
85. Wilby, P.R. The Role of Organic Matrices in Post-Mortem Phosphatization of Soft-Tissues. *Kaupla Darm. Beitr. Naturg.* **1993**, *2*, 99–113.
86. Hubert, B.; Álvaro, J.J.; Chen, J.Y. Microbially mediated phosphatization in the Neoproterozoic Doushantuo Lagerstätte, South China. *Bull. Soc. Geol. Fr.* **2005**, *176*, 355–361. [CrossRef]
87. Dornbos, S.Q. Phosphatization through the Phanerozoic. In *Taphonomy*; Springer: Dordrecht, The Netherlands, 2010; Volume 32, pp. 435–456. [CrossRef]
88. Martill, D.M. The Medusa effect: Instantaneous fossilization. *Geol. Today* **1989**, *5*, 201–205. [CrossRef]

Review

The Rising of Paleontology in China: A Century-Long Road

Zhonghe Zhou

Institute of Vertebrate Paleontology and Paleoanthropology, Chinese Academy of Sciences, 142 Xizhimenwai Dajie, Beijing 100044, China; zhouzhonghe@ivpp.ac.cn

Simple Summary: A brief account of the history of Chinese paleontology for about one century is provided with some perspectives for its future development. The development of Chinese paleontology is closely related to its social-ecological background as well as its connection to the outside world. On the other hand, the rising of the Chinese paleontology also benefitted from its rich fossil resources as well as the integration with other biological and geological disciplines and the use of new technologies.

Abstract: In this paper, the history of paleontology in China from 1920 to 2020 is divided into three major stages, i.e., 1920–1949, 1949–1978, and 1979–2020. As one of the first scientific disciplines to have earned international fame in China, the development of Chinese paleontology benefitted from international collaborations and China's rich resources. Since 1978, China's socio-economic development and its open-door policy to the outside world have also played a key role in the growth of Chinese paleontology. In the 21st century, thanks to constant funding from the government and the rise of the younger generation of paleontologists, Chinese paleontology is expected to make even more contributions to the integration of paleontology with both biological and geological research projects by taking advantage of new technologies and China's rich paleontological resources.

Keywords: paleontology; China; history; 20th century; 21th century

Citation: Zhou, Z. The Rising of Paleontology in China: A Century-Long Road. *Biology* **2022**, *11*, 1104. https://doi.org/10.3390/biology11081104

Academic Editor: Raymond Louis Bernor

Received: 13 June 2022
Accepted: 14 July 2022
Published: 25 July 2022

Publisher's Note: MDPI stays neutral with regard to jurisdictional claims in published maps and institutional affiliations.

Copyright: © 2022 by the author. Licensee MDPI, Basel, Switzerland. This article is an open access article distributed under the terms and conditions of the Creative Commons Attribution (CC BY) license (https://creativecommons.org/licenses/by/4.0/).

1. Introduction

This paper aims to provide a brief account of the history of the development of paleontology in China. The year 1920 is chosen as an important starting point for Chinese paleontology. During the past century from 1920 to 2020, Chinese paleontology has gone through several difficult periods and was only able to usher in the rapid growth seen over the past several decades thanks to the continuous efforts of the first-generation paleontologists in the 20th century to the youngest generation in the 21th century. In a chronological approach, this paper also intends to reflect the major characteristics of the discipline at various stages of the history with a brief introduction of the major achievements and an overview of how Chinese paleontologists have benefitted from international collaborations and the social-economic changes in China, focusing in particular on those since 1978. Finally, this paper discusses a few brief predictions regarding the future directions and challenges for Chinese paleontology.

2. Before 1920—The Collecting and Studying of Chinese Fossils by Foreigners

In China, fossils have been recognized as the remains of ancient animals, plants and fungi as well as seen as evidence of changes of the Earth's environment for over two thousand years. However, they had never been studied by Chinese scholars as a scientific subject until modern sciences (e.g., geology, paleontology) began to be introduced to China during the 19th century. The first scientific collection of Chinese fossils was made by western explorers, missionaries, geographers, and geologists, particularly gaining momentum since the opium war in 1940, and then studied by western paleontologists from Europe and the Americas. For instance, L. de Korunck published on two brachiopods from China in

1943, and R. Owen published the first paper on Chinese mammal fossils [1]. E. Koken and M. Schlosser published books on Chinese fossil mammals in 1885 and 1903, respectively, which represent the earliest systematic studies on Chinese vertebrate fossils [2,3].

Beginning in the 1860s, French missionary P.A. David started to collect fossil fish from the Mesozoic lake deposits in western Liaoning. These fossils were later named as *Lycoptera* and are regarded as one of the typical elements of the Jehol Biota, which is now best known for producing many exceptional feathered dinosaurs, birds, and mammals. American geologist and explorer R. Pumpelly was one of the first western scientists to carry out geological surveys in China. He collected many fossils from China from 1863~1865 and proposed the lacustrine facies of the Chinese loess [4]. German scientist F. Richthofen made extensive geographic and geological explorations in China during 1868 and 1872. He also collected many invertebrate fossils as well as recorded their stratigraphic data in the majority of Chinese provinces.

During the late 19th century and early 20th century, western explorers and scholars also made many scientific expeditions to China. Swedish geographer and explorer S.A. Hedin made several expeditions to western China starting in 1890, and the ancient city of Loulan (Kroraina) in Xinjiang was one of his major discoveries during his voyage across the Taklamakan Desert. Also notable is the Central Asian Expeditions organized by the American Museum of Natural History in 1916 and 1919, which resulted in the discovery of many fossil mammals in Inner Mongolia. In the early 20th century, Russian geologists collected many dinosaur fossils from Heilongjiang, northeastern China, including some duck-billed dinosaurs.

Before 1920, there were hardly any Chinese scientists known mainly as a paleontologist. The first Chinese scientist who published his research on paleontology was Rongguang Qi, who was among the first group of teenagers sent to America by the Qing Imperial Dynasty government in 1871. He was trained in geology and mineral resources. In 1910, he published his study on invertebrates and plants from Hebei, North China.

The first attempt to teach college-level geology in China started in 1909 at the Imperial University of Peking (known as Peking University since 1912). In 1914, Wenjiang Ding (V.K. Ting; 1887–1936), a 1911 graduate of the University of Glasgow, started to teach China's first college-level course in paleontology at the Geological Institute established in 1913. This produced the first generation of domestically trained geologists, including a few paleontologists who were then hired by the National Geological Survey of China that was officially founded in 1916 with Ding as its first director. The National Geological Survey of China has also been generally regarded as the first national scientific institution with an international reputation in Chinese history. Among many of his tremendous achievements, Ding also made contributions to many pioneering works in Chinese paleontology and stratigraphy. He discovered the first Devonian fish in Yunnan in 1913 and pioneered the study of fossil plants in China.

In summary, upon entering the 20th century, China had gone through major social changes, and its door began to open to the outside world. In 1911, the Imperial Qing Dynasty was ended and replaced by the Republic of China. A few more young Chinese were sent to Western countries to learn science and technology. The early 20th century was the preparation stage for the early development of paleontology in China [5].

3. From 1920 to 1949—Founding Stages of Chinese Paleontology

1920 was a turning point for Chinese paleontology. By the invitation of Wenjiang Ding, two prominent western-trained geologists and paleontologists joined Peking University. One of them was A.W. Grabau from Columbia University. Grabau not only taught at Peking University, but was also the Chief Paleontologist at the Chinese Geological Survey (Figure 1). The second was Siguang Li who graduated from Birmingham University with a master's degree in 1919 (Ph.D. degree in 1931). During 1920–1937, Grabau, Li, and their colleagues mentored and trained many of the first generation of geologists and paleontologists in China who would later become the leading figures in the majority of disciplines in Chinese geology and paleontology. It is worthy to note that Grabau has been also well known for proposing the Jehol Fauna and Jehol Series in western Liaoning, northeastern China [6,7]. He also named the Changxing limestone in Zhejiang Province, where two GSSPs (Global Boundary Stratotype Sections and Points) were later established, including the boundary between the Permian and Triassic.

Figure 1. Founders of Chinese geology and paleontology gathering at Amadeus W. Grabau's home in Beijing in 1935. Front row, from the left: Hongzhao Zhang, Wenjiang Ding, Amadeus W. Grabau, Wenhao Weng, and Pierre Teilhard de Chardin. Middle row, from the left: Zhongjian Yang, Zanheng Zhou, Jiarong Xie, Guangxi Xu, Yunzhu Sun, Xiechou Tan, Shaowen Wang, Zanxun Yin, and Fuli Yuan. Back row, from the left: Zuolin He, Hengshen Wang, Zhuquan Wang, Yuelun Wang, Huanwen Zhu, Rongseng Ji, and Jianchu Sun (credit to IVPP).

Many of the first-generation Chinese paleontologists had been trained overseas and earned their Ph.D. degrees during the 1920s–1940s before they returned to China to pioneer their research fields. Among them were the invertebrate paleontologists Yunzhu Sun and Senxun Yue, vertebrate paleontologist Zhongjian Yang (Chung-Chien Young), and paleobotanist Xingjian Si, who received their Ph.D. degrees in Germany in 1927, 1936, 1928, and 1933, respectively. Zanxun Yin, geologist and paleontologist, and Wenzhong Pei, paleoanthropologist and archaeologist, earned their Ph.D. degrees from France in 1931 and 1937, respectively. Invertebrate paleontologists Jianzhang Yu and Hongzhen Wang received their Ph.D. degrees from England in 1935 and 1947. Zunyi Yang, an invertebrate paleontologist, received his Ph.D. degree from the USA in 1939. They have all became the backbones of Chinese paleontology after they returned to China.

It is notable that scientific expeditions organized by organizations of Western countries continued to be active in China during the 1920s and 1930s. For instance, the Central Asiatic Expeditions led by R.C. Andrews of the American Museum of Natural History during 1922 and 1930 resulted in the discovery of many dinosaur skeletons and dinosaur eggs in Inner Mongolia as well as many Mesozoic and Cenozoic mammals, including some of the earliest placental mammals. However, with the appearance of the first generation of Chinese paleontologists, international collaborations for the study of Chinese paleontology and stratigraphy became more and more prominent and productive. For instance, the Sino-Swedish Expedition which carried out multi-discipline scientific research in north and northwest China from 1927–1935 was co-led by S. Anders and Fuli Yuan, a geologist from Peking University, and many fossil reptiles and mammal fossils were collected during these field excursions. Zhongjian Yang and Wenzhong Pei joined the Central Asiatic Expedition in 1930.

One of the most successful collaborations between the first generation of Chinese paleontologists and their Western colleagues is probably the excavation of the fossil humans in Choukoudian (Figure 2) as well as the study on Cenozoic vertebrates and their stratigraphy. The Swedish geologist J.G. Andersson was invited by the Chinese government to act as consultant for the mining industry from 1914 to 1924. He discovered the Choukoudian site in 1918. During 1921–1922, Andersson assigned the Austrian paleontologist O. Zdansky to excavate at Choukoudian, which resulted in the discovery of two human teeth in addition to an abundance of mammal fossils. In 1927, with the support from the Rockefeller Foundation, D. Black from the Beijing Union Medical College Hospital (BUMCH) was able to continue the excavation of Choukoudian and study the Cenozoic cave deposits based on an agreement with the Geological Survey of China. In 1929, Wenzhong Pei led the excavation in Choukoudian and discovered the first skull of the Peking Man (*Homo erectus*), which drew some international attention for Chinese paleontology and paleoanthropology.

Figure 2. Paleontologists in Choukoudian, Beijing in 1934. From the left: Wenzhong Pei, Shiguang Li, Pierre Teilhard de Chardin, Meinian Bian, Zhongjian Yang, and George B. Barbour (credit to IVPP).

The collaboration between Black and the National Geological Survey of China also resulted in the founding of the Cenozoic Research Laboratory of the Geological Survey of China in 1929, which was the predecessor of the Institute of Vertebrate Paleontology and Paleoanthropology (IVPP). Black was appointed as the honorary head of the lab with Zhongjian Yang as the deputy head and P. T. de Chardin as the adviser. After Black died in 1925, his position was filled by German paleontologist F. Weidenreich. Clearly, the Cenozoic laboratory was designed and intended to be an international research unit.

In 1922, with the support of Grabau and Anderson, Ding founded the first Chinese paleontological journal *Palaeontologia Sinica*, and he became the chief editor of this journal for nearly 15 years. The journal published only English and German papers in the beginning, but later published Chinese papers as well. The majority of the important Chinese paleontological and stratigraphic studies during the 1920s and 1930 were published in this journal.

In 1928, the National Geological Survey of China formally established its Paleontology Laboratory. In the same year, the National Research Institute of Geology was founded in Nanjing with its own Stratigraphy and Paleontology Laboratory, representing the growth of paleontological research being conducted in China.

The paleontological and stratigraphic studies carried out by Chinese paleontologists started to grow in number quickly during the 1920s and 1930s. To list a few examples, Xichou Tan excavated dinosaurs in Laiyang, Shandong Province in 1923 and collected many dinosaur bones, some of which were later published and named as a duck-billed dinosaur *Tanius sinensis* [8]. In addition, a large number of fish, insect, and plant fossils were also collected. In 1923, Zanheng Zhou, a graduate from the National Research Institute of Geology in 1916 who also later trained in Sweden, published his study on the fossil plants from Shandong Province [9], representing the first paleobotanic paper by a Chinese paleontologist.

The book "Contributions to Cambrian Faunas of North China" written by Yunzhu Sun [10] marked the first paleontological book by a Chinese paleontologist. Some domestically educated paleontologists also did great work in paleontology and biostratigraphy. For instance, Yazhen Zhao, a graduate from Peking University was hired by the National Geological Survey of China in 1923. During his short but glorious career, he published several classic books on the study of brachiopods and stratigraphy before he was tragically killed by a bandit while in the field in 1929 at the age of 31.

Siguang Li published his classic monography "Fusulinidae of North China" in 1927 [11]. In 1927, Zhongjian Yang published the book "Fossil Rodents from North China" [12] based on his doctoral dissertation, which was the first monography on Chinese vertebrate paleontology. Zhongjian Yang became the father of Chinese vertebrate paleontology. His earlier work focused on Zhoukoudian and other Late Cenozoic mammalian faunas of northern China. After 1938, his main research interest shifted to Mesozoic dinosaurs and synapsids. His collaborations with P. Teilhard de Chardin were very successful and contributed greatly to the establishment of the geochronology and the stratigraphic succession of the Tertiary and Quaternary periods in China.

Xingjian Si (Hsing-Chien Sze) came back to China after he was awarded a Ph.D. degree for his research on Chinese fossil plants [13] and contributed greatly to the study of Paleozoic and Mesozoic plants in China, which earned him international fame.

Thanks to the efforts of Zhongjian Yang and Yunzhu Sun, the Paleontological Society of China was founded in 1929. Yunzhu Sun elected the first president, and he was elected the vice chairman of the International Paleontological Association in 1948.

Although the 1920s and 1930s witnessed the rapid growth of paleontology in China, starting in 1937 China entered the Second Sino-Japanese War that ended in 1945 and then the Chinese civil war that ended in 1949, which resulted in the founding of the People's Republic of China. Paleontological study during these years became difficult. The National Geological Survey of China moved to Yunnan. Peking University also moved to Yunnan to merge with Tsinghua University and Nankai University, and together they became the National Southwestern Associated University. At this new conglomerate University, students continued to be taught geology and paleontology, and many of the graduates later became the some of the most important figures in Chinese paleontology (e.g., Hongzhen Wang, Enzhi Mu, Dongsheng Liu, Zhiwei Gu, Yichun Hao etc.). As a 1937 graduate from Peking University, Yanhao Lu, a Cambrian and trilobite expert, first worked in the National Southwestern Associated University and then moved to the National Geological Survey of China. In reality, Chinese paleontological study had not completely stopped. For instance, Zhongjian Yang continued his work on the Lufeng dinosaur fauna in Yunnan during the trying years from 1938 to 1945 and had discovered the Jurassic dinosaur *Lufengosaurus* Fauna [14]. *Lufengosaurus* also represents China's first dinosaur with a mounted and complete skeleton. In 1940, Young was appointed the head of the Laboratory of Vertebrate Palaeontology and the honorary head of the Cenozoic Department of the National Geological Survey of China. In addition to the work he did in Yunnan, he also carried out fieldwork in Northwestern China. During the same time, Zanxun Yin and Yanhao Lu studied the Paleozoic biostratigraphy in southwestern China.

In sum, after the fall of Qing Dynasty in the 1911 and before the Japanese invasion in 1937, China had witnessed a relatively fast period of economic growth and the construction of social infrastructure including railroads, communications etc. despite civil war and political instability. As a result, Chinese paleontology had experienced a rapid growth in both the number of paleontologists and paleontological research. Their research covered nearly all major animal phyla and plant divisions. Furthermore, their studies not only included taxonomic and biostratigraphic work, but also included discussions on the evolution of some specific taxonomic groups. Chinese paleontologists had also established a preliminary stratigraphic framework spanning from the Precambrian to the Quaternary. It is also notable that many of the first-generation paleontologists had some background in studying overseas or benefitted in some way from collaborating with some of the talented western paleontologists. However, despite the persistence of dedicated scientists, overall, the wars and social turmoil had slowed down the sometimes-fast development of Chinese paleontology.

4. From 1949 to 1977—The Expanding Stage of Chinese Paleontology

The founding of the People's Republic of China in 1949 had two immediate impacts: first, the newfound social stability and the secondly, the cessation of academic exchange with western countries. Learning from and modeling after the former Soviet Union became a trend. China's connection with countries other than the former Soviet Union was limited. For instance, Zhang Miman (Meemann Chang) spent a year visiting the Swedish Museum of Natural History in 1966 (Figure 3), but did not return there to receive her Ph.D. degree from Stockholm University until 1982.

Figure 3. Miman Zhang and her colleagues in Stockholm, Sweden in 1966. Front row, front the left: Erik Stensiö, Miman Zhang, E. Mark-Kurik. Back row, from the left: Hans-Peter Schultze, Tor Örvig, Hans Bjerring, Gareth Nelson, Ray Thorsteinsson, Erik Jarvik, and Hans Jessen (credit to IVPP).

To reboot its economy and social development following the civil war, the new Chinse government established the Chinese Academy of Sciences in 1950 which includes many different research institutes. Siguang Li was appointed the director of the Nanjing Institute of Paleontology in 1950, and the institute was officially founded in 1951, which also includes the Department of Vertebrate Paleontology in Beijing. The institute was based on staff merged from the paleontological group of the Institute of Geology, the Central Research Academy, the department of paleontology and the Cenozoic Research Laboratory (Beijing) of the National Geological Survey of China. In 1953, the Department of Vertebrate Paleontology, based in Beijing, was separated the from the Nanjing Institute of Geology and Paleontology and became affiliated directly with the Chinese Academy of Sciences. In 1957, the department was upgraded to an institute and assigned its current name, the

Institute of Vertebrate Paleontology and Paleoanthropology (IVPP), in 1960 with Zhongjian Yang as its first director.

In 1959, the Nanjing Institute of Paleontology was renamed as the Nanjing Institute of Geology and Paleontology (NIGPAS) and has since focused on the study of invertebrate paleontology, paleobotany, and stratigraphy, in order to be distinct from the IVPP, but both are still affiliated with the Chinese Academy of Sciences (CAS) today. The two research institutes of the CAS have since become the major forces in paleontological study in China. It is also noteworthy that the two institutes were also able to accept graduate students and helped to educate many of the outstanding young paleontologists of the next generation.

In addition to the positions available at the two research institutes of the CAS (NIGPAS and IVPP), many other paleontologists were mainly employed at universities such as Peking University, Nanjing University, Northwest University, Beijing Institute of Geology (now the China University of Geoscience in Beijing and Wuhan), Beijing Institute of Mining and Technology (now the China University of Mining and Technology), Changchun Institute of Geology (now merged into Jilin University), and Chengdu Institute of Geology (now the Chengdu University of Technology) etc. They have educated many students in paleontology and stratigraphy, in addition to conducting extensive studies on invertebrates, plants and stratigraphy.

In addition, the Geological Academy of Sciences affiliated to the Ministry of Geology and other institutes affiliated to the later established Ministry of Oil have also recruited many staff members who are working on paleontology and stratigraphy, mainly to fulfil the tasks of geological surveying and exploring for geological resources.

In 1953, the journal *Acta Paleontologica Sinica* was established, with Zunyi Yang as chair of the editorial board. Minzhen Zhou, a 1950 Ph.D. from Lehigh University in the USA, helped create the journal *Vertebrata PalAsiatica* in 1956, with Zhongjian Yang as the chair of the editorial board. In 1957, Zunyi Yang and Yichun Hao, graduates from the National Southwestern Associated University, published the first textbook on paleontology for Chinese students. In the same year, Yichun Hao also published the first Chinese textbook on micropaleontology.

During this time, the extensive geological surveys conducted due to the nationwide need to domestically locate oil, coal, and other mineral resources stimulated a rapid growth in the number of paleontologists around the country. Micropaleontology was quickly developed, largely in response to the practical demand for experts in the field. While academic exchange between China and western countries had nearly stopped, a new generation of students were sent to the former Soviet Union to study paleontology. Among the graduates were Yichun Hao, Pinxian Wang, and Miman Zhang (Meemann Chang). Chinese paleontologists had been asked to edit books on categories of fossils from various regions of China and made progress on the study of fossils and stratigraphy of nearly all Phanerozoic ages of China. For instance, dinosaur excavations at Shandong, Sichuan, Inner Mongolia and Xingjiang have produced many previously unknown taxa from the Jurassic and Cretaceous. In addition, large scale expeditions were organized to investigate the geology of Tibet.

Despite the general practical need of the geological surveys to explore mineral resources, there was still some scientific consideration for the paleontological development in China thanks to the enduring visions of the first generation of Chinese paleontologists (Figure 4). For instance, Zhongjian Yang had outlined the focus and directions of the IVPP as "four origins and two deposits" in 1955. The four origins include the origins of vertebrates, tetrapods, and mammals, as well as humans, primates and cultures from the perspective of biology. The two deposits represent the fossil bearing red beds in southern China and the soil-like deposits in northern China from the perspective of geology. In 1958, Yang further proposed the goals of "three gaps to be filled" for the IVPP, i.e., the evolutionary gap, regional gap, and the stratigraphic gap. In the 1960s, the discovery of human fossils had been remarkable, including some new specimens of *Homo erectus*, and *Homo sapiens*.

Figure 4. Group photo of officials elected at the first National Congress of the Chinese Paleontological Society in 1956. Front row, front the left: Senxun Yue, Zanxun Yin, Xiaohe Zhou, Zhongjian Yang, Yunzhu Sun, Xingjian Si, Jinke Zhao, and Zunyi Yang. Back row, from the left: Hongzhen Wang, Zhiwei Gu, Yuanren Qu, Shicheng Huo, Ren Xu, Minzhen Zhou, Yu Wang, Enzhi Mu, and Longqing Chang (credit to NIGPAS). The Chinese characters means the photo was taken at the first National Congress of the Chinese Paleontological Society on June 16th, 1956.

International collaboration between China and the former Soviet Union is not only limited to the education of Chinese students by experts from the former Soviet Union. In 1959, in adherence to an agreement between the Academies of Sciences of the two countries, a joint expedition to Central Asia was formed (Figure 5), and was led on the Chinese side by Zhou Minzhen from the IVPP. Unfortunately, the joint expedition did not last long and was ended in 1960 due to the deterioration of the political relationship between the two countries.

Figure 5. Field camp of the Sino–Soviet Union joint expedition in Inner Mongolia in 1959 (credit to IVPP).

During the Cultural Revolution (1966–1976), much like many other scientific disciplines, Chinese paleontology was seriously disrupted. In particular, the Chinese paleontological community was nearly completely isolated from the outside world. Despite the turmoil, field excavations continued to produce some exciting finds, and the empirical paleontological and stratigraphic work carried out during that time filled many gaps in our understanding of both the evolution of life and the stratigraphic record. In addition, micropaleontology witnessed a stage of rapid development due to its application in surveying and exploring for geological resources.

In summation, with the second generation of paleontologists entering the mainstage during the period from 1949 to 1977, the Chinese paleontological community was greatly expanded. On one hand, this was under the guidance of the first generation of paleontologists and their international vision. On the other hand, it was also impacted by the influence of the former Soviet Union and the growth of Chinese paleontology had been closely related to the national industrial requirement for extensive geological surveys and prospecting with a focus on the collection of fossils from across the country spanning from the Precambrian to the Quaternary. A preliminary biostratigraphic frame was established, and the study of the Chinese paleontologists covered nearly all the taxonomic categories of paleontology. Overall, Chinese paleontology had generally been geology-oriented, although the biological significance of some fossils had also been discussed to a degree.

5. From 1977 to 2020

In 1977, the National College Entrance Exam was reinstated after it had been abandoned for ten years and as a result many more talented students were able to study paleontology in college. A new generation of graduate from Chinese universities were able to be recruited into the NIGPAS, IVPP, and other research institutes as well universities, and they remain the backbone of Chinese paleontology today.

With the beginning of the open-door policy in 1978, the connection between China and western countries was able to be re-established. While some senior paleontologists were able to communicate with their international colleagues and had more and more visits from both sides, some Chinese students were able to pursue their degrees in western universities. Meanwhile, new ideas and the most recent developments including scientific methods in western countries such as cladistics and paleoecology began to be introduced into China [15,16].

Since the 1980s, Chinese paleontologists have been frequently invited to participate in the editing of the Treatise on Invertebrate Paleontology. In addition, Chinese paleontologists have published many books about their systematic work on various groups such as trilobites, brachiopods, graptolites, corals, bivalves, cephalopod, insects, conchostrans etc.

Starting in the late 1990s, Chinese paleontology began to gradually enter its "golden age". First, some of the western educated students came back to China and were awarded enough funding for their research to establish their own research labs. Secondly, many domestically educated students graduated from colleges and joined research institutes such as the NIGPAS and IVPP. The third generation of paleontologists, or the "reform and open generation", had the advantage of having a better grasp of the English language and were able to more easily learn new technique and methods. It is notable that some of the Chinese students chose to stay in western countries after earning their Ph.D. degrees continued their paleontological career, and they helped to foster collaborations between China and various western countries in order to both educate students and sponsor further collaborative projects. More international meetings were hosted in China and more Chinese paleontologists were able to participate in meetings outside China.

Starting in 2000, the scale of funding from the government for basic scientific research began to be increased. The National Natural Science Foundation, which was founded in 1986, has been growing steadily and has become the main source of support for Chinese paleontologists since the 1990s. For instance, over two dozen of the third and fourth generation of paleontologists have been supported by the Distinguished Young Scientist Fund, and paleontologists from the NIGPAS, IVPP, China University of Geosciences, and Northwest University have also been awarded the Innovation Research Group Fund to ensure that their research is being supported in the long term.

In addition to the funding from the NSFC, Chinese paleontologists had a chance to be awarded large grants from other government agencies to help organize a large group of paleontologists to work together on a major scientific endeavor. For instance, the Special Funds for Basic Research of China ("973" projects) by the Ministry of Sciences and Technology provided grants amounting to twenty million Chinese yuan for a five year's project [17]. Since 2000, several paleontologists from the NIGPAS, IVPP, and the University of Geosciences (Wuhan, Beijing) have been awarded this grant. In addition, Chinese paleontologists have recently been able to secure other major grant, i.e., the Strategic Priority Research Program from the Chinese Academy of Sciences, to investigate the coevolution of life and paleoenvironment during major critical intervals of Earth's history. Most recently, a 10-year multidiscipline project was awarded to a paleontologist by the NSFC. It is intended to study the Mesozoic terrestrial biota and its tectonic background from the perspective of Earth system science, and it has drawn participants from paleontology, geochemistry, geophysics, and sedimentology. Furthermore, even more paleontologists have been invited to participate in several other multidiscipline projects, e.g., the new Tibetan Plateau Expeditions.

The state key laboratory administrated by the Ministry of Sciences and Technology is another way to support basic scientific research in China. The State Key Laboratory of Paleobiology and Stratigraphy, based at the Nanjing Institute of Geology and paleontology, and the State Key Laboratory of Biogeology and Environmental Geology, based at the Chinese University of Geosciences were established in 2002 and 2012, respectively. These labs are annually fiscally supported not only in scientific research and collaboration, but also in the acquisition of new equipment.

Due to the increasing funding capacity available to the promising young generation of paleontologists, Chinese paleontology has been growing at an unprecedent rate, and their research scope and production have been greatly expanded, spanning from traditional paleontology and stratigraphy, to paleobiology and some newly developed fields. Since 2000, some students and postdocs from western countries started to join the Chinese institutions and universities, which have further fostered the international collaborations between China and its international community.

In traditional paleontology and stratigraphy, Chinese paleontologists have greatly extended their scope of excavations of fossils to nearly all areas of China, spanning from Precambrian to the Quaternary. As a result, the rate of discovery of new taxa in various biological groups has significantly increased. For instance, the rate of commonly accepted dinosaur species named from China now outnumbers that of any other country. Starting 2015, supported by the "Special Research Program of Basic Science and Technology of the Ministry of Science and Technology", Chinese paleontologists launched an ambitious project to publish a complete series of *Palaeovertebrata Sinica*, which comprises three volumes and 23 fascicles on the taxonomy of all the vertebrate fossil species (nearly 10 thousand) in China. Currently, 13 fascicles have been published, and several more are nearly finished.

Among many of the discoveries, there are several world famous Lagerstätte are probably best known to the international paleontological community, i.e., the Neoprotozoic Wengan Biota (580 my) in Guizhou, Early Cambrian Chengjiang Biota (520 my) in Yunnan, the late Jurassic Yanliao Biota (160 my), and the Early Cretaceous Jehol Biota (125 my) in western Liaoning and neighboring areas, which have produced many world-known exceptionally preserved fossils bearing a lot evolutionary significant data on the early evolution of early life including the evolution of animals [18–21], origin of birds, and regarding the early evolution of mammals, birds, pterosaurs, and flowering plants etc. [22–25]. In addition, these discoveries have also drawn great attention from the media and public, hence increasing the profile of Chinese paleontology in popular scientific communities. In 2001, *Science* magazine published a special issue introducing the highlights of the discoveries and research done in Chinese paleontology.

Thanks to sufficient funding and the great efforts put into field investigations, many other Cambrian fauna have also been discovered, including the Neoprotozoic Lantian Biota (600 my) in Anhui [26] and the Miaohe Biota in Hubei, the terminal Ediacaran Shibantan Biota of Yangtze Gorges, South China [27], the Edicarian fauna in Guizhou [18,28], the Early

Cambrian Burgess Shale-type fossil Lagerstätte the Qingjiang Biota [29], and the Middle Cambrian Burgess Shale-type fossil Lagerstätte the Lingyi Biota [30] (Figure 6), etc. Newly discovered Silurian fishes from Chongqing have produced many exceptionally preserved articulated jawed fished, among many other discoveries. Devonian plants from Yunnan and other regions produced unknown vascular plants [31]. The Permian Pompeii (about 300 my) preserved remarkable plants from coal in Wuhai, Inner Mongolia [32]. Many new species of Middle Triassic marine reptiles and fishes often represented by complete skeletons, including the earliest turtle, have also been discovered in Guizhou and Yunnan provinces [33]. The Miocene Hezheng Fauna has also produced abundant and diverse mammals and birds, sometimes with exceptional preservation of soft tissues and gut contents [34]. Lastly, the Miocene Zhangpu Biota from Fujian represents another amber treasure trove, recording a diverse fauna and flora of the tropical forest ecosystem [35].

Figure 6. Reconstruction of the latest discovered Middle Cambrian Lagerstätte Lingyi Biota in Shandong (credit to Fangcheng Zhao).

Chinese paleontologists have also made significant contributions to several of the major evolutionary issues, such as the Cambrian explosive radiation and its background [36], the origin of animals [37], the study on the Great Ordovician Radiation and the end-Ordovician mass extinction [38,39], the study of early vertebrates [40], the P-T extinctions [41], the biodiversity changes based on big data [42], the origin of birds and their flight [23,43], the early evolution of Mesozoic mammals including middle ears [22,44], the Mesozoic insect–plant coevolution [45], sexual selection [46,47], the interaction of the Cenozoic biota and flora with the uprising of the Tibet Plateau [48,49], the evolution and dispersal of modern and archaic humans in China, etc. [50–52].

In the study of stratigraphy, Chinese paleontologists have now witnessed the ratification of 11 global boundary stratotype sections and points (GSSP, or the so-called golden spikes) in China, including the GSSP of the boundary between the Paleozoic and Mesozoic in Zhejiang Province [53,54]. From the study of terrestrial strata, the Paleogene and Neogene biochronologic frameworks in China were preliminarily established [55]. New integrative stratigraphy and timescales for 13 geological periods (Ediacaran–Quaternary) in China were published in the special issue of SCIENCE CHINA Earth Sciences co-edited by Shen and Rong [56]. This research summarized the latest advances in stratigraphy and timescale and discussed the correlation among different blocks in China with international timescales.

High-resolution stratigraphic studies have also been helpful in oil and gas explorations as well as marine geology. Chinese paleontologists, particularly invertebrate paleontologists and micropaleontologists, continue to have work in close collaboration with oil companies for the prospecting of oil and gas resources in which the Silurian division by graptolites has played a key role [57].

Paleogeographic, paleoecological, and paleoenvironmental research has also been active. For instance, paleogeographic and paleoecological reconstruction has been useful in studying the controversial rising history of the Tibetan Plateau [49,58], and the paleobiogeographic and paleogeographic evolution of blocks in the Qinghai–Tibet Plateau [59].

It is notable that the geochronological progress in China has made a great contribution to the establishment of stratigraphic frame, which is critical for the discussion of various geological and evolutionary questions. In particular, the dating of the volcanic ashes interbedded in the terrestrial sediments provided a rare chance for the precise correlation of deposits as well as the fossils in them [60–63]. Furthermore, the precise dating of the fossil-bearing deposits enabled us to better relate the evolution of the Jehol Biota to the major tectonic background [64,65]. Exciting dating results have been obtained from deposits ranging from the Edicaran to the Quaternary, thus providing a solid ground for discussing the interactions between the evolution of different forms of life and their geological and paleoenvironmental background.

Geobiology, derived from the study of geomicrobiology, has also been developing quickly in China [66], and shows a promising future as it can better bridge the geological and biological processes in Earth's history by taking advantage of the latest advances in the study of geochemistry and molecular biology. It is also noteworthy that an institute called the Institute of Geo-Biology was founded in Beijing in 1940, with P. Teilhard de Chardin as the honorary president and zoologist P. Leroy as the director. It was based on the collections and laboratories of the Huangho-Paiho Museum (now the Tianjin Natural History Museum) founded by F. Licent in 1915.

The application of the modern genomic technique to the study of fossils has quickened the development of the discipline of molecular paleobiology, particularly regarding the study of fossil DNA. In the past decade, Chinese paleontologists have succeeded in extracting the DNA sequences from bones of various fossils of *Homo sapiens*, as well that of Denisovan from sediments [67,68]. The rapid progress in this area has also been well connected with the study of archaeology and will certainly make a big impact on the study of the early history of modern humans in China as well as the cultural exchanges between various human populations in the past 100 ka [69,70]. In addition, the extraction of fossil DNA has been increasingly often used in the study of Quaternary fossil mammals such as the Giant Panda.

The study of molecular paleobiology is not limited to the study of fossil DNA. In fact, the study of fossil proteins in deep time has also shown a promising future. A recently published work on the paleoproteomics of the mysterious primate *Gigantopithecus*, dating back 1.9 million years, provided some interesting phylogenetic information [71]. It is also notable that Chinese paleontologists have also been working hard on the identification and detailed analysis of keratin from fossil feathers of dinosaurs and birds from the Jurassic and Cretaceous [72].

The rich paleontological and stratigraphic resources that remain to be further investigated and the unique geological history of the Chinese continent provide numerous chances for the future development of Chinese paleontology, particularly when paleontology is regarded as an integrated part of the Earth System. For instance, the collision between the Indian plate and Eurasian plate and the subsequent rise of the Tibetan Plateau in the Cenozoic has been one of the major tectonic events that has made an important impact on the geography, climate, and history of China. The study of the evolution of life, including human evolution in the Tibetan Plateau against its unique geological background, is expected to produce more exciting multidiscipline results. The subduction of the paleo-Pacific plate towards the Eastern Asia during the late Mesozoic in East Asia has been regarded as related to the evolution of the Yanliao and Jehol Biota, and the study of the control of deep tectonic activities on the surface geological processes and its impact on the evolution of terrestrial life has now drawn attention from paleontologists, geologists, and geochemists.

Technical applications, such as synchrotron, have also played a key role in the study of paleontology in China [73–75]. Several Chinese institutes and paleontological labs in universities have now been equipped with high resolution CT in addition to SEM, TEM, etc. The Synchrotron facilities in Shanghai, Beijing, and Hefei as well as those in Taiwan have also been used by Chinese paleontologists, bringing forward more opportunities and research directions for the younger generation of paleontologists [76–78].

The NIGPA has constructed an impressive Geobiodiversity Database (GBDB) thanks to the longstanding support of the State Key Laboratory of Paleobiology and Stratigraphy and other major projects. Recently, a high-resolution summary of Cambrian to Early-Triassic marine invertebrate biodiversity curves with an imputed temporal resolution of 26 ± 14.9 thousand years was published based on the Geobiodiversity Database (GBDB), which used quantitative data from 11,000 marine fossil species collected from more than 3000 stratigraphic sections in China [42].

Benefitted by the rapid economic growth and the need for popular science education, many new museums of natural history or geology have been constructed in China, which have created more chances for the employment of graduates of paleontology. While the NIGPAS and IVPP remain the two biggest paleontological centers in China, a few universities, e.g., the Chinese University of Geosciences, Peking University, Nanjing University, Northwest University, Lanzhou University, and the Geological Academy of Sciences remain to have a strong paleontological program. New paleontological programs have been established and are growing in several other universities, such as the Yunnan University, Sun Yat-Sen University, Capital Normal University in Beijing, Shenyang Normal University in Liaoning, Hefei University of Technology in Anhui, and Lingyi University in Shandong, etc. In addition, paleontologists are also active in some of the biological institutes, e.g., the Xishuangbanna Tropical Botanical Garden, Institute of Botany of the CAS, and many provincial museums.

Despite the remarkable progress made during the past three decades, the advancement of Chinese paleontology is also held back by several challenges. For instance, the illegal collecting and marketing of vertebrate fossils remain unsolved issues, while scientific collecting often meets great difficulty mainly due to the immature administrative management of fossil resources and lack of local interest.

In sum, due to over 40 years of reform and the open-door policy of China in addition to the rapid growth of the nation's economy, paleontology in China has largely merged into the global paleontological community and became a major force that constantly produces exciting new discoveries of fossils that arouse public interest. Furthermore, it has contributed important evidence to ongoing evolutionary research, which has added to our understanding of the tree of life in deep time. The integration of paleontology with biological and geological sciences has indisputably proved its importance.

6. Future Directions and Challenges

The new generations of paleontologists in China, who have benefited from the success of their predecessors against the background of digital age and several decades of the fast growing of Chinese economy, seem to be more optimistic and confident in using new technology and methods in paleontological studies. In addition, they are more willing and prepared to participate in multidisciplinary research that often involves a background in geochemistry as well as in molecular and developmental biology. More urgent global changes also encourage them to pay more attention to the interactions between life evolution and paleoenvironment during deep time.

During the digital age, more attention has been paid to the construction of Big Data and the application of AI technology to paleontological study. Paleontological data will probably be better integrated with geochronological and geochemical data, which may help produce more interesting results regarding paleobiodiversity and paleogeography, as well as the impact of paleoenviroment, on the evolution of life on Earth.

Considering the population of China, in the future, there will be even more museums or universities with paleontology programs to increase the employment chances for paleontology students and arouse more public interest in this discipline. Young paleontologists can find their positions either in the department of Earth sciences or biology. Their expertise in the history of the Earth and the evolution of life on Earth will be a necessity not only for students of Earth sciences and biological majors, but also for any students who may be interested in the history of the Earth and life, or the theory of evolution (Figure 7).

Figure 7. Group photo of participants of the conference celebrating the 70th anniversary of the Nanjing Institute of Geology and Paleontology in Nanjing in 2021 (credit to NIGPAS). The Chinese characters means a warm congratulation to the 70th anniversary of the Nanjing Institute of Geology and Paleontology of the Chinese Academy of Sciences.

With the growing need for science communication in order the increase the scientific literacy of the public, Chinese paleontologists should be optimistic and enthusiastic about the great potential of paleontology in attracting both kids and adults. It is important to note that when natural sciences are further integrated with liberal sciences, the knowledge gained in paleontology and evolutionary biology will be beneficial to a wide range of readers and future scientists.

However, challenges remain in Chinese paleontology. The practical philosophy of Chinese culture has hindered the development of basic scientific research. Some of the students chose paleontology because they saw it as an advantageous career rather than out of their personal interest. The NIGPAS and IVPP, as CAS's two institutes, seem to be under some pressure to focus on institutional organized projects, rather than small individual projects born out of pure curiosity.

Due to historical and cultural reasons, it seems that Chinese paleontologists are shy to propose new hypotheses, attempt contribute more to evolutionary theory from paleontological evidence, or to try to integrate paleontological data with biological data. Cross-discipline interaction is never easy, and it only happens when the scientific culture is suitable for scientists to focus on purely scientific pursuits. Yu recently provided an interesting account of the social, cultural, and disciplinary factors that influenced the reception and appropriation of Darwinism by China's first-generation paleontologists [79]. With the development of young generations of paleontologists in the new century, it is hoped that Chinese paleontology will continue to make more contributions to world's paleontological community. This is accomplished not only by making more exciting discoveries, but also by engaging in new studies that enrich our understanding of the evolutionary mechanisms of life on Earth and provide more clues to our understanding of the impact of global changes in biodiversity on human evolution.

Funding: This research was funded by National Natural Science Foundation of China grant number 42288201.

Institutional Review Board Statement: Not applicable.

Informed Consent Statement: Not applicable.

Data Availability Statement: Not applicable.

Acknowledgments: I wish to thank Mary H. Schweitzer and Ferhat Kaya for their kind invitation to contribute to the special issue "Paleontology in the 21st Century". Yinmai O'Connor helped improve the English. I also thank Renbin Zhan for providing help with the figures. In addition, three referees provided many valuable suggestions that helped improve the manuscript.

Conflicts of Interest: The authors declare no conflict of interest.

References

1. Owen, R. On Fossil Remains of Mammals found in China. *Quart. J. Geol. Soc.* **1870**, *26*, 417–434. [CrossRef]
2. Koken, E. Urber fossile Säugetiere aus China. *Palaeont. Abhandl.* **1885**, *3*, 31–114.
3. Schlosser, M. Die fossilen Säugetiere Chinas nebst einer Ordontographie der recenten Antilopen. *Abhandl. Bayer. Akad. Wissensch.* **1903**, *22*, 1–221.
4. Richthofen, B.F. II.—On the mode of origin of the loess. *Geol. Mag.* **1882**, *9*, 293–305. [CrossRef]
5. Chinese Paleontological Society. *History of Chinese Paleontology*; University of Science and Technology Press: Hefei, China, 2014; pp. 1–328. (In Chinese)
6. Grabau, A.W. Cretaceous Mollusca from north China. *Bull. Geol. Surv. China* **1923**, *5*, 183–198.
7. Grabau, A.W. *Stratigraphy of China. Part 2 Mesozoic*; Geological Survey of China: Beijing, China, 1928; pp. 1–774.
8. Wiman, C. Die Kreide-Dinosaurier aus Shantung. *Palaeontol. Sin. C* **1929**, *6*, 1–67.
9. Chow, T. A preliminary note on some younger Mesozoic plants from Shangtung. *Bull. Geol. Surv. China* **1923**, *5*, 81–141. (In Chinese with English Summary)
10. Sun, Y. Contributions to the Cambrian Faunas of North China. *Palaeontol. Sin.* **1924**, *1*, 1–109.
11. Lee, J.S. Fusulinidae of North China. *Palaeontol. Sin. B* **1927**, *4*, 1–123.
12. Young, C.C. Fossile Nagetiere aus Nord-China. *Palaeontol. Sin. C* **1927**, *5*, 1–82.
13. Sze, X. *Beiträge zur liasischen Flora von China*; Academia Sinica—Memoirs of the National Research Institute of Geology; The National Research Institute of Geology: Shangai, China, 1931; Volume 12, pp. 1–85.
14. Young, C.C. Preliminary notes on the Lufeng vertebrate fossils. *Bull. Geol. Soc. China* **1940**, *20*, 235–239.
15. Zhou, M.; Zhang, M.; Yu, X. *Selected Papers in Cladistics*; Science Press: Beijing, China, 1983; pp. 1–209. (In Chinese)
16. Rong, J.; Fang, Z.; Wu, T. *Selected Papers of Theoretical Palaeontology*; Nanjing University Press: Nanjing, China, 1990; pp. 1–447. (In Chinese)
17. Rong, J.; Fang, Z.; Zhou, Z.; Zhan, R.; Wang, X.; Yuan, X. *Originations, Radiations and Biodiversity Changes—Evidences from the Chinese Fossil Record*; Science Press: Beijing, China, 2006; pp. 1–962.
18. Zhang, Y.; Yuan, X.; Yin, L. Interpreting late Precambrian microfossils. *Science* **1998**, *282*, 1783. [CrossRef]
19. Shu, D.; Luo, H.; Conway Morris, S.; Zhang, X.; Hu, S.; Chen, L.; Han, J.; Zhu, M.; Li, Y.; Chen, L. Lower Cambrian vertebrates from south China. *Nature* **1999**, *402*, 42–46. [CrossRef]
20. Hou, X.; Aldridge, R.J.; Bergström, J.; Siveter, D.J.; Siveter, D.J.; Feng, X. *The Cambrian Fossils of Chengjiang, China: The Flowering of Early Animal Life*; Blackwell: Oxford, UK, 2004; Volume 233, pp. 1–233.
21. Xiao, S.; Muscente, A.D.; Chen, L.; Zhou, C.; Schiffbauer, J.D.; Wood, A.D.; Polys, N.F.; Yuan, X. The Weng'an biota and the Ediacaran radiation of multicellular eukaryotes. *Natl. Sci. Rev.* **2014**, *1*, 498–520. [CrossRef]
22. Meng, J. Mesozoic mammals of China: Implications for phylogeny and early evolution of mammals. *Natl. Sci. Rev.* **2014**, *1*, 521–542. [CrossRef]
23. Xu, X.; Zhou, Z.; Dudley, R.; Mackem, S.; Chuong, C.M.; Erickson, G.M.; Varricchio, D.J. An integrative approach to understanding bird origins. *Science* **2014**, *346*, 1253293. [CrossRef]
24. Zhou, Z. The Jehol Biota, an Early Cretaceous terrestrial Lagerstätte: New discoveries and implications. *Natl. Sci. Rev.* **2014**, *1*, 543–559. [CrossRef]
25. Chiappe, L.M.; Meng, Q. *Birds of Stone: Chinese Avian Fossils from the Age of Dinosaurs*; Johns Hopkins University Press: Baltimore, MD, USA, 2016; pp. 1–304.
26. Yuan, X.; Chen, Z.; Xiao, S.; Zhou, C.; Hua, H. An early Ediacaran assemblage of macroscopic and morphologically differentiated eukaryotes. *Nature* **2011**, *470*, 390–393. [CrossRef]
27. Chen, Z.; Zhou, C.; Yuan, X.; Xiao, S. Death march of a segmented and trilobate bilaterian elucidates early animal evolution. *Nature* **2019**, *573*, 412–415. [CrossRef]
28. Xiao, S.; Zhang, Y.; Knoll, A.H. Three-dimensional preservation of algae and animal embryos in a Neoproterozoic phosphorite. *Nature* **1998**, *391*, 553–558. [CrossRef]

29. Fu, D.; Tong, G.; Dai, T.; Liu, W.; Yang, Y.; Zhang, Y.; Cui, L.; Li, L.; Yun, H.; Wu, Y.; et al. The Qingjiang biota—a Burgess Shale–type fossil Lagerstätte from the early Cambrian of South China. *Science* **2019**, *363*, 1338–1342. [CrossRef] [PubMed]

30. Sun, Z.; Zhao, F.; Zeng, H.; Luo, C.; Van Iten, H.; Zhu, M. The middle Cambrian Linyi Lagerstatte from the North China Craton: A new window on the Cambrian evolutionary fauna. *Natl. Sci. Rev.* **2022**, *9*, nwac069. [CrossRef] [PubMed]

31. Hao, S.; Xue, J. *The Early Devonian Posongchong Flora of Yunnan*; Science Press: Beijing, China, 2013; pp. 1–366.

32. Wang, J.; Pfefferkorn, H.W.; Zhang, Y.; Feng, Z. Permian vegetational Pompeii from Inner Mongolia and its implications for landscape paleoecology and paleobiogeography of Cathaysia. *Proc. Natl. Acad. Sci. USA* **2012**, *109*, 4927–4932. [CrossRef] [PubMed]

33. Li, C.; Fraser, N.C.; Rieppel, O.; Wu, X. A Triassic stem turtle with an edentulous beak. *Nature* **2018**, *560*, 476–479. [CrossRef] [PubMed]

34. Li, Z.; Stidham, T.A.; Zheng, X.; Wang, Y.; Zhao, T.; Deng, T.; Zhou, Z. Early evolution of diurnal habits in owls (Aves, Strigiformes) documented by a new and exquisitely preserved Miocene owl fossil from China. *Proc. Natl. Acad. Sci. USA* **2022**, *119*, e2119217119. [CrossRef] [PubMed]

35. Wang, B.; Shi, G.; Xu, C.; Spicer, R.A.; Perrichot, V.; Schmidt, A.R.; Feldberg, K.; Heinrichs, J.; Chény, C.; Pang, H.; et al. The mid-Miocene Zhangpu biota reveals an outstandingly rich rainforest biome in East Asia. *Sci. Adv.* **2021**, *7*, eabg0625. [CrossRef] [PubMed]

36. He, T.; Zhu, M.; Mills, B.J.; Wynn, P.M.; Zhuravlev, A.Y.; Tostevin, R.; von Strandmann, P.E.P.; Yang, A.; Poulton, S.W.; Hields, G.A. Possible links between extreme oxygen perturbations and the Cambrian radiation of animals. *Nat. Geosci.* **2019**, *12*, 468–474. [CrossRef] [PubMed]

37. Yin, Z.; Vargas, K.; Cunningham, J.; Bengtson, S.; Zhu, M.; Marone, F.; Donoghue, P. The early Ediacaran Caveasphaera foreshadows the evolutionary origin of animal-like embryology. *Curr. Biol.* **2019**, *29*, 4307–4314. [CrossRef]

38. Ling, M.; Zhan, R.; Wang, G.; Wang, Y.; Amelin, Y.; Tang, P.; Liu, J.; Jin, J.; Huang, B.; Wu, R.; et al. An extremely brief end Ordovician mass extinction linked to abrupt onset of glaciation. *Solid Earth Sci.* **2019**, *4*, 190–198. [CrossRef]

39. Wang, G.; Zhan, R.; Percival, I.G. The end-Ordovician mass extinction: A single-pulse event? *Earth Sci. Rev.* **2019**, *192*, 15–33. [CrossRef]

40. Zhu, M.; Yu, X.; Ahlberg, P.E.; Choo, B.; Lu, J.; Qiao, T.; Qu, Q.; Zhao, W.; Jia, L.; Lu, J.; et al. A Silurian placoderm with osteichthyan-like marginal jaw bones. *Nature* **2013**, *502*, 188–193. [CrossRef]

41. Shen, S.; Crowley, J.L.; Wang, Y.; Bowring, S.A.; Erwin, D.H.; Sadler, P.M.; Cao, C.; Rothman, D.H.; Henderson, C.M.; Ramezani, J.; et al. Calibrating the end-Permian mass extinction. *Science* **2011**, *334*, 1367–1372. [CrossRef] [PubMed]

42. Fan, J.; Shen, S.; Erwin, D.H.; Sadler, P.M.; MacLeod, N.; Cheng, Q.; Hou, X.; Yang, J.; Wang, X.; Wang, Y.; et al. A high-resolution summary of Cambrian to Early Triassic marine invertebrate biodiversity. *Science* **2020**, *367*, 272–277. [CrossRef] [PubMed]

43. Wang, M.; O'Connor, J.K.; Xu, X.; Zhou, Z. A new Jurassic scansoriopterygid and the loss of membranous wings in theropod dinosaurs. *Nature* **2019**, *569*, 256–259. [CrossRef] [PubMed]

44. Mao, F.; Liu, C.; Chase, M.H.; Smith, A.K.; Meng, J. Exploring ancestral phenotypes and evolutionary development of the mammalian middle ear based on Early Cretaceous Jehol mammals. *Proc. Natl. Acad. Sci. USA* **2020**, *8*, nwaa188. [CrossRef]

45. Ren, D.; Gao, T.; Shih, C.; Yao, Y.; Wang, Y. *Rhythms of Insect Evolution: Evidence from the Jurassic and Cretaceous in Northern China*; Whiley-Blackwell: Hoboken, NJ, USA, 2019; pp. 1–728.

46. Wang, M.; O'Connor, J.K.; Zhao, T.; Pan, Y.; Zheng, X.; Wang, X.; Zhou, Z. An Early Cretaceous enantiornithine bird with a pintail. *Curr. Biol.* **2021**, *31*, 4845–4852. [CrossRef]

47. Wang, S.; Ye, J.; Meng, J.; Li, C.; Costeur, L.; Mennecart, B.; Zhang, C.; Zhang, J.; Aiglstorfer, M.; Wang, Y.; et al. Sexual selection promotes giraffoid head-neck evolution and ecological adaptation. *Science* **2022**, *376*, eabl8316. [CrossRef]

48. Deng, T.; Ding, L. Paleoaltimetry reconstructions of the Tibetan Plateau: Progress and contradictions. *Natl. Sci. Rev.* **2015**, *2*, 417–437. [CrossRef]

49. Su, T.; Spicer, R.A.; Wu, F.; Farnsworth, A.; Huang, J.; Del Rio, C.; Deng, T.; Ding, L.; Deng, W.; Huang, Y.; et al. A Middle Eocene lowland humid subtropical "Shangri-La" ecosystem in central Tibet. *Proc. Natl. Acad. Sci. USA* **2020**, *117*, 32989–32995. [CrossRef]

50. Li, Z.; Wu, X.; Zhou, L.; Liu, W.; Gao, X.; Nian, X.; Trinkaus, E. Late Pleistocene archaic human crania from Xuchang, China. *Science* **2017**, *355*, 969–972. [CrossRef]

51. Wu, X.; Pei, S.; Cai, Y.; Tong, H.; Li, Q.; Dong, Z.; Sheng, J.; Jin, Z.; Ma, D.; Xing, S.; et al. Archaic human remains from Hualongdong, China, and Middle Pleistocene human continuity and variation. *Proc. Natl. Acad. Sci. USA* **2019**, *116*, 9820–9824. [CrossRef] [PubMed]

52. Ni, X.; Ji, Q.; Wu, W.; Shao, Q.; Ji, Y.; Zhang, C.; Liang, L.; Ge, J.; Guo, Z.; Li, J.; et al. Massive cranium from Harbin in northeastern China establishes a new Middle Pleistocene human lineage. *The Innovation* **2021**, *2*, 100130. [CrossRef] [PubMed]

53. Yin, H.; Zhang, K.; Tong, J.; Yang, Z.; Wu, S. The global stratotype section and point (GSSP) of the Permian-Triassic boundary. *Episodes* **2001**, *24*, 102–114.

54. Zhan, R.; Zhang, Y. *Stratigraphic "Golden Spikes": Critical Points in the Evolution of the Earth*; Phoenix Science Press: Nanjing, China, 2022; pp. 1–707. (In Chinese)

55. Qiu, Z.; Qiu, Z.; Deng, T.; Li, C.; Zhang, Z.; Wang, B.; Wang, X. Neogene land mammal stages/ages of China. In *Fossil Mammals of Asia*; University Press: Columbia, SC, USA, 2013; pp. 29–90.

56. Shen, S.; Rong, J. Preface: New advances in the integrative stratigraphy and timescale of China. *Sci. China Earth Sci.* **2019**, *62*, 1–6. [CrossRef]

57. Chen, X.; Wang, H.; Zhao, Q.; Qiu, Z.; Chen, J.; Ni, H. *Late Ordovician to Early Silurian Shale Gas Bearing Strata from the Yangtze Region, China*; Zhejiang University Press: Hangzhou, China, 2021; pp. 1–266. (In Chinese)

58. Deng, T.; Li, Q.; Tseng, Z.J.; Takeuchi, G.T.; Wang, Y.; Xie, G.; Wang, S.; Hou, S.; Wang, X. Locomotive implication of a Pliocene three-toed horse skeleton from Tibet and its paleo-altimetry significance. *Proc. Natl. Acad. Sci. USA* **2012**, *109*, 7374–7378. [CrossRef]

59. Zhang, Y.; Zhai, Q.; Fan, J.; Song, P.; Qie, W. Editorial preface to special issue: From prototethys to neotethys: Deep time paleobiogeographic and paleogeographic evolution of blocks in the Qinghai-Tibet Plateau. *Palaeogeogr. Palaeoclimatol. Palaeoecol.* **2022**, *599*, 111046. [CrossRef]

60. Swisher, C.C.; Wang, Y.; Wang, X.; Xu, X.; Wang, Y. Cretaceous age for the feathered dinosaurs of Liaoning, China. *Nature* **1999**, *400*, 58–61. [CrossRef]

61. Zhu, M.; Zhang, J.; Yang, A. Integrated Ediacaran (Sinian) chronostratigraphy of South China. *Palaeogeogr. Palaeoclimatol. Palaeoecol.* **2007**, *254*, 7–61. [CrossRef]

62. He, H.; Wang, X.; Zhou, Z.; Wang, F.; Boven, A.; Shi, G.; Zhu, R. Timing of the Jiufotang Formation (Jehol Group) in Liaoning, northeastern China and its implications. *Geophys. Res. Lett.* **2004**, *31*, L019790. [CrossRef]

63. Zhong, Y.; Huyskens, M.H.; Yin, Q.; Wang, Y.; Ma, Q.; Xu, Y. High-precision geochronological constraints on the duration of 'Dinosaur Pompeii' and the Yixian Formation. *Natl. Sci. Rev.* **2021**, *8*, nwab063. [CrossRef]

64. Zhu, R.; Zhou, Z.; Meng, Q. Destruction of the North China Craton and its influence on surface geology and terrestrial biotas. *Chin. Sci. Bull.* **2020**, *65*, 2954. [CrossRef]

65. Zhou, Z.; Meng, Q.; Zhu, R.; Wang, M. Spatiotemporal evolution of the Jehol Biota: Responses to the North China craton destruction in the Early Cretaceous. *Proc. Natl. Acad. Sci. USA* **2021**, *118*, e2107859118. [CrossRef] [PubMed]

66. Xie, S.; Yin, H. Progress and perspective on frontiers of geobiology. *Sci. China Earth Sci.* **2014**, *57*, 855–868. [CrossRef]

67. Fu, Q.; Meyer, M.; Gao, X.; Stenzel, U.; Burbano, H.A.; Kelso, J.; Pääbo, S. DNA analysis of an early modern human from Tianyuan Cave, China. *Proc. Natl. Acad. Sci. USA* **2013**, *110*, 2223–2227. [CrossRef]

68. Zhang, D.; Xia, H.; Chen, F.; Li, B.; Slon, V.; Cheng, T.; Yang, R.; Jacobs, Z.; Dai, Q.; Massilani, D.; et al. Denisovan DNA in late Pleistocene sediments from Baishiya Karst Cave on the Tibetan plateau. *Science* **2020**, *370*, 584–587. [CrossRef]

69. Yang, M.A.; Fan, X.; Sun, B.; Chen, C.; Lang, J.; Ko, Y.C.; Tsang, C.; Chiu, H.; Wang, T.; Bao, Q.; et al. Ancient DNA indicates human population shifts and admixture in northern and southern China. *Science* **2020**, *369*, 282–288. [CrossRef]

70. Zhang, P.; Zhang, X.; Zhang, X.; Gao, X.; Huerta-Sanchez, E.; Zwyns, N. Denisovans and *Homo sapiens* on the Tibetan Plateau: Dispersals and adaptations. *Trends Ecol. Evol.* **2021**, *37*, 257–267. [CrossRef]

71. Welker, F.; Ramos-Madrigal, J.; Kuhlwilm, M.; Liao, W.; Gutenbrunner, P.; de Manuel, M.; Samodova, D.; Mackie, M.; Allentoft, M.E.; Bacon, A.M.; et al. Enamel proteome shows that *Gigantopithecus* was an early diverging pongine. *Nature* **2019**, *576*, 262–265. [CrossRef]

72. Pan, Y.; Zheng, W.; Sawyer, R.H.; Pennington, M.W.; Zheng, X.; Wang, X.; Wang, M.; Hu, L.; O'Connor, J.; Zhao, T.; et al. The molecular evolution of feathers with direct evidence from fossils. *Proc. Natl. Acad. Sci. USA* **2019**, *116*, 3018–3023. [CrossRef]

73. Chen, J.; Bottjer, D.J.; Davidson, E.H.; Dornbos, S.Q.; Gao, X.; Yang, Y.; Li, J.; Li, G.; Wang, X.; Xian, D.; et al. Phosphatized polar lobe-forming embryos from the Precambrian of southwest China. *Science* **2006**, *312*, 1644–1646. [CrossRef]

74. Gai, Z.; Donoghue, P.C.; Zhu, M.; Janvier, P.; Stampanoni, M. Fossil jawless fish from China foreshadows early jawed vertebrate anatomy. *Nature* **2011**, *476*, 324–327. [CrossRef] [PubMed]

75. Ni, X.; Gebo, D.L.; Dagosto, M.; Meng, J.; Tafforeau, P.; Flynn, J.J.; Beard, K.C. The oldest known primate skeleton and early haplorhine evolution. *Nature* **2013**, *498*, 60–64. [CrossRef] [PubMed]

76. Zhang, X.; Pratt, B.R. The first stalk-eyed phosphatocopine crustacean from the Lower Cambrian of China. *Curr. Biol.* **2012**, *22*, 2149–2154. [CrossRef] [PubMed]

77. Xing, L.; McKellar, R.C.; Xu, X.; Li, G.; Bai, M.; Persons IV, W.S.; Miyashita, T.; Benton, M.; Zhang, J.; Wolfe, A.P.; et al. A feathered dinosaur tail with primitive plumage trapped in mid-Cretaceous amber. *Curr. Biol.* **2016**, *26*, 3352–3360. [CrossRef] [PubMed]

78. Ma, X.; Wang, G.; Wang, M. Impact of Chinese palaeontology on evolutionary research. *Philos. Trans. R. Soc. B* **2022**, *377*, 20210029. [CrossRef]

79. Yu, X. Chinese paleontology and the reception of Darwinism in early twentieth century. *Stud. Hist. Philos. Sci. Part C Stud. Hist. Philos. Biol. Biomed. Sci.* **2017**, *66*, 46–54. [CrossRef]

MDPI

Commentary

A Plea for a New Synthesis: From Twentieth-Century Paleobiology to Twenty-First-Century Paleontology and Back Again

Marco Tamborini

Department of Philosophy, Technische Universität Darmstadt, Marktplatz 15 (Residenzschloss), 64283 Darmstadt, Germany; marco.tamborini@tu-darmstadt.de

Simple Summary: This article examines the relationship between twentieth- and twenty-first-century paleobiology. After summarizing the disciplinary problem of paleontology in the mid-twentieth century, I focus on five representative research topics in contemporary paleontology. In doing so, I outline twenty-first-century paleontology as a science that seeks more data, more technology, and more integration. At the end of the paper, I highlight a possible new paleobiological revolution: it would give paleontologists strong political representation to deal with issues such as expanded evolutionary synthesis, conservation of the Earth's environment, and global climate change.

Abstract: In this paper, I will briefly discuss the elements of novelty and continuity between twentieth-century paleobiology and twenty-first-century paleontology. First, I will outline the heated debate over the disciplinary status of paleontology in the mid-twentieth century. Second, I will analyze the main theoretical issue behind this debate by considering two prominent case studies within the broader paleobiology agenda. Third, I will turn to twenty-first century paleontology and address five representative research topics. In doing so, I will characterize twenty-first century paleontology as a science that strives for more data, more technology, and more integration. Finally, I will outline what twenty-first-century paleontology might inherit from twentieth-century paleobiology: the pursuit of and plea for a new synthesis that could lead to a second paleobiological revolution. Following in the footsteps of the paleobiological revolution of the 1960s and 1970s, the paleobiological revolution of the twenty-first century would enable paleontologists to gain strong political representation and argue with a decisive voice at the "high table" on issues such as the expanded evolutionary synthesis, the conservation of Earth's environment, and global climate change.

Keywords: paleontology; paleobiology; history and philosophy of paleontology; twenty-first-century paleontology; paleobiological revolution; technoscience and global issues

Citation: Tamborini, M. A Plea for a New Synthesis: From Twentieth-Century Paleobiology to Twenty-First-Century Paleontology and Back Again. *Biology* **2022**, *11*, 1120. https://doi.org/10.3390/biology11081120

Academic Editors: Mary H. Schweitzer and Ferhat Kaya

Received: 27 June 2022
Accepted: 24 July 2022
Published: 26 July 2022

Publisher's Note: MDPI stays neutral with regard to jurisdictional claims in published maps and institutional affiliations.

Copyright: © 2022 by the author. Licensee MDPI, Basel, Switzerland. This article is an open access article distributed under the terms and conditions of the Creative Commons Attribution (CC BY) license (https://creativecommons.org/licenses/by/4.0/).

1. Introduction

At the beginning of his book *Tempo and Mode in Evolution* (1944), North American paleontologist George Gaylord Simpson (1902–1984) provided an interesting picture of the difficulties and lack of interaction between the paleontological and biological communities during the middle of the twentieth century. He wrote, "not long ago [...] geneticists said that paleontology had no further contributions to make to biology, that [...] it was a subject too purely descriptive to merit the name "science". The paleontologist, they believed, is like a man who undertakes to study the principles of the internal combustion engine by standing on a street corner and watching the motor cars whiz by" [1].

The same opinion was shared by prominent biologists. For instance, British geneticist John Maynard Smith (1920–2004) noted that that during the 1960s there were continuous frictions between paleontologists and biologists. As he recalled in 1984, "[at] that time, the attitude of populations genetics to any paleontologist rash enough to offer a contribution to evolutionary theory has been to tell him go away and find another fossil, and not to bother the grownups" [2]. However, in the same piece, Maynard Smith added two lines, in effect showing the different perception and value of paleontology. He admitted, "Palaeontology has been too long absent from the high table. Welcome back" [2].

In welcoming paleontology to the table that matters, the geneticist paved the way for some historical and theoretical questions: What happened between the 1960s and 1980s that was so imprinted as to make his judgment change? Does twenty-first-century paleontology still sit at the high-table? And how does the scientific agenda of today's paleontology relate to its scientific past and future?

In this paper, I will briefly address these questions, in effect reflecting on the elements of novelty and continuity between twentieth-century paleobiology and twenty-first-century paleontology. First, I will expose the heated debate about the disciplinary status of paleontology during the mid-twentieth century. Second, I will analyze the main theoretical issue behind this debate by looking at two prominent case studies within the broader paleobiological agenda. Third, I will move to twenty-first-century paleontology and address five representative research topics. By doing so, I will characterize twenty-first-century paleontology as a science that seeks *more data, more technology, and more integration.* Moreover, I will I focus on the notion of the "second digital revolution" as it occurs in paleontology. In the conclusion, I will state what twenty-first-century paleontology might take from twentieth-century paleobiology: the aspiration of and plea for a new synthesis that may lead to a second paleobiological revolution.

2. Paleobiology vs. Paleontology

"We are paleontologists, so we need a name to contrast ourselves with all you folks who study modern organisms in human or ecological time. You therefore become neontologists" [3]. With his classical provocative and quite "parochialism" jargon, North American paleontologist Stephan Jay Gould (1941–2002) called attention to two important elements that characterized the methodology and disciplinary status of paleontology throughout the twentieth century. First, Gould insisted that a different name was necessary to differentiate two quite different approaches to evolution. According to Gould, the term 'paleontology' was too weak and too historically laden to achieve the aim of promoting paleontologists as genuine biologists and not as geologists. Together with several US colleagues, such as Niles Eldredge, David Raup, Jack Sepkoski, Steven Stanley, and others, he chose the term "paleobiology" as the perfect name to describe paleontological interests in evolutionary biology [4,5]. Paleobiology had its heyday between 1970 and 1985. This name though goes back to Austrian paleontologist and biologist Othenio Abel (1875–1946). He coined the term "paleobiology" to emphasize the biological meaning of this discipline. Paleobiology, so Abel, was able to investigate evolutionary mechanisms, in effect providing the deep-time perspective to evolutionary study [6–8]. Hence, paleontology should be considered a biological sub-discipline and not practiced in geological departments.

Second, Gould contrasted between evolutionary investigations of living organisms (what he called neontological analyses) and the paleo(bio)ontological ones, which study what happened in the deep past. Indeed, Gould's statement derived from the empirical results he himself and other paleontologists were obtaining during the 1960s and 1970s. These scientists were keen to present their discipline as a rigorous investigation of evolutionary mechanisms. By uncovering deep-time patterns and processes, paleontologists were indeed able to contribute to and expand the evolutionary mechanisms set during the modern synthesis of evolution. Gould coupled this theoretical aim with a broader idea of science, paleontology, and evolutionary time [5,9–11].

As noted, the second ground behind Gould's provocative statement was his emphasis on the importance of deep-time investigations as defining characteristic of paleontology. Although the discovery and conquest of deep time was a classical argument for the importance of paleontology since the seventeenth century [12–16], this dimension acquired extra (biological) value in the twentieth century. In 1985, Gould published a paper in the journal *Paleobiology*, arguing that "Nature's discontinuities occur at different scales of time or tiers". He distinguished three distinct temporal tiers. The first tier includes events at the ecological movement; the second tier encompasses events which occur during "millions of years in "normal" geological time"; whereas the most exciting subject in paleontology lies in our recognition that one of our best-recognized and most puzzling phenomena, mass extinction, is not merely more and quicker of the same, but a third distinct tier with rules and principles of its own [17].

Successively, he stated that "whatever accumulates at the first tier is sufficiently reversed, undone, or overridden by processes of the higher tiers" [17]. For instance, mass extinction occurs at the third tier. It "works by different rules and may undo whatever the lower tiers had accumulated" [17]. Evolution is an extremely hierarchical phenomenon. Hence, Gould noted that only paleobiologists are able to investigate what happened in the second and third tier; whereas "neontologists" study "modern organisms in human or ecological time" [17].

Gould's idea was that evolutionary time could be seen as a system of distinct tiers—and the problem of transpacific evolution requires an explicit study of their interaction. Darwinian tradition leads us to deny this kind of structuring, to view time as a continuous, and to seek the source of causality at all scales in observable events and processes at smallest [9,18].

To put it simply, Gould and colleagues meant to "(1) make paleontology more theoretical and less descriptive; (2) introduce models and quantitative analysis into paleontological methodology; (3) import ideas and techniques from other disciplines (especially biology) into paleontology; (4) emphasize the evolutionary implications of the fossil record" [5]. Or, as philosopher Derek Turner put it in seven slogans, "(1) Paleontology has more to contribute to biology than to geology; (2) Study fossils in bulk—individual specimens don't tell you much about evolution; (3) Paleontology needs theories; (4) If you can't experiment, then simulate; (5) don't assume that the fossil record is incomplete; analyze the incompleteness; (6) resist reductionism; (7) don't shy away from raising big questions about evolution" [19]. This does not mean, though, that all these features were literally invented by paleobiologists, but rather that Gould and colleagues put these elements at the center of their research programs.

Following these starting points, Gould asserted that paleontology was a legitimate biological discipline able to uncover patterns and mechanisms which could be found at the second and tier tiers of time [20]. For instance, at the second tier, Gould individuated phenomena which occur primary within a punctuated equilibrium-pattern (as he had formulated years earlier together with Niles Eldredge [21]), whereas the third tier is dominated by mass extinction phenomena. These patterns and mechanisms complete, expand, and in part revise the neo-Darwinian picture of evolution.

Nine months after his article "The Paradox of the first Tier" was submitted to *Paleobiology*, Gould wrote to the journal's editor, Jack Sepkoski, explaining the necessity of an interaction between different and autonomous layers of the evolutionary theory:

Hierarchy, as here discussed in its genealogical context, is an 'internalistic' theory about evolution dynamics. And we need to formulate it properly if we are to tackle this internal dynamic with the other great mover of life's patterns—the externals of geological history especially mass extinctions, that so impact life's history . . . in other words, all the data that you and your colleagues are treating in such new and exciting ways. Hierarchy confronts the geological dynamic, and we will not get it right until we reformulate both sides. Gould to Sepkoski, 13 August 1985 in [5].

In this letter, Gould clearly affirmed the importance of reformulating the hierarchical model of evolution in the light of the momentous and exciting research techniques used in the investigation of mass extinctions. The reformulation Gould had in mind, however, concerned the entire paleontological discipline. He suggested a "nomothetic and idiographic" approach to the fossil record based upon David Raup and Sepkoski's studies on the structure of the mass extinction [22–24].

The Kantian philosopher Wilhelm Windelband (1848–1915) coined the nomothetic vs. idiographic distinction. On May 4, 1894, he gave his rectoral address on the methodological differences between History and Natural Science at the Kaiser-Wilhelms-Universität Strasburg. During that address, he distinguished historical from natural sciences by focusing on the "formal character of their cognitive goals" [25]. Nomothetic disciplines aimed at formulating general laws and general judgments, while the ideographic sciences merely collected historical facts. The former were sciences of laws, the latter sciences of events: "the former teach what always is, the latter what once was" [25]. This distinction was not about the contents of the two sciences; but rather it was about how scientists produce knowledge. That means that the main difference between nomothetic and ideographical disciplines was methodological. With his programmatic statements expressed publicly in a paper, Gould intended to bridge the methodological gap between the natural and bio-historical sciences [20].

To accomplish the reformulation, Gould promoted a methodological synthesis. He clearly suggested this in commenting on Sepkoski's study on Phanerozoic diversity [26,27]. Gould affirmed that, "Here we see an interesting and fruitful interaction of nomothetics and ideographics. The form of the model remains nomothetic—the "real" pattern arises as an interaction between two general curves of the same form, but with different parameters. Ideographic factors determine the parameters and then enter as boundary conditions into a nomothetic model" [20].

The ideographic factors mentioned by Gould derive from Sepkoski's famous Compendium. It gave the required data a mathematical treatment of data. This in turn made visible what is invisible: the structure and development of the Phanerozoic diversity [28,29].

Hence, the contrast between paleontology and paleobiology was mainly based on disciplinary and methodological issues. Gould and colleagues first created a disciplinary space in which to insert their research agenda (they named their approach "paleobiology"). Successively, they sought to a possible methodological and disciplinary synthesis, in effect avoiding possible dichotomies. Famously and quoting German philosopher Immanuel Kant, Gould wrote "with all biology and no geology, paleontology is empty; but with geology alone, it is blind" [20]. Furthermore, by calling for a synthesis, another theoretical issue was tackled by twentieth-century paleobiology: the possible over- and under-determination of paleontological explanations.

3. Over- and Under-Determination

It is quite difficult to find well-preserved fossils that immediately resemble living organisms that can also be exhibited, used, and taken at face value. In fact, once an organism dies, it is subjected to the several taphonomic processes. These processes destroy and change the features of the original organism. For instance, it is rare to find fossils with their soft parts preserved. There are two ways to practically address this problem: (1) focus on structures that are more frequently preserved; (2) focus on exceptionally well-preserved sites. These provide more selective observations that can be essential, but have their limitations when studying rapid changes during evolution (I thank an anonymous reviewer who pointed this out to me. See [30,31])

The imperfect and incomplete nature of the paleontological record gave several paleontologists cause to reflect on the epistemic aspect of the fossil record. As a result, paleontologists came up with practices intended to overcome the incomplete nature of the records of the past. In addition to working with imperfect and incomplete data, paleontologists must face another difficulty: the so-called over and underdetermination issue.

As philosopher of science Carol E. Cleland has pointed out, historical sciences, such as paleontology, are subjected to the asymmetry of overdetermination [32,33]. They are in the same condition as the investigator who is trying to reconstruct what, exactly, shattered a window starting from the traces on the floor. Let us imagine that three different people throw different objects at the same window at the same time. In that case, "the breaking of the window is overdetermined by numerous sub collections of shards of glass lying on the kitchen floor. That overdetermination of earlier facts by later traces occurs whenever a window breaks" [34].

Following Turner, we can develop the thought experiment a bit further. "The owners of the house sweep up the shards, throw the baseball in the bin, and eventually repair the window. A few weeks later, the only traces of the event that remain are a few shards of glass under the refrigerator. The housecleaning and repair are examples of what Sober (1988, 3) calls *information-destroying processes*" [34].

Let us follow Turner again in this line of reasoning. Let us assume that a future investigator discovers glass shards on the kitchen floor. He will then ask what kind of shards they are—are they pieces of a glass, a window, a vase, etc.? "Even if the historical investigator recognizes the traces for what they are", noted Turner, "rival hypotheses about earlier events and processes will often be underdetermined by the available traces. After studying the shards under the refrigerator, the historical investigator will be completely confused: The evidence does not permit her to discriminate at all between incompatible opposing hypotheses (window vs. wine glass, football vs. baseball, etc.) In other words, she confronts a local underdetermination problem" [34].

Therefore, the present event (a broken window, or, in paleobiology, a mass extinction) over- or under-determines its possible causes. Replaying the tape of time, we cannot be sure to identify the correct sequence of cause–effect, since there are many possible causal chains backwards from the local event.

Hence, first, the record of the past is always imperfect and incomplete, and second, we are not able to state whether our imperfect and incomplete data overdetermine or underdetermine the phenomena paleontologist would like to investigate.

Given these two issues, how can paleontologists bring out patterns and mechanisms that might expand evolutionary theory? Are these two issues not a death kiss for paleontology? And more broadly, what is paleontological business about? To answer these questions, I will briefly recall some research results Gould himself put in the middle of his agenda.

One main topic of twentieth-century paleobiology was the debate about the importance and dynamics of mass extinction [5,35]. One central research result was the famous paper (and graph) representing the periodicity in mass extinctions within geological time. It was used by Gould to indicate a classical phenomenon which happens on the third temporal tier. This representation has many interesting peculiarities, which are related to the notion of paleobiological data [16,29,36,37].

Due to Alvarez's team discovery (1978) of the iridium anomaly in rocks formation at K–T boundary, mass extinction became the hot topic of paleontology in the 1980s. One of the main questions during those years was concerning the number and intensity of the mass extinctions.

This issue was resolved in two famous papers written by David Raup and Jack Sepkoski. Using Sepkoski's database (1982), Raup and Sepkoski identified the number of mass extinctions and the periodicity of this phenomenon. Sepkoski and Raup plotted a huge number of fossil marine families against geological time and found that five mass extinctions clearly occurred in the history of the Earth as "statistically distinct from the background extinction levels". These "five extinction events are seen as sharp drops in standing diversity", and therefore they were easily detected. A mass extinction is thus an event in which a "large number of organisms have disappeared over relatively short time" [22,23].

This case study is emblematic since it shows the *highly integrative practice* used in twentieth-century paleobiology. As for the seminal investigation of morphogenesis conducted by David Raup and Adolf Seilacher during the 1960s and 1970s [11,38], also in this case paleobiologists sought to integrate data and technologies to produce possible coherent scenarios of the past. Since a direct access to the deep past is impossible, scientists technically recreate possible scenarios which tell the scientists that something happened in the past. Hence, to overcome the issue of over- and underdetermination, paleobiologists stretched and elaborated their data with the help of technology (such as computer, electron microscope, databases, etc.). The key methodological insight was indeed *the fruitful combination of science and technology*. This implied that phenomena such as mass-extinctions or a life-like display of extinct specimens in a museum's hall depend both on the correct use of technological devices and on the interplay between these devices and theories. Hence, paleobiology can be seen as a phenomena-lead discipline. These "investigations are such because the relevance of evidence turns on relationships between phenomena, the hypotheses pertaining to them" [39] and, I would add, the technologies used to stage deep time. Shades of glass on the ground or fossil excavated in a particular region are evidence if these can be used into this integrative process [40].

4. Twenty-First-Century Paleontology: More Data, More Technology, and More Integration

What is left of the paleobiological research agenda? As I will detail in the conclusion, twentieth-century paleobiology shares with twenty-first-century paleontology the needs for integration and synthesis. In this section, I would like to briefly single out five promising research topics of current paleontology and connect them with twentieth-century paleobiology. I would like to characterize the research of twenty-first-century paleontology with a slogan: more data, more technology, and more integration.

The first topic I chose is about the emergence of paleocolor as a testable field of enquiry. To achieve this aim, it is important to have more clean data. In addition to the publication of studies on feather taphonomy, scientists are teaming up to use ion beam scanning electron microscopy to investigate the preserved melanin pigments. By integrating more data with more technologies, paleocolor can provided precious insights into the behavior and ecology of extinct organisms [34,41–48].

Second, as noted, one main issue paleontologists have to face is the lack of appropriate data. The clarification around the preservability of organic chemicals in fossils, and more broadly, the mechanisms behind taphonomy, is another key research program of twenty-first-century paleontology. Also in this case, the choice and use of appropriate technologies (such as mass spectrometry methodology, appropriate databases, etc.) is essential to obtain more data and thus pose new questions on phylogenetic hypotheses [49–54].

Third, another promising topic of twenty-first-century paleontology is given by the intersection between morphology, evolutionary theory, and various technologies. The intersection between robotics and morphology provides one compelling example. Scientists are working together with engineers to model and construct bio-inspired robots to understand evolution. One example above all (for another example published recently, see [55]) is given by research on the morphology of *Orobates pabsti*. This is a four-legged vertebrate organism that went extinct about 300 million years ago. The study of the morphology of this well-preserved specimen is very important because it could offer valuable insights into the evolution of terrestrial vertebrates. *Orobates* are an early evolution of the lineage that led to amniotes. These made the vertebrate transition to land by becoming independent of open water during the early stages of development. Thus, studying and understanding how these species were able to transition from water to land is essential to better understand one of the major transitions in vertebrate evolution.

To perform this research, scientists designed the OroBOT robot. It was designed to account for the locomotion dynamics of *Orobates*. The OroBot was built in collaboration with bioengineers at the École Polytechnique Fédérale de Lausanne (EPFL) in Lausanne. The spine of the OroBot was segmented into eight joints: two for the neck, four for the trunk, and two for the tail. The feet consisted of three passive and compliant joints. The designed parts of OroBOT were made of polyamide plastic material and created by selective laser sintering. Working in vivo with these robots, Nyakatura's team was able to reach conclusions about the complex form-function that characterized the vertebrates' transition to land [56–59].

The same logic is behind the creation of a chewing machine, an artificial mechanical chewing machine, used to study the micro-wear process and the possible correlation with diet and the rate of tooth wear. As the authors noted, "the aim was a simplified system that would allow changing various factors in mastication, for example different standardized feeds, to test whether expectations based on a simplistic interpretation of microscopic and macroscopic wear could be confirmed. The aim of this study was to produce microwear features seen in nature, i.e., pits and scratches, and to quantify macroscopic wear with real teeth and diets, in order to achieve a detailed picture of the wear process and its components" [60]. As in the case of the OroBot, the use of technology and various machines here also brings the investigation of the past (micro-wear process) close to engineering [59,61].

This synergy between technology and paleontological research has also taken different forms in the present day. Among them, one is virtual paleontology [62–67]. This leads to a second digital revolution in paleontology. This second digital revolution is signed by the passage from bio-robotics, or nature-inspired robotics, to robotics-inspired biology. This transition implies bridging of the gap between technology and nature. Form changes should now be studied through in vivo investigations (such as the classical anatomical dissection), in silico (as, for example, through CT scanners or computer simulations), and eventually again in a hybrid and highly integrated in vivo–silico–robotic environment through bio-robotics [61]. The full integration of these methodological levels would help illustrate the structural interplay of elements that characterizes form change.

The fourth hot topic of twenty-first-century paleontology is deeply rooted in the twentieth-century paleobiology. As noted in the introduction, paleobiology was launched to reorient the paleontological agenda toward broader evolutionary problems. This was the main reason for the paleobiological revolution of the 1960s and 1970s—think of Eldredge and Gould's call for punctuated equilibrium. The same search for new evolutionary mechanisms pervaded paleobiological research during the first encounters with the emerging evo-devo community. Or rather, the evolution of evo-devo was shaped by paleobiological questions from the beginning, and vice versa. In fact, at the 1981 Dahlem conference on "Evolution and Development", which will be considered as the grounding meeting of evolutionary developmental biology as an autonomous evolutionary discipline, biologists and paleontologists discussed together the relationship between evolution and development. For instance, in the working group on "The Role of Development in Macroevolutionary Change" biologists, such as Jim Murray, Pere Alberch, Brian Goodwin, Gunther Wagner, Tony Hoffman, and David Wake, defined with paleontologists Stephen Gould, Adolf Seilacher, and David Raup the role of constrains in evolution [11,68–70].

The same line of continuity and interaction denotes current paleontology, which even formalizes the paleontological contribution to evo-devo as Paleo-Evo-Devo. Therefore, the paleontology of the twenty-first century is also characterized by the search for a strong integration in this fourth case [71–79].

The fifth topic I have singled out is also anchored in the twentieth-century paleobiological agenda: the discussion of abiotic and biotic factors in evolution. Recall that I opened this paper with a quote from Maynard Smith. He welcomed the return of paleontology to the high table as a result of Raup and Sepkoski's findings on mass extinction and the consequent focus on biotic elements in evolution. This research, in turn, had emerged from the study of the dynamics of Phanerozoic marine paleodiversity (and the proposed model analyses for studying diversity) and the Red Queen model proposed by Leigh Van Valen [5].

Twenty-first-century paleontology is capitalizing on this line of research. At the same time, paleontologists are trying to synthetize biotic and abiotic factors. For instance, Mike Benton notes, "The realization that the Red Queen and Court Jester models may be scale-dependent, and that evolution may be pluralistic, opens opportunities for dialog" [80]. He goes on asserting, "methods are shared by paleontologists and neontologists, and this allows direct communication on the patterns and processes of macroevolution" [80]. Fortelius and colleagues are also proposing "a tentative synthesis, characterized by interdependence between physical forcing and biotic interactions" [81].

Along these lines, paleontologist Tyler Faith and his colleagues emphasized the key methodological elements of current hominin evolution. First, paleoanthropology, as paleontology, is all about data and time scales (or as Gould put it tiers): "There is no universally 'correct' scale of observation, but to address questions linking ecology and evolution, the scales of the processes of interest must align with the scales of the available data" [82]. Second, paleoanthropology is desperately in need of more theory (recall Turner's third point for describing paleobiology), or rather of a new hypothetic-deductive methodology [83]: "Incorporating a stronger theoretical framework into the agenda of hominin paleoecology will allow researchers to reverse the typical direction of inference (i.e., from data to hypothesis) by generating theoretically informed predictions that are tested with the data, and then determining if a hypothesis should be modified or rejected" [82]. By granting a new methodology, a new balance may emerge "between inferring evolutionary narratives from the data and testing process-based hypotheses using those data" [82]. Hence, again, twenty-first-century paleo-research is about more data, more technology, and more integration.

5. Conclusions: A Plea for a New Synthesis

What do the five topics just discussed tell us about the elements of continuity between twentieth-century paleobiology and twenty-first-century paleontology? First, both enterprises called paleontologists' attention to a *cooperative effort* to understand the (deep) past. As paleontological data are always imperfect and incomplete, paleontologists should omnivorously assimilate every method to successfully and opportunistically work on and with deep time [40,84,85]. Molecular approaches to the deep past are therefore supplementary not antagonistic to classical paleontological analyses of forms [86]. Sometimes, they allow paleontologists to see things better or, at least, in a different fine grade (e.g., [87]). How, however, can paleontologists confidently use data, names, technology, and knowledge outside their field of application if they do not know the correct boundaries of application of such tools in their own? Critical integration of different methods can help mediate and hopefully overcome technical issues. This should be paired with inter- and multidisciplinary programs to enable students and scholars to fruitfully learn and apply different methods—that was one of insights of the paleobiological revolution of the 1960s and 1970s.

Indeed, this implies a return to the sprit embodied by the paleobiological revolution of the mid twentieth-century. At the end, what Gould and colleagues were asking for was a genuine synthesis of knowledge. They asked for blurring the disciplinary borders between natural and historical sciences as well as between science and technology. This spirit should be put at the center of twenty-first-century paleontology – this agenda is also perused by other technoscientific disciplines such as biorobotics, synthetic biology, nano(bio)technology, etc. See, for instance, [59,61,88–91]. This enterprise should be characterized by a cooperative model of knowledge production. Instead of supporting a never-ending disciplinary struggle to define what paleontology is (or is not), to draw a sharp line between paleontology and neontology, or between morphological and molecular phylogenetics, scholars should work together and ideally share their data and method to address new challenges, as has happened during periods of major theoretical transitions in the history of paleontology [5,29,92]. The results provided by Perri et al. [93] are a clear example of the continuum of approaches that characterizes paleontology and biology.

A future task for philosophers and historians of paleontology would be to understand the various practices and reasons for the paleontological plea for synthesis and integration in the last century. In other words, the analysis should focus on what (and why) paleontologists sought (and are seeking) collaboration with neontologists and promote the circulation of knowledge and technologies [94] (an important starting point would be the analysis of how paleontologists and neontologists are publishing together or taking part in joint conference. I thank one referee for this point. See, for instance, the papers gathered in the special issue "Crossing the Palaeontological-Ecological Gap" published in *Methods in Ecology and Evolution* 2016, 7. Furthermore, it is perhaps worth noting that Wolfgang Kiessling was the first paleontologist to play a major role in the Intergovernmental Panel on Climate Change (IPCC) report on climate change, working on "Climate Change 2021: Impacts, Adaptation and Vulnerability", thus bridging the gap between the two communities - I thank one referee for this. See also [35,95–97]).

This point will make visible the role of the so-called invisible technicians, i.e., all those who work (silently and without much recognition) on the production of paleontological knowledge. In addition to the flood of data and the transformation of the paleontologist into a data scientist (and vice versa), *field and preparatory work needs to be recognized as it remains at the heart of paleontology today* [98–100]. Today, more than ever, the classic metaphor of the earth as an archive of knowledge and data is still valid. In fact, the greatest amount of data is still buried in the field. On the importance of field work for paleontology [101–106]. This focus will create an awareness of the new social hierarchies in twenty-first century paleontology that result from the massive use of technology.

Hence, and to conclude, if constructively interpreted and read through the recent history of paleontology, the five topics singled out may provide some potentialities for a new synthesis between genetic and morphological approaches to the evolution of forms in the attempt to overcome the limits, issues, and problems of phylogenetic reconstructions. The major implication would be a broader reflection on what paleontology might become in light of recent technical and molecular revolutions. This would imply a second paleobiological revolution. In fact, as David Sepkoski put it, "the paleobiological "revolution" was more like a political contest in which one group perceives itself to be disenfranchised and agitates for greater representation in government than a contest of lofty ideas. The 1970s was a period of revolution in paleontology because paleobiologists saw themselves, and described what they were doing, as revolutionary" [5]. Echoing the paleobiological revolution of the 1960s and 1970s, a twenty-first-century paleobiological revolution would enable paleontologists to gain strong political representation and to argue with a decisive voice at "the high table" on topics such as the extended evolutionary synthesis, the Anthropocene, conservation of Earth's environment, and global climate change.

Funding: This research received no external funding.

Institutional Review Board Statement: Not applicable.

Informed Consent Statement: Not applicable.

Data Availability Statement: Not applicable.

Acknowledgments: I thank the guest editors of the Special Issue and the three anonymous referees for their valuable feedback.

Conflicts of Interest: The author declares no conflict of interest.

References

1. Simpson, G.G. *Tempo and Mode in Evolution*; Columbia University Press: New York, NY, USA, 1944.
2. Smith, J.M. Palaeontology at the High Table. *Nature* **1984**, *309*, 401–402. [CrossRef]
3. Gould, S.J. *The Structure of Evolutionary Theory*; Harvard University Press: Cambridge, MA, USA, 2002.
4. Sepkoski, D.; Ruse, M. (Eds.) *The Paleobiological Revolution Essays on the Growth of Modern Paleontology*; University of Chicago Press: Chicago, IL, USA, 2009.
5. Sepkoski, D. *Rereading the Fossil Record: The Growth of Paleobiology as an Evolutionary Discipline*; University of Chicago Press: Chicago, IL, USA, 2012.
6. Abel, O. Paläontologie und Paläozoologie. In *Die Kultur der Gegenwart*; Hinneberg, P., Ed.; B. G. Teubner: Berlin, Germany; Leipzig, Germany, 1914.
7. Tamborini, M. "If the Americans Can Do It, So Can We": How Dinosaur Bones Shaped German Paleontology. *Hist. Sci.* **2016**, *54*, 225–256. [CrossRef]

8.   Rieppel, O. Othenio Abel: The Rise and Decline of Paleobiology in German Paleontology. *Hist. Biol.* **2013**, *25*, 313–325. [CrossRef]
9.   Dresow, M.W. Before Hierarchy: The Rise and Fall of Stephen Jay Gould's First Macroevolutionary Synthesis. *Hist. Philos. Life Sci.* **2017**, *39*, 6. [CrossRef] [PubMed]
10.   Dresow, M. Macroevolution Evolving: Punctuated Equilibria and the Roots of Stephen Jay Gould's Second Macroevolutionary Synthesis. *Stud. Hist. Philos. Biol. Biomed. Sci.* **2019**, *75*, 15–23. [CrossRef]
11.   Tamborini, M. The Architecture of Evolution: The Science of Form. In *Twentieth-Century Evolutionary Biology*; University of Pittsburgh Press: Pittsburgh, PA, USA, 2022.
12.   Rossi, P. *The Dark Abyss of Time*; The University of Chicago Press: Chicago, IL, USA, 1984.
13.   Rudwick, M.J.S. *Bursting the Limits of Time: The Reconstruction of Geohistory in the Age of Revolution*; The University of Chicago Press: Chicago, IL, USA, 2005.
14.   Rudwick, M.J.S. *Worlds before Adam: The Reconstruction of Geohistory in the Age of Reform*; The University of Chicago Press: Chicago, IL, USA, 2008.
15.   Tamborini, M. "From the Known to the Unknown or Backwards": Visualization and Conceptualization of Paleontological Time in Nineteenth Century Paleontology. In *The Fascination with Unknown Time*; Baumbach, S., Henningsen, L., Oschema, K., Eds.; Palgrave: London, UK, 2017; pp. 115–140.
16.   Tamborini, M. Data and Visualizations of Deep Time. In *Handbook of the Historiography of the Earth and Environmental Sciences*; Springer: Berlin/Heidelberg, Germany, In print.
17.   Gould, S.J. The Paradox of the First Tier: An Agenda for Paleobiology. *Paleobiology* **1985**, *11*, 2–12. [CrossRef]
18.   Eldredge, N.; Pievani, T.; Serrelli, E.; Tëmkin, I. *Evolutionary Theory: A Hierarchical Perspective*; University of Chicago Press: Chicago, IL, USA, 2016.
19.   Turner, D. *Paleontology a Philosophical Introduction*; Cambridge University Press: Cambridge, UK, 2011.
20.   Gould, S.J. The Promise of Paleontology as a Nomothetic, Evolutionary Discipline. *Paleobiology* **1980**, *6*, 96–118. [CrossRef]
21.   Eldredge, N.; Gould, S.J. Punctuated Equilibria: An Alternative to Phyletic Gradualism. In *Models in Paleobiology*; Schopf, T.J.M., Ed.; Freeman, Cooper and Co.: San Francisco, CA, USA, 1972.
22.   Raup, D.M.; Sepkoski, J.J., Jr. Periodicity of Extinctions in the Geologic Past. *Proc. Natl. Acad. Sci. USA* **1984**, *81*, 801–805. [CrossRef]
23.   Raup, D.M.; Sepkoski, J.J., Jr. Mass Extinctions in the Marine Fossil Record. *Science* **1982**, *215*, 1501–1503. [CrossRef]
24.   Ebbighausen, R.; Korn, D. Paleontology as a Circumstantial Evidence Lawsuit. *Hist. Biol.* **2013**, *25*, 283–295. [CrossRef]
25.   Windelband, W. *Geschichte und Naturwissenschaft. Rede zum Antritt des. Rectorats der Kaiser–Wilhelm–Universität Strassburg, Geh. Am. 1. Mai 1894*; Heitz: Strassburg, France, 1900.
26.   Sepkoski, J.J., Jr. A Kinetic Model of Phanerozoic Taxonomic Diversity. I. Analysis of Marine Orders. *Paleobiology* **1978**, *4*, 223–251. [CrossRef]
27.   Sepkoski, J.J., Jr.; Bambach, R.K.; Raup, D.M.; Valentine, J.W. Phanerozoic Marine Diversity and The Fossil Record. *Nature* **1981**, *293*, 435–437. [CrossRef]
28.   Sepkoski, J.J., Jr. *A Compendium of Fossil Marine Families*; Milwaukee Public Museum: Milwaukee, WI, USA, 1982.
29.   Sepkoski, D.; Tamborini, M. "An Image of Science": Cameralism, Statistics, and the Visual Language of Natural History in the Nineteenth Century. *Hist. Stud. Nat. Sci.* **2018**, *48*, 56–109. [CrossRef]
30.   Donoghue, P.C.; Yang, Z. The Evolution of Methods for Establishing Evolutionary Timescales. *Philos. Trans. R. Soc. B Biol. Sci.* **2016**, *371*, 20160020. [CrossRef]
31.   Patzkowsky, M.E.; Holland, S.M. *Stratigraphic Paleobiology*; University of Chicago Press: Chicago, IL, USA, 2012.
32.   Cleland, C.E. Methodological and Epistemic Differences between Historical Science and Experimental Science. *Philos. Sci.* **2002**, *69*, 447–451. [CrossRef]
33.   Cleland, C.E. Prediction and Explanation in Historical Natural Science. *Br. J. Philos. Sci.* **2011**, *62*, 551–582. [CrossRef]
34.   Turner, D. *Making Prehistory: Historical Science and the Scientific Realism Debate*; Cambridge University Press: Cambridge, UK, 2007.
35.   Sepkoski, D. *Catastrophic Thinking: Extinction and the Value of Diversity from Darwin to the Anthropocene*; University of Chicago Press: Chicago, IL, USA, 2020.
36.   Tamborini, M. Paleontology and Darwin's Theory of Evolution: The Subversive Role of Statistics at the End of the 19th Century. *J. Hist. Biol.* **2015**, *48*, 575–612. [CrossRef]
37.   Sepkoski, D. The Database before the Computer? *Osiris* **2017**, *32*, 175–201. [CrossRef]
38.   Raup, D.M.; Seilacher, A. Fossil Foraging Behavior: Computer Simulation. *Science* **1969**, *166*, 994–995. [CrossRef]
39.   Currie, A. Mass Extinctions as Major Transitions. *Biol. Philos.* **2019**, *34*, 1–24. [CrossRef]
40.   Tamborini, M. Technoscientific Approaches to Deep Time. *Stud. Hist. Philos. Sci. Part A* **2020**, *79*, 57–67. [CrossRef]
41.   McNamara, M.E.; Briggs, D.E.; Orr, P.J.; Field, D.J.; Wang, Z. Experimental Maturation of Feathers: Implications for Reconstructions of Fossil Feather Colour. *Biol. Lett.* **2013**, *9*, 20130184. [CrossRef] [PubMed]
42.   Smithwick, F.; Vinther, J. Palaeocolour: A History and State of the Art. *Evol. Feathers* **2020**, 185–211.
43.   Zhang, F.; Kearns, S.L.; Orr, P.J.; Benton, M.J.; Zhou, Z.; Johnson, D.; Xu, X.; Wang, X. Fossilized Melanosomes and the Colour of Cretaceous Dinosaurs and Birds. *Nature* **2010**, *463*, 1075–1078. [CrossRef] [PubMed]
44.   Benton, M.J. A Colourful View of the Origin of Dinosaur Feathers. *Nature* **2022**, *604*, 630–631. [CrossRef] [PubMed]
45.   Dance, A. Prehistoric Animals, in Living Color. *Proc. Natl. Acad. Sci. USA* **2016**, *113*, 8552–8556. [CrossRef] [PubMed]

46. Negro, J.J.; Finlayson, C.; Galván, I. Melanins in Fossil Animals: Is It Possible to Infer Life History Traits from the Coloration of Extinct Species? *Int. J. Mol. Sci.* **2018**, *19*, 230. [CrossRef] [PubMed]
47. Vinther, J.; Briggs, D.E.; Prum, R.O.; Saranathan, V. The Colour of Fossil Feathers. *Biol. Lett.* **2008**, *4*, 522–525. [CrossRef]
48. Turner, D.D. A Second Look at the Colors of the Dinosaurs. *Stud. Hist. Philos. Sci. Part A* **2016**, *55*, 60–68. [CrossRef]
49. Heingård, M.; Sjövall, P.; Schultz, B.P.; Sylvestersen, R.L.; Lindgren, J. Preservation and Taphonomy of Fossil Insects from the Earliest Eocene of Denmark. *Biology* **2022**, *11*, 395.
50. Botfalvai, G.; Kocsis, L.; Szabó, M.; Király, E.; Sebe, K. Preliminarily Report on Rare Earth Element Taphonomy of a Miocene Mixed Age Fossil Vertebrate Assemblage (Pécs-Danitzpuszta, Mecsek Mts., Hungary): Uptake Mechanism and Possible Separation of Palaeocommunities. *Hist. Biol.* **2022**, 1–20. [CrossRef]
51. Ullmann, P.V.; Macauley, K.; Ash, R.D.; Shoup, B.; Scannella, J.B. Taphonomic and Diagenetic Pathways to Protein Preservation, Part I: The Case of Tyrannosaurus Rex Specimen MOR 1125. *Biology* **2021**, *10*, 1193. [CrossRef]
52. Allison, P.A.; Bottjer, D.J. Taphonomy: Bias and Process through Time. In *Taphonomy*; Springer: Berlin/Heidelberg, Germany, 2011; pp. 1–17.
53. Gupta, N.S.; Briggs, D.E. Taphonomy of Animal Organic Skeletons through Time. In *Taphonomy*; Springer: Berlin/Heidelberg, Germany, 2011; pp. 199–221.
54. Baets, K.D.; Huntley, J.W.; Klompmaker, A.A.; Schiffbauer, J.D.; Muscente, A.D. The Fossil Record of Parasitism: Its Extent and Taphonomic Constraints. In *The Evolution and Fossil Record of Parasitism*; Springer: Berlin/Heidelberg, Germany, 2021; pp. 1–50.
55. Peterman, D.J.; Ritterbush, K.A. Resurrecting Extinct Cephalopods with Biomimetic Robots to Explore Hydrodynamic Stability, Maneuverability, and Physical Constraints on Life Habits. *Sci. Rep.* **2022**, *12*, 11287. [CrossRef] [PubMed]
56. Nyakatura, J.A. Early Primate Evolution: Insights into the Functional Significance of Grasping from Motion Analyses of Extant Mammals. *Biol. J. Linn. Soc.* **2019**, *127*, 611–631. [CrossRef]
57. Nyakatura, J.A.; Melo, K.; Horvat, T.; Karakasiliotis, K.; Allen, V.R.; Andikfar, A.; Andrada, E.; Arnold, P.; Lauströer, J.; Hutchinson, J.R. Reverse-Engineering the Locomotion of a Stem Amniote. *Nature* **2019**, *565*, 351. [CrossRef]
58. Nyakatura, J. Learning to Move on Land. *Science* **2016**, *353*, 120–121. [CrossRef]
59. Tamborini, M. *Entgrenzung. Die Biologisierung der Technik und die Technisierung der Biologie*; Meiner: Hamburg, Germany, 2022.
60. Karme, A.; Rannikko, J.; Kallonen, A.; Clauss, M.; Fortelius, M. Mechanical Modelling of Tooth Wear. *J. R. Soc. Interface* **2016**, *13*, 20160399. [CrossRef] [PubMed]
61. Tamborini, M. The Material Turn in The Study of Form: From Bio-Inspired Robots to Robotics-Inspired Morphology. *Perspect. Sci.* **2021**, *29*, 643–665. [CrossRef]
62. Sutton, M.; Rahman, I.; Garwood, R. Virtual Paleontology—An Overview. *Paleontol. Soc. Pap.* **2016**, *22*, 1–20. [CrossRef]
63. Rahman, I.A.; Smith, S.Y. Virtual Paleontology: Computer-Aided Analysis of Fossil form and Function. *J. Paleontol.* **2014**, *88*, 633–635. [CrossRef]
64. Cunningham, J.A.; Rahman, I.A.; Lautenschlager, S.; Rayfield, E.J.; Donoghue, P.C. A Virtual World of Paleontology. *Trends Ecol. Evol.* **2014**, *29*, 347–357. [CrossRef]
65. Cirilli, O.; Melchionna, M.; Serio, C.; Bernor, R.L.; Bukhsianidze, M.; Lordkipanidze, D.; Rook, L.; Profico, A.; Raia, P. Target Deformation of the Equus Stenonis Holotype Skull: A Virtual Reconstruction. *Front. Earth Sci.* **2020**, *8*, 247. [CrossRef]
66. Pandolfi, L.; Raia, P.; Fortuny, J.; Rook, L. Evolving Virtual and Computational Paleontology. *Front. Earth Sci.* **2020**, *8*, 591813. [CrossRef]
67. Lautenschlager, S.; Rücklin, M. Beyond the Print—Virtual Paleontology in Science Publishing, Outreach, and Education. *J. Paleontol.* **2014**, *88*, 727–734. [CrossRef] [PubMed]
68. Tamborini, M. Challenging the Adaptationist Paradigm: Morphogenesis, Constraints, and Constructions. *J. Hist. Biol.* **2020**, *53*, 269–294. [CrossRef]
69. Love, A. Conceptual Change and Evolutionary Developmental Biology. In *Conceptual Change in Biology. Scientific and Philosophical Perspectives on Evolution and Development*; Love, A., Ed.; Springer: Berlin/Heidelberg, Germany, 2015; pp. 1–54.
70. Love, A. Morphological and Paleontological Perspectives for a History of Evo-Devo. In *From Embryology to Evo-Devo: A History of Developmental Evolution*; Laubichler, M.D., Maienschein, J., Eds.; MIT Press: Cambridge, MA, USA, 2007; pp. 267–307.
71. Chipman, A.D.; Edgecombe, G.D. Developing an Integrated Understanding of the Evolution of Arthropod Segmentation Using Fossils and Evo-Devo. *Proc. R. Soc. B* **2019**, *286*, 20191881. [CrossRef]
72. Davidson, E.H.; Erwin, D.H. Gene Regulatory Networks and the Evolution of Animal Body Plans. *Science* **2006**, *311*, 796–800. [CrossRef] [PubMed]
73. Haug, C.; Haug, J.T. Methods and Practices in Paleo-Evo-Devo. In *Evolutionary Developmental Biology: A Reference Guide*; Springer: Cham, Switzerland, 2021; pp. 1151–1164.
74. Hlusko, L.J.; Sage, R.D.; Mahaney, M.C. Modularity in the Mammalian Dentition: Mice and Monkeys Share a Common Dental Genetic Architecture. *J. Exp. Zool. Part. B Mol. Dev. Evol.* **2011**, *316*, 21–49. [CrossRef] [PubMed]
75. Jernvall, J.; Jung, H. Genotype, Phenotype, and Developmental Biology of Molar Tooth Characters. *Am. J. Phys. Anthropol. Off. Publ. Am. Assoc. Phys. Anthropol.* **2000**, *113*, 171–190. [CrossRef]
76. Raff, R.A. Written in Stone: Fossils, Genes and Evo–Devo. *Nat. Rev. Genet.* **2007**, *8*, 911–920. [CrossRef]
77. Tomescu, A.M.; Rothwell, G.W. Fossils and Plant Evolution: Structural Fingerprints and Modularity in the Evo-Devo Paradigm. *EvoDevo* **2022**, *13*, 1–19. [CrossRef] [PubMed]

78. Hlusko, L.J. Integrating the Genotype and Phenotype in Hominid Paleontology. *Proc. Natl. Acad. Sci. USA* **2004**, *101*, 2653–2657. [CrossRef] [PubMed]

79. Urdy, S.; Wilson, L.A.B.; Haug, J.T.; Sánchez-Villagra, M.R. On the Unique Perspective of Paleontology in the Study of Developmental Evolution and Biases. *Biol. Theory* **2013**, *8*, 293–311. [CrossRef]

80. Benton, M.J. The Red Queen and the Court Jester: Species Diversity and the Role of Biotic and Abiotic Factors through Time. *Science* **2009**, *323*, 728–732. [CrossRef] [PubMed]

81. Fortelius, M.; Eronen, J.T.; Kaya, F.; Tang, H.; Raia, P.; Puolamäki, K. Evolution of Neogene Mammals in Eurasia: Environmental Forcing and Biotic Interactions. *Annu. Rev. Earth Planet. Sci.* **2014**, *42*, 579–604. [CrossRef]

82. Faith, J.T.; Du, A.; Behrensmeyer, A.K.; Davies, B.; Patterson, D.B.; Rowan, J.; Wood, B. Rethinking the Ecological Drivers of Hominin Evolution. *Trends Ecol. Evol.* **2021**, *36*, 797–807. [CrossRef]

83. Carrier, M. *Wissenschaftstheorie zur Einführung*; Junius Verlag: Hamburg, Germany, 2019.

84. Currie, A. *Rock, Bone, and Ruin: An. Optimist's Guide to the Historical Sciences*; MIT Press: Cambridge, MA, USA, 2018.

85. Currie, A. Marsupial Lions and Methodological Omnivory: Function, Success and Reconstruction in Paleobiology. *Biol. Philos.* **2015**, *30*, 187–209. [CrossRef]

86. Barido-Sottani, J.; Pohle, A.; de Baets, K.; Murdock, D.; Warnock, R.C.M. Putting the F in FBD Analyses: Tree Constraints or Morphological Data? *bioRxiv* **2022**. [CrossRef]

87. Rook, L. The Plio-Pleistocene Old World Canis (Xenocyon) Ex Gr. Falconeri. *Boll. Della Soc. Paleontol. Ital.* **1994**, *33*, 71–82.

88. Nordmann, A. Collapse of Distance: Epistemic Strategies of Science and Technoscience. *Dan. Yearb. Philos.* **2006**, *41*, 7–34. [CrossRef]

89. Köchy, K.; Norwig, M.; Hofmeister, G. (Eds.) *Nanobiotechnologien: Philosophische, Anthropologische und Ethische Fragen*; Alber: Freiburg, Germany, 2008; ISBN 978-3-495-48347-3.

90. Gutmann, M. *Leben und Form. Zur Technischen Form des. Wissens vom Lebendigen*; Springer: Wiesbaden, Germany, 2017.

91. Tamborini, M. Technische Form und Konstruktion. *Dtsch. Z. für Philos.* **2020**, *68*, 712–733. [CrossRef]

92. Rudwick, M.J.S. *Earth's Deep History: How It Was Discovered and Why It Matters*; University of Chicago Press: Chicago, IL, USA, 2016.

93. Perri, A.R.; Mitchell, K.J.; Mouton, A.; Alvarez-Carretero, S.; Hulme-Beaman, A.; Haile, J.; Jamieson, A.; Meachen, J.; Lin, A.T.; Schubert, B.W. Dire Wolves Were the Last of an Ancient New World Canid Lineage. *Nature* **2021**, *591*, 87–91. [CrossRef]

94. Tamborini, M. The Circulation of Morphological Knowledge: Understanding 'Form' across Disciplines in the Twentieth and Twenty-First Centuries. *Isis* **2022**, 4.

95. Kiessling, W.; Raja, N.B.; Roden, V.J.; Turvey, S.T.; Saupe, E.E. Addressing Priority Questions of Conservation Science with Palaeontological Data. *Philos. Trans. R. Soc. B* **2019**, *374*, 20190222. [CrossRef]

96. Dietl, G.P.; Flessa, K.W. *Conservation Paleobiology: Science and Practice*; University of Chicago Press: Chicago, IL, USA, 2017; ISBN 0-226-50686-X.

97. Dietl, G.P.; Flessa, K.W. Conservation Paleobiology: Putting the Dead to Work. *Trends Ecol. Evol.* **2011**, *26*, 30–37. [CrossRef] [PubMed]

98. Wylie, C.D. *Preparing Dinosaurs: The Work Behind the Scenes*; MIT Press: Cambridge, MA, USA, 2021.

99. Wylie, C.D. Trust in Technicians in Paleontology Laboratories. *Sci. Technol. Hum. Values* **2018**, *43*, 324–348. [CrossRef]

100. Wylie, C.D. Preparation in Action: Paleontological Skill and the Role of the Fossil Preparator. In Proceedings of the Methods in Fossil Preparation: Proceedings of the First Annual Fossil Preparation and Collections Symposium, Petrified Forest National Park, AZ, USA, April 2008; pp. 3–12.

101. Heumann, I.; Stoecker, H.; Tamborini, M.; Vennen, M. *Dinosaurierfragmente: Zur Geschichte der Tendaguru-Expedition und Ihrer Objekte, 1906–2017*; Wallstein Verlag: Göttingen, Germany, 2018; ISBN 3-8353-4305-X.

102. Rieppel, L. *Assembling the Dinosaur*; Harvard University Press: Cambridge, MA, USA, 2019; ISBN 0-674-24033-2.

103. Tamborini, M.; Vennen, M. Disruptions and Changing Habits: The Case of the Tendaguru Expedition. *Mus. Hist. J.* **2017**, *10*, 183–199. [CrossRef]

104. Nieuwland, I. *American Dinosaur Abroad: A Cultural History of Carnegie's Plaster Diplodocus*; University of Pittsburgh Press: Pittsburgh, PA, USA, 2019; ISBN 0-8229-8666-3.

105. Podgorny, I. La Prueba Asesinda. El Trabajo de Campo y Los Métodos de Registro En La Arquelogía de Lod Inicios Del Siglo XX. In *Saberes Locales, Ensayos Sobre Historia de la Ciencia*; López, C.B., Gorbach, F., Eds.; El Colegio de Michoacán: Zamora de Hidalgo, Mexico, 2008.

106. Podgorny, I. Towards a Bureaucratic History of Archaeology. A Preliminary Essay. In *Historiographical Approaches to Past Archaeological Research. Berlin Studies of the Ancient World*; Eberhardt, G., Link, F., Eds.; E-Topois: Berlin, Germany, 2015; Volume 32.

MDPI

St. Alban-Anlage 66

4052 Basel

Switzerland

Tel. +41 61 683 77 34

Fax +41 61 302 89 18

www.mdpi.com

Biology Editorial Office

E-mail: biology@mdpi.com

www.mdpi.com/journal/biology

www.ingramcontent.com/pod-product-compliance
Lightning Source LLC
LaVergne TN
LVHW071624170726
843515LV00009B/2474